PARALLELOGRAM

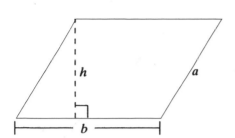

Perimeter: $P = 2a + 2b$
Area: $A = bh$

CIRCLE

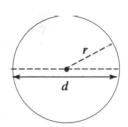

Circumference: $C = \pi d$
$C = 2\pi r$
Area: $A = \pi r^2$

RECTANGULAR SOLID

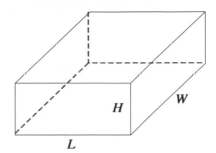

Volume: $V = LWH$
Surface Area: $A = 2HW + 2LW + 2LH$

CUBE

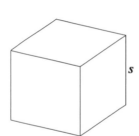

Volume: $V = s^3$
Surface Area: $A = 6s^2$

CONE

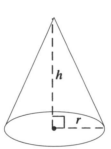

Volume: $V = \dfrac{1}{3}\pi r^2 h$
Lateral Surface Area: $A = \pi r \sqrt{r^2 + h^2}$

RIGHT CIRCULAR CYLINDER

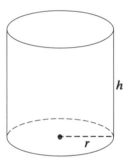

Volume: $V = \pi r^2 h$
Surface Area:
$A = 2\pi rh + 2\pi r^2$

OTHER FORMULAS

Distance: $d = rt$ (r = rate, t = time)

Temperature: $F = \dfrac{9}{5}C + 32$ $C = \dfrac{5}{9}(F - 32)$

Simple Interest: $I = Prt$
(P = principal, r = annual interest rate, t = time in years)

Compound Interest: $A = P\left(1 + \dfrac{r}{n}\right)^{nt}$

(P = principal, r = annual interest rate, t = time in years, n = number of compoundings per year)

Companion Website
http://www.prenhall.com/martin-gay

BEGINNING ALGEBRA

BEGINNING ALGEBRA

Third Edition

K. ELAYN MARTIN-GAY

University of New Orleans

Prentice Hall

Upper Saddle River, New Jersey 07458

Library of Congress Cataloging-in-Publication Data

Martin-Gay, K. Elayn
   Beginning algebra/K. Elayn Martin-Gay.—3rd ed.
    p. cm.
   Includes index.
   ISBN 0-13-086763-2 (alk. paper)
   1. Algebra.   I. Title.
QA152.2.M367 2001
512.9—dc21

00-035951
CIP

Executive Acquisition Editor: Karin E. Wagner
Editor in Chief: Christine Hoag
Project Manager: Ann Marie Jones
Assistant Vice President of Production and Manufacturing: David W. Riccardi
Executive Managing Editor: Kathleen Schiaparelli
Senior Managing Editor: Linda Mihatov Behrens
Project Management: Elm Street Publishing Services, Inc.
Manufacturing Buyer: Alan Fischer
Manufacturing Manager: Trudy Pisciotti
Senior Marketing Manager: Eilish Collins Main
Marketing Assistant: Dan Auld
Director of Marketing: John Tweeddale
Development Editors: Tony Palermino/Emily Keaton
Editor in Chief, Development: Carol Trueheart
Associate Editor, Mathematics/Statistics Media: Audra J. Walsh
Editorial Assistant/Supplements Editor: Kate Marks
Art Director: Maureen Eide
Assistant to the Art Director: John Christiana
Interior Designer: Donna Wickes
Cover Designer: Joseph Sengotta
Art Editor: Grace Hazeldine
Art Manager: Gus Vibal
Director of Creative Services: Paul Belfanti
Photo Researcher: Kathy Ringrose
Cover Photo: Alex Demyan Photographs
Art Studio: Academy Artworks
Compositor: Preparé Inc., Italy

© 2001 by Prentice-Hall, Inc.
Upper Saddle River, NJ 07458

Printed in the United States of America
10 9 8 7 6 5

ISBN 0-13-086763-2

Prentice-Hall International (UK) Limited, *London*
Prentice-Hall of Australia Pty. Limited, *Sydney*
Prentice-Hall Canada, Inc., *Toronto*
Prentice-Hall Hispanoamericana, S.A., *Mexico*
Prentice-Hall of India Private Limited, *New Delhi*
Prentice-Hall of Japan, Inc., *Tokyo*
Pearson Education Asia, Pte. Ltd.
Editora Prentice-Hall do Brasil, Ltda., *Rio de Janeiro*

To my mother, Barbara M. Miller,
and her husband, Leo Miller,
and to the memory
of my father, Robert J. Martin

CONTENTS

Preface   xi

1 REVIEW OF REAL NUMBERS   2

1.1   Tips for Success in Mathematics   4
1.2   Symbols and Sets of Numbers   8
1.3   Fractions   17
1.4   Introduction to Variable Expressions and Equations   25
1.5   Adding Real Numbers   34
1.6   Subtracting Real Numbers   41
1.7   Multiplying and Dividing Real Numbers   48
1.8   Properties of Real Numbers   56
1.9   Reading Graphs   61
      Chapter 1 Project: Creating and Interpreting Graphs   68
      Chapter 1 Vocabulary Check   68
      Chapter 1 Highlights   69
      Chapter 1 Review   73
      Chapter 1 Test   76

2 EQUATIONS, INEQUALITIES, AND PROBLEM SOLVING   78

2.1   Simplifying Algebraic Expressions   80
2.2   The Addition Property of Equality   86
2.3   The Multiplication Property of Equality   93
2.4   Solving Linear Equations   99
2.5   An Introduction to Problem Solving   107
2.6   Formulas and Problem Solving   116
2.7   Percent and Problem Solving   124
2.8   Further Problem Solving   133
2.9   Solving Linear Inequalities   140
      Chapter 2 Project: Developing a Budget   150
      Chapter 2 Vocabulary Check   151

Chapter 2 Highlights   *151*
Chapter 2 Review   *156*
Chapter 2 Test   *158*
Chapter 2 Cumulative Review   *159*

3   GRAPHING   160

3.1   The Rectangular Coordinate System   162
3.2   Graphing Linear Equations   173
3.3   Intercepts   182
3.4   Slope   189
3.5   Graphing Linear Inequalities   202
Chapter 3 Project: Financial Analysis   *208*
Chapter 3 Vocabulary Check   *208*
Chapter 3 Highlights   *209*
Chapter 3 Review   *212*
Chapter 3 Test   *214*
Chapter 3 Cumulative Review   *215*

4   EXPONENTS AND POLYNOMIALS   216

4.1   Exponents   218
4.2   Adding and Subtracting Polynomials   228
4.3   Multiplying Polynomials   237
4.4   Special Products   241
4.5   Negative Exponents and Scientific Notation   247
4.6   Division of Polynomials   256
Chapter 4 Project: Modeling with Polynomials   *262*
Chapter 4 Vocabulary Check   *263*
Chapter 4 Highlights   *264*
Chapter 4 Review   *266*
Chapter 4 Test   *269*
Chapter 4 Cumulative Review   *270*

5   FACTORING POLYNOMIALS   272

5.1   The Greatest Common Factor and Factoring by Grouping   274
5.2   Factoring Trinomials of the Form $x^2 + bx + c$ 280
5.3   Factoring Trinomials of the Form $ax^2 + bx + c$ 286
5.4   Factoring Binomials   294
5.5   Choosing a Factoring Strategy   300
5.6   Solving Quadratic Equations by Factoring   305
5.7   Quadratic Equations and Problem Solving   313
Chapter 5 Project: Choosing Among Building Options   *322*
Chapter 5 Vocabulary Check   *322*
Chapter 5 Highlights   *323*
Chapter 5 Review   *326*
Chapter 5 Test   *327*
Chapter 5 Cumulative Review   *328*

6 RATIONAL EXPRESSIONS   330

6.1   Simplifying Rational Expressions   332
6.2   Multiplying and Dividing Rational Expressions   339
6.3   Adding and Subtracting Rational Expressions with Common Denominators and Least Common Denominator   346
6.4   Adding and Subtracting Rational Expressions with Unlike Denominators   353
6.5   Simplifying Complex Fractions   359
6.6   Solving Equations Containing Rational Expressions   365
6.7   Ratio and Proportion   373
6.8   Rational Equations and Problem Solving   379
      Chapter 6 Project: Comparing Dosage Formulas   387
      Chapter 6 Vocabulary Check   388
      Chapter 6 Highlights   389
      Chapter 6 Review   393
      Chapter 6 Test   395
      Chapter 6 Cumulative Review   396

7 FURTHER GRAPHING   398

7.1   The Slope-Intercept Form   400
7.2   The Point-Slope Form   406
7.3   Graphing Nonlinear Equations   414
7.4   Functions   421
      Chapter 7 Project: Matching Descriptions of Linear Data to Their Equations and Graphs   430
      Chapter 7 Vocabulary Check   431
      Chapter 7 Highlights   432
      Chapter 7 Review   433
      Chapter 7 Test   435
      Chapter 7 Cumulative Review   436

8 SOLVING SYSTEMS OF LINEAR EQUATIONS   438

8.1   Solving Systems of Linear Equations by Graphing   440
8.2   Solving Systems of Linear Equations by Substitution   447
8.3   Solving Systems of Linear Equations by Addition   453
8.4   Systems of Linear Equations and Problem Solving   459
8.5   Systems of Linear Inequalities   468
      Chapter 8 Project: Analyzing the Courses of Ships   472
      Chapter 8 Vocabulary Check   473
      Chapter 8 Highlights   474
      Chapter 8 Review   477
      Chapter 8 Test   478
      Chapter 8 Cumulative Review   479

9 ROOTS AND RADICALS   480

9.1 Introduction to Radicals   482
9.2 Simplifying Radicals   488
9.3 Adding and Subtracting Radicals   494
9.4 Multiplying and Dividing Radicals   497
9.5 Solving Equations Containing Radicals   504
9.6 Radical Equations and Problem Solving   509
9.7 Rational Exponents   516
   Chapter 9 Project: Investigating the Dimensions of Cylinders   520
   Chapter 9 Vocabulary Check   521
   Chapter 9 Highlights   522
   Chapter 9 Review   525
   Chapter 9 Test   526
   Chapter 9 Cumulative Review   527

10 SOLVING QUADRATIC EQUATIONS   528

10.1 Solving Quadratic Equations by the Square Root Method   530
10.2 Solving Quadratic Equations by Completing the Square   534
10.3 Solving Quadratic Equations by the Quadratic Formula   539
10.4 Summary of Methods for Solving Quadratic Equations and Problem Solving   546
10.5 Complex Solutions of Quadratic Equations   550
10.6 Graphing Quadratic Equations   555
   Chapter 10 Project: Modeling a Physical Situation   562
   Chapter 10 Vocabulary Check   563
   Chapter 10 Highlights   564
   Chapter 10 Review   567
   Chapter 10 Test   569
   Chapter 10 Cumulative Review   570

APPENDICES

A. Operations on Decimals   573
B. Review of Angles, Lines, and Special Triangles   575
C. Review of Geometric Figures   583
D. Mean, Median, and Mode   587
E. Table of Squares and Square Roots   591
F. Review of Volume and Surface Area   593

ANSWERS TO SELECTED EXERCISES   A1

INDEX   I-1

PHOTO CREDITS   P-1

PREFACE

ABOUT THE BOOK

Beginning Algebra, Third Edition was written to provide a **solid foundation in algebra** for students who might have had no previous experience in algebra. Specific care has been taken to ensure that students have the most **up-to-date and relevant** text preparation for their next mathematics course, as well as to help students to succeed in nonmathematical courses that require a grasp of algebraic fundamentals. I have tried to achieve this by writing a user-friendly text that is keyed to objectives and contains many worked-out examples. The basic concepts of graphing are introduced early, and problem solving techniques, real-life and real-data applications, data interpretation, appropriate use of technology, mental mathematics, number sense, critical thinking, decision-making, and geometric concepts are emphasized and integrated throughout the book.

The many factors that contributed to the success of the first two editions have been retained. In preparing this edition, I considered the comments and suggestions of colleagues throughout the country, students, and many users of the prior editions. The AMATYC Crossroads in Mathematics: Standards for Introductory College Mathematics before Calculus and the MAA and NCTM standards (plus Addenda), together with advances in technology, also influenced the writing of this text.

Beginning Algebra, Third Edition is **part of a series of texts** that can include *Basic College Mathematics, Prealgebra, Third Edition, Intermediate Algebra, Third Edition*, or *Intermediate Algebra: A Graphing Approach, Second Edition*, and *Beginning and Intermediate Algebra, Second Edition*, a combined algebra text. Throughout the series, pedagogical features are designed to develop student proficiency in algebra and problem solving, and to prepare students for future courses.

KEY PEDAGOGICAL FEATURES IN THE THIRD EDITION

Readability and Connections   I have tried to make the writing style as clear as possible while still retaining the mathematical integrity of the content. When a new topic is presented, an effort has been made to **relate the new ideas to those that students**

may already know. Constant reinforcement and connections within problem solving strategies, data interpretation, geometry, patterns, graphs, and situations from everyday life can help students gradually master both new and old information.

Problem Solving Process   This is formally introduced in Chapter 2 with a **new four-step process that is integrated throughout the text**. The four steps are Understand, Translate, Solve, and Interpret. The repeated use of these steps throughout the text in a variety of examples shows their wide applicability. Reinforcing the steps can increase students' confidence in tackling problems.

Applications and Connections   Every effort was made to include as many accessible, interesting, and relevant real-life applications as possible throughout the text in both worked-out examples and exercise sets. The applications **strengthen students' understanding of mathematics in the real world** and help to motivate students. They show connections to a wide range of fields including agriculture, allied health, art, astronomy, automotive ownership, aviation, biology, business, chemistry, communication, computer technology, construction, consumer affairs, demographics, earth science, education, entertainment, environmental issues, finance and economics, food service, geography, government, history, hobbies, labor and career issues, life science, medicine, music, nutrition, physics, political science, population, recreation, sports, technology, transportation, travel, weather, and important related mathematical areas such as geometry and statistics. (See the Index of Applications on page xxi.) Many of the applications are based on **recent and interesting real-life data**. Sources for data include newspapers, magazines, government publications, publicly held companies, special interest groups, research organizations, and reference books. Opportunities for obtaining your own real data are also included.

Helpful Hints   Helpful Hints, formerly Reminders, contain practical advice on applying mathematical concepts. These are found throughout the text and **strategically placed** where students are most likely to need immediate reinforcement. They are highlighted in a box for quick reference and, as appropriate, an indicator line is used to precisely identify the particular part of a problem or concept being discussed. For instance, see pages 96 and 408.

Visual Reinforcement of Concepts   The text contains numerous graphics, models, and illustrations to visually clarify and reinforce concepts. These include **new and updated** bar graphs, circle graphs in two and three dimensions, line graphs, calculator screens, application illustrations, photographs, and geometric figures. There are now **over 1,000 figures**.

Real World Chapter Openers   The new two-page chapter opener focuses on how math is used in a specific career, provides links to the World Wide Web, and references a "Spotlight on Decision Making" feature within the chapter for further exploration of the **career and the relevance of algebra**. For example, look at the opener for Chapter 8. The opening pages also contain a list of section titles, and an introduction to the mathematics to be studied together with mathematical connections to previous chapters in the text.

Student Resource Icons   At the beginning of each section, videotape, tutorial software CD Rom, Student Solutions Manual, and Study Guide icons are displayed. These icons help reinforce that these learning aids are available should students wish to use them to review concepts and skills at their own pace. These items have **direct correlation to the text** and emphasize the text's methods of solution.

Chapter Highlights   Found at the end of each chapter, the Chapter Highlights contain key definitions, concepts, *and* examples to **help students understand and retain** what they have learned.

Chapter Project   This feature occurs at the end of each chapter, often serving as a chapter wrap-up. For **individual or group completion**, the multi-part Chapter Project, usually hands-on or data based, allows students to problem solve, make interpretations, and to think and write about algebra.

Functional Use of Color and New Design   Elements of this text are highlighted with color or design to make it easier for students to read and study. Special care has been taken to use color within solutions to examples or in the art to **help clarify, distinguish, or connect concepts**. For example, look at pages 190 and 191 in Section 3.4.

EXERCISE SETS

Each text section ends with an exercise set, usually divided into two parts. Both parts contain graded exercises. The **first part is carefully keyed** to at least one worked example in the text. Once a student has gained confidence in a skill, **the second part contains exercises not keyed to examples**. Exercises and examples marked with a video icon (✎) have been worked out step-by-step by the author in the videos that accompany this text.

Throughout the text exercises there is an emphasis on data and graphical interpretation via tables, charts, and graphs. The ability to interpret data and read and create a variety of types of graphs is developed gradually so students become comfortable with it. Similarly, throughout the text there is integration of geometric concepts, such as perimeter and area. Exercises and examples marked with a geometry icon (△) have been identified for convenience.

Each exercise set contains one or more of the following features.

Spotlight on Decision Making   These unique **new, specially designed applications** help students develop their decision-making and problem solving abilities, skills useful in mathematics and in life. Appropriately placed before an exercise set begins, students have an opportunity to immediately practice and reinforce basic algebraic concepts found in the accompanying section in relevant, accessible contexts. There is an emphasis on workplace or job-related career situations (such as the decisions of a small business owner in Section 3.1, a physical therapist in Section 7.2, or a registered nurse in Section 8.5) as well as decision-making in general (such as choosing a homeowner's insurance policy in Section 2.8 or choosing a credit card in Section 5.5 or deciding when to plant flower bulbs in Section 10.6).

Mental Mathematics   These problems are found at the beginning of many exercise sets. They are mental warm-ups that **reinforce concepts** found in the accompanying section and increase students' confidence before they tackle an exercise set. By relying on their own mental skills, students increase not only their confidence in themselves, but also their number sense and estimation ability.

Writing Exercises   These exercises now found in almost every exercise set are marked with the icon (✎). They require students to **assimilate information** and provide a written response to explain concepts or justify their thinking. Guidelines recommended by the American Mathematical Association of Two Year Colleges (AMATYC) and other professional groups recommend incorporating writing in mathematics courses to reinforce concepts. Writing opportunities also occur within features such as Spotlight on Decision Making and Chapter Projects.

Data and Graphical Interpretation   Throughout the text there is an emphasis on data interpretation in exercises via tables, bar charts, line graphs, or circle graphs. The ability to interpret data and read and create a variety of graphs is **developed gradually** so students become comfortable with it. In addition, there is an appendix on mean, median, and mode together with exercises.

Calculator Explorations and Exercises   These optional explorations offer guided instruction, through examples and exercises, on the proper use of **scientific and graphing calculators or computer graphing utilities as tools in the mathematical problem-solving process**. Placed appropriately throughout the text, these explorations reinforce concepts or motivate discovery learning.

Additional exercises building on the skills developed in the Explorations may be found in exercise sets throughout the text, and are marked with the icon ▦ for scientific calculator use and with the icon ▦ for graphing calculator use.

Review Exercises   These exercises occur in each exercise set (except for those in Chapter 1). These problems are **keyed to earlier sections** and review concepts learned earlier in the text that are needed in the next section or in the next chapter. These exercises show the **links between earlier topics and later material**.

A Look Ahead   These exercises occur at the end of some exercise sets. This section contains examples and problems similar to those found in a subsequent algebra course. "A Look Ahead" is presented as **a natural extension of the material** and contains an example followed by advanced exercises.

In addition to the approximately 5,500 exercises within chapters, exercises may also be found in the Vocabulary Checks, Chapter Reviews, Chapter Tests, as Cumulative Reviews.

Vocabulary Checks   Vocabulary checks, **new to this edition**, provide an opportunity for students to become more familiar with the use of mathematical terms as they strengthen verbal skills.

Chapter Review and Chapter Test   The end of each chapter contains a review of topics introduced in the chapter. The review problems are keyed to sections. The chapter test is not keyed to sections.

Cumulative Review   Each chapter after the first contains a **cumulative review of all chapters beginning with the first** up through the chapter at hand. Each problem contained in the cumulative review is actually an earlier worked example in the text that is referenced in the back of the book along with the answer. Students who need to see a complete worked-out solution, with explanation, can do so by turning to the appropriate example in the text.

KEY CONTENT FEATURES IN THE THIRD EDITION

Overview   This new edition retains many of the factors that have contributed to its success. Even so, **every section of the text was carefully re-examined**. Throughout the new edition you will find numerous new applications, examples, and many real-life applications and exercises. For example, look at Sections 1.9, 2.5, or 7.2. Some sections have internal re-organization to better clarify and enhance the presentation.

Increased Integration of Geometry Concepts   In addition to the traditional topics in beginning algebra courses, this text contains a strong emphasis on problem solving, and geometric concepts are integrated throughout. The geometry concepts presented are those most important to a students' understanding of algebra, and I have included **many applications and exercises** devoted to this topic. These are marked with the icon △. Also, geometric figures, a review of angles, lines, and special triangles, as well as a *new* review of volume and surface area are covered in the appendices. The inside front cover provides a quick reference of geometric formulas.

Review of Real Numbers   Chapter 1 has been streamlined and refreshed for greater efficiency and relevance. Former Sections 1.3 and 1.4 were merged to form new Section 1.4 for a smoother, more efficient flow. Chapter 1 now begins with Study Tips for Success in Mathematics (Section 1.1). **New applications** and real data enhance the chapter, especially in the reading graphs section.

Early and Intuitive Introduction to Graphing   As bar and line graphs are gradually introduced in Chapters 1 and 2, an emphasis is placed on the notion of paired data. This leads naturally to the concepts of ordered pair and the rectangular coordinate system introduced in Chapter 3. Chapter 3 is devoted to graphing and concepts of graphing linear equations such as slope and intercepts. **These concepts are reinforced throughout exercise sets** in subsequent chapters, helping prepare students for more work with equations in Chapter 7.

Chapter 3 has been updated, and the overall emphasis was to **better reinforce key concepts.** Reviewers have been pleased. Following user recommendations, a few of the changes are: Section 3.1 contains scattergrams of real data. Section 3.2 contains a new example and exercises on graphing and interpreting linear equations that model real data. Section 3.4 contains a new example and exercises interpreting slope as a rate of change. As usual, exercise sets progress gradually from easier to more difficult exercises.

Increased Attention to Problem Solving   Building on the strengths of the prior editions, a special emphasis and strong commitment is given to contemporary, accessible, and practical applications of algebra. **Real data** was drawn from a variety of sources including internet sources, magazines, newspapers, government publications, and reference books. **New Spotlight on Decision Making exercises and a new four-step problem solving process are incorporated throughout** to focus on helping to build students problem-solving skills.

Increased Opportunities for Using Technology   Optional explorations for a calculator or graphing calculator (or graphing utility such as Texas Instruments Interactive), are integrated appropriately **throughout the text** in Calculator Explorations features and in exercises marked with a calculator icon. The Martin-Gay companion website includes links to internet sites to allow opportunities for finding data and researching potential mathematically related careers branching from the chapter openers.

New Examples   Detailed step-by-step examples were added, deleted, replaced, or u dated as needed. Many of these reflect real life. **Examples are used in two ways**. Oft there are numbered, formal examples, and occasionally an example or applicatio used to introduce a topic or informally discuss the topic.

New Exercises   A significant amount of time was spent on the exercise sets. New exercises and examples **help address a wide range of student learning styles and abilities**. The text now includes the following types of exercises: spotlight on decision making exercises, mental math, computational exercises, real-life applications, writing exercises, multi-part exercises, review exercises, a look ahead exercises, optional calculator or graphing calculator exercises, data analysis from tables and graphs, vocabulary checks, and projects for individual or group assignment.

Enhanced Supplements Package   The new Third Edition is supported by a wealth of supplements designed for **added effectiveness and efficiency**. New items include the MathPro 4.0 Explorer tutorial software together with a unique video clip feature, a new computerized testing system TestGen-EQ, and an expanded and improved Martin-Gay companion website. Some highlights in print materials include the addition of teaching tips in the Annotated Instructor's Edition, and an expanded Instructor's Resource Manual with Tests including additional exercises and short group activities in a ready-to-use format. Please see the list of supplements for descriptions.

On-Line Options for Distance Learning

For maximum convenience, Prentice Hall offers on-line interactivity and delivery options for a variety of distance learning needs. Instructors may access or adopt these in conjunction with this text, *Beginning Algebra*.

Companion Website
Visit http://www.prenhall.com/martin-gay
The companion website includes basic distance learning access to provide links to the text's Real World Activities, career related sites referenced in the chapter opening pages and a selection of on-line self quizzes. E-mail is available. For quick reference, the inside front cover of this text also lists the companion website URL.

WebCT
WebCT includes distance learning access to content found in the Martin-Gay Companion Website plus more. WebCT provides tools to create, manage, and use on-line course materials. Save time and take advantage of items such as on-line help, communication tools, and access to instructor and student manuals. Your college may already have WebCT's software installed on their server or your may choose to download it should you decide on this option. Contact your local Prentice Hall sales representative for details or a preview.

For a *complete* computer-based internet course ...
Prentice Hall Interactive Math
Visit http://www.prenhall.com/interactive math

Prentice Hall Interactive math is an exciting, proven choice to help students succeed in math. Created for a computer-based course, it provides the effective teaching philosophy of K. Elayn Martin-Gay in an Internet-based course format. Interactive Math, Intermediate Algebra, takes advantage of state-of-the-art technology to provide highly flexible and user-friendly course management tools and an engaging, highly interactive student learning program that easily accommodates the variety of learning styles and broad spectrum of students presented by the typical intermediate algebra class. Personalized learning includes reading, writing, watching video clips, and exploring concepts through interactive questions and activities. Contact your local Prentice Hall sales representative for details.

SUPPLEMENTS FOR THE INSTRUCTOR

Printed Supplements

Annotated Instructor's Edition (ISBN 0-13-086764-0)

- Answers to exercises on the same text page or in Graphing Answer Section
- Graphing Answer section contains answers to exercises requiring graphical solutions, chapter projects, and Spotlight on Decision Making exercises
- Teaching Tips throughout the text placed at key points in the margin, found in places where students historically need extra help together with ideas on how to help students through these concepts, as well as placed appropriately to provide ideas for expanding upon a certain concept, other ways to present a concept, or ideas for classroom activities

Instructor's Solutions Manual (ISBN 0-13-087208-3)

- Detailed step-by-step solutions to even-numbered section exercises
- Solutions to every Spotlight on Decision Making exercise
- Solutions to every Calculator Exploration exercise
- Solutions to every Chapter Test and Chapter Review exercise
- Solution methods reflect those emphasized in the textbook

Instructor's Resource Manual with Tests (ISBN 0-13-087207-5)

- Notes to the Instructor that includes an introduction to Interactive Learning, Interpreting Graphs and Data, Alternative Assessment, Using Technology and Helping Students Succeed
- Eight Chapter Tests per chapter (5 free response, 3 multiple choice)
- Two Cumulative Review Tests (one free response, one multiple choice)
- Eight Final Exams (4 free response, 4 multiple choice)
- Twenty additional exercises per section for added test exercises or worksheets, if needed
- Group Activities (on average of two per chapter; providing short group activities in a convenient ready-to-use handout format)
- Answers to all items

Media Supplements

TestGen-EQ CD-Rom (Windows/Macintosh) (ISBN 0-13-088072-8)

- Algorithmically driven, text specific testing program
- Networkable for administering tests and capturing grades on-line
- Edit or add your own questions to create a nearly unlimited number of tests and worksheets
- Use the new "Function Plotter" to create graphs
- Tests can be easily exported to HTML so they can be posted to the Web for student practice

Computerized Tutorial Software Course Management Tools
MathPro 4.0 Explorer Network CD-Rom (ISBN 0-13-088073-6)

- Enables instructors to create either customized or algorithmically generated practice tests from any section of a chapter, or a test of random items

- Includes an e-mail function for network users, enabling instructors to send a message to a specific student or to an entire group
- Network based reports and summaries for a class or student and for cumulative or selected scores are available

Companion Website: http://www.prenhall.com/martin-gay

- Create a customized online syllabus with Syllabus Manager
- Assign Internet-based Real World Activities wherein students find and retrieve real data for use in guided problem solving.
- Assign quizzes or monitor student self quizzes by having students e-mail results, such as true/false reading quizzes or vocabulary check quizzes
- Destination links provide additional opportunities to explore related sites

SUPPLEMENTS FOR THE STUDENT

Printed Supplements

Student Solutions Manual (ISBN 0-13-087209-1)

- Detailed step-by-step solutions to odd-numbered section exercises
- Solutions to every (odd and even) Mental Math exercise
- Solutions to odd-numbered Calculator Exploration exercises
- Solutions to every (odd and even) exercise found in the Chapter Reviews and Chapter Tests
- Solution methods reflect those emphasized in the textbook
- Ask your bookstore about ordering

Study Guide (ISBN 0-13-087200-8)

- Additional step-by-step worked out examples and exercises
- Practice tests and final examination
- Includes Study Skills and Note-taking suggestions
- Solutions to all exercises, tests, and final examination
- Solution methods reflect those emphasized in the text
- Ask your bookstore about ordering

How to Study Mathematics

- Have your instructor contact the local Prentice Hall sales representative

Math on the Internet: A Student's Guide

- Have your instructor contact the local Prentice Hall sales representative

Prentice Hall/New York Times, Theme of the Times Newspaper Supplement

- Have your instructor contact the local Prentice Hall sales representative

Media Supplements

Computerized Tutorial Software
MathPro 4.0 Explorer Network CD-Rom (ISBN 0-13-088073-6)
MathPro 4.0 Explorer Student CD-Rom (ISBN 0-13-088079-5)

- Keyed to each section of the text for text-specific tutorial exercises and instruction

- Warm-up exercises and graded Practice Problems
- Video clips, providing a problem (similar to the one being attempted) being explained and worked out on the board
- Explorations, allowing explorations of concepts associated with objectives in more detail
- Algorithmically generated exercises, and includes bookmark, on-line help, glossary, and summary of scores for the exercises tried
- Have your instructor contact the local Prentice Hall sales representative—also available for home use

Videotape Series (ISBN 0-13-088074-4)

- Written and presented by textbook author K. Elayn Martin-Gay
- Keyed to each section of the text
- Presentation and step-by-step solutions to exercises from each section of the text. Exercises that are worked in the videos are marked with a video icon ✎.
- Key concepts are explained

Companion Website: www.prenhall.com/martin-gay

- Offers Warm-ups, Real World Activities, True/False Reading Quizzes, Chapter Quizzes, and Vocabulary Check Quizzes
- Option to e-mail results to your instructor
- Destination links provide additional opportunities to explore other related sites, such as those mentioned in this text's chapter opening pages

ACKNOWLEDGMENTS

First, as usual, I would like to thank my husband, Clayton, for his constant encouragement. I would also like to thank my children, Eric and Bryan, for continuing to eat my burnt meals. Thankfully, they have started to cook a little themselves.

I would also like to thank my extended family for their invaluable help and wonderful sense of humor. Their contributions are too numerous to list. They are Rod and Karen Pasch; Peter, Michael, Christopher, Matthew, and Jessica Callac; Stuart, Earline, Melissa, Mandy, and Bailey Martin; Mark, Sabrina, and Madison Martin; Leo, Barbara, Aaron, and Andrea Miller; and Jewett Gay.

A special thank you to all the users of the first and second editions of this text and for their suggestions for improvements that were incorporated into the third edition. I would also like to thank the following reviewers for their input and suggestions:

Beth Andrews, *Craven Community College*

Ken Araujo, *Florence-Darlington Technical College*

Daniel Cronin, *New Hampshire Technical Institute*

Karen Estes, *St. Petersburg Junior College*

Ann Flamm, *Reading Area Community College*

Teresa Hasenauer, *Indian River Community College*

Jennifer Laveglia, *Bellevue Community College*

Doug Mace, *Baker College*

James Matovina, *Community College of Southern Nevada*

Chris McNally, *Tallahassee Community College*
Gabrielle Michaelis, *Manatee Community College*
Linda Mudge, *Illinois Valley Community College*
Janet Peart, *Oakland Community College*
Bernard Piña, *New Mexico State University—Dona Ana Branch*
Susan Sabrio, *Texas A & M University—Kingsville*
Peggy Tibbs, *Arkansas Tech University*
Bettie Truitt, *Black Hawk College*
Mary Vachon, *San Joaquin Delta College*
Pat Velicky, *Florence-Darlington Technical College*
Bill Witte, *Montgomery College*
Kathleen Wolf, *Kent State University*

There were many people who helped me develop this text and I will attempt to thank some of them here. Cheryl Cantwell was invaluable for contributing to the overall accuracy of this text. Emily Keaton was also invaluable for her many suggestions and contributions during the development and writing of this text. I thank Tony Palermino for his many suggestions during the editing process and a special thank you to Ann Marie Jones, my project manager. She kept us all organized and on task. Ingrid Mount at Elm Street Publishing Services provided guidance throughout the production process. I very much appreciated the writers, formatters, and accuracy checkers of the supplements, including Cheryl Cantwell, Mark Serebransky, and Penny Arnold. I thank Terri Bittner, Carrie Green, Cindy Trimble, Jeff Rector, and Teri Lovelace at Laurel Technical Services for all their work on some of the supplements, and providing a thorough accuracy check. Lastly, a special thank you to my editor Karin Wagner for her support and assistance throughout the development and production of this text and to all the staff at Prentice Hall: Chris Hoag, Linda Behrens, Alan Fischer, Maureen Eide, Grace Hazeldine, Gus Vibal, Audra Walsh, Kate Marks, Eilish Main, Daniel Auld, Elise Schneider, Stephanie Szolusha, John Tweedale, Paul Corey, and Tim Bozik.

K. Elayn Martin-Gay

ABOUT THE AUTHOR

K. Elayn Martin-Gay has taught mathematics at the University of New Orleans for more than 20 years. Her numerous teaching awards include the local University Alumni Association's Award for Excellence in Teaching, and Outstanding Developmental Educator at University of New Orleans, presented by the Louisiana Association of Developmental Educators.

Prior to writing textbooks, K. Elayn Martin-Gay developed an acclaimed series of lecture videos to support developmental mathematics students in their quest for success. These highly successful videos originally served as the foundation material for her texts. Today the tapes specifically support each book in the Martin-Gay series.

Elayn is the author of over nine published textbooks as well as multimedia interactive mathematics, all specializing in developmental mathematics courses such as basic mathematics, prealgebra, beginning and intermediate algebra. She has provided author participation across the broadest range of materials: textbook, videos, tutorial software, and Interactive Math courseware. All the components are designed to work together. This offers an opportunity of various combinations for an integrated teaching and learning package offering great consistency and comfort for the student.

APPLICATIONS INDEX

Agriculture
  beef production, 236
  cattle on South Dakota farms, 533
  durum yield, 405
  farms in U.S., 321
  small farms, 479
  soybean yield in Nebraska, 533
Animals
  heartworm preventive for dog, 429
Architecture
  missing dimension in plans, 386
  wall length on blueprint, 377
Astronomy
  apparent magnitude, 17
  Comet Hale-Bopp, 255
  distance between Earth and the sun, 255
  fireball volume, 123
  galaxies in Virgo constellation, 268
  Julian day number, 364
  moons per planet, 113
  planetary alignment, 352
  reflected light from moon to Earth, 256
Automobiles
  average miles per gallon, 201
  car dealerships, 181
  compact car expenses, 201
  electric vehicles, 412
  engine displacement, 225
  gasoline price per gallon, 170
  operating costs, 212
  pickup truck expenses, 201
  rentals, 112, 114
  skidding and speed of car, 515
  speeds of, 381–82, 385, 394

Business
  air conditioning decision, 427
  annual revenues, 164–65
  apparel/accessory stores, 413
  average salaries, 15
  breaking even, 140
  building depreciation, 413
  business travelers in hotels, 157
  cable TV subscribers, 215
  cell phone subscribers, 75
  coaching company team, 383
  computer depreciation, 413
  computer desk manufacturing cost, 395
  computer desk production, 172
  computer value, 168–69
  consumer research, 68
  conveyor belts, 384
  daily sales prediction, 413
  decrease in employees, 158
  defective bulbs, 396

  delivery service costs, 493
  diameters of Easter egg openings, 149
  diameters of washers, 149
  Disney's top animated films, 62–63
  fax machine manufacturing cost, 338
  Home Depot stores, 549
  home office growth, 534
  hotel/motel room rate, 267
  hourly wage, 172
  long-distance phone calls, 34, 196-97
  machine screws, 15
  mail-order, 169
  manufacturing unit costs, 321
  morning edition newspapers, 458
  negative net income, 40, 55
  net income, 76, 549
  production line upgrading, 544
  profits on product, 413
  quantity pricing, 212
  revenue for Home Depot stores, 180–81
  sales prediction, 409–10
  straight-line depreciation, 241
  supermarket services/products, 131
  Target stores, 172
  time budgeting, 472
  travel time for manager, 385
  U.S. labor force, 67
  Wal-Mart sales, 550
  work rates, 380, 384, 385, 395

Chemistry
  acid solution, 139, 216, 468, 478
  alcohol solution, 467
  copper alloy, 139
  hydrochloric acid solution, 467
  percent for lab experiment, 135–36
  saline solution, 464–65, 467, 527
  weed killer, 378
Construction
  bathroom molding, 73
  building options, 322
  diagonal brace, 514
  ladder for gutter installation, 524
  landscaping, 318
  length/width of den, 314–15
  pitch of roof, 200, 201
  railroad track expansion joints, 516
  storage bin braces, 497
  underground pipelines, 514
  warehouse roof, 487
  water trough, 497
  wooden deck installation, 513

Distance. See Time and distance

Education
  ACT Assessment scores, 131
  algebra problem, 352
  algebra test scores, 149
  college attendance costs, 128
  community college enrollment, 131
  degrees attained, 214
  dental hygiene curriculum, 131
  elementary school teachers, 181
  extra credit project, 91
  floor space for students, 378
  high school graduates, 280
  institutions of higher learning, 15, 170
  LAN administration, 32
  public school teacher's salary, 72
  students per computer, 67
Electricity
  coulombs, 255
  kilowatt-hour charges, 61–62, 269
  photons of light emitted by bulb, 268
  wind-generated, 378
Electronics
  total resistance, 364
Entertainment
  broadcast television stations, 458
  CD sales, 280
  CD shipments, 452
  cost of recordable compact discs, 375–76
  "Grease" tickets, 461–62
  linear data description matched to
    equations/graphs, 431
  movie ticket sales, 215
  music cassette sales, 172
  orchestra section seating, 477
  top-grossing concert tours, 262
  total revenue for CDs, 338
Environment
  curbside plastics recycling, 16
  trail chart, 24

Finance and economics
  amounts in savings/checking accounts,
    466
  anniversary reception, 148
  bankruptcy cases, 269
  Barbie doll percent increase, 131
  best buys, 378, 396, 397
  birthday party tickets, 139
  bounced checks fees, 131
  budget development, 150
  car rental daily fee/mileage charge, 468
  cellular phone usage, 110
  charity income sources, 158
  child care center selection, 377
  clothes budget for women/girls, 130

  compound interest, 321
  consulting fees, 394
  consumer spending on media, 262–63
  cost of feeding family, 413
  credit card offers, 303
  credit card solicitation, 131
  crude oil production, 406
  estate distribution, 114
  financial analysis, 208
  fishery products, 447
  fuel oil consumption, 452
  home maintenance spending, 126, 129
  homeowner's insurance, 138
  income tax preparation, 128
  interest rate on investment, 139, 550
  interest rate on loan, 34
  investment amounts, 137, 139, 149, 158,
    396, 472
  investments at each rate, 478
  labor/material costs, 468
  merchandise trade balance, 75
  mineral production, 569
  minimum wage, 427
  money in account, 520
  mortgage payments, 170
  number of one/five dollar bills, 479
  percent increase, 130
  platinum price, 568
  plumbing estimate, 112
  pottery sale prices, 467
  price decrease/sale price, 130
  price of CDs/music cassette tapes, 465,
    466
  price of computer disks/notebooks, 465
  retirement party, 149
  sales commissions, 149, 158
  savings account balance, 16
  silver price, 568
  Social Security payments, 212
  spaghetti supper, 467
  stamp price mixture, 466
  stock investments, 37, 39, 75, 158, 466
  tin price trend, 549
  tourism budgets, 114
  train fares, 466
  value of coins, 466
  wedding budget, 146–47
  women earning bigger paychecks, 378
Food
  Cajun coffee, 139
  calories in cereal, 377
  calories in Eagle Brand Milk, 378
  cereal cost comparison, 376–77
  coffee bean blend, 467
  deli charges, 478

hot dogs/buns purchase, 352
iceberg lettuce, 131
nut mix, 139, 378, 467
nutrition labeling, 132
pepper hotness, 130
pizza size, 123
trail mix, 139

Geometry
  angles
    complementary, 47, 92, 357, 371, 467
    measures of, 111-12, 113, 114
    supplementary, 47, 92, 357, 371, 467
  area
    circle, 227
    largest square, 293
    parallelogram, 227
    plane figure, 73
    rectangle, 23, 34, 73, 227, 235, 239, 240,
      246, 344, 357, 364, 393, 502
    shaded regions, 279
    square, 235, 241, 246, 247, 344
    trapezoid, 34
    triangle, 23, 241, 246, 255, 394
  circle
    radius of, 319, 533
    rationalizing denominator, 502
    relationship between radius and cir-
      cumference, 413
    unknown part of, 23, 73
  cube
    side length, 493
    volume, 502
  cylinder
    dimensions of, 520-21
    radius of, 509
    volume of, 124, 227
  Golden Ratio, 478, 550
  golden rectangle, 115
  length of sides
    of isosceles triangle, 465
    of pentagon, 106
    of triangle, 106
  parallelogram
    area of, 124
    base and height of, 319
    height of, 261
    measure of angles of, 15, 112
  perimeter
    of plane figure, 24, 73
    of polygon, 496
    of quadrilateral, 234
    of rectangle, 34, 73, 85, 181, 357, 393,
      496
    of square, 351
    of trapezoid, 181, 351
    of triangle, 34, 85, 234, 394
  plane
    distances between points in, 511-12
  polygons
    diagonals of, 319
  quadrilateral
    lengths of sides of, 319
    measures of angles of, 93, 114
  rectangle
    dimensions of, 133-34, 138, 269, 319,
      321, 327, 328, 465, 467-68, 550
    length of, 118, 149, 215, 280, 318, 526
    width of, 121, 122, 261, 318
  rectangular box
    surface area of, 267
  rectangular solid
    length of, 261
    volume of, 124
  right triangle
    length of hypotenuse of, 510, 527
    length of leg of, 510
    length of long leg of, 327
    length of shorter leg of, 320
    length of unknown leg of, 526
    unknown sides of, 513
  similar triangles
    lengths of missing sides of, 384, 386
    lengths of sides of, 383
    missing lengths of, 395, 396, 527
    unknown lengths of, 385, 493
  sphere
    circumference of, 123
    rationalizing denominator for, 503
  square
    length of sides of, 318

side of, 320
square-based pyramid
  length of side of base of, 509
  volume of, 515
surface area
  of cube, 226
  of rectangular solid, 236
trapezoid
  area of, 122
  base and height of, 318
triangle
  base of, 327
  dimensions of, 138, 317
  height and base of, 550
  height of, 122, 304, 320
  length of base of, 328
  length of base and height of, 397
  length of shortest side of, 138
  length of side of, 139
  length of unknown side of, 526, 555
  lengths of legs of, 320
  lengths of sides of, 319, 320
  maximum lengths of sides of, 149
  measure of angle for, 93, 115
  measure of unknown angle for, 74
  sum of measures of angles of, 15
  unknown length of, 515
volume
  of cube, 226, 227, 241, 254

Law/Law enforcement
  forensics, 336
  mineral rights, 253

Medicine
  antibiotic dosage, 98
  antibiotic solution, 139
  blood cholesterol levels, 148
  blood pressure reading, 471
  body-mass index, 338
  child's medicine dosages, 338, 387-88
  cosmetic surgeons, 201
  farsighted/nearsighted Americans, 131
  fluid intake record, 93
  gum tissue pocket depth, 44
  hepatitis B vaccine, 131
  HMO enrollments, 149
  medical assistants, 178-79
  octuplets born, 74
  over-the-counter drug use, 130
  physician visits, 236
  smoking and pulse rate, 64-65
  toenail growth, 113
  treadmill treatment, 411

Miscellaneous
  age epigram, 387
  antifreeze solution, 139
  base/height of sail, 315-16
  beam lengths, 112, 235
  birth years, 47
  casino gaming area, 343
  coins in pay phone, 158
  compass height, 312
  computer desk, 92
  Diophantus' age at death, 386-87
  Eiffel Tower height, 157
  Empire State Building floor space, 114
  eye blinking, 113
  height of rocket, 312
  highest price paid for diamond, 255
  home phone usage, 128
  international phone calls, 113
  juveniles starting fires, 66
  kite height, 514
  latitude and longitude, 67
  length of time passed, 313-14, 319, 320,
    327, 328
  lobster traps, 24
  machine processing time, 394
  measurement conversions, 345
  molding length, 235
  negative card game score, 47
  phoneless households, 157
  pipe filling tank, 396
  pipes filling pond, 395
  popular names for cities/towns, 106
  pump filling tank, 385, 386
  sewer pipes, 200
  side of gold cube, 487
  side of TV picture tube, 533
  storage pool pipe, 385

string lengths, 92, 114
suitcase dimension, 304
teenagers using Internet, 65
telephone switchboard, 321
tourist destinations, 66
vacationers in U.S., 214
walking speeds, 462-63
water seal on deck, 158
wheel chair ramps, 200
width of swimming pool, 157
wire lengths, 114, 514
work rate, 386, 394

Physics
  fireball height and return to ground,
    561
  parabolic path of water, 562-63
  rocket height/time it strikes ground, 546
  velocity of fallen object, 514
  velocity of watermelon, 512
Politics
  governors per party, 114
  gubernatorial salaries, 112
  mayoral election, 92
  representatives per party, 114
  senate race, 93
  senators in each party, 109-10, 215
  votes per candidate, 113
Population
  annual growth rate, 520
  California, 268
  Minnesota, 172
  uninsured, 378
  United States, 255
  world, 255

Real estate
  agency choices, 197
  rental increases, 75

Science
  carbon isotope, 255
  cephalic index, 338
  density of object, 34
  fish in tank, 122
  flying fish, 123
  forensics, 429
  fresh water distribution, 24
  hydrogen atom, 255
  light travel, 123
  radius of Earth, 268
  robot dimensions/traveling rate, 122
  space probe, 113
  speed of sound conversion, 345
  volcanic eruption, 123
Sports
  advertising, 65-66
  America's Cup, 321
  average rushing yards, 378
  baseball attendance, 16
  basketball scores, 113
  bowling score, 149
  cycling, 158, 384, 385
  discus throwing records, 130
  diving length of time, 548, 549
  fastest lap speeds, 387
  football yards gained/lost, 16, 47, 76
  games won by Super Bowl winner, 170
  girls in high school athletics, 181
  golf scores, 40, 75, 158
  heights for NBA team, 149
  hockey points scored, 466
  jogging, 384, 478
  Ryder Cup championship, 113
  skateboard competition, 114
  Summer Olympics gold medals, 112
  WNBA highest scorers, 466
  women's javelin throw, 24
  world land speed record, 345

Technology
  computers in households, 158, 533
  Internet users, 200
  personal computers and household in-
    come, 294
Temperature
  daily lows, 40, 46
  equation relating Celsius/Fahrenheit,
    406
  equation relating Kelvin scale/Celsius
    scale, 406

Fahrenheit/Celsius interconversions,
  118-19, 122, 123, 124, 149, 157
freezing point/boiling point of water, 15
ground, 561
in scientific notation, 255
Winter Olympics, 76
Time and distance
  auto ferry trips, 122
  average speed on level road, 138
  baseball diamond, 486
  bicycle time/time walking, 468
  bicycling trip, 134-35
  boat speed in still water, 394, 396
  boat travel, 320, 384
  bullet train record speed, 123
  car overtaking bus, 138
  car's speeds, 392, 467
  charity 10K race, 157
  cruise control, 123
  current rate, 467
  distance across pond, 514
  distance between actors, 526
  distance formula, 364
  distance seen from height, 515
  distance to Disneyland, 138
  diving time, 568, 569
  driving and car phone range, 139
  driving distance apart, 516
  elevation, 40
  falling object, 33, 515
  fisherman rowing, 385
  free-fall time, 568
  glacier movement, 117, 123
  hang glider flight, 122
  height of dropped object, 230, 234, 269,
    299
  hiking times, 139, 140
  hyena pursuing giraffe, 387
  jet liner vertical change, 47
  lightning travel, 123
  light travel, 255, 256
  limousine travel, 122
  luggage hitting ground, 436
  planes from starting point, 138
  rate of SR-71, 122
  river current/speed of ship in still water,
    478
  river flow, 255
  rocket height, 236
  rowing, 140, 467
  ship courses, 472-73
  space plane travel, 122
  speeding before getting stopped, 140
  speed of dropped rock, 413
  speed of wind and speed of plane, 467
  speed per hour, 34
  speeds of eastbound/westbound trains,
    468
  surveying distance across lake, 511, 570
  swimming distance, 493
  time of free-fall, 533, 549
  trains catching up with each other, 158
  trains traveling in opposite directions,
    158
  travel time, 123
  vacation trip, 122
  variation in elevation, 44, 47, 74
  walking speeds, 570
Transportation
  airline departures, 236
  barge speed in still water, 385
  bridge lengths, 92
  "bullet" trains, 200
  plane rate in still air, 385
  radius of curvature, 486, 515
  road grade, 196, 200
  speeds of car and motorcycle, 386
  traffic fatalities in Florida, 236
  truck flatland/mountain rate, 384
  truck rental, 63-64

Weather
  Fujita Scale for tornadoes, 148
  rainfall intensity, 338
  sunrise time, 424-25, 428
  sunset time, 435
Weight
  elephant on Pluto, 377
  satellite on Mars, 377

Highlights of *Beginning Algebra,* *Third Edition*

Beginning Algebra, Third Edition has been written and designed to help you succeed in this course. Specific care has been taken to ensure you have the most up-to-date and relevant text features to provide you with a solid foundation in algebra, as many accessible real-world applications as possible, and to prepare you for future courses.

Get Motivated!

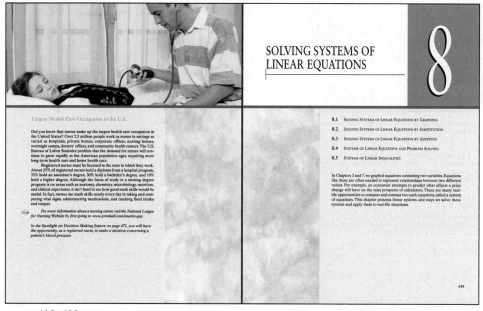

pages 438, 439

◀ **REAL-WORLD CHAP-TER OPENERS**

New Real-World Chapter Openers focus on how algebraic concepts relate to the world around you so that you see the relevance and practical applications of algebra in daily life.

They also provide links to the World Wide Web and reference a *Spotlight on Decision Making* feature within the chapter for further exploration.

SPOTLIGHT ON ▶ DECISION-MAKING

These unique new applications encourage you to develop your decision-making and problem solving abilities, and develop life skills, primarily using workplace or career-related situations.

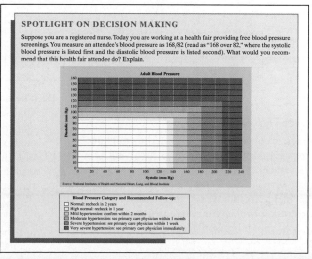

page 471

Become a Confident Problem-Solver!

A goal of this text is to help you develop problem-solving abilities.

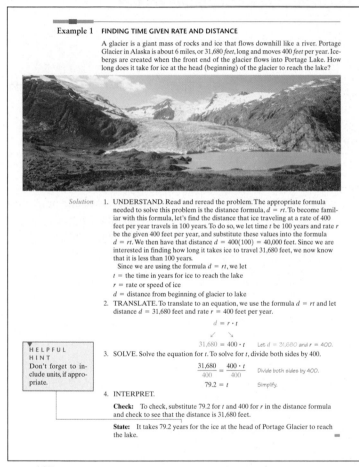

Example 1   **FINDING TIME GIVEN RATE AND DISTANCE**

A glacier is a giant mass of rocks and ice that flows downhill like a river. Portage Glacier in Alaska is about 6 miles, or 31,680 *feet*, long and moves 400 *feet* per year. Icebergs are created when the front end of the glacier flows into Portage Lake. How long does it take for ice at the head (beginning) of the glacier to reach the lake?

Solution   1. UNDERSTAND. Read and reread the problem. The appropriate formula needed to solve this problem is the distance formula, $d = rt$. To become familiar with this formula, let's find the distance that ice traveling at a rate of 400 feet per year travels in 100 years. To do so, we let time t be 100 years and rate r be the given 400 feet per year, and substitute these values into the formula $d = rt$. We then have that distance $d = 400(100) = 40,000$ feet. Since we are interested in finding how long it takes ice to travel 31,680 feet, we now know that it is less than 100 years.

    Since we are using the formula $d = rt$, we let

t = the time in years for ice to reach the lake
r = rate or speed of ice
d = distance from beginning of glacier to lake

2. TRANSLATE. To translate to an equation, we use the formula $d = rt$ and let distance $d = 31,680$ feet and rate $r = 400$ feet per year.

$$d = r \cdot t$$

$$31,680 = 400 \cdot t \quad \text{Let } d = 31,680 \text{ and } r = 400.$$

3. SOLVE. Solve the equation for t. To solve for t, divide both sides by 400.

$$\frac{31,680}{400} = \frac{400 \cdot t}{400} \quad \text{Divide both sides by 400.}$$

$$79.2 = t \quad \text{Simplify.}$$

HELPFUL HINT
Don't forget to include units, if appropriate.

4. INTERPRET.

   Check: To check, substitute 79.2 for t and 400 for r in the distance formula and check to see that the distance is 31,680 feet.

   State: It takes 79.2 years for the ice at the head of Portage Glacier to reach the lake.

page 117

◀ **GENERAL STRATEGY FOR PROBLEM-SOLVING**

Save time by having a plan. This text's organization can help you. Note the outlined problem-solving steps, *Understand, Translate, Solve,* and *Interpret.*

   Problem-solving is introduced early and emphasized and integrated throughout the book. The author provides patient explanations and illustrates how to apply the problem-solving procedure to the in-text examples.

GEOMETRY ▶

Geometric concepts are integrated throughout the text. Examples and exercises involving geometric concepts are identified with a triangle icon.

   The inside front cover of this text contains *Geometric Formulas* for convenient reference, as well as appendices on geometry.

15. The flag of Equatorial Guinea contains an isosceles triangle. (Recall that an isosceles triangle contains two angles with the same measure.) If the measure of the third angle of the triangle is 30° more than twice the measure of either of the other two angles, find the measure of each angle of the triangle. (*Hint:* Recall that the sum of the measures of the angles of a triangle is 180°.)

page 112

Get Involved!

Real-world applications in this textbook will help to reinforce your problem-solving skills, and show you how algebra is connected to a wide range of fields like consumer affairs, sports, and business. You will be asked to evaluate and interpret real data in graphs, tables, and in context to solve applications. See also the Index of Applications located on page xxi for a quick way to locate those in your areas of interest.

INTERESTING, RELEVANT, AND PRACTICAL REAL-WORLD APPLICATIONS

Accessible applications reinforce concepts needed for success in this course, and relate those concepts to everyday life.

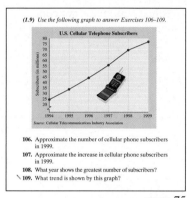

(1.9) *Use the following graph to answer Exercises 106–109.*

U.S. Cellular Telephone Subscribers

Source: Cellular Telecommunications Industry Association

106. Approximate the number of cellular phone subscribers in 1999.
107. Approximate the increase in cellular phone subscribers in 1999.
108. What year shows the greatest number of subscribers?
109. What trend is shown by this graph?

page 75

18. The table shows the number of regular-season NFL football games won by the winner of the Super Bowl for the years shown. (*Source:* National Football League)

Year	Regular-Season Games Won by Super Bowl Winner
1994	12
1995	13
1996	12
1997	13
1998	12
1999	14
2000	13

a. Write each paired data as an ordered pair of the form (year, games won).
b. Draw a grid such as the one in Example 2 and create a scatter diagram of the paired data.

page 170

▶ INTERESTING REAL DATA

Real-world applications include those based on real data.

29. Tony Hawk landed the first 900 in skateboard competition history during the X Games "Best Trick" competition in 1999. A 900 means that he rotated 900° in the air with his skateboard. If there are 360° in a single rotation, how many rotations were there?

page 114

For additional Chapter Projects, visit the Real World Activities Website by going to http://www.prenhall.com/martin-gay.

3 CHAPTER PROJECT

Financial Analysis

Investment analysts investigate a company's sales, net profit, debt, and assets to decide whether investing in it is a wise choice. One way to analyze this data is to graph it and look for trends over time. Another way is to find algebraically the rate at which the data change over time.

The table below gives the net profits in millions of dollars for the leading U.S. businesses in the pharmaceutical industry for the years 1997 and 1998. In this project, you will analyze the performances of these companies and, based on this information alone, make an investment recommendation. This project may be completed by working in groups or individually.

PHARMACEUTICAL INDUSTRY NET PROFITS (IN MILLIONS OF DOLLARS)

Company	1997	1998
Abbott Laboratories	2094.5	2333.2
American Home Products Corp.	2043.1	2474.3
Bristol Myers Squibb Co.	3205.0	3141.0
Eli Lilly & Co.	−385.1	2097.9
Johnson & Johnson	3303.0	3059.0
Merck & Co. Inc.	4614.1	5248.2
Pfizer Inc.	2213.0	3351.0
Pharmacia & Upjohn Inc.	323.0	691.0
Schering Plough Corp.	1444.0	1756.0
Warner Lambert Co.	869.5	1254.0

(Source: Disclosure Incorporated)

1. Scan the table. Did any of the companies have a loss during the years shown? If so, which company and when? What does this mean?
2. Write the data for each company as two ordered pairs of the form (year, net profit). Assuming that the trends in net profit are linear, use graph paper to graph the line represented by the ordered pairs for each company. Describe the trend shown by each graph.
3. Find the slope of the line for each company.
4. Which of the lines, if any, have positive slopes? What does that mean in this context? Which of the lines, if any, have negative slopes? What does that mean in this context?
5. Of these pharmaceutical companies, which one(s) would you recommend as an investment choice? Why?
6. Do you think it is wise to make a decision after looking at only 2 years of net profits? What other factors do you think should be taken into consideration when making an investment choice?
7. (Optional) Use financial magazines, company annual reports, or online investing information to find net profit information for two different years for two to four companies in the same industry. Analyze the net profits and make an investment recommendation.

◄ CHAPTER PROJECT

New *Chapter Projects* for individuals or groups provide chapter wrap-up and extend the chapter's concepts in a multi-part application.

In addition, references to optional alternative *Real World Activities* are given. This **internet option** invites you to find and retrieve real data for use in solving problems.

page 208

Be Confident!

Several features of this text can be helpful in building your confidence and mathematical competence. As you study, also notice the connections the author makes to relate new material to ideas that you may already know.

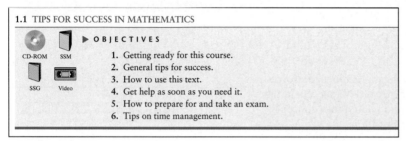

1.1 TIPS FOR SUCCESS IN MATHEMATICS

▶ **OBJECTIVES**
1. Getting ready for this course.
2. General tips for success.
3. How to use this text.
4. Get help as soon as you need it.
5. How to prepare for and take an exam.
6. Tips on time management.

CD-ROM SSM SSG Video

page 4

◀ **TIPS FOR SUCCESS**
New coverage of study skills in Section 1.1 reinforces this important component to success in this course.

MENTAL MATH ▶

Mental Math warm-up exercises reinforce concepts found in the accompanying section and can increase your confidence before beginning an exercise set.

MENTAL MATH

Answer the following true or false.

1. The graph of $x = 2$ is a horizontal line.
2. All lines have an x-intercept *and* a y-intercept.
3. The graph of $y = 4x$ contains the point $(0, 0)$.
4. The graph of $x + y = 5$ has an x-intercept of $(5, 0)$ and a y-intercept of $(0, 5)$.

page 187

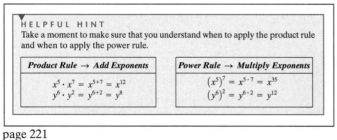

HELPFUL HINT
Take a moment to make sure that you understand when to apply the product rule and when to apply the power rule.

Product Rule → Add Exponents	*Power Rule → Multiply Exponents*
$x^5 \cdot x^7 = x^{5+7} = x^{12}$ $y^6 \cdot y^2 = y^{6+2} = y^8$	$(x^5)^7 = x^{5 \cdot 7} = x^{35}$ $(y^6)^2 = y^{6 \cdot 2} = y^{12}$

page 221

◀ **HELPFUL HINTS**
Found throughout the text, these contain practical advice on applying mathematical concepts. They are strategically placed where you are most likely to need immediate reinforcement.

VOCABULARY CHECKS ▶
New *Vocabulary Checks* allow you to write your answers to questions about chapter content and strengthen verbal skills.

CHAPTER 4 VOCABULARY CHECK

Fill in each blank with one of the words or phrases listed below.

term   coefficient   monomial   binomial   trinomial
polynomials   degree of a term   degree of a polynomial   FOIL

1. A _____ is a number or the product of numbers and variables raised to powers.
2. The _____ method may be used when multiplying two binomials.
3. A polynomial with exactly 3 terms is called a _____.
4. The _____ is the greatest degree of any term of the polynomial.
5. A polynomial with exactly 2 terms is called a _____.
6. The _____ of a term is its numerical factor.
7. The _____ is the sum of the exponents on the variables in the term.
8. A polynomial with exactly 1 term is called a _____.
9. Monomials, binomials, and trinomials are all examples of _____.

page 263

Visualize It!

The third edition increases emphasis on visualization. Graphing is introduced early and intuitively. Knowing how to read and use graphs is a valuable skill in the workplace as well as in this and other courses.

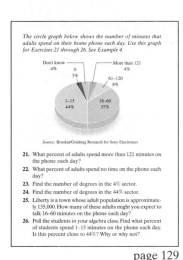

The circle graph below shows the number of minutes that adults spend on their home phone each day. Use this graph for Exercises 21 through 26. See Example 4.

Don't know 4%
0 3%
More than 121 4%
61–120 8%
1–15 44%
16–60 37%

Source: Bruskin/Goldring Research for Sony Electronics

21. What percent of adults spend more than 121 minutes on the phone each day?
22. What percent of adults spend no time on the phone each day?
23. Find the number of degrees in the 4% sector.
24. Find the number of degrees in the 44% sector.
25. Liberty is a town whose adult population is approximately 135,000. How many of these adults might you expect to talk 16–60 minutes on the phone each day?
26. Poll the students in your algebra class. Find what percent of students spend 1–15 minutes on the phone each day. Is this percent close to 44%? Why or why not?

page 129

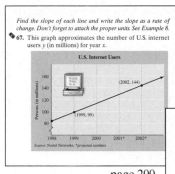

Find the slope of each line and write the slope as a rate of change. Don't forget to attach the proper units. See Example 8.

67. This graph approximates the number of U.S. internet users y (in millions) for year x.

U.S. Internet Users

(2002, 144)
(1999, 99)

Source: Nortel Networks, *projected numbers

page 200

◀ **VISUALIZATION OF TOPICS**

Many illustrations, models, photographs, tables, charts, and graphs provide visual reinforcement of concepts and opportunities for data interpretation.

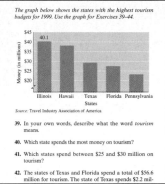

The graph below shows the states with the highest tourism budgets for 1999. Use the graph for Exercises 39–44.

40.1

Illinois   Hawaii   Texas   Florida   Pennsylvania
States

Source: Travel Industry Association of America

39. In your own words, describe what the word *tourism* means.
40. Which state spends the most money on tourism?
41. Which states spend between $25 and $30 million on tourism?
42. The states of Texas and Florida spend a total of $56.6 million for tourism. The state of Texas spends $2.2 mil-

page 114

SCIENTIFIC AND GRAPHING CALCULATOR EXPLORATIONS

Enhanced *Explorations* contain examples and exercises to **reinforce** concepts, help **interpret** graphs, or **motivate** discovery learning. Scientific and graphing utility exercises can also be found in exercise sets.

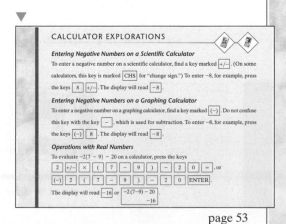

CALCULATOR EXPLORATIONS

Entering Negative Numbers on a Scientific Calculator

To enter a negative number on a scientific calculator, find a key marked $+/-$. (On some calculators, this key is marked CHS for "change sign.") To enter -8, for example, press the keys 8 $+/-$. The display will read -8.

Entering Negative Numbers on a Graphing Calculator

To enter a negative number on a graphing calculator, find a key marked $(-)$. Do not confuse this key with the key $-$, which is used for subtraction. To enter -8, for example, press the keys $(-)$ 8. The display will read -8.

Operations with Real Numbers

To evaluate $-2(7-9) - 20$ on a calculator, press the keys

2 $+/-$ $\times$ $($ 7 $-$ 9 $)$ $-$ 2 0 $=$, or

$(-)$ 2 $($ 7 $-$ 9 $)$ $-$ 2 0 ENTER.

The display will read -16 or $-2(7-9) - 20$.
-16

page 53

GRAPHING CALCULATOR EXPLORATIONS

It is possible to use a grapher and sketch the graph of more than one equation on the same set of axes. This feature can be used to confirm our findings from Section 3.2 when we learned that the graph of an equation written in the form $y = mx + b$ has a y-intercept of b. For example, graph the equations $y = \frac{2}{5}x$, $y = \frac{2}{5}x + 7$, and $y = \frac{2}{5}x - 4$ on the same set of axes. To do so, press the $Y =$ key and enter the equations on the first three lines.

$$Y_1 = \left(\frac{2}{5}\right)x$$

$$Y_2 = \left(\frac{2}{5}\right)x + 7$$

$$Y_3 = \left(\frac{2}{5}\right)x - 4$$

The screen should look like:

$y_2 = \frac{2}{5}x + 7$ $y_1 = \frac{2}{5}x$
$y_3 = \frac{2}{5}x - 4$

Notice that all three graphs appear to have the same positive slope. The graph of $y = \frac{2}{5}x + 7$ is the graph of $y = \frac{2}{5}x$ moved 7 units upward with a y-intercept of 7. Also, the graph of $y = \frac{2}{5}x - 4$ is the graph of $y = \frac{2}{5}x$ moved 4 units downward with a y-intercept of -4.

Graph the equations on the same set of axes. Describe the similarities and differences in their graphs.

1. $y = 3.8x$, $y = 3.8x - 3$, $y = 3.8x + 9$
2. $y = -4.9x$, $y = -4.9x + 1$, $y = -4.9x + 8$
3. $y = \frac{1}{4}x$; $y = \frac{1}{4}x + 5$, $y = \frac{1}{4}x - 8$
4. $y = -\frac{3}{4}x$, $y = -\frac{3}{4}x - 5$, $y = -\frac{3}{4}x + 6$

page 198

Discover the Best Supplemental Resource for You!
Integrated Learning Program

All of the components of the Martin-Gay supplemental resources fit together to help you learn and understand algebra. Use these student resources based on your personal learning style to enhance what you learn from your instructor and textbook.

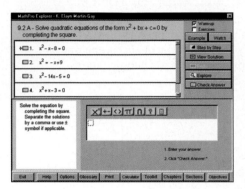

◀ MATHPRO EXPLORER 4.0

This **interactive** tutorial software is developed around the content and concepts of *Beginning Algebra*. It provides:

- virtually unlimited practice problems with immediate feedback.
- video clips
- step-by-step solutions
- exploratory activities
- on-line help
- summary of progress

Available on CD-ROM

LECTURE VIDEO SERIES BY ▶ K. ELAYN MARTIN-GAY

Hosted by the award-winning teacher and author of *Beginning Algebra,* these videos cover each objective in every chapter section as a supplementary review. Problems are taken directly from the text. Look for the video icon next to examples and exercises in this text.

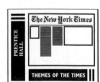

WWW.PRENHALL.COM/MARTIN-GAY COMPANION WEBSITE!

The website offers warm-ups, real world activities, reading quizzes, vocabulary quizzes, and chapter quizzes, that you can send e-mail to your instructor, and links to explore related sites such as those noted in this text's chapter-opening pages.

ALSO AVAILABLE

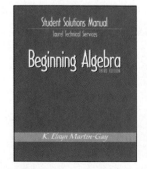

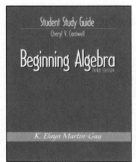

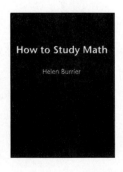

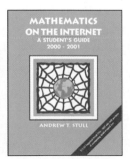

The New York Times/ Themes of the Times
Newspaper-format supplement

BEGINNING ALGEBRA

A Demanded Technical Position

Computers are everywhere: banks, real estate offices, churches, post offices, factories, corporations, grocery stores, libraries, homes, and schools. With so many computers in use in today's society, it has become necessary to network, or connect together, computers to share resources such as printers or file storage. A popular computer network for linking computers in the same building is the Local Area Network (LAN). According to the Unicom Institute of Technology, LAN administrators are among the technical workers most in demand in today's job market.

LAN administrators are generally responsible for designing and configuring the network, adding printers or other devices to the network as necessary, installing networked application software, maintaining system files, ensuring network security, and troubleshooting problems on the network. LAN administrators enjoy solving problems, communicate effectively, and read and understand technical material.

 For more information about the field of computer networking, or a number of other information technology (IT) professions, visit the Planet IT Website by going to http://www.prenhall.com/martin-gay.

In the Spotlight on Decision Making feature on page 32, you will have the opportunity to make a decision concerning the feasibility of a proposed computer network as a LAN administrator.

REVIEW OF REAL NUMBERS

1.1 TIPS FOR SUCCESS IN MATHEMATICS

1.2 SYMBOLS AND SETS OF NUMBERS

1.3 FRACTIONS

1.4 INTRODUCTION TO VARIABLE EXPRESSIONS AND EQUATIONS

1.5 ADDING REAL NUMBERS

1.6 SUBTRACTING REAL NUMBERS

1.7 MULTIPLYING AND DIVIDING REAL NUMBERS

1.8 PROPERTIES OF REAL NUMBERS

1.9 READING GRAPHS

The power of mathematics is its flexibility. We apply numbers to almost every aspect of our lives, from an ordinary trip to the grocery store to a rocket launched into space. The power of algebra is its generality. Using letters to represent numbers, we tie together the trip to the grocery store and the launched rocket.

In this chapter we review the basic symbols and words—the language— of arithmetic and introduce using variables in place of numbers. This is our starting place in the study of algebra.

1.1 TIPS FOR SUCCESS IN MATHEMATICS

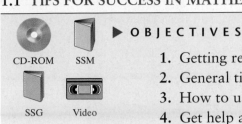

CD-ROM   SSM

SSG   Video

▶ **O B J E C T I V E S**

1. Getting ready for this course.
2. General tips for success.
3. How to use this text.
4. Get help as soon as you need it.
5. How to prepare for and take an exam.
6. Tips on time management.

Before reading this section, remember that your instructor is your best source for information. Please see your instructor for any additional help or information.

1   Now that you have decided to take this course, remember that a positive attitude will make all the difference in the world. Your belief that you can succeed is just as important as your commitment to this course. Make sure that you are ready for this course by having the time and positive attitude that it takes to succeed.

Next, make sure that you have scheduled your math course at a time that will give you the best chance for success. For example, if you are also working, you may want to check with your employer to make sure that your work hours will not conflict with your course schedule.

Now you are ready for your first class period. Double-check your schedule and allow yourself extra time to arrive in case of traffic or in case you have trouble locating your classroom. Make sure that you bring at least your textbook, paper, and a writing instrument with you. Are you required to have a lab manual, graph paper, calculator, or some other supply besides this text? If so, bring this material with you also.

2   Below are some general tips that will increase your chance for success in a mathematics class. Many of these tips will also help you in other courses you may be registered for.

Exchange names and phone numbers with at least one other person in class. This contact person can be a great help in case you miss the class assignment or want to discuss math concepts or exercises that you find difficult.

Choose to attend all class periods. If possible, sit near the front of the classroom. This way, you will see and hear the presentation better. It may also be easier for you to participate in classroom activities.

Do your homework. You've probably heard the phrase "practice makes perfect" in relation to music and sports. It also applies to mathematics. You will find that the more time you spend solving mathematics problems, the easier the process becomes. Be sure to block enough time to complete your assignments before the next class period.

Check your work. Review the steps you made while working a problem. Learn to check your answers in the original problems. You may also compare your answers to the answers to selected exercises listed in the back of the book. If you have made a mistake, figure out what went wrong. Then correct your mistake. If you can't find your mistake, don't erase your work or throw it away. Bring your work to your instructor, a tutor in a math lab, or a classmate. Someone can help you find where you had trouble only if they have your work to look at.

Learn from your mistakes. Everyone, even your instructor, makes mistakes. (That definitely includes me—Elayn Martin-Gay. You usually don't see my mistakes because many other people doublecheck my work in this text. If I make a mistake on a videotape, it is edited out so that you are not confused by it.) Use your mistakes to learn and to become a better math student. The key is finding and understanding your mistakes. Was your mistake a careless mistake or did you make it because you can't read your own "math" writing? If so, try to work more slowly or write more neatly and make a conscious effort to carefully check your work. Did you make a mistake because you don't understand a concept? Take the time to review the concept or ask questions to better understand the concept.

Know how to get help if you need it. It's OK to ask for help. In fact, it's a good idea to ask for help whenever there is something that you don't understand. Make sure you know when your instructor has office hours and how to find his or her office. Find out if math tutoring services are available on your campus. Check out the hours, location, and requirements of the tutoring service. Know whether videotapes or software are available and how to access those resources.

Organize your class materials, including homework assignments, graded quizzes and tests, and notes from your class or lab. All of these items will make valuable references throughout your course and as you study for upcoming tests and your final exam. Make sure that you can locate any of these materials when you need them.

Read your textbook before class. Reading a mathematics textbook is unlike entertainment reading such as reading a newspaper. Your pace will be much slower. It is helpful to have a pencil and paper with you when you read. Try to work out examples on your own as you encounter them in your text. You may also write down any questions that you want to ask in class. I know that when you read a mathematics textbook, sometimes some of the information in a section will still be unclear. But once you hear a lecture or watch a video on that section, you will understand it much more easily than if you had not read your text.

Don't be afraid to ask questions. From experience, I can tell you that you are not the only person in class with questions. Other students are normally grateful that someone has spoken up.

Hand in assignments on time. This way you can be sure that you will not lose points needlessly for being late. Show every step of a solution and be neat and organized. Also be sure that you understand which problems are assigned for homework. You can always doublecheck this assignment with another student in your class.

3 There are many helpful resources that are available to you in this text. It is important that you become familiar with and use these resources. This should increase your chances for success in this course.
For example:

- If you need help in a particular section, check at the beginning of the section to see what videotapes or software are available. These resources are usually available to you in a tutorial lab, resource center, or library.
- Many of the exercises in this text are referenced by an example(s). Use this referencing in case you have trouble completing an assignment from the exercise set.

- Make sure that you understand the meaning of the icons that are beside many exercises. The video icon ✎ tells you that the corresponding exercise may be viewed on the videotape that corresponds with that section. The pencil icon ✎ tells you that this exercise is a writing exercise in which you should answer in complete sentences. The calculator icons are placed by exercises that can be worked more efficiently with the use of a scientific calculator ⊞ or a graphing calculator. ⊞

- There are many opportunities at the end of each chapter to help you understand the concepts of the chapter.

  Vocabulary Checks provide a vocabulary self-check to make sure that you know the vocabulary in that chapter.

  Highlights contain chapter summaries with examples.

  Chapter Review contains additional exercises that are keyed to sections of the chapter.

  Chapter Test is a sample test to help you prepare for an exam.

  Cumulative Review is a review consisting of material from the beginning of the book to the end of the particular chapter.

4 If you have trouble completing assignments or understanding the mathematics, get help as soon as you need it! This tip is presented as an objective on its own because it is *so* important. In mathematics, usually the material presented in one section builds on your understanding of the previous section. What does this mean? It means that if you don't understand the concepts covered during a class period, there is a good chance that you will not understand the concepts covered during the next class period. If this happens to you, get help as soon as you can.

Where can you get help? Many suggestions have been made in this section on where to get help and now it is up to you to do it. Try your instructor, a tutor center, or math lab, or you may want to form a study group with fellow classmates. If you do decide to see your instructor or go to a tutor center, make sure that you have a neat notebook and be ready with your questions.

5 Make sure that you allow yourself plenty of time to prepare for a test. If you think that you are a little math anxious, it may be that you are not preparing for a test in a way that will ensure success. The way that you prepare for a test in mathematics is important.

To prepare for a test,

1. Review your previous homework assignments.
2. Review any notes from class and section level quizzes you may have taken. (If this is a final exam, review chapter tests you have taken also.)
3. Review concepts and definitions by reading the Highlights at the end of each chapter.
4. Practice working exercises by completing the Chapter Review found at the end of each chapter. (If this is a final exam, work a Cumulative Review. There is one found at the end of each chapter (except Chapter 1). Choose the review found at the end of the latest chapter that you have covered in your course.)

Don't stop here!

5. It is important that you place yourself in conditions similar to test conditions to see how you will perform. In other words, once you feel that you know the material, get out a few blank sheets of paper and take a sample test. There is a Chapter Test available at the end of each chapter, or you can work selected

problems from the Chapter Review, or your instructor may provide you with a review sheet. During this sample test, do not use your notes or your textbook. Then check your sample test. If you are not satisfied with the results, study the areas that you are weak in and try again.

6. On the day of the test, allow yourself plenty of time to arrive where you will be taking your exam.

When taking your test,

1. Read the directions on the test carefully.
2. Read each problem carefully as you take your test. Make sure that you answer the question asked.
3. Watch your time and pace yourself so that you may attempt each problem on your test.
4. If you have time, check your work and answers.
5. Do not turn your test in early. If you have extra time, spend it double-checking your work.

6 As a college student, you know the demands that classes, homework, work, and family place on your time. Some days you probably wonder how you'll ever get everything done. One key to managing your time is developing a schedule. Here are some hints for making a schedule:

1. Make a list of all of your weekly commitments for the term. Include classes, work, regular meetings, extracurricular activities, etc. You may also find it helpful to list such things as doing laundry, regular workouts, grocery shopping, etc.
2. Next, estimate the time needed for each item on the list. Also make a note of how often you will need to do each item. Don't forget to include time estimates for reading, studying, and homework you do outside of your classes. You may want to ask your instructor for help estimating the time needed for this item.
3. In the exercise set below, you are asked to block out a typical week on the schedule grid given. Start with items with fixed time slots, like classes and work.
4. Next, include the items on your list with flexible time slots. Think carefully about how best to schedule some items such as study time.
5. Don't fill up every time slot on the schedule. Remember that you need to allow time for eating, sleeping, and relaxing! You should also allow a little extra time in case things take longer than planned.
6. If you find that your weekly schedule is too full for you to handle, you may need to make some changes in your workload, class load, or in other areas of your life. You may want to talk to your advisor, manager or supervisor at work, or someone in your college's academic counseling center for help with such decisions.

Exercise Set 1.1

1. What is your instructor's name?
2. What are your instructor's office location and office hours?
3. What is the best way to contact your instructor?
4. What does this icon ✎ mean?
5. What does this icon ◆ mean?

6. Do you have the name and contact information of at least one other student in class?
7. Will your instructor allow you to use a calculator in this class?
8. Are videotapes and/or tutorial software available to you?

9. Is there a tutoring service available? If so, what are its hours?

10. Have you attempted this course before? If so write down ways that you may improve your chances of success during this attempt.

11. List some steps that you may take in case you begin having trouble understanding the material or completing an assignment.

12. Read or reread objective ⁶ and fill out the schedule grid below.

	Monday	Tuesday	Wednesday	Thursday	Friday	Saturday	Sunday
7:00 A.M.							
8:00 A.M.							
9:00 A.M.							
10:00 A.M.							
11:00 A.M.							
12:00 P.M.							
1:00 P.M.							
2:00 P.M.							
3:00 P.M.							
4:00 P.M.							
5:00 P.M.							
6:00 P.M.							
7:00 P.M.							
8:00 P.M.							
9:00 P.M.							

1.2   SYMBOLS AND SETS OF NUMBERS

CD-ROM   SSM

SSG   Video

▶ **OBJECTIVES**

1. Use the number line to order numbers.
2. Translate sentences into mathematical statements.
3. Identify natural numbers, whole numbers, integers, rational numbers, irrational numbers, and real numbers.
4. Find the absolute value of a real number.

1   We begin with a review of the set of natural numbers and the set of whole numbers and how we use symbols to compare these numbers. A **set** is a collection of objects, each of which is called a **member** or **element** of the set. A pair of brace symbols { } encloses the list of elements and is translated as "the set of" or "the set containing."

NATURAL NUMBERS

The set of **natural numbers** is $\{1, 2, 3, 4, 5, 6, \ldots\}$.

WHOLE NUMBERS

The set of **whole numbers** is $\{0, 1, 2, 3, 4, \dots\}$.

The three dots (an ellipsis) at the end of the list of elements of a set means that the list continues in the same manner indefinitely.

These numbers can be pictured on a **number line**. We will use the number line often to help us visualize distance and relationships between numbers. Visualizing mathematical concepts is an important skill and tool, and later we will develop and explore other visualizing tools.

To draw a number line, first draw a line. Choose a point on the line and label it 0. To the right of 0, label any other point 1. Being careful to use the same distance as from 0 to 1, mark off equally spaced distances. Label these points 2, 3, 4, 5, and so on. Since the whole numbers continue indefinitely, it is not possible to show every whole number on the number line. The arrow at the right end of the line indicates that the pattern continues indefinitely.

Picturing whole numbers on a number line helps us to see the order of the numbers. Symbols can be used to describe concisely in writing the order that we see.

The **equal symbol** $=$ means "is equal to."

The symbol $\neq$ means "is not equal to."

These symbols may be used to form a **mathematical statement**. The statement might be true or it might be false. The two statements below are both true.

$2 = 2$ states that "two is equal to two"

$2 \neq 6$ states that "two is not equal to six"

$2 > 0$ or $0 < 2$

If two numbers are not equal, then one number is larger than the other. The symbol $>$ means "is greater than." The symbol $<$ means "is less than." For example,

$2 > 0$ states that "two is greater than zero"

$3 < 5$ states that "three is less than five"

$3 < 5$

On the number line, we see that a number **to the right of** another number is **larger**. Similarly, a number **to the left of** another number is smaller. For example, 3 is to the left of 5 on the number line, which means that 3 is less than 5, or $3 < 5$. Similarly, 2 is to the right of 0 on the number line, which means 2 is greater than 0, or $2 > 0$. Since 0 is to the left of 2, we can also say that 0 is less than 2, or $0 < 2$.

The symbols $\neq$, $<$, and $>$ are called **inequality symbols**.

> **HELPFUL HINT**
> Notice that $2 > 0$ has exactly the same meaning as $0 < 2$. Switching the order of the numbers and reversing the "direction of the inequality symbol" does not change the meaning of the statement.
>
> $5 > 3$ has the same meaning as $3 < 5$.
>
> Also notice that, when the statement is true, the inequality arrow points to the smaller number.

Example 1 Insert $<$, $>$, or $=$ in the space between the paired numbers to make each statement true.

        a. 2   3   **b.** 7   4   **c.** 72   27

Solution **a.** $2 < 3$ since 2 is to the left of 3 on the number line.
        b. $7 > 4$ since 7 is to the right of 4 on the number line.
        c. $72 > 27$ since 72 is to the right of 27 on the number line.

Two other symbols are used to compare numbers. The symbol $\leq$ means "is less than or equal to." The symbol $\geq$ means "is greater than or equal to." For example,

$$7 \leq 10 \text{ states that "seven is less than or equal to ten"}$$

This statement is true since $7 < 10$ is true. If either $7 < 10$ or $7 = 10$ is true, then $7 \leq 10$ is true.

$$3 \geq 3 \text{ states that "three is greater than or equal to three"}$$

This statement is true since $3 = 3$ is true. If either $3 > 3$ or $3 = 3$ is true, then $3 \geq 3$ is true.

The statement $6 \geq 10$ is false since neither $6 > 10$ nor $6 = 10$ is true.

The symbols $\leq$ and $\geq$ are also called **inequality symbols**.

Example 2 Tell whether each statement is true or false.

        a. $8 \geq 8$ **b.** $8 \leq 8$ **c.** $23 \leq 0$ **d.** $23 \geq 0$

Solution **a.** True, since $8 = 8$ is true.   **b.** True, since $8 = 8$ is true.
        c. False, since neither $23 < 0$ nor $23 = 0$ is true.   **d.** True, since $23 > 0$ is true.

2   Now, let's use the symbols discussed above to translate sentences into mathematical statements.

Example 3 Translate each sentence into a mathematical statement.

        a. Nine is less than or equal to eleven.
        b. Eight is greater than one.
        c. Three is not equal to four.

Solution **a.**

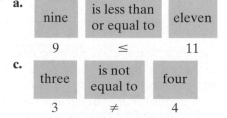

        b.

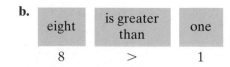

3   Whole numbers are not sufficient to describe many situations in the real world. For example, quantities smaller than zero must sometimes be represented, such as temperatures less than 0 degrees.

We can picture numbers less than zero on the number line as follows:

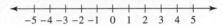

$$-5 \ -4 \ -3 \ -2 \ -1 \quad 0 \quad 1 \quad 2 \quad 3 \quad 4 \quad 5$$

Numbers less than 0 are to the left of 0 and are labeled $-1, -2, -3$, and so on. A $-$ sign, such as the one in -1, tells us that the number is to the left of 0 on the number line. In words, -1 is read "negative one." A $+$ sign or no sign tells us that a number lies to the right of 0 on the number line. For example, 3 and $+3$ both mean positive three.

The numbers we have pictured are called the set of **integers**. Integers to the left of 0 are called **negative integers**; integers to the right of 0 are called **positive integers**. The integer 0 is neither positive nor negative.

INTEGERS

The set of **integers** is $\{\ldots, -3, -2, -1, 0, 1, 2, 3, \ldots\}$.

Notice the ellipses (three dots) to the left and to the right of the list for the integers. This indicates that the positive integers and the negative integers continue indefinitely.

Example 4   Use an integer to express the number in the following. "Pole of Inaccessibility, Antarctica, is the coldest location in the world, with an average annual temperature of 72 degrees below zero." (*Source: The Guinness Book of Records*)

Solution   The integer -72 represents 72 degrees below zero.

A problem with integers in real-life settings arises when quantities are smaller than some integer but greater than the next smallest integer. On the number line, these quantities may be visualized by points between integers. Some of these quantities between integers can be represented as a quotient of integers. For example,

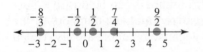

The point on the number line halfway between 0 and 1 can be represented by $\frac{1}{2}$, a quotient of integers.

The point on the number line halfway between 0 and -1 can be represented by $-\frac{1}{2}$. Other quotients of integers and their graphs are shown.

The set numbers, each of which can be represented as a quotient of integers, is called the set of **rational numbers**. Notice that every integer is also a rational number since each integer can be expressed as a quotient of integers. For example, the integer 5 is also a rational number since $5 = \frac{5}{1}$.

RATIONAL NUMBERS

The set of **rational numbers** is the set of all numbers that can be expressed as a quotient of integers.

The number line also contains points that cannot be expressed as quotients of integers. These numbers are called **irrational numbers** because they cannot be represented by rational numbers. For example, $\sqrt{2}$ and π are irrational numbers.

IRRATIONAL NUMBERS

The set of **irrational numbers** is the set of all numbers that correspond to points on the number line but that are not rational numbers. That is, an irrational number is a number that cannot be expressed as a quotient of integers.

Rational numbers and irrational numbers can be written as decimal numbers. The decimal equivalent of a rational number will either terminate or repeat in a pattern. For example, upon dividing we find that

$$\frac{3}{4} = 0.75 \text{ (decimal number terminates or ends) and}$$

$$\frac{2}{3} = 0.66666\ldots \text{ (decimal number repeats in a pattern)}$$

The decimal representation of an irrational number will neither terminate nor repeat. (For further review of decimals, see the appendix.)

The set of numbers, each of which corresponds to a point on the number line, is called the set of **real numbers**. One and only one point on the number line corresponds to each real number.

REAL NUMBERS

The set of **real numbers** is the set of all numbers each of which corresponds to a point on the number line.

On the following number line, we see that real numbers can be positive, negative, or 0. Numbers to the left of 0 are called **negative numbers**; numbers to the right of 0 are called **positive numbers**. Positive and negative numbers are also called **signed numbers**.

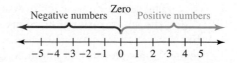

Several different sets of numbers have been discussed in this section. The following diagram shows the relationships among these sets of real numbers.

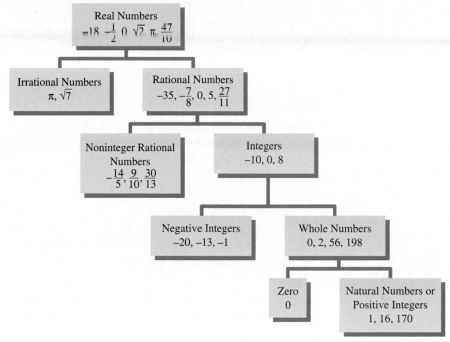

Common Sets of Numbers

Example 5 Given the set $\left\{-2, 0, \frac{1}{4}, 112, -3, 11, \sqrt{2}\right\}$, list the numbers in this set that belong to the set of:

  a. Natural numbers   **b.** Whole numbers   **c.** Integers
  d. Rational numbers   **e.** Irrational numbers   **f.** Real numbers

Solution   **a.** The natural numbers are 11 and 112.
          b. The whole numbers are 0, 11, and 112.
          c. The integers are $-3, -2, 0, 11,$ and 112.
          d. Recall that integers are rational numbers also. The rational numbers are $-3, -2,$ $0, \frac{1}{4}, 11,$ and 112.
          e. The irrational number is $\sqrt{2}$.
          f. The real numbers are all numbers in the given set.   ▬

We can now extend the meaning and use of inequality symbols such as $<$ and $>$ to apply to all real numbers.

ORDER PROPERTY FOR REAL NUMBERS

Given any two real numbers a and b, $a < b$ if a is to the left of b on the number line. Similarly, $a > b$ if a is to the right of b on the number line.

Example 6 Insert $<$, $>$, or $=$ in the appropriate space to make the statement true.

    a. -1 0   **b.** 7 $\frac{14}{2}$ **c.** -5 -6

Solution   **a.** $-1 < 0$ since -1 is to the left of 0 on the number line.
    b. $7 = \frac{14}{2}$ since $\frac{14}{2}$ simplifies to 7.
    c. $-5 > -6$ since -5 is to the right of -6 on the number line.

4 The number line not only gives us a picture of the real numbers, it also helps us visualize the distance between numbers. The distance between a real number a and 0 is given a special name called the **absolute value** of a. "The absolute value of a" is written in symbols as $|a|$.

> ### ABSOLUTE VALUE
>
> The absolute value of a real number a, denoted by $|a|$, is the distance between a and 0 on a number line.

For example, $|3| = 3$ and $|-3| = 3$ since both 3 and -3 are a distance of 3 units from 0 on the number line.

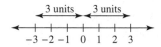

> **HELPFUL HINT**
> Since $|a|$ is a distance, $|a|$ is always either positive or 0, never negative. That is, **for any real number a, $|a| \geq 0$.**

Example 7 Find the absolute value of each number.

    a. $|4|$ **b.** $|-5|$ **c.** $|0|$

Solution   **a.** $|4| = 4$ since 4 is 4 units from 0 on the number line.
    b. $|-5| = 5$ since -5 is 5 units from 0 on the number line.
    c. $|0| = 0$ since 0 is 0 units from 0 on the number line.

Example 8 Insert $<$, $>$, or $=$ in the appropriate space to make the statement true.

    a. $|0|$ 2   **b.** $|-5|$ 5   **c.** $|-3|$ $|-2|$ **d.** $|5|$ $|6|$ **e.** $|-7|$ $|6|$

Solution   **a.** $|0| < 2$ since $|0| = 0$ and $0 < 2$.
    b. $|-5| = 5$.
    c. $|-3| > |-2|$ since $3 > 2$.
    d. $|5| < |6|$ since $5 < 6$.
    e. $|-7| > |6|$ since $7 > 6$.

SPOTLIGHT ON DECISION MAKING

Suppose you are a quality control engineering technician in a factory that makes machine screws. You have just helped to install programmable machinery on the production line that measures the length of each screw. If a screw's length is greater than 4.05 centimeters or less than or equal to 3.98 centimeters, the machinery is programmed to discard the screw. To check that the machinery works properly, you test six screws with known lengths. The results of the test are displayed. Is the new machinery working properly? Explain.

TEST RESULTS

Test Screw	Actual Length of Test Screw (cm)	Machine Action on Test Screw
A	4.03	Accept
B	3.96	Reject
C	4.05	Accept
D	4.08	Reject
E	3.98	Reject
F	4.01	Accept

Exercise Set 1.2

Insert <, >, or = in the appropriate space to make the statement true. See Example 1.

1. 4   10
2. 8   5
3. 7   3
4. 9   15
5. 6.26   6.26
6. 2.13   1.13
7. 0   7
8. 20   0

9. The freezing point of water is 32° Fahrenheit. The boiling point of water is 212° Fahrenheit. Write an inequality statement using < or > comparing the numbers 32 and 212.

10. The freezing point of water is 0° Celsius. The boiling point of water is 100° Celsius. Write an inequality statement using < or > comparing the numbers 0 and 100.

11. The average salary in the United States for an experienced registered nurse is $44,300. The average salary for a drafter is $34,611. Write an inequality statement using < or > comparing the numbers 44,300 and 34,611. (*Source*: U.S. Department of Labor)

12. The state of New York is home to 312 institutions of higher learning. California claims a total of 384 colleges and universities. Write an inequality statement using < or > comparing the numbers 312 and 384. (*Source*: U.S. Department of Education)

Are the following statements true or false? See Example 2.

13. $11 \leq 11$
14. $4 \geq 7$
15. $10 > 11$
16. $17 > 16$
17. $3 + 8 \geq 3(8)$
18. $8 \cdot 8 \leq 8 \cdot 7$
19. $7 > 0$
20. $4 < 7$

△ **21.** An angle measuring 30° is shown and an angle measuring 45° is shown. Use the inequality symbol $\leq$ or $\geq$ to write a statement comparing the numbers 30 and 45.

△ **22.** The sum of the measures of the angles of a triangle is 180°. The sum of the measures of the angles of a parallelogram is 360°. Use the inequality symbol $\leq$ or $\geq$ to write a statement comparing the numbers 360 and 180.

Write each sentence as a mathematical statement. See Example 3.

23. Eight is less than twelve.
24. Fifteen is greater than five.
25. Five is greater than or equal to four.
26. Negative ten is less than or equal to thirty-seven.
27. Fifteen is not equal to negative two.
28. Negative seven is not equal to seven.

Use integers to represent the values in each statement. See Example 4.

29. Driskill Mountain, in Louisiana, has an altitude of 535 feet. New Orleans, Louisiana lies 8 feet below sea level. (*Source:* U.S. Geological Survey)

30. During a Green Bay Packers football game, the team gained 23 yards and then lost 12 yards on consecutive plays.

31. From 1997 to 1998, Chicago Cubs home attendance decreased by 433,853. (*Source:* The Baseball Archive)

32. From 1997 to 1998, Boston Red Sox home attendance increased by 108,559. (*Source:* The Baseball Archive)

33. Aaron Miller deposited $350 in his savings account. He later withdrew $126.

34. Aris Peña was deep-sea diving. During her dive, she ascended 30 feet and later descended 50 feet.

The graph below is called a bar graph. This particular graph shows the number of communities in the United States that have curbside recycling programs for plastics. Each bar represents a different year and the height of the bar represents the number of communities with curbside plastics recycling.

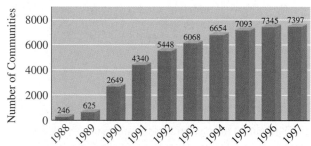

Source: American Plastics Council

35. In which year was the number of communities with curbside plastics recycling the lowest?

36. What is the highest number of communities with curbside plastics recycling?

37. In what years was the number of curbside plastic recycling programs less than 5000?

38. In what years was the number of curbside plastic recycling programs greater than 7000?

39. Write an inequality statement using $\leq$ or $\geq$ comparing the number of communities with curbside plastic recycling for 1993 and 1990.

40. Do you notice any trends shown by this bar graph?

Tell which set or sets each number belongs to: natural numbers, whole numbers, integers, rational numbers, irrational numbers, and real numbers. See Example 5.

41. 0

42. $\frac{1}{4}$

43. -2

44. $-\frac{1}{2}$

45. 6

46. 5

47. $\frac{2}{3}$

48. $\sqrt{3}$

49. $-\sqrt{5}$

50. $-1\frac{5}{6}$

Tell whether each statement is true or false.

51. Every rational number is also an integer.

52. Every negative number is also a rational number.

53. Every natural number is positive.

54. Every rational number is also a real number.

55. 0 is a real number.

56. Every real number is also a rational number.

57. Every whole number is an integer.

58. $\frac{1}{2}$ is an integer.

59. A number can be both rational and irrational.

60. Every whole number is positive.

Insert $<$, $>$, or $=$ in the appropriate space to make a true statement. See Examples 6 through 8.

61. -10 -100

62. -200 -20

63. 32 5.2

64. 7 -7

65. $\frac{18}{3}$ $\frac{24}{3}$

66. $\frac{8}{2}$ $\frac{12}{3}$

67. -51 -50

68. $|-20|$ -200

69. $|-5|$ -4

70. 0 $|0|$

71. $|-1|$ $|1|$

72. $\left|\frac{2}{5}\right|$ $\left|-\frac{2}{5}\right|$

73. $|-2|$ $|-3|$

74. -500 $|-50|$

75. $|0|$ $|-8|$

76. $|-12|$ $\frac{24}{2}$

The apparent magnitude of a star is the measure of its brightness as seen by someone on Earth. The smaller the apparent magnitude, the brighter the star. Use the apparent magnitudes in the table to answer Exercises 77–82.

Star	Apparent Magnitude	Star	Apparent Magnitude
Arcturus	−0.04	Spica	0.98
Sirius	−1.46	Rigel	0.12
Vega	0.03	Regulus	1.35
Antares	0.96	Canopus	−0.72
Sun	−26.7	Hadar	0.61

Source: Norton's 2000.0: Star Atlas and Reference Handbook, 18th ed., Longman Group, UK, 1989

77. The apparent magnitude of the sun is −26.7. The apparent magnitude of the star Arcturus is −0.04. Write an inequality statement comparing the numbers −0.04 and −26.7.

78. The apparent magnitude of Antares is 0.96. The apparent magnitude of Spica is 0.98. Write an inequality statement comparing the numbers 0.96 and 0.98.

79. Which is brighter, the sun or Arcturus?

80. Which is dimmer, Antares or Spica?

81. Which star listed is the brightest?

82. Which star listed is the dimmest?

Rewrite the following inequalities so that the inequality symbol points in the opposite direction and the resulting statement has the same meaning as the given one.

83. $25 \geq 20$ **84.** $-13 \leq 13$

85. $0 < 6$ **86.** $5 > 3$

87. $-10 > -12$ **88.** $-4 < -2$

89. In your own words, explain how to find the absolute value of a number.

90. Give an example of a real-life situation that can be described with integers but not with whole numbers.

1.3 FRACTIONS

CD-ROM   SSM

SSG   Video

▶ **OBJECTIVES**

1. Write fractions in simplest form.
2. Multiply and divide fractions.
3. Add and subtract fractions.

1

A quotient of two numbers such as $\frac{2}{9}$ is called a **fraction**. In the fraction $\frac{2}{9}$, the top number, 2, is called the **numerator** and the bottom number, 9, is called the **denominator**.

A fraction may be used to refer to part of a whole. For example, $\frac{2}{9}$ of the circle below is shaded. The denominator 9 tells us how many equal parts the whole circle is divided into and the numerator 2 tells us how many equal parts are shaded.

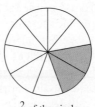

$\frac{2}{9}$ of the circle
is shaded.

To simplify fractions, we can factor the numerator and the denominator. In the statement $3 \cdot 5 = 15$, 3 and 5 are called **factors** and 15 is the **product**. (The raised dot symbol indicates multiplication.)

$$\underset{\underset{\text{factor}}{\uparrow}}{3} \quad \cdot \quad \underset{\underset{\text{factor}}{\uparrow}}{5} \quad = \quad \underset{\underset{\text{product}}{\uparrow}}{15}$$

To **factor** 15 means to write it as a product. The number 15 can be factored as $3 \cdot 5$ or as $1 \cdot 15$.

A fraction is said to be **simplified** or in **lowest terms** when the numerator and the denominator have no factors in common other than 1. For example, the fraction $\frac{5}{11}$ is in lowest terms since 5 and 11 have no common factors other than 1.

To help us simplify fractions, we write the numerator and the denominator as a product of **prime numbers**.

PRIME NUMBER

A prime number is a whole number, other than 1, whose only factors are 1 and itself. The first few prime numbers are

2, 3, 5, 7, 11, 13, 17, 19, 23, 29, and so on.

Example 1 Write each of the following numbers as a product of primes.

a. 40 **b.** 63

Solution **a.** First, write 40 as the product of any two whole numbers, other than 1.

$$40 = 4 \cdot 10$$

Next, factor each of these numbers. Continue this process until all of the factors are prime numbers.

$$40 = 4 \cdot 10$$
$$= 2 \cdot 2 \cdot 2 \cdot 5$$

All the factors are now prime numbers. Then 40 written as a product of primes is

$$40 = 2 \cdot 2 \cdot 2 \cdot 5$$

b. $63 = 9 \cdot 7$
$$= 3 \cdot 3 \cdot 7$$

To use prime factors to write a fraction in lowest terms, apply the fundamental principle of fractions.

FUNDAMENTAL PRINCIPLE OF FRACTIONS

If $\frac{a}{b}$ is a fraction and c is a nonzero real number, then

$$\frac{a \cdot c}{b \cdot c} = \frac{a}{b}$$

Example 2 Write each fraction in lowest terms.

a. $\dfrac{42}{49}$ **b.** $\dfrac{11}{27}$ **c.** $\dfrac{88}{20}$

Solution **a.** Write the numerator and the denominator as products of primes; then apply the fundamental principle to the common factor 7.

$$\frac{42}{49} = \frac{2 \cdot 3 \cdot \boxed{7}}{7 \cdot \boxed{7}} = \frac{2 \cdot 3}{7} = \frac{6}{7}$$

b. $\dfrac{11}{27} = \dfrac{11}{3 \cdot 3 \cdot 3}$

There are no common factors other than 1, so $\frac{11}{27}$ is already in lowest terms.

c. $\dfrac{88}{20} = \dfrac{\boxed{2} \cdot \boxed{2} \cdot 2 \cdot 11}{\boxed{2} \cdot \boxed{2} \cdot 5} = \dfrac{22}{5}$

2 To multiply two fractions, multiply numerator times numerator to obtain the numerator of the product; multiply denominator times denominator to obtain the denominator of the product.

MULTIPLYING FRACTIONS

$$\frac{a}{b} \cdot \frac{c}{d} = \frac{a \cdot c}{b \cdot d}, \qquad \text{if } b \neq 0 \text{ and } d \neq 0$$

Example 3 Find the product of $\dfrac{2}{15}$ and $\dfrac{5}{13}$. Write the product in lowest terms.

Solution $\dfrac{2}{15} \cdot \dfrac{5}{13} = \dfrac{2 \cdot 5}{15 \cdot 13}$ Multiply numerators.
Multiply denominators.

Next, simplify the product by dividing the numerator and the denominator by any common factors.

$$= \frac{2 \cdot \boxed{5}}{3 \cdot \boxed{5} \cdot 13}$$

$$= \frac{2}{39}$$

Before dividing fractions, we first define **reciprocals**. Two fractions are reciprocals of each other if their product is 1. For example $\frac{2}{3}$ and $\frac{3}{2}$ are reciprocals since $\frac{2}{3} \cdot \frac{3}{2} = 1$. Also, the reciprocal of 5 is $\frac{1}{5}$ since $5 \cdot \frac{1}{5} = \frac{5}{1} \cdot \frac{1}{5} = 1$.

To divide fractions, multiply the first fraction by the reciprocal of the second fraction.

DIVIDING FRACTIONS

$$\frac{a}{b} \div \frac{c}{d} = \frac{a}{b} \cdot \frac{d}{c}, \qquad \text{if } b \neq 0, d \neq 0, \text{and } c \neq 0$$

Example 4 Find each quotient. Write all answers in lowest terms.

a. $\dfrac{4}{5} \div \dfrac{5}{16}$ **b.** $\dfrac{7}{10} \div 14$ **c.** $\dfrac{3}{8} \div \dfrac{3}{10}$

Solution   **a.** $\dfrac{4}{5} \div \dfrac{5}{16} = \dfrac{4}{5} \cdot \dfrac{16}{5} = \dfrac{4 \cdot 16}{5 \cdot 5} = \dfrac{64}{25}$

       b. $\dfrac{7}{10} \div 14 = \dfrac{7}{10} \div \dfrac{14}{1} = \dfrac{7}{10} \cdot \dfrac{1}{14} = \dfrac{7 \cdot 1}{2 \cdot 5 \cdot 2 \cdot 7} = \dfrac{1}{20}.$

       c. $\dfrac{3}{8} \div \dfrac{3}{10} = \dfrac{3}{8} \cdot \dfrac{10}{3} = \dfrac{3 \cdot 2 \cdot 5}{2 \cdot 2 \cdot 2 \cdot 3} = \dfrac{5}{4}$

3   To add or subtract fractions with the same denominator, combine numerators and place the sum or difference over the common denominator.

ADDING AND SUBTRACTING FRACTIONS WITH THE SAME DENOMINATOR

$$\frac{a}{b} + \frac{c}{b} = \frac{a + c}{b}, \qquad \text{if } b \neq 0$$

$$\frac{a}{b} - \frac{c}{b} = \frac{a - c}{b}, \qquad \text{if } b \neq 0$$

Example 5 Add or subtract as indicated. Write each result in lowest terms.

a. $\dfrac{2}{7} + \dfrac{4}{7}$ **b.** $\dfrac{3}{10} + \dfrac{2}{10}$ **c.** $\dfrac{9}{7} - \dfrac{2}{7}$ **d.** $\dfrac{5}{3} - \dfrac{1}{3}$

Solution   **a.** $\dfrac{2}{7} + \dfrac{4}{7} = \dfrac{2 + 4}{7} = \dfrac{6}{7}$

       b. $\dfrac{3}{10} + \dfrac{2}{10} = \dfrac{3 + 2}{10} = \dfrac{5}{10} = \dfrac{5}{2 \cdot 5} = \dfrac{1}{2}$

       c. $\dfrac{9}{7} - \dfrac{2}{7} = \dfrac{9 - 2}{7} = \dfrac{7}{7} = 1$

       d. $\dfrac{5}{3} - \dfrac{1}{3} = \dfrac{5 - 1}{3} = \dfrac{4}{3}$

To add or subtract fractions without the same denominator, first write the fractions as **equivalent fractions** with a common denominator. Equivalent fractions are

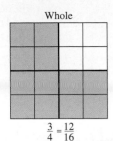

Whole

$$\frac{3}{4} = \frac{12}{16}$$

fractions that represent the same quantity. For example, $\frac{3}{4}$ and $\frac{12}{16}$ are equivalent fractions since they represent the same portion of a whole, as the diagram shows. Count the larger squares and the shaded portion is $\frac{3}{4}$. Count the smaller squares and the shaded portion is $\frac{12}{16}$. Thus, $\frac{3}{4} = \frac{12}{16}$.

We can write equivalent fractions by multiplying a given fraction by 1, as shown in the next example. Multiplying a fraction by 1 does not change the value of the fraction.

Example 6 Write $\frac{2}{5}$ as an equivalent fraction with a denominator of 20.

Solution Since $5 \cdot 4 = 20$, multiply the fraction by $\frac{4}{4}$. Multiplying by $\frac{4}{4} = 1$ does not change the value of the fraction.

Multiply by $\frac{4}{4}$ or 1.

$$\frac{2}{5} = \frac{2}{5} \cdot \frac{4}{4} = \frac{2 \cdot 4}{5 \cdot 4} = \frac{8}{20}$$

Example 7 Add or subtract as indicated. Write each answer in lowest terms.

a. $\dfrac{2}{5} + \dfrac{1}{4}$ **b.** $\dfrac{1}{2} + \dfrac{17}{22} - \dfrac{2}{11}$ **c.** $3\dfrac{1}{6} - 1\dfrac{11}{12}$

Solution **a.** Fractions must have a common denominator before they can be added or subtracted. Since 20 is the smallest number that both 5 and 4 divide into evenly, 20 is the **least common denominator**. Write both fractions as equivalent fractions with denominators of 20. Since

$$\frac{2}{5} \cdot \frac{4}{4} = \frac{2 \cdot 4}{5 \cdot 4} = \frac{8}{20} \quad \text{and} \quad \frac{1}{4} \cdot \frac{5}{5} = \frac{1 \cdot 5}{4 \cdot 5} = \frac{5}{20}$$

then

$$\frac{2}{5} + \frac{1}{4} = \frac{8}{20} + \frac{5}{20} = \frac{13}{20}$$

b. The least common denominator for denominators 2, 22, and 11 is 22. First, write each fraction as an equivalent fraction with a denominator of 22. Then add or subtract from left to right.

$$\frac{1}{2} = \frac{1}{2} \cdot \frac{11}{11} = \frac{11}{22}, \qquad \frac{17}{22} = \frac{17}{22}, \qquad \text{and} \qquad \frac{2}{11} = \frac{2}{11} \cdot \frac{2}{2} = \frac{4}{22}$$

Then

$$\frac{1}{2} + \frac{17}{22} - \frac{2}{11} = \frac{11}{22} + \frac{17}{22} - \frac{4}{22} = \frac{24}{22} = \frac{12}{11}$$

c. To find $3\frac{1}{6} - 1\frac{11}{12}$, first rewrite each mixed number as follows:

$$3\frac{1}{6} = 3 + \frac{1}{6} = \frac{18}{6} + \frac{1}{6} = \frac{19}{6}$$

$$1\frac{11}{12} = 1 + \frac{11}{12} = \frac{12}{12} + \frac{11}{12} = \frac{23}{12}$$

Then

$$3\frac{1}{6} - 1\frac{11}{12} = \frac{19}{6} - \frac{23}{12} = \frac{38}{12} - \frac{23}{12} = \frac{15}{12} = \frac{5}{4} \text{ or } 1\frac{1}{4}$$

HELPFUL HINT

Notice that we wrote the mixed number $3\frac{1}{6}$ as $\frac{19}{6}$ in Example 7c. Recall a short-cut process for writing a mixed number as an improper fraction:

$$3\frac{1}{6} = \frac{6 \cdot 3 + 1}{6} = \frac{19}{6}$$

SPOTLIGHT ON DECISION MAKING

Suppose you are fishing on a freshwater lake in Canada. You catch a whitefish weighing $14\frac{5}{32}$ pounds. According to the International Game Fish Association, the world's record for largest lake whitefish ever caught is $14\frac{3}{8}$ pounds. Did you set a new world's record? Explain. By how much did you beat or miss the existing world record?

MENTAL MATH

Represent the shaded part of each geometric figure by a fraction.

1.

2.

3.

4.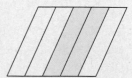

For Exercises 5 and 6, fill in the blank.

5. In the fraction $\frac{3}{5}$, 3 is called the _____ and 5 is called the _____.

6. The reciprocal of $\frac{7}{11}$ is ____.

Exercise Set 1.3

Write each number as a product of primes. See Example 1.

1. 33   **2.** 60   **5.** 20   **6.** 56

3. 98   **4.** 27   **7.** 75   **8.** 32

                                          9. 45   **10.** 24

Write the fraction in lowest terms. See Example 2.

11. $\dfrac{2}{4}$

12. $\dfrac{3}{6}$

 13. $\dfrac{10}{15}$

14. $\dfrac{15}{20}$

15. $\dfrac{3}{7}$

16. $\dfrac{5}{9}$

17. $\dfrac{18}{30}$

18. $\dfrac{42}{45}$

Multiply or divide as indicated. Write the answer in lowest terms. See Examples 3 and 4.

19. $\dfrac{1}{2} \cdot \dfrac{3}{4}$

20. $\dfrac{10}{6} \cdot \dfrac{3}{5}$

21. $\dfrac{2}{3} \cdot \dfrac{3}{4}$

22. $\dfrac{7}{8} \cdot \dfrac{3}{21}$

23. $\dfrac{1}{2} \div \dfrac{7}{12}$

24. $\dfrac{7}{12} \div \dfrac{1}{2}$

25. $\dfrac{3}{4} \div \dfrac{1}{20}$

26. $\dfrac{3}{5} \div \dfrac{9}{10}$

27. $\dfrac{7}{10} \cdot \dfrac{5}{21}$

28. $\dfrac{3}{35} \cdot \dfrac{10}{63}$

29. $2\dfrac{7}{9} \cdot \dfrac{1}{3}$

30. $\dfrac{1}{4} \cdot 5\dfrac{5}{6}$

The area of a plane figure is a measure of the amount of surface of the figure. Find the area of each figure below. (The area of a rectangle is the product of its length and width. The area of a triangle is $\frac{1}{2}$ the product of its base and height.)

△ **31.**

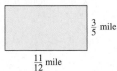

$\dfrac{3}{5}$ mile, $\dfrac{11}{12}$ mile

△ **32.**

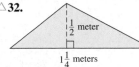

$1\dfrac{1}{2}$ meter, $1\dfrac{1}{4}$ meters

Add or subtract as indicated. Write the answer in lowest terms. See Example 5.

33. $\dfrac{4}{5} - \dfrac{1}{5}$

34. $\dfrac{6}{7} - \dfrac{1}{7}$

35. $\dfrac{4}{5} + \dfrac{1}{5}$

36. $\dfrac{6}{7} + \dfrac{1}{7}$

37. $\dfrac{17}{21} - \dfrac{10}{21}$

38. $\dfrac{18}{35} - \dfrac{11}{35}$

39. $\dfrac{23}{105} + \dfrac{4}{105}$

40. $\dfrac{13}{132} + \dfrac{35}{132}$

Write each fraction as an equivalent fraction with the given denominator. See Example 6.

41. $\dfrac{7}{10}$ with a denominator of 30

42. $\dfrac{2}{3}$ with a denominator of 9

43. $\dfrac{2}{9}$ with a denominator of 18

44. $\dfrac{8}{7}$ with a denominator of 56

45. $\dfrac{4}{5}$ with a denominator of 20

46. $\dfrac{4}{5}$ with a denominator of 25

Add or subtract as indicated. Write the answer in lowest terms. See Example 7.

47. $\dfrac{2}{3} + \dfrac{3}{7}$

48. $\dfrac{3}{4} + \dfrac{1}{6}$

49. $2\dfrac{13}{15} - 1\dfrac{1}{5}$

50. $5\dfrac{2}{9} - 3\dfrac{1}{6}$

51. $\dfrac{5}{22} - \dfrac{5}{33}$

52. $\dfrac{7}{10} - \dfrac{8}{15}$

53. $\dfrac{12}{5} - 1$

54. $2 - \dfrac{3}{8}$

Each circle below represents a whole, or 1. Determine the unknown part of the circle.

55.

$\dfrac{3}{10}$, $\dfrac{5}{10}$, ?

56.

$\dfrac{3}{11}$, $\dfrac{2}{11}$, ?

57.

?, $\dfrac{1}{4}$, $\dfrac{3}{8}$

58.

$\dfrac{5}{12}$, $\dfrac{1}{3}$, $\dfrac{1}{6}$, ?

59.

$\dfrac{1}{2}$, ?, $\dfrac{2}{9}$, $\dfrac{1}{6}$

60.

?, $\dfrac{3}{5}$, $\dfrac{1}{10}$

Perform the following operations. Write answers in lowest terms.

61. $\dfrac{10}{21} + \dfrac{5}{21}$

62. $\dfrac{11}{35} + \dfrac{3}{35}$

63. $\dfrac{10}{3} - \dfrac{5}{21}$

64. $\dfrac{11}{7} - \dfrac{3}{35}$

65. $\frac{2}{3} \cdot \frac{3}{5}$

66. $\frac{2}{3} \div \frac{3}{4}$

67. $\frac{3}{4} \div \frac{7}{12}$

68. $\frac{3}{5} + \frac{2}{3}$

69. $\frac{5}{12} + \frac{4}{12}$

70. $\frac{2}{7} + \frac{4}{7}$

71. $5 + \frac{2}{3}$

72. $7 + \frac{1}{10}$

73. $\frac{7}{8} \div 3\frac{1}{4}$

74. $3 \div \frac{3}{4}$

75. $\frac{7}{18} \div \frac{14}{36}$

76. $4\frac{3}{7} \div \frac{31}{7}$

77. $\frac{23}{105} - \frac{2}{105}$

78. $\frac{57}{132} - \frac{13}{132}$

79. $1\frac{1}{2} + 3\frac{2}{3}$

80. $2\frac{3}{5} + 4\frac{7}{10}$

81. $\frac{2}{3} - \frac{5}{9} + \frac{5}{6}$

82. $\frac{8}{11} - \frac{1}{4} + \frac{1}{2}$

The perimeter of a plane figure is the total distance around the figure. Find the perimeter of each figure below.

△ **83.**

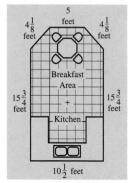

△ **84.**

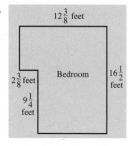

85. In 1988, Petra Felke of East Germany set an Olympic record in the women's javelin throw of $74\frac{17}{25}$ meters. Later that year she set a new world record for the javelin throw of 80 meters. By how much did her world record beat her Olympic record? (*Source: World Almanac and Book of Facts, 1999*)

86. In March 1999, a proposal to increase the size of rectangular escape vents in lobster traps was brought before the Maine state legislature. The proposed change would add $\frac{1}{16}$ of an inch to the current vent height of $1\frac{7}{8}$ inches. What would be the new vent height under the proposal? (*Source: The Boston Sunday Globe, April 4, 1999*)

87. In your own words, explain how to add two fractions with different denominators.

88. In your own words, explain how to multiply two fractions.

The following trail chart is given to visitors at the Lakeview Forest Preserve.

Trail Name	Distance (miles)
Robin Path	$3\frac{1}{2}$
Red Falls	$5\frac{1}{2}$
Green Way	$2\frac{1}{8}$
Autumn Walk	$1\frac{3}{4}$

89. How much longer is Red Falls Trail than Green Way Trail?

90. Find the total distance traveled by someone who hiked along all four trails.

Most of the water on Earth is in the form of oceans. Only a small part is fresh water. The graph below is called a circle graph or pie chart. This particular circle graph shows the distribution of fresh water.

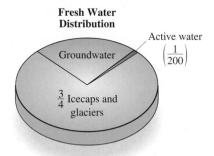

Fresh Water Distribution

91. What fractional part of fresh water is icecaps and glaciers?

92. What fractional part of fresh water is active water?

93. What fractional part of fresh water is groundwater?

94. What fractional part of fresh water is groundwater or icecaps and glaciers?

1.4 INTRODUCTION TO VARIABLE EXPRESSIONS AND EQUATIONS

CD-ROM   SSM

SSG   Video

▶ **O B J E C T I V E S**

1. Define and use exponents and the order of operations.
2. Evaluate algebraic expressions, given replacement values for variables.
3. Determine whether a number is a solution of a given equation.
4. Translate phrases into expressions and sentences into equations.

1

2 cm

Volume is $(2 \cdot 2 \cdot 2)$ cubic centimeters.

Frequently in algebra, products occur that contain repeated multiplication of the same factor. For example, the volume of a cube whose sides each measure 2 centimeters is $(2 \cdot 2 \cdot 2)$ cubic centimeters. We may use **exponential notation** to write such products in a more compact form. For example,

$$2 \cdot 2 \cdot 2 \quad \textit{may be written as} \quad 2^3.$$

The 2 in 2^3 is called the **base**; it is the repeated factor. The 3 in 2^3 is called the **exponent** and is the number of times the base is used as a factor. The expression 2^3 is called an **exponential expression**.

$$2^3 = 2 \cdot 2 \cdot 2 = 8$$

exponent

base ⎯ 2 is a factor 3 times

Example 1   Evaluate the following:

   a. 3^2 [read as "3 squared" or as "3 to second power"]
   b. 5^3 [read as "5 cubed" or as "5 to the third power"]
   c. 2^4 [read as "2 to the fourth power"]
   d. 7^1 **e.** $\left(\dfrac{3}{7}\right)^2$

Solution   **a.** $3^2 = 3 \cdot 3 = 9$
   b. $5^3 = 5 \cdot 5 \cdot 5 = 125$
   c. $2^4 = 2 \cdot 2 \cdot 2 \cdot 2 = 16$
   d. $7^1 = 7$
   e. $\left(\dfrac{3}{7}\right)^2 = \left(\dfrac{3}{7}\right)\left(\dfrac{3}{7}\right) = \dfrac{9}{49}$

> **H E L P F U L   H I N T**
> $2^3 \neq 2 \cdot 3$ since 2^3 indicates repeated **multiplication** of the same factor.
> $$2^3 = 2 \cdot 2 \cdot 2 = 8, \text{ whereas } 2 \cdot 3 = 6.$$

Using symbols for mathematical operations is a great convenience. The more operation symbols presented in an expression, the more careful we must be when performing the indicated operation. For example, in the expression $2 + 3 \cdot 7$, do we add first or multiply first? To eliminate confusion, **grouping symbols** are used.

Examples of grouping symbols are parentheses ( ), brackets [ ], braces { }, and the fraction bar. If we wish $2 + 3 \cdot 7$ to be simplified by adding first, we enclose $2 + 3$ in parentheses.

$$(2 + 3) \cdot 7 = 5 \cdot 7 = 35$$

If we wish to multiply first, $3 \cdot 7$ may be enclosed in parentheses.

$$2 + (3 \cdot 7) = 2 + 21 = 23$$

To eliminate confusion when no grouping symbols are present, use the following agreed upon order of operations.

ORDER OF OPERATIONS

Simplify expressions using the following order. If grouping symbols such as parentheses are present, simplify expressions within those first, starting with the innermost set. If fraction bars are present, simplify the numerator and the denominator separately.

1. Evaluate exponential expressions.
2. Perform multiplications or divisions in order from left to right.
3. Perform additions or subtractions in order from left to right.

Now simplify $2 + 3 \cdot 7$. There are no grouping symbols and no exponents, so we multiply and then add.

$$2 + 3 \cdot 7 = 2 + 21 \qquad \text{Multiply.}$$
$$= 23 \qquad \text{Add.}$$

Example 2 Simplify each expression.

a. $6 \div 3 + 5^2$ **b.** $\dfrac{2(12 + 3)}{|-15|}$ **c.** $3 \cdot 10 - 7 \div 7$ **d.** $3 \cdot 4^2$ **e.** $\dfrac{3}{2} \cdot \dfrac{1}{2} - \dfrac{1}{2}$

Solution **a.** Evaluate 5^2 first.

$$6 \div 3 + 5^2 = 6 \div 3 + 25$$

Next divide, then add.

$$6 \div 3 + 25 = 2 + 25 \qquad \text{Divide.}$$
$$= 27 \qquad \text{Add.}$$

b. First, simplify the numerator and the denominator separately.

$$\frac{2(12 + 3)}{|-15|} = \frac{2(15)}{15} \qquad \text{Simplify numerator and denominator separately.}$$
$$= \frac{30}{15}$$
$$= 2 \qquad \text{Simplify.}$$

c. Multiply and divide from left to right. Then subtract.

$$3 \cdot 10 - 7 \div 7 = 30 - 1$$
$$= 29 \qquad \text{Subtract.}$$

d. In this example, only the 4 is squared. The factor of 3 is not part of the base because no grouping symbol includes it as part of the base.

$$3 \cdot 4^2 = 3 \cdot 16 \qquad \text{Evaluate the exponential expression.}$$
$$= 48 \qquad \text{Multiply.}$$

e. The order of operations applies to operations with fractions in exactly the same way as it applies to operations with whole numbers.

$$\frac{3}{2} \cdot \frac{1}{2} - \frac{1}{2} = \frac{3}{4} - \frac{1}{2} \qquad \text{Multiply.}$$

$$= \frac{3}{4} - \frac{2}{4} \qquad \text{The least common denominator is 4.}$$

$$= \frac{1}{4} \qquad \text{Subtract.}$$

HELPFUL HINT

Be careful when evaluating an exponential expression. In $3 \cdot 4^2$, the exponent 2 applies only to the base 4. In $(3 \cdot 4)^2$, we multiply first because of parentheses, so the exponent 2 applies to the product $3 \cdot 4$.

$$3 \cdot 4^2 = 3 \cdot 16 = 48 \qquad (3 \cdot 4)^2 = (12)^2 = 144$$

Expressions that include many grouping symbols can be confusing. When simplifying these expressions, keep in mind that grouping symbols separate the expression into distinct parts. Each is then simplified separately.

Example 3 Simplify $\dfrac{3 + |4 - 3| + 2^2}{6 - 3}$.

Solution The fraction bar serves as a grouping symbol and separates the numerator and denominator. Simplify each separately. Also, the absolute value bars here serve as a grouping symbol. We begin in the numerator by simplifying within the absolute value bars.

$$\frac{3 + |4 - 3| + 2^2}{6 - 3} = \frac{3 + |1| + 2^2}{6 - 3} \qquad \begin{array}{l}\text{Simplify the expression}\\\text{inside the absolute value}\\\text{bars.}\end{array}$$

$$= \frac{3 + 1 + 2^2}{3} \qquad \begin{array}{l}\text{Find the absolute value}\\\text{and simplify the denomi-}\\\text{nator.}\end{array}$$

$$= \frac{3 + 1 + 4}{3} \qquad \begin{array}{l}\text{Evaluate the exponential}\\\text{expression.}\end{array}$$

$$= \frac{8}{3} \qquad \text{Simplify the numerator.}$$

Example 4   Simplify $3[4(5 + 2) - 10]$.

Solution   Notice that both parentheses and brackets are used as grouping symbols. Start with the innermost set of grouping symbols.

$$\begin{aligned} 3[4(5 + 2) - 10] &= 3[4(7) - 10] &&\text{Simplify the expression in parentheses.} \\ &= 3[28 - 10] &&\text{Multiply 4 and 7.} \\ &= 3[18] &&\text{Subtract inside the brackets.} \\ &= 54 &&\text{Multiply.} \end{aligned}$$

Example 5   Simplify $\dfrac{8 + 2 \cdot 3}{2^2 - 1}$.

Solution $\dfrac{8 + 2 \cdot 3}{2^2 - 1} = \dfrac{8 + 6}{4 - 1} = \dfrac{14}{3}$

2   In algebra, we use symbols, usually letters such as x, y, or z, to represent unknown numbers. A symbol that is used to represent a number is called a **variable**. An **algebraic expression** is a collection of numbers, variables, operation symbols, and grouping symbols. For example,

$$2x, \qquad -3, \qquad 2x + 10, \qquad 5(p^2 + 1), \qquad \text{and} \qquad \frac{3y^2 - 6y + 1}{5}$$

are algebraic expressions. The expression $2x$ means $2 \cdot x$. Also, $5(p^2 + 1)$ means $5 \cdot (p^2 + 1)$ and $3y^2$ means $3 \cdot y^2$. If we give a specific value to a variable, we can **evaluate an algebraic expression**. To evaluate an algebraic expression means to find its numerical value once we know the values of the variables.

Algebraic expressions often occur during problem solving. For example, the expression

$$16t^2$$

gives the distance in feet (neglecting air resistance) that an object will fall in t seconds. (See Exercise 63 in this section.)

Example 6   Evaluate each expression if $x = 3$ and $y = 2$.

a. $2x - y$ **b.** $\dfrac{3x}{2y}$ **c.** $\dfrac{x}{y} + \dfrac{y}{2}$ **d.** $x^2 - y^2$

Solution   **a.** Replace x with 3 and y with 2.

$$2x - y = 2(3) - 2 \qquad \text{Let } x = 3 \text{ and } y = 2.$$
$$= 6 - 2 \qquad \text{Multiply.}$$
$$= 4 \qquad \text{Subtract.}$$

b. $\dfrac{3x}{2y} = \dfrac{3 \cdot 3}{2 \cdot 2} = \dfrac{9}{4}$ Let $x = 3$ and $y = 2.$

c. Replace x with 3 and y with 2. Then simplify.

$$\frac{x}{y} + \frac{y}{2} = \frac{3}{2} + \frac{2}{2} = \frac{5}{2}$$

d. Replace x with 3 and y with 2.

$$x^2 - y^2 = 3^2 - 2^2 = 9 - 4 = 5$$

3   Many times a problem-solving situation is modeled by an equation. An **equation** is a mathematical statement that two expressions have equal value. The equal symbol "=" is used to equate the two expressions. For example, $3 + 2 = 5$, $7x = 35$, $\dfrac{2(x - 1)}{3} = 0$, and $I = PRT$ are all equations.

> **HELPFUL HINT**
> An equation contains the equal symbol "=". An algebraic expression does not.

When an equation contains a variable, deciding which values of the variable make an equation a true statement is called **solving** an equation for the variable. A **solution** of an equation is a value for the variable that makes the equation true. For example, 3 is a solution of the equation $x + 4 = 7$, because if x is replaced with 3 the statement is true.

$$x + 4 = 7$$
$$\downarrow$$
$$3 + 4 = 7 \qquad \text{Replace } x \text{ with 3.}$$
$$7 = 7 \qquad \text{True.}$$

Similarly, 1 is not a solution of the equation $x + 4 = 7$, because $1 + 4 = 7$ is **not** a true statement.

Example 7   Decide whether 2 is a solution of $3x + 10 = 8x$.

Solution   Replace x with 2 and see if a true statement results.

$$3x + 10 = 8x \qquad \text{Original equation}$$
$$3(2) + 10 \stackrel{?}{=} 8(2) \qquad \text{Replace } x \text{ with 2.}$$
$$6 + 10 \stackrel{?}{=} 16 \qquad \text{Simplify each side.}$$
$$16 = 16 \qquad \text{True.}$$

Since we arrived at a true statement after replacing x with 2 and simplifying both sides of the equation, 2 is a solution of the equation.

4 Now that we know how to represent an unknown number by a variable, let's practice translating phrases into algebraic expressions and sentences into equations. Oftentimes solving problems involves the ability to translate word phrases and sentences into symbols. Below is a list of some key words and phrases to help us translate.

Addition (+)	*Subtraction* (−)	*Multiplication* (·)	*Division* (÷)	*Equality* (=)
Sum	Difference of	Product	Quotient	Equals
Plus	Minus	Times	Divide	Gives
Added to	Subtracted from	Multiply	Into	Is/was/ should be
More than	Less than	Twice	Ratio	Yields
Increased by	Decreased by	Of	Divided by	Amounts to
Total	Less			Represents Is the same as

Example 8 Write an algebraic expression that represents each phrase. Let the variable x represent the unknown number.

  a. The sum of a number and 3
  b. The product of 3 and a number
  c. Twice a number
  d. 10 decreased by a number
  e. 5 times a number increased by 7

Solution **a.** $x + 3$ since "sum" means to add
  b. $3 \cdot x$ and $3x$ are both ways to denote the product of 3 and x
  c. $2 \cdot x$ or $2x$
  d. $10 - x$ because "decreased by" means to subtract
  e. $\underbrace{5x}_{\substack{5 \text{ times} \\ \text{a number}}} + 7$

> **HELPFUL HINT**
> Make sure you understand the difference when translating phrases containing "decreased by," "subtracted from," and "less than."
>
> | Phrase | Translation | |
> |---|---|---|
> | A number decreased by 10 | $x - 10$ | Notice the order. |
> | A number subtracted from 10 | $10 - x$ | |
> | 10 less than a number | $x - 10$ | |

Now let's practice translating sentences into equations.

Example 9 Write each sentence as an equation. Let x represent the unknown number.

  a. The quotient of 15 and a number is 4.
  b. Three subtracted from 12 is a number.
  c. Four times a number added to 17 is 21.

Solution   **a.** In words:

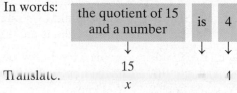

Translate:

$$\frac{15}{x} = 4$$

b. In words:

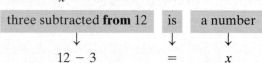

Translate: $12 - 3$ $=$ x

Care must be taken when the operation is subtraction. The expression $3 - 12$ would be incorrect. Notice that $3 - 12 \neq 12 - 3$.

c. In words:

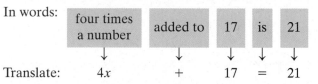

Translate: $4x$ $+$ 17 $=$ 21

CALCULATOR EXPLORATIONS

Exponents

To evaluate exponential expressions on a scientific calculator, find the key marked $\boxed{y^x}$ or

$\boxed{\wedge}$. To evaluate, for example, 3^5, press the following keys: $\boxed{3}$ $\boxed{y^x}$ (or $\boxed{\wedge}$) $\boxed{5}$ $\boxed{=}$ (or

$\boxed{\text{ENTER}}$). The display should read $\boxed{\qquad 243}$ or $\boxed{\begin{array}{l} 3\text{^}5 \\ \qquad 243 \end{array}}$.

Order of Operations

Although most calculators follow the order of operations, parentheses must sometimes be inserted when evaluating an expression. For example, to find the value of $\frac{5}{12 - 7}$, press the keys

$\boxed{5}$ $\boxed{\div}$ $\boxed{(}$ $\boxed{1}$ $\boxed{2}$ $\boxed{-}$ $\boxed{7}$ $\boxed{)}$ $\boxed{=}$ (or $\boxed{\text{ENTER}}$).

The display should read $\boxed{\qquad 1}$ or $\boxed{\begin{array}{l} 5/(12 - 7) \\ \qquad\qquad 1 \end{array}}$.

Use a calculator to evaluate each expression.

1. 5^3 **2.** 7^4

3. 9^5 **4.** 8^6

5. $2(20 - 5)$ **6.** $3(14 - 7) + 21$

7. $24(862 - 455) + 89$ **8.** $99 + (401 + 962)$

SPOTLIGHT ON DECISION MAKING

Suppose you are a local area network (LAN) administrator for a small college and you are configuring a new LAN for the mathematics department. The department would like a network of 20 computers so that each user can transmit data over the network at a speed of 0.25 megabits per second. The collective speed for a LAN is given by the expression rn, where r is the data transmission speed needed by each of the n computers on the LAN. You know that the network will drastically lose its efficiency if the collective speed of the network exceeds 8 megabits per second. Decide whether the LAN requested by the math department will operate efficiently. Explain your reasoning.

MENTAL MATH

Fill in the blank with add, subtract, multiply, or divide.

1. To simplify the expression $1 + 3 \cdot 6$, first _____.
2. To simplify the expression $(1 + 3) \cdot 6$, first _____.
3. To simplify the expression $(20 - 4) \cdot 2$, first _____.
4. To simplify the expression $20 - 4 \div 2$, first _____.

Exercise Set 1.4

Evaluate. See Example 1.

1. 3^5
2. 5^3
3. 3^3
4. 4^4
5. 1^5
6. 1^8
7. 5^1
8. 8^1
9. $\left(\dfrac{1}{5}\right)^3$
10. $\left(\dfrac{6}{11}\right)^2$
11. $\left(\dfrac{2}{3}\right)^4$
12. $\left(\dfrac{1}{2}\right)^5$
13. 7^2
14. 9^2
15. 4^2
16. 2^4
17. $(1.2)^2$
18. $(0.07)^2$

Simplify each expression. See Examples 2 through 5.

19. $5 + 6 \cdot 2$
20. $8 + 5 \cdot 3$
21. $4 \cdot 8 - 6 \cdot 2$
22. $12 \cdot 5 - 3 \cdot 6$
23. $2(8 - 3)$
24. $5(6 - 2)$
25. $2 + (5 - 2) + 4^2$
26. $6 - 2 \cdot 2 + 2^5$
27. $5 \cdot 3^2$
28. $2 \cdot 5^2$

29. $\dfrac{1}{4} \cdot \dfrac{2}{3} - \dfrac{1}{6}$
30. $\dfrac{3}{4} \cdot \dfrac{1}{2} + \dfrac{2}{3}$
31. $\dfrac{6 - 4}{9 - 2}$
32. $\dfrac{8 - 5}{24 - 20}$
33. $2[5 + 2(8 - 3)]$
34. $3[4 + 3(6 - 4)]$
35. $\dfrac{19 - 3 \cdot 5}{6 - 4}$
36. $\dfrac{4 \cdot 3 + 2}{4 + 3 \cdot 2}$
37. $\dfrac{|6 - 2| + 3}{8 + 2 \cdot 5}$
38. $\dfrac{15 - |3 - 1|}{12 - 3 \cdot 2}$
39. $\dfrac{3 + 3(5 + 3)}{3^2 + 1}$
40. $\dfrac{3 + 6(8 - 5)}{4^2 + 2}$
41. $\dfrac{6 + |8 - 2| + 3^2}{18 - 3}$
42. $\dfrac{16 + |13 - 5| + 4^2}{17 - 5}$

43. Are parentheses necessary in the expression $2 + (3 \cdot 5)$? Explain your answer.

44. Are parentheses necessary in the expression $(2 + 3) \cdot 5$? Explain your answer.

For Exercises 45 and 46, match each expression in the first column with its value in the second column.

45. **a.** $(6 + 2) \cdot (5 + 3)$ — 19
b. $(6 + 2) \cdot 5 + 3$ — 22
c. $6 + 2 \cdot 5 + 3$ — 64
d. $6 + 2 \cdot (5 + 3)$ — 43

46. **a.** $(1 + 4) \cdot 6 - 3$ — 15
b. $1 + 4 \cdot (6 - 3)$ — 13
c. $1 + 4 \cdot 6 - 3$ — 27
d. $(1 + 4) \cdot (6 - 3)$ — 22

Evaluate each expression when $x = 1$, $y = 3$, and $z = 5$. See Example 6.

47. $3y$

48. $4x$

49. $\dfrac{z}{5x}$

50. $\dfrac{y}{2z}$

51. $3x - 2$

52. $6y - 8$

53. $|2x + 3y|$

54. $|5z - 2y|$

55. $5y^2$

56. $2z^2$

Evaluate each expression if $x = 12$, $y = 8$, and $z = 4$. See Example 6.

57. $\dfrac{x}{z} + 3y$

58. $\dfrac{y}{z} + 8x$

59. $x^2 - 3y + x$

60. $y^2 - 3x + y$

61. $\dfrac{x^2 + z}{y^2 + 2z}$

62. $\dfrac{y^2 + x}{x^2 + 3y}$

Neglecting air resistance, the expression $16t^2$ gives the distance in feet an object will fall in t seconds.

63. Complete the chart below. To evaluate $16t^2$, remember to first find t^2, then multiply by 16.

Time t (in seconds)	Distance $16t^2$ (in feet)
1	
2	
3	
4	

64. Does an object fall the same distance *during* each second? Why or why not? (See Exercise 63.)

Decide whether the given number is a solution of the given equation. See Example 7.

65. Is 5 a solution of $3x - 6 = 9$?

66. Is 6 a solution of $2x + 7 = 3x$?

67. Is 0 a solution of $2x + 6 = 5x - 1$?

68. Is 2 a solution of $4x + 2 = x + 8$?

69. Is 8 a solution of $2x - 5 = 5$?

70. Is 6 a solution of $3x - 10 = 8$?

71. Is 2 a solution of $x + 6 = x + 6$?

72. Is 10 a solution of $x + 6 = x + 6$?

73. Is 0 a solution of $x = 5x + 15$?

74. Is 1 a solution of $4 = 1 - x$?

Write each phrase as an algebraic expression. Let x represent the unknown number. See Example 8.

75. Fifteen more than a number

76. One-half times a number

77. Five subtracted from a number

78. The quotient of a number and 9

79. Three times a number increased by 22

80. The product of 8 and a number

Write each sentence as an equation. Use x to represent any unknown number. See Example 9.

81. One increased by two equals the quotient of nine and three.

82. Four subtracted from eight is equal to two squared.

83. Three is not equal to four divided by two.

84. The difference of sixteen and four is greater than ten.

85. The sum of 5 and a number is 20.

86. Twice a number is 17.

87. Thirteen minus three times a number is 13.

88. Seven subtracted from a number is 0.

89. The quotient of 12 and a number is $\dfrac{1}{2}$.

90. The sum of 8 and twice a number is 42.

91. In your own words, explain the difference between an expression and an equation.

92. Determine whether each is an expression or an equation.
a. $3x^2 - 26$
b. $3x^2 - 26 = 1$
c. $2x - 5 = 7x - 5$
d. $9y + x - 8$

Solve the following.

△ **93.** The perimeter of a figure is the distance around the figure. The expression $2l + 2w$ represents the perimeter of a rectangle when l is its length and w is its width. Find the perimeter of the following rectangle by substituting 8 for l and 6 for w.

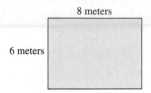

8 meters

6 meters

△ **94.** The expression $a + b + c$ represents the perimeter of a triangle when a, b, and c are the lengths of its sides. Find the perimeter of the following triangle.

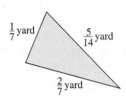

$\frac{1}{7}$ yard $\frac{5}{14}$ yard

$\frac{2}{7}$ yard

△ **95.** The area of a figure is the total enclosed surface of the figure. Area is measured in square units. The expression lw represents the area of a rectangle when l is its length and w is its width. Find the area of the following rectangular-shaped lot.

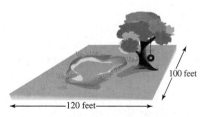

100 feet

← 120 feet →

△ **96.** A trapezoid is a four-sided figure with exactly one pair of parallel sides. The expression $\frac{1}{2}h(B + b)$ represents its area, when B and b are the lengths of the two parallel sides and h is the height between these sides. Find the area if $B = 15$ inches, $b = 7$ inches, and $h = 5$ inches.

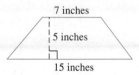

7 inches

5 inches

15 inches

97. The expression $\frac{I}{PT}$ represents the rate of interest being charged if a loan of P dollars for T years required I dollars in interest to be paid. Find the interest rate if a $650 loan for 3 years to buy a used IBM personal computer requires $126.75 in interest to be paid.

98. The expression $\frac{d}{t}$ represents the average speed r in miles per hour if a distance of d miles is traveled in t hours. Find the rate to the nearest whole number if the distance between Dallas, Texas, and Kaw City, Oklahoma, is 432 miles, and it takes Peter Callac 8.5 hours to drive the distance.

99. Sprint Communications Company offers a long-distance telephone plan called Sprint Sense AnyTime that charges $4.95 per month and $0.10 per minute of calling. The expression $4.95 + 0.10m$ represents the monthly long-distance bill for a customer who makes m minutes of long-distance calling on this plan. Find the monthly bill for a customer who makes 228 minutes of long-distance calls on the Sprint Sense AnyTime plan.

100. In forensics, the density of a substance is used to help identify it. The expression $\frac{M}{V}$ represents the density of an object with a mass of M grams and a volume of V milliliters. Find the density of an object having a mass of 29.76 grams and a volume of 12 milliliters.

1.5 ADDING REAL NUMBERS

CD-ROM   SSM

SSG   Video

▶ **OBJECTIVES**

1. Add real numbers with the same sign.
2. Add real numbers with unlike signs.
3. Solve problems that involve addition of real numbers.
4. Find the opposite of a number.

1

Real numbers can be added, subtracted, multiplied, divided, and raised to powers, just as whole numbers can. We use the number line to help picture the addition of real numbers.

Example 1   Add: 3 + 2

Solution   We start at 0 on a number line, and draw an arrow representing 3. This arrow is three units long and points to the right since 3 is positive. From the tip of this arrow, we draw another arrow representing 2. The number below the tip of this arrow is the sum, 5.

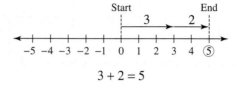

$$3 + 2 = 5$$

Example 2   Add: −1 + (−2)

Solution   We start at 0 on a number line, and draw an arrow representing −1. This arrow is one unit long and points to the left since −1 is negative. From the tip of this arrow, we draw another arrow representing −2. The number below the tip of this arrow is the sum, −3.

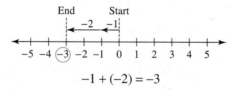

$$-1 + (-2) = -3$$

Thinking of signed numbers as money earned or lost might help make addition more meaningful. Earnings can be thought of as positive numbers. If $1 is earned and later another $3 is earned, the total amount earned is $4. In other words, 1 + 3 = 4.

On the other hand, losses can be thought of as negative numbers. If $1 is lost and later another $3 is lost, a total of $4 is lost. In other words, (−1) + (−3) = −4.

Using a number line each time we add two numbers can be time consuming. Instead, we can notice patterns in the previous examples and write rules for adding signed numbers. When adding two numbers with the same sign, notice that the sign of the sum is the same as the sign of the addends.

ADDING TWO NUMBERS WITH THE SAME SIGN

Add their absolute values. Use their common sign as the sign of the sum.

Example 3   Add.

a. −3 + (−7)   **b.** −1 + (−20)   **c.** −2 + (−10)

Solution   Notice that each time, we are adding numbers with the same sign.

a. −3 + (−7) = −10 ← Add their absolute values: 3 + 7 = 10.
                           └── Use their common sign.

b. $-1 + (-20) = -21$ ← *Add their absolute values: 1 + 20 = 21.*
              └─── *Common sign.*

c. $-2 + (-10) = -12$ ← *Add their absolute values.*
              └─── *Common sign.*

2   Adding numbers whose signs are not the same can also be pictured on a number line.

Example 4   Add: $-4 + 6$

Solution

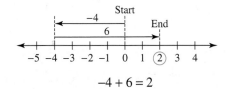

$$-4 + 6 = 2$$

       Using temperature as an example, if the thermometer registers 4 degrees below 0 degrees and then rises 6 degrees, the new temperature is 2 degrees above 0 degrees. Thus, it is reasonable that $-4 + 6 = 2$.

       Once again, we can observe a pattern: when adding two numbers with different signs, the sign of the sum is the same as the sign of the addend whose absolute value is larger.

ADDING TWO NUMBERS WITH DIFFERENT SIGNS

Subtract the smaller absolute value from the larger absolute value. Use the sign of the number whose absolute value is larger as the sign of the sum.

Example 5   Add.

     a. $3 + (-7)$ **b.** $-2 + 10$ **c.** $0.2 + (-0.5)$

Solution   Notice that each time, we are adding numbers with the different signs.

     a. $3 + (-7) = -4$ ← *Subtract their absolute values: 7 − 3 = 4.*
                   └─── *The negative number, −7, has the larger absolute value so the sum is negative.*

     b. $-2 + 10 = 8$ ← *Subtract their absolute values: 10 − 2 = 8.*
                   └─── *The positive number, 10, has the larger absolute value so the sum is understood positive.*

     c. $0.2 + (-0.5) = -0.3$ ← *Subtract their absolute values: 0.5 − 0.2 = 0.3.*
                   └─── *The negative number, −0.5, has the larger absolute value so the sum is negative.*

Example 6   Add.

     a. $-8 + (-11)$ **b.** $-5 + 35$ **c.** $0.6 + (-1.1)$

     d. $-\dfrac{7}{10} + \left(-\dfrac{1}{10}\right)$ **e.** $11.4 + (-4.7)$ **f.** $-\dfrac{3}{8} + \dfrac{2}{5}$

Solution
a. $-8 + (-11) = -19$ Same sign. Add absolute values and use the common sign.

b. $-5 + 35 = 30$ Different signs. Subtract absolute values and use the sign of the number with the larger absolute value. Different signs.

c. $0.6 + (-1.1) = -0.5$

d. $-\dfrac{7}{10} + \left(-\dfrac{1}{10}\right) = -\dfrac{8}{10} = -\dfrac{4}{5}$ Same sign.

e. $11.4 + (-4.7) = 6.7$

f. $-\dfrac{3}{8} + \dfrac{2}{5} = -\dfrac{15}{40} + \dfrac{16}{40} = \dfrac{1}{40}$

> **HELPFUL HINT**
> Don't forget that a common denominator is needed when adding or subtracting fractions. The common denominator here is 40.

Example 7   Add.

a. $3 + (-7) + (-8)$ **b.** $[7 + (-10)] + [-2 + (-4)]$

Solution   **a.** Perform the additions from left to right.

$$3 + (-7) + (-8) = -4 + (-8)$$ Adding numbers with different signs.
$$= -12$$ Adding numbers with like signs.

b. Simplify inside brackets first.

$$[7 + (-10)] + [-2 + (-4)] = [-3] + [-6]$$
$$= -9$$ Add.

> **HELPFUL HINT**
> Don't forget that brackets are grouping symbols. We simplify within them first.

---

3   Positive and negative numbers are often used in everyday life. Stock market returns show gains and losses as positive and negative numbers. Temperatures in cold climates often dip into the negative range, commonly referred to as "below zero" temperatures. Bank statements report deposits and withdrawals as positive and negative numbers.

Example 8   **FINDING THE GAIN OR LOSS OF A STOCK**

During a three-day period, a share of Electronic's International stock recorded the following gains and losses:

Monday	Tuesday	Wednesday
a gain of $2	a loss of $1	a loss of $3

Find the overall gain or loss for the stock for the three days.

Solution   Gains can be represented by positive numbers. Losses can be represented by negative numbers. The overall gain or loss is the sum of the gains and losses.

In words: gain   plus   loss   plus   loss

Translate: $2 \quad + \quad (-1) \quad + \quad (-3) = -2$

The overall loss is $2.

$$\frac{4}{}$$ To help us subtract real numbers in the next section, we first review the concept of opposites. The graph of 4 and −4 is shown on the number line below.

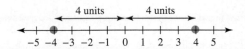

Notice that 4 and −4 lie on opposite sides of 0, and each is 4 units away from 0.

This relationship between −4 and +4 is an important one. Such numbers are known as **opposites** or **additive inverses** of each other.

OPPOSITES OR ADDITIVE INVERSES

Two numbers that are the same distance from 0 but lie on opposite sides of 0 are called opposites or additive inverses of each other.

Example 9   Find the opposite or additive inverse of each number.

   a. 5   **b.** −6   **c.** $\dfrac{1}{2}$ **d.** −4.5

Solution   **a.** The opposite of 5 is −5. Notice that 5 and −5 are on opposite sides of 0 when plotted on a number line and are equal distances away.
   b. The opposite of −6 is 6.
   c. The opposite of $\dfrac{1}{2}$ is $-\dfrac{1}{2}$.
   d. The opposite of −4.5 is 4.5.

We use the symbol "−" to represent the phrase "the opposite of" or "the additive inverse of." In general, if a is a number, we write the opposite or additive inverse of a as $-a$. We know that the opposite of −3 is 3. Notice that this translates as

the opposite of	−3	is	3
↓	↓	↓	↓
−	(**−3**)	=	**3**

This is true in general.

If a is a number, then $-(-a) = a$.

Example 10   Simplify each expression.

   a. $-(-10)$ **b.** $-\left(-\dfrac{1}{2}\right)$ **c.** $-(-2x)$ **d.** $-|-6|$

Solution   **a.** $-(-10) = 10$ **b.** $-\left(-\dfrac{1}{2}\right) = \dfrac{1}{2}$ **c.** $-(-2x) = 2x$
   d. Since $|-6| = 6$, then $-|-6| = -6$.

Let's discover another characteristic about opposites. Notice that the sum of a number and its opposite is 0.

$$10 + (-10) = 0$$
$$-3 + 3 = 0$$
$$\frac{1}{2} + \left(-\frac{1}{2}\right) = 0$$

In general, we can write the following:

The sum of a number a and its opposite $-a$ is 0.

$$a + (-a) = 0$$

Notice that this means that the opposite of 0 is then 0 since $0 + 0 = 0$.

SPOTLIGHT ON DECISION MAKING

Suppose you own stock in XYZ Corp and like to follow the stock's price. You see the following information on a stock ticker indicating the current price of XYZ stock followed by the change in price from its previous closing price. You decide that you would like to make a trade. Should you sell some of the XYZ stock you already own or should you buy additional XYZ stock? Explain your reasoning. What other factors would you want to consider?

XYZ 63.625 -7.125   PCX 46.875 +3.375   JCP 5

MENTAL MATH

Tell whether the sum is a positive number, a negative number, or 0. Do not actually find the sum.

1. $-80 + (-127)$

2. $-162 + 164$

3. $-162 + 162$

4. $-1.26 + (-8.3)$

5. $-3.68 + 0.27$

6. $-\frac{2}{3} + \frac{2}{3}$

Exercise Set 1.5

Add. See Examples 1 through 7

1. $6 + 3$

2. $9 + (-12)$

3. $-6 + (-8)$

4. $-6 + (-14)$

5. $8 + (-7)$

6. $6 + (-4)$

7. $-14 + 2$

8. $-10 + 5$

9. $-2 + (-3)$

10. $-7 + (-4)$

11. $-9 + (-3)$

12. $7 + (-5)$

13. $-7 + 3$

14. $-5 + 9$

15. $10 + (-3)$

16. $8 + (-6)$

17. $5 + (-7)$

18. $3 + (-6)$

19. $-16 + 16$

20. $23 + (-23)$

21. $27 + (-46)$

22. $53 + (-37)$

23. $-18 + 49$

24. $-26 + 14$

25. $-33 + (-14)$

26. $-18 + (-26)$

27. $6.3 + (-8.4)$

28. $9.2 + (-11.4)$

29. $|-8| + (-16)$

30. $|-6| + (-61)$

31. $117 + (-79)$

32. $144 + (-88)$

33. $-9.6 + (-3.5)$

34. $-6.7 + (-7.6)$

35. $-\dfrac{3}{8} + \dfrac{5}{8}$

36. $-\dfrac{5}{12} + \dfrac{7}{12}$

37. $-\dfrac{7}{16} + \dfrac{1}{4}$

38. $-\dfrac{5}{9} + \dfrac{1}{3}$

39. $-\dfrac{7}{10} + \left(-\dfrac{3}{5}\right)$

40. $-\dfrac{5}{6} + \left(-\dfrac{2}{3}\right)$

41. $-15 + 9 + (-2)$

42. $-9 + 15 + (-5)$

43. $-21 + (-16) + (-22)$

44. $-18 + (-6) + (-40)$

45. $-23 + 16 + (-2)$

46. $-14 + (-3) + 11$

47. $|5 + (-10)|$

48. $|7 + (-17)|$

49. $6 + (-4) + 9$

50. $8 + (-2) + 7$

51. $[-17 + (-4)] + [-12 + 15]$

52. $[-2 + (-7)] + [-11 + 22]$

53. $|9 + (-12)| + |-16|$

54. $|43 + (-73)| + |-20|$

55. $-1.3 + [0.5 + (-0.3) + 0.4]$

56. $-3.7 + [0.1 + (-0.6) + 8.1]$

The following bar graph shows the daily low temperatures for a week in Sioux Falls, South Dakota.

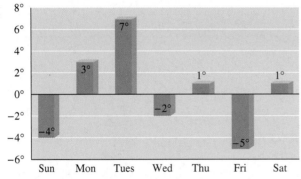

57. On what day of the week was the graphed temperature the highest?

58. On what day of the week was the graphed temperature the lowest?

59. What is the highest temperature shown on the graph?

60. What is the lowest temperature shown on the graph?

61. Find the average daily low temperature for Sunday through Thursday. (*Hint*: To find the average of the five temperatures, find their sum and divide by 5.)

62. Find the average daily low temperature for Tuesday through Thursday.

Solve. See Example 8.

63. The low temperature in Anoka, Minnesota, was $-15°$ last night. During the day it rose only $9°$. Find the high temperature for the day.

64. On January 2, 1943, the temperature was $-4°$ at 7:30 a.m. in Spearfish, South Dakota. Incredibly, it got $49°$ warmer in the next 2 minutes. To what temperature did it rise by 7:32?

65. The lowest elevation on Earth is -1312 feet (that is, 1312 feet below sea level) at the Dead Sea. If you are standing 658 feet above the Dead Sea, what is your elevation? (*Source*: National Geographic Society)

66. The lowest point in Africa is -512 feet at Lake Assal in Djibouti. If you are standing at a point 658 feet above Lake Assal, what is your elevation? (*Source*: Microsoft Encarta)

67. A negative net income results when a company's expenses are more than the money it brings in. Polaroid Corporation had net incomes of $-\$41.1$ million, $-\$126.7$ million, and $-\$51.0$ million in 1996, 1997, and 1998, respectively. What was Polaroid's total net income for these three years? (*Source*: Polaroid Corporation)

68. Apple Computer had net incomes of $-\$816$ million, $-\$1045$ million, and $\$309$ million in 1996, 1997, and 1998, respectively. What was Apple's total net income for these three years? (*Source*: Apple Computer, Inc.)

69. In golf, scores that are under par for the entire round are shown as negative scores; positive scores are shown for scores that are over par, and 0 is par. During the 1999 LPGA Sara Lee Classic, winner Meg Mallon had scores of -6, -7, and -4. What was her overall score? (*Source*: Ladies Professional Golf Association)

70. During the 1999 PGA Masters Tournament, winner Jose Maria Olazabal of Spain had scores of -2, -6, $+1$, and -1. What was his overall score? (*Source*: Professional Golf Association)

Find each additive inverse or opposite. See Example 9.

71. 6   **72.** 4

73. −2   **74.** −8

75. 0   **76.** $-\dfrac{1}{4}$

77. |−6| **78.** |−11|

79. In your own words, explain how to find the opposite of a number.

80. In your own words, explain why 0 is the only number that is its own opposite.

Simplify each of the following. See Example 10.

81. −|−2| **82.** −(−3)

83. −|0| **84.** $\left|-\dfrac{2}{3}\right|$

85. $-\left|-\dfrac{2}{3}\right|$ **86.** −(−7)

87. Explain why adding a negative number to another negative number always gives a negative sum.

88. When a positive and a negative number are added, sometimes the sum is positive, sometimes it is zero, and sometimes it is negative. Explain why this happens.

If a is a positive number and b is a negative number, fill in the blanks with the words positive or negative.

89. −*a* is a _____.

90. −*b* is a _____.

91. *a* + *a* is a _____.

92. *b* + *b* is a _____.

Decide whether the given number is a solution of the given equation.

93. Is −4 a solution of $x + 9 = 5$?

94. Is 10 a solution of $7 = -x + 3$?

95. Is −1 a solution of $y + (-3) = -7$?

96. Is −6 a solution of $1 = y + 7$?

1.6   SUBTRACTING REAL NUMBERS

CD-ROM   SSM

SSG   Video

▶ **OBJECTIVES**

  1. Subtract real numbers.
  2. Add and subtract real numbers.
  3. Evaluate algebraic expressions using real numbers.
  4. Solve problems that involve subtraction of real numbers.

1   Now that addition of signed numbers has been discussed, we can explore subtraction. We know that $9 - 7 = 2$. Notice that $9 + (-7) = 2$, also. This means that

$$9 - 7 = 9 + (-7)$$

Notice that the difference of 9 and 7 is the same as the sum of 9 and the opposite of 7. In general, we have the following.

SUBTRACTING TWO REAL NUMBERS

If *a* and *b* are real numbers, then $a - b = a + (-b)$.

In other words, to find the difference of two numbers, add the first number to the opposite of the second number.

Example 1   Subtract.

  a. −13 − 4   **b.** 5 − (−6)   **c.** 3 − 6   **d.** −1 − (−7)

Solution

a. $\overset{\text{add}}{\overbrace{-13 - 4 = -13 + (-4)}}$ Add −13 to the opposite of +4, which is −4.

$\underset{\text{opposite}}{}$

$= -17$

b. $\overset{\text{add}}{\overbrace{5 - (-6) = 5 + (6)}}$ Add 5 to the opposite of −6, which is 6.

$\underset{\text{opposite}}{}$

$= 11$

c. $3 - 6 = 3 + (-6)$ Add 3 to the opposite of 6, which is −6.

$ = -3$

d. $-1 - (-7) = -1 + (7) = 6$

HELPFUL HINT

Study the patterns indicated.

No change — Change to addition.
— Change to opposite.

$$5 - 11 = 5 + (-11) = -6$$

$$-3 - 4 = -3 + (-4) = -7$$

$$7 - (-1) = 7 + (1) = 8$$

Example 2   Subtract.

a. $5.3 - (-4.6)$ **b.** $-\dfrac{3}{10} - \dfrac{5}{10}$ **c.** $-\dfrac{2}{3} - \left(-\dfrac{4}{5}\right)$

Solution   **a.** $5.3 - (-4.6) = 5.3 + (4.6) = 9.9$

b. $-\dfrac{3}{10} - \dfrac{5}{10} = -\dfrac{3}{10} + \left(-\dfrac{5}{10}\right) = -\dfrac{8}{10} = -\dfrac{4}{5}$

c. $-\dfrac{2}{3} - \left(-\dfrac{4}{5}\right) = -\dfrac{2}{3} + \left(\dfrac{4}{5}\right) = -\dfrac{10}{15} + \dfrac{12}{15} = \dfrac{2}{15}$ The common denominator is 15.

Example 3   Subtract 8 from −4.

Solution   Be careful when interpreting this: The order of numbers in subtraction is important. 8 is to be subtracted **from** −4.

$$-4 - 8 = -4 + (-8) = -12$$

---

2   If an expression contains additions and subtractions, just write the subtractions as equivalent additions. Then simplify from left to right.

Example 4   Simplify:

$$-14 - 8 + 10 - (-6)$$

Solution $-14 - 8 + 10 - (-6) = -14 + (-8) + 10 + 6 = -6$ ▬

When an expression contains parentheses and brackets, remember the order of operations. Start with the innermost set of parentheses or brackets and work your way outward.

Example 5   Simplify each expression.

  a. $-3 + [(-2 - 5) - 2]$ **b.** $2^3 - |10| + [-6 - (-5)]$

Solution   **a.** Start with the innermost sets of parentheses. Rewrite $-2 - 5$ as a sum.

$$\begin{aligned}
-3 + [(-2 - 5) - 2] &= -3 + [(-2 + (-5)) - 2] \\
&= -3 + [(-7) - 2] & \text{Add: } -2 + (-5). \\
&= -3 + [-7 + (-2)] & \text{Write } -7 - 2 \text{ as a sum.} \\
&= -3 + [-9] & \text{Add.} \\
&= -12 & \text{Add.}
\end{aligned}$$

  b. Start simplifying the expression inside the brackets by writing $-6 - (-5)$ as a sum.

$$\begin{aligned}
2^3 - |10| + [-6 - (-5)] &= 2^3 - |10| + [-6 + 5] \\
&= 2^3 - |10| + [-1] & \text{Add.} \\
&= 8 - 10 + (-1) & \text{Evaluate } 2^3 \text{ and } |10|. \\
&= 8 + (-10) + (-1) & \text{Write } 8 - 10 \text{ as a sum.} \\
&= -2 + (-1) & \text{Add.} \\
&= -3 & \text{Add.}
\end{aligned}$$ ▬

3   Knowing how to evaluate expressions for given replacement values is helpful when checking solutions of equations and when solving problems whose unknowns satisfy given expressions. The next example illustrates this.

Example 6   Find the value of each expression when $x = 2$ and $y = -5$.

  a. $\dfrac{x - y}{12 + x}$ **b.** $x^2 - y$

Solution   **a.** Replace x with 2 and y with -5. Be sure to put parentheses around -5 to separate signs. Then simplify the resulting expression.

$$\frac{x - y}{12 + x} = \frac{2 - (-5)}{12 + 2} = \frac{2 + 5}{14} = \frac{7}{14} = \frac{1}{2}$$

  b. Replace the x with 2 and y with -5 and simplify.

$$x^2 - y = 2^2 - (-5) = 4 - (-5) = 4 + 5 = 9$$ ▬

4   One use of positive and negative numbers is in recording altitudes above and below sea level, as shown in the next example.

Example 7 **FINDING THE VARIATION IN ELEVATION**

The lowest point in North America is in Death Valley, at an elevation of 282 feet below sea level. Nearby, Mount Whitney reaches 14,494 feet, the highest point in the United States outside Alaska. How much of a variation in elevation is there between these two extremes?

Solution To find the variation in elevation between the two heights, find the difference of the high point and the low point.

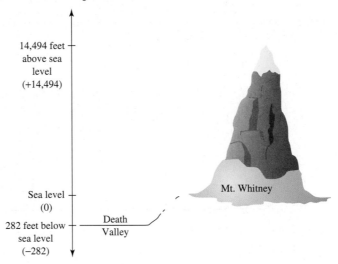

14,494 feet above sea level (+14,494)

Sea level (0)

282 feet below sea level (−282)

Death Valley

Mt. Whitney

In words: | high point | minus | low point |

Translate: 14,494 − (−282) = 14,494 + 282
= 14,776 feet

Thus, the variation in elevation is 14,776 feet.

SPOTLIGHT ON DECISION MAKING

Suppose you are a dental hygienist. As part of a new patient assessment, you measure the depth of the gum tissue pocket around the patient's teeth with a dental probe and record the results. If these pockets deepen over time, this could indicate a problem with gum health or be an indication of gum disease. Now, a year later, you measure the patient's gum tissue pocket depth again to compare to the initial measurements. Based on these findings, would you alert the dentist to a possible problem with the health of the patient's gums? Explain.

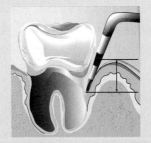

DENTAL CHART

	Gum Tissue Pocket Depth (millimeters)					
Tooth:	22	23	24	25	26	27
Initial	2	3	3	2	4	2
Current	2	2	4	5	6	5

A knowledge of geometric concepts is needed by many professionals, such as doctors, carpenters, electronic technicians, gardeners, machinists, and pilots, just to name a few. With this in mind, we review the geometric concepts of **complementary** and **supplementary angles**.

COMPLEMENTARY AND SUPPLEMENTARY ANGLES

Two angles are **complementary** if their sum is 90°.

$$x + y = 90°$$

Two angles are **supplementary** if their sum is 180°.

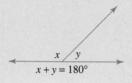

$$x + y = 180°$$

△ **Example 8** Find each unknown complementary or supplementary angle.

a.

b.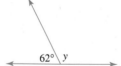

Solution　**a.** These angles are complementary, so their sum is 90°. This means that x is 90° − 38°.

$$x = 90° - 38° = 52°$$

b. These angles are supplementary, so their sum is 180°. This means that y is 180° − 62°.

$$y = 180° - 62° = 118°$$

Exercise Set 1.6

Subtract. See Examples 1 through 5.

1. $-6 - 4$

2. $-12 - 8$

3. $4 - 9$

4. $8 - 11$

5. $16 - (-3)$

6. $12 - (-5)$

7. $\dfrac{1}{2} - \dfrac{1}{3}$

8. $\dfrac{3}{4} - \dfrac{7}{8}$

9. $-16 - (-18)$

10. $-20 - (-48)$

11. $-6 - 5$

12. $-8 - 4$

13. $7 - (-4)$

14. $3 - (-6)$

15. $-6 - (-11)$

16. $-4 - (-16)$

17. $16 - (-21)$

18. $15 - (-33)$

19. $9.7 - 16.1$

20. $8.3 - 11.2$

21. $-44 - 27$

22. $-36 - 51$

23. $-21 - (-21)$

24. $-17 - (-17)$

25. $-2.6 - (-6.7)$

26. $-6.1 - (-5.3)$

27. $-\dfrac{3}{11} - \left(-\dfrac{5}{11}\right)$

28. $-\dfrac{4}{7} - \left(-\dfrac{1}{7}\right)$

29. $-\dfrac{1}{6} - \dfrac{3}{4}$

30. $-\dfrac{1}{10} - \dfrac{7}{8}$

31. $8.3 - (-0.62)$

32. $4.3 - (-0.87)$

Perform the operation. See Example 3.

33. Subtract -5 from 8.

34. Subtract 3 from -2.

35. Subtract -1 from -6.

36. Subtract 17 from 1.

37. Subtract 8 from 7.

38. Subtract 9 from -4.

39. Decrease -8 by 15.

40. Decrease 11 by -14.

41. In your own words, explain why $5 - 8$ simplifies to a negative number.

42. Explain why $6 - 11$ is the same as $6 + (-11)$.

Simplify each expression. (Remember the order of operations.) See Examples 4 and 5.

43. $-10 - (-8) + (-4) - 20$

44. $-16 - (-3) + (-11) - 14$

45. $5 - 9 + (-4) - 8 - 8$

46. $7 - 12 + (-5) - 2 + (-2)$

47. $-6 - (2 - 11)$

48. $-9 - (3 - 8)$

49. $3^3 - 8 \cdot 9$

50. $2^3 - 6 \cdot 3$

51. $2 - 3(8 - 6)$

52. $4 - 6(7 - 3)$

53. $(3 - 6) + 4^2$

54. $(2 - 3) + 5^2$

55. $-2 + [(8 - 11) - (-2 - 9)]$

56. $-5 + [(4 - 15) - (-6) - 8]$

57. $|-3| + 2^2 + [-4 - (-6)]$

58. $|-2| + 6^2 + (-3 - 8)$

Evaluate each expression when $x = -5$, $y = 4$, and $t = 10$. See Example 6.

59. $x - y$

60. $y - x$

61. $|x| + 2t - 8y$

62. $|x + t - 7y|$

63. $\dfrac{9 - x}{y + 6}$

64. $\dfrac{15 - x}{y + 2}$

65. $y^2 - x$

66. $t^2 - x$

67. $\dfrac{|x - (-10)|}{2t}$

68. $\dfrac{|5y - x|}{6t}$

The following bar graph shows each month's average daily low temperature in degrees Fahrenheit for Fairbanks, Alaska. Use the graph to answer Exercises 69–74. See Example 7.

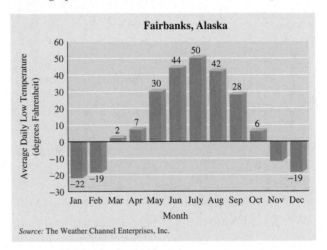

Source: The Weather Channel Enterprises, Inc.

69. In which month is the average daily low temperature the lowest? What is the average daily low temperature for this month?

70. In which month is the average daily low temperature the highest? What is the average daily low temperature for this month?

71. The average daily low temperature for November is 18 degrees less than the average daily low temperature for October. What is the average daily low temperature for November?

72. Which months have the same average daily low temperature?

73. How many degrees warmer is May's average low temperature than February's average low temperature?

74. What is the difference between the average daily low temperatures for July and January?

Solve. See Example 7.

75. Within 24 hours in 1916, the temperature in Browning, Montana, fell from 44 degrees to -56 degrees. How large a drop in temperature was this?

76. Much of New Orleans is just barely above sea level. If George descends 12 feet from an elevation of 5 feet above sea level, what is his new elevation?

77. In a series of plays, the San Francisco 49ers gain 2 yards, lose 5 yards, and then lose another 20 yards. What is their total gain or loss of yardage?

78. In some card games, it is possible to have a negative score. Lavonne Schultz currently has a score of 15 points. She then loses 24 points. What is her new score?

79. Aristotle died in the year −322 (or 322 B.C.). When was he born, if he was 62 years old when he died?

80. Augustus Caesar died in A.D. 14 in his 77th year. When was he born?

81. Tyson Industries stock posted a loss of $1\frac{5}{8}$ points yesterday. If it drops another $\frac{3}{4}$ points today, find its overall change for the two days.

82. A commercial jet liner hits an air pocket and drops 250 feet. After climbing 120 feet, it drops another 178 feet. What is its overall vertical change?

83. The highest point in South America is Mount Aconcagua, Argentina, at an elevation of 22,834 feet. The lowest point is Valdes Peninsula, Argentina, at 131 feet below sea level. How much higher is Mount Aconcagua than Valdes Peninsula? (*Source*: National Geographic Society)

84. The lowest altitude in Antarctica is the Bentley Subglacial Trench at 8327 feet below sea level. The highest altitude is Vinson Massif at an elevation of 16,864 feet above sea level. What is the difference between these altitudes? (*Source*: National Geographic Society)

Find each unknown complementary or supplementary angle. See Example 8.

△ **85.**

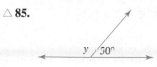

△ **86.**

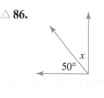

△ **87.**

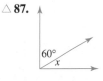

△ **88.**

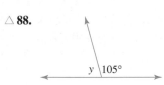

Decide whether the given number is a solution of the given equation.

89. Is −4 a solution of $x - 9 = 5$?

90. Is 3 a solution of $x - 10 = -7$?

91. Is −2 a solution of $-x + 6 = -x - 1$?

92. Is −10 a solution of $-x - 6 = -x - 1$?

93. Is 2 a solution of $-x - 13 = -15$?

94. Is 5 a solution of $4 = 1 - x$?

If a is a positive number and b is a negative number, determine whether each statement is true or false.

95. $a - b$ is always a positive number.

96. $b - a$ is always a negative number.

97. $|b| - |a|$ is always a positive number.

98. $|b - a|$ is always a positive number.

Without calculating, determine whether each answer is positive or negative. Then use a calculator to find the exact difference.

99. 56,875 − 87,262   **100.** 4.362 − 7.0086

1.7 MULTIPLYING AND DIVIDING REAL NUMBERS

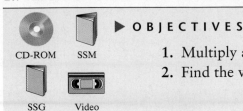

CD-ROM   SSM

SSG   Video

▶ **O B J E C T I V E S**

1. Multiply and divide real numbers.
2. Find the values of algebraic expressions.

1   In this section, we discover patterns for multiplying and dividing real numbers. To discover sign rules for multiplication, recall that multiplication is repeated addition. Thus $3 \cdot 2$ means that 2 is an addend 3 times. That is,

$$2 + 2 + 2 = 3 \cdot 2$$

which equals 6. Similarly, $3 \cdot (-2)$ means -2 is an addend 3 times. That is,

$$(-2) + (-2) + (-2) = 3 \cdot (-2)$$

Since $(-2) + (-2) + (-2) = -6$, then $3 \cdot (-2) = -6$. This suggests that the product of a positive number and a negative number is a negative number.

What about the product of two negative numbers? To find out, consider the following pattern.

Factor decreases by 1 each time

$$\left. \begin{array}{l} -3 \cdot 2 = -6 \\ -3 \cdot 1 = -3 \\ -3 \cdot 0 = 0 \end{array} \right\} \quad \text{Product increases by 3 each time.}$$

This pattern continues as

Factor decreases by 1 each time

$$\left. \begin{array}{l} -3 \cdot -1 = 3 \\ -3 \cdot -2 = 6 \end{array} \right\} \quad \text{Product increases by 3 each time.}$$

This suggests that the product of two negative numbers is a positive number.

> **MULTIPLYING REAL NUMBERS**
>
> 1. The product of two numbers with the *same* sign is a positive number.
> 2. The product of two numbers with *different* signs is a negative number.

Example 1   Find the product.

a. $(-6)(4)$ **b.** $2(-1)$ **c.** $(-5)(-10)$

Solution   **a.** $(-6)(4) = -24$ **b.** $2(-1) = -2$ **c.** $(-5)(-10) = 50$ ■

We know that every whole number multiplied by zero equals zero. This remains true for signed numbers.

> **ZERO AS A FACTOR**
>
> If b is a real number, then $b \cdot 0 = 0$. Also, $0 \cdot b = 0$.

Example 2 Perform the indicated operations.

a. $(7)(0)(-6)$ b. $(-2)(-3)(-4)$ c. $(-1)(5)(-9)$ d. $(-4)(-11) - (5)(-2)$

Solution a. By the order of operations, we multiply from left to right. Notice that, because one of the factors is 0, the product is 0.

$$(7)(0)(-6) = 0(-6) = 0$$

b. Multiply two factors at a time, from left to right.

$$(-2)(-3)(-4) = (6)(-4) \quad \text{Multiply } (-2)(-3).$$
$$= -24$$

c. Multiply from left to right.

$$(-1)(5)(-9) = (-5)(-9) \quad \text{Multiply } (-1)(5).$$
$$= 45$$

d. Follow the rules for order of operation.

$$(-4)(-11) - (5)(-2) = 44 - (-10) \quad \text{Find each products.}$$
$$= 44 + 10 \quad \text{Add 44 to the opposite of } -10.$$
$$= 54 \quad \text{Add.}$$ ▬

Multiplying signed decimals or fractions is carried out exactly the same way as multiplying by integers.

Example 3 Find each product.

a. $(-1.2)(0.05)$ b. $\dfrac{2}{3} \cdot -\dfrac{7}{10}$

Solution a. The product of two numbers with different signs is negative.

$$(-1.2)(0.05) = -\big[(1.2)(0.05)\big]$$
$$= -0.06$$

b. $\dfrac{2}{3} \cdot -\dfrac{7}{10} = -\dfrac{2 \cdot 7}{3 \cdot 10} = -\dfrac{2 \cdot 7}{3 \cdot 2 \cdot 5} = -\dfrac{7}{15}$ ▬

Now that we know how to multiply positive and negative numbers, let's see how we find the values of $(-4)^2$ and -4^2, for example. Although these two expressions look similar, the difference between the two is the parentheses. In $(-4)^2$, the parentheses tell us that the base, or repeated factor, is -4. In -4^2, only 4 is the base. Thus,

$$(-4)^2 = (-4)(-4) = 16 \quad \text{The base is } -4.$$
$$-4^2 = -(4 \cdot 4) = -16 \quad \text{The base is 4.}$$

Example 4   Evaluate.

    a. $(-2)^3$ **b.** -2^3 **c.** $(-3)^2$ **d.** -3^2

Solution   **a.** $(-2)^3 = (-2)(-2)(-2) = -8$ The base is -2.
          b. $-2^3 = -(2 \cdot 2 \cdot 2) = -8$ The base is 2.
          c. $(-3)^2 = (-3)(-3) = 9$ The base is -3.
          d. $-3^2 = -(3 \cdot 3) = -9$ The base is 3.

HELPFUL HINT

Be careful when identifying the base of an exponential expression.

$$(-3)^2 \qquad\qquad\qquad\qquad -3^2$$
$$\text{Base is } -3 \qquad\qquad\qquad \text{Base is } 3$$
$$(-3)^2 = (-3)(-3) = 9 \qquad -3^2 = -(3 \cdot 3) = -9$$

Just as every difference of two numbers $a - b$ can be written as the sum $a + (-b)$, so too every quotient of two numbers can be written as a product. For example, the quotient $6 \div 3$ can be written as $6 \cdot \frac{1}{3}$. Recall that the pair of numbers 3 and $\frac{1}{3}$ has a special relationship. Their product is 1 and they are called reciprocals or **multiplicative inverses** of each other.

RECIPROCALS OR MULTIPLICATIVE INVERSES

Two numbers whose product is 1 are called reciprocals or multiplicative inverses of each other.

Notice that **0 has no multiplicative inverse** since 0 multiplied by any number is never 1 but always 0.

Example 5   Find the reciprocal of each number.

    a. 22   **b.** $\dfrac{3}{16}$ **c.** -10 **d.** $-\dfrac{9}{13}$

Solution   **a.** The reciprocal of 22 is $\frac{1}{22}$ since $22 \cdot \frac{1}{22} = 1$.
          b. The reciprocal of $\frac{3}{16}$ is $\frac{16}{3}$ since $\frac{3}{16} \cdot \frac{16}{3} = 1$.
          c. The reciprocal of -10 is $-\frac{1}{10}$.
          d. The reciprocal of $-\frac{9}{13}$ is $-\frac{13}{9}$.

We may now write a quotient as an equivalent product.

QUOTIENT OF TWO REAL NUMBERS

If a and b are real numbers and b is not 0, then

$$\frac{a}{b} = a \cdot \frac{1}{b}$$

In other words, the quotient of two real numbers is the product of the first number and the multiplicative inverse or reciprocal of the second number.

Example 6 Use the definition of the quotient of two numbers to find each quotient.

a. $-18 \div 3$ **b.** $\dfrac{-14}{-2}$ **c.** $\dfrac{20}{-4}$

Solution **a.** $-18 \div 3 = -18 \cdot \dfrac{1}{3} = -6$ **b.** $\dfrac{-14}{-2} = -14 \cdot -\dfrac{1}{2} = 7$

c. $\dfrac{20}{-4} = 20 \cdot -\dfrac{1}{4} = -5$

Since the quotient $a \div b$ can be written as the product $a \cdot \frac{1}{b}$, it follows that sign patterns for dividing two real numbers are the same as sign patterns for multiplying two real numbers.

MULTIPLYING AND DIVIDING REAL NUMBERS

1. The product or quotient of two numbers with the *same* sign is a positive number.
2. The product or quotient of two numbers with *different* signs is a negative number.

Example 7 Find each quotient.

a. $\dfrac{-24}{-4}$ **b.** $\dfrac{-36}{3}$ **c.** $\dfrac{2}{3} \div \left(-\dfrac{5}{4}\right)$

Solution **a.** $\dfrac{-24}{-4} = 6$ **b.** $\dfrac{-36}{3} = -12$ **c.** $\dfrac{2}{3} \div \left(-\dfrac{5}{4}\right) = \dfrac{2}{3} \cdot \left(-\dfrac{4}{5}\right) = -\dfrac{8}{15}$

The definition of the quotient of two real numbers does not allow for division by 0 because 0 does not have a multiplicative inverse. There is no number we can multiply 0 by to get 1. How then do we interpret $\frac{3}{0}$? We say that division by 0 is not allowed or not defined and that $\frac{3}{0}$ does not represent a real number. The denominator of a fraction can never be 0.

Can the numerator of a fraction be 0? Can we divide 0 by a number? Yes. For example,

$$\frac{0}{3} = 0 \cdot \frac{1}{3} = 0$$

In general, the quotient of 0 and any nonzero number is 0.

ZERO AS A DIVISOR OR DIVIDEND

1. The quotient of any nonzero real number and 0 is undefined. In symbols, if $a \neq 0$, $\dfrac{a}{0}$ is **undefined**.
2. The quotient of 0 and any real number except 0 is 0. In symbols, if $a \neq 0$, $\dfrac{0}{a} = 0$.

Example 8 Perform the indicated operations.

  a. $\dfrac{1}{0}$ **b.** $\dfrac{0}{-3}$ **c.** $\dfrac{0(-8)}{2}$

Solution **a.** $\dfrac{1}{0}$ is undefined   **b.** $\dfrac{0}{-3} = 0$ **c.** $\dfrac{0(-8)}{2} = \dfrac{0}{2} = 0$

Notice that $\dfrac{12}{-2} = -6, -\dfrac{12}{2} = -6,$ and $\dfrac{-12}{2} = -6.$ This means that

$$\dfrac{12}{-2} = -\dfrac{12}{2} = \dfrac{-12}{2}$$

In words, a single negative sign in a fraction can be written in the denominator, in the numerator, or in front of the fraction without changing the value of the fraction. Thus,

$$\dfrac{1}{-7} = \dfrac{-1}{7} = -\dfrac{1}{7}$$

In general, if a and b are real numbers, $b \neq 0$, $\dfrac{a}{-b} = \dfrac{-a}{b} = -\dfrac{a}{b}$.

Examples combining basic arithmetic operations along with the principles of order of operations help us to review these concepts.

Example 9 Simplify each expression.

  a. $\dfrac{(-12)(-3) + 3}{-7 - (-2)}$ **b.** $\dfrac{2(-3)^2 - 20}{-5 + 4}$

Solution **a.** First, simplify the numerator and denominator separately, then divide.

$$\dfrac{(-12)(-3) + 3}{-7 - (-2)} = \dfrac{36 + 3}{-7 + 2}$$
$$= \dfrac{39}{-5} \text{ or } -\dfrac{39}{5}$$

b. Simplify the numerator and denominator separately, then divide.

$$\dfrac{2(-3)^2 - 20}{-5 + 4} = \dfrac{2 \cdot 9 - 20}{-5 + 4} = \dfrac{18 - 20}{-5 + 4} = \dfrac{-2}{-1} = 2$$

2 Using what we have learned about multiplying and dividing real numbers, we continue to practice evaluating algebraic expressions.

Example 10 If $x = -2$ and $y = -4$, evaluate each expression.

  a. $5x - y$ **b.** $x^3 - y^2$ **c.** $\dfrac{3x}{2y}$

Solution   **a.** Replace x with -2 and y with -4 and simplify.

$$5x - y = 5(-2) - (-4) = -10 - (-4) = -10 + 4 = -6$$

b. Replace x with -2 and y with -4.

$$x^3 - y^2 = (-2)^3 - (-4)^2 \qquad \text{Substitute the given values for the variables.}$$
$$= -8 - (16) \qquad \text{Evaluate exponential expressions.}$$
$$= -8 + (-16) \qquad \text{Write as a sum.}$$
$$= -24 \qquad \text{Add.}$$

c. Replace x with -2 and y with -4 and simplify.

$$\frac{3x}{2y} = \frac{3(-2)}{2(-4)} = \frac{-6}{-8} = \frac{3}{4}$$

CALCULATOR EXPLORATIONS

Entering Negative Numbers on a Scientific Calculator

To enter a negative number on a scientific calculator, find a key marked $\boxed{+/-}$. (On some calculators, this key is marked $\boxed{\text{CHS}}$ for "change sign.") To enter -8, for example, press the keys $\boxed{8}$ $\boxed{+/-}$. The display will read $\boxed{-8}$.

Entering Negative Numbers on a Graphing Calculator

To enter a negative number on a graphing calculator, find a key marked $\boxed{(-)}$. Do not confuse this key with the key $\boxed{-}$, which is used for subtraction. To enter -8, for example, press the keys $\boxed{(-)}$ $\boxed{8}$. The display will read $\boxed{-8}$.

Operations with Real Numbers

To evaluate $-2(7 - 9) - 20$ on a calculator, press the keys

$\boxed{2}$ $\boxed{+/-}$ $\boxed{\times}$ $\boxed{(}$ $\boxed{7}$ $\boxed{-}$ $\boxed{9}$ $\boxed{)}$ $\boxed{-}$ $\boxed{2}$ $\boxed{0}$ $\boxed{=}$, or

$\boxed{(-)}$ $\boxed{2}$ $\boxed{(}$ $\boxed{7}$ $\boxed{-}$ $\boxed{9}$ $\boxed{)}$ $\boxed{-}$ $\boxed{2}$ $\boxed{0}$ $\boxed{\text{ENTER}}$.

The display will read $\boxed{-16}$ or $\boxed{\begin{array}{c}-2(7-9)-20 \\ -16\end{array}}$.

Use a calculator to simplify each expression.

1. $-38(26 - 27)$
2. $-59(-8) + 1726$
3. $134 + 25(68 - 91)$
4. $45(32) - 8(218)$
5. $\dfrac{-50(294)}{175 - 265}$
6. $\dfrac{-444 - 444.8}{-181 - 324}$
7. $9^5 - 4550$
8. $5^8 - 6259$
9. $(-125)^2$ (Be careful.)
10. -125^2 (Be careful.)

MENTAL MATH

Answer the following with positive or negative.

1. The product of two negative numbers is a _____ number.
2. The quotient of two negative numbers is a _____ number.
3. The quotient of a positive number and a negative number is a _____ number.
4. The product of a positive number and a negative number is a _____ number.
5. The reciprocal of a positive number is a _____ number.
6. The opposite of a positive number is a _____ number.

Exercise Set 1.7

Multiply. See Examples 1 through 3.

1. $-6(4)$
2. $-8(5)$
3. $2(-1)$
4. $7(-4)$
5. $-5(-10)$
6. $-6(-11)$
7. $-3 \cdot 4$
8. $-2 \cdot 8$
9. $-6(-7)$
10. $-6(-9)$
11. $2(-9)$
12. $3(-5)$
13. $-\dfrac{1}{2}\left(-\dfrac{3}{5}\right)$
14. $-\dfrac{1}{8}\left(-\dfrac{1}{3}\right)$
15. $-\dfrac{3}{4}\left(-\dfrac{8}{9}\right)$
16. $-\dfrac{5}{6}\left(-\dfrac{3}{10}\right)$
17. $5(-1.4)$
18. $6(-2.5)$
19. $-0.2(-0.7)$
20. $-0.5(-0.3)$
21. $-10(80)$
22. $-20(60)$
23. $4(-7)$
24. $5(-9)$
25. $(-5)(-5)$
26. $(-7)(-7)$
27. $\dfrac{2}{3}\left(-\dfrac{4}{9}\right)$
28. $\dfrac{2}{7}\left(-\dfrac{2}{11}\right)$
29. $-11(11)$
30. $-12(12)$
31. $-\dfrac{20}{25}\left(\dfrac{5}{16}\right)$
32. $-\dfrac{25}{36}\left(\dfrac{6}{15}\right)$
33. $-2.1(-0.4)$
34. $-1.3(-0.6)$
35. $(-1)(2)(-3)(-5)$
36. $(-2)(-3)(-4)(-2)$
37. $(2)(-1)(-3)(5)(3)$
38. $(3)(-5)(-2)(-1)(-2)$

Decide whether each statement is true or false.

39. The product of three negative integers is negative.
40. The product of three positive integers is positive.
41. The product of four negative integers is negative.
42. The product of four positive integers is positive.

Evaluate. See Example 4.

43. $(-2)^4$
44. -2^4
45. -1^5
46. $(-1)^5$
47. $(-5)^2$
48. -5^2
49. -7^2
50. $(-7)^2$

Find each reciprocal or multiplicative inverse. See Example 5.

51. 9
52. 100
53. $\dfrac{2}{3}$
54. $\dfrac{1}{7}$
55. -14
56. -8
57. $-\dfrac{3}{11}$
58. $-\dfrac{6}{13}$
59. 0.2
60. 1.5
61. $\dfrac{1}{-6.3}$
62. $\dfrac{1}{-8.9}$

Divide. See Examples 6 through 8.

63. $\dfrac{18}{-2}$
64. $\dfrac{20}{-10}$
65. $\dfrac{-16}{-4}$
66. $\dfrac{-18}{-6}$
67. $\dfrac{-48}{12}$
68. $\dfrac{-60}{5}$
69. $\dfrac{0}{-4}$
70. $\dfrac{0}{-9}$
71. $-\dfrac{15}{3}$
72. $-\dfrac{24}{8}$

73. $\dfrac{5}{0}$

74. $\dfrac{3}{0}$

75. $\dfrac{-12}{-4}$

76. $\dfrac{-45}{-9}$

77. $\dfrac{30}{-2}$

70. $\dfrac{14}{-2}$

79. $\dfrac{6}{7} \div \left(-\dfrac{1}{3}\right)$

80. $\dfrac{4}{5} \div \left(-\dfrac{1}{2}\right)$

81. $-\dfrac{5}{9} \div \left(-\dfrac{3}{4}\right)$

82. $-\dfrac{1}{10} \div \left(-\dfrac{8}{11}\right)$

83. $-\dfrac{4}{9} \div \dfrac{4}{9}$

84. $-\dfrac{5}{12} \div \dfrac{5}{12}$

Simplify. See Example 9.

85. $\dfrac{-9(-3)}{-6}$

86. $\dfrac{-6(-3)}{-4}$

87. $\dfrac{12}{9 - 12}$

88. $\dfrac{-15}{1 - 4}$

89. $\dfrac{-6^2 + 4}{-2}$

90. $\dfrac{3^2 + 4}{5}$

91. $\dfrac{8 + (-4)^2}{4 - 12}$

92. $\dfrac{6 + (-2)^2}{4 - 9}$

93. $\dfrac{22 + (3)(-2)}{-5 - 2}$

94. $\dfrac{-20 + (-4)(3)}{1 - 5}$

95. $\dfrac{-3 - 5^2}{2(-7)}$

96. $\dfrac{-2 - 4^2}{3(-6)}$

97. $\dfrac{6 - 2(-3)}{4 - 3(-2)}$

98. $\dfrac{8 - 3(-2)}{2 - 5(-4)}$

99. $\dfrac{-3 - 2(-9)}{-15 - 3(-4)}$

100. $\dfrac{-4 - 8(-2)}{-9 - 2(-3)}$

101. $\dfrac{|5 - 9| + |10 - 15|}{|2(-3)|}$

102. $\dfrac{|-3 + 6| + |-2 + 7|}{|-2 \cdot 2|}$

If $x = -5$ and $y = -3$, evaluate each expression. See Example 10.

103. $3x + 2y$

104. $4x + 5y$

105. $2x^2 - y^2$

106. $x^2 - 2y^2$

107. $x^3 + 3y$

108. $y^3 + 3x$

109. $\dfrac{2x - 5}{y - 2}$

110. $\dfrac{2y - 12}{x - 4}$

111. $\dfrac{6 - y}{x - 4}$

112. $\dfrac{4 - 2x}{y + 3}$

113. Amazon.com is an Internet bookseller. At the end of 1998, Amazon posted a net income of −$124.5 million. If

this continued, what would Amazon's income be after four years? (*Source:* Amazon.com, Inc.)

114. Union Pacific provides rail transportation services. At the end of 1998, Union Pacific posted a net income of −$633 million. If this continued, what would Union Pacific's income be after three years? (*Source:* Union Pacific Corp.)

115. Explain why the product of an even number of negative numbers is a positive number.

116. If a and b are any real numbers, is the statement $a \cdot b = b \cdot a$ always true? Why or why not?

117. Find any real numbers that are their own reciprocal.

118. Explain why 0 has no reciprocal.

If q is a negative number, r is a negative number, and t is a positive number, determine whether each expression simplifies to a positive or negative number If it is not possible to determine, state so.

119. $\dfrac{q}{r \cdot t}$

120. $q^2 \cdot r \cdot t$

121. $q + t$

122. $t + r$

123. $t(q + r)$

124. $r(q - t)$

Write each of the following as an expression and evaluate.

125. The sum of −2 and the quotient of −15 and 3

126. The sum of 1 and the product of −8 and −5

127. Twice the sum of −5 and −3

128. 7 subtracted from the quotient of 0 and 5

Decide whether the given number is a solution of the given equation.

129. Is 7 a solution of $-5x = -35$?

130. Is −4 a solution of $2x = x - 1$?

131. Is −20 a solution of $\dfrac{x}{-10} = 2$?

132. Is −3 a solution of $\dfrac{45}{x} = -15$?

133. Is 5 a solution of $-3x - 5 = -20$?

134. Is −4 a solution of $2x + 4 = x + 8$?

1.8 PROPERTIES OF REAL NUMBERS

CD-ROM   SSM

SSG   Video

▶ **OBJECTIVES**

1. Use the commutative and associative properties.
2. Use the distributive property.
3. Use the identity and inverse properties.

1 In this section we give names to properties of real numbers with which we are already familiar. Throughout this section, the variables a, b, and c represent real numbers.

We know that order does not matter when adding numbers. For example, we know that $7 + 5$ is the same as $5 + 7$. This property is given a special name—the **commutative property of addition**. We also know that order does not matter when multiplying numbers. For example, we know that $-5(6) = 6(-5)$. This property means that multiplication is commutative also and is called the **commutative property of multiplication**.

COMMUTATIVE PROPERTIES

Addition: $a + b = b + a$

Multiplication: $a \cdot b = b \cdot a$

These properties state that the *order* in which any two real numbers are added or multiplied does not change their sum or product. For example, if we let $a = 3$ and $b = 5$, then the commutative properties guarantee that

$$3 + 5 = 5 + 3 \quad \text{and} \quad 3 \cdot 5 = 5 \cdot 3$$

> **HELPFUL HINT**
> Is subtraction also commutative? Try an example. Is $3 - 2 = 2 - 3$? **No!** The left side of this statement equals 1; the right side equals -1. There is no commutative property of subtraction. Similarly, there is no commutative property for division. For example, $10 \div 2$ does not equal $2 \div 10$.

Example 1 Use a commutative property to complete each statement.

a. $x + 5 = $ _____ **b.** $3 \cdot x = $ _____

Solution **a.** $x + 5 = 5 + x$ By the commutative property of addition
b. $3 \cdot x = x \cdot 3$ By the commutative property of multiplication

Let's now discuss grouping numbers. We know that when we add three numbers, the way in which they are grouped or associated does not change their sum. For example, we know that $2 + (3 + 4) = 2 + 7 = 9$. This result is the same if we group the numbers differently. In other words, $(2 + 3) + 4 = 5 + 4 = 9$, also. Thus, $2 + (3 + 4) = (2 + 3) + 4$. This property is called the **associative property of addition**.

We also know that changing the grouping of numbers when multiplying does not change their product. For example, $2 \cdot (3 \cdot 4) = (2 \cdot 3) \cdot 4$ (check it). This is the **associative property of multiplication**.

ASSOCIATIVE PROPERTIES

Addition:	$(a + b) + c = a + (b + c)$
Multiplication:	$(a \cdot b) \cdot c = a \cdot (b \cdot c)$

These properties state that the way in which three numbers are *grouped* does not change their sum or their product.

Example 2   Use an associative property to complete each statement.

a. $5 + (4 + 6) = $ _____   **b.** $(-1 \cdot 2) \cdot 5 = $ _____

Solution   **a.** $5 + (4 + 6) = (5 + 4) + 6$ By the associative property of addition
b. $(-1 \cdot 2) \cdot 5 = -1 \cdot (2 \cdot 5)$ By the associative property of multiplication   ∎

HELPFUL HINT
Remember the difference between the commutative properties and the associative properties. The commutative properties have to do with the *order* of numbers, and the associative properties have to do with the *grouping* of numbers.

Let's now illustrate how these properties can help us simplify expressions.

Example 3   Simplify each expression.

a. $10 + (x + 12)$ **b.** $-3(7x)$

Solution   **a.** $10 + (x + 12) = 10 + (12 + x)$ By the commutative property of addition
$= (10 + 12) + x$ By the associative property of addition
$= 22 + x$ Add.
b. $-3(7x) = (-3 \cdot 7)x$ By the associative property of multiplication
$= -21x$ Multiply.   ∎

---

2   The **distributive property of multiplication over addition** is used repeatedly throughout algebra. It is useful because it allows us to write a product as a sum or a sum as a product.

We know that $7(2 + 4) = 7(6) = 42$. Compare that with $7(2) + 7(4) = 14 + 28 = 42$. Since both original expressions equal 42, they must equal each other, or

$$7(2 + 4) = 7(2) + 7(4)$$

This is an example of the distributive property. The product on the left side of the equal sign is equal to the sum on the right side. We can think of the 7 as being distributed to each number inside the parentheses.

DISTRIBUTIVE PROPERTY OF MULTIPLICATION OVER ADDITION

$$a(b + c) = ab + ac$$

Since multiplication is commutative, this property can also be written as

$$(b + c)a = ba + ca$$

The distributive property can also be extended to more than two numbers inside the parentheses. For example,

$$3(x + y + z) = 3(x) + 3(y) + 3(z)$$
$$= 3x + 3y + 3z$$

Since we define subtraction in terms of addition, the distributive property is also true for subtraction. For example

$$2(x - y) = 2(x) - 2(y)$$
$$= 2x - 2y$$

Example 4   Use the distributive property to write each expression without parentheses. Then simplify the result.

    a. $2(x + y)$ **b.** $-5(-3 + 2z)$ **c.** $5(x + 3y - z)$
    d. $-1(2 - y)$ **e.** $-(3 + x - w)$ **f.** $4(3x + 7) + 10$

Solution   **a.** $2(x + y) = 2 \cdot x + 2 \cdot y$
                $= 2x + 2y$

      b. $-5(-3 + 2z) = -5(-3) + (-5)(2z)$
                       $= 15 - 10z$

> **HELPFUL HINT**
> Notice in part **e** that $-(3 + x - w)$ is first rewritten as $-1(3 + x - w)$.

      c. $5(x + 3y - z) = 5(x) + 5(3y) - 5(z)$
                    $= 5x + 15y - 5z$

      d. $-1(2 - y) = (-1)(2) - (-1)(y)$
                 $= -2 + y$

      e. $-(3 + x - w) = -1(3 + x - w)$
                     $= (-1)(3) + (-1)(x) - (-1)(w)$
                     $= -3 - x + w$

      f. $4(3x + 7) + 10 = 4(3x) + 4(7) + 10$ *Apply the distributive property.*
                       $= 12x + 28 + 10$ *Multiply.*
                       $= 12x + 38$ *Add.*

The distributive property can also be used to write a sum as a product.

Example 5   Use the distributive property to write each sum as a product.

    a. $8 \cdot 2 + 8 \cdot x$ **b.** $7s + 7t$

Solution   **a.** $8 \cdot 2 + 8 \cdot x = 8(2 + x)$ **b.** $7s + 7t = 7(s + t)$

3 Next, we look at the **identity properties**.

The number 0 is called the identity for addition because when 0 is added to any real number, the result is the same real number. In other words, the *identity* of the real number is not changed.

The number 1 is called the identity for multiplication because when a real number is multiplied by 1, the result is the same real number. In other words, the *identity* of the real number is not changed.

IDENTITIES FOR ADDITION AND MULTIPLICATION

0 is the identity element for addition.

$$a + 0 = a \quad \text{and} \quad 0 + a = a$$

1 is the identity element for multiplication.

$$a \cdot 1 = a \quad \text{and} \quad 1 \cdot a = a$$

Notice that 0 is the *only* number that can be added to any real number with the result that the sum is the same real number. Also, 1 is the *only* number that can be multiplied by any real number with the result that the product is the same real number.

Additive inverses or **opposites** were introduced in Section 1.3. Two numbers are called additive inverses or opposites if their sum is 0. The additive inverse or opposite of 6 is -6 because $6 + (-6) = 0$. The additive inverse or opposite of -5 is 5 because $-5 + 5 = 0$.

Reciprocals or **multiplicative inverses** were introduced in Section 1.2. Two nonzero numbers are called reciprocals or multiplicative inverses if their product is 1. The reciprocal or multiplicative inverse of $\frac{2}{3}$ is $\frac{3}{2}$ because $\frac{2}{3} \cdot \frac{3}{2} = 1$. Likewise, the reciprocal of -5 is $-\frac{1}{5}$ because $-5\left(-\frac{1}{5}\right) = 1$.

ADDITIVE OR MULTIPLICATIVE INVERSES

The numbers a and $-a$ are additive inverses or opposites of each other because their sum is 0; that is,

$$a + (-a) = 0$$

The numbers b and $\frac{1}{b}$ (for $b \neq 0$) are reciprocals or multiplicative inverses of each other because their product is 1; that is,

$$b \cdot \frac{1}{b} = 1$$

Example 6 Name the property illustrated by each true statement.

Solution
a. $3 \cdot y = y \cdot 3$ Commutative property of multiplication (order changed)
b. $(x + 7) + 9 = x + (7 + 9)$ Associative property of addition (grouping changed)
c. $(b + 0) + 3 = b + 3$ Identity element for addition

d. $2 \cdot (z \cdot 5) = 2 \cdot (5 \cdot z)$ Commutative property of multiplication (order changed)

e. $-2 \cdot \left(-\dfrac{1}{2}\right) = 1$ Multiplicative inverse property

f. $-2 + 2 = 0$ Additive inverse property

g. $-6 \cdot (y \cdot 2) = (-6 \cdot 2) \cdot y$ Commutative and associative properties of multiplication (order and grouping changed)

Exercise Set 1.8

Use a commutative property to complete each statement. See Examples 1 and 3.

1. $x + 16 =$ _____   **2.** $4 + y =$ _____

3. $-4 \cdot y =$ _____   **4.** $-2 \cdot x =$ _____

5. $xy =$ _____   **6.** $ab =$ _____

7. $2x + 13 =$ _____   **8.** $19 + 3y =$ _____

Use an associative property to complete each statement. See Examples 2 and 3.

9. $(xy) \cdot z =$ _____   **10.** $3 \cdot (xy) =$ _____

11. $2 + (a + b) =$ _____   **12.** $(y + 4) + z =$ _____

13. $4 \cdot (ab) =$ _____   **14.** $(-3y) \cdot z =$ _____

15. $(a + b) + c =$ _____   **16.** $6 + (r + s) =$ _____

Use the commutative and associative properties to simplify each expression. See Example 3.

17. $8 + (9 + b)$ **18.** $(r + 3) + 11$

19. $4(6y)$ **20.** $2(42x)$

21. $\dfrac{1}{5}(5y)$ **22.** $\dfrac{1}{8}(8z)$

23. $(13 + a) + 13$ **24.** $7 + (x + 4)$

25. $-9(8x)$ **26.** $-3(12y)$

27. $\dfrac{3}{4}\left(\dfrac{4}{3}s\right)$ **28.** $\dfrac{2}{7}\left(\dfrac{7}{2}r\right)$

29. Write an example that shows that division is not commutative.

30. Write an example that shows that subtraction is not commutative.

Use the distributive property to write each expression without parentheses. Then simplify the result. See Example 4.

31. $4(x + y)$ **32.** $7(a + b)$

33. $9(x - 6)$ **34.** $11(y - 4)$

35. $2(3x + 5)$ **36.** $5(7 + 8y)$

37. $7(4x - 3)$ **38.** $3(8x - 1)$

39. $3(6 + x)$ **40.** $2(x + 5)$

41. $-2(y - z)$ **42.** $-3(z - y)$

43. $-7(3y + 5)$ **44.** $-5(2r + 11)$

45. $5(x + 4m + 2)$ **46.** $8(3y + z - 6)$

47. $-4(1 - 2m + n)$ **48.** $-4(4 + 2p + 5)$

49. $-(5x + 2)$ **50.** $-(9r + 5)$

51. $-(r - 3 - 7p)$ **52.** $-(q - 2 + 6r)$

53. $\dfrac{1}{2}(6x + 8)$ **54.** $\dfrac{1}{4}(4x - 2)$

55. $-\dfrac{1}{3}(3x - 9y)$ **56.** $-\dfrac{1}{5}(10a - 25b)$

57. $3(2r + 5) - 7$ **58.** $10(4s + 6) - 40$

59. $-9(4x + 8) + 2$ **60.** $-11(5x + 3) + 10$

61. $-4(4x + 5) - 5$ **62.** $-6(2x + 1) - 1$

Use the distributive property to write each sum as a product. See Example 5.

63. $4 \cdot 1 + 4 \cdot y$ **64.** $14 \cdot z + 14 \cdot 5$

65. $11x + 11y$ **66.** $9a + 9b$

67. $(-1) \cdot 5 + (-1) \cdot x$ **68.** $(-3)a + (-3)b$

69. $30a + 30b$ **70.** $25x + 25y$

Name the properties illustrated by each true statement. See Example 6.

71. $3 \cdot 5 = 5 \cdot 3$

72. $4(3 + 8) = 4 \cdot 3 + 4 \cdot 8$

73. $2 + (x + 5) = (2 + x) + 5$

74. $(x + 9) + 3 = (9 + x) + 3$

75. $9(3 + 7) = 9 \cdot 3 + 9 \cdot 7$

76. $1 \cdot 9 = 9$

77. $(4 \cdot y) \cdot 9 = 4 \cdot (y \cdot 9)$

78. $6 \cdot \dfrac{1}{6} = 1$

79. $0 + 6 = 6$

80. $(a + 9) + 6 = a + (9 + 6)$

81. $-4(y + 7) = -4 \cdot y + (-4) \cdot 7$

82. $(11 + r) + 8 = (r + 11) + 8$

83. $-4 \cdot (8 \cdot 3) = (8 \cdot -4) \cdot 3$

84. $r + 0 = r$

Fill in the table with the opposite (additive inverse), and the reciprocal (multiplicative inverse). Assume that the value of each expression is not 0.

	Expression	Opposite	Reciprocal
85.	8		
86.	$-\frac{2}{3}$		
87.	x		
88.	$4y$		
89.	$2x$		
90.	$-7x$		

Determine which pairs of actions are commutative.

91. "taking a test" and "studying for the test"

92. "putting on your shoes" and "putting on your socks"

93. "putting on your left shoe" and "putting on your right shoe"

94. "reading the sports section" and "reading the comics section"

95. Explain why 0 is called the identity element for addition.

96. Explain why 1 is called the identity element for multiplication.

1.9 READING GRAPHS

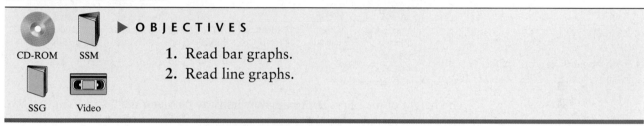

CD-ROM   SSM   SSG   Video

▶ **OBJECTIVES**

1. Read bar graphs.
2. Read line graphs.

In today's world, where the exchange of information must be fast and entertaining, graphs are becoming increasingly popular. They provide a quick way of making comparisons, drawing conclusions, and approximating quantities.

1 A **bar graph** consists of a series of bars arranged vertically or horizontally. The bar graph in Example 1 shows a comparison of the rates charged by selected electricity companies. The names of the companies are listed horizontally and a bar is shown for each company. Corresponding to the height of the bar for each company is a number along a vertical axis. These vertical numbers are cents charged for each kilowatt-hour of electricity used.

◈ **Example 1** The following bar graph shows the cents charged per kilowatt-hour for selected electricity companies.

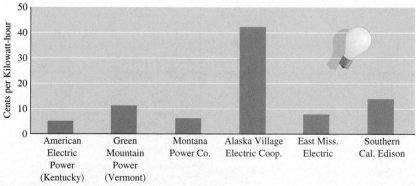

Source: Electric Company Listed

a. Which company charges the highest rate?

b. Which company charges the lowest rate?

c. Approximate the electricity rate charged by the first four companies listed.

d. Approximate the difference in the rates charged by the companies in parts (a) and (b).

Solution **a.** The tallest bar corresponds to the company that charges the highest rate. Alaska Village Electric Cooperative charges the highest rate.

b. The shortest bar corresponds to the company that charges the lowest rate. American Electric Power in Kentucky charges the lowest rate.

c. To approximate the rate charged by American Electric Power, we go to the top of the bar that corresponds to this company. From the top of the bar, we move horizontally to the left until the vertical axis is reached.

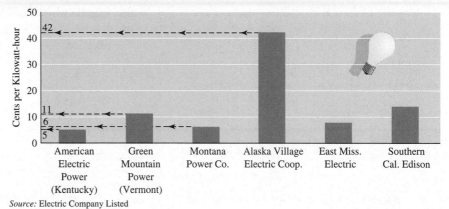

Source: Electric Company Listed

The height of the bar is approximately halfway between the 0 and 10 marks. We therefore conclude that

American Electric Power charges approximately 5¢ per kilowatt-hour.
Green Mountain Power charges approximately 11¢ per kilowatt-hour.
Montana Power Co. charges approximately 6¢ per kilowatt-hour.
Alaska Village Electric charges approximately 42¢ per kilowatt-hour.

d. The difference in rates for Alaska Village Electric Cooperative and American Electric Power is approximately 42¢ − 5¢ or 37¢.

Example 2 The following bar graph shows Disney's top animated films and the amount of money they generated at theaters.

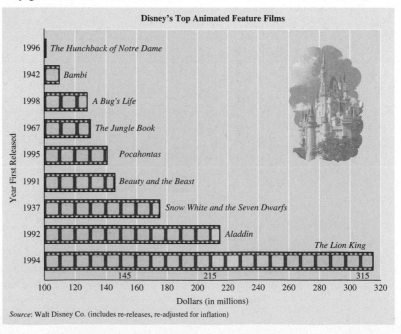

Source: Walt Disney Co. (includes re-releases, re-adjusted for inflation)

a. Find the film shown that generated the most income for Disney and approximate the income.

b. How much more money did the film *Aladdin* make than the film *Beauty and the Beast*?

Solution   **a.** Since these bars are arranged horizontally, we look for the longest bar, which is the bar representing the film *The Lion King*. To approximate the income from this film, we move from the right edge of this bar vertically downward to the dollars axis. This film generated approximately 315 million dollars, or $315,000,000, the most income for Disney.

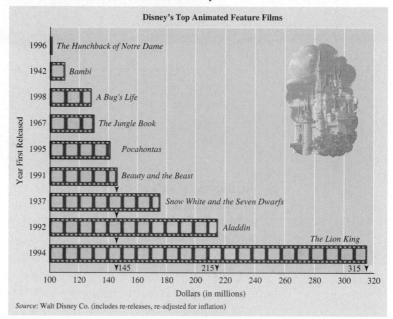

Disney's Top Animated Feature Films

Source: Walt Disney Co. (includes re-releases, re-adjusted for inflation)

b. *Aladdin* generated approximately 215 million dollars. *Beauty and the Beast* generated approximately 145 million dollars. To find how much more money *Aladdin* generated than *Beauty and the Beast*, we subtract 215 − 145 = 70 million dollars, or $70,000,000.

2   A **line graph** consists of a series of points connected by a line. The graph in Example 3 is a line graph.

Example 3   The line graph below shows the relationship between the distance driven in a 14-foot U-Haul truck in one day and the total cost of renting this truck for that day. Notice that the horizontal axis is labeled Distance and the vertical axis is labeled Total Cost.

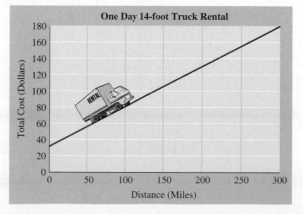

One Day 14-foot Truck Rental

a. Find the total cost of renting the truck if 100 miles are driven.
b. Find the number of miles driven if the total cost of renting is $140.

Solution **a.** Find the number 100 on the horizontal scale and move vertically upward until the line is reached. From this point on the line, we move horizontally to the left until the vertical scale is reached. We find that the total cost of renting the truck if 100 miles are driven is approximately $80.

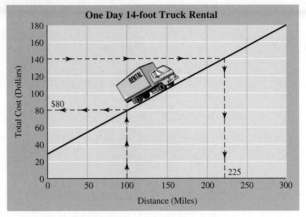

One Day 14-foot Truck Rental

b. We find the number 140 on the vertical scale and move horizontally to the right until the line is reached. From this point on the line, we move vertically downward until the horizontal scale is reached. We find that the truck is driven approximately 225 miles.

From the previous example, we can see that graphing provides a quick way to approximate quantities. In Chapter 6 we show how we can use equations to find exact answers to the questions posed in Example 3. The next graph is another example of a line graph. It is also sometimes called a **broken line graph**.

◆ **Example 4** The line graph shows the relationship between time spent smoking a cigarette and pulse rate. Time is recorded along the horizontal axis in minutes, with 0 minutes being the moment a smoker lights a cigarette. Pulse is recorded along the vertical axis in heartbeats per minute.

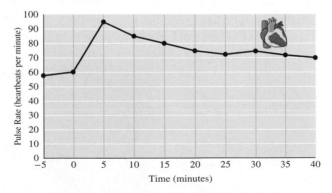

a. What is the pulse rate 15 minutes after lighting a cigarette?
b. When is the pulse rate the lowest?
c. When does the pulse rate show the greatest change?

Solution **a.** We locate the number 15 along the time axis and move vertically upward until the line is reached. From this point on the line, we move horizontally to the left until the pulse rate axis is reached. Reading the number of beats per minute, we find that the pulse rate is 80 beats per minute 15 minutes after lighting a cigarette.

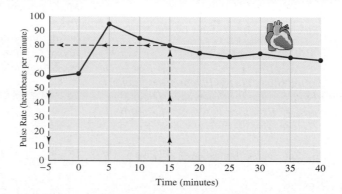

b. We find the lowest point of the line graph, which represents the lowest pulse rate. From this point, we move vertically downward to the time axis. We find that the pulse rate is the lowest at −5 minutes, which means 5 minutes *before* lighting a cigarette.

c. The pulse rate shows the greatest change during the 5 minutes between 0 and 5. Notice that the line graph is *steepest* between 0 and 5 minutes.

Exercise Set 1.9

The following bar graph shows the number of teenagers expected to use the Internet for the years shown. Use this graph to answer Exercises 1–4. See Example 1.

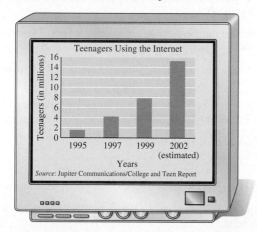

1. Approximate the number of teenagers expected to use the Internet in 1999.

2. Approximate the number of teenagers who use the Internet in 1995.

3. What year shows the greatest *increase* in number of teenagers using the Internet?

4. How many more teenagers are expected to use the Internet in 2002 than in 1999?

The following bar graph shows the amounts of money used by major pro sports for advertising in a recent year. Use this graph to answer Exercises 5–10. See Example 2.

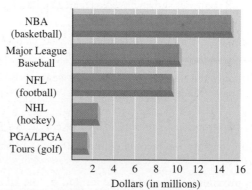

Source: "Competitive Media Reporting," *USA Today*, 6/9/97

5. Which major pro sport used the least amount of money for advertising?

6. Which major pro sport used the greatest amount of money for advertising?

7. Which major pro sports spent over $10,000,000 in advertising?

8. Which major pro sports spent under $5,000,000 in advertising?

9. Estimate the amount of money spent by the NBA for advertising.

10. Estimate the amount of money spent by the NHL for advertising.

The following bar graph shows the top 10 tourist destinations and the number of tourists that visit each country per year. Use this graph to answer Exercises 11–16. See Examples 1 and 2.

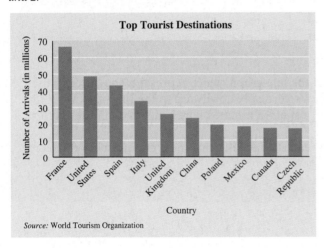

Source: World Tourism Organization

11. Which country is the most popular tourist destination?

12. Which country shown is the least popular tourist destination?

13. Which countries have more than 30 million tourists per year?

14. Which countries shown have less than 20 million tourists per year?

15. Estimate the number of tourists per year whose destination is Italy.

16. Estimate the number of tourists per year whose destination is France.

Use the bar graph in Example 2 to answer Exercises 17–22.

17. Approximate the income generated by the film *Pocahontas*.

18. Approximate the income generated by the film *The Hunchback of Notre Dame*.

19. Before 1990, which Disney film generated the most income?

20. After 1990, which Disney film generated the most income?

21. Why do you think that the Disney film *The Little Mermaid* is not shown on this graph?

22. How much less money did the film *The Hunchback of Notre Dame* generate than *Pocahontas*?

Many fires are deliberately set. An increasing number of those arrested for arson are juveniles (age 17 and under). The following line graph shows the percent of deliberately set fires started by juveniles. Use this graph to answer Exercises 23–30. See Examples 3 and 4.

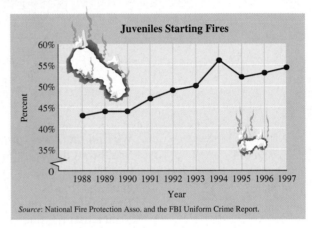

Source: National Fire Protection Asso. and the FBI Uniform Crime Report.

23. What year shows the highest percent of arson fires started by juveniles?

24. What year since 1990 shows a decrease in the percent of fires started by juveniles?

25. Name two consecutive years where the percent appears to remain the same.

26. What year shows the lowest percent of arson fires started by juveniles?

27. Estimate the percent of arson fires started by juveniles in 1997.

28. Estimate the percent of arson fires started by juveniles in 1992.

29. What year shows the greatest increase in the percent of fires started by juveniles?

30. What trend do you notice from this graph?

Use the line graph in Example 4 to answer Exercises 31–34.

31. Approximate the pulse rate 5 minutes before lighting a cigarette.

32. Approximate the pulse rate 10 minutes after lighting a cigarette.

33. Find the difference in pulse rate between 5 minutes before and 10 minutes after lighting a cigarette.

34. When is the pulse rate less than 60 heartbeats per minute?

The line graph below shows the number of students per computer in U.S. public schools. Use this graph for Exercises 35–39. See Examples 3 and 4.

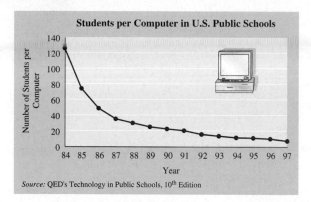

35. Approximate the number of students per computer in 1991.
36. Approximate the number of students per computer in 1997.
37. During what year was the greatest decrease in number of students per computer?
38. What was the first year that the number of students per computer fell below 20?
39. Discuss any trends shown by this line graph.

The special bar graph shown in the next column is called a double bar graph. This double bar graph is used to compare men and women in the U.S. labor force per year . Use this graph for Exercises 40–48.

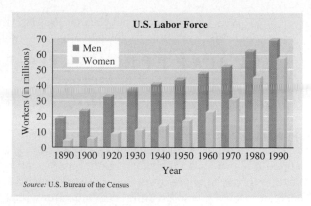

Source: U.S. Bureau of the Census

40. Estimate the number of men in the workforce in 1890.
41. Estimate the number of women in the workforce in 1890.
42. Estimate the number of women in the workforce in 1990.
43. Estimate the number of men in the workforce in 1990.
44. Give the first year that the number of men in the workforce rose above 20 million.
45. Give the first year that the number of women in the workforce rose above 20 million.
46. Estimate the difference in the number of men and women in the workforce in 1940.
47. Estimate the difference in the number of men and women in the workforce in 1990.
48. Discuss any trends shown by this graph.

Geographic locations can be described by a gridwork of lines called latitudes and longitudes, as shown below. For example, the location of Houston, Texas, can be described by latitude 30° north and longitude 95° west. Use the map shown to answer Exercises 49–52.

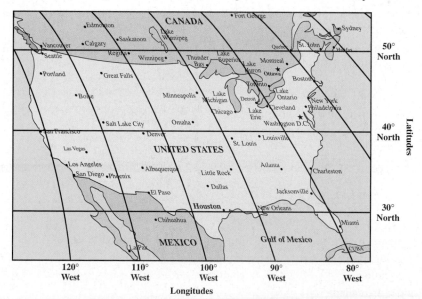

49. Using latitude and longitude, describe the location of New Orleans, Louisiana.
50. Using latitude and longitude, describe the location of Denver, Colorado.
51. Use an atlas and describe the location of your hometown.
52. Give another name for 0° latitude.

1

For additional Chapter Projects, visit the Real World Activities Website by going to http://www.prenhall.com/martin-gay.

CHAPTER PROJECT

Creating and Interpreting Graphs

Companies often rely on market research to investigate the types of consumers who buy their products and their competitors' products. One way of gathering this type of data is through the use of surveys. The raw data collected from surveys can be difficult to interpret without some organization. Graphs organize data visually. They also allow a user to interpret the data quickly.

In this project, you will conduct a brief survey and create tables and graphs to represent the results. This project may be completed by working in groups or individually.

1. Begin by conducting this survey with fellow students, either in this class or another class.
2. For each survey question, tally the results for each response category. Present the results in a table.
3. For each survey question, find the fraction of the total number of responses that fall in each response category.
4. For each survey question, decide which type of graph would best represent the data: bar graph, line graph, or circle graph. Then create an appropriate graph for each set of responses to the questions.
5. Use the data for the two Internet questions and the two newspaper questions to record the increase or decrease in the number of people who participated in these activities. Complete the table in the next column.
6. Study the tables and graphs. What may you conclude from them? What do they tell you about your survey respondents? Write a paragraph summarizing your findings.

Activity	Increase or Decrease from 3 Years Ago
Using the Internet more than 2 hours per week	
Reading a newspaper at least once per week	

SURVEY

What is your age?

Under 20   20s   30s   40s   50s   60 and older

What is your gender?

Female   Male

Do you currently use the Internet more than 2 hours per week?

Yes   No

Three years ago, did you use the Internet more than 2 hours per week?

Yes   No

Do you currently read a newspaper at least once per week?

Yes   No

Three years ago, did you read a newspaper at least once per week?

Yes   No

CHAPTER 1 VOCABULARY CHECK

Fill in each blank with one of the words or phrases listed below.

set	inequality symbols	opposites	absolute value	numerator
denominator	grouping symbols	exponent	base	reciprocals
variable	equation	solution		

1. The symbols $\neq, <,$ and $>$ are called _____ .
2. A mathematical statement that two expressions are equal is called an _____ .
3. The _____ of a number is the distance between that number and 0 on the number line.
4. A symbol used to represent a number is called a _____ .

5. Two numbers that are the same distance from 0 but lie on opposite sides of 0 are called _____.

6. The number in a fraction above the fraction bar is called the _____.

7. A _____ of an equation is a value for the variable that makes the equation a true statement.

8. Two numbers whose product is 1 are called _____.

9. In 2^3, the 2 is called the _____ and the 3 is called the _____.

10. The number in a fraction below the fraction bar is called the _____.

11. Parentheses and brackets are examples of _____.

12. A _____ is a collection of objects.

CHAPTER 1 HIGHLIGHTS

DEFINITIONS AND CONCEPTS	EXAMPLES

Section 1.2 Symbols and Sets of Numbers

A **set** is a collection of objects, called **elements**, enclosed in braces.

$\{a, c, e\}$

Natural Numbers: $\{1, 2, 3, 4, \dots\}$

Whole Numbers: $\{0, 1, 2, 3, 4, \dots\}$

Integers: $\{\dots, -3, -2, -1, 0, 1, 2, 3, \dots\}$

Rational Numbers: {real numbers that can be expressed as a quotient of integers}

Irrational Numbers: {real numbers that cannot be expressed as a quotient of integers}

Real Numbers: {all numbers that correspond to a point on the number line}

Given the set $\{-3.4, \sqrt{3}, 0, \frac{2}{3}, 5, -4\}$, list the numbers that belong to the set of

Natural numbers 5

Whole numbers 0, 5

Integers $-4, 0, 5$

Rational numbers $-3.4, 0, \frac{2}{3}, 5, -4$

Irrational numbers $\sqrt{3}$

Real numbers $-3.4, \sqrt{3}, 0, \frac{2}{3}, 5, -4$

A line used to picture numbers is called a **number line**.

The **absolute value** of a real number a, denoted by $|a|$, is the distance between a and 0 on the number line.

$|5| = 5 \qquad |0| = 0 \qquad |-2| = 2$

Symbols:
= is equal to
≠ is not equal to
> is greater than
< is less than
≤ is less than or equal to
≥ is greater than or equal to

$-7 = -7$

$3 \neq -3$

$4 > 1$

$1 < 4$

$6 \leq 6$

$18 \geq -\dfrac{1}{3}$

Order Property for Real Numbers

For any two real numbers a and b, a is less than b if a is to the left of b on the number line.

$-3 < 0 \qquad 0 > -3 \qquad 0 < 2.5 \qquad 2.5 > 0$

Section 1.3 Fractions

A quotient of two integers is called a **fraction**. The **numerator** of a fraction is the top number. The **denominator** of a fraction is the bottom number.

$\dfrac{13}{17} \begin{array}{l} \leftarrow \text{ numerator} \\ \leftarrow \text{ denominator} \end{array}$

If $a \cdot b = c$, then a and b are **factors** and c is the **product**.

$\underset{\text{factor}}{7} \quad \cdot \quad \underset{\text{factor}}{9} \quad = \quad \underset{\text{product}}{63}$ *(continued)*

DEFINITIONS AND CONCEPTS	EXAMPLES

Section 1.3 Fractions

A fraction is in **lowest terms** when the numerator and the denominator have no factors in common other than 1.

To write a fraction in lowest terms, factor the numerator and the denominator; then apply the fundamental property.

$\frac{13}{17}$ is in lowest terms.

Write in lowest terms.

$$\frac{6}{14} = \frac{2 \cdot 3}{2 \cdot 7} = \frac{3}{7}$$

Two fractions are **reciprocals** if their product is 1. The reciprocal of $\frac{a}{b}$ is $\frac{b}{a}$.

The reciprocal of

$$\frac{6}{25} \text{ is } \frac{25}{6}$$

To multiply fractions, numerator times numerator is the numerator of the product and denominator times denominator is the denominator of the product.

To divide fractions, multiply the first fraction by the reciprocal of the second fraction.

To add fractions with the same denominator, add the numerators and place the sum over the common denominator.

To subtract fractions with the same denominator, subtract the numerators and place the difference over the common denominator.

Fractions that represent the same quantity are called **equivalent fractions**.

Perform the indicated operations.

$$\frac{2}{5} \cdot \frac{3}{7} = \frac{6}{35}$$

$$\frac{5}{9} \div \frac{2}{7} = \frac{5}{9} \cdot \frac{7}{2} = \frac{35}{18}$$

$$\frac{5}{11} + \frac{3}{11} = \frac{8}{11}$$

$$\frac{13}{15} - \frac{3}{15} = \frac{10}{15} = \frac{2}{3}$$

$$\frac{1}{5} = \frac{1 \cdot 4}{5 \cdot 4} = \frac{4}{20}$$

$\frac{1}{5}$ and $\frac{4}{20}$ are equivalent fractions.

Section 1.4 Introduction to Variable Expressions and Equations

The expression a^n is an **exponential expression**. The number a is called the **base**; it is the repeated factor. The number n is called the **exponent**; it is the number of times that the base is a factor.

$$4^3 = 4 \cdot 4 \cdot 4 = 64$$
$$7^2 = 7 \cdot 7 = 49$$

Order of Operations

Simplify expressions in the following order. If grouping symbols are present, simplify expressions within those first, starting with the innermost set. Also, simplify the numerator and the denominator of a fraction separately.

1. Simplify exponential expressions.
2. Multiply or divide in order from left to right.
3. Add or subtract in order from left to right.

$$\frac{8^2 + 5(7 - 3)}{3 \cdot 7} = \frac{8^2 + 5(4)}{21}$$
$$= \frac{64 + 5(4)}{21}$$
$$= \frac{64 + 20}{21}$$
$$= \frac{84}{21}$$
$$= 4$$

A symbol used to represent a number is called a **variable**.

Examples of variables are:
$$q, x, z$$

An **algebraic expression** is a collection of numbers, variables, operation symbols, and grouping symbols.

Examples of algebraic expressions are:
$$5x, 2(y - 6), \frac{q^2 - 3q + 1}{6}$$

To evaluate an algebraic expression containing a variable, substitute a given number for the variable and simplify.

Evaluate $x^2 - y^2$ if $x = 5$ and $y = 3$.
$$x^2 - y^2 = (5)^2 - 3^2$$
$$= 25 - 9$$
$$= 16 \qquad \text{(continued)}$$

DEFINITIONS AND CONCEPTS	EXAMPLES

Section 1.4 Introduction to Variable Expressions and Equations

A mathematical statement that two expressions are equal is called an **equation**.	Equations: $$3x - 9 = 20$$ $$A = \pi r^2$$
A **solution** or **root** of an equation is a value for the variable that makes the equation a true statement.	Determine whether 4 is a solution of $5x + 7 = 27$. $$5x + 7 = 27$$ $$5(4) + 7 = 27$$ $$20 + 7 = 27$$ $$27 = 27 \quad \text{True}$$ 4 is a solution.

Section 1.5 Adding Real Numbers

To Add Two Numbers with the Same Sign **1.** Add their absolute values. **2.** Use their common sign as the sign of the sum.	Add. $$10 + 7 = 17$$ $$-3 + (-8) = -11$$
To Add Two Numbers with Different Signs **1.** Subtract their absolute values. **2.** Use the sign of the number whose absolute value is larger as the sign of the sum.	$$-25 + 5 = -20$$ $$14 + (-9) = 5$$
Two numbers that are the same distance from 0 but lie on opposite sides of 0 are called **opposites** or **additive inverses**. The opposite of a number a is denoted by $-a$.	The opposite of -7 is 7. The opposite of 123 is -123.
The sum of a number a and its opposite, $-a$, is 0. $$a + (-a) = 0$$ If a is a number, then $-(-a) = a$.	$$-4 + 4 = 0$$ $$12 + (-12) = 0$$ $$-(-8) = 8$$ $$-(-14) = 14$$

Section 1.6 Subtracting Real Numbers

To subtract two numbers a and b, add the first number a to the opposite of the second number b. $$a - b = a + (-b)$$	Subtract. $$3 - (-44) = 3 + 44 = 47$$ $$-5 - 22 = -5 + (-22) = -27$$ $$-30 - (-30) = -30 + 30 = 0$$

Section 1.7 Multiplying and Dividing Real Numbers

Quotient of two real numbers $$\frac{a}{b} = a \cdot \frac{1}{b}$$	Multiply or divide. $$\frac{42}{2} = 42 \cdot \frac{1}{2} = 21$$

(continued)

DEFINITIONS AND CONCEPTS	EXAMPLES

Section 1.7 Multiplying and Dividing Real Numbers

Multiplying and Dividing Real Numbers

The product or quotient of two numbers with the same sign is a positive number. The product or quotient of two numbers with different signs is a negative number.

$$7 \cdot 8 = 56 \qquad -7 \cdot (-8) = 56$$
$$-2 \cdot 4 = -8 \qquad 2 \cdot (-4) = -8$$
$$\frac{90}{10} = 9 \qquad \frac{-90}{-10} = 9$$

Products and Quotients Involving Zero

The product of 0 and any number is 0.

$$b \cdot 0 = 0 \quad \text{and} \quad 0 \cdot b = 0$$

The quotient of a nonzero number and 0 is undefined.

$$\frac{b}{0} \text{ is undefined.}$$

The quotient of 0 and any nonzero number is 0.

$$\frac{0}{b} = 0$$

$$\frac{42}{-6} = -7 \qquad \frac{-42}{6} = -7$$
$$-4 \cdot 0 = 0 \qquad 0 \cdot \left(-\frac{3}{4}\right) = 0$$
$$\frac{-85}{0} \text{ is undefined.}$$
$$\frac{0}{18} = 0 \qquad \frac{0}{-47} = 0$$

Section 1.8 Properties of Real Numbers

Commutative Properties

Addition: $a + b = b + a$

Multiplication: $a \cdot b = b \cdot a$

Associative Properties

Addition: $(a + b) + c = a + (b + c)$

Multiplication: $(a \cdot b) \cdot c = a \cdot (b \cdot c)$

Two numbers whose product is 1 are called **multiplicative inverses** or

reciprocals. The reciprocal of a nonzero number a is $\frac{1}{a}$ because $a \cdot \frac{1}{a} = 1$.

Distributive Property $a(b + c) = a \cdot b + a \cdot c$

Identities $\begin{aligned} a + 0 &= a \qquad 0 + a = a \\ a \cdot 1 &= a \qquad 1 \cdot a = a \end{aligned}$

Inverses

Addition or opposite: $a + (-a) = 0$

Multiplication or reciprocal: $b \cdot \frac{1}{b} = 1$

$$3 + (-7) = -7 + 3$$
$$-8 \cdot 5 = 5 \cdot (-8)$$
$$(5 + 10) + 20 = 5 + (10 + 20)$$
$$(-3 \cdot 2) \cdot 11 = -3 \cdot (2 \cdot 11)$$

The reciprocal of 3 is $\frac{1}{3}$.

The reciprocal of $-\frac{2}{5}$ is $-\frac{5}{2}$.

$$5(6 + 10) = 5 \cdot 6 + 5 \cdot 10$$
$$-2(3 + x) = -2 \cdot 3 + (-2)(x)$$
$$5 + 0 = 5 \qquad 0 + (-2) = -2$$
$$-14 \cdot 1 = -14 \qquad 1 \cdot 27 = 27$$

$$7 + (-7) = 0$$

$$3 \cdot \frac{1}{3} = 1$$

Section 1.9 Reading Graphs

To find the value on the vertical axis representing a location on a graph, move horizontally from the location on the graph until the vertical axis is reached. To find the value on the horizontal axis representing a location on a graph, move vertically from the location on the graph until the horizontal axis is reached.

The broken line graph to the right shows the average public classroom teachers' salaries for the school year ending in the years shown.

Estimate the average public teacher's salary for the school year ending in 1998.

Find the earliest year that the average salary rose above $37,000.

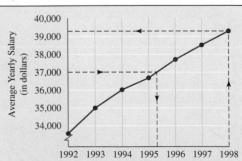

Source: U.S. Bureau of the Census, *Statistical Abstract of the United States: 1999* (114th edition) Washington, D.C., 1999

CHAPTER 1 REVIEW

(1.2) *Insert* <, >, *or* = *in the appropriate space to make the following statements true.*

1. 8   10

2. 7   2

3. −4   −5

4. $\dfrac{12}{2}$ −8

5. $|-7|$ $|-8|$

6. $|-9|$ −9

7. $-|-1|$ −1

8. $|-14|$ $-(-14)$

9. 1.2   1.02

10. $-\dfrac{3}{2}$ $-\dfrac{3}{4}$

Translate each statement into symbols.

11. Four is greater than or equal to negative three.

12. Six is not equal to five.

13. 0.03 is less than 0.3.

14. Lions and hyenas were featured in the Disney film *The Lion King*. For short distances, lions can run at a rate of 50 miles per hour whereas hyenas can run at a rate of 40 miles per hour. Write an inequality statement comparing the numbers 50 and 40.

Given the following sets of numbers, list the numbers in each set that also belong to the set of:

  a. Natural numbers   **b.** Whole numbers

  c. Integers   **d.** Rational numbers

  e. Irrational numbers   **f.** Real numbers

15. $\left\{-6, 0, 1, 1\dfrac{1}{2}, 3, \pi, 9.62\right\}$

16. $\left\{-3, -1.6, 2, 5, \dfrac{11}{2}, 15.1, \sqrt{5}, 2\pi\right\}$

The following chart shows the gains and losses in dollars of Density Oil and Gas stock for a particular week.

Day	Gain or Loss in Dollars
Monday	+1
Tuesday	−2
Wednesday	+5
Thursday	+1
Friday	−4

17. Which day showed the greatest loss?

18. Which day showed the greatest gain?

(1.3) *Write the number as a product of prime factors.*

19. 36   **20.** 120

Perform the indicated operations. Write results in lowest terms.

21. $\dfrac{8}{15} \cdot \dfrac{27}{30}$

22. $\dfrac{7}{8} \div \dfrac{21}{32}$

23. $\dfrac{7}{15} + \dfrac{5}{6}$

24. $\dfrac{3}{4} - \dfrac{3}{20}$

25. $2\dfrac{3}{4} + 6\dfrac{5}{8}$

26. $7\dfrac{1}{6} - 2\dfrac{2}{3}$

27. $5 \div \dfrac{1}{3}$

28. $2 \cdot 8\dfrac{3}{4}$

29. Determine the unknown part of the given circle.

Find the area and the perimeter of each figure.

△ **30.**

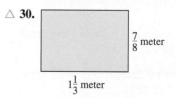

$\dfrac{7}{8}$ meter

$1\dfrac{1}{3}$ meter

△ **31.**

$\dfrac{5}{11}$ in.

$\dfrac{3}{11}$ in.

$\dfrac{3}{11}$ in.

$\dfrac{5}{11}$ in.

△ **32.** A trim carpenter needs a piece of quarter round molding $6\dfrac{1}{8}$ feet long for a bathroom. She finds a piece $7\dfrac{1}{2}$ feet long. How long a piece does she need to cut from the $7\dfrac{1}{2}$-foot-long molding in order to use it in the bathroom?

In December 1998, Nkem Chukwu gave birth to the world's first surviving octuplets in Houston, Texas. The following chart gives the octuplets' birthweights. The babies are listed in order of birth.

Baby's Name	Gender	Birthweight (pounds)
Ebuka	girl	$1\frac{1}{2}$
Chidi	girl	$1\frac{11}{16}$
Echerem	girl	$1\frac{3}{4}$
Chima	girl	$1\frac{5}{8}$
Odera	girl	$\frac{11}{16}$
Ikem	boy	$1\frac{1}{8}$
Jioke	boy	$1\frac{13}{16}$
Gorom	girl	$1\frac{1}{8}$

Source: Texas Children's Hospital, Houston, Texas

33. What was the total weight of the boy octuplets?

34. What was the total weight of the girl octuplets?

35. Find the combined weight of all eight octuplets.

36. Which baby weighed the most?

37. Which baby weighed the least?

38. How much more did the heaviest baby weigh than the lightest baby?

39. By March 1999, Chima weighed $5\frac{1}{2}$ pounds. How much weight had she gained since birth?

40. By March 1999, Ikem weighed $4\frac{5}{32}$ pounds. How much weight had he gained since birth?

(1.4) Simplify each expression.

41. 2^4

42. 5^2

43. $\left(\frac{2}{7}\right)^2$

44. $\left(\frac{3}{4}\right)^3$

45. $6 \cdot 3^2 + 2 \cdot 8$

46. $68 - 5 \cdot 2^3$

47. $3(1 + 2 \cdot 5) + 4$

48. $8 + 3(2 \cdot 6 - 1)$

49. $\dfrac{4 + |6 - 2| + 8^2}{4 + 6 \cdot 4}$

50. $5[3(2 + 5) - 5]$

Translate each word statement to symbols.

51. The difference of twenty and twelve is equal to the product of two and four.

52. The quotient of nine and two is greater than negative five.

Evaluate each expression if $x = 6$, $y = 2$, and $z = 8$.

53. $2x + 3y$

54. $x(y + 2z)$

55. $\dfrac{x}{y} + \dfrac{z}{2y}$

56. $x^2 - 3y^2$

△ **57.** The expression $180 - a - b$ represents the measure of the unknown angle of the given triangle. Replace a with 37 and b with 80 to find the measure of the unknown angle.

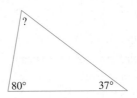

Decide whether the given number is a solution to the given equation.

58. Is $x = 3$ a solution of $7x - 3 = 18$?

59. Is $x = 1$ a solution of $3x^2 + 4 = x - 1$?

(1.5) Find the additive inverse or the opposite.

60. -9

61. $\dfrac{2}{3}$

62. $|-2|$

63. $-|-7|$

Find the following sums.

64. $-15 + 4$

65. $-6 + (-11)$

66. $\dfrac{1}{16} + \left(-\dfrac{1}{4}\right)$

67. $-8 + |-3|$

68. $-4.6 + (-9.3)$

69. $-2.8 + 6.7$

70. The lowest elevation in North America is -282 feet at Death Valley in California. If you are standing at a point 728 feet above Death Valley, what is your elevation? (*Source*: National Geographic Society)

(1.6) Perform the indicated operations.

71. $6 - 20$

72. $-3.1 - 8.4$

73. $-6 - (-11)$

74. $4 - 15$

75. $-21 - 16 + 3(8 - 2)$

76. $\dfrac{11 - (-9) + 6(8 - 2)}{2 + 3 \cdot 4}$

If x = 3, y = −6, and z = −9, evaluate each expression.

77. $2x^2 - y + z$

78. $\dfrac{y - x + 5x}{2x}$

79. At the beginning of the week the price of Density Oil and Gas stock from Exercises 17 and 18 is $50 per share. Find the price of a share of stock at the end of the week.

80. The expression $E - I$ represents a country's merchandise trade balance if the country has exports worth E dollars and imports worth I dollars. Find the merchandise trade balance for a country with exports worth $412 billion and imports worth $536 billion.

(1.7) Find the multiplicative inverse or reciprocal.

81. -6

82. $\dfrac{3}{5}$

Simplify each expression.

83. $6(-8)$

84. $(-2)(-14)$

85. $\dfrac{-18}{-6}$

86. $\dfrac{42}{-3}$

87. $-3(-6)(-2)$

88. $(-4)(-3)(0)(-6)$

89. $\dfrac{4(-3) + (-8)}{2 + (-2)}$

90. $\dfrac{3(-2)^2 - 5}{-14}$

91. $\dfrac{-6}{0}$

92. $\dfrac{0}{-2}$

93. During the 1999 LPGA Sara Lee Classic, Michelle McGann had scores of −9, −7, and +1 in three rounds of golf. Find her average score per round. (*Source:* Ladies Professional Golf Association)

94. During the 1999 PGA Masters Tournament, Bob Estes had scores of −1, 0, −3, and 0 in four rounds of golf. Find his average score per round. (*Source:* Professional Golf Association)

(1.8) Name the property illustrated.

95. $-6 + 5 = 5 + (-6)$

96. $6 \cdot 1 = 6$

97. $3(8 - 5) = 3 \cdot 8 + 3 \cdot (-5)$

98. $4 + (-4) = 0$

99. $2 + (3 + 9) = (2 + 3) + 9$

100. $2 \cdot 8 = 8 \cdot 2$

101. $6(8 + 5) = 6 \cdot 8 + 6 \cdot 5$

102. $(3 \cdot 8) \cdot 4 = 3 \cdot (8 \cdot 4)$

103. $4 \cdot \dfrac{1}{4} = 1$

104. $8 + 0 = 8$

105. $4(8 + 3) = 4(3 + 8)$

(1.9) Use the following graph to answer Exercises 106–109.

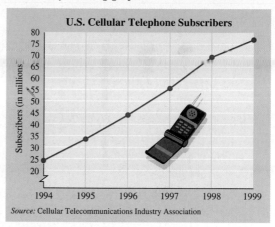

U.S. Cellular Telephone Subscribers

Source: Cellular Telecommunications Industry Association

106. Approximate the number of cellular phone subscribers in 1999.

107. Approximate the increase in cellular phone subscribers in 1999.

108. What year shows the greatest number of subscribers?

109. What trend is shown by this graph?

The following bar graph shows the average annual percent increase in rent from June 1997 to June 1998 in selected major metropolitan areas.

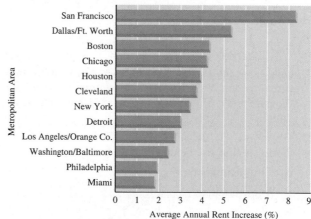

Source: U.S. Department of Labor

110. Which metropolitan area had the greatest increase in annual rent? Approximate the average annual rent increase in this area.

111. Which metropolitan area had the least increase in annual rent? Approximate the average annual rent increase in this area.

112. For this period, the national average increase in rent was 3.2%. Which metropolitan areas had rental increases below the national average?

113. Which metropolitan areas had rental increases above the national average (see Exercise 112)?

CHAPTER 1 TEST

Translate the statement into symbols.

1. The absolute value of negative seven is greater than five.

2. The sum of nine and five is greater than or equal to four.

Simplify the expression.

3. $-13 + 8$

4. $-13 - (-2)$

5. $6 \cdot 3 - 8 \cdot 4$

6. $(13)(-3)$

7. $(-6)(-2)$

8. $\dfrac{|-16|}{-8}$

9. $\dfrac{-8}{0}$

10. $\dfrac{|-6| + 2}{5 - 6}$

11. $\dfrac{1}{2} - \dfrac{5}{6}$

12. $-1\dfrac{1}{8} + 5\dfrac{3}{4}$

13. $-\dfrac{3}{5} + \dfrac{15}{8}$

14. $3(-4)^2 - 80$

15. $6[5 + 2(3 - 8) - 3]$

16. $\dfrac{-12 + 3 \cdot 8}{4}$

17. $\dfrac{(-2)(0)(-3)}{-6}$

Insert $<$, $>$, or $=$ in the appropriate space to make each of the following statements true.

18. $-3 \quad -7$

19. $4 \quad -8$

20. $|-3| \quad 2$

21. $|-2| \quad -1 - (-3)$

22. In the state of Massachusetts, there are 2221 licensed child care centers and 10,993 licensed home-based child care providers. Write an inequality statement comparing the numbers 2221 and 10,993. (*Source*: Children's Foundation)

23. Given $\left\{-5, -1, 0, \frac{1}{4}, 1, 7, 11.6, \sqrt{7}, 3\pi\right\}$, list the numbers in this set that also belong to the set of:

    a. Natural numbers

    b. Whole numbers

    c. Integers

    d. Rational numbers

    e. Irrational numbers

    f. Real numbers

If $x = 6$, $y = -2$, and $z = -3$, evaluate each expression.

24. $x^2 + y^2$

25. $x + yz$

26. $2 + 3x - y$

27. $\dfrac{y + z - 1}{x}$

Identify the property illustrated by each expression.

28. $8 + (9 + 3) = (8 + 9) + 3$

29. $6 \cdot 8 = 8 \cdot 6$

30. $-6(2 + 4) = -6 \cdot 2 + (-6) \cdot 4$

31. $\dfrac{1}{6}(6) = 1$

32. Find the opposite of -9.

33. Find the reciprocal of $-\dfrac{1}{3}$.

The New Orleans Saints were 22 yards from the goal when the following series of gains and losses occurred.

Gains and Losses in Yards	
First Down	5
Second Down	-10
Third Down	-2
Fourth Down	29

34. During which down did the greatest loss of yardage occur?

35. Was a touchdown scored?

36. The temperature at the Winter Olympics was a frigid 14 degrees below zero in the morning, but by noon it had risen 31 degrees. What was the temperature at noon?

37. United HealthCare is a health insurance provider. It had net incomes of $356 million, $460 million, and −$166 million in 1996, 1997, and 1998, respectively. What was United HealthCare's total net income for these three years? (*Source*: United HealthCare Corp.)

38. Jean Avarez decided to sell 280 shares of stock, which decreased in value by $1.50 per share yesterday. How much money did she lose?

Intel is a semiconductor manufacturer that makes almost one-third of the world's computer chips. (You may have seen the slogan "Intel Inside" in commercials on television.) The line graph below shows Intel's net revenues in billions of dollars. Use this figure to answer the questions below.

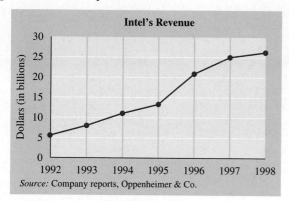

Source: Company reports, Oppenheimer & Co.

39. Estimate Intel's revenue in 1993.
40. Estimate Intel's revenue in 1997.
41. Find the increase in Intel's revenue from 1993 to 1995.
42. What year shows the greatest increase in revenue?

The following bar graph shows the top steel-producing states ranked by total tons of raw steel produced in 1997.

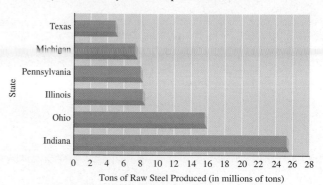

Source: American Iron & Steel Institute

43. Which state was the top steel producer? Approximate the amount of raw steel it produced.

44. Which of the top steel producing states produced the least steel? Approximate the amount of raw steel it produced.

45. Approximate the amount of raw steel produced by Ohio.

46. Approximately how much more steel was produced in Pennsylvania than Texas?

Educational Opportunities

Over 3 million teachers and nearly 1 million teacher aides are at work in this country. While teachers develop and execute lesson plans to help students learn and apply concepts in a variety of subjects, teacher aides provide instructional and clerical support for teachers.

All states require their teachers to have at least a bachelor's degree. Requirements for teacher aides vary from a high school diploma to some college training. Working as a teacher aide provides an excellent opportunity to advance to classroom teaching positions. In fact, according to a survey conducted by the National Education Association, half of all teacher aides aspire to become classroom teachers. Teachers and teacher aides should enjoy working with children, communicate well, and have a solid understanding of topics including science, mathematics, and English. Educators and support staff need good math skills not only for teaching math, but also for figuring grades and other recordkeeping.

    For more information about teachers and teacher aides, visit the National Education Association Website by first going to www.prenhall.com/martin-gay.

In the Spotlight on Decision Making feature on page 91, you will have the opportunity to make a decision involving grades as a teacher aide.

EQUATIONS, INEQUALITIES, AND PROBLEM SOLVING

2.1 SIMPLIFYING ALGEBRAIC EXPRESSIONS

2.2 THE ADDITION PROPERTY OF EQUALITY

2.3 THE MULTIPLICATION PROPERTY OF EQUALITY

2.4 SOLVING LINEAR EQUATIONS

2.5 AN INTRODUCTION TO PROBLEM SOLVING

2.6 FORMULAS AND PROBLEM SOLVING

2.7 PERCENT AND PROBLEM SOLVING

2.8 FURTHER PROBLEM SOLVING

2.9 SOLVING LINEAR INEQUALITIES

Much of mathematics relates to deciding which statements are true and which are false. When a statement, such as an equation, contains variables, it is usually not possible to decide whether the equation is true or false until the variable has been replaced by a value. For example, the statement $x + 7 = 15$ is an equation stating that the sum $x + 7$ has the same value as 15. Is this statement true or false? It is false for some values of x and true for just one value of x, namely 8. Our purpose in this chapter is to learn ways of deciding which values make an equation or an inequality true.

2.1 SIMPLIFYING ALGEBRAIC EXPRESSIONS

▶ **OBJECTIVES**

1. Identify terms, like terms, and unlike terms.
2. Combine like terms.
3. Use the distributive property to remove parentheses.
4. Write word phrases as algebraic expressions.

CD-ROM   SSM   SSG   Video

As we explore in this section, an expression such as $3x + 2x$ is not as simple as possible, because—even without replacing x by a value—we can perform the indicated addition.

1   Before we practice simplifying expressions, some new language of algebra is presented. A **term** is a number or the product of a number and variables raised to powers.

> **Terms**
>
> $$-y, \quad 2x^3, \quad -5, \quad 3xz^2, \quad \frac{2}{y}, \quad 0.8z$$

The **numerical coefficient** of a term is the numerical factor. The numerical coefficient of $3x$ is 3. Recall that $3x$ means $3 \cdot x$.

Term	Numerical Coefficient	
$3x$	3	
$\dfrac{y^3}{5}$	$\dfrac{1}{5}$	since $\dfrac{y^3}{5}$ means $\dfrac{1}{5} \cdot y^3$
$0.7ab^3c^5$	0.7	
z	1	
$-y$	-1	
-5	-5	

> **HELPFUL HINT**
> The term $-y$ means $-1y$ and thus has a numerical coefficient of -1. The term z means $1z$ and thus has a numerical coefficient of 1.

Example 1   Identify the numerical coefficient in each term.

a. $-3y$ **b.** $22z^4$ **c.** y **d.** $-x$ **e.** $\dfrac{x}{7}$

Solution   **a.** The numerical coefficient of $-3y$ is -3.
b. The numerical coefficient of $22z^4$ is 22.
c. The numerical coefficient of y is 1, since y is $1y$.
d. The numerical coefficient of $-x$ is -1, since $-x$ is $-1x$.

e. The numerical coefficient of $\dfrac{x}{7}$ is $\dfrac{1}{7}$, since $\dfrac{x}{7}$ is $\dfrac{1}{7} \cdot x$.   ▬

Terms with the same variables raised to exactly the same powers are called **like terms**. Terms that aren't like terms are called **unlike terms**.

Like Terms	Unlike Terms	
$3x, 2x$	$5x, 5x^2$	Why? Same variable x, but different powers x and x^2
$-6x^2y, 2x^2y, 4x^2y$	$7y, 3z, 8x^2$	Why? Different variables
$2ab^2c^3, ac^3b^2$	$6abc^3, 6ab^2$	Why? Different variables and different powers

▼
HELPFUL HINT
In like terms, each variable and its exponent must match exactly, but these factors don't need to be in the same order.

$$2x^2y \text{ and } 3yx^2 \text{ are like terms.}$$

Example 2   Determine whether the terms are like or unlike.

a. $2x, 3x^2$ **b.** $4x^2y, x^2y, -2x^2y$ **c.** $-2yz, -3zy$ **d.** $-x^4, x^4$

Solution **a.** Unlike terms, since the exponents on x are not the same.
b. Like terms, since each variable and its exponent match.
c. Like terms, since $zy = yz$ by the commutative property.
d. Like terms.   ▬

2   An algebraic expression containing the sum or difference of like terms can be simplified by applying the distributive property. For example, by the distributive property, we rewrite the sum of the like terms $3x + 2x$ as

$$3x + 2x = (3 + 2)x = 5x$$

Also,

$$-y^2 + 5y^2 = (-1 + 5)y^2 = 4y^2$$

Simplifying the sum or difference of like terms is called **combining like terms**.

◈ **Example 3**   Simplify each expression by combining like terms.

a. $7x - 3x$ **b.** $10y^2 + y^2$ **c.** $8x^2 + 2x - 3x$

Solution **a.** $7x - 3x = (7 - 3)x = 4x$
b. $10y^2 + y^2 = (10 + 1)y^2 = 11y^2$
c. $8x^2 + 2x - 3x = 8x^2 + (2 - 3)x = 8x^2 - x$ ▬

Example 4   Simplify each expression by combining like terms.

   a. $2x + 3x + 5 + 2$ **b.** $-5a - 3 + a + 2$
   c. $4y - 3y^2$ **d.** $2.3x + 5x - 6$

Solution   Use the distributive property to combine the numerical coefficients of like terms.

   a. $2x + 3x + 5 + 2 = (2 + 3)x + (5 + 2)$
   $$= 5x + 7$$
   b. $-5a - 3 + a + 2 = -5a + 1a + (-3 + 2)$
   $$= (-5 + 1)a + (-3 + 2)$$
   $$= -4a - 1$$
   c. $4y - 3y^2$ These two terms cannot be combined because they are unlike terms.
   d. $2.3x + 5x - 6 = (2.3 + 5)x - 6$
   $$= 7.3x - 6$$

The examples above suggest the following:

COMBINING LIKE TERMS

To **combine like terms**, add the numerical coefficients and multiply the result by the common variable factors.

---

3   Simplifying expressions makes frequent use of the distributive property to remove parentheses.

Example 5   Find each product by using the distributive property to remove parentheses.

   a. $5(x + 2)$ **b.** $-2(y + 0.3z - 1)$ **c.** $-(x + y - 2z + 6)$

Solution   **a.** $5(x + 2) = 5 \cdot x + 5 \cdot 2$ Apply the distributive property.
   $$= 5x + 10$$ Multiply.

   b. $-2(y + 0.3z - 1) = -2(y) + (-2)(0.3z) + (-2)(-1)$ Apply the distributive property.
   $$= -2y - 0.6z + 2$$ Multiply.
   c. $-(x + y - 2z + 6) = -1(x + y - 2z + 6)$ Distribute -1 over each term.
   $$= -1(x) - 1(y) - 1(-2z) - 1(6)$$
   $$= -x - y + 2z - 6$$

HELPFUL HINT
If a "−" sign precedes parentheses, the sign of each term inside the parentheses is changed when the distributive property is applied to remove parentheses.

Examples:

   $-(2x + 1) = -2x - 1$ $-(-5x + y - z) = 5x - y + z$

   $-(x - 2y) = -x + 2y$ $-(-3x - 4y - 1) = 3x + 4y + 1$

When simplifying an expression containing parentheses, we often use the distributive property first to remove parentheses and then again to combine any like terms.

Example 6   Simplify the following expressions.

a. $3(2x - 5) + 1$ **b.** $8 - (7x + 2) + 3x$ **c.** $-2(4x + 7) - (3x - 1)$

Solution   **a.** $3(2x - 5) + 1 = 6x - 15 + 1$ Apply the distributive property.
                          $= 6x - 14$ Combine like terms.
      b. $8 - (7x + 2) + 3x = 8 - 7x - 2 + 3x$ Apply the distributive property.
                          $= -7x + 3x + 8 - 2$
                          $= -4x + 6$ Combine like terms.
      c. $-2(4x + 7) - (3x - 1) = -8x - 14 - 3x + 1$ Apply the distributive property.
                          $= -11x - 13$ Combine like terms.   ▮

Example 7   Subtract $4x - 2$ from $2x - 3$.

Solution   "Subtract $4x - 2$ **from** $2x - 3$" translates to $(2x - 3) - (4x - 2)$. Next, simplify the algebraic expression.

$$(2x - 3) - (4x - 2) = 2x - 3 - 4x + 2$$ Apply the distributive property.
$$= -2x - 1$$ Combine like terms.   ▮

4   Next, we practice writing word phrases as algebraic expressions.

Example 8   Write the following phrases as algebraic expressions and simplify if possible. Let x represent the unknown number.

a. Twice a number, added to 6
b. The difference of a number and 4, divided by 7
c. Five added to 3 times the sum of a number and 1
d. The sum of twice a number, 3 times the number, and 5 times the number

Solution   **a.** In words:

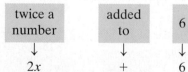

b. In words:

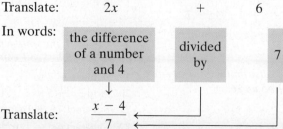

c. In words:

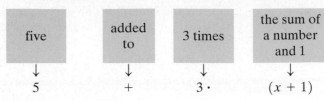

five	added to	3 times	the sum of a number and 1
↓	↓	↓	↓
5	+	3 ·	$(x + 1)$

Translate:

Next, we simplify this expression.

$$5 + 3(x + 1) = 5 + 3x + 3 \qquad \text{Use the distributive property.}$$
$$= 8 + 3x \qquad \text{Combine like terms.}$$

d. The phrase "the sum of" means that we add.

In words:

twice a number	added to	3 times the number	added to	5 times the number
↓	↓	↓	↓	↓
$2x$	+	$3x$	+	$5x$

Translate:

Now let's simplify.

$$2x + 3x + 5x = 10x \qquad \text{Combine like terms.}$$

MENTAL MATH

Identify the numerical coefficient of each term. See Example 1.

1. $-7y$

2. $3x$

3. x

4. $-y$

5. $17x^2y$

6. $1.2xyz$

Indicate whether the following lists of terms are like or unlike. See Example 2.

7. $5y, -y$

8. $-2x^2y, 6xy$

9. $2z, 3z^2$

10. $ab^2, -7ab^2$

11. $8wz, \frac{1}{7}zw$

12. $7.4p^3q^2, 6.2p^3q^2r$

Exercise Set 2.1

Simplify each expression by combining any like terms. See Examples 3 and 4.

1. $7y + 8y$

2. $3x + 2x$

3. $8w - w + 6w$

4. $c - 7c + 2c$

5. $3b - 5 - 10b - 4$

6. $6g + 5 - 3g - 7$

7. $m - 4m + 2m - 6$

8. $a + 3a - 2 - 7a$

Simplify each expression. First use the distributive property to remove any parentheses. See Examples 5 and 6.

9. $5(y - 4)$

10. $7(r - 3)$

11. $7(d - 3) + 10$

12. $9(z + 7) - 15$

13. $-(3x - 2y + 1)$

14. $-(y + 5z - 7)$

15. $5(x + 2) - (3x - 4)$

16. $4(2x - 3) - 2(x + 1)$

17. In your own words, explain how to combine like terms.

18. Do like terms contain the same numerical coefficients? Explain your answer.

Write each of the following as an algebraic expression. Simplify if possible. See Example 7.

19. Add $6x + 7$ to $4x - 10$.

20. Add $3y - 5$ to $y + 16$.

21. Subtract $7x + 1$ from $3x - 8$.

22. Subtract $4x - 7$ from $12 + x$.

Simplify each expression.

23. $7x^2 + 8x^2 - 10x^2$ **24.** $8x + x - 11x$

25. $6x - 5x + x - 3 + 2x$

26. $8h + 13h - 6 + 7h - h$

27. $-5 + 8(x - 6)$ **28.** $-6 + 5(r - 10)$

29. $5g - 3 - 5 - 5g$ **30.** $8p + 4 - 8p - 15$

31. $6.2x - 4 + x - 1.2$ **32.** $7.9y - 0.7 - y + 0.2$

33. $2k - k - 6$ **34.** $7c - 8 - c$

35. $0.5(m + 2) + 0.4m$ **36.** $0.2(k + 8) - 0.1k$

37. $-4(3y - 4)$ **38.** $-3(2x + 5)$

39. $3(2x - 5) - 5(x - 4)$ **40.** $2(6x - 1) - (x - 7)$

41. $3.4m - 4 - 3.4m - 7$

42. $2.8w - 0.9 - 0.5 - 2.8w$

43. $6x + 0.5 - 4.3x - 0.4x + 3$

44. $0.4y - 6.7 + y - 0.3 - 2.6y$

45. $-2(3x - 4) + 7x - 6$ **46.** $8y - 2 - 3(y + 4)$

47. $-9x + 4x + 18 - 10x$ **48.** $5y - 14 + 7y - 20y$

49. $5k - (3k - 10)$ **50.** $-11c - (4 - 2c)$

51. $(3x + 4) - (6x - 1)$ **52.** $(8 - 5y) - (4 + 3y)$

Write each of the following phrases as an algebraic expression and simplify if possible. Let x represent the unknown number See Example 8.

53. Twice a number, decreased by four

54. The difference of a number and two, divided by five

55. Three-fourths of a number, increased by twelve

56. Eight more than triple a number

57. The sum of 5 times a number and -2, added to 7 times a number

58. The sum of 3 times a number and 10, **subtracted from** 9 times a number

59. Subtract $5m - 6$ from $m - 9$.

60. Subtract $m - 3$ from $2m - 6$.

61. Eight times the sum of a number and six

62. Five, subtracted from four times a number

63. Double a number, minus the sum of the number and ten

64. Half a number, minus the product of the number and eight

65. Seven, multiplied by the quotient of a number and six

66. The product of a number and ten, less twenty

67. The sum of 2, three times a number, -9, and four times a number

68. The sum of twice a number, -1, five times a number, and -12

△ **69.** Recall that the perimeter of a figure is the total distance around the figure. Given the following rectangle, express

the perimeter as an algebraic expression containing the variable x.

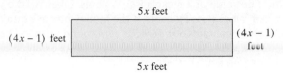

5x feet

$(4x - 1)$ feet $(4x - 1)$ feet

5x feet

△ **70.** Given the following triangle, express its perimeter as an algebraic expression containing the variable x.

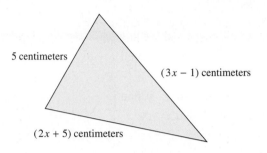

5 centimeters

$(3x - 1)$ centimeters

$(2x + 5)$ centimeters

Given the following, determine whether each scale is balanced or not.

1 cone balances 1 cube

1 cylinder balances 2 cubes

71.

72.

73.

74.

75. To convert from feet to inches, we multiply by 12. For example, the number of inches in 2 feet is $12 \cdot 2$ inches. If one board has a length of $(x + 2)$ *feet* and a second board has a length of $(3x - 1)$ *inches*, express their total length in inches as an algebraic expression.

76. The value of 7 nickels is $5 \cdot 7$ cents. Likewise, the value of x nickels is $5x$ cents. If the money box in a drink machine contains x *nickels*, $3x$ *dimes*, and $(30x - 1)$ *quarters*, express their total value in cents as an algebraic expression.

REVIEW EXERCISES

Evaluate the following expressions for the given values. See Section 1.7.

77. If $x = -1$ and $y = 3$, find $y - x^2$.

78. If $g = 0$ and $h = -4$, find $gh - h^2$.

79. If $a = 2$ and $b = -5$, find $a - b^2$.

80. If $x = -3$, find $x^3 - x^2 + 4$.

81. If $y = -5$ and $z = 0$, find $yz - y^2$.

82. If $x = -2$, find $x^3 - x^2 - x$.

A Look Ahead

Example

Simplify $-3xy + 2x^2y - (2xy - 1)$.

Solution:

$-3xy + 2x^2y - (2xy - 1)$
$= -3xy + 2x^2y - 2xy + 1 = -5xy + 2x^2y + 1$

Simplify each expression.

83. $5b^2c^3 + 8b^3c^2 - 7b^3c^2$

84. $4m^4p^2 + m^4p^2 - 5m^2p^4$

85. $3x - (2x^2 - 6x) + 7x^2$

86. $9y^2 - (6xy^2 - 5y^2) - 8xy^2$

87. $-(2x^2y + 3z) + 3z - 5x^2y$

88. $-(7c^3d - 8c) - 5c - 4c^3d$

2.2 THE ADDITION PROPERTY OF EQUALITY

CD-ROM

SSM

SSG

Video

▶ **OBJECTIVES**

1. Define linear equation in one variable and equivalent equations.
2. Use the addition property of equality to solve linear equations.
3. Write word phrases as algebraic expressions.

1 Recall from Section 1.4 that an equation is a statement that two expressions have the same value. Also, a value of the variable that makes an equation a true statement is called a solution or root of the equation. The process of finding the solution of an equation is called **solving** the equation for the variable. In this section we concentrate on solving **linear equations** in one variable.

> **LINEAR EQUATION IN ONE VARIABLE**
>
> A **linear equation in one variable** can be written in the form
>
> $$ax + b = c$$
>
> where a, b, and c are real numbers and $a \neq 0$.

Evaluating a linear equation for a given value of the variable, as we did in Section 1.4, can tell us whether that value is a solution, but we can't rely on evaluating an equation as our method of solving it.

Instead, to solve a linear equation in x, we write a series of simpler equations, all *equivalent* to the original equation, so that the final equation has the form

$$x = \textbf{number} \quad \text{or} \quad \textbf{number} = x$$

Equivalent equations are equations that have the same solution. This means that the "number" above is the solution to the original equation.

2

The first property of equality that helps us write simpler equivalent equations is the **addition property of equality**.

ADDITION PROPERTY OF EQUALITY

If a, b, and c are real numbers, then

$$a = b \quad \text{and} \quad a + c = b + c$$

are equivalent equations.

This property guarantees that adding the same number to both sides of an equation does not change the solution of the equation. Since subtraction is defined in terms of addition, we may also **subtract the same number from both sides** without changing the solution.

A good way to picture a true equation is as a balanced scale. Since it is balanced, each side of the scale weighs the same amount.

$x - 2$ 5

If the same weight is added to or subtracted from each side, the scale remains balanced.

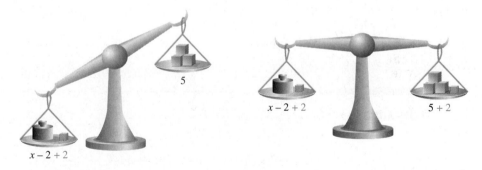

5

$x - 2 + 2$

$x - 2 + 2$ $5 + 2$

We use the addition property of equality to write equivalent equations until the variable is by itself on one side of the equation, and the equation looks like "$x = $ number" or "number $= x$."

Example 1   Solve $x - 7 = 10$ for x.

Solution   To solve for x, we want x alone on one side of the equation. To do this, we add 7 to both sides of the equation.

$$x - 7 = 10$$
$$x - 7 + 7 = 10 + 7 \qquad \text{Add 7 to both sides.}$$
$$x = 17 \qquad \text{Simplify.}$$

The solution of the equation $x = 17$ is obviously 17. Since we are writing equivalent equations, the solution of the equation $x - 7 = 10$ is also 17.

To check, replace x with 17 in the original equation.

$$x - 7 = 10$$
$$17 - 7 \stackrel{?}{=} 10 \qquad \text{Replace } x \text{ with 17 in the original equation.}$$
$$10 = 10 \qquad \text{True.}$$

Since the statement is true, 17 is the solution.

Example 2 Solve $y + 0.6 = -1.0$ for y.

Solution To get y alone on one side of the equation, subtract 0.6 from both sides of the equation.

$$y + 0.6 = -1.0$$
$$y + 0.6 - 0.6 = -1.0 - 0.6 \qquad \text{Subtract 0.6 from both sides.}$$
$$y = -1.6 \qquad \text{Combine like terms.}$$

To check the proposed solution, -1.6, replace y with -1.6 in the original equation.

Check

$$y + 0.6 = -1.0$$
$$-1.6 + 0.6 \stackrel{?}{=} -1.0 \qquad \text{Replace } y \text{ with } -1.6 \text{ in the original equation.}$$
$$-1.0 = -1.0 \qquad \text{True.}$$

The solution is -1.6.

Example 3 Solve: $\dfrac{1}{2} = x - \dfrac{3}{4}$

Solution To get x alone, we add $\dfrac{3}{4}$ to both sides.

$$\frac{1}{2} = x - \frac{3}{4}$$

$$\frac{1}{2} + \frac{3}{4} = x - \frac{3}{4} + \frac{3}{4} \qquad \text{Add } \frac{3}{4} \text{ to both sides.}$$

$$\frac{1}{2} \cdot \frac{2}{2} + \frac{3}{4} = x \qquad \text{The LCD is 4.}$$

$$\frac{2}{4} + \frac{3}{4} = x \qquad \text{Add the fractions.}$$

$$\frac{5}{4} = x$$

Check

$$\frac{1}{2} = x - \frac{3}{4}$$ Original equation.

$$\frac{1}{2} \overset{?}{=} \frac{5}{4} - \frac{3}{4}$$ Replace x with $\frac{5}{4}$.

$$\frac{1}{2} \overset{?}{=} \frac{2}{4}$$ Subtract.

$$\frac{1}{2} = \frac{1}{2}$$ True.

The solution is $\frac{5}{4}$.

HELPFUL HINT

We may solve an equation so that the variable is alone on *either* side of the equation. For example, $\frac{5}{4} = x$ is equivalent to $x = \frac{5}{4}$.

Example 4   Solve $5t - 5 = 6t + 2$ for t.

Solution   To solve for t, we first want all terms containing t on one side of the equation and all other terms on the other side of the equation. To do this, first subtract $5t$ from both sides of the equation.

$$5t - 5 = 6t + 2$$
$$5t - 5 - 5t = 6t + 2 - 5t$$ Subtract $5t$ from both sides.
$$-5 = t + 2$$ Combine like terms.

Next, subtract 2 from both sides and the variable t will be isolated.

$$-5 = t + 2$$
$$-5 - 2 = t + 2 - 2$$ Subtract 2 from both sides.
$$-7 = t$$

Check the solution, -7, in the original equation. The solution is -7.

Many times, it is best to simplify one or both sides of an equation before applying the addition property of equality.

Example 5   Solve: $2x + 3x - 5 + 7 = 10x + 3 - 6x - 4$

Solution   First we simplify both sides of the equation.

$$2x + 3x - 5 + 7 = 10x + 3 - 6x - 4$$
$$5x + 2 = 4x - 1$$ Combine like terms on each side of the equation.

Next, we want all terms with a variable on one side of the equation and all numbers on the other side.

$$5x + 2 - 4x = 4x - 1 - 4x$$ Subtract $4x$ from both sides.
$$x + 2 = -1$$ Combine like terms.
$$x + 2 - 2 = -1 - 2$$ Subtract 2 from both sides to get x alone.
$$x = -3$$ Combine like terms.

Check
$$2x + 3x - 5 + 7 = 10x + 3 - 6x - 4 \qquad \text{Original equation.}$$
$$2(-3) + 3(-3) - 5 + 7 \stackrel{?}{=} 10(-3) + 3 - 6(-3) - 4 \qquad \text{Replace } x \text{ with } -3.$$
$$-6 - 9 - 5 + 7 \stackrel{?}{=} -30 + 3 + 18 - 4 \qquad \text{Multiply.}$$
$$-13 = -13 \qquad \text{True.}$$

The solution is -3.

If an equation contains parentheses, we use the distributive property to remove them, as before. Then we combine any like terms.

Example 6 Solve: $6(2a - 1) - (11a + 6) = 7$

Solution
$$6(2a - 1) - 1(11a + 6) = 7$$
$$6(2a) + 6(-1) - 1(11a) - 1(6) = 7 \qquad \text{Apply the distributive property.}$$
$$12a - 6 - 11a - 6 = 7 \qquad \text{Multiply.}$$
$$a - 12 = 7 \qquad \text{Combine like terms.}$$
$$a - 12 + 12 = 7 + 12 \qquad \text{Add 12 to both sides.}$$
$$a = 19 \qquad \text{Simplify.}$$

Check Check by replacing a with 19 in the original equation.

Example 7 Solve: $3 - x = 7$

Solution First we subtract 3 from both sides.

$$3 - x = 7$$
$$3 - x - 3 = 7 - 3 \qquad \text{Subtract 3 from both sides.}$$
$$-x = 4 \qquad \text{Simplify.}$$

We have not yet solved for x since x is not alone. However, this equation does say that the opposite of x is 4. If the opposite of x is 4, then x is the opposite of 4, or $x = -4$. If $-x = 4$, then $x = -4$.

Check
$$3 - x = 7 \qquad \text{Original equation.}$$
$$3 - (-4) \stackrel{?}{=} 7 \qquad \text{Replace } x \text{ with } -4.$$
$$3 + 4 \stackrel{?}{=} 7 \qquad \text{Add.}$$
$$7 = 7 \qquad \text{True.}$$

The solution is -4.

3 Next, we practice writing algebraic expressions.

Example 8
a. The sum of two numbers is 8. If one number is 3, find the other number.
b. The sum of two numbers is 8. If one number is x, write an expression representing the other number.

Solution   **a.** If the sum of two numbers is 8 and one number is 3, we find the other number by subtracting 3 from 8. The other number is $8 - 3$ or 5.

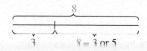

b. If the sum of two numbers is 8 and one number is x, we find the other number by subtracting x from 8. The other number is represented by $8 - x$.

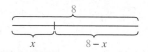

SPOTLIGHT ON DECISION MAKING

Suppose you are a teacher's aide in Mr. Mankato's eighth-grade science class. You are responsible for recording grades and totaling scores. After students took Test 4, Mr. Mankato decided to allow anyone who scored 75 or less on that test the opportunity to do an extra credit project. While preparing a list of students eligible for the extra credit project, you found that one of the entries had gotten smudged and was illegible. Should Demetra Brown be included in the list of students eligible for the extra credit project? Explain.

GRADE BOOK

Teacher: *Mr. Mankato*
Period: *4th*

Student	#1	#2	#3	#4	Total
Brown, Demetra	92	86	94	✸	344
Mankad, Rahul	88	89	91	73	341
Mendoza, Carlos	91	85	83	80	339
Roberge, Ann	93	84	92	74	343

MENTAL MATH

Solve each equation mentally. See Examples 1 and 2.

1. $x + 4 = 6$

2. $x + 7 = 10$

3. $n + 18 = 30$

4. $z + 22 = 40$

5. $b - 11 = 6$

6. $d - 16 = 5$

Exercise Set 2.2

Solve each equation. Check each solution. See Examples 1 through 3.

1. $x + 7 = 10$

2. $x + 14 = 25$

3. $x - 2 = -4$

4. $y - 9 = 1$

5. $3 + x = -11$

6. $8 + z = -8$

7. $r - 8.6 = -8.1$

8. $t - 9.2 = -6.8$

9. $\frac{1}{3} + f = \frac{3}{4}$

10. $c + \frac{1}{6} = \frac{3}{8}$

11. $5b - 0.7 = 6b$

12. $9x + 5.5 = 10x$

13. $7x - 3 = 6x$

14. $18x - 9 = 19x$

15. In your own words, explain what is meant by the solution of an equation.

16. In your own words, explain how to check a solution of an equation.

Solve each equation. Don't forget to first simplify each side of the equation, if possible. Check each solution. See Examples 4 through 7.

17. $7x + 2x = 8x - 3$

18. $3n + 2n = 7 + 4n$

19. $2y + 10 = 5y - 4y$

20. $4x - 4 = 10x - 7x$

21. $3x - 6 = 2x + 5$

22. $7y + 2 = 6y + 2$

23. $5x - \dfrac{1}{6} = 6x - \dfrac{5}{6}$

24. $2x + \dfrac{1}{8} = x - \dfrac{3}{8}$

25. $8y + 2 - 6y = 3 + y - 10$

26. $4p - 11 - p = 2 + 2p - 20$

27. $13x - 9 + 2x - 5 = 12x - 1 + 2x$

28. $15x + 20 - 10x - 9 = 25x + 8 - 21x - 7$

29. $-6.5 - 4x - 1.6 - 3x = -6x + 9.8$

30. $-1.4 - 7x - 3.6 - 2x = -8x + 4.4$

31. $\dfrac{3}{8}x - \dfrac{1}{6} = -\dfrac{5}{8}x - \dfrac{2}{3}$

32. $\dfrac{2}{5}x - \dfrac{1}{12} = -\dfrac{3}{5}x - \dfrac{3}{4}$

33. $2(x - 4) = x + 3$

34. $3(y + 7) = 2y - 5$

35. $7(6 + w) = 6(2 + w)$

36. $6(5 + c) = 5(c - 4)$

37. $10 - (2x - 4) = 7 - 3x$

38. $15 - (6 - 7k) = 2 + 6k$

39. $-5(n - 2) = 8 - 4n$

40. $-4(z - 3) = 2 - 3z$

41. $-3\left(x - \dfrac{1}{4}\right) = -4x$

42. $-2\left(x - \dfrac{1}{7}\right) = -3x$

43. $3(n - 5) - (6 - 2n) = 4n$

44. $5(3 + z) - (8z + 9) = -4z$

45. $-2(x + 6) + 3(2x - 5) = 3(x - 4) + 10$

46. $-5(x + 1) + 4(2x - 3) = 2(x + 2) - 8$

47. $7(m - 2) - 6(m + 1) = -20$

48. $-4(x - 1) - 5(2 - x) = -6$

49. $0.8t + 0.2(t - 0.4) = 1.75$

50. $0.6v + 0.4(0.3 + v) = 2.34$

See Example 8.

51. Two numbers have a sum of 20. If one number is p, express the other number in terms of p.

52. Two numbers have a sum of 13. If one number is y, express the other number in terms of y.

53. A 10-foot board is cut into two pieces. If one piece is x feet long, express the other length in terms of x.

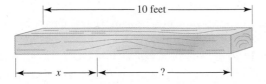

54. A 5-foot piece of string is cut into two pieces. If one piece is x feet long, express the other length in terms of x.

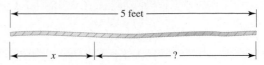

△ **55.** Two angles are *supplementary* if their sum is 180°. If one angle measures $x°$, express the measure of its supplement in terms of x.

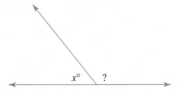

△ **56.** Two angles are *complementary* if their sum is 90°. If one angle measures $x°$, express the measure of its complement in terms of x.

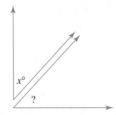

57. In a mayoral election, April Catarella received 284 more votes than Charles Pecot. If Charles received n votes, how many votes did April receive?

58. The length of the top of a computer desk is $1\frac{1}{2}$ feet longer than its width. If its width measures m feet, express its length as an algebraic expression in m.

59. The Verrazano-Narrows Bridge in New York City is the longest suspension bridge in North America. The Golden Gate Bridge in San Francisco is 60 feet shorter than the Verrazano-Narrows Bridge. If the length of the Verrazano-Narrows Bridge is m feet, express the length of the Golden Gate Bridge as an algebraic expression in m. (*Source:* World Almanac, 2000)

60. In a recent U.S. Senate race in Maine, Susan M. Collins received 30,898 more votes than Joseph E. Brennan. If Joseph received n votes, how many did Susan receive? (*Source:* Voter News Service)

△ **61.** The sum of the angles of a triangle is 180°. If one angle of a triangle measures $x°$ and a second angle measures $(2x + 7)°$, express the measure of the third angle in terms of x. Simplify the expression.

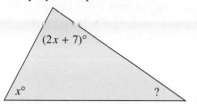

△ **62.** A quadrilateral is a four-sided figure like the one shown below whose angle sum is 360°. If one angle measures $x°$, a second angle measures $3x°$, and a third angle measures $5x°$, express the measure of the fourth angle in terms of x. Simplify the expression.

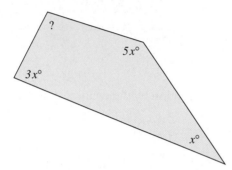

63. A nurse's aide recorded the following fluid intakes for a patient on her night shift: 200 ml, 150 ml, 400 ml. If the patient's doctor requested that a total of 1000 ml of fluid be taken by the patient overnight, how much more fluid must the nurse give the patient? To solve this problem, solve the equation $200 + 150 + 400 + x = 1000$.

64. Let $x = 1$ and then $x = 2$ in the equation $x + 5 = x + 6$. Is either number a solution? How many solutions do you think this equation has? Explain your answer.

65. Let $x = 1$ and then $x = 2$ in the equation $x + 3 = x + 3$. Is either number a solution? How many solutions do you think this equation has? Explain your answer.

Use a calculator to determine whether the given value is a solution of the given equation.

66. $1.23x - 0.06 = 2.6x - 0.1285$; $x = 0.05$

67. $8.13 + 5.85y = 20.05y - 8.91$; $y = 1.2$

68. $3(a + 4.6) = 5a + 2.5$; $a = 6.3$

69. $7(z - 1.7) + 9.5 = 5(z + 3.2) - 9.2$; $z = 4.8$

REVIEW EXERCISES

Find the reciprocal or multiplicative inverse of each. See Section 1.7.

70. $\dfrac{5}{8}$ **71.** $\dfrac{7}{6}$ **72.** 2

73. 5 **74.** $-\dfrac{1}{9}$ **75.** $-\dfrac{3}{5}$

Perform each indicated operation and simplify. See Section 1.7.

76. $\dfrac{3x}{3}$ **77.** $\dfrac{-2y}{-2}$ **78.** $-5\left(-\dfrac{1}{5}y\right)$

79. $7\left(\dfrac{1}{7}r\right)$ **80.** $\dfrac{3}{5}\left(\dfrac{5}{3}x\right)$ **81.** $\dfrac{9}{2}\left(\dfrac{2}{9}x\right)$

2.3   THE MULTIPLICATION PROPERTY OF EQUALITY

CD-ROM   SSM

SSG   Video

▶ **O B J E C T I V E S**

1. Use the multiplication property of equality to solve linear equations.
2. Use both the addition and multiplication properties of equality to solve linear equations.
3. Write word phrases as algebraic expressions.

1   As useful as the addition property of equality is, it cannot help us solve every type of linear equation in one variable. For example, adding or subtracting a value on both sides of the equation does not help solve

$$\frac{5}{2}x = 15.$$

Instead, we apply another important property of equality, the **multiplication property of equality**.

MULTIPLICATION PROPERTY OF EQUALITY

If a, b, and c are real numbers and $c \neq 0$, then

$$a = b \quad \text{and} \quad ac = bc$$

are equivalent equations.

This property guarantees that multiplying both sides of an equation by the same nonzero number does not change the solution of the equation. Since division is defined in terms of multiplication, we may also **divide both sides of the equation by the same nonzero number** without changing the solution.

Example 1   Solve for x: $\dfrac{5}{2} x = 15$.

Solution   To get x alone, multiply both sides of the equation by the reciprocal of $\dfrac{5}{2}$, which is $\dfrac{2}{5}$.

$$\frac{5}{2} x = 1.5$$

$$\frac{2}{5} \cdot \frac{5}{2} x = \frac{2}{5} \cdot 15 \qquad \text{Multiply both sides by } \frac{2}{5}.$$

$$\left(\frac{2}{5} \cdot \frac{5}{2} \right) x = \frac{2}{5} \cdot 15 \qquad \text{Apply the associative property.}$$

$$1x = 6 \qquad \text{Simplify.}$$

or

$$x = 6$$

Check   Replace x with 6 in the original equation.

$$\frac{5}{2} x = 15 \qquad \text{Original equation.}$$

$$\frac{5}{2} (6) \overset{?}{=} 15 \qquad \text{Replace } x \text{ with 6.}$$

$$15 = 15 \qquad \text{True.}$$

The solution is 6.

In the equation $\dfrac{5}{2} x = 15$, $\dfrac{5}{2}$ is the coefficient of x. When the coefficient of x is a *fraction*, we will get x alone by multiplying by the reciprocal. When the coefficient of x is an integer or a decimal, it is usually more convenient to divide both sides by the coefficient. (Dividing by a number is, of course, the same as multiplying by the reciprocal of the number.)

Example 2   Solve: $-3x = 33$

Solution   Recall that $-3x$ means $-3 \cdot x$. To get x alone, we divide both sides by the coefficient of x, that is, -3.

$$-3x = 33$$

$$\frac{-3x}{-3} = \frac{33}{-3} \qquad \text{Divide both sides by } -3.$$

$$1x = -11 \qquad \text{Simplify.}$$

$$x = -11$$

Check
$$-3x = 33 \qquad \text{Original equation.}$$

$$-3(-11) \overset{?}{=} 33 \qquad \text{Replace } x \text{ with } -11.$$

$$33 = 33 \qquad \text{True.}$$

The solution is -11.

Example 3   Solve: $\dfrac{y}{7} = 20$

Solution   Recall that $\dfrac{y}{7} = \dfrac{1}{7}y$. To get y alone, we multiply both sides of the equation by 7, the reciprocal of $\dfrac{1}{7}$.

$$\frac{y}{7} = 20$$

$$\frac{1}{7}y = 20$$

$$7 \cdot \frac{1}{7}y = 7 \cdot 20 \qquad \text{Multiply both sides by 7.}$$

$$1y = 140 \qquad \text{Simplify.}$$

$$y = 140$$

Check
$$\frac{y}{7} = 20 \qquad \text{Original equation.}$$

$$\frac{140}{7} \overset{?}{=} 20 \qquad \text{Replace } y \text{ with 140.}$$

$$20 = 20 \qquad \text{True.}$$

The solution is 140.

Example 4   Solve: $3.1x = 4.96$

Solution
$$3.1x = 4.96$$

$$\frac{3.1x}{3.1} = \frac{4.96}{3.1} \qquad \text{Divide both sides by 3.1.}$$

$$1x = 1.6 \qquad \text{Simplify.}$$

$$x = 1.6$$

Check   Check by replacing x with 1.6 in the original equation. The solution is 1.6.

Example 5 Solve: $-\dfrac{2}{3}x = -\dfrac{5}{2}$

Solution To get x alone, we multiply both sides of the equation by $-\dfrac{3}{2}$, the reciprocal of the coefficient of x.

> **HELPFUL HINT**
> Don't forget to multiply *both sides* by $-\dfrac{3}{2}$.

$$-\frac{2}{3}x = -\frac{5}{2}$$

$$-\frac{3}{2} \cdot -\frac{2}{3}x = -\frac{3}{2} \cdot -\frac{5}{2} \qquad \text{Multiply both sides by } -\frac{3}{2},$$

$$\text{the reciprocal of } -\frac{2}{3}.$$

$$x = \frac{15}{4} \qquad \text{Simplify.}$$

Check Check by replacing x with $\dfrac{15}{4}$ in the original equation. The solution is $\dfrac{15}{4}$. ▬

2 We are now ready to combine the skills learned in the last section with the skills learned from this section to solve equations by applying more than one property.

Example 6 Solve: $-z - 4 = 6$

Solution First, get $-z$, the term containing the variable alone on one side. To do so, add 4 to both sides of the equation.

$$-z - 4 + 4 = 6 + 4 \qquad \text{Add 4 to both sides.}$$

$$-z = 10 \qquad \text{Simplify.}$$

Next, recall that $-z$ means $-1 \cdot z$. To get z alone, either multiply or divide both sides of the equation by -1. In this example, we divide.

$$-z = 10$$

$$\frac{-z}{-1} = \frac{10}{-1} \qquad \text{Divide both sides by the coefficient } -1.$$

$$z = -10 \qquad \text{Simplify.}$$

Check To check, replace z with -10 in the original equation. The solution is -10. ▬

Example 7 Solve: $12a - 8a = 10 + 2a - 13 - 7$

Solution First, simplify both sides of the equation by combining like terms.

$$12a - 8a = 10 + 2a - 13 - 7$$

$$4a = 2a - 10 \qquad \text{Combine like terms.}$$

To get all terms containing a variable on one side, subtract $2a$ from both sides.

$$4a - 2a = 2a - 10 - 2a \qquad \text{Subtract } 2a \text{ from both sides.}$$

$$2a = -10 \qquad \text{Simplify.}$$

$$\frac{2a}{2} = \frac{-10}{2} \qquad \text{Divide both sides by 2.}$$

$$a = -5 \qquad \text{Simplify.}$$

Check Check by replacing a with -5 in the original equation. The solution is -5. ▬

3   Next, we continue to sharpen our problem-solving skills by writing algebraic expressions.

Example 8   If x is the first of three consecutive integers, express the sum of the three integers in terms of x. Simplify if possible.

Solution   An example of three consecutive integers is

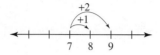

The second consecutive integer is always 1 more than the first, and the third consecutive integer is 2 more than the first. If x is the first of three consecutive integers, the three consecutive integers are

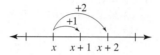

Their sum is

In words:

first integer	+	second integer	+	third integer

Translate: x + $(x + 1)$ + $(x + 2)$

which simplifies to $3x + 3$.

Below are examples of consecutive even and odd integers.

Even integers:

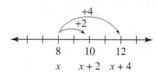

Odd integers:

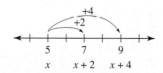

> **HELPFUL HINT**
> If x is an odd integer, then $x + 2$ is the next odd integer. This 2 simply means that odd integers are always 2 units from each other.
>
>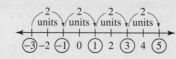

MENTAL MATH

Solve each equation mentally. See Examples 2 and 3.

1. $3a = 27$

2. $9c = 54$

3. $5b = 10$

4. $7t = 14$

5. $6x = -30$

6. $8r = -64$

Exercise Set 2.3

Solve each equation. Check each solution. See Examples 1 through 5.

1. $-5x = 20$

2. $7x = 49$

3. $3x = 0$

4. $2x = 0$

5. $-x = -12$

6. $-y = 8$

7. $\dfrac{2}{3}x = -8$

8. $\dfrac{3}{4}n = -15$

9. $\dfrac{1}{6}d = \dfrac{1}{2}$

10. $\dfrac{1}{8}v = \dfrac{1}{4}$

11. $\dfrac{a}{2} = 1$

12. $\dfrac{d}{15} = 2$

13. $\dfrac{k}{-7} = 0$

14. $\dfrac{f}{-5} = 0$

15. $1.7x = 10.71$

16. $8.5y = 18.7$

17. $42 = 7x$

18. $81 = 3x$

19. $4.4 = -0.8x$

20. $6.3 = -0.6x$

21. $-\dfrac{3}{7}p = -2$

22. $-\dfrac{4}{5}r = -5$

23. $-\dfrac{4}{3}x = 12$

24. $-\dfrac{10}{3}x = 30$

Solve each equation. Check each solution. See Examples 6 and 7.

25. $2x - 4 = 16$

26. $3x - 1 = 26$

27. $-x + 2 = 22$

28. $-x + 4 = -24$

29. $6a + 3 = 3$

30. $8t + 5 = 5$

31. $6x + 10 = -20$

32. $-10y + 15 = 5$

33. $5 - 0.3k = 5$

34. $2 + 0.4p = 2$

35. $-2x + \dfrac{1}{2} = \dfrac{7}{2}$

36. $-3n - \dfrac{1}{3} = \dfrac{8}{3}$

37. $\dfrac{x}{3} + 2 = -5$

38. $\dfrac{b}{4} - 1 = -7$

39. $10 = 2x - 1$

40. $12 = 3j - 4$

41. $6z - 8 - z + 3 = 0$

42. $4a + 1 + a - 11 = 0$

43. $10 - 3x - 6 - 9x = 7$

44. $12x + 30 + 8x - 6 = 10$

45. $1 = 0.4x - 0.6x - 5$

46. $19 = 0.4x - 0.9x - 6$

47. $z - 5z = 7z - 9 - z$

48. $t - 6t = -13 + t - 3t$

49. $0.4x - 0.6x - 5 = 1$

50. $0.4x - 0.9x - 6 = 19$

51. $6 - 2x + 8 = 10$

52. $-5 - 6y + 6 = 19$

53. $-3a + 6 + 5a = 7a - 8a$

54. $4b - 8 - b = 10b - 3b$

55. The equation $3x + 6 = 2x + 10 + x - 4$ is true for all real numbers. Substitute a few real numbers for x to see that this is so and then try solving the equation.

56. The equation $6x + 2 - 2x = 4x + 1$ has no solution. Try solving this equation for x and see what happens.

57. From the results of Exercises 55 and 56, when do you think an equation has all real numbers as its solution set?

58. From the results of Exercises 55 and 56, when do you think an equation has no solution?

Write each algebraic expression described. Simplify if possible. See Example 8.

59. If x represents the first of two consecutive odd integers, express the sum of the two integers in terms of x.

60. If x is the first of four consecutive even integers, write their sum as an algebraic expression in x.

61. If x is the first of three consecutive integers, express the sum of the first integer and the third integer as an algebraic expression containing the variable x.

62. If x is the first of two consecutive integers, express the sum of 20 and the second consecutive integer as an algebraic expression containing the variable x.

63. A licensed practical nurse is instructed to give a patient 2100 milligrams of an antibiotic over a period of 36 hours. If the antibiotic is to be given every 4 hours starting immediately, how much antibiotic should be given in each dose? To answer this question, solve the equation $9x = 2100$.

Solve each equation.

64. $-3.6x = 10.62$

65. $4.95y = -31.185$

66. $7x - 5.06 = -4.92$

67. $0.06y + 2.63 = 2.5562$

REVIEW EXERCISES

Simplify each expression. See Section 2.1.

68. $5x + 2(x - 6)$

69. $-7y + 2y - 3(y + 1)$

70. $6(2z + 4) + 20$

71. $-(3a - 3) + 2a - 6$

72. $-(r - 1) + r$

73. $8(z - 6) + 7z - 1$

Insert $<$, $>$, or $=$ in the appropriate space to make each statement true. See Sections 1.2 and 1.7.

74. $(-3)^2$ ___ -3^2

75. $(-2)^4$ ___ -2^4

76. $(-2)^3$ ___ -2^3

77. $(-4)^3$ ___ -4^3

78. $-|-6|$ ___ 6

79. $-|-0.7|$ ___ -0.7

2.4 SOLVING LINEAR EQUATIONS

CD-ROM   SSM

SSG   Video

▶ **OBJECTIVES**

1. Apply the general strategy for solving a linear equation.
2. Solve equations containing fractions.
3. Solve equations containing decimals.
4. Recognize identities and equations with no solution.
5. Write sentences as equations and solve.

1 We now present a general strategy for solving linear equations. One new piece of strategy is a suggestion to "clear an equation of fractions" as a first step. Doing so makes the equation more manageable, since operating on integers is more convenient than operating on fractions.

SOLVING LINEAR EQUATIONS IN ONE VARIABLE

Step 1. Multiply on both sides to clear the equation of fractions if they occur.

Step 2. Use the distributive property to remove parentheses if they occur.

Step 3. Simplify each side of the equation by combining like terms.

Step 4. Get all variable terms on one side and all numbers on the other side by using the addition property of equality.

Step 5. Get the variable alone by using the multiplication property of equality.

Step 6. Check the solution by substituting it into the original equation.

Example 1 Solve: $4(2x - 3) + 7 = 3x + 5$

Solution There are no fractions, so we begin with Step 2.

$$4(2x - 3) + 7 = 3x + 5$$

Step 2. $8x - 12 + 7 = 3x + 5$ Apply the distributive property.

Step 3. $8x - 5 = 3x + 5$ Combine like terms.

Step 4.   Get all variable terms on the same side of the equation by subtracting $3x$ from both sides, then adding 5 to both sides.

$$8x - 5 - 3x = 3x + 5 - 3x$$ Subtract $3x$ from both sides.

$$5x - 5 = 5$$ Simplify.

$$5x - 5 + 5 = 5 + 5$$ Add 5 to both sides.

$$5x = 10$$ Simplify.

Step 5. Use the multiplication property of equality to get x alone.

$$\frac{5x}{5} = \frac{10}{5} \qquad \text{Divide both sides by 5.}$$

$$x = 2 \qquad \text{Simplify.}$$

Step 6. Check.

$$4(2x - 3) + 7 = 3x + 5 \qquad \text{Original equation}$$
$$4[2(2) - 3] + 7 \stackrel{?}{=} 3(2) + 5 \qquad \text{Replace } x \text{ with 2.}$$
$$4(4 - 3) + 7 \stackrel{?}{=} 6 + 5$$
$$4(1) + 7 \stackrel{?}{=} 11$$
$$4 + 7 \stackrel{?}{=} 11$$
$$11 = 11 \qquad \text{True.}$$

The solution is 2.

> **H E L P F U L   H I N T**
> When checking solutions, use the original written equation.

Example 2   Solve: $8(2 - t) = -5t$

Solution   First, we apply the distributive property.

$$8(2 - t) = -5t$$

Step 2. $16 - 8t = -5t$ Use the distributive property.

Step 4. $16 - 8t + 8t = -5t + 8t$ To get variable terms on one side, add $8t$ to both sides.

$$16 = 3t \qquad \text{Combine like terms.}$$

Step 5. $\dfrac{16}{3} = \dfrac{3t}{3}$ Divide both sides by 3.

$$\frac{16}{3} = t \qquad \text{Simplify.}$$

Step 6.   Check.

$$8(2 - t) = -5t \qquad \text{Original equation}$$

$$8\left(2 - \frac{16}{3}\right) \stackrel{?}{=} -5\left(\frac{16}{3}\right) \qquad \text{Replace } t \text{ with } \frac{16}{3}.$$

$$8\left(\frac{6}{3} - \frac{16}{3}\right) \stackrel{?}{=} -\frac{80}{3} \qquad \text{The LCD is 3.}$$

$$8\left(-\frac{10}{3}\right) \stackrel{?}{=} -\frac{80}{3} \qquad \text{Subtract fractions.}$$

$$-\frac{80}{3} = -\frac{80}{3} \qquad \text{True.}$$

The solution is $\dfrac{16}{3}$.

2   If an equation contains fractions, we can clear the equation of fractions by multiplying both sides by the LCD of all denominators. By doing this, we avoid working with time-consuming fractions.

Example 3   Solve: $\dfrac{x}{2} - 1 = \dfrac{2}{3}x - 3$

Solution   We begin by clearing fractions. To do this, we multiply both sides of the equation by the LCD of 2 and 3, which is 6.

$$\frac{x}{2} - 1 = \frac{2}{3}x - 3$$

Step 1. $6\left(\dfrac{x}{2} - 1\right) = 6\left(\dfrac{2}{3}x - 3\right)$ Multiply both sides by the LCD, 6.

Step 2. $6\left(\dfrac{x}{2}\right) - 6(1) = 6\left(\dfrac{2}{3}x\right) - 6(3)$ Apply the distributive property.

> **HELPFUL HINT**
> Don't forget to multiply *each* term by the LCD.

$$3x - 6 = 4x - 18 \qquad \text{Simplify.}$$

There are no longer grouping symbols and no like terms on either side of the equation, so we continue with Step 4.

$$3x - 6 = 4x - 18$$

Step 4. $3x - 6 - 3x = 4x - 18 - 3x$ To get variable terms on one side, subtract $3x$ from both sides.
$-6 = x - 18$ Simplify.
$-6 + 18 = x - 18 + 18$ Add 18 to both sides.
$12 = x$ Simplify.

Step 5.   The variable is now alone, so there is no need to apply the multiplication property of equality.

Step 6.   Check.

$$\frac{x}{2} - 1 = \frac{2}{3}x - 3 \qquad \text{Original equation}$$

$$\frac{12}{2} - 1 \stackrel{?}{=} \frac{2}{3} \cdot 12 - 3 \qquad \text{Replace } x \text{ with 12.}$$

$$6 - 1 \stackrel{?}{=} 8 - 3 \qquad \text{Simplify.}$$

$$5 = 5 \qquad \text{True.}$$

The solution is 12.

Example 4   Solve: $\dfrac{2(a + 3)}{3} = 6a + 2$

Solution   We clear the equation of fractions first.

$$\frac{2(a + 3)}{3} = 6a + 2$$

Step 1. $3 \cdot \dfrac{2(a + 3)}{3} = 3(6a + 2)$ Clear the fraction by multiplying both sides by the LCD, 3.

Step 2. Next, we use the distributive property and remove parentheses.

$2a + 6 = 18a + 6$ Apply the distributive property.

Step 4. $2a + 6 - 6 = 18a + 6 - 6$ Subtract 6 from both sides.

$2a = 18a$

$2a - 18a = 18a - 18a$ Subtract 18a from both sides.

$-16a = 0$

Step 5. $\dfrac{-16a}{-16} = \dfrac{0}{-16}$ Divide both sides by −16.

$a = 0$ Write the fraction in simplest form.

Step 6. To check, replace a with 0 in the original equation. The solution is 0. ▄

3

When solving a problem about money, you may need to solve an equation containing decimals. If you choose, you may multiply to clear the equation of decimals.

◆ **Example 5**   Solve: $0.25x + 0.10(x - 3) = 0.05(22)$

Solution   First we clear this equation of decimals by multiplying both sides of the equation by 100. Recall that multiplying a decimal number by 100 has the effect of moving the decimal point 2 places to the right.

$$0.25x + 0.10(x - 3) = 0.05(22)$$

Step 1. $0.25x + 0.10(x - 3) = 0.05(22)$ Multiply both sides by 100.

$25x + 10(x - 3) = 5(22)$

Step 2. $25x + 10x - 30 = 110$ Apply the distributive property.

Step 3. $35x - 30 = 110$ Combine like terms.

Step 4. $35x - 30 + 30 = 110 + 30$ Add 30 to both sides.

$35x = 140$ Combine like terms.

Step 5. $\dfrac{35x}{35} = \dfrac{140}{35}$ Divide both sides by 35.

$x = 4$

Step 6. To check, replace x with 4 in the original equation. The solution is 4. ▄

4

So far, each equation that we have solved has had a single solution. However, not every equation in one variable has a single solution. Some equations have no solution, while others have an infinite number of solutions. For example,

$$x + 5 = x + 7$$

has no solution since no matter which **real number** we replace x with, the equation is false.

real number + 5 = same **real number** + 7   **FALSE**

On the other hand,

$$x + 6 = x + 6$$

has infinitely many solutions since x can be replaced by any real number and the equation is always true.

real number $+$ 6 $=$ same real number $+$ 6   **TRUE**

The equation $x + 6 = x + 6$ is called an **identity**. The next few examples illustrate special equations like these.

Example 6   Solve: $-2(x - 5) + 10 = -3(x + 2) + x$

Solution
$$-2(x - 5) + 10 = -3(x + 2) + x$$
$$-2x + 10 + 10 = -3x - 6 + x \qquad \text{Apply the distributive property on both sides.}$$
$$-2x + 20 = -2x - 6 \qquad \text{Combine like terms.}$$
$$-2x + 20 + 2x = -2x - 6 + 2x \qquad \text{Add } 2x \text{ to both sides.}$$
$$20 = -6 \qquad \text{Combine like terms.}$$

The final equation contains no variable terms, and there is no value for x that makes $20 = -6$ a true equation. We conclude that there is **no solution** to this equation. ■

Example 7   Solve: $3(x - 4) = 3x - 12$

Solution
$$3(x - 4) = 3x - 12$$
$$3x - 12 = 3x - 12 \qquad \text{Apply the distributive property.}$$

The left side of the equation is now identical to the right side. Every real number may be substituted for x and a true statement will result. We arrive at the same conclusion if we continue.

$$3x - 12 = 3x - 12$$
$$3x - 12 + 12 = 3x - 12 + 12 \qquad \text{Add 12 to both sides.}$$
$$3x = 3x \qquad \text{Combine like terms.}$$
$$3x - 3x = 3x - 3x \qquad \text{Subtract } 3x \text{ from both sides.}$$
$$0 = 0$$

Again, one side of the equation is identical to the other side. Thus, $3(x - 4) = 3x - 12$ is an **identity** and **every real number** is a solution. ■

5   We can apply our equation-solving skills to solving problems written in words. Many times, writing an equation that describes or models a problem involves a direct translation from a word sentence to an equation.

Example 8   **FINDING AN UNKNOWN NUMBER**

Twice a number, added to seven, is the same as three subtracted from the number. Find the number.

Solution   Translate the sentence into an equation and solve.

In words:	twice a number	added to	seven	is the same as	three subtracted from the number
Translate:	$2x$	$+$	7	$=$	$x - 3$

HELPFUL HINT

When checking solutions, go back to the original stated problem, rather than to your equation in case errors have been made in translating to an equation.

To solve, begin by subtracting x on both sides to isolate the variable term.

$$2x + 7 = x - 3$$

$$2x + 7 - x = x - 3 - x \qquad \text{Subtract } x \text{ from both sides.}$$

$$x + 7 = -3 \qquad \text{Combine like terms.}$$

$$x + 7 - 7 = -3 - 7 \qquad \text{Subtract 7 from both sides.}$$

$$x = -10 \qquad \text{Combine like terms.}$$

Check the solution in the problem as it was originally stated. To do so, replace "number" in the sentence with -10. Twice "-10" added to 7 is the same as 3 subtracted from "-10."

$$2(-10) + 7 = -10 - 3$$

$$-13 = -13$$

The unknown number is -10.

CALCULATOR EXPLORATIONS

Checking Equations

We can use a calculator to check possible solutions of equations. To do this, replace the variable by the possible solution and evaluate both sides of the equation separately.

Equation: $3x - 4 = 2(x + 6)$ Solution: $x = 16$

$$3x - 4 = 2(x + 6) \qquad \text{Original equation}$$

$$3(16) - 4 \stackrel{?}{=} 2(16 + 6) \qquad \text{Replace } x \text{ with 16.}$$

Now evaluate each side with your calculator.

Evaluate left side: | 3 | × | 16 | − | 4 | = | or | ENTER | Display: | 44 |

or

$$3 * 16 - 4$$
$$44$$

Evaluate right side: | 2 | (| 16 | + | 6 |) | = | or | ENTER | Display: | 44 |

or

$$2(16 + 6)$$
$$44$$

Since the left side equals the right side, the equation checks.

Use a calculator to check the possible solutions to each equation.

1. $2x = 48 + 6x; \quad x = -12$
2. $-3x - 7 = 3x - 1; \quad x = -1$
3. $5x - 2.6 = 2(x + 0.8); \quad x = 4.4$
4. $-1.6x - 3.9 = -6.9x - 25.6; \quad x = 5$
5. $\dfrac{564x}{4} = 200x - 11(649); \quad x = 121$
6. $20(x - 39) = 5x - 432; \quad x = 23.2$

Exercise Set 2.4

Solve each equation. See Examples 1 and 2.

1. $-2(3x - 4) = 2x$

2. $-(5x - 1) = 9$

3. $4(2n - 1) = (6n + 4) + 1$

4. $3(4y + 2) = 2(1 + 6y) + 8$

5. $5(2x - 1) - 2(3x) = 1$

6. $3(2 - 5x) + 4(6x) = 12$

7. $6(x - 3) + 10 = -8$

8. $-4(2 + n) + 9 = 1$

Solve each equation. See Examples 3 through 5.

9. $\frac{3}{4}x - \frac{1}{2} = 1$

10. $\frac{2}{3}x + \frac{5}{3} = \frac{5}{3}$

11. $x + \frac{5}{4} = \frac{3}{4}x$

12. $\frac{7}{8}x + \frac{1}{4} = \frac{3}{4}x$

13. $\frac{x}{2} - 1 = \frac{x}{5} + 2$

14. $\frac{x}{5} - 2 = \frac{x}{3}$

15. $\frac{6(3 - z)}{5} = -z$

16. $\frac{4(5 - w)}{3} = -w$

17. $\frac{2(x + 1)}{4} = 3x - 2$

18. $\frac{3(y + 3)}{5} = 2y + 6$

19. $.50x + .15(70) = .25(142)$

20. $.40x + .06(30) = .20(49)$

21. $.12(y - 6) + .06y = .08y - .07(10)$

22. $.60(z - 300) + .05z = .70z - .41(500)$

Solve each equation. See Examples 6 and 7.

23. $5x - 5 = 2(x + 1) + 3x - 7$

24. $3(2x - 1) + 5 = 6x + 2$

25. $\frac{x}{4} + 1 = \frac{x}{4}$

26. $\frac{x}{3} - 2 = \frac{x}{3}$

27. $3x - 7 = 3(x + 1)$

28. $2(x - 5) = 2x + 10$

29. Explain the difference between simplifying an expression and solving an equation.

30. When solving an equation, if the final equivalent equation is $0 = 5$, what can we conclude? If the final equivalent equation is $-2 = -2$, what can we conclude?

31. On your own, construct an equation for which every real number is a solution.

32. On your own, construct an equation that has no solution.

Solve each equation.

33. $4x + 3 = 2x + 11$

34. $6y - 8 = 3y + 7$

35. $-2y - 10 = 5y + 18$

36. $7n + 5 = 10n - 10$

37. $.6x - .1 = .5x + .2$

38. $.2x - .1 = .6x - 2.1$

39. $2y + 2 = y$

40. $7y + 4 = -3$

41. $3(5c - 1) - 2 = 13c + 3$

42. $4(3t + 4) - 20 = 3 + 5t$

43. $x + \frac{7}{6} = 2x - \frac{7}{6}$

44. $\frac{5}{2}x - 1 = x + \frac{1}{4}$

45. $2(x - 5) = 7 + 2x$

46. $-3(1 - 3x) = 9x - 3$

47. $\frac{2(z + 3)}{3} = 5 - z$

48. $\frac{3(w + 2)}{4} = 2w + 3$

49. $\frac{4(y - 1)}{5} = -3y$

50. $\frac{5(1 - x)}{6} = -4x$

51. $8 - 2(a - 1) = 7 + a$

52. $5 - 6(2 + b) = b - 14$

53. $2(x + 3) - 5 = 5x - 3(1 + x)$

54. $4(2 + x) + 1 = 7x - 3(x - 2)$

55. $\frac{5x - 7}{3} = x$

56. $\frac{7n + 3}{5} = -n$

57. $\frac{9 + 5v}{2} = 2v - 4$

58. $\frac{6 - c}{2} = 5c - 8$

59. $-3(t - 5) + 2t = 5t - 4$

60. $-(4a - 7) - 5a = 10 + a$

61. $.02(6t - 3) = .12(t - 2) + .18$

62. $.03(2m + 7) = .06(5 + m) - .09$

63. $.06 - .01(x + 1) = -.02(2 - x)$

64. $-.01(5x + 4) = .04 - .01(x + 4)$

65. $\frac{3(x - 5)}{2} = \frac{2(x + 5)}{3}$

66. $\frac{5(x - 1)}{4} = \frac{3(x + 1)}{2}$

67. $1000(7x - 10) = 50(412 + 100x)$

68. $10{,}000(x + 4) = 100(16 + 7x)$

69. $.035x + 5.112 = .010x + 5.107$

70. $.127x - 2.685 = .027x - 2.38$

Write each of the following as equations. Then solve. See Example 8.

71. The sum of twice a number and $\frac{1}{5}$ is equal to the difference between three times a number and $\frac{4}{5}$. Find the number.

72. The sum of four times a number and $\frac{2}{3}$ is equal to the difference of five times the number and $\frac{5}{6}$. Find the number.

73. The sum of twice a number and 7 is equal to the sum of a number and 6. Find the number.

74. The difference of three times a number and 1 is the same as twice a number. Find the number.

75. Three times a number, minus 6, is equal to two times a number, plus 8. Find the number.

76. The sum of 4 times a number and -2 is equal to the sum of 5 times a number and -2. Find the number.

77. One-third of a number is five-sixths. Find the number.

78. Seven-eighths of a number is one-half. Find the number.

79. The difference of a number and four is twice the number. Find the number.

80. The sum of double a number and six is four times the number. Find the number.

81. If the quotient of a number and 4 is added to $\frac{1}{2}$, the result is $\frac{3}{4}$. Find the number.

82. If $\frac{3}{4}$ is added to three times a number, the result is $\frac{1}{2}$ subtracted from twice the number. Find the number.

△ **83.** The perimeter of a geometric figure is the sum of the lengths of its sides. If the perimeter of the following pentagon (five-sided figure) is 28 centimeters, find the length of each side.

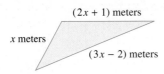

△ **84.** The perimeter of the following triangle is 35 meters. Find the length of each side.

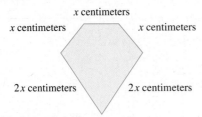

85. Five times a number subtracted from ten is triple the number. Find the number.

86. Nine is equal to ten subtracted from double a number. Find the number.

The graph below is called a three-dimensional bar graph. It shows the five most common names of cities, towns, or villages in the United States.

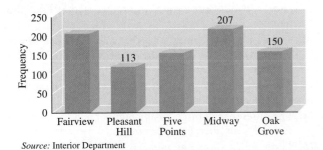

Source: Interior Department

87. What is the most popular name of a city, town, or village in the United States?

88. How many more cities, towns, or villages are named Oak Grove than named Pleasant Hill?

89. Let x represent "the number of towns, cities, or villages named Five Points" and use the information given to determine the unknown number. "The number of towns, cities, or villages named Five Points" added to 55 is equal to twice "the number of towns, cities, or villages named Five Points" minus 90. Check your answer by noticing the height of the bar representing Five Points. Is your answer reasonable?

90. Let x represent "the number of towns, cities, or villages named Fairview" and use the information given to determine the unknown number. Three times "the number of towns, cities, or villages named Fairview" added to 24 is equal to 168 subtracted from 4 times "the number of towns, cities, or villages named Fairview." Check your answer by noticing the height of the bar representing Fairview. Is your answer reasonable?

REVIEW EXERCISES

Evaluate. See Section 1.7.

91. $|2^3 - 3^2| - |5 - 7|$

92. $|5^2 - 2^2| + |9 \div (-3)|$

93. $\dfrac{5}{4 + 3 \cdot 7}$

94. $\dfrac{8}{24 - 8 \cdot 2}$

See Section 2.1.

△ **95.** A plot of land is in the shape of a triangle. If one side is x meters, a second side is $(2x - 3)$ meters and a third side is $(3x - 5)$ meters, express the perimeter of the lot as a simplified expression in x.

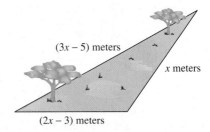

96. A portion of a board has length x feet. The other part has length $(7x - 9)$ feet. Express the total length of the board as a simplified expression in x.

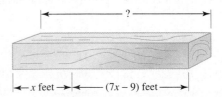

A Look Ahead

Example

Solve $t(t + 4) = t^2 - 2t + 10$.

Solution:
$$t(t + 4) = t^2 - 2t + 10$$
$$t^2 + 4t = t^2 - 2t + 10$$
$$t^2 + 4t - t^2 = t^2 - 2t + 10 - t^2$$
$$4t = -2t + 10$$
$$4t + 2t = -2t + 10 + 2t$$
$$6t = 10$$
$$\frac{6t}{6} = \frac{10}{6}$$
$$t = \frac{5}{3}$$

Solve each equation.

97. $x(x - 3) = x^2 + 5x + 7$

98. $t^2 - 6t = t(8 + t)$

99. $2z(z + 6) = 2z^2 + 12z - 8$

100. $y^2 - 4y + 10 = y(y - 5)$

101. $n(3 + n) = n^2 + 4n$

102. $3c^2 - 8c + 2 = c(3c - 8)$

2.5   AN INTRODUCTION TO PROBLEM SOLVING

CD-ROM

SSM

SSG

Video

▶ **OBJECTIVE**

1. Apply the steps for problem solving.

1   In previous sections, you practiced writing word phrases and sentences as algebraic expressions and equations to help prepare for problem solving. We now use these translations to help write equations that model a problem. The problem-solving steps given next may be helpful.

> **GENERAL STRATEGY FOR PROBLEM SOLVING**
>
> 1. UNDERSTAND the problem. During this step, become comfortable with the problem. Some ways of doing this are:
>   Read and reread the problem.
>   Choose a variable to represent the unknown.
>   Construct a drawing.
>   Propose a solution and check. Pay careful attention to how you check your proposed solution. This will help when writing an equation to model the problem.
> 2. TRANSLATE the problem into an equation.
> 3. SOLVE the equation.
> 4. INTERPRET the results: *Check* the proposed solution in the stated problem and *state* your conclusion.

Much of problem solving involves a direct translation from a sentence to an equation.

Example 1   **FINDING AN UNKNOWN NUMBER**

Twice the sum of a number and 4 is the same as four times the number decreased by 12. Find the number.

Solution   1. UNDERSTAND. Read and reread the problem. If we let

$$x = \text{the unknown number, then}$$

"the sum of a number and 4" translates to "$x + 4$" and "four times the number" translates to "$4x$."

2. TRANSLATE.

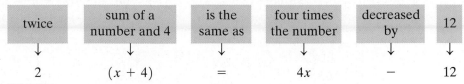

twice	sum of a number and 4	is the same as	four times the number	decreased by	12
↓	↓	↓	↓	↓	↓
2	$(x + 4)$	=	$4x$	−	12

3. SOLVE.

$$2(x + 4) = 4x - 12$$
$$2x + 8 = 4x - 12 \qquad \text{Apply the distributive property.}$$
$$2x + 8 - 4x = 4x - 12 - 4x \qquad \text{Subtract } 4x \text{ from both sides.}$$
$$-2x + 8 = -12$$
$$-2x + 8 - 8 = -12 - 8 \qquad \text{Subtract 8 from both sides.}$$
$$-2x = -20$$
$$\frac{-2x}{-2} = \frac{-20}{-2} \qquad \text{Divide both sides by } -2.$$
$$x = 10$$

4. INTERPRET.

Check: Check this solution in the problem as it was originally stated. To do so, replace "number" with 10. Twice the sum of "10" and 4 is 28, which is the same as 4 times "10" decreased by 12.

State: The number is 10.

Example 2   **FINDING THE LENGTH OF A BOARD**

A 10-foot board is to be cut into two pieces so that the longer piece is 4 times the shorter. Find the length of each piece.

Solution   1. UNDERSTAND the problem. To do so, read and reread the problem. You may also want to propose a solution. For example, if 3 feet represents the length of the shorter piece, then $4(3) = 12$ feet is the length of the longer piece, since it is 4 times the length of the shorter piece. This guess gives a total board length of 3 feet + 12 feet = 15 feet, too long. However, the purpose of proposing a solution is not to guess correctly, but to help better understand the problem and how to model it.

Since the length of the longer piece is given in terms of the length of the shorter piece, let's let

$$x = \text{length of shorter piece, then}$$
$$4x = \text{length of longer piece}$$

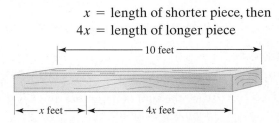

2. TRANSLATE the problem. First, we write the equation in words.

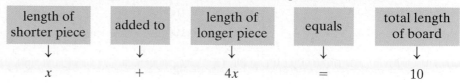

length of shorter piece	added to	length of longer piece	equals	total length of board
↓	↓	↓	↓	↓
x	$+$	$4x$	$=$	10

3. SOLVE.

$$x + 4x = 10$$

$$5x = 10 \qquad \text{Combine like terms.}$$

$$\frac{5x}{5} = \frac{10}{5} \qquad \text{Divide both sides by 5.}$$

$$x = 2$$

4. INTERPRET.

Check: Check the solution in the stated problem. If the shorter piece of board is 2 feet, the longer piece is $4 \cdot (2 \text{ feet}) = 8$ feet and the sum of the two pieces is 2 feet $+$ 8 feet $=$ 10 feet.

State: The shorter piece of board is 2 feet and the longer piece of board is 8 feet.

HELPFUL HINT
Make sure that units are included in your answer, if appropriate.

Example 3   **FINDING THE NUMBER OF REPUBLICAN AND DEMOCRATIC SENATORS**

In a recent year, Congress had 8 more Republican senators than Democratic. If the total number of senators is 100, how many senators of each party were there?

Solution   1. UNDERSTAND the problem. Read and reread the problem. Let's suppose that there are 40 Democratic senators. Since there are 8 more Republicans than Democrats, there must be $40 + 8 = 48$ Republicans. The total number of Democrats and Republicans is then $40 + 48 = 88$. This is incorrect since the total should be 100, but we now have a better understanding of the problem.
   Since the number of Republican senators is given in terms of the number of Democratic senators, let's let

$$x = \text{number of Democrats, then}$$

$$x + 8 = \text{number of Republicans}$$

2. TRANSLATE the problem. First, we write the equation in words.

number of Democrats	added to	number of Republicans	equals	100
↓	↓	↓	↓	↓
x	$+$	$(x + 8)$	$=$	100

3. SOLVE.

$$x + (x + 8) = 100$$
$$2x + 8 = 100 \qquad \text{Combine like terms.}$$
$$2x + 8 - 8 = 100 - 8 \qquad \text{Subtract 8 from both sides.}$$
$$2x = 92$$
$$\frac{2x}{2} = \frac{92}{2} \qquad \text{Divide both sides by 2.}$$
$$x = 46$$

4. INTERPRET.

Check: If there are 46 Democratic senators, then there are $46 + 8 = 54$ Republican senators. The total number of senators is then $46 + 54 = 100$. The results check.

State: There were 46 Democratic and 54 Republican senators.

Example 4 **CALCULATING CELLULAR PHONE USAGE**

A local cellular phone company charges Mike and Elaine Shubert $50 per month and $0.36 per minute of phone use in their usage category. If they were charged $99.68 for a month's cellular phone use, determine the number of whole minutes of phone use.

Solution 1. UNDERSTAND. Read and reread the problem. Let's propose that the Shuberts use the phone for 70 minutes. Pay careful attention as to how we calculate their bill. For 70 minutes of use, their phone bill will be $50 plus $0.36 per minute of use. This is $50 + 0.36(70) = 75.20, less than $99.68. We now understand the problem and know that the number of minutes is greater than 70.

If we let

$$x = \text{number of minutes, then}$$
$$0.36x = \text{charge per minute of phone use}$$

2. TRANSLATE.

$50	added to	minute charge	is equal to	$99.68
↓	↓	↓	↓	↓
50	+	$0.36x$	=	99.68

3. SOLVE.

$$50 + 0.36x = 99.68$$
$$50 + 0.36x - 50 = 99.68 - 50 \qquad \text{Subtract 50 from both sides.}$$
$$0.36x = 49.68 \qquad \text{Simplify.}$$
$$\frac{0.36x}{0.36} = \frac{49.68}{0.36} \qquad \text{Divide both sides by 0.36.}$$
$$x = 138 \qquad \text{Simplify.}$$

4. INTERPRET.

Check: If the Shuberts spend 138 minutes on their cellular phone, their bill is $50 + \$0.36(138) = \$99.68$.

State: The Shuberts spent 138 minutes on their cellular phone this month.

△ **Example 5**   **FINDING ANGLE MEASURES**

If the two walls of the Vietnam Veterans Memorial in Washington D.C. were connected, an isosceles triangle would be formed. The measure of the third angle is 97.5° more than the measure of either of the other two equal angles. Find the measure of the third angle. (*Source:* National Park Service)

Solution   1. **UNDERSTAND.** Read and reread the problem. We then draw a diagram (recall that an isosceles triangle has two angles with the same measure) and let

$$x = \text{degree measure of one angle}$$
$$x = \text{degree measure of the second equal angle}$$
$$x + 97.5 = \text{degree measure of the third angle}$$

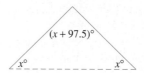

2. **TRANSLATE.** Recall that the sum of the measures of the angles of a triangle equals 180.

measure of first angle		measure of second angle		measure of third angle		equals		180
↓		↓		↓		↓		↓
x	$+$	x		$+(x + 97.5)$		$=$		180

3. **SOLVE.**

$$x + x + (x + 97.5) = 180$$
$$3x + 97.5 = 180 \qquad \text{Combine like terms.}$$
$$3x + 97.5 - 97.5 = 180 - 97.5 \qquad \text{Subtract 97.5 from both sides.}$$
$$3x = 82.5$$
$$\frac{3x}{3} = \frac{82.5}{3} \qquad \text{Divide both sides by 3.}$$
$$x = 27.5$$

4. **INTERPRET.**

Check: If $x = 27.5$, then the measure of the third angle is $x + 97.5 = 125$. The sum of the angles is then $27.5 + 27.5 + 125 = 180$, the correct sum.

State: The third angle measures 125°.*
(*This is rounded to the nearest whole degree. The two walls actually meet at an angle of 125 degrees 12 minutes.)

Exercise Set 2.5

Solve. See Example 1.

1. The sum of twice a number and $\frac{1}{5}$ is equal to the difference between three times the number and $\frac{4}{5}$. Find the number.

2. The sum of four times a number and $\frac{2}{3}$ is equal to the difference of five times the number and $\frac{5}{6}$. Find the number.

3. Twice the difference of a number and 8 is equal to three times the sum of the number and 3. Find the number.

4. Five times the sum of a number and -1 is the same as 6 times the number. Find the number.

5. The product of twice a number and three is the same as the difference of five times the number and $\frac{3}{4}$. Find the number.

6. If the difference of a number and four is doubled, the result is $\frac{1}{4}$ less than the number. Find the number.

7. If the sum of a number and five is tripled, the result is one less than twice the number. Find the number.

8. Twice the sum of a number and six equals three times the sum of the number and four. Find the number.

Solve. See Examples 2 through 5.

9. The governor of Washington state makes about twice as much money as the governor of Nebraska. If the total of their salaries is $195,000, find the salary of each.

10. In the 1996 Summer Olympics, the United States Team won 24 more gold medals than the German Team. If the total number of gold medals for both is 64, find the number of gold medals that each team won. (*Source: World Almanac*, 2000)

11. A 40-inch board is to be cut into three pieces so that the second piece is twice as long as the first piece and the third piece is 5 times as long as the first piece. If x represents the length of the first piece, find the lengths of all three pieces.

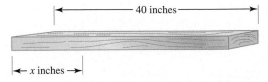

12. A 21-foot beam is to be divided so that the longer piece is 1 foot more than 3 times the shorter piece. If x represents the length of the shorter piece, find the lengths of both pieces.

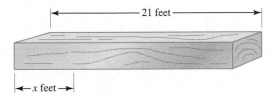

13. A car rental agency advertised renting a Buick Century for $24.95 per day and $0.29 per mile. If you rent this car for 2 days, how many whole miles can you drive on a $100 budget?

14. A plumber gave an estimate for the renovation of a kitchen. Her hourly pay is $27 per hour and the plumber's parts will cost $80. If her total estimate is $404, how many hours does she expect this job to take?

△ 15. The flag of Equatorial Guinea contains an isosceles triangle. (Recall that an isosceles triangle contains two angles with the same measure.) If the measure of the third angle of the triangle is 30° more than twice the measure of either of the other two angles, find the measure of each angle of the triangle. (*Hint:* Recall that the sum of the measures of the angles of a triangle is 180°.)

△ 16. The flag of Brazil contains a parallelogram. One angle of the parallelogram is 15° less than twice the measure of the angle next to it. Find the measure of each angle of the parallelogram. (*Hint:* Recall that opposite angles of a parallelogram have the same measure and that the sum of the measures of the angles is 360°.)

17. In a recent election in Florida for a seat in the United States House of Representatives, Corrine Brown received 13,288 more votes than Bill Randall. If the total number of votes was 119,436, find the number of votes for each candidate.

18. In a recent election in Texas for a seat in the United States House of Representatives, Max Sandlin received 25,557 more votes than opponent Dennis Boerner. If the total number of votes was 135,821, find the number of votes for each candidate. (*Source:* Voter News Service)

19. Two angles are supplementary if their sum is 180°. One angle measures three times the measure of a smaller angle. If x represents the measure of the smaller angle and these two angles are supplementary, find the measure of each angle.

20. Two angles are complementary if their sum is 90°. Given the measures of the complementary angles shown, find the measure of each angle.

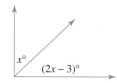

21. To make an international telephone call, you need the code for the country you are calling. The codes for Belgium, France, and Spain are three consecutive integers whose sum is 99. Find the code for each country. (*Source: The World Almanac and Book of Facts,* 1999)

22. To make an international telephone call, you need the code for the country you are calling. The codes for Mali Republic, Côte d'Ivoire, and Niger are three consecutive odd integers whose sum is 675. Find the code for each country.

23. Human fingernails grow much faster than human toenails. Healthy fingernails grow at an average rate of 0.8 inches per year. This is about four times as fast as toenails grow. About how fast do human toenails grow?

24. The human eye blinks once every 5 seconds on average. How many times does the average eye blink in one day? In one year?

25. On December 7, 1995, a probe launched from the robot explorer called Galileo entered the atmosphere of Jupiter at 100,000 miles per hour. The diameter of the probe is 19 inches less than twice its height. If the sum of

the height and the diameter is 83 inches, find each dimension.

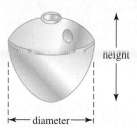

26. Over the past few years the satellite Voyager II has passed by the planets Saturn, Uranus, and Neptune, continually updating information about these planets, including the number of moons for each. Uranus is now believed to have 7 more moons than Neptune. Also, Saturn is now believed to have 2 more than twice the number of moons of Neptune. If the total number of moons for these planets is 41, find the number of moons for each planet. (*Source:* National Space Science Data Center)

27. In 1999, the San Antonio Spurs won the NBA championship against the New York Knicks. The two teams' final scores for the last game of the championship were two consecutive integers whose sum was 155. Find each final score. (*Source: USA Today,* 6/28/99)

28. In September 1999, Team U.S.A. defeated Team Europe by 1 point to win the Ryder Cup, a golf championship. If the total of these two scores is 28, find each team score.

29. Tony Hawk landed the first 900 in skateboard competition history during the X Games "Best Trick" competition in 1999. A 900 means that he rotated 900° in the air with his skateboard. If there are 360° in a single rotation, how many rotations were there?

30. The Pentagon Building in Washington, D.C., is the headquarters for the U.S. Department of Defense. The Pentagon is also the world's largest office building in terms of ground space with a floor area of over 6.5 million square feet. This is three times the floor area of the Empire State Building. About how much floor space does the Empire State Building have? Round to the nearest tenth.

31. After a recent election, there were 14 more Republican governors than Democratic governors in the United States. In addition, 2 of the 50 state governors were neither Republican nor Democrat. How many Democrats and how many Republicans held governor's offices after the 1998 election? (*Source: The World Almanac and Book of Facts*, 1999)

32. In a recent year, the total number of Democrats and Republicans in the U.S. House of Representatives was 434. There were 12 more Republicans than Democrats. Find the number of representatives from each party.

33. A 17-foot piece of string is cut into two pieces so that one piece is 2 feet longer than twice the shorter piece. If the shorter piece is x feet long, find the lengths of both pieces.

34. An 18-foot wire is to be cut so that the longer piece is 5 times longer than the shorter piece. Find the length of each piece.

△ **35.** The measures of the angles of a triangle are 3 consecutive even integers. Find the measure of each angle.

△ **36.** A quadrilateral is a polygon with 4 sides. The sum of the measures of the 4 angles in a quadrilateral is 360°. If the measures of the angles of a quadrilateral are consecutive odd integers, find the measures.

37. Enterprise Car Rental charges a daily rate of $34 plus $0.20 per mile. Suppose that you rent a car for a day and your bill (before taxes) is $104. How many miles did you drive?

38. A woman's $15,000 estate is to be divided so that her husband receives twice as much as her son. If x represents the amount of money that her son receives, find the amount of money that her husband receives and the amount of money that her son receives.

The graph below shows the states with the highest tourism budgets for 1999. Use the graph for Exercises 39–44.

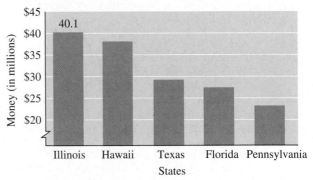

Source: Travel Industry Association of America

39. In your own words, describe what the word *tourism* means.

40. Which state spends the most money on tourism?

41. Which states spend between $25 and $30 million on tourism?

42. The states of Texas and Florida spend a total of $56.6 million for tourism. The state of Texas spends $2.2 million more than the state of Florida. Find the amount that each state spends on tourism.

43. The states of Hawaii and Pennsylvania spend a total of $60.9 million for tourism. The state of Hawaii spends $8.1 million less than twice the amount of money that the state of Pennsylvania spends. Find the amount that each state spends on tourism.

44. Compare the heights of the bars in the graph with your results of Exercises 42 and 43. Are your answers reasonable?

△ **45.** The golden rectangle is a rectangle whose length is approximately 1.6 times its width. The early Greeks thought that a rectangle with these dimensions was the most pleasing to the eye and examples of the golden rectangle are found in many early works of art. For example, the Parthenon in Athens contains many examples of golden rectangles. Mike Hallahan would like to plant a rectangular garden in the shape of a golden rectangle. If he has 78 feet of fencing available, find the dimensions of the garden.

The length-width rectangle approximates the golden rectangle as well as the width-height rectangle.

46. Dr. Dorothy Smith gave the students in her geometry class at the University of New Orleans the following question. Is it possible to construct a triangle such that the second angle of the triangle has a measure that is twice the measure of the first angle and the measure of the third angle is 3 times the measure of the first? If so, find the measure of each angle. (*Hint:* Recall that the sum of the measures of the angles of a triangle is 180°.)

47. Give an example of how you recently solved a problem using mathematics.

48. In your own words, explain why a solution of a word problem should be checked using the original wording of the problem and not the equation written from the wording.

Recall from Exercise 45 that the golden rectangle is a rectangle whose length is approximately 1.6 times its width.

△ **49.** It is thought that for about 75% of adults, a rectangle in the shape of the golden rectangle is the most pleasing to the eye. Draw 3 rectangles, one in the shape of the golden rectangle, and poll your class. Do the results agree with the percentage given above?

△ **50.** Examples of golden rectangles can be found today in architecture and manufacturing packaging. Find an example of a golden rectangle in your home. A few suggestions: the front face of a book, the floor of a room, the front of a box of food.

△ **51.** Measure the dimensions of each rectangle and decide which one best approximates the shape of the golden rectangle.

a.

b.

c.

REVIEW EXERCISES

Perform the indicated operations. See Sections 1.5 and 1.6.

52. $3 + (-7)$

53. $-2 + (-8)$

54. $4 - 10$

55. $-11 + 2$

56. $-5 - (-1)$

57. $-12 - 3$

Translate each sentence into an equation. See Sections 1.4 and 2.4.

58. Half of the difference of a number and one is thirty-seven.

59. Five times the opposite of a number is the number plus sixty.

60. If three times the sum of a number and 2 is divided by 5, the quotient is 0.

61. If the sum of a number and 9 is subtracted from 50, the result is 0.

2.6 FORMULAS AND PROBLEM SOLVING

CD-ROM SSM SSG Video

▶ **OBJECTIVES**

1. Use formulas to solve problems.
2. Solve a formula or equation for one of its variables.

1 An equation that describes a known relationship among quantities, such as distance, time, volume, weight, and money is called a **formula**. These quantities are represented by letters and are thus variables of the formula. Here are some common formulas and their meanings.

$A = lw$
Area of a rectangle = length · width

$I = PRT$
Simple Interest = Principal · Rate · Time

$P = a + b + c$
Perimeter of a triangle = side a + side b + side c

$d = rt$
distance = rate · time

$V = lwh$
Volume of a rectangular solid = length · width · height

$F = \left(\dfrac{9}{5}\right)C + 32$

degrees Fahrenheit = $\left(\dfrac{9}{5}\right)$ · degrees Celsius + 32

Formulas are valuable tools because they allow us to calculate measurements as long as we know certain other measurements. For example, if we know we traveled a distance of 100 miles at a rate of 40 miles per hour, we can replace the variables d and r in the formula $d = rt$ and find our time, t.

$$d = rt \qquad \text{Formula.}$$
$$100 = 40t \qquad \text{Replace } d \text{ with 100 and } r \text{ with 40.}$$

This is a linear equation in one variable, t. To solve for t, divide both sides of the equation by 40.

$$\frac{100}{40} = \frac{40t}{40} \qquad \text{Divide both sides by 40.}$$
$$\frac{5}{2} = t \qquad \text{Simplify.}$$

The time traveled is $\frac{5}{2}$ hours or $2\frac{1}{2}$ hours.

In this section we solve problems that can be modeled by known formulas. We use the same problem-solving steps that were introduced in the previous section. These steps have been slightly revised to include formulas.

Example 1   FINDING TIME GIVEN RATE AND DISTANCE

A glacier is a giant mass of rocks and ice that flows downhill like a river. Portage Glacier in Alaska is about 6 miles, or 31,680 *feet*, long and moves 400 *feet* per year. Icebergs are created when the front end of the glacier flows into Portage Lake. How long does it take for ice at the head (beginning) of the glacier to reach the lake?

Solution   1. UNDERSTAND. Read and reread the problem. The appropriate formula needed to solve this problem is the distance formula, $d = rt$. To become familiar with this formula, let's find the distance that ice traveling at a rate of 400 feet per year travels in 100 years. To do so, we let time t be 100 years and rate r be the given 400 feet per year, and substitute these values into the formula $d = rt$. We then have that distance $d = 400(100) = 40,000$ feet. Since we are interested in finding how long it takes ice to travel 31,680 feet, we now know that it is less than 100 years.

Since we are using the formula $d = rt$, we let

t = the time in years for ice to reach the lake

r = rate or speed of ice

d = distance from beginning of glacier to lake

2. TRANSLATE. To translate to an equation, we use the formula $d = rt$ and let distance $d = 31,680$ feet and rate $r = 400$ feet per year.

$$d = r \cdot t$$

$$31,680 = 400 \cdot t \qquad \text{Let } d = 31,680 \text{ and } r = 400.$$

3. SOLVE. Solve the equation for t. To solve for t, divide both sides by 400.

$$\frac{31,680}{400} = \frac{400 \cdot t}{400} \qquad \text{Divide both sides by 400.}$$

$$79.2 = t \qquad \text{Simplify.}$$

4. INTERPRET.

Check: To check, substitute 79.2 for t and 400 for r in the distance formula and check to see that the distance is 31,680 feet.

State: It takes 79.2 years for the ice at the head of Portage Glacier to reach the lake.

> **HELPFUL HINT**
> Don't forget to include units, if appropriate.

△ Example 2 CALCULATING THE LENGTH OF A GARDEN

Charles Pecot can afford enough fencing to enclose a rectangular garden with a perimeter of 140 feet. If the width of his garden must be 30 feet, find the length.

Solution

l

$w = 30$ feet

1. UNDERSTAND. Read and reread the problem. The formula needed to solve this problem is the formula for the perimeter of a rectangle, $P = 2l + 2w$. Before continuing, let's become familar with this formula.

   l = the length of the rectangular garden

   w = the width of the rectangular garden

   P = perimeter of the garden

2. TRANSLATE. To translate to an equation, we use the formula $P = 2l + 2w$ and let perimeter P = 140 feet and width w = 30 feet.

$$P = 2l + 2w \qquad \text{Let } P = 140 \text{ and } w = 30.$$
$$140 = 2l + 2(30)$$

3. SOLVE.

$$
\begin{aligned}
140 &= 2l + 2(30) \\
140 &= 2l + 60 &&\text{Multiply 2(30).} \\
140 - 60 &= 2l + 60 - 60 &&\text{Subtract 60 from both sides.} \\
80 &= 2l &&\text{Combine like terms.} \\
40 &= l &&\text{Divide both sides by 2.}
\end{aligned}
$$

4. INTERPRET.

   Check: Substitute 40 for l and 30 for w in the perimeter formula and check to see that the perimeter is 140 feet.

   State: The length of the rectangular garden is 40 feet.

Example 3 FINDING AN EQUIVALENT TEMPERATURE

The average maximum temperature for January in Algerias, Algeria, is 59° Fahrenheit. Find the equivalent temperature in degrees Celsius.

Solution

1. UNDERSTAND. Read and reread the problem. A formula that can be used to solve this problem is the formula for converting degrees Celsius to degrees Fahrenheit, $F = \frac{9}{5}C + 32$. Before continuing, become familiar with this formula. Using this formula, we let

   C = temperature in degrees Celsius, and

   F = temperature in degrees Fahrenheit.

2. TRANSLATE. To translate to an equation, we use the formula $F = \frac{9}{5}C + 32$ and let degrees Fahrenheit F = 59.

$$\text{Formula:} \qquad F = \frac{9}{5}C + 32$$

$$\text{Substitute:} \qquad 59 = \frac{9}{5}C + 32 \qquad \text{Let } F = 59.$$

3. SOLVE.

$$59 = \frac{9}{5}C + 32$$

$$59 - 32 = \frac{9}{5}C + 32 - 32 \qquad \text{Subtract 32 from both sides.}$$

$$27 = \frac{9}{5}C \qquad\qquad \text{Combine like terms.}$$

$$\frac{5}{9} \cdot 27 = \frac{5}{9} \cdot \frac{9}{5}C \qquad \text{Multiply both sides by } \tfrac{5}{9}.$$

$$15 = C \qquad\qquad \text{Simplify.}$$

4. INTERPRET.

Check: To check, replace C with 15 and F with 59 in the formula and see that a true statement results.

State: Thus, 59° Fahrenheit is equivalent to 15° Celsius.

2 We say that the formula $F = \frac{9}{5}C + 32$ is solved for F because F is alone on one side of the equation and the other side of the equation contains no F's. Suppose that we need to convert many Fahrenheit temperatures to equivalent degrees Celsius. In this case, it is easier to perform this task by solving the formula $F = \frac{9}{5}C + 32$ for C. (See Example 7.) For this reason, it is important to be able to solve an equation for any one of its specified variables. For example, the formula $d = rt$ is solved for d in terms of r and t. We can also solve $d = rt$ for t in terms of d and r. To solve for t, divide both sides of the equation by r.

$$d = rt$$

$$\frac{d}{r} = \frac{rt}{r} \qquad \text{Divide both sides by } r.$$

$$\frac{d}{r} = t \qquad \text{Simplify.}$$

To solve a formula or an equation for a specified variable, we use the same steps as for solving a linear equation. These steps are listed next.

SOLVING EQUATIONS FOR A SPECIFIED VARIABLE

Step 1. Multiply on both sides to clear the equation of fractions if they occur.

Step 2. Use the distributive property to remove parentheses if they occur.

Step 3. Simplify each side of the equation by combining like terms.

Step 4. Get all terms containing the specified variable on one side and all other terms on the other side by using the addition property of equality.

Step 5. Get the specified variable alone by using the multiplication property of equality.

△ **Example 4**   Solve $V = lwh$ for l.

Solution   This formula is used to find the volume of a box. To solve for l, divide both sides by wh.

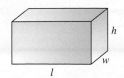

$$V = lwh$$

$$\frac{V}{wh} = \frac{lwh}{wh} \qquad \text{Divide both sides by } wh.$$

$$\frac{V}{wh} = l \qquad \text{Simplify.}$$

Since we have l alone on one side of the equation, we have solved for l in terms of V, w, and h. Remember that it does not matter on which side of the equation we isolate the variable. ▬

Example 5   Solve $y = mx + b$ for x.

Solution   The term containing the variable we are solving for, mx, is on the right side of the equation. Get mx alone by subtracting b from both sides.

$$y = mx + b$$

$$y - b = mx + b - b \qquad \text{Subtract } b \text{ from both sides.}$$

$$y - b = mx \qquad \text{Combine like terms.}$$

Next, solve for x by dividing both sides by m.

$$\frac{y - b}{m} = \frac{mx}{m}$$

$$\frac{y - b}{m} = x \qquad \text{Simplify.} \quad ▬$$

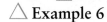

△ **Example 6**   Solve $P = 2l + 2w$ for w.

Solution   This formula relates the perimeter of a rectangle to its length and width. Find the term containing the variable w. To get this term, $2w$, alone subtract $2l$ from both sides.

$$P = 2l + 2w$$

$$P - 2l = 2l + 2w - 2l \qquad \text{Subtract } 2l \text{ from both sides.}$$

$$P - 2l = 2w \qquad \text{Combine like terms.}$$

$$\frac{P - 2l}{2} = \frac{2w}{2} \qquad \text{Divide both sides by 2.}$$

$$\frac{P - 2l}{2} = w \qquad \text{Simplify.}$$

The next example has an equation containing a fraction. We will first clear the equation of fractions and then solve for the specified variable. ▬

Example 7   Solve $F = \frac{9}{5}C + 32$ for C.

Solution

$$F = \frac{9}{5}C + 32$$

$$5(F) = 5\left(\frac{9}{5}C + 32\right) \qquad \text{Clear the fraction by multiplying both sides by the LCD.}$$

$$5F = 9C + 160 \qquad \text{Distribute the 5.}$$

$$5F - 160 = 9C + 160 - 160 \qquad \text{To get the term containing the variable } C \text{ alone, subtract 160 from both sides.}$$

$$5F - 160 = 9C \qquad \text{Combine like terms.}$$

$$\frac{5F - 160}{9} = \frac{9C}{9} \qquad \text{Divide both sides by 9.}$$

$$\frac{5F - 160}{9} = C \qquad \text{Simplify.}$$

Exercise Set 2.6

Substitute the given values into each given formula and solve for the unknown variable. If necessary, round to one decimal place. See Examples 1 through 3.

△ **1.** $A = bh$; $A = 45, b = 15$
(Area of a parallelogram)

2. $d = rt$; $d = 195, t = 3$
(Distance formula)

△ **3.** $S = 4lw + 2wh$; $S = 102, l = 7, w = 3$
(Surface area of a special rectangular box)

△ **4.** $V = lwh$; $l = 14, w = 8, h = 3$
(Volume of a rectangular box)

△ **5.** $A = \frac{1}{2}h(B + b)$; $A = 180, B = 11, b = 7$

(Area of a trapezoid)

△ **6.** $A = \frac{1}{2}h(B + b)$; $A = 60, B = 7, b = 3$

(Area of a trapezoid)

7. $P = a + b + c$; $P = 30, a = 8, b = 10$
△ (Perimeter of a triangle)

△ **8.** $V = \frac{1}{3}Ah$; $V = 45, h = 5$

(Volume of a pyramid)

9. $C = 2\pi r$; $C = 15.7$ (use the approximation 3.14 or a
△ calculator approximation for π)
(Circumference of a circle)

10. $A = \pi r^2$; $r = 4.5$ (use the approximation 3.14 or a
△ calculator approximation for π)
(Area of a circle)

11. $I = PRT$; $I = 3750, P = 25,000, R = 0.05$
(Simple interest formula)

12. $I = PRT$; $I = 1,056,000, R = 0.055, T = 6$
(Simple interest formula)

13. $V = \frac{1}{3}\pi r^2 h$; $V = 565.2, r = 6$ (use the calculator
△ approximation for π)
(Volume of a cone)

14. $V = \frac{4}{3}\pi r^3$; $r = 3$ (use a calculator approximation
△ for π)
(Volume of a sphere)

Solve each formula for the specified variable. See Examples 4 through 7.

15. $f = 5gh$ for h △ **16.** $C = 2\pi r$ for r

17. $V = LWH$ for W **18.** $T = mnr$ for n

19. $3x + y = 7$ for y **20.** $-x + y = 13$ for y

21. $A = P + PRT$ for R **22.** $A = P + PRT$ for T

23. $V = \frac{1}{3}Ah$ for A **24.** $D = \frac{1}{4}fk$ for k

25. $P = a + b + c$ for a

26. $PR = s_1 + s_2 + s_3 + s_4$ for s_3

27. $S = 2\pi rh + 2\pi r^2$ for h

△ **28.** $S = 4lw + 2wh$ for h

Solve. See Examples 1 through 3.

△ **29.** The world's largest sign for Coca-Cola is located in Arica, Chile. The rectangular sign has a length of 400 feet and has an area of 52,400 square feet. Find the width of the sign. (*Source:* Fabulous Facts about Coca-Cola, Atlanta, GA)

△ **30.** The length of a rectangular garden is 6 meters. If 21 meters of fencing are required to fence the garden, find its width.

6 meters

31. A limousine built in 1968 for the president cost $500,000 and weighed 5.5 tons. This Lincoln Continental Executive could travel at 50 miles per hour with all of its tires shot away. At this rate, how long would it take to travel from Charleston, Virginia, to Washington, D.C., a distance of 375 miles?

32. The SR-71 is a top secret spy plane. It is capable of traveling from Rochester, New York, to San Francisco, California, a distance of approximately 3000 miles, in $1\frac{1}{2}$ hours. Find the rate of the SR-71.

33. Convert Nome, Alaska's 14°F high temperature to Celsius.

34. Convert Paris, France's low temperature of −5°C to Fahrenheit.

△ **35.** Piranha fish require 1.5 cubic feet of water per fish to maintain a healthy environment. Find the maximum number of piranhas you could put in a tank measuring 8 feet by 3 feet by 6 feet.

△ **36.** Find how many goldfish you can put in a cylindrical tank whose diameter is 8 meters and whose height is 3 meters, if each goldfish needs 2 cubic meters of water.

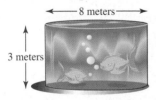

8 meters

3 meters

37. The Cat is a high-speed catamaran auto ferry that operates between Bar Harbor, Maine, and Yarmouth, Nova Scotia. The Cat can make the trip in about $2\frac{1}{2}$ hours at a speed of 55 mph. About how far apart are Bar Harbor and Yarmouth? (*Source:* Bay Ferries)

38. A family is planning their vacation to Disney World. They will drive from a small town outside New Orleans, Louisiana, to Orlando, Florida, a distance of 700 miles. They plan to average a rate of 55 miles per hour. How long will this trip take?

△ **39.** A lawn is in the shape of a trapezoid with a height of 60 feet and bases of 70 feet and 130 feet. How many bags of

fertilizer must be purchased to cover the lawn if each bag covers 4000 square feet?

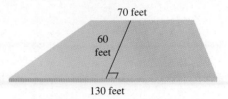

70 feet

60 feet

130 feet

△ **40.** If the area of a right-triangularly shaped sail is 20 square feet and its base is 5 feet, find the height of the sail.

?

5 feet

41. The X-30 is a "space plane" that skims the edge of space at 4000 miles per hour. Neglecting altitude, if the circumference of the Earth is approximately 25,000 miles, how long will it take for the X-30 to travel around the Earth?

42. In the United States, a notable hang glider flight was a 303-mile, $8\frac{1}{2}$ hour flight from New Mexico to Kansas. What was the average rate during this flight?

The Dante II is a spider-like robot that is used to map the depths of an active Alaskan volcano.

△ **43.** The dimensions of the Dante II are 10 feet long by 8 feet wide by 10 feet high. Find the volume of the smallest box needed to store this robot.

44. The Dante II traveled 600 feet into an active Alaskan volcano in $3\frac{1}{3}$ hours. Find the traveling rate of Dante II in feet per minute. (*Hint:* First convert $3\frac{1}{3}$ hours to minutes.)

△ **45.** Maria's Pizza sells one 16-inch cheese pizza or two 10-inch cheese pizzas for $9.99. Determine which size gives more pizza.

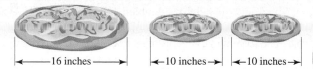

←————16 inches————→ ←10 inches→ ←10 inches→

△ **46.** Find how much rope is needed to wrap around the Earth at the equator, if the radius of the Earth is 4000 miles. (*Hint:* Use 3.14 for π and the formula for circumference.)

47. A Japanese "bullet" train set a new world record for train speed at 552 kilometers per hour during a manned test run on the Yamanashi Maglev Test Line in April 1999. The Yamanashi Maglev Test Line is 42.8 kilometers long. How many *minutes* would a test run on the Yamanashi Line last at this record-setting speed? Round to the nearest hundredth of a minute. (*Source:* Japan Railways Central Co.)

48. In 1983, the Hawaiian volcano Kilauea began erupting in a series of episodes still occurring at the time of this writing. At times, the lava flows advanced at speeds of up to 0.5 kilometer per hour. In 1983 and 1984 lava flows destroyed 16 homes in the Royal Gardens subdivision, about 6 km away from the eruption site. Roughly how long did it take the lava to reach Royal Gardens? (*Source:* U.S. Geological Survey Hawaiian Volcano Observatory)

49. Find how long it takes Tran Nguyen to drive 135 miles on I-10 if he merges onto I-10 at 10 A.M. and drives nonstop with his cruise control set on 60 mph.

50. Beaumont, Texas, is about 150 miles from Toledo Bend. If Leo Miller leaves Beaumont at 4 A.M. and averages 45 mph, when should he arrive at Toledo Bend?

51. Dry ice is a name given to solidified carbon dioxide. At −78.5° Celsius it changes directly from a solid to a gas. Convert this temperature to Fahrenheit.

52. Lightning bolts can reach a temperature of 50,000° Fahrenheit. Convert this temperature to Celsius.

53. The distance from the sun to the Earth is approximately 93,000,000 miles. If light travels at a rate of 186,000 miles per second, how long does it take light from the sun to reach us?

54. Light travels at a rate of 186,000 miles per second. If our moon is 238,860 miles from the Earth, how long does it take light from the moon to reach us? (Round to the nearest tenth of a second.)

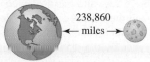

238,860
←— miles —→

55. On July 16, 1994, the Shoemaker-Levy 9 comet collided
△ with Jupiter. The impact of the largest fragment of the comet, a massive chunk of rock and ice, created a fireball with a radius of 2000 miles. Find the volume of this spherical fireball. (Use a calculator approximation for π. Round to the nearest whole cubic mile.)

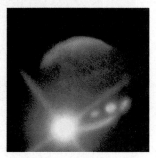

Color image made by the Hubble Space Telescope of the impact site of Fragment G of the comet.

56. The fireball from the largest fragment of the comet (see Exercise 55) immediately collapsed, as it was pulled down by gravity. As it fell, it cooled to approximately −350°F. Convert this temperature to Celsius.

57. Bolts of lightning can travel at 270,000 miles per second. How many times can a lightning bolt travel around the world in one second? (See Exercise 46. Round to the nearest tenth.)

58. A glacier is a giant mass of rocks and ice that flows downhill like a river. Exit Glacier, near Seward, Alaska, moves at a rate of 20 inches a day. Find the distance in feet the glacier moves in a year. (Assume 365 days a year. Round to 2 decimal places.)

59. Flying fish do not *actually* fly, but glide. They have been known to travel a distance of 1300 feet at a rate of 20 miles per hour. How many seconds did it take to travel this distance? (*Hint:* First convert miles per hour to feet per second. Recall that 1 mile = 5280 feet. Round to the nearest tenth of a second.)

60. Stalactites join stalagmites to form columns. A column found at Natural Bridge Caverns near San Antonio, Texas, rises 15 feet and has a *diameter* of only 2 inches. Find the volume of this column in cubic inches. (*Hint:* Use the formula for volume of a cylinder and use a calculator approximation for π.)

61. Find the temperature at which the Celsius measurement and Fahrenheit measurement are the same number.

62. Normal room temperature is about 78°F. Convert this temperature to Celsius.

63. The formula $V = LWH$ is used to find the volume of a box. If the length of a box is doubled, the width is doubled, and the height is doubled, how does this affect the volume?

64. The formula $A = bh$ is used to find the area of a parallelogram. If the base of a parallelogram is doubled and its height is doubled, how does this affect the area?

REVIEW EXERCISES

Write the following phrases as algebraic expressions. See Section 2.1.

65. Nine divided by the sum of a number and 5

66. Half the product of a number and five

67. Three times the sum of a number and four

68. One-third of the quotient of a number and six

69. Double the sum of ten and four times a number

70. Twice a number divided by three times the number

71. Triple the difference of a number and twelve

72. A number minus the sum of the number and six

2.7 PERCENT AND PROBLEM SOLVING

CD-ROM   SSM

SSG   Video

▶ **O B J E C T I V E S**

1. Write percents as decimals and decimals as percents.
2. Find percents of numbers.
3. Read and interpret graphs containing percents.
4. Solve percent equations.
5. Solve problems containing percents.

1 Much of today's statistics is given in terms of percent: a basketball player's free throw percent, current interest rates, stock market trends, and nutrition labeling, just to name a few. In this section, we explore percent and applications involving percents.

If 37 percent of all households in the United States have a home computer, what does this percent mean? It means that 37 households out of every 100 households have a home computer. In other words,

the word **percent** means **per hundred** so that

37 **percent** means 37 **per hundred** or $\dfrac{37}{100}$.

We use the symbol % to denote percent, and we can write a percent as a fraction or a decimal.

$$37\% = \frac{37}{100} = 0.37$$

$$8\% = \frac{8}{100} = 0.08$$

$$100\% = \frac{100}{100} = 1.00$$

This suggests the following:

WRITING A PERCENT AS A DECIMAL

Drop the percent symbol and move the decimal point two places to the left.

Example 1   Write each percent as a decimal.

     a. 35%   **b.** 89.5%   **c.** 150%

Solution   **a.** 35% = 0.35   **b.** 89.5% = 0.895   **c.** 150% = 1.5

To write a decimal as a percent, we reverse the procedure above.

WRITING A DECIMAL AS A PERCENT

Move the decimal point two places to the right and attach the percent symbol, %.

Example 2   Write each number as a percent.

     a. 0.73   **b.** 1.39   **c.** $\dfrac{1}{4}$

Solution   **a.** 0.73 = 73%   **b.** 1.39 = 139%

           c. First, write $\frac{1}{4}$ as a decimal.

$$\frac{1}{4} = 0.25 = 25\%.$$

2   To find a percent of a number, recall that the word "of" means multiply.

◈ Example 3   Find 72% of 200.

Solution   To find 72% of 200, we multiply.

$$
\begin{aligned}
72\% \text{ of } 200 &= 72\%(200) \\
&= 0.72(200) &&\text{Write 72\% as 0.72.} \\
&= 144. &&\text{Multiply.}
\end{aligned}
$$

Thus, 72% of 200 is 144.

3   As mentioned earlier, percents are often used in statistics. Recall that the graph below is called a circle graph or a pie chart. The circle or pie represents a whole, or 100%. Each circle is divided into sectors (shaped like pieces of a pie) which represent various parts of the whole 100%.

Example 4   The circle graph below shows how much money homeowners in the United States spend annually on maintaining their homes. Use this graph to answer the questions below.

Yearly Home Maintenance in the U.S.

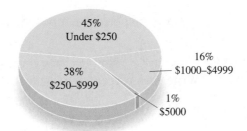

a. What percent of homeowners spend under $250 on yearly home maintenance?
b. What percent of homeowners spend less than $1000 per year on maintenance?
c. How many of the 22,000 homeowners in a town called Fairview might we expect to spend under $250 a year on home maintenance?
d. Find the number of degrees in the 16% sector.

Solution   a. From the circle graph, we see that 45% of homeowners spend under $250 per year on home maintenance.
b. From the circle graph, we know that 45% of homeowners spend under $250 per year and 38% of homeowners spend $250–$999 per year, so that the sum 45% + 38% or 83% of homeowners spend less than $1000 per year.
c. Since 45% of homeowners spend under $250 per year on maintenance, we find 45% of 22,000.

$$45\% \text{ of } 22{,}000 = 0.45(22{,}000)$$
$$= 9900$$

We might then expect that 9900 homeowners in Fairview spend under $250 per year on home maintenance.
d. To find the number of degrees in the 16% sector, recall that the number of degrees around a circle is 360°. Thus, to find the number of degrees in the 16% sector, we find 16% of 360°.

$$16\% \text{ of } 360° = 0.16(360°)$$
$$= 57.6°$$

The 16% sector contains 57.6°.

4   Next, we practice writing sentences as percent equations.

Example 5 The number 63 is what percent of 72?

Solution 1. UNDERSTAND. Read and reread the problem. Next, let's guess a solution. Suppose we guess that 63 is 80% of 72. We may check our guess by finding 80% of 72. 80% of 72 = 0.80(72) = 57.6. Close, but not 63. At this point, though, we have a better understanding of the problem, we know the correct answer is close to and greater than 80%, and we know how to check our proposed solution later.

Let x = the unknown percent.

2. TRANSLATE. Recall that "is" means "equals" and "of" signifies multiplying. Translate the sentence directly.

In words:

The number 63	is	what percent	of	72	?

Translate:

| 63 | = | x | $\cdot$ | 72 | |

3. SOLVE.

$$63 = 72x$$
$$0.875 = x \qquad \text{Divide both sides by 72.}$$
$$87.5\% = x \qquad \text{Write as a percent.}$$

4. INTERPRET.
  Check: Check by verifying that 87.5% of 72 is 63.
  State: the results. The number 63 is 87.5% of 72.

Example 6 The number 120 is 15% of what number?

Solution 1. UNDERSTAND. Read and reread the problem. Guess a solution and check your guess.

Let x = the unknown number.

2. TRANSLATE.

In words:

The number 120	is	15%	of	what number	?

Translate:

| 120 | = | 15% | $\cdot$ | x | |

3. SOLVE.

$$120 = 0.15x$$
$$800 = x \qquad \text{Divide both sides by 0.15.}$$

4. INTERPRET.
  Check: Check the proposed solution of 800 by finding 15% of 800 and verifying that the result is 120.
  State: Thus, 120 is 15% of 800.

---

5 Percent increase or percent decrease is a common way to describe how some measurement has increased or decreased. For example, crime increased by 8%, teachers received a 5.5% increase in salary, or a company decreased its employees by 10%. The next example is a review of percent increase.

Example 7 CALCULATING THE COST OF ATTENDING COLLEGE

The cost of attending a public college rose from $5324 in 1990 to $8086 in 2000. Find the percent increase. (*Source:* U.S. Department of Education. *Note:* These costs include room and board.)

Solution

1. UNDERSTAND. Read and reread the problem. Let's guess that the percent increase is 20%. To see if this is the case, we find 20% of $5324 to find the *increase* in cost. Then we add this increase to $5324 to find the *new cost*. In other words, 20%($5324) = 0.20($5324) = $1064.80, the *increase* in cost. The new cost then would be $5324 + $1064.80 = $6388.80, less than the actual new cost of $8086. We now know that the increase is greater than 20% and we know how to check our proposed solution.

   Let x = the percent increase.

2. TRANSLATE. First, find the **increase**, and then the **percent increase**. The increase in cost is found by:

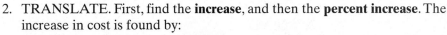

In words: increase = new cost − old cost or

Translate: increase = $8086   − $5324
                     = $2762

Next, find the percent increase. The percent increase or percent decrease is always a percent of the original number or in this case, the old cost.

In words: | increase | is | what percent increase | of | old cost |

Translate: $2762   = x · $5324

3. SOLVE.

$$2762 = 5324x$$
$$0.519 \approx x \quad \text{Divide both sides by 5324 and}$$
$$ \quad \text{round to 3 decimal places.}$$
$$51.9\% \approx x \quad \text{Write as a percent.}$$

4. INTERPRET.

   Check: Check the proposed solution.

   State: The percent increase in cost is approximately 51.9%.

SPOTLIGHT ON DECISION MAKING

Suppose you are a personal income tax preparer. Your clients, Jose and Felicia Fernandez, are filing jointly Form 1040 as their individual income tax return. You know that medical expenses may be written off as an itemized deduction if the expenses exceed 7.5% of their adjusted gross income. Furthermore, only the portion of medical expenses that exceed 7.5% of their adjusted gross income can be deducted. Is the Fernandez family eligible to deduct their medical expenses? Explain.

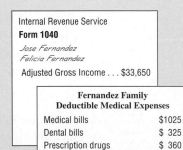

Internal Revenue Service
Form 1040
Jose Fernandez
Felicia Fernandez
Adjusted Gross Income . . . $33,650

Fernandez Family Deductible Medical Expenses	
Medical bills	$1025
Dental bills	$ 325
Prescription drugs	$ 360
Medical Insurance premiums	$1200

MENTAL MATH

Tell whether the percent labels in the circle graphs are correct.

1.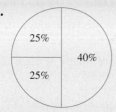
25% 40% 25%

2.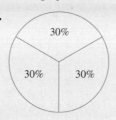
30% 30% 30%

3.
25% 25% 25% 25%

4.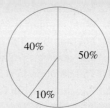
40% 50% 10%

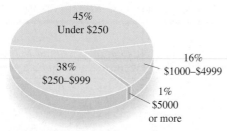

Exercise Set 2.7

Write each percent as a decimal. See Example 1.

1. 120%
2. 73%
3. 22.5%
4. 4.2%
5. 0.12%
6. 0.86%

Write each number as a percent. See Example 2.

7. 0.75
8. 0.3
9. 2
10. 5.1
11. $\frac{1}{8}$
12. $\frac{3}{5}$

Use the home maintenance graph to answer Exercises 13 through 20. See Example 4.

Yearly Home Maintenance in the U.S.

45% Under $250

38% $250–$999

16% $1000–$4999

1% $5000 or more

Source: Census Bureau

13. What percent of homeowners spend $250–$999 on yearly home maintenance?

14. What percent of homeowners spend $5000 or more on yearly home maintenance?

15. What percent of homeowners spend $250–$4999 on yearly home maintenance?

16. What percent of homeowners spend $250 or more on yearly home maintenance?

17. Find the number of degrees in the 38% sector.

18. Find the number of degrees in the 1% sector.

19. How many homeowners in your town might you expect to spend $250–$999 on yearly home maintenance?

20. How many homeowners in your town might you expect to spend $5000 or more on yearly home maintenance?

The circle graph below shows the number of minutes that adults spend on their home phone each day. Use this graph for Exercises 21 through 26. See Example 4.

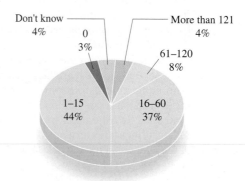

Don't know 4%
0 3%
More than 121 4%
61–120 8%
1–15 44%
16–60 37%

Source: Bruskin/Goldring Research for Sony Electronics

21. What percent of adults spend more than 121 minutes on the phone each day?

22. What percent of adults spend no time on the phone each day?

23. Find the number of degrees in the 4% sector.

24. Find the number of degrees in the 44% sector.

25. Liberty is a town whose adult population is approximately 135,000. How many of these adults might you expect to talk 16–60 minutes on the phone each day?

26. Poll the students in your algebra class. Find what percent of students spend 1–15 minutes on the phone each day. Is this percent close to 44%? Why or why not?

Solve the following. See Examples 3, 5, and 6.

27. What number is 16% of 70?
28. What number is 88% of 1000?
29. The number 28.6 is what percent of 52?
30. The number 87.2 is what percent of 436?
31. The number 45 is 25% of what number?
32. The number 126 is 35% of what number?
33. Find 23% of 20.
34. Find 140% of 86.
35. The number 40 is 80% of what number?
36. The number 56.25 is 45% of what number?
37. The number 144 is what percent of 480?
38. The number 42 is what percent of 35?

Solve. See Example 7. Many applications in this exercise set may be solved more efficiently with the use of a calculator.

39. Dillard's advertised a 25% off sale. If a London Fog coat originally sold for $156, find the decrease and the sale price.
40. Time Saver increased the price of a $0.75 cola by 15%. Find the increase and the new price.
41. At this writing, the women's world record for throwing a discus is held by Gabriele Reinsch of East Germany. Her throw was 252 feet. The men's record is held by Jürgen Schult also of East Germany. His throw was 3.7% shorter than Gabriele's. Find the length of his throw. (Round to the nearest whole foot.) (*Source: Guiness Book of Records,* 1999. *Note:* The women's discus weighs 2 lb 3 oz while the men's weighs 4 lb 8 oz.)
42. Scoville units are used to measure the hotness of a pepper. An alkaloid, capsaicin, is the ingredient that makes a pepper hot and liquid chromatography measures the amount of capsaicin in parts per million. The jalapeno measures around 5000 Scoville units, while the hottest pepper, the habanero, measures around 3000% as hot as the measure of the jalapeno. Find the measure of the habanero pepper.

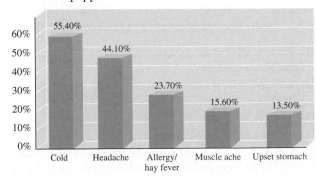

Source: Simmons Market Research Bureau

The graph at the bottom of the previous column shows the percent of people in a survey who have used various over-the-counter drugs in a twelve-month period. Use this graph for Exercises 43–48.

43. What percent of those surveyed used over-the-counter drugs to combat the common cold?
44. What percent of those surveyed used over-the-counter drugs to combat an upset stomach?
45. If 230 people were surveyed, how many of these used over-the-counter drugs for allergies?
46. The city of Chattanooga has a population of approximately 152,000. How many of these people would you expect to have used over-the-counter drugs for relief of a headache?
47. Do the percents shown in the graph have a sum of 100%? Why or why not?
48. Survey your algebra class and find what percent of the class has used over-the-counter drugs for each of the categories listed. Draw a bar graph of the results
49. A recent study showed that 26% of men have dozed off at their place of work. If you currently employ 121 men, how many of these men might you expect to have dozed off at work? (*Source:* Better Sleep Council)
50. A recent study showed that women and girls spend 41% of the household clothes budget. If a family spent $2000 last year on clothing, how much might have been spent on clothing for women and girls? (*Source:* The Interep Radio Store)

Fill in the percent column in each table. The first entry has been done for you. See Examples 1 through 3.

51. **WOMEN IN THE U.S. ARMED FORCES, 1999**

Service	Number of Women	Percent of Total (round to the nearest whole percent)
Army	58,408	$\frac{58,408}{186,697} \approx 31\%$
Navy	50,287	
Marines	9696	
Air Force	64,427	
Coast Guard	3879	
Total	186,697	

(*Source:* U.S. Dept. of Defense)

52. U.S. PUBLIC SCHOOLS WITH COMPUTERS, 1999

Type of School	Number of Schools	Percent of Total (round to the nearest whole percent)
Elementary	52,276	$\frac{52{,}276}{84{,}002} \approx 62\%$
Junior High	14,419	
Senior High	17,307	
Total	84,002	

(*Source:* Quality Education Data, Inc.)

53. The hepatitis B vaccine, given as part of the standard regimen of childhood vaccinations, is required for public-school admission in 42 states. Each year about 1800 of the nearly 1,200,000 people who receive the shot have a serious adverse event. What percent of hepatitis B vaccine recipients report a serious adverse event? (*Source:* Vaccine Adverse Event Reporting System)

54. At schools that offer an accredited dental hygiene program, the curriculum requires 1948 clock hours of course and lab work on average. This includes a total of 585 hours of supervised clinical dental hygiene instruction. What percent of dental hygiene students' curriculum is normally devoted to supervised clinical instruction? Round to the nearest whole percent. (*Source:* The American Dental Hygienists' Association)

55. Approximately 10.4 million students are enrolled in community colleges. This represents 44% of all U.S. undergraduates. How many U.S. undergraduates are there in all? Round to the nearest tenth of a million. (*Source:* American Association of Community Colleges)

56. In 1998, a total of 916 million people were solicited by credit card issuers. This represents a 276% increase since 1992. How many people were solicited in 1992? Round to the nearest whole million. (*Source: USA Today,* 5/12/99)

57. Iceberg lettuce is grown and shipped to stores for about 40 cents a head, and consumers purchase it for about 70 cents a head. Find the percent increase.

58. The lettuce consumption per capita in 1968 was about 21.5 pounds, and in 1997 the consumption rose to 24.3 pounds. Find the percent increase. (Round to the nearest tenth of a percent.)

59. During a recent 5-year period, bank fees for bounced checks have risen from $12.62 to $15.65. Find the percent increase. (Round to the nearest whole percent.) If inflation over the same period has been 16%, do you think the increase in bank fees is fair? (*Source:* Federal Reserve)

60. During a recent 5-year period, the bank fee for depositing a bad check has risen from $5.38 to $6.08. Find the percent increase. (Round to the nearest whole percent.) Given the inflation rate of 16%, do you think that this increase in bank fees is fair?

61. The first Barbie doll was introduced in March 1959 and cost $3. This same 1959 Barbie doll now costs up to $5000. Find the percent increase rounded to the nearest whole percent.

62. The ACT Assessment is a college entrance exam taken by about 60% of college-bound students. The national average score was 20.7 in 1993 and rose to 21.0 in 1998. Find the percent increase. (Round to the nearest hundredth of a percent.)

The double bar graph below shows selected services and products and the percent of supermarkets offering each.

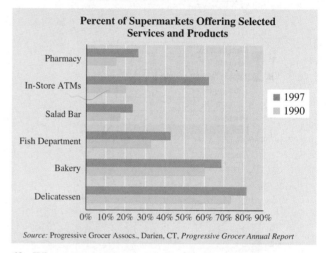

63. What percent of supermarkets had a pharmacy in 1990?

64. What percent of supermarkets had a bakery in 1997?

65. Suppose that there are 50 supermarkets in your city. How many of these 50 supermarkets might have offered customers a salad bar in 1997?

66. How many more supermarkets (in percent) offered in-store ATMs in 1997 than in 1990?

67. Do you notice any trends shown in this graph?

68. Approximately 60% of all Americans are farsighted, and about 30% are nearsighted. In a group of 450 Americans, how many people would you expect to be (a) farsighted? (b) neither farsighted nor nearsighted?

69. The number of crimes reported in Baltimore, Maryland, was 38,831 in the first half of 1997 and 34,611 in the first half of 1998. Find the percent decrease in the number of reported crimes in Baltimore from 1997 to 1998. Round to the nearest whole percent. (*Source:* Federal Bureau of Investigation, *Uniform Crime Reports*, January–June 1998)

Standardized nutrition labels like the ones below have been displayed on food items since 1994. The percent column on the right shows the percent of daily values based on a 2000-calorie diet shown at the bottom of the label. For example, a serving of this food contains 4 grams of total fat, where the recommended daily fat based on a 2000-calorie diet is 65 grams of fat. This means that $\frac{4}{65}$ or approximately 6% (as shown) of your daily recommended fat is taken in by eating a serving of this food.

Nutrition Facts

Serving Size 18 crackers (31g)
Servings Per Container About 9

Amount Per Serving

Calories 130	Calories from Fat 35

	% Daily Value*
Total Fat 4g	**6%**
Saturated Fat 0.5g	**3%**
Polyunsaturated Fat 0g	
Monounsaturated Fat 1.5g	
Cholesterol 0mg	**0%**
Sodium 230mg	**x**
Total Carbohydrate 23g	**y**
Dietary Fiber 2g	**8%**
Sugars 3g	
Protein 2g	

Vitamin A 0%	•	Vitamin C 0%
Calcium 2%	•	Iron 6%

*Percent Daily Values are based on a 2,000 calorie diet. Your daily values may be higher or lower depending on your calorie needs.

		Calories:	2,000	2,500
Total Fat	Less than		65g	80g
Sat Fat	Less than		20g	25g
Cholesterol	Less than		300mg	300mg
Sodium	Less than		2400mg	2400mg
Total Carbohydrate			300g	375g
Dietary Fiber			25g	30g

70. Based on a 2000-calorie diet, what percent of daily values of sodium is contained in a serving of this food? In other words, find *x*. (Round to the nearest tenth of a percent.)

71. Based on a 2000-calorie diet, what percent of daily values of total carbohydrate is contained in a serving of this food? In other words, find *y*. (Round to the nearest tenth of a percent.)

72. Notice on the nutrition label that one serving of this food contains 130 calories and 35 of these calories are from fat. Find the percent of calories from fat. (Round to the nearest tenth of a percent.) It is recommended that no more than 30% of calorie intake come from fat. Does this food satisfy this recommendation?

Below is a nutrition label for a particular food.

NUTRITIONAL INFORMATION PER SERVING

Serving Size: 9.8 oz.   **Servings Per Container: 1**

Calories.................280	Polyunsaturated Fat..........1g
Protein...................12g	Saturated Fat...............3g
Carbohydrate.............45g	Cholesterol..............20mg
Fat.......................6g	Sodium................520mg
Percent of Calories from Fat...... ?	Potassium.............220mg

73. If fat contains approximately 9 calories per gram, find the percent of calories from fat in one serving of this food. (Round to the nearest tenth of a percent.)

74. If protein contains approximately 4 calories per gram, find the percent of calories from protein from one serving of this food. (Round to the nearest tenth of a percent.)

75. Find a food that contains more than 30% of its calories per serving from fat. Analyze the nutrition label and verify that the percents shown are correct.

76. The table shows where lightning strikes. Use this table to draw a circle graph or pie chart of this information.

Fields, Ballparks	*Under trees*	*Bodies of water*	*Golf courses*	*Near heavy equipment*	*Telephone poles*	*Other*
28%	17%	13%	4%	6%	1%	31%

REVIEW EXERCISES

Evaluate the following expressions for the given values. See Section 1.7.

77. $2a + b - c$; $a = 5, b = -1$, and $c = 3$

78. $-3a + 2c - b$; $a = -2, b = 6$, and $c = -7$

79. $4ab - 3bc$; $a = -5, b = -8$, and $c = 2$

80. $ab + 6bc$; $a = 0, b = -1$, and $c = 9$

81. $n^2 - m^2$; $n = -3$ and $m = -8$

82. $2n^2 + 3m^2$; $n = -2$ and $m = 7$

2.8 FURTHER PROBLEM SOLVING

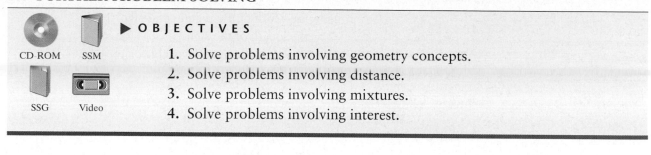

CD·ROM SSM

SSG Video

▶ **OBJECTIVES**

1. Solve problems involving geometry concepts.
2. Solve problems involving distance.
3. Solve problems involving mixtures.
4. Solve problems involving interest.

This section is devoted to solving problems in the categories listed. The same problem-solving steps used in previous sections are also followed in this section. They are listed below for review.

GENERAL STRATEGY FOR PROBLEM SOLVING

1. UNDERSTAND the problem. During this step, become comfortable with the problem. Some ways of doing this are:
   Read and reread the problem.
   Choose a variable to represent the unknown.
   Construct a drawing.
   Propose a solution and check. Pay careful attention to how you check your proposed solution. This will help writing an equation to model the problem.
2. TRANSLATE the problem into an equation.
3. SOLVE the equation.
4. INTERPRET the results: *Check* the proposed solution in the stated problem and *state* your conclusion.

1   The first example involves geometry concepts.

△ **Example 1**   **FINDING ROAD SIGN DIMENSIONS**

The length of a rectangular road sign is 2 feet less than three times its width. Find the dimensions if the perimeter is 28 feet.

WESTVIEW   92
RIVERSIDE   162
SUNDALE   205

Solution

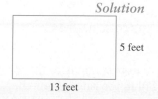

5 feet

13 feet

1. UNDERSTAND. Read and reread the problem. Recall that the formula for the perimeter of a rectangle is $P = 2l + 2w$. Draw a rectangle and guess the solution. If the width of the rectangular sign is 5 feet, its length is 2 feet less than 3 times the width or 3(5 feet) − 2 feet = 13 feet. The perimeter P of the rectangle is then 2(13 feet) + 2(5 feet) = 36 feet, too much. We now know that the width is less than 5 feet.

Let

w = the width of the rectangular sign; then

$3w - 2$ = the length of the sign.

Draw a rectangle and label it with the assigned variables.

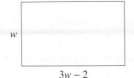

2. TRANSLATE.

Formula: $P = 2l$ $+ \; 2w$ or

Substitute: $28 = 2(3w - 2) + \; 2w.$

3. SOLVE.

$$28 = 2(3w - 2) + 2w$$
$$28 = 6w - 4 + 2w \qquad \text{\textit{Apply the distributive property.}}$$
$$28 = 8w - 4$$
$$28 + 4 = 8w - 4 + 4 \qquad \text{\textit{Add 4 to both sides.}}$$
$$32 = 8w$$
$$\frac{32}{8} = \frac{8w}{8} \qquad \text{\textit{Divide both sides by 8.}}$$
$$4 = w$$

4. INTERPRET.

Check: If the width of the sign is 4 feet, the length of the sign is 3(4 feet) $-$ 2 feet = 10 feet. This gives a perimeter of P = 2(4 feet) + 2(10 feet) = 28 feet, the correct perimeter.

State: The width of the sign is 4 feet and the length of the sign is 10 feet.

2 The next example involves distance.

Example 2   FINDING TIME GIVEN RATE AND DISTANCE

Marie Antonio, a bicycling enthusiast, rode her 10-speed at an average speed of 18 miles per hour on level roads and then slowed down to an average of 10 miles per hour on the hilly roads of the trip. If she covered a distance of 98 miles, how long did the entire trip take if traveling the level roads took the same time as traveling the hilly roads?

Solution   1. UNDERSTAND the problem. To do so, read and reread the problem. The formula $d = r \cdot t$ is needed. At this time, let's guess a solution. Suppose that she spent 2 hours traveling on the level roads. This means that she also spent 2 hours traveling on the hilly roads, since the times spent were the same. What is her total distance? Her distance on the level road is rate $\cdot$ time = 18(2) = 36 miles. Her distance on the hilly roads is rate $\cdot$ time = 10(2) = 20 miles. This gives a total distance of 36 miles + 20 miles = 56 miles, not the correct distance of 98 miles. Remember that the purpose of guessing a solution is not to guess correctly (although this may happen) but to help better understand the problem and how to model it with an equation.

We are looking for the length of the entire trip, so we begin by letting

x = the time spent on level roads.

Because the same amount of time is spent on hilly roads, then also

x = the time spent on hilly roads.

2. TRANSLATE. To help us translate to an equation, we now summarize the information from the problem on the following chart. Fill in the rates given, the

variables used to represent the times, and use the formula $d = r \cdot t$ to fill in the distance column.

	Rate · Time = Distance		
LEVEL	18	x	18x
HILLY	10	x	10x

Since the entire trip covered 98 miles, we have that

In words:

total distance	=	level distance	+	hilly distance

Translate:

98	=	18x	+	10x

3. SOLVE.

$$98 = 28x \qquad \text{Add like terms.}$$

$$\frac{98}{28} = \frac{28x}{28} \qquad \text{Divide both sides by 28.}$$

$$3.5 = x$$

4. INTERPRET the results.

   Check: Recall that x represents the time spent on the level portion of the trip and also the time spent on the hilly portion. If Marie rides for 3.5 hours at 18 mph, her distance is $18(3.5) = 63$ miles. If Marie rides for 3.5 hours at 10 mph, her distance is $10(3.5) = 35$ miles. The total distance is 63 miles + 35 miles = 98 miles, the required distance.

   State: The time of the entire trip is then 3.5 hours + 3.5 hours or 7 hours. ■

---

3 Mixture problems involve two or more different quantities being combined to form a new mixture. These applications range from Dow Chemical's need to form a chemical mixture of a required strength to Planter's Peanut Company's need to find the correct mixture of peanuts and cashews, given taste and price constraints.

Example 3 **CALCULATING PERCENT FOR A LAB EXPERIMENT**

A chemist working on his doctoral degree at Massachusetts Institute of Technology needs 12 liters of a 50% acid solution for a lab experiment. The stockroom has only 40% and 70% solutions. How much of each solution should be mixed together to form 12 liters of a 50% solution?

Solution
1. UNDERSTAND. First, read and reread the problem a few times. Next, guess a solution. Suppose that we need 7 liters of the 40% solution. Then we need $12 - 7 = 5$ liters of the 70% solution. To see if this is indeed the solution, find the amount of pure acid in 7 liters of the 40% solution, in 5 liters of the 70% solution, and in 12 liters of a 50% solution, the required amount and strength.

number of liters	×	acid strength	=	amount of pure acid
7 liters	×	40%	=	7(0.40) or 2.8 liters
5 liters	×	70%	=	5(0.70) or 3.5 liters
12 liters	×	50%	=	12(0.50) or 6 liters

Since 2.8 liters + 3.5 liters = 6.3 liters and not 6, our guess is incorrect, but we have gained some valuable insight into how to model and check this problem. Let

$$x = \text{number of liters of 40\% solution; then}$$

$$12 - x = \text{number of liters of 70\% solution.}$$

2. TRANSLATE. To help us translate to an equation, the following table summarizes the information given. Recall that the amount of acid in each solution is found by multiplying the acid strength of each solution by the number of liters.

	*No. of liters · Strength = *Acid* Amount of Acid*		
40% SOLUTION	x	40%	$0.40x$
70% SOLUTION	$12 - x$	70%	$0.70(12 - x)$
50% SOLUTION NEEDED	12	50%	$0.50(12)$

The amount of acid in the final solution is the sum of the amounts of acid in the two beginning solutions.

In words:

acid in 40% solution	+	acid in 70% solution	=	acid in 50% mixture

Translate: $\quad 0.40x \quad + \ 0.70(12 - x) = \quad 0.50(12)$

3. SOLVE.

$$0.40x + 0.70(12 - x) = 0.50(12)$$

$$0.4x + 8.4 - 0.7x = 6 \qquad \text{Apply the distributive property.}$$

$$-0.3x + 8.4 = 6 \qquad \text{Combine like terms.}$$

$$-0.3x = -2.4 \qquad \text{Subtract 8.4 from both sides.}$$

$$x = 8 \qquad \text{Divide both sides by } -0.3.$$

4. INTERPRET.

Check: To check, recall how we checked our guess.

State: If 8 liters of the 40% solution are mixed with $12 - 8$ or 4 liters of the 70% solution, the result is 12 liters of a 50% solution.

4 The next example is an investment problem.

Example 4 FINDING THE INVESTMENT AMOUNT

Rajiv Puri invested part of his $20,000 inheritance in a mutual funds account that pays 7% simple interest yearly and the rest in a certificate of deposit that pays 9% simple interest yearly. At the end of one year, Rajiv's investments earned $1550. Find the amount he invested at each rate.

Solution
1. UNDERSTAND: Read and reread the problem. Next, guess a solution. Suppose that Rajiv invested $8000 in the 7% fund and the rest, $12,000, in the fund paying 9%. To check, find his interest after one year. Recall the formula, $I = PRT$, so the interest from the 7% fund = $8000(0.07)(1) = $560. The interest from the 9% fund = $12,000(0.09)(1) = $1080. The sum of the interests is $560 + $1080 = $1640. Our guess is incorrect, since the sum of the interests is not $1550, but we now have a better understanding of the problem.

   Let

$$x = \text{amount of money in the account paying 7\%. The rest of the money is \$20,000 less } x \text{ or}$$
$$20{,}000 - x = \text{amount of money in the account paying 9\%.}$$

2. TRANSLATE. We apply the simple interest formula $I = PRT$ and organize our information in the following chart. Since there are two different rates of interest and two different amounts invested, we apply the formula twice.

	Principal ·	Rate ·	Time =	Interest
7% FUND	x	0.07	1	$x(0.07)(1)$ or $0.07x$
9% FUND	$20{,}000 - x$	0.09	1	$(20{,}000 - x)(0.09)(1)$ or $0.09(20{,}000 - x)$
TOTAL	20,000			1550

The total interest earned, $1550, is the sum of the interest earned at 7% and the interest earned at 9%.

In words:

interest at 7%	+	interest at 9%	=	total interest

Translate: $\quad 0.07x \quad + \quad 0.09(20{,}000 - x) \quad = \quad 1550$

3. SOLVE.

$$0.07x + 0.09(20{,}000 - x) = 1550$$
$$0.07x + 1800 - 0.09x = 1550 \qquad \text{Apply the distributive property.}$$
$$1800 - 0.02x = 1550 \qquad \text{Combine like terms.}$$
$$-0.02x = -250 \qquad \text{Subtract 1800 from both sides.}$$
$$x = 12{,}500 \qquad \text{Divide both sides by } -0.02.$$

4. INTERPRET.

   Check: If $x = 12{,}500$, then $20{,}000 - x = 20{,}000 - 12{,}500$ or 7500. These solutions are reasonable, since their sum is $20,000 as required. The annual interest on $12,500 at 7% is $875; the annual interest on $7500 at 9% is $675, and $875 + $675 = $1550.

   State: The amount invested at 7% is $12,500. The amount invested at 9% is $7500.

SPOTLIGHT ON DECISION MAKING

Suppose you have just purchased a house. You are required by your mortgage company to buy homeowner's insurance. In choosing a policy, you must also choose a deductible level. A deductible is how much you, the homeowner, pay to repair or replace a loss before the insurance company starts paying. Typically, deductibles start at $250. However, other levels such as $500 or $1000 are also available. One way to save money when purchasing your homeowner's policy is to raise the deductible amount. However, doing so means you will have to pay more out of pocket in case of theft or fire. Study the accompanying chart. Which amount of deductible would you choose? Why?

Increase Your Deductible to:	
$500	save up to 12%
$1000	save up to 24%
$2500	save up to 30%
$5000	save up to 37%

on the cost of a homeowner's policy, depending on your insurance company.

Source: Insurance Information Institute

Exercise Set 2.8

Solve each word problem. See Example 1.

△ **1.** An architect designs a rectangular flower garden such that the width is exactly two-thirds of the length. If 260 feet of antique picket fencing are to be used, find the dimensions of the garden.

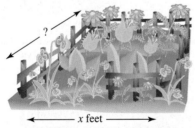

△ **2.** If the length of a rectangular parking lot is 10 meters less than twice its width, and the perimeter is 400 meters, find the length of the parking lot.

△ **3.** A flower bed is in the shape of a triangle with one side twice the length of the shortest side, and the third side is 30 feet more than the length of the shortest side. Find the dimensions if the perimeter is 102 feet.

△ **4.** The perimeter of a yield sign in the shape of an isosceles triangle is 22 feet. If the shortest side is 2 feet less than the other two sides, find the length of the shortest side. (*Hint:* An isosceles triangle has two sides the same length.)

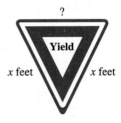

Solve. See Example 2.

5. A jet plane traveling at 500 mph overtakes a propeller plane traveling at 200 mph that had a 2-hour head start. How far from the starting point are the planes?

6. How long will it take a bus traveling at 60 miles per hour to overtake a car traveling at 40 mph if the car had a 1.5-hour head start?

7. The Jones family drove to Disneyland at 50 miles per hour and returned on the same route at 40 mph. Find the distance to Disneyland if the total driving time was 7.2 hours.

8. A bus traveled on a level road for 3 hours at an average speed 20 miles per hour faster than it traveled on a winding road. The time spent on the winding road was 4 hours. Find the average speed on the level road if the entire trip was 305 miles.

Solve. See Example 3.

9. How much pure acid should be mixed with 2 gallons of a 40% acid solution in order to get a 70% acid solution?

10. How many cubic centimeters of a 25% antibiotic solution should be added to 10 cubic centimeters of a 60% antibiotic solution in order to get a 30% antibiotic solution?

11. Planter's Peanut Company wants to mix 20 pounds of peanuts worth $3 a pound with cashews worth $5 a pound in order to make an experimental mix worth $3.50 a pound. How many pounds of cashews should be added to the peanuts?

12. Community Coffee Company wants a new flavor of Cajun coffee. How many pounds of coffee worth $7 a pound should be added to 14 pounds of coffee worth $4 a pound to get a mixture worth $5 a pound?

13. Is it possible to mix a 30% antifreeze solution with a 50% antifreeze solution and obtain a 70% antifreeze solution? Why or why not?

14. A trail mix is made by combining peanuts worth $3 a pound, raisins worth $2 a pound, and M & M's worth $4 a pound. Would it make good business sense to sell the trail mix for $1.98 a pound? Why or why not?

15. Zoya Lon invested part of her $25,000 advance at 8% annual simple interest and the rest at 9% annual simple interest. If her total yearly interest from both accounts was $2135, find the amount invested at each rate.

16. Karen Waugthaul invested some money at 9% annual simple interest and $250 more than that amount at 10% annual simple interest. If her total yearly interest was $101, how much was invested at each rate?

17. Sam Mathius invested part of his $10,000 bonus in a fund that paid an 11% profit and invested the rest in stock that suffered a 4% loss. Find the amount of each investment if his overall net profit was $650.

18. Bruce Blossum invested a sum of money at 10% annual simple interest and invested twice that amount at 12% annual simple interest. If his total yearly income from both investments was $2890, how much was invested at each rate?

Solve.

△ 19. The perimeter of an equilateral triangle is 7 inches more than the perimeter of a square, and the side of the triangle is 5 inches longer than the side of the square. Find the side of the triangle.

△ 20. A square animal pen and a pen shaped like an equilateral triangle have equal perimeters. Find the length of the sides of each pen if the sides of the triangular pen are fifteen less than twice a side of the square pen. (*Hint:* An equilateral triangle has three sides the same length.)

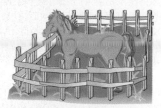

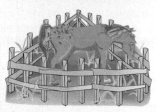

21. How can $54,000 be invested, part at 8% annual simple interest and the remainder at 10% annual simple interest, so that the interest earned by the two accounts will be equal?

22. Kathleen and Cade Williams leave simultaneously from the same point hiking in opposite directions, Kathleen walking at 4 miles per hour and Cade at 5 mph. How long can they talk on their walkie-talkies if the walkie-talkies have a 20-mile radius?

23. Ms. Mills invested her $20,000 bonus in two accounts. She took a 4% loss on one investment and made a 12% profit on another investment, but ended up breaking even. How much was invested in each account?

24. Alan and Dave Schaferkötter leave from the same point driving in opposite directions, Alan driving at 55 miles per hour and Dave at 65 mph. Alan has a one-hour head start. How long will they be able to talk on their car phones if the phones have a 250-mile range?

25. If $3000 is invested at 6% annual simple interest, how much should be invested at 9% annual simple interest so that the total yearly income from both investments is $585?

26. How much of an alloy that is 20% copper should be mixed with 200 ounces of an alloy that is 50% copper in order to get an alloy that is 30% copper?

27. Trudy Waterbury, a financial planner, invested a certain amount of money at 9% annual simple interest, twice that amount at 10% annual simple interest, and three times that amount at 11% annual simple interest. Find the amount invested at each rate if her total yearly income from the investments was $2790.

28. How much water should be added to 30 gallons of a solution that is 70% antifreeze in order to get a mixture that is 60% antifreeze?

29. April Thrower spent $32.25 to take her daughter's birthday party guests to the movies. Adult tickets cost $5.75 and children tickets cost $3.00. If 8 persons were at the party, how many adult tickets were bought?

30. Nedra and Latonya Dominguez are 12 miles apart hiking toward each other. How long will it take them to meet if Nedra walks at 3 miles per hour and Latonya walks 1 mph faster?

 31. Two hikers are 11 miles apart and walking toward each other. They meet in 2 hours. Find the rate of each hiker if one hiker walks 1.1 miles per hour faster than the other.

32. On a 255-mile trip, Gary Alessandrini traveled at an average speed of 70 miles per hour, got a speeding ticket, and then traveled at 60 mph for the remainder of the trip. If the entire trip took 4.5 hours and the speeding ticket stop took 30 minutes, how long did Gary speed before getting stopped?

33. Mark Martin can row upstream at 5 miles per hour and downstream at 11 mph. If Mark starts rowing upstream until he gets tired and then rows downstream to his starting point, how far did Mark row if the entire trip took 4 hours?

*To "break even" in a manufacturing business, revenue R (income) **must equal** the cost C of production, or R = C.*

34. The cost C to produce x number of skateboards is given by $C = 100 + 20x$. The skateboards are sold wholesale for $24 each, so revenue R is given by $R = 24x$. Find how many skateboards the manufacturer needs to produce and sell to break even. (*Hint:* Set the expression for R equal to the expression for C, then solve for x.)

35. The revenue R from selling x number of computer boards is given by $R = 60x$, and the cost C of producing them is given by $C = 50x + 5000$. Find how many boards must be sold to break even. Find how much money is needed to produce the break-even number of boards.

36. The cost C of producing x number of paperback books is given by $C = 4.50x + 2400$. Income R from these books is given by $R = 7.50x$. Find how many books should be produced and sold to break even.

37. Find the break-even quantity for a company that makes x number of computer monitors at a cost C given by $C = 870 + 70x$ and receives revenue R given by $R = 105x$.

38. Problems 34 through 37 involve finding the break-even point for manufacturing. Discuss what happens if a company makes and sells fewer products than the break-even point. Discuss what happens if more products than the break-even point are made and sold.

REVIEW EXERCISES

Perform the indicated operations. See Section 1.6.

39. $3 + (-7)$

40. $(-2) + (-8)$

41. $\dfrac{3}{4} - \dfrac{3}{16}$

42. $-11 + 2.9$

43. $-5 - (-1)$

44. $-12 - 3$

Place $<, >,$ or $=$ in the appropriate space to make each a true statement. See Sections 1.1 and 1.7.

45. $-5 \qquad -7$

46. $\dfrac{12}{3} \qquad 2^2$

47. $|-5| \qquad -(-5)$

48. $-3^3 \qquad (-3)^3$

2.9 SOLVING LINEAR INEQUALITIES

CD-ROM   SSM

SSG   Video

▶ **OBJECTIVES**

1. Define linear inequality in one variable.
2. Graph solution sets on a number line.
3. Solve linear inequalities.
4. Solve compound inequalities.
5. Solve inequality applications.

1   In Chapter 1, we reviewed these inequality symbols and their meanings:

< means "is less than"   ≤ means "is less than or equal to"
> means "is greater than"   ≥ means "is greater than or equal to"

A linear inequality is similar to a linear equation except that the equality symbol is replaced with an inequality symbol.

Equations	Inequalities
$x = 3$	$x \leq 3$
$5n - 6 = 14$	$5n - 6 > 14$
$12 = 7 - 3y$	$12 \leq 7 - 3y$
$\dfrac{x}{4} - 6 = 1$	$\dfrac{x}{4} - 6 > 1$

LINEAR INEQUALITY IN ONE VARIABLE

A **linear inequality in one variable** is an inequality that can be written in the form

$$ax + b < c$$

where a, b, and c are real numbers and a is not 0.

This definition and all other definitions, properties, and steps in this section also hold true for the inequality symbols, $>$, $\geq$, and $\leq$.

2   A **solution of an inequality** is a value of the variable that makes the inequality a true statement. The solution set is the set of all solutions. In the inequality $x < 3$, replacing x with any number less than 3, that is, to the left of 3 on the number line, makes the resulting inequality true. This means that any number less than 3 is a solution of the inequality $x < 3$. Since there are infinitely many such numbers, we cannot list all the solutions of the inequality. We *can* use set notation and write

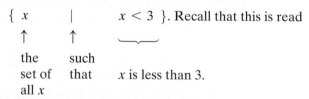

$\{\ x\ \ |\ \ x < 3\ \}$. Recall that this is read

the set of all x such that x is less than 3.

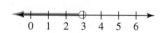

We can also picture the solution set on a number line. To do so, shade the portion of the number line corresponding to numbers less than 3.

Recall that all the numbers less than 3 lie to the left of 3 on the number line. An open circle about the point representing 3 indicates that 3 **is not** a solution of the inequality: 3 **is not** less than 3. The shaded arrow indicates that the solutions of $x < 3$ continue indefinitely to the left of 3.

Picturing the solutions of an inequality on a number line is called **graphing** the solutions or graphing the inequality, and the picture is called the **graph** of the inequality.

To graph $x \leq 3$, shade the numbers to the left of 3 and place a closed circle on the point representing 3. The closed circle indicates that 3 **is** a solution: 3 **is** less than or equal to 3.

Example 1   Graph $x \geq -1$.

Solution   We place a closed circle at -1 since the inequality symbol is $\geq$ and -1 is greater than or equal to -1. Then we shade to the right of -1.

Inequalities containing one inequality symbol are called **simple inequalities**, while inequalities containing two inequality symbols are called **compound inequalities**. A compound inequality is really two simple inequalities in one. The compound inequality

$$3 < x < 5 \qquad \text{means} \qquad 3 < x \text{ and } x < 5$$

This can be read "x is greater than 3 and less than 5."

A solution of a compound inequality is a value that is a solution of both of the simple inequalities that make up the compound inequality. For example,

$$4\frac{1}{2} \text{ is a solution of } 3 < x < 5 \text{ since } 3 < 4\frac{1}{2} \text{ and } 4\frac{1}{2} < 5.$$

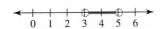

To graph $3 < x < 5$, place open circles at both 3 and 5 and shade between.

Example 2   Graph $2 < x \leq 4$.

Solution   Graph all numbers greater than 2 and less than or equal to 4. Place an open circle at 2, a closed circle at 4, and shade between.

3   When solutions of a linear inequality are not immediately obvious, they are found through a process similar to the one used to solve a linear equation. Our goal is to get the variable alone, and we use properties of inequality similar to properties of equality.

ADDITION PROPERTY OF INEQUALITY

If a, b, and c are real numbers, then

$$a < b \qquad \text{and} \qquad a + c < b + c$$

are equivalent inequalities.

This property also holds true for subtracting values, since subtraction is defined in terms of addition. In other words, adding or subtracting the same quantity from both sides of an inequality does not change the solution of the inequality.

Example 3   Solve $x + 4 \leq -6$ for x. Graph the solution set.

Solution   To solve for x, subtract 4 from both sides of the inequality.

$x + 4 \leq -6$	Original inequality.
$x + 4 - 4 \leq -6 - 4$	Subtract 4 from both sides.
$x \leq -10$	Simplify.

The solution set is $\{x \mid x \leq -10\}$.

HELPFUL HINT

Notice that any number less than or equal to −10 is a solution to $x \leq -10$. For example, solutions include

$$-10, \quad -200, \quad -11\frac{1}{2}, \quad -7\pi, \quad -\sqrt{130}, \quad -50.3$$

An important difference between linear equations and linear inequalities is shown when we multiply or divide both sides of an inequality by a nonzero real number. For example, start with the true statement $6 < 8$ and multiply both sides by 2. As we see below, the resulting inequality is also true.

$$6 < 8 \qquad \text{True.}$$
$$2(6) < 2(8) \qquad \text{Multiply both sides by 2.}$$
$$12 < 16 \qquad \text{True.}$$

But if we start with the same true statement $6 < 8$ and multiply both sides by −2, the resulting inequality is not a true statement.

$$6 < 8 \qquad \text{True.}$$
$$-2(6) < -2(8) \qquad \text{Multiply both sides by −2.}$$
$$-12 < -16 \qquad \text{False.}$$

Notice, however, that if we reverse the direction of the inequality symbol, the resulting inequality is true.

$$-12 < -16 \qquad \text{False.}$$
$$-12 > -16 \qquad \text{True.}$$

This demonstrates the multiplication property of inequality.

MULTIPLICATION PROPERTY OF INEQUALITY

1. If a, b, and c are real numbers, and c is **positive**, then

$$a < b \qquad \text{and} \qquad ac < bc$$

are equivalent inequalities.

2. If a, b, and c are real numbers, and c is **negative**, then

$$a < b \qquad \text{and} \qquad ac > bc$$

are equivalent inequalities.

Because division is defined in terms of multiplication, this property also holds true when dividing both sides of an inequality by a nonzero number: If we multiply or divide both sides of an inequality by a negative number, **the direction of the inequality sign must be reversed for the inequalities to remain equivalent**.

HELPFUL HINT

Whenever both sides of an inequality are multiplied or divided by a negative number, the direction of the inequality symbol **must be** reversed to form an equivalent inequality.

Example 4   Solve $-2x \leq -4$. Graph the solution set.

Solution   Remember to reverse the direction of the inequality symbol when dividing by a negative number.

> **HELPFUL HINT**
> Don't forget to reverse the direction of the inequality sign.

$$-2x \leq -4$$
$$\frac{-2x}{-2} \geq \frac{-4}{-2} \quad \text{Divide both sides by } -2 \text{ and reverse the inequality sign.}$$
$$x \geq 2 \quad \text{Simplify.}$$

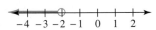

The solution set $\{x \mid x \geq 2\}$ is graphed as shown.   ▬

Example 5   Solve $2x < -4$, and graph the solution set.

Solution
> **HELPFUL HINT**
> Do not reverse the inequality sign.

$$2x < -4$$
$$\frac{2x}{2} < \frac{-4}{2} \quad \text{Divide both sides by 2.} \\ \text{Do not reverse the inequality sign.}$$
$$x < -2 \quad \text{Simplify.}$$

The graph of $\{x \mid x < -2\}$ is shown.   ▬

The following steps may be helpful when solving inequalities. Notice that these steps are similar to the ones given in Section 2.4 for solving equations.

> **SOLVING LINEAR INEQUALITIES IN ONE VARIABLE**
>
> **Step 1.** Clear the inequality of fractions by multiplying both sides of the inequality by the lowest common denominator (LCD) of all fractions in the inequality.
> **Step 2.** Remove grouping symbols such as parentheses by using the distributive property.
> **Step 3.** Simplify each side of the inequality by combining like terms.
> **Step 4.** Write the inequality with variable terms on one side and numbers on the other side by using the addition property of inequality.
> **Step 5.** Get the variable alone by using the multiplication property of inequality.

Don't forget that if both sides of an inequality are multiplied or divided by a negative number, the direction of the inequality sign must be reversed.

Example 6   Solve $-4x + 7 \geq -9$, and graph the solution set.

Solution
$$-4x + 7 \geq -9$$
$$-4x + 7 - 7 \geq -9 - 7 \quad \text{Subtract 7 from both sides.}$$
$$-4x \geq -16 \quad \text{Simplify.}$$

$$\frac{-4x}{-4} \leq \frac{-16}{-4}$$ Divide both sides by −4 and reverse the direction of the inequality sign.

$$x \leq 4$$ Simplify.

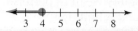

The graph of $\{x \mid x \leq 4\}$ is shown.

Example 7   Solve $2x + 7 \leq x - 11$, and graph the solution set.

Solution

$$2x + 7 \leq x - 11$$

$$2x + 7 - x \leq x - 11 - x$$ Subtract x from both sides.

$$x + 7 \leq -11$$ Combine like terms.

$$x + 7 - 7 \leq -11 - 7$$ Subtract 7 from both sides.

$$x \leq -18$$ Combine like terms.

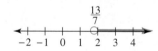

The graph of the solution set $\{x \mid x \leq -18\}$ is shown.

Example 8   Solve $-5x + 7 < 2(x - 3)$, and graph the solution set.

Solution

$$-5x + 7 < 2(x - 3)$$

$$-5x + 7 < 2x - 6$$ Apply the distributive property.

$$-5x + 7 - 2x < 2x - 6 - 2x$$ Subtract $2x$ from both sides.

$$-7x + 7 < -6$$ Combine like terms.

$$-7x + 7 - 7 < -6 - 7$$ Subtract 7 from both sides.

$$-7x < -13$$ Combine like terms.

$$\frac{-7x}{-7} > \frac{-13}{-7}$$ Divide both sides by −7 and reverse the direction of the inequality sign.

$$x > \frac{13}{7}$$ Simplify.

The graph of the solution set $\left\{x \mid x > \frac{13}{7}\right\}$ is shown.

Example 9   Solve $2(x - 3) - 5 \leq 3(x + 2) - 18$, and graph the solution set.

Solution

$$2(x - 3) - 5 \leq 3(x + 2) - 18$$

$$2x - 6 - 5 \leq 3x + 6 - 18$$ Apply the distributive property.

$$2x - 11 \leq 3x - 12$$ Combine like terms.

$$-x - 11 \leq -12$$ Subtract $3x$ from both sides.

$$-x \leq -1$$ Add 11 to both sides.

$$\frac{-x}{-1} \geq \frac{-1}{-1}$$ Divide both sides by −1 and reverse the direction of the inequality sign.

$$x \geq 1$$ Simplify.

The graph of the solution set $\{x \mid x \geq 1\}$ is shown.

4

When we solve a simple inequality, we isolate the variable on one side of the inequality. When we solve a compound inequality, we isolate the variable in the middle part of the inequality. Also, when solving a compound inequality, we must perform the same operation to all **three** parts of the inequality: left, middle, and right.

Example 10   Solve $-1 \le 2x - 3 < 5$, and graph the solution set.

Solution

$$-1 \le 2x - 3 < 5$$
$$-1 + 3 \le 2x - 3 + 3 < 5 + 3 \qquad \text{Add 3 to all three parts.}$$
$$2 \le 2x < 8 \qquad \text{Combine like terms.}$$
$$\frac{2}{2} \le \frac{2x}{2} < \frac{8}{2} \qquad \text{Divide all three parts by 2.}$$
$$1 \le x < 4 \qquad \text{Simplify.}$$

The solution set $\{x \mid 1 \le x < 4\}$ is graphed.

Example 11   Solve $3 \le \dfrac{3x}{2} + 4 \le 5$, and graph the solution set.

Solution

$$3 \le \frac{3x}{2} + 4 \le 5$$

$$2(3) \le 2\left(\frac{3x}{2} + 4\right) \le 2(5) \qquad \begin{array}{l}\text{Multiply all three parts by 2 to}\\ \text{clear the fraction.}\end{array}$$

$$6 \le 3x + 8 \le 10 \qquad \text{Distribute.}$$
$$-2 \le 3x \le 2 \qquad \text{Subtract 8 from all three parts.}$$
$$\frac{-2}{3} \le \frac{3x}{3} \le \frac{2}{3} \qquad \text{Divide all three parts by 3.}$$
$$-\frac{2}{3} \le x \le \frac{2}{3} \qquad \text{Simplify.}$$

The graph of the solution set $\left\{x \mid -\frac{2}{3} \le x \le \frac{2}{3}\right\}$ is shown.

5

Problems containing words such as "at least," "at most," "between," "no more than," and "no less than" usually indicate that an inequality should be solved instead of an equation. In solving applications involving linear inequalities, use the same procedure you use to solve applications involving linear equations.

Example 12   **STAYING WITHIN BUDGET**

Marie Chase and Jonathan Edwards are having their wedding reception at the Gallery reception hall. They may spend at most $2000 for the reception. If the reception hall charges a $100 cleanup fee plus $28 per person, find the greatest number of people that they can invite and still stay within their budget.

Solution   1. UNDERSTAND. Read and reread the problem. Next, guess a solution. If 50 people attend the reception, the cost is $100 + $28(50) = $100 + $1400 = $1500. Let x = the number of people who attend the reception.

2. TRANSLATE.

   In words:

cleanup fee	+	cost per person	must be less than or equal to	$2000
100	+	28x	≤	2000

   Translate:

3. SOLVE.

$$100 + 28x \leq 2000$$
$$28x \leq 1900 \qquad \text{Subtract 100 from both sides.}$$
$$x \leq 67\frac{6}{7} \qquad \text{Divide both sides by 28.}$$

4. INTERPRET.

   Check: Since x represents the number of people, we round down to the nearest whole, or 67. Notice that if 67 people attend, the cost is $100 + $28(67) = $1976. If 68 people attend, the cost is $100 + $28(68) = $2004, which is more than the given $2000.

   State: Marie Chase and Jonathan Edwards can invite at most 67 people to the reception.

MENTAL MATH

Solve each of the following inequalities.

1. $5x > 10$ **2.** $4x < 20$ **3.** $2x \geq 16$ **4.** $9x \leq 63$

Exercise Set 2.9

Graph each inequality on a number line. See Examples 1 and 2.

1. $x \leq -1$

2. $y < 0$

3. $x < \dfrac{1}{2}$

4. $z \leq -\dfrac{2}{3}$

5. $-1 < x < 3$

6. $2 \leq y \leq 3$

7. $0 \leq y < 2$

8. $-1 \leq x \leq 4$

Solve each inequality and graph the solution set. See Examples 3 through 5.

9. $2x < -6$

10. $3x > -9$

11. $x - 2 \geq -7$

12. $x + 4 \leq 1$

13. $-8x \leq 16$

14. $-5x < 20$

Solve, each inequality and graph the solution set. See Examples 6 and 7.

15. $3x - 5 > 2x - 8$

16. $3 - 7x \geq 10 - 8x$

17. $4x - 1 \leq 5x - 2x$

18. $7x + 3 < 9x - 3x$

Solve each inequality and graph the solution set. See Examples 8 and 9.

19. $x - 7 < 3(x + 1)$

20. $3x + 9 \geq 5(x - 1)$

21. $-6x + 2 \geq 2(5 - x)$

22. $-7x + 4 > 3(4 - x)$

23. $4(3x - 1) \leq 5(2x - 4)$

24. $3(5x - 4) \leq 4(3x - 2)$

25. $3(x + 2) - 6 > -2(x - 3) + 14$

26. $7(x - 2) + x \leq -4(5 - x) - 12$

Solve each inequality, and then graph the solution set. See Examples 10 and 11.

27. $-3 < 3x < 6$

28. $-5 < 2x < -2$

29. $2 \leq 3x - 10 \leq 5$

30. $4 \leq 5x - 6 \leq 19$

31. $-4 < 2(x - 3) < 4$

32. $0 < 4(x + 5) < 8$

33. Explain how solving a linear inequality is similar to solving a linear equation.

34. Explain how solving a linear inequality is different from solving a linear equation.

Solve the following inequalities. Graph each solution set.

35. $-2x \leq -40$

36. $-7x > 21$

37. $-9 + x > 7$

38. $y - 4 \leq 1$

39. $3x - 7 < 6x + 2$

40. $2x - 1 \geq 4x - 5$

41. $5x - 7x \leq x + 2$

42. $4 - x < 8x + 2x$

43. $\frac{3}{4}x > 2$

44. $\frac{5}{6}x \geq -8$

45. $3(x - 5) < 2(2x - 1)$

46. $5(x + 4) < 4(2x + 3)$

47. $4(2x + 1) > 4$

48. $6(2 - x) \geq 12$

49. $-5x + 4 \leq -4(x - 1)$

50. $-6x + 2 < -3(x + 4)$

51. $-2 < 3x - 5 < 7$

52. $1 < 4 + 2x \leq 7$

53. $-2(x - 4) - 3x < -(4x + 1) + 2x$

54. $-5(1 - x) + x \leq -(6 - 2x) + 6$

55. $-3x + 6 \geq 2x + 6$

56. $-(x - 4) < 4$

57. $-6 < 3(x - 2) < 8$

58. $-5 \leq 2(x + 4) < 8$

Solve the following. See Example 12.

59. High blood cholesterol levels increase the risk of heart disease in adults. Doctors recommend that total blood cholesterol be less than 200 milligrams per deciliter. Total cholesterol levels from 200 up to 240 milligrams per deciliter are considered borderline. Any total cholesterol reading above 240 milligrams per deciliter is considered high.

Letting x represent a patient's total blood cholesterol level, write a series of three inequalities that describe the ranges corresponding to recommended, borderline, and high levels of total blood cholesterol.

60. T. Theodore Fujita created the Fujita Scale (or F-Scale), which uses ratings from 0 to 5 to classify tornadoes. An F-0 tornado has wind speeds between 40 and 72 mph, inclusive. The winds in an F-1 tornado range from 73 to 112 mph. In an F-2 tornado, winds are from 113 to 157 mph. An F-3 tornado has wind speeds ranging from 158 to 206 mph. Wind speeds in an F-4 tornado are clocked at 207 to 260 mph. The most violent tornadoes are ranked at F-5, with wind speeds of at least 261 mph. (*Source:* National Weather Service)

Letting y represent a tornado's wind speed, write a series of six inequalities that describe the wind speed ranges corresponding to each Fujita Scale rank.

61. Six more than twice a number is greater than negative fourteen. Find all numbers that make this statement true.

62. Five times a number increased by one is less than or equal to ten. Find all such numbers.

63. Dennis and Nancy Wood are celebrating their 30th wedding anniversary by having a reception at Tiffany Oaks reception hall. They have budgeted $3000 for their reception. If the reception hall charges a $50.00 cleanup fee plus $34 per person, find the greatest number of people that they may invite and still stay within their budget.

64. A surprise retirement party is being planned for Pratep Puri. A total of $860 has been collected for the event, which is to be held at a local reception hall. This reception hall charges a cleanup fee of $40 and $15 per person for drinks and light snacks. Find the greatest number of people that may be invited and still stay within $860.

△ **65.** The perimeter of a rectangle is to be no greater than 100 centimeters and the width must be 15 centimeters. Find the maximum length of the rectangle.

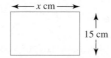

△ **66.** One side of a triangle is four times as long as another side, and the third side is 12 inches long. If the perimeter can be no longer than 87 inches, find the maximum lengths of the other two sides.

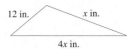

67. The temperatures in Ohio range from $-39°C$ to $45°C$. Use a compound inequality to convert these temperatures to Fahrenheit temperatures. (*Hint:* Use $C = \frac{5}{9}(F - 32)$.)

68. Mario Lipco has scores of 85, 95, and 92 on his algebra tests. Use a compound inequality to find the range of scores he can make on his final exam in order to receive an A in the course. The final exam counts as three tests, and an A is received if the final course average is from 90 to 100. (*Hint:* The average of a list of numbers is their sum divided by the number of numbers in the list.)

69. The formula $C = 3.14d$ can be used to approximate the circumference of a circle given its diameter. Waldo Manufacturing manufactures and sells a certain washer with an outside circumference of 3 centimeters. The company has decided that a washer whose actual circumference is in the interval $2.9 \leq C \leq 3.1$ centimeters is acceptable. Use a compound inequality and find the corresponding interval for diameters of these washers. (Round to 3 decimal places.)

70. Bunnie Supplies manufactures plastic Easter eggs that open. The company has determined that if the circumference of the opening of each part of the egg is in the interval $118 \leq C \leq 122$ millimeters, the eggs will open and close comfortably. Use a compound inequality and find the corresponding interval for diameters of these openings. (Round to 2 decimal places.)

71. Twice a number increased by one is between negative five and seven. Find all such numbers.

72. Half a number decreased by four is between two and three. Find all such numbers.

73. A financial planner has a client with $15,000 to invest. If he invests $10,000 in a certificate of deposit paying 11% annual simple interest, at what rate does the remainder of the money need to be invested so that the two investments together yield at least $1600 in yearly interest?

74. Alex earns $600 per month plus 4% of all his sales over $1000. Find the minimum sales that will allow Alex to earn at least $3000 per month.

75. Ben Holladay bowled 146 and 201 in his first two games. What must he bowl in his third game to have an average of at least 180?

76. On an NBA team the two forwards measure 6′8″ and 6′6″ and the two guards measure 6′0″ and 5′9″ tall. How tall a center should they hire if they wish to have a starting team average height of at least 6′5″?

REVIEW EXERCISES

Evaluate the following. See Section 1.3.

77. $(2)^3$

78. $(3)^3$

79. $(1)^{12}$

80. 0^5

81. $\left(\dfrac{4}{7}\right)^2$

82. $\left(\dfrac{2}{3}\right)^3$

This broken line graph shows the enrollment of people (members) in a Health Maintenance Organization (HMO). The height of each dot corresponds to the number of members (in millions).

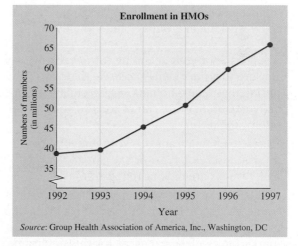

Source: Group Health Association of America, Inc., Washington, DC

83. How many people were enrolled in Health Maintenance Organizations in 1995?

84. How many people were enrolled in Health Maintenance Organizations in 1997?

85. Which year shows the greatest increase in number of members?

86. In what years were there over 50,000,000 members of HMOs?

A Look Ahead

Example

Solve $x(x - 6) > x^2 - 5x + 6$, and then graph the solution.

Solution:

$$x(x - 6) > x^2 - 5x + 6$$
$$x^2 - 6x > x^2 - 5x + 6$$
$$x^2 - 6x - x^2 > x^2 - 5x + 6 - x^2$$
$$-6x > -5x + 6$$
$$-x > 6$$
$$\frac{-x}{-1} < \frac{6}{-1}$$
$$x < -6$$

The solution set $\{x \mid x < -6\}$ is graphed as shown.

$$\xleftarrow{\qquad} \begin{array}{ccccccc} | & \oplus & | & | & | & | & | \\ -7 & -6 & -5 & -4 & -3 & -2 & -1 \end{array} \xrightarrow{\qquad}$$

Solve each inequality and then graph the solution set.

87. $x(x + 4) > x^2 - 2x + 6$

88. $x(x - 3) \geq x^2 - 5x - 8$

89. $x^2 + 6x - 10 < x(x - 10)$

90. $x^2 - 4x + 8 < x(x + 8)$

91. $x(2x - 3) \leq 2x^2 - 5x$

92. $x(4x + 1) < 4x^2 - 3x$

For additional Chapter Projects, visit the Real World Activities Website by going to http://www.prenhall.com/martin-gay.

2 CHAPTER PROJECT

Developing a Budget

Whether you are rolling in dough or pinching every penny, a budget can help put you in control of your personal finances. Listing your income lets you see where your money comes from, and by tracking your expenses you can analyze your spending practices. Putting these together in a budget helps you make informed decisions about what you spend.

In this project, you will have the opportunity to develop a personal budget. This project may be completed by working in groups or individually.

1. Keep track of your spending for a week or two. Be sure to write down *everything* that you spend during that time—no matter how small—from a daily newspaper or a can of soda to your tuition bill for the term.

2. Decide on a timeframe for your budget: a month, a term, a school year, or the entire year. How many weeks are in your timeframe?

3. Make a list of all of your income for your timeframe of choice. Be sure to include sources like grants or scholarships, portion of savings earmarked for college expenses for this timeframe, take-home pay from a job, gifts, financial support from family, and interest. Total your income from all sources.

4. Make a list of all of your expected expenses for your timeframe of choice. Be sure to include major expenses such as car payments, auto and health insurance, room and board or rent/mortgage payment, and tuition. Try to estimate more variable expenses like telephone and utilities, books and supplies, groceries and/or restaurant meals, entertainment, transportation or parking, personal care, clothing, dues and/or lab fees, etc. Don't forget to take into account small or irregular purchases. Use the expense record you made in Question 1 to help you gauge your levels of spending and identify spending categories. Total your expected expenses in all categories.

5. In the equation $I = E + wx$, I represents your income from Question 3, E represents your expected expenses from Question 4, w represents the number of weeks in your timeframe of choice (from Question 2), and x represents the weekly shortage or surplus in your budget. If x is positive, this is the extra amount you have in your budget each week to save or spend as "pocket money." If x is negative, this is the amount by which you fall short each week. To keep your budget balanced, you should try to reduce nonessential expenses by this amount each week.

Substitute for I, E, and w in the equation and solve for x. Interpret your result.

6. If you have a weekly shortage, which expenses could you try to reduce to balance your budget? If you have a weekly surplus, what would you do with this extra money?

CHAPTER 2 VOCABULARY CHECK

Fill in each blank with one of the words or phrases listed below.

like terms   numerical coefficient   linear equation in one variable

equivalent equations   formula   compound inequalities

linear inequality in one variable   percent

1. Terms with the same variables raised to exactly the same powers are called _____.
2. A _____ can be written in the form $ax + b = c$.
3. Equations that have the same solution are called _____.
4. Inequalities containing two inequality symbols are called _____.
5. An equation that describes a known relationship among quantities is called a _____.
6. The word _____ means per hundred.
7. A _____ can be written in the form $ax + b < c, (\text{or} >, \leq, \geq)$.
8. The _____ of a term is its numerical factor.

CHAPTER 2 HIGHLIGHTS

DEFINITIONS AND CONCEPTS	EXAMPLES

Section 2.1 Simplifying Algebraic Expressions

The **numerical coefficient** of a **term** is its numerical factor.

Term	Numerical Coefficient
$-7y$	-7
x	1
$\dfrac{1}{5}a^2b$	$\dfrac{1}{5}$

Terms with the same variables raised to exactly the same powers are **like terms**.

Like Terms	Unlike Terms
$12x, -x$	$3y, 3y^2$
$-2xy, 5yx$	$7a^2b, -2ab^2$

To combine like terms, add the numerical coefficients and multiply the result by the common variable factor.

$$9y + 3y = 12y$$
$$-4z^2 + 5z^2 - 6z^2 = -5z^2$$

To remove parentheses, apply the distributive property.

$$-4(x + 7) + 10(3x - 1)$$
$$= -4x - 28 + 30x - 10$$
$$= 26x - 38$$

Section 2.2 The Addition Property of Equality

A **linear equation in one variable** can be written in the form $ax + b = c$ where $a, b,$ and c are real numbers and $a \neq 0$.

Linear Equations
$$-3x + 7 = 2$$
$$3(x - 1) = -8(x + 5) + 4$$

Equivalent equations are equations that have the same solution.

$x - 7 = 10$ and $x = 17$ are equivalent equations.

Addition Property of Equality

Adding the same number to or subtracting the same number from both sides of an equation does not change its solution.

$$y + 9 = 3$$
$$y + 9 - 9 = 3 - 9$$
$$y = -6 \quad \text{(continued)}$$

DEFINITIONS AND CONCEPTS	EXAMPLES

Section 2.3 The Multiplication Property of Equality

Multiplication Property of Equality

Multiplying both sides or dividing both sides of an equation by the same nonzero number does not change its solution.

$$\frac{2}{3}a = 18$$

$$\frac{3}{2}\left(\frac{2}{3}a\right) = \frac{3}{2}(18)$$

$$a = 27$$

Section 2.4 Solving Linear Equations

To Solve Linear Equations

Solve: $\dfrac{5(-2x + 9)}{6} + 3 = \dfrac{1}{2}$

1. Clear the equation of fractions.

   1. $6 \cdot \dfrac{5(-2x + 9)}{6} + 6 \cdot 3 = 6 \cdot \dfrac{1}{2}$

   $$5(-2x + 9) + 18 = 3$$

2. Remove any grouping symbols such as parentheses.

   2. $-10x + 45 + 18 = 3$ Distributive property.

3. Simplify each side by combining like terms.

   3. $-10x + 63 = 3$ Combine like terms.

4. Write variable terms on one side and numbers on the other side using the addition property of equality.

   4. $-10x + 63 - 63 = 3 - 63$ Subtract 63.
   $$-10x = -60$$

5. Get the variable alone using the multiplication property of equality.

   5. $\dfrac{-10x}{-10} = \dfrac{-60}{-10}$ Divide by -10.
   $$x = 6$$

6. Check by substituting in the original equation.

   6. $\dfrac{5(-2x + 9)}{6} + 3 = \dfrac{1}{2}$

   $$\frac{5(-2 \cdot 6 + 9)}{6} + 3 \stackrel{?}{=} \frac{1}{2}$$

   $$\frac{5(-3)}{6} + 3 \stackrel{?}{=} \frac{1}{2}$$

   $$-\frac{5}{2} + \frac{6}{2} \stackrel{?}{=} \frac{1}{2}$$

   $$\frac{1}{2} = \frac{1}{2} \quad \text{True.}$$

Section 2.5 An Introduction to Problem Solving

Problem-Solving Steps

The height of the Hudson volcano in Chili is twice the height of the Kiska volcano in the Aleutian Islands. If the sum of their heights is 12,870 feet, find the height of each.

1. UNDERSTAND the problem.

1. Read and reread the problem. Guess a solution and check your guess.
   Let x be the height of the Kiska volcano. Then $2x$ is the height of the Hudson volcano.

x $2x$

Kiska   Hudson   *(continued)*

DEFINITIONS AND CONCEPTS	EXAMPLES

Section 2.5 An Introduction to Problem Solving

Problem-Solving Steps

2. TRANSLATE the problem.

2. In words:

height of Kiska	added to	height of Hudson	is	12,870

Translate: x $+$ $2x$ $=$ $12{,}870$

3. SOLVE.

3.

$$x + 2x = 12{,}870$$
$$3x = 12{,}870$$
$$x = 4290$$

4. INTERPRET the results.

4. *Check:* If x is 4290 then $2x$ is $2(4290)$ or 8580. Their sum is $4290 + 8580$ or 12,870, the required amount.

State: Kiska volcano is 4290 feet high and Hudson volcano is 8580 feet high.

Section 2.6 Formulas and Problem Solving

An equation that describes a known relationship among quantities is called a **formula**.

To solve a formula for a specified variable, use the same steps as for solving a linear equation. Treat the specified variable as the only variable of the equation.

Formulas

$A = lw$ (area of a rectangle)

$I = PRT$ (simple interest)

Solve $P = 2l + 2w$ for l.

$$P = 2l + 2w$$
$$P - 2w = 2l + 2w - 2w \quad \text{Subtract } 2w.$$
$$P - 2w = 2l$$
$$\frac{P - 2w}{2} = \frac{2l}{2} \quad \text{Divide by 2.}$$
$$\frac{P - 2w}{2} = l \quad \text{Simplify.}$$

If all values for the variables in a formula are known except for one, this unknown value may be found by substituting in the known values and solving.

If $d = 182$ miles and $r = 52$ miles per hour in the formula $d = r \cdot t$, find t.

$$d = r \cdot t$$
$$182 = 52 \cdot t \quad \text{Let } d = 182 \text{ and } r = 52.$$
$$3.5 = t$$

The time is 3.5 hours.

Section 2.7 Percent and Problem Solving

The word **percent** means **per hundred**. The symbol % is used to denote percent.

$$49\% = \frac{49}{100}, \qquad 1\% = \frac{1}{100}$$

To write a percent as a decimal, drop the percent symbol and move the decimal point two places to the left.

$$85.\% = 0.85, \qquad 3.5\% = 0.035$$

To write a decimal as a percent, move the decimal point two places to the right and attach the percent symbol, %.

$$0.35 = 35\%, \quad 10.1 = 1010\%$$

$$\frac{1}{8} = 0.125 = 12.5\%$$

(continued)

DEFINITIONS AND CONCEPTS	EXAMPLES

Section 2.7 Percent and Problem Solving

Use the same problem-solving steps to solve a problem containing percents.	32% of what number is 36.8?
1. UNDERSTAND.	1. Read and reread. Guess a solution and check. Let x = the unknown number.
2. TRANSLATE.	2. In words:

32%	of	what number	is	36.8

Translate: 32%   · x =   36.8

3. SOLVE.	3. *Solve* $32\% \cdot x = 36.8$

$$0.32x = 36.8$$
$$\frac{0.32x}{0.32} = \frac{36.8}{.32} \quad \text{Divide by .32.}$$
$$x = 115 \quad \text{Simplify.}$$

4. INTERPRET.	4. *Check:* 32% of 115 is $0.32(115) = 36.8$.
	State: The unknown number is 115.

Section 2.8 Further Problem Solving

	How many liters of a 20% acid solution must be mixed with a 50% acid solution in order to obtain 12 liters of a 30% solution?
1. UNDERSTAND.	1. Read and reread. Guess a solution and check. Let x = number of liters of 20% solution. Then $12 - x$ = number of liters of 50% solution.
2. TRANSLATE.	2.

	No. of liters ·	*Acid* Strength =	Amount of Acid
20% SOLUTION	x	20%	$0.20x$
50% SOLUTION	$12 - x$	50%	$0.50(12 - x)$
30% SOLUTION NEEDED	12	30%	$0.30(12)$

In words:

acid in 20% solution	+	acid in 50% solution	=	acid in 30% solution

Translate: $0.20x$ $+ 0.50(12 - x) =$ $0.30(12)$

3. SOLVE.	3. Solve $0.20x + 0.50(12 - x) = 0.30(12)$

$$0.20x + 6 - 0.50x = 3.6 \quad \text{Apply the distributive property.}$$
$$-0.30x + 6 = 3.6$$
$$-0.30x = -2.4 \quad \text{Subtract 6.}$$
$$x = 8 \quad \text{Divide by } -0.30.$$

4. INTERPRET.	4. *Check*, then *state*. If 8 liters of a 20% acid solution are mixed with $12 - 8$ or 4 liters of a 50% acid solution, the result is 12 liters of a 30% solution. (*continued*)

DEFINITIONS AND CONCEPTS	EXAMPLES

Section 2.9 Solving Linear Inequalities

A linear inequality in one variable is an inequality that can be written in one of the forms:

$$ax + b < c \qquad ax + b \leq c$$
$$ax + b > c \qquad ax + b \geq c$$

where a, b, and c are real numbers and a is not 0.

Linear Inequalities

$$2x + 3 < 6 \qquad\qquad 5(x - 6) \geq 10$$

$$\frac{x - 2}{5} > \frac{5x + 7}{2} \qquad \frac{-(x + 8)}{9} \leq \frac{-2x}{11}$$

Addition Property of Inequality

Adding the same number to or subtracting the same number from both sides of an inequality does not change the solutions.

$$y + 4 \leq -1$$
$$y + 4 - 4 \leq -1 - 4 \qquad \text{Subtract 4.}$$
$$y \leq -5$$

 (number line −6 −5 −4 −3 −2 −1 0 1 2)

Multiplication Property of Inequality

Multiplying or dividing both sides of an inequality by the same *positive number* does not change its solutions.

$$\frac{1}{3}x > -2$$
$$3\left(\frac{1}{3}x\right) > 3 \cdot -2 \qquad \text{Multiply by 3.}$$
$$x > -6$$

(number line −6 −4 −2 0 2)

Multiplying or dividing both sides of an inequality by the same **negative number and reversing the direction of the inequality sign** does not change its solutions.

$$-2x \leq 4$$
$$\frac{-2x}{-2} \geq \frac{4}{-2} \qquad \text{Divide by −2, reverse inequality sign.}$$
$$x \geq -2$$

(number line −3 −2 −1 0 1 2)

To Solve Linear Inequalities

1. Clear the equation of fractions.

2. Remove grouping symbols.

3. Simplify each side by combining like terms.

4. Write variable terms on one side and numbers on the other side using the addition property of inequality.

5. Get the variable alone using the multiplication property of inequality.

Solve: $3(x + 2) \leq -2 + 8$

1. No fractions to clear. $3(x + 2) \leq -2 + 8$

2. $3x + 6 \leq -2 + 8$ \qquad Distributive property.

3. $3x + 6 \leq 6$ \qquad Combine like terms.

4. $3x + 6 - 6 \leq 6 - 6$ \qquad Subtract 6.

   $3x \leq 0$

5. $\dfrac{3x}{3} \leq \dfrac{0}{3}$ \qquad Divide by 3.

   $x \leq 0$

(number line −2 −1 0 1 2)

Inequalities containing two inequality symbols are called **compound inequalities**.

Compound Inequalities

$$-2 < x < 6$$
$$5 \leq 3(x - 6) < \frac{20}{3}$$

(continued)

DEFINITIONS AND CONCEPTS	EXAMPLES

Section 2.9 Solving Linear Inequalities

To solve a compound inequality, isolate the variable in the middle part of the inequality. Perform the same operation to all three parts of the inequality: left, middle, right.

Solve: one only $-2 < 3x + 1 < 7$

$$-2 - 1 < 3x + 1 - 1 < 7 - 1 \quad \text{Subtract 1.}$$

$$-3 < 3x < 6$$

$$\frac{-3}{3} < \frac{3x}{3} < \frac{6}{3} \quad \text{Divide by 3.}$$

$$-1 < x < 2$$

CHAPTER 2 REVIEW

(2.1) *Simplify the following expressions.*

1. $5x - x + 2x$

2. $0.2z - 4.6x - 7.4z$

3. $\frac{1}{2}x + 3 + \frac{7}{2}x - 5$

4. $\frac{4}{5}y + 1 + \frac{6}{5}y + 2$

5. $2(n - 4) + n - 10$

6. $3(w + 2) - (12 - w)$

7. Subtract $7x - 2$ from $x + 5$.

8. Subtract $1.4y - 3$ from $y - 0.7$.

Write each of the following as algebraic expressions.

9. Three times a number decreased by 7

10. Twice the sum of a number and 2.8 added to 3 times a number

(2.2) *Solve the following.*

11. $8x + 4 = 9x$

12. $5y - 3 = 6y$

13. $3x - 5 = 4x + 1$

14. $2x - 6 = x - 6$

15. $4(x + 3) = 3(1 + x)$

16. $6(3 + n) = 5(n - 1)$

Use the addition property to fill in the blank so that the middle equation simplifies to the last equation.

17. $\quad\quad x - 5 = 3$
$x - 5 + \underline{\quad} = 3 + \underline{\quad}$
$\quad\quad x = 8$

18. $\quad\quad x + 9 = -2$
$x + 9 - \underline{\quad} = -2 - \underline{\quad}$
$\quad\quad x = -11$

Write each as an algebraic expression.

19. The sum of two numbers is 10. If one number is x, express the other number in terms of x.

20. Mandy is 5 inches taller than Melissa. If x inches represents the height of Mandy, express Melissa's height in terms of x.

△ **21.** If one angle measures $(x + 5)°$, express the measure of its supplement in terms of x.

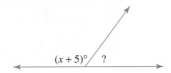

(2.3) *Solve each equation.*

22. $\frac{3}{4}x = -9$

23. $\frac{x}{6} = \frac{2}{3}$

24. $-3x + 1 = 19$

25. $5x + 25 = 20$

26. $5x + x = 9 + 4x - 1 + 6$

27. $-y + 4y = 7 - y - 3 - 8$

28. Express the sum of three even consecutive integers as an expression in x. Let x be the first even integer.

(2.4) *Solve the following.*

29. $\frac{2}{7}x - \frac{5}{7} = 1$

30. $\frac{5}{3}x + 4 = \frac{2}{3}x$

31. $-(5x + 1) = -7x + 3$

32. $-4(2x + 1) = -5x + 5$

33. $-6(2x - 5) = -3(9 + 4x)$

34. $3(8y - 1) = 6(5 + 4y)$

35. $\frac{3(2 - z)}{5} = z$

36. $\frac{4(n + 2)}{5} = -n$

37. $5(2n - 3) - 1 = 4(6 + 2n)$

38. $-2(4y - 3) + 4 = 3(5 - y)$

39. $9z - z + 1 = 6(z - 1) + 7$

40. $5t - 3 - t = 3(t + 4) - 15$

41. $-n + 10 = 2(3n - 5)$

42. $-9 - 5a = 3(6a - 1)$

43. $\frac{5(c + 1)}{6} = 2c - 3$

44. $\frac{2(8 - a)}{3} = 4 - 4a$

45. $200(70x - 3560) = -179(150x - 19,300)$

46. $1.72y - .04y = 0.42$

Solve.

47. The quotient of a number and 3 is the same as the difference of the number and two. Find the number.

48. Double the sum of a number and six is the opposite of the number. Find the number.

(2.5) Solve each of the following.

49. The height of the Eiffel Tower is 68 feet more than three times a side of its square base. If the sum of these two dimensions is 1380 feet, find the height of the Eiffel Tower.

50. A 12-foot board is to be divided into two pieces so that one piece is twice as long as the other. If x represents the length of the shorter piece, find the length of each piece.

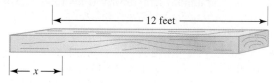

51. One area code in Ohio is 34 more than three times another area code used in Ohio. If the sum of these area codes is 1262, find the two area codes.

52. Find three consecutive even integers whose sum is negative 114.

(2.6) Substitute the given values into the given formulas and solve for the unknown variable.

△**53.** $P = 2l + 2w; P = 46, l = 14$

△**54.** $V = lwh; V = 192, l = 8, w = 6$

Solve each of the following for the indicated variable.

55. $y = mx + b$ for m

56. $r = vst - 9$ for s

57. $2y - 5x = 7$ for x

58. $3x - 6y = -2$ for y

△**59.** $C = \pi D$ for π

△**60.** $C = 2\pi r$ for π

△**61.** A swimming pool holds 900 cubic meters of water. If its length is 20 meters and its height is 3 meters, find its width.

62. The highest temperature on record in Rome, Italy, is 104° Fahrenheit. Convert this temperature to Celsius.

63. A charity 10K race is given annually to benefit a local hospice organization. How long will it take to run/walk a 10K race (10 kilometers or 10,000 meters) if your average pace is 125 **meters** per minute?

(2.7) Solve.

64. Find 12% of 250.   **65.** Find 110% of 85.

66. The number 9 is what percent of 45?

67. The number 59.5 is what percent of 85?

68. The number 137.5 is 125% of what number?

69. The number 768 is 60% of what number?

70. The state of Mississippi has the highest phoneless rate in the United States, 12.6% of households. If a city in Mississippi has 50,000 households, how many of these would you expect to be phoneless?

The graph below shows how business travelers relax when in their hotel rooms.

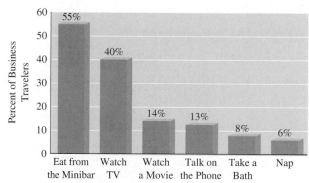

Source: USA Today, 1994.

71. What percent of business travelers surveyed relax by taking a nap?

72. What is the most popular way to relax according to the survey?

73. If a hotel in New York currently has 300 business travelers, how many might you expect to relax by watching TV?

74. Do the percents in the graph above have a sum of 100%? Why or why not?

75. The number of employees at Arnold's Box Manufacturers just decreased from 210 to 180. Find the percent decrease. Round to the nearest tenth of a percent.

(2.8) *Solve each of the following.*

76. A $50,000 retirement pension is to be invested into two accounts: a money market fund that pays 8.5% and a certificate of deposit that pays 10.5%. How much should be invested at each rate in order to provide a yearly interest income of $4550?

77. A pay phone is holding its maximum number of 500 coins consisting of nickels, dimes, and quarters. The number of quarters is twice the number of dimes. If the value of all the coins is $88.00, how many nickels were in the pay phone?

78. How long will it take an Amtrak passenger train to catch up to a freight train if their speeds are 60 and 45 miles per hour and the freight train had an hour and a half head start?

79. Florian Rousseau, from Italy, won a gold medal in cycling during the 1996 Summer Olympics. Suppose he rides a bicycle up a mountain trail at 8 miles per hour and down the same trail at 12 mph. Find the round-trip distance traveled if the total travel time was 5 hours.

(2.9) *Solve and graph the solution of each of the following inequalities.*

80. $x \leq -2$

81. $x > 0$

82. $-1 < x < 1$

83. $0.5 \leq y < 1.5$

84. $-2x \geq -20$

85. $-3x > 12$

86. $5x - 7 > 8x + 5$

87. $x + 4 \geq 6x - 16$

88. $2 \leq 3x - 4 < 6$

89. $-3 < 4x - 1 < 2$

90. $-2(x - 5) > 2(3x - 2)$ **91.** $4(2x - 5) \leq 5x - 1$

92. Tina earns $175 per week plus a 5% commission on all her sales. Find the minimum amount of sales to ensure that she earns at least $300 per week.

93. Ellen Catarella shot rounds of 76, 82, and 79 golfing. What must she shoot on her next round so that her average will be below 80?

CHAPTER 2 TEST

Simplify each of the following expressions.

1. $2y - 6 - y - 4$

2. $2.7x + 6.1 + 3.2x - 4.9$

3. $4(x - 2) - 3(2x - 6)$

4. $-5(y + 1) + 2(3 - 5y)$

Solve each of the following equations.

5. $-\dfrac{4}{5}x = 4$

6. $4(n - 5) = -(4 - 2n)$

7. $5y - 7 + y = -(y + 3y)$

8. $4z + 1 - z = 1 + z$

9. $\dfrac{2(x + 6)}{3} = x - 5$

10. $\dfrac{4(y - 1)}{5} = 2y + 3$

11. $\dfrac{1}{2} - x + \dfrac{3}{2} = x - 4$

12. $\dfrac{1}{3}(y + 3) = 4y$

13. $-0.3(x - 4) + x = 0.5(3 - x)$

14. $-4(a + 1) - 3a = -7(2a - 3)$

Solve each of the following applications.

15. A number increased by two-thirds of the number is 35. Find the number.

△ **16.** A gallon of water seal covers 200 square feet. How many gallons are needed to paint two coats of water seal on a deck that measures 20 feet by 35 feet?

35 feet   20 feet

17. Sedric Angell invested an amount of money in Amoxil stock that earned an annual 10% return, and then he invested twice the original amount in IBM stock that earned an annual 12% return. If his total return from both investments was $2890, find how much he invested in each stock.

18. Two trains leave Los Angeles simultaneously traveling on the same track in opposite directions at speeds of 50 and 64 miles per hour. How long will it take before they are 285 miles apart?

19. Find the value of x if $y = -14$, $m = -2$, and $b = -2$ in the formula $y = mx + b$.

Solve each of the following equations for the indicated variable.

△ **20.** $V = \pi r^2 h$ for h

21. $3x - 4y = 10$ for y

Solve and graph each of the following inequalities.

22. $3x - 5 > 7x + 3$

23. $x + 6 > 4x - 6$

24. $-2 < 3x + 1 < 8$

25. $0 < 4x - 7 < 9$

26. $\dfrac{2(5x + 1)}{3} > 2$

The following graph shows the source of income for charities.

81.3%
Individuals

6.7%
Bequests

7.3%
Foundations

4.7%
Corporations

27. What percent of charity income comes from individuals?

28. If the total annual income for charities is $126.2 billion, find the amount that comes from corporations.

29. Find the number of degrees in the Bequests sector.

Use the following double bar graph for Exercises 30–32.

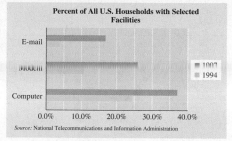

Percent of All U.S. Households with Selected Facilities

E-mail

Modem

Computer

0.0%   10.0%   20.0%   30.0%   40.0%

■ 1997
■ 1994

Source: National Telecommunications and Information Administration

30. What percent of U.S. households had e-mail in 1997?

31. How many more households (in percent) had computers in 1997 than in 1994?

32. In a city with 23,000 households, how many of these would you expect to have modems?

CHAPTER 2 CUMULATIVE REVIEW

1. Given the set $\left\{-2, 0, \frac{1}{4}, 112, -3, 11, \sqrt{2}\right\}$, list the numbers in this set that belong to the set of:
  a. Natural numbers
  b. Whole numbers
  c. Integers
  d. Rational numbers
  e. Irrational numbers
  f. Real numbers

2. Find the absolute value of each number.
  a. $|4|$ **b.** $|-5|$ **c.** $|0|$

3. Write each of the following numbers as a product of primes.
  a. 40   **b.** 63

4. Write $\frac{2}{5}$ as an equivalent fraction with a denominator of 20.

5. Simplify $3[4(5 + 2) - 10]$.

6. Decide whether 2 is a solution of $3x + 10 = 8x$.

Add.

7. $-1 + (-2)$ **8.** $-4 + 6$

9. Simplify each expression.
  a. $-(-10)$ **b.** $-\left(-\frac{1}{2}\right)$
  c. $-(-2x)$ **d.** $-|-6|$

10. Subtract.
  a. $5.3 - (-4.6)$ **b.** $-\frac{3}{10} - \frac{5}{10}$
  c. $-\frac{2}{3} - \left(-\frac{4}{5}\right)$

11. Find each unknown complementary or supplementary angle.
  a.   **b.**

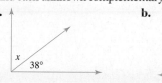

x
38°

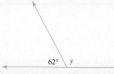

62° y

12. Find each product.
  a. $(-1.2)(0.05)$ **b.** $\frac{2}{3} \cdot -\frac{7}{10}$

13. Find each quotient.
  a. $\frac{-24}{-4}$ **b.** $\frac{-36}{3}$
  c. $\frac{2}{3} \div \left(-\frac{5}{4}\right)$

14. Use a commutative property to complete each statement.
  a. $x + 5 = $ _____
  b. $3 \cdot x = $ _____

15. Use the distributive property to write each sum as a product.
  a. $8 \cdot 2 + 8 \cdot x$
  b. $7s + 7t$

16. Subtract $4x - 2$ from $2x - 3$.

Solve.

17. $\frac{1}{2} = x - \frac{3}{4}$

18. $6(2a - 1) - (11a + 6) = 7$

19. $\frac{y}{7} = 20$

20. $4(2x - 3) + 7 = 3x + 5$

21. Twice the sum of a number and 4 is the same as four times the number decreased by 12. Find the number.

22. Solve $V = lwh$ for l.

23. Write each percent as a decimal.
  a. 35% **b.** 89.5%
  c. 150%

24. The number 63 is what percent of 72?

25. Solve $x + 4 \le -6$ for x. Graph the solution set.

No Small Part of the U.S. Economy

Did you know that approximately 99% of all businesses in this country are small businesses, as classified by the U.S. Small Business Administration? Small businesses employ 53% of the workforce and provide 64% of all new jobs in this country. It's also true that small businesses make 47% of all sales. It's easy to see that small businesses play no small role in the U.S. economy.

Have you ever thought of starting your own business? Most small business owners started their businesses because they wanted to be their own bosses, wanted more freedom in their work, and wanted financial independence. Starting and running a business takes motivation, discipline, and hard work. It also takes solid math skills for tasks such as raising money to start the business, managing the business' finances, and pricing and marketing products or services.

 For more information about starting, financing, or expanding your own business, visit the U.S. Small Business Administration Website by first going to www.prenhall.com/martin-gay.

In the Spotlight on Decision Making feature on page 169, you will have the chance to consider a new business opportunity for a small business owner.

GRAPHING

3.1   THE RECTANGULAR COORDINATE SYSTEM

3.2   GRAPHING LINEAR EQUATIONS

3.3   INTERCEPTS

3.4   SLOPE

3.5   GRAPHING LINEAR INEQUALITIES

In the previous chapter we learned to solve and graph the solutions of linear equations and inequalities in one variable. Now we define and present techniques for solving and graphing linear equations and inequalities in two variables.

3.1   THE RECTANGULAR COORDINATE SYSTEM

CD-ROM   SSM

SSG   Video

▶ **O B J E C T I V E S**

1. Define the rectangular coordinate system and plot ordered pairs of numbers.
2. Graph paired data to create a scattergram.
3. Determine whether an ordered pair is a solution of an equation in two variables.
4. Find the missing coordinate of an ordered pair solution, given one coordinate of the pair.

1

In Section 1.9, we learned how to read graphs. Example 4 in Section 1.9 presented the graph below showing the relationship between time spent smoking a cigarette and pulse rate. Notice in this graph that there are two numbers associated with each point of the graph. For example, we discussed earlier that 15 minutes after "lighting up," the pulse rate is 80 beats per minute. If we agree to write the time first and the pulse rate second, we can say there is a point on the graph corresponding to the **ordered pair** of numbers $(15, 80)$. A few more ordered pairs are listed alongside their corresponding points.

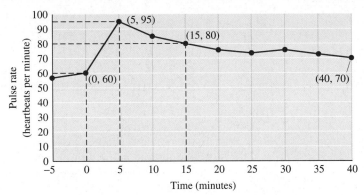

In general, we use this same ordered pair idea to describe the location of a point in a plane (such as a piece of paper). We start with a horizontal and a vertical axis. Each axis is a number line, and for the sake of consistency we construct our axes to intersect at the 0 coordinate of both. This point of intersection is called the **origin**. Notice that these two number lines or axes divide the plane into four regions called **quadrants**. The quadrants are usually numbered with Roman numerals as shown. The axes are not considered to be in any quadrant.

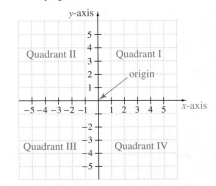

It is helpful to label axes, so we label the horizontal axis the **x-axis** and the vertical axis the **y-axis**. We call the system described above the **rectangular coordinate system**.

Just as with the pulse rate graph, we can then describe the locations of points by ordered pairs of numbers. We list the horizontal **x-axis** measurement first and the vertical **y-axis** measurement second.

To plot or graph the point corresponding to the ordered pair

$$(a, b)$$

we start at the origin. We then move a units left or right (right if a is positive, left if a is negative). From there, we move b units up or down (up if b is positive, down if b is negative). For example, to plot the point corresponding to the ordered pair $(3, 2)$, we start at the origin, move 3 units right, and from there move 2 units up. (See the figure below.) The x-value, 3, is also called the **x-coordinate** and the y-value, 2, is also called the **y-coordinate**. From now on, we will call the point with coordinates $(3, 2)$ simply the point $(3, 2)$. The point $(-2, 5)$ is graphed below also.

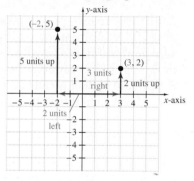

Does the order in which the coordinates are listed matter? Yes! Notice that the point corresponding to the ordered pair $(2, 3)$ is in a different location than the point corresponding to $(3, 2)$. These two ordered pairs of numbers describe two different points of the plane.

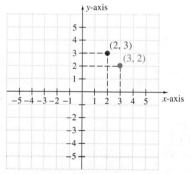

> **HELPFUL HINT**
> Don't forget that **each ordered pair corresponds to exactly one point in the plane and that each point in the plane corresponds to exactly one ordered pair.**

Example 1 On a single coordinate system, plot each ordered pair. State in which quadrant, if any, each point lies.

a. $(5, 3)$ **b.** $(-5, 3)$ **c.** $(-2, -4)$ **d.** $(1, -2)$

e. $(0, 0)$ **f.** $(0, 2)$ **g.** $(-5, 0)$ **h.** $\left(0, -5\frac{1}{2}\right)$

Solution   Point $(5, 3)$ lies in quadrant I.
Point $(-5, 3)$ lies in quadrant II.
Point $(-2, -4)$ lies in quadrant III.
Point $(1, -2)$ lies in quadrant IV.

Points $(0, 0)$, $(0, 2)$, $(-5, 0)$, and $\left(0, -5\frac{1}{2}\right)$

lie on axes, so they are not in any quadrant.

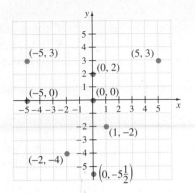

From Example 1, notice that the *y*-coordinate of any point on the *x*-axis is 0. For example, the point $(-5, 0)$ lies on the *x*-axis. Also, the *x*-coordinate of any point on the *y*-axis is 0. For example, the point $(0, 2)$ lies on the *y*-axis.

2   Data that can be represented as an ordered pair is called **paired data**. Many types of data collected from the real world are paired data. For instance, the annual measurement of a child's height can be written as an ordered pair of the form (year, height in inches) and is paired data. The graph of paired data as points in the rectangular coordinate system is called a **scatter diagram**. Scatter diagrams can be used to look for patterns and trends in paired data.

Example 2   The table gives the annual revenues for Wal-Mart Stores for the years shown. (*Source:* Wal-Mart Stores, Inc.)

Year	Wal-Mart Revenue (in billions of dollars)
1993	56
1994	68
1995	83
1996	95
1997	106
1998	121
1999	139

a. Write this paired data as a set of ordered pairs of the form (year, revenue in billions of dollars).
b. Create a scatter diagram of the paired data.
c. What trend in the paired data does the scatter diagram show?

Solution   a. The ordered pairs are $(1993, 56)$, $(1994, 68)$, $(1995, 83)$, $(1996, 95)$, $(1997, 106)$, $(1998, 121)$, and $(1999, 139)$.
b. We begin by plotting the ordered pairs. Because the *x*-coordinate in each ordered pair is a year, we label the *x*-axis "Year" and mark the horizontal axis with the years given. Then we label the *y*-axis or vertical axis "Wal-Mart Revenue (in billions of dollars)." It is convenient to mark the vertical axis in multiples of 20, starting with 0.

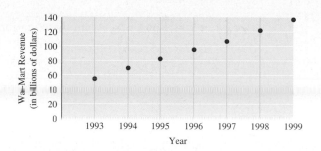

c. The scatter diagram shows that Wal-Mart revenue steadily increased over the years 1993–1999.

3   Let's see how we can use ordered pairs to record solutions of equations containing two variables. An equation in one variable such as $x + 1 = 5$ has one solution, which is 4: the number 4 is the value of the variable x that makes the equation true.

An equation in two variables, such as $2x + y = 8$, has solutions consisting of two values, one for x and one for y. For example, $x = 3$ and $y = 2$ is a solution of $2x + y = 8$ because, if x is replaced with 3 and y with 2, we get a true statement.

$$2x + y = 8$$
$$2(3) + 2 = 8$$
$$8 = 8 \qquad \text{True.}$$

The solution $x = 3$ and $y = 2$ can be written as $(3, 2)$, an **ordered pair** of numbers. The first number, 3, is the x-value and the second number, 2, is the y-value.

In general, an ordered pair is a **solution** of an equation in two variables if replacing the variables by the values of the ordered pair results in a true statement.

Example 3   Determine whether each ordered pair is a solution of the equation $x - 2y = 6$.

a. $(6, 0)$ **b.** $(0, 3)$ **c.** $\left(1, -\dfrac{5}{2}\right)$

Solution   **a.** Let $x = 6$ and $y = 0$ in the equation $x - 2y = 6$.

$$x - 2y = 6$$
$$6 - 2(0) = 6 \qquad \text{Replace } x \text{ with 6 and } y \text{ with 0.}$$
$$6 - 0 = 6 \qquad \text{Simplify.}$$
$$6 = 6 \qquad \text{True.}$$

$(6, 0)$ is a solution, since $6 = 6$ is a true statement.

b. Let $x = 0$ and $y = 3$.

$$x - 2y = 6$$
$$0 - 2(3) = 6 \qquad \text{Replace } x \text{ with 0 and } y \text{ with 3.}$$
$$0 - 6 = 6$$
$$-6 = 6 \qquad \text{False.}$$

$(0, 3)$ is *not* a solution, since $-6 = 6$ is a false statement.

c. Let $x = 1$ and $y = -\dfrac{5}{2}$ in the equation.

$$x - 2y = 6$$

$$1 - 2\left(-\frac{5}{2}\right) = 6 \qquad \text{Replace } x \text{ with 1 and } y \text{ with } -\frac{5}{2}.$$

$$1 + 5 = 6$$

$$6 = 6 \qquad \text{True.}$$

$\left(1, -\dfrac{5}{2}\right)$ is a solution, since $6 = 6$ is a true statement.

4   If one value of an ordered pair solution of an equation is known, the other value can be determined. To find the unknown value, replace one variable in the equation by its known value. Doing so results in an equation with just one variable that can be solved for the variable using the methods of Chapter 2.

Example 4   Complete the following ordered pair solutions for the equation $3x + y = 12$.

a. $(0, \ )$ **b.** $(\ , 6)$ **c.** $(-1, \ )$

Solution   **a.** In the ordered pair $(0, \ )$, the x-value is 0. Let $x = 0$ in the equation and solve for y.

$$3x + y = 12$$

$$3(0) + y = 12 \qquad \text{Replace } x \text{ with O.}$$

$$0 + y = 12$$

$$y = 12$$

The completed ordered pair is $(0, 12)$.

b. In the ordered pair $(\ , 6)$, the y-value is 6. Let $y = 6$ in the equation and solve for x.

$$3x + y = 12$$

$$3x + 6 = 12 \qquad \text{Replace } y \text{ with 6.}$$

$$3x = 6 \qquad \text{Subtract 6 from both sides.}$$

$$x = 2 \qquad \text{Divide both sides by 3.}$$

The ordered pair is $(2, 6)$.

c. In the ordered pair $(-1, \ )$, the x-value is -1. Let $x = -1$ in the equation and solve for y.

$$3x + y = 12$$

$$3(-1) + y = 12 \qquad \text{Replace } x \text{ with } -1.$$

$$-3 + y = 12$$

$$y = 15 \qquad \text{Add 3 to both sides.}$$

The ordered pair is $(-1, 15)$.

Solutions of equations in two variables can also be recorded in a **table of values**, as shown in the next example.

Example 5 Complete the table for the equation $y = 3x$.

	x	y
a.	-1	
b.		0
c.		-9

Solution **a.** Replace x with -1 in the equation and solve for y.

$$y = 3x$$
$$y = 3(-1) \quad \text{Let } x = -1.$$
$$y = -3$$

The ordered pair is $(-1, -3)$.

b. Replace y with 0 in the equation and solve for x.

$$y = 3x$$
$$0 = 3x \quad \text{Let } y = 0.$$
$$0 = x \quad \text{Divide both sides by 3.}$$

The completed ordered pair is $(0, 0)$.

c. Replace y with -9 in the equation and solve for x.

x	y
-1	-3
0	0
-3	-9

$$y = 3x$$
$$-9 = 3x \quad \text{Let } y = -9.$$
$$-3 = x \quad \text{Divide both sides by 3.}$$

The completed ordered pair is $(-3, -9)$. The completed table is shown to the left.

Example 6 Complete the table for the equation $y = 3$.

x	y
-2	
0	
-5	

Solution The equation $y = 3$ is the same as $0x + y = 3$. Replace x with -2 and we have $0(-2) + y = 3$ or $y = 3$. Notice that no matter what value we replace x by, y always equals 3. The completed table is:

x	y
-2	3
0	3
-5	3

By now, you have noticed that equations in two variables often have more than one solution. We discuss this more in the next section.

A table showing ordered pair solutions may be written vertically or horizontally as shown in the next example.

Example 7   **FINDING THE VALUE OF A COMPUTER**

A small business purchased a computer for $2000. The business predicts that the computer will be used for 5 years and the value in dollars y of the computer in x years is $y = -300x + 2000$. Complete the table.

x	0	1	2	3	4	5
y						

Solution   To find the value of y when x is 0, replace x with 0 in the equation. We use this same procedure to find y when x is 1 and when x is 2.

When x = 0,
$y = -300x + 2000$
$y = -300 \cdot 0 + 2000$
$y = 0 + 2000$
$y = 2000$

When x = 1,
$y = -300x + 2000$
$y = -300 \cdot 1 + 2000$
$y = -300 + 2000$
$y = 1700$

When x = 2,
$y = -300x + 2000$
$y = -300 \cdot 2 + 2000$
$y = -600 + 2000$
$y = 1400$

We have the ordered pairs (0, 2000), (1, 1700), and (2, 1400). This means that in 0 years the value of the computer is $2000, in 1 year the value of the computer is $1700, and in 2 years the value is $1400. Complete the table of values.

When x = 3,
$y = -300x + 2000$
$y = -300 \cdot 3 + 2000$
$y = -900 + 2000$
$y = 1100$

When x = 4,
$y = -300x + 2000$
$y = -300 \cdot 4 + 2000$
$y = -1200 + 2000$
$y = 800$

When x = 5,
$y = -300x + 2000$
$y = -300 \cdot 5 + 2000$
$y = -1500 + 2000$
$y = 500$

The completed table is

x	0	1	2	3	4	5
y	2000	1700	1400	1100	800	500

The ordered pair solutions recorded in the completed table for the example above are graphed on the following page. Notice that the graph gives a visual picture of the decrease in value of the computer.

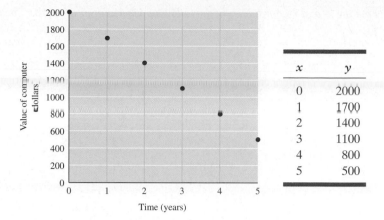

x	y
0	2000
1	1700
2	1400
3	1100
4	800
5	500

SPOTLIGHT ON DECISION MAKING

Suppose you own a small mail-order business. One of your employees brings you this graph showing the expected retail revenue from on-line shopping sales on the Internet. Should you consider expanding your business to include taking orders over the Internet? Explain your reasoning. What other factors would you want to consider?

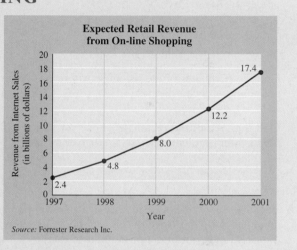

Source: Forrester Research Inc.

MENTAL MATH

Give two ordered pair solutions for each of the following linear equations.

1. $x + y = 10$ **2.** $x + y = 6$ **3.** $x = 3$ **4.** $y = -2$

Exercise Set 3.1

Plot the ordered pairs. State in which quadrant, if any, each point lies. See Example 1.

1. $(1, 5)$ **2.** $(-5, -2)$

3. $(-3, 0)$ **4.** $(0, -1)$

5. $(2, -4)$ **6.** $\left(-1, 4\frac{1}{2}\right)$

7. $\left(4\frac{3}{4}, 0\right)$ **8.** $\left(0, \frac{7}{8}\right)$

9. $(0, 0)$ **10.** $(5, 0)$

11. $(0, 4)$ **12.** $(-3, -3)$

Find the x- and y-coordinates of the following labeled points.

13.

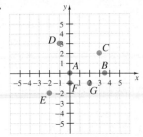

14.

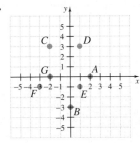

△ **15.** Find the perimeter of the rectangle whose vertices are the points with coordinates $(-1, 5)$, $(3, 5)$, $(3, -4)$, and $(-1, -4)$.

△ **16.** Find the area of the rectangle whose vertices are the points with coordinates $(5, 2)$, $(5, -6)$, $(0, -6)$, and $(0, 2)$.

Solve. See Example 2.

17. The table shows the average price of a gallon of regular unleaded gasoline (in dollars) for the years shown. (*Source:* Energy Information Administration)

Year	Price per Gallon of Unleaded Gasoline (in dollars)
1991	1.14
1992	1.13
1993	1.11
1994	1.11
1995	1.15
1996	1.23
1997	1.23
1998	1.06
1999	1.17

  a. Write each paired data as an ordered pair of the form (year, gasoline price).

  b. Draw a grid such as the one in Example 2 and create a scatter diagram of the paired data.

18. The table shows the number of regular-season NFL football games won by the winner of the Super Bowl for the years shown. (*Source:* National Football League)

Year	Regular-Season Games Won by Super Bowl Winner
1994	12
1995	13
1996	12
1997	13
1998	12
1999	14
2000	13

  a. Write each paired data as an ordered pair of the form (year, games won).

  b. Draw a grid such as the one in Example 2 and create a scatter diagram of the paired data.

↖ **19.** The table shows the average monthly mortgage payment made by Americans during the years shown. (*Source:* National Association of REALTORS®)

Year	Average Monthly Mortgage Payment (in dollars)
1994	578
1995	613
1996	654
1997	675
1998	717

  a. Write each paired data as an ordered pair of the form (year, mortgage payment).

  b. Draw a grid such as the one in Example 2 and create a scatter diagram of the paired data.

  c. What trend in the paired data does the scatter diagram show?

↖ **20.** The table shows the number of institutions of higher education in the United States for the years shown. (*Source:* U.S. Department of Education)

Year	Number of Institutions of Higher Learning
1909	951
1919	1041
1929	1409
1939	1708
1949	1851
1959	2008
1969	2525
1979	3152
1989	3535
1999	3800

a. Write each paired data as an ordered pair of the form (year, number of institutions).

b. Draw a grid such as the one in Example 2 and create a scatter diagram of the paired data.

c. What trend in the paired data does the scatter diagram show?

Determine whether each ordered pair is a solution of the given linear equation. See Example 3.

21. $2x + y = 7$; $(3, 1), (7, 0), (0, 7)$

22. $x - y = 6$; $(5, -1), (7, 1), (0, -6)$

23. $y = -5x$; $(-1, -5), (0, 0), (2, -10)$

24. $x = 2y$; $(0, 0), (2, 1), (-2, -1)$

25. $x = 5$; $(4, 5), (5, 4), (5, 0)$

26. $y = 2$; $(-2, 2), (2, 2), (0, 2)$

27. $x + 2y = 9$; $(5, 2), (0, 9)$

28. $3x + y = 8$; $(2, 3), (0, 8)$

29. $2x - y = 11$; $(3, -4), (9, 8)$

30. $x - 4y = 14$; $(2, -3), (14, 6)$

31. $x = \frac{1}{3} y$; $(0, 0), (3, 9)$ **32.** $y = -\frac{1}{2} x$; $(0, 0), (4, 2)$

33. $y = -2$; $(-2, -2), (5, -2)$

34. $x = 4$; $(4, 0), (4, 4)$

Complete each ordered pair so that it is a solution of the given linear equation. See Examples 4 through 6.

35. $x - 4y = 4$; $(, -2), (4,)$

36. $x - 5y = -1$; $(, -2), (4,)$

37. $3x + y = 9$; $(0,), (, 0)$

38. $x + 5y = 15$; $(0,), (, 0)$

39. $y = -7$; $(11,), (, -7)$

40. $x = \frac{1}{2}$; $(, 0), \left(\frac{1}{2}, \right)$

Complete the table of values for each given linear equation; then plot each solution. Use a single coordinate system for each equation. See Examples 4 through 6.

41. $x + 3y = 6$

x	y
0	
	0
	1

42. $2x + y = 4$

x	y
0	
	0
	2

43. $2x - y = 12$

x	y
0	
	-2
-3	

44. $-5x + y = 10$

x	y
	0
	5
2	

45. $2x + 7y = 5$

x	y
0	
	0
	1

46. $x - 6y = 3$

x	y
0	
1	
	-1

47. $x = 3$

x	y
	0
	-0.5
	$\frac{1}{4}$

48. $y = -1$

x	y
-2	
0	
-1	

49. $x = -5y$

x	y
	0
	1
10	

50. $y = -3x$

x	y
0	
-2	
	9

51. Discuss any similarities in the graphs of the ordered pair solutions for Exercises 41–50.

52. Explain why equations in two variables have more than one solution.

Solve. See Example 7.

53. The cost in dollars y of producing x computer desks is given by $y = 80x + 5000$.

  a. Complete the following table and graph the results.

x	100	200	300
y			

  b. Find the number of computer desks that can be produced for $8600. (*Hint:* Find x when $y = 8600$.)

54. The hourly wage y of an employee at a certain production company is given by $y = 0.25x + 9$ where x is the number of units produced in an hour.

  a. Complete the table and graph the results.

x	0	1	5	10
y				

  b. Find the number of units that must be produced each hour to earn an hourly wage of $12.25. (*Hint:* Find x when $y = 12.25$.)

55. The population density y of Minnesota (in people per square mile of land) from 1920 through 1990 is given by $y = 0.364x + 21.939$. In the equation, x represents the number of years after 1900. (*Source:* Based on data from the U.S. Bureau of the Census)

  a. Complete the table.

x	20	65	90
y			

  b. Find the year in which the population density was approximately 50 people per square mile. (*Hint:* Find x when $y = 50$ and round to the nearest whole number.)

56. The percentage y of recorded music sales that were in cassette format from 1991 through 1998 is given by $y = -6.09x + 55.99$. In the equation, x represents the number of years after 1991. (*Source:* Based on data from the Recording Industry Association of America)

  a. Complete the table.

x	1	3	5
y			

  b. Find the year in which approximately 31.6% of recorded music sales were cassettes. (*Hint:* Find x when $y = 31.6$ and round to the nearest whole number.)

The graph below shows the number of Target stores for each year. Use this graph to answer Exercises 57–60.

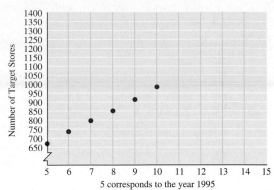

Source: USA Today 1/14/99

57. The ordered pair $(5, 670)$ is a point of the graph. Write a sentence describing the meaning of this ordered pair.

58. The ordered pair $(10, 984)$ is a point of the graph. Write a sentence describing the meaning of this ordered pair.

59. Estimate the increase in Target stores for years 6, 7, and 8.

60. Use a straightedge or ruler and this graph to predict the number of Target stores in the year 2005.

61. When is the graph of the ordered pair (a, b) the same as the graph of the ordered pair (b, a)?

62. In your own words, describe how to plot an ordered pair.

Determine the quadrant or quadrants in which the points described below lie.

63. The first coordinate is positive and the second coordinate is negative.

64. Both coordinates are negative.

65. The first coordinate is negative.

66. The second coordinate is positive.

REVIEW EXERCISES

Solve each equation for y. See Section 2.4.

67. $x + y = 5$

68. $x - y = 3$

69. $2x + 4y = 5$

70. $5x + 2y = 7$

71. $10x = -5y$

72. $4y = -8x$

73. $x - 3y = 6$

74. $2x - 9y = -20$

3.2 GRAPHING LINEAR EQUATIONS

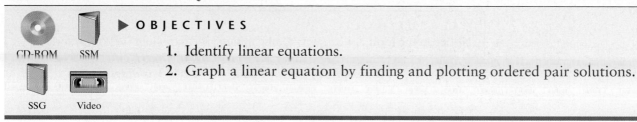

CD-ROM   SSM

SSG   Video

▶ O B J E C T I V E S

1. Identify linear equations.
2. Graph a linear equation by finding and plotting ordered pair solutions.

1

In the previous section, we found that equations in two variables may have more than one solution. For example, both $(6, 0)$ and $(2, -2)$ are solutions of the equation $x - 2y = 6$. In fact, this equation has an infinite number of solutions. Other solutions include $(0, -3)$, $(4, -1)$, and $(-2, -4)$. If we graph these solutions, notice that a pattern appears.

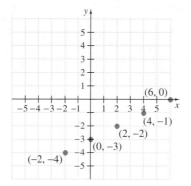

These solutions all appear to lie on the same line, which has been filled in below. It can be shown that every ordered pair solution of the equation corresponds to a point on this line, and every point on this line corresponds to an ordered pair solution. Thus, we say that this line is the graph of the equation $x - 2y = 6$.

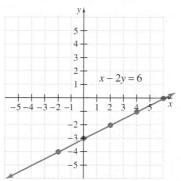

The equation $x - 2y = 6$ is called a **linear equation in two variables** and **the graph of every linear equation in two variables is a line.**

LINEAR EQUATION IN TWO VARIABLES

A linear equation in two variables is an equation that can be written in the form

$$Ax + By = C$$

where A, B, and C are real numbers and A and B are not both 0.

The form $Ax + By = C$ is called **standard form**.

HELPFUL HINT

Notice in the form $Ax + By = C$, the understood exponent on both x and y is 1.

Examples of Linear Equations in Two Variables

$$2x + y = 8 \qquad -2x = 7y \qquad y = \frac{1}{3}x + 2 \qquad y = 7$$

Before we graph linear equations in two variables, let's practice identifying these equations.

Example 1　Identify the linear equations in two variables.

　　　a. $x - 1.5y = -1.6$ 　　**b.** $y = -2x$ 　　**c.** $x + y^2 = 9$ 　　**d.** $x = 5$

Solution　**a.** This is a linear equation in two variables because it is written in the form
　　　$Ax + By = C$ with $A = 1$, $B = -1.5$, and $C = -1.6$.
　　b. This is a linear equation in two variables because it can be written in the form
　　　$Ax + By = C$.

$$y = -2x$$
$$2x + y = 0 \qquad \text{Add } 2x \text{ to both sides.}$$

　　c. This is *not* a linear equation in two variables because y is squared.
　　d. This is a linear equation in two variables because it can be written in the form
　　　$Ax + By = C$.

$$x = 5$$
$$x + 0y = 5 \qquad \text{Add } 0 \cdot y.$$

2

From geometry, we know that a straight line is determined by just two points. Graphing a linear equation in two variables, then, requires that we find just two of its infinitely many solutions. Once we do so, we plot the solution points and draw the line connecting the points. Usually, we find a third solution as well, as a check.

Example 2　Graph the linear equation $2x + y = 5$.

Solution　Find three ordered pair solutions of $2x + y = 5$. To do this, choose a value for one variable, x or y, and solve for the other variable. For example, let $x = 1$. Then $2x + y = 5$ becomes

$$2x + y = 5$$
$$2(\mathbf{1}) + y = 5 \qquad \text{Replace } x \text{ with 1.}$$
$$2 + y = 5 \qquad \text{Multiply.}$$
$$y = \mathbf{3} \qquad \text{Subtract 2 from both sides.}$$

Since $y = 3$ when $x = 1$, the ordered pair $(1, 3)$ is a solution of $2x + y = 5$. Next, let $x = 0$.

$$2x + y = 5$$
$$2(\mathbf{0}) + y = 5 \qquad \text{Replace } x \text{ with 0.}$$
$$0 + y = 5$$
$$y = \mathbf{5}$$

The ordered pair $(0, 5)$ is a second solution.

The two solutions found so far allow us to draw the straight line that is the graph of all solutions of $2x + y = 5$. However, we find a third ordered pair as a check. Let $y = -1$.

$$2x + y = 5$$
$$2x + (-1) = 5 \qquad \text{Replace } y \text{ with } -1.$$
$$2x - 1 = 5$$
$$2x = 6 \qquad \text{Add 1 to both sides.}$$
$$x = 3 \qquad \text{Divide both sides by 2.}$$

The third solution is $(3, -1)$. These three ordered pair solutions are listed in table form as shown. The graph of $2x + y = 5$ is the line through the three points.

x	y
1	3
0	5
3	-1

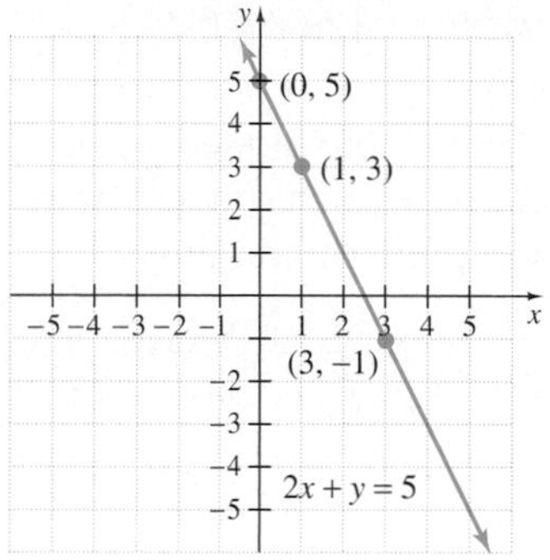

HELPFUL HINT
How does a third ordered pair solution above help us check our work? If the three ordered pair solutions do not lie along the same straight line when graphed, either a calculation is wrong or a mistake was made when graphing.

Example 3 Graph the linear equation $-5x + 3y = 15$.

Solution Find three ordered pair solutions of $-5x + 3y = 15$.

Let $x = 0$.

$$-5x + 3y = 15$$
$$-5 \cdot 0 + 3y = 15$$
$$0 + 3y = 15$$
$$3y = 15$$
$$y = 5$$

Let $y = 0$.

$$-5x + 3y = 15$$
$$-5x + 3 \cdot 0 = 15$$
$$-5x + 0 = 15$$
$$-5x = 15$$
$$x = -3$$

Let $x = -2$.

$$-5x + 3y = 15$$
$$-5(-2) + 3y = 15$$
$$10 + 3y = 15$$
$$3y = 5$$
$$y = \frac{5}{3}$$

The ordered pairs are $(0, 5)$, $(-3, 0)$, and $\left(-2, \frac{5}{3}\right)$. The graph of $-5x + 3y = 15$ is the line through the three points.

x	y
0	5
-3	0
-2	$\frac{5}{3} = 1\frac{2}{3}$

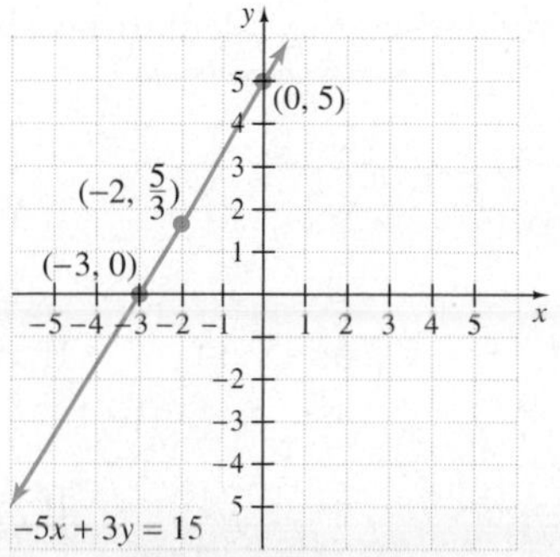

Example 4   Graph the linear equation $y = 3x$.

Solution   To graph this linear equation, we find three ordered pair solutions. Since this equation is solved for y, choose three x values.

x	y
2	6
0	0
−1	−3

If $x = 2$, $y = 3 \cdot 2 = 6$.

If $x = 0$, $y = 3 \cdot 0 = 0$.

If $x = -1$, $y = 3 \cdot -1 = -3$.

Next, graph the ordered pair solutions listed in the table above and draw a line through the plotted points. The line is the graph of $y = 3x$. Every point on the graph represents an ordered pair solution of the equation and every ordered pair solution is a point on this line.

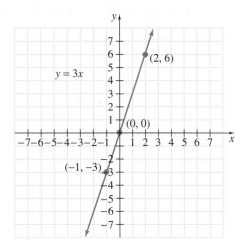

Example 5   Graph the linear equation $y = -\frac{1}{3}x$.

Solution   Find three ordered pair solutions, graph the solutions, and draw a line through the plotted solutions. To avoid fractions, choose x values that are multiples of 3 to substitute in the equation. When a multiple of 3 is multiplied by $-\frac{1}{3}$, the result is an integer. See the calculations shown to the right of the table below.

x	y
6	−2
0	0
−3	1

If $x = 6$, then $y = -\dfrac{1}{3} \cdot 6 = -2$.

If $x = 0$, then $y = -\dfrac{1}{3} \cdot 0 = 0$.

If $x = -3$, then $y = -\dfrac{1}{3} \cdot -3 = 1$.

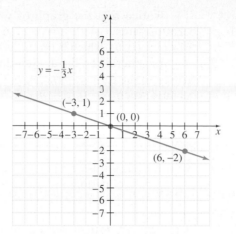

Let's compare the graphs in Examples 4 and 5. The graph of $y = 3x$ tilts upward (as we follow the line from left to right) and the graph of $y = -\frac{1}{3}x$ tilts downward (as we follow the line from left to right). Also notice that both lines go through the origin or that $(0, 0)$ is an ordered pair solution of both equations. In general, the graphs of $y = 3x$ and $y = -\frac{1}{3}x$ are of the form $y = mx$ where m is a constant. The graph of an equation in this form always goes through the origin $(0, 0)$ because when x is 0, $y = mx$ becomes $y = m \cdot 0 = 0$.

Example 6 Graph the linear equation $y = 3x + 6$ and compare this graph with the graph of $y = 3x$ in Example 4.

Solution Find ordered pair solutions, graph the solutions, and draw a line through the plotted solutions. We choose x values and substitute in the equation $y = 3x + 6$.

x	y
-3	-3
0	6
1	9

If $x = -3$, then $y = 3(-3) + 6 = -3$.
If $x = 0$, then $y = 3(0) + 6 = 6$.
If $x = 1$, then $y = 3(1) + 6 = 9$.

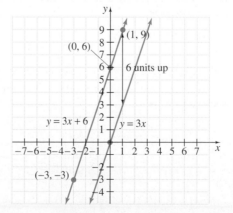

The most startling similarity is that both graphs appear to have the same upward tilt as we move from left to right. Also, the graph of $y = 3x$ crosses the y-axis at the origin, while the graph of $y = 3x + 6$ crosses the y-axis at 6. In fact, the graph of $y = 3x + 6$ is the same as the graph of $y = 3x$ moved vertically upward 6 units.

Notice that the graph of $y = 3x + 6$ crosses the y-axis at 6. This happens because when $x = 0$, $y = 3x + 6$ becomes $y = 3 \cdot 0 + 6 = 6$. The graph contains the point $(0, 6)$, which is on the y-axis.

In general, if a linear equation in two variables is solved for y, we say that it is written in the form $y = mx + b$. The graph of this equation contains the point $(0, b)$ because when $x = 0$, $y = mx + b$ is $y = m \cdot 0 + b = b$.

> The graph of $y = mx + b$ crosses the y-axis at b.

We will review this again in Section 7.1.

Linear equations are often used to model real data as seen in the next example.

Example 7   ESTIMATING THE NUMBER OF MEDICAL ASSISTANTS

One of the occupations expected to have the most growth in the next few years is medical assistant. The number of people y (in thousands) employed as medical assistants in the United States can be estimated by the linear equation $y = 11x + 217$, where x is the number of years after the year 1995. (*Source:* based on data from the Bureau of Labor Statistics)

Graph the equation and use the graph to predict the number of medical assistants in the year 2005.

Solution   To graph $y = 11x + 217$, choose x-values and substitute in the equation.

x	y
0	217
2	239
7	294

If $x = 0$, then $y = 11(0) + 217 = 217$.

If $x = 2$, then $y = 11(2) + 217 = 239$.

If $x = 7$, then $y = 11(7) + 217 = 294$.

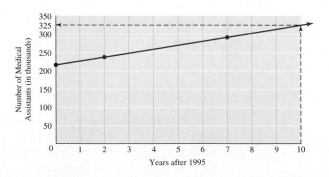

To use the graph to predict the number of medical assistants in the year 2005, we need to find the y-coordinate that corresponds to $x = 10$. (10 years after 1995 is the year 2005.) To do so, find 10 on the x-axis. Move vertically upward to the graphed line and then horizontally to the left. We approximate the number on the y-axis to be 325. Thus in the year 2005, we predict that there will be 325 thousand medical assistants. (The actual value, using 10 for x, is 327.)

GRAPHING CALCULATOR EXPLORATIONS

In this section, we begin an optional study of graphing calculators and graphing software packages for computers. These graphers use the same point plotting technique that was introduced in this section. The advantage of this graphing technology is, of course, that graphing calculators and computers can find and plot ordered pair solutions much faster than we can. Note, however, that the features described in these boxes may not be available on all graphing calculators.

The rectangular screen where a portion of the rectangular coordinate system is displayed is called a **window**. We call it a **standard window** for graphing when both the x- and y-axes show coordinates between -10 and 10. This information is often displayed in the window menu on a graphing calculator as

$$\text{Xmin} = -10$$
$$\text{Xmax} = 10$$
$$\text{Xscl} = 1 \qquad \textit{The scale on the x-axis is one unit per tick mark.}$$
$$\text{Ymin} = -10$$
$$\text{Ymax} = 10$$
$$\text{Yscl} = 1 \qquad \textit{The scale on the y-axis is one unit per tick mark.}$$

To use a graphing calculator to graph the equation $y = 2x + 3$, press the $\boxed{\text{Y=}}$ key and enter the keystrokes $\boxed{2}$ $\boxed{x}$ $\boxed{+}$ $\boxed{3}$. The top row should now read $Y_1 = 2x + 3$. Next press the $\boxed{\text{GRAPH}}$ key, and the display should look like this:

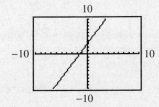

Use a standard window and graph the following linear equations. (Unless otherwise stated, use a standard window when graphing.)

1. $y = -3x + 7$

2. $y = -x + 5$

3. $y = 2.5x - 7.9$

4. $y = -1.3x + 5.2$

5. $y = -\dfrac{3}{10}x + \dfrac{32}{5}$

6. $y = \dfrac{2}{9}x - \dfrac{22}{3}$

Exercise Set 3.2

Determine whether each equation is a linear equation in two variables. See Example 1.

1. $-x = 3y + 10$

2. $y = x - 15$

3. $x = y$

4. $x = y^3$

5. $x^2 + 2y = 0$

6. $0.01x - 0.2y = 8.8$

7. $y = -1$

8. $x = 25$

Graph each linear equation. See Examples 2 through 5.

9. $x + y = 4$

10. $x + y = 7$

11. $x - y = -2$

12. $-x + y = 6$

13. $x - 2y = 6$

14. $-x + 5y = 5$

15. $y = 6x + 3$

16. $y = -2x + 7$

17. $x - 2y = -6$

18. $-x + 2y = 5$

19. $y = 6x$

20. $x = -2y$

21. $3y - 10 = 5x$

22. $-2x + 7 = 2y$

23. $x + 3y = 9$

24. $2x + y = -2$

25. $y - x = -1$

26. $x - y = 5$

27. $x = -3y$

28. $y = -x$

29. $5x - y = 10$

30. $7x - y = 2$

31. $y = \dfrac{1}{2}x + 2$

32. $y = -\dfrac{1}{5}x - 1$

Graph each pair of linear equations on the same set of axes. Discuss how the graphs are similar and how they are different. See Example 6.

33. $y = 5x; y = 5x + 4$

34. $y = 2x; y = 2x + 5$

35. $y = -2x; y = -2x - 3$

36. $y = x; y = x - 7$

37. $y = \dfrac{1}{2}x; y = \dfrac{1}{2}x + 2$

38. $y = -\dfrac{1}{4}x; y = -\dfrac{1}{4}x + 3$

The graph of $y = 5x$ is below as well as Figures a–d. For Exercises 39 through 42, match each equation with its graph. Hint: Recall that if an equation is written in the form $y = mx + b$, its graph crosses the y-axis at b.

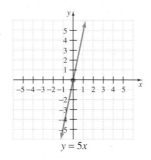

$y = 5x$

a.

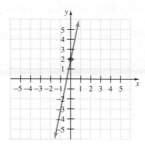

b.

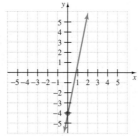

c.

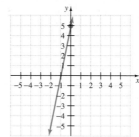

d.

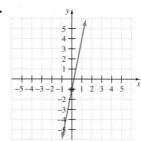

39. $y = 5x + 5$

40. $y = 5x - 4$

41. $y = 5x - 1$

42. $y = 5x + 2$

Solve. See Example 7.

43. The revenue y (in billions of dollars) for Home Depot stores during the years 1996 through 1999 is given by the equation $y = 5x + 15$, where x is the number of years after 1996. Graph this equation and use the graph to predict the revenue for Home Depot Stores in the year 2006. (*Source:* Based on data from The Home Depot Inc.)

44. The number of girls y (in thousands) participating in high school athletic programs in the United States during the years 1993 through 1997 is approximated by the equation $y = 120x + 2004$, where x is the number of years after 1993. Graph this equation and use the graph to predict the number of girls in U.S. high school athletic programs in the year 2010. (*Source:* Based on data from the National Federation of State High School Associations)

45. A fast-growing occupation in the next few years is elementary school teacher. The number of people y (in thousands) employed as elementary teachers in the United States can be estimated by the linear equation $y = 20x + 1539$, where x is the number of years after the year 2000. Graph this equation and use the graph to predict the number of elementary teachers in the year 2008. (*Source:* Based on data from the Bureau of Labor Statistics)

46. The number y of franchised new car dealerships in the United States during the years 1994 through 1997 is given by $y = -50x + 22,850$, where x is the number of years after 1994. Graph this equation and use the graph to predict the number of new car dealerships in the year 2010. (*Source:* Based on data from National Automobile Dealers Association)

Write each statement as an equation in two variables. Then graph the equation.

47. The y-value is 5 more than the x-value.

48. The y-value is twice the x-value.

49. Two times the x-value added to three times the y-value is 6.

50. Five times the x-value added to twice the y-value is -10.

51. Explain how to find ordered pair solutions of linear equations in two variables.

△ **52.** The perimeter of the trapezoid below is 22 centimeters. Write a linear equation in two variables for the perimeter. Find y if x is 3 cm.

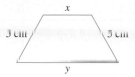

53. If (a, b) is an ordered pair solution of $x + y = 5$, is (b, a) also a solution? Explain why or why not.

△ **54.** The perimeter of the rectangle below is 50 miles. Write a linear equation in two variables for this perimeter. Use this equation to find x when y is 20.

55. Graph the nonlinear equation $y = x^2$ by completing the table shown. Plot the ordered pairs and connect them with a smooth curve.

x	y
0	
1	
-1	
2	
-2	

56. Graph the nonlinear equation $y = |x|$ by completing the table shown. Plot the ordered pairs and connect them. This curve is "V" shaped.

$$y = |x|$$

x	y
0	
1	
-1	
2	
-2	

REVIEW EXERCISES

△ **57.** The coordinates of three vertices of a rectangle are $(-2, 5)$, $(4, 5)$, and $(-2, -1)$. Find the coordinates of the fourth vertex. See Section 3.1.

△ **58.** The coordinates of two vertices of a square are $(-3, -1)$ and $(2, -1)$. Find the coordinates of two pairs of points possible for the third and fourth vertices. See Section 3.1.

Solve the following equations. See Section 2.4.

59. $3(x - 2) + 5x = 6x - 16$

60. $5 + 7(x + 1) = 12 + 10x$

61. $3x + \dfrac{2}{5} = \dfrac{1}{10}$

62. $\dfrac{1}{6} + 2x = \dfrac{2}{3}$

Complete each table. See Section 3.1.

63. $x - y = -3$

x	y
0	
	0

64. $y - x = 5$

x	y
0	
	0

65. $y = 2x$

x	y
0	
	0

66. $x = -3y$

x	y
0	
	0

3.3 INTERCEPTS

CD-ROM   SSM

SSG   Video

▶ **OBJECTIVES**

1. Identify intercepts of a graph.
2. Graph a linear equation by finding and plotting intercepts.
3. Identify and graph vertical and horizontal lines.

1

In this section, we graph linear equations in two variables by identifying intercepts. For example, the graph of $y = 4x - 8$ is shown below. Notice that this graph crosses the y-axis at the point $(0, -8)$. This point is called the **y-intercept**. Likewise, the graph crosses the x-axis at $(2, 0)$, and this point is called the **x-intercept**.

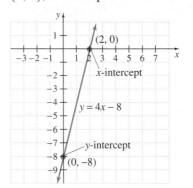

Example 1   Identify the x- and y-intercepts.

a.

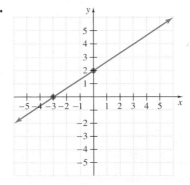

b.

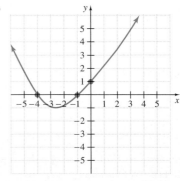

c.

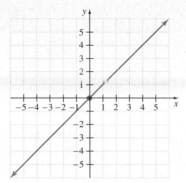

d.

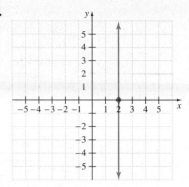

e.

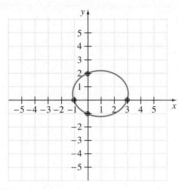

Solution   **a.** The graph crosses the x-axis at -3, so the x-intercept is $(-3, 0)$. The graph crosses the y-axis at 2, so the y-intercept is $(0, 2)$.
   b. The graph crosses the x-axis at -4 and -1, so the x-intercepts are $(-4, 0)$ and $(-1, 0)$. The graph crosses the y-axis at 1, so the y-intercept is $(0, 1)$.
   c. The x-intercept and the y-intercept are both $(0, 0)$.
   d. The x-intercept is $(2, 0)$. There is no y-intercept.
   e. The x-intercepts are $(-1, 0)$ and $(3, 0)$. The y-intercepts are $(0, -1)$, and $(0, 2)$.

2

Given an equation of a line, intercepts are usually easy to find since one coordinate is 0.
   One way to find the y-intercept of a line, given its equation, is to let $x = 0$, since a point on the y-axis has an x-coordinate of 0. To find the x-intercept of a line, let $y = 0$, since a point on the x-axis has a y-coordinate of 0.

FINDING *X*- AND *Y*-INTERCEPTS

To find the x-intercept, let $y = 0$ and solve for x.
To find the y-intercept, let $x = 0$ and solve for y.

Example 2   Graph $x - 3y = 6$ by finding and plotting intercepts.

Solution   Let $y = 0$ to find the x-intercept and let $x = 0$ to find the y-intercept.

$$
\begin{array}{ll}
\text{Let } y = 0 & \text{Let } x = 0 \\
x - 3y = 6 & x - 3y = 6 \\
x - 3(0) = 6 & 0 - 3y = 6 \\
x - 0 = 6 & -3y = 6 \\
x = 6 & y = -2
\end{array}
$$

The x-intercept is $(6, 0)$ and the y-intercept is $(0, -2)$. We find a third ordered pair solution to check our work. If we let $y = -1$, then $x = 3$. Plot the points $(6, 0)$, $(0, -2)$, and $(3, -1)$. The graph of $x - 3y = 6$ is the line drawn through these points, as shown.

x	y
6	0
0	-2
3	-1

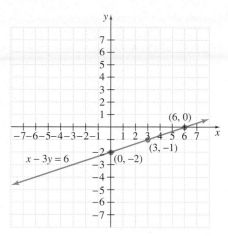

Example 3   Graph $x = -2y$ by plotting intercepts.

Solution   Let $y = 0$ to find the x-intercept and $x = 0$ to find the y-intercept.

$$\text{Let } y = 0 \qquad\qquad \text{Let } x = 0$$
$$x = -2y \qquad\qquad\qquad x = -2y$$
$$x = -2(0) \qquad\qquad\quad 0 = -2y$$
$$x = 0 \qquad\qquad\qquad\quad 0 = y$$

Both the x-intercept and y-intercept are $(0, 0)$. In other words, when $x = 0$, then $y = 0$, which gives the ordered pair $(0, 0)$. Also, when $y = 0$, then $x = 0$, which gives the same ordered pair $(0, 0)$. This happens when the graph passes through the origin. Since two points are needed to determine a line, we must find at least one more ordered pair that satisfies $x = -2y$. Let $y = -1$ to find a second ordered pair solution and let $y = 1$ as a checkpoint.

$$\text{Let } y = -1 \qquad\qquad \text{Let } y = 1$$
$$x = -2(-1) \qquad\qquad\quad x = -2(1)$$
$$x = 2 \qquad\qquad\qquad\quad x = -2$$

The ordered pairs are $(0, 0)$, $(2, -1)$, and $(-2, 1)$. Plot these points to graph $x = -2y$.

x	y
0	0
2	-1
-2	1

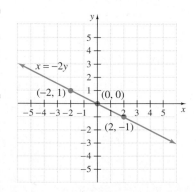

Example 4   Graph $4x = 3y - 9$.

Solution   Find the x- and y-intercepts, and then choose $x = 2$ to find a third checkpoint.

Let $y = 0$	Let $x = 0$	Let $x = 2$
$4x = 3(0) - 9$	$4 \cdot 0 = 3y - 9$	$4(2) = 3y - 9$
$4x = -9$	$9 = 3y$	$8 = 3y - 9$
Solve for x.	Solve for y.	Solve for y.
$x = -\dfrac{9}{4}$ or $-2\dfrac{1}{4}$	$3 = y$	$17 = 3y$
		$\dfrac{17}{3} = y$ or $y = 5\dfrac{2}{3}$

The ordered pairs are $\left(-2\dfrac{1}{4}, 0\right)$, $(0, 3)$, and $\left(2, 5\dfrac{2}{3}\right)$. The equation $4x = 3y - 9$ is graphed as follows.

x	y
$-2\frac{1}{4}$	0
0	3
2	$5\frac{2}{3}$

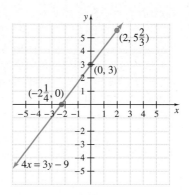

3   The equation $x = c$, where c is a real number constant, is a linear equation in two variables because it can be written in the form $x + 0y = c$. The graph of this equation is a vertical line as shown in the next example.

Example 5   Graph $x = 2$.

Solution   The equation $x = 2$ can be written as $x + 0y = 2$. For any y-value chosen, notice that x is 2. No other value for x satisfies $x + 0y = 2$. Any ordered pair whose x-coordinate is 2 is a solution of $x + 0y = 2$. We will use the ordered pair solutions $(2, 3)$, $(2, 0)$, and $(2, -3)$ to graph $x = 2$.

x	y
2	3
2	0
2	-3

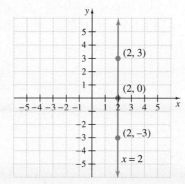

The graph is a vertical line with x-intercept $(2, 0)$. Note that this graph has no y-intercept because x is never 0.

VERTICAL LINES

The graph of $x = c$, where c is a real number, is a vertical line with x-intercept $(c, 0)$.

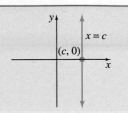

Example 6   Graph $y = -3$.

Solution   The equation $y = -3$ can be written as $0x + y = -3$. For any x-value chosen, y is -3. If we choose $4, 1$, and -2 as x-values, the ordered pair solutions are $(4, -3)$, $(1, -3)$, and $(-2, -3)$. Use these ordered pairs to graph $y = -3$. The graph is a horizontal line with y-intercept $(0, -3)$ and no x-intercept.

x	y
4	−3
1	−3
−2	−3

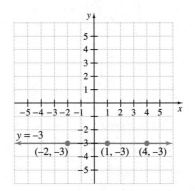

HORIZONTAL LINES

The graph of $y = c$, where c is a real number, is a horizontal line with y-intercept $(0, c)$.

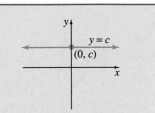

GRAPHING CALCULATOR EXPLORATIONS

You may have noticed that to use the $\boxed{Y=}$ key on a grapher to graph an equation, the equation must be solved for y. For example, to graph $2x + 3y = 7$, we solve this equation for y.

$$2x + 3y = 7$$
$$3y = -2x + 7 \qquad \text{Subtract } 2x \text{ from both sides.}$$
$$\frac{3y}{3} = -\frac{2x}{3} + \frac{7}{3} \qquad \text{Divide both sides by 3.}$$
$$y = -\frac{2}{3}x + \frac{7}{3} \qquad \text{Simplify.}$$

To graph $2x + 3y = 7$ or $y = -\dfrac{2}{3}x + \dfrac{7}{3}$, press the $\boxed{Y=}$ key and enter

$$Y_1 = -\frac{2}{3}x + \frac{7}{3}$$

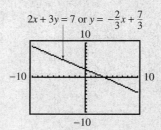

Graph each linear equation.

1. $x = 3.78y$ **2.** $-2.61y = x$ **3.** $3x + 7y = 21$

4. $-4x + 6y = 12$ **5.** $-2.2x + 6.8y = 15.5$ **6.** $5.9x - 0.8y = -10.4$

MENTAL MATH

Answer the following true or false.

1. The graph of $x = 2$ is a horizontal line.

2. All lines have an x-intercept *and* a y-intercept.

3. The graph of $y = 4x$ contains the point $(0, 0)$.

4. The graph of $x + y = 5$ has an x-intercept of $(5, 0)$ and a y-intercept of $(0, 5)$.

Exercise Set 3.3

Identify the intercepts. See Example 1.

1.

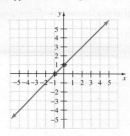

2.

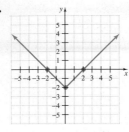

3.

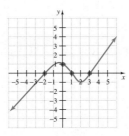

4.

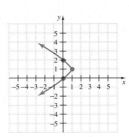

5.

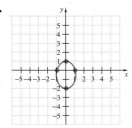

6.

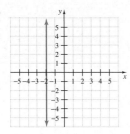

7. What is the greatest number of intercepts for a line?

8. What is the least number of intercepts for a line?

9. What is the least number of intercepts for a circle?

10. What is the greatest number of intercepts for a circle?

Graph each linear equation by finding x- and y-intercepts. See Examples 2 through 4.

11. $x - y = 3$ **12.** $x - y = -4$

13. $x = 5y$ **14.** $2x = y$

15. $-x + 2y = 6$ **16.** $x - 2y = -8$

17. $2x - 4y = 8$ **18.** $2x + 3y = 6$

Graph each linear equation. See Examples 5 and 6.

19. $x = -1$ **20.** $y = 5$

21. $y = 0$ **22.** $x = 0$

23. $y + 7 = 0$ **24.** $x - 2 = 0$

Graph each linear equation.

25. $x + 2y = 8$ **26.** $x - 3y = 3$

27. $x - 7 = 3y$ **28.** $y - 3x = 2$

29. $x = -3$ **30.** $y = 3$

31. $3x + 5y = 7$ **32.** $3x - 2y = 5$

33. $x = y$ **34.** $x = -y$

35. $x + 8y = 8$ **36.** $x - 3y = 9$

37. $5 = 6x - y$ **38.** $4 = x - 3y$

39. $-x + 10y = 11$ **40.** $-x + 9 = -y$

41. $y = 1$ **42.** $x = 1$

43. $x = 2y$ **44.** $y = -2x$

45. $x + 3 = 0$ **46.** $y - 6 = 0$

47. $x = 4y - \dfrac{1}{3}$ **48.** $y = -3x + \dfrac{3}{4}$

49. $2x + 3y = 6$ **50.** $4x + y = 5$

For Exercises 51 through 56, match each equation with its graph.

A.

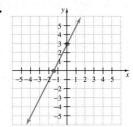

B.

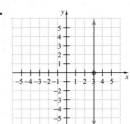

C.

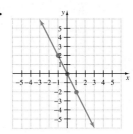

D.

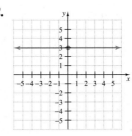

E.

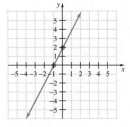

F.

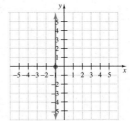

51. $y = 3$ **52.** $y = 2x + 2$

53. $x = -1$ **54.** $x = 3$

55. $y = 2x + 3$ **56.** $y = -2x$

57. The production supervisor at Alexandra's Office Products finds that it takes 3 hours to manufacture a particular office chair and 6 hours to manufacture an office desk. A total of 1200 hours is available to produce office chairs and desks of this style. The linear equation that models this situation is $3x + 6y = 1200$, where x represents the number of chairs produced and y the number of desks manufactured.

   a. Complete the ordered pair solution $(0, \ )$ of this equation. Describe the manufacturing situation that corresponds to this solution.

   b. Complete the ordered pair solution $(\ , 0)$ of this equation. Describe the manufacturing situation that corresponds to this solution.

   c. Use the ordered pairs found above and graph the equation $3x + 6y = 1200$.

   d. If 50 desks are manufactured, find the greatest number of chairs that they can make.

58. U.S. farm expenses y for livestock feed (in billions of dollars) can be modeled by the linear equation $y = 1.2x + 23.6$, where x represents the number of years after 1995. (*Source:* Based on data from the National Agricultural Statistics Service)

   a. Find the y-intercept of this equation.

   b. What does this y-intercept mean?

59. The number of music cassettes y (in millions) shipped to retailers in the United States can be modeled by the equation $y = -22.5x + 467.0$, where x represents the number of years after 1987. (*Source:* Based on data from the Recording Industry Association of America)

   a. Find the x-intercept of this equation. (Round to the nearest tenth.)

   b. What does this x-intercept mean?

60. Discuss whether a vertical line ever has a y-intercept.

61. Explain why it is a good idea to use three points to graph a linear equation.

62. Discuss whether a horizontal line ever has an x-intercept.

63. Explain how to find intercepts.

Two lines in the same plane that do not intersect are called **parallel lines.**

△ **64.** Draw a line parallel to the line $x = 5$ that intersects the x-axis at $(1, 0)$. What is the equation of this line?

△ **65.** Draw a line parallel to the line $y = -1$ that intersects the y-axis at $(0, -4)$. What is the equation of this line?

REVIEW EXERCISES

Simplify.

66. $\dfrac{-6 - 3}{2 - 8}$ **67.** $\dfrac{4 - 5}{-1 - 0}$

68. $\dfrac{-8 - (-2)}{-3 - (-2)}$ **69.** $\dfrac{12 - 3}{10 - 9}$

70. $\dfrac{0 - 6}{5 - 0}$ **71.** $\dfrac{2 - 2}{3 - 5}$

3.4 SLOPE

CD-ROM   SSM

SSG   Video

▶ **OBJECTIVES**

1. Find the slope of a line given two points of the line.
2. Find the slopes of horizontal and vertical lines.
3. Compare the shapes of parallel and perpendicular lines.
4. Solve problems of slope.

1 Thus far, much of this chapter has been devoted to graphing lines. You have probably noticed by now that a key feature of a line is its slant or steepness. In mathematics, the slant or steepness of a line is formally known as its **slope.** We measure the slope of a line by the ratio of vertical change to the corresponding horizontal change as we move along the line.

On the line on the next page, for example, suppose that we begin at the point $(1, 2)$ and move to the point $(4, 6)$. The vertical change is the change in *y*-coordinates: $6 - 2$ or 4 units. The corresponding horizontal change is the change in *x*-coordinates: $4 - 1 = 3$ units. The ratio of these changes is

$$\text{slope} = \frac{\text{change in } y \text{ (vertical change)}}{\text{change in } x \text{ (horizontal change)}} = \frac{4}{3}$$

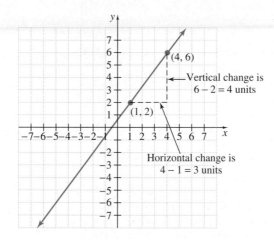

The slope of this line, then, is $\frac{4}{3}$: for every 4 units of change in *y*-coordinates, there is a corresponding change of 3 units in *x*-coordinates.

> ▼ **HELPFUL HINT**
> It makes no difference what two points of a line are chosen to find its slope. The slope of a line is the same everywhere on the line.

To find the slope of a line, then, choose two points of the line. Label the two *x*-coordinates of two points, x_1 and x_2 (read "*x* sub one" and "*x* sub two"), and label the corresponding *y*-coordinates y_1 and y_2.

The vertical change or **rise** between these points is the difference in the *y*-coordinates: $y_2 - y_1$. The horizontal change or **run** between the points is the difference of the *x*-coordinates: $x_2 - x_1$. The slope of the line is the ratio of $y_2 - y_1$ to $x_2 - x_1$, and we traditionally use the letter *m* to denote slope.

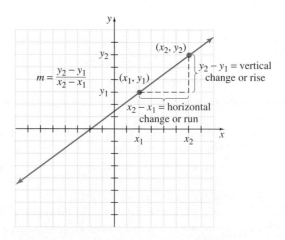

SLOPE OF A LINE

The slope m of the line containing the points (x_1, y_1) and (x_2, y_2) is given by

$$m = \frac{\text{rise}}{\text{run}} = \frac{\text{change in } y}{\text{change in } x} = \frac{y_2 - y_1}{x_2 - x_1}, \qquad \text{as long as } x_2 \neq x_1$$

Example 1   Find the slope of the line through $(-1, 5)$ and $(2, -3)$. Graph the line.

Solution   If we let (x_1, y_1) be $(-1, 5)$, then $x_1 = -1$ and $y_1 = 5$. Also, let (x_2, y_2) be point $(2, -3)$ so that $x_2 = 2$ and $y_2 = -3$. Then, by the definition of slope,

$$m = \frac{y_2 - y_1}{x_2 - x_1}$$

$$= \frac{-3 - 5}{2 - (-1)}$$

$$= \frac{-8}{3} = -\frac{8}{3}$$

The slope of the line is $-\dfrac{8}{3}$.

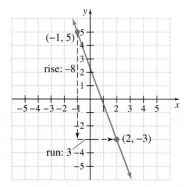

> **HELPFUL HINT**
> When finding slope, it makes no difference which point is identified as (x_1, y_1) and which is identified as (x_2, y_2). Just remember that whatever y-value is first in the numerator, its corresponding x-value is first in the denominator. Another way to calculate the slope in Example 1 is:
>
> $$m = \frac{y_2 - y_1}{x_2 - x_1} = \frac{5 - (-3)}{-1 - 2} = \frac{8}{-3} \quad \text{or} \quad -\frac{8}{3} \qquad \leftarrow \text{Same slope as found in Example 1.}$$

Example 2   Find the slope of the line through $(-1, -2)$ and $(2, 4)$. Graph the line.

Solution

$$m = \frac{\overset{\text{y-value}}{-2} - 4}{\underset{\text{corresponding x-value}}{-1} - 2} = \frac{-6}{-3} = 2$$

The slope is the same if we begin with the other *y*-value.

$$m = \frac{\overset{\text{y-value}}{4} - (-2)}{\underset{\text{corresponding x-value}}{2} - (-1)} = \frac{6}{3} = 2$$

The slope is 2.

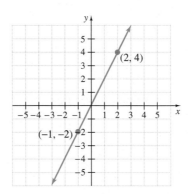

Notice that the slope of the line in Example 1 is negative, whereas the slope of the line in Example 2 is positive. Let your eye follow the line with negative slope from left to right and notice that the line "goes down." Following the line with positive slope from left to right, notice that the line "goes up." This is true in general.

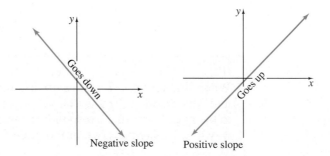

Negative slope   Positive slope

2   If a line tilts upward from left to right, its slope is positive. If a line tilts downward from left to right, its slope is negative. Let's now find the slopes of two special lines, horizontal and vertical lines.

Example 3   Find the slope of the line $y = -1$.

Solution   Recall that $y = -1$ is a horizontal line with y-intercept $(0, -1)$. To find the slope, find two ordered pair solutions of $y = -1$. Solutions of $y = -1$ must have a y-value of -1. Let's use points $(2, -1)$ and $(-3, -1)$, which are on the line.

$$m = \frac{y_2 - y_1}{x_2 - x_1} = \frac{-1 - (-1)}{-3 - 2} = \frac{0}{-5} = 0$$

The slope of the line $y = -1$ is 0.

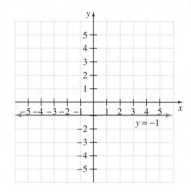

Any two points of a horizontal line will have the same y-values. This means that the y-values will always have a difference of 0 for all horizontal lines. Thus, **all horizontal lines have a slope 0.**

Example 4   Find the slope of the line $x = 5$.

Solution   Recall that the graph of $x = 5$ is a vertical line with x-intercept $(5, 0)$.

To find the slope, find two ordered pair solutions of $x = 5$. Solutions of $x = 5$ must have an x-value of 5. Let's use points $(5, 0)$ and $(5, 4)$, which are on the line.

$$m = \frac{y_2 - y_1}{x_2 - x_1} = \frac{4 - 0}{5 - 5} = \frac{4}{0}$$

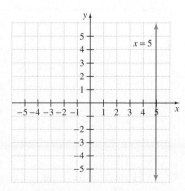

Since $\frac{4}{0}$ is undefined, we say the slope of the vertical line $x = 5$ is undefined.

Any two points of a vertical line will have the same x-values. This means that the x-values will always have a difference of 0 for all vertical lines. Thus **all vertical lines have undefined slope.**

> ▼ **HELPFUL HINT**
> Slope of 0 and undefined slope are not the same. Vertical lines have undefined slope or no slope, while horizontal lines have a slope of 0.

Here is a general review of slope.

SUMMARY OF SLOPE

Slope m of the line through (x_1, y_1) and (x_2, y_2) is given by the equation

$$m = \frac{y_2 - y_1}{x_2 - x_1}.$$

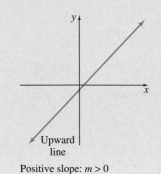

Upward line

Positive slope: $m > 0$

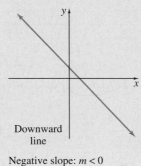

Downward line

Negative slope: $m < 0$

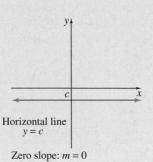

Horizontal line
$y = c$

Zero slope: $m = 0$

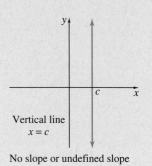

Vertical line
$x = c$

No slope or undefined slope

3 Two lines in the same plane are **parallel** if they do not intersect. Slopes of lines can help us determine whether lines are parallel. Parallel lines have the same steepness, so it follows that they have the same slope.

PARALLEL LINES

Nonvertical parallel lines have the same slope.

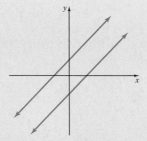

How do the slopes of perpendicular lines compare? Two lines that intersect at right angles are said to be **perpendicular**. The product of the slopes of two perpendicular lines is -1.

PERPENDICULAR LINES

If the product of the slopes of two lines is -1, the lines are perpendicular.

(Two nonvertical lines are perpendicular if the slopes of one is the negative reciprocal of the slope of the other.

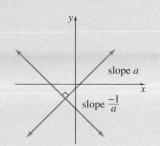

HELPFUL HINT
Here are examples of numbers that are negative (opposite) reciprocals.

Number	Negative Reciprocal	Their product is -1.
$\frac{2}{3}$	$-\frac{3}{2}$	$\frac{2}{3} \cdot -\frac{3}{2} = -\frac{6}{6} = -1$
-5 or $-\frac{5}{1}$	$\frac{1}{5}$	$-5 \cdot \frac{1}{5} = -\frac{5}{5} = -1$

△ **Example 5**   Is the line passing through the points $(-6, 0)$ and $(-2, 3)$ parallel to the line passing through the points $(5, 4)$ and $(9, 7)$?

Solution   To see if these lines are parallel, we find and compare slopes. The line passing through the points $(-6, 0)$ and $(-2, 3)$ has slope

$$m = \frac{3 - 0}{-2 - (-6)} = \frac{3}{4}$$

The line passing through the points $(5, 4)$ and $(9, 7)$ has slope

$$m = \frac{7 - 4}{9 - 5} = \frac{3}{4}$$

Since the slopes are the same, these lines are parallel.

△ **Example 6**   Find the slope of a line perpendicular to the line passing through the points $(-1, 7)$ and $(2, 2)$.

Solution   First, let's find the slope of the line through $(-1, 7)$ and $(2, 2)$. This line has slope

$$m = \frac{2 - 7}{2 - (-1)} = \frac{-5}{3}$$

The slope of every line perpendicular to the given line has a slope equal to the negative reciprocal of

$$-\frac{5}{3} \quad \text{or} \quad -\left(-\frac{3}{5}\right) = \frac{3}{5}$$

$$\underset{\text{negative}}{\uparrow} \quad \underset{\text{reciprocal}}{\uparrow}$$

The slope of a line perpendicular to the given line has slope $\frac{3}{5}$. ▬

4   There are many real-world applications of slope. For example, the pitch of a roof, used by builders and architects, is its slope. The pitch of the roof on the left is $\frac{7}{10}$ $\left(\frac{\text{rise}}{\text{run}}\right)$. This means that the roof rises vertically 7 feet for every horizontal 10 feet.

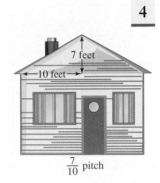

The grade of a road is its slope written as a percent. A 7% grade, as shown below, means that the road rises (or falls) 7 feet for every horizontal 100 feet. (Recall that $7\% = \frac{7}{100}$.)

$\frac{7}{10}$ pitch

$\frac{7}{100} = 7\%$ grade   7 feet   100 feet

Example 7   **FINDING THE GRADE OF A ROAD**

At one part of the road to the summit of Pikes Peak, the road rises 15 feet for a horizontal distance of 250 feet. Find the grade of the road.

Solution   Recall that the grade of a road is its slope written as a percent.

$$\text{grade} = \frac{\text{rise}}{\text{run}} = \frac{15}{250} = 0.06 = 6\%$$

15 feet   250 feet

The grade is 6%. ▬

Slope can also be interpreted as a rate of change. In other words, slope tells us how fast y is changing with respect to x.

Example 8   **FINDING THE SLOPE OF A LINE**

The following graph shows the cost y (in cents) of an in-state long-distance telephone call in Massachusetts where x is the length of the call in minutes. Find the slope of the line and attach the proper units for the rate of change.

Solution   Use (2, 48) and (5, 81) to calculate slope.

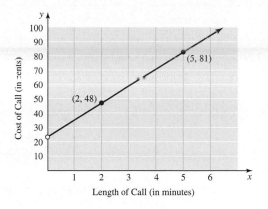

$$m = \frac{81 - 48}{5 - 2} = \frac{33}{3} = \frac{11}{1} \frac{\text{cents}}{\text{minute}}$$

This means that the rate of change of a phone call is 11 cents per 1 minute or the cost of the phone call increases 11 cents per minute.

SPOTLIGHT ON DECISION MAKING

Suppose you own a house that you are trying to sell. You want a real estate agency to handle the sale of your house. You begin by scanning your local newspaper to find likely candidates, and spot these ads:

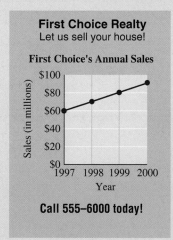

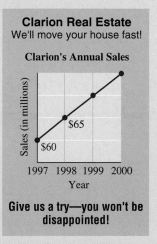

Which real estate agency would you want to check into first? Explain. What other factors would you want to consider?

GRAPHING CALCULATOR EXPLORATIONS

It is possible to use a grapher and sketch the graph of more than one equation on the same set of axes. This feature can be used to confirm our findings from Section 3.2 when we learned that the graph of an equation written in the form $y = mx + b$ has a y-intercept of b. For example, graph the equations $y = \frac{2}{5}x$, $y = \frac{2}{5}x + 7$, and $y = \frac{2}{5}x - 4$ on the same set of axes. To do so, press the $\boxed{Y=}$ key and enter the equations on the first three lines.

$$Y_1 = \left(\frac{2}{5}\right)x$$

$$Y_2 = \left(\frac{2}{5}\right)x + 7$$

$$Y_3 = \left(\frac{2}{5}\right)x - 4$$

The screen should look like:

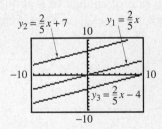

Notice that all three graphs appear to have the same positive slope. The graph of $y = \frac{2}{5}x + 7$ is the graph of $y = \frac{2}{5}x$ moved 7 units upward with a y-intercept of 7. Also, the graph of $y = \frac{2}{5}x - 4$ is the graph of $y = \frac{2}{5}x$ moved 4 units downward with a y-intercept of -4.

Graph the equations on the same set of axes. Describe the similarities and differences in their graphs.

1. $y = 3.8x$, $y = 3.8x - 3$, $y = 3.8x + 9$ **2.** $y = -4.9x$, $y = -4.9x + 1$, $y = -4.9x + 8$

3. $y = \frac{1}{4}x$; $y = \frac{1}{4}x + 5$, $y = \frac{1}{4}x - 8$ **4.** $y = -\frac{3}{4}x$, $y = -\frac{3}{4}x - 5$, $y = -\frac{3}{4}x + 6$

MENTAL MATH

Decide whether a line with the given slope is upward, downward, horizontal, or vertical.

1. $m = \frac{7}{6}$ **2.** $m = -3$ **3.** $m = 0$ **4.** m is undefined.

Exercise Set 3.4

Find the slope of each line if it exists. See Example 1.

1.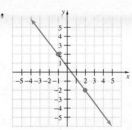

2.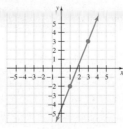

3.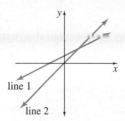

4.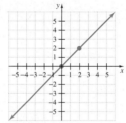

5.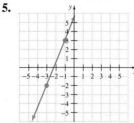

6.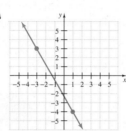

Find the slope of the line that goes through the given points. See Examples 1 and 2.

7. $(0, 0)$ and $(7, 8)$

8. $(-1, 5)$ and $(0, 0)$

9. $(-1, 5)$ and $(6, -2)$

10. $(-1, 9)$ and $(-3, 4)$

11. $(1, 4)$ and $(5, 3)$

12. $(3, 1)$ and $(2, 6)$

13. $(-4, 3)$ and $(-4, 5)$

14. $(6, -6)$ and $(6, 2)$

15. $(-2, 8)$ and $(1, 6)$

16. $(4, -3)$ and $(2, 2)$

17. $(1, 0)$ and $(1, 1)$

18. $(0, 13)$ and $(-4, 13)$

19. $(5, 1)$ and $(-2, 1)$

20. $(5, 4)$ and $(0, 5)$

For each graph, determine which line has the greater slope.

21.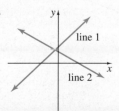

22.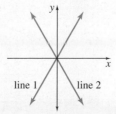

23.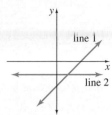

24.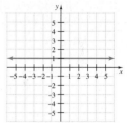

Match each line with its slope.

A. $m = 0$ **B.** no slope **C.** $m = 3$

D. $m = 1$ **E.** $m = -\dfrac{1}{2}$ **F.** $m = -\dfrac{3}{4}$

25.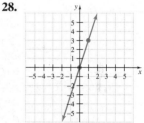

26.

27.

28.

29.

30.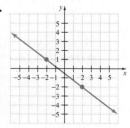

Find the slope of each line. See Examples 3 and 4.

31. $x = 6$

32. $y = 4$

33. $y = -4$

34. $x = 2$

35. $x = -3$

36. $x = -11$

37. $y = 0$

38. $x = 0$

△ *Find the slope of the line* ***a.*** *parallel and* ***b.*** *perpendicular to the line through each pair of points. See Examples 5 and 6.*

39. $(-3, -3)$ and $(0, 0)$ **40.** $(6, -2)$ and $(1, 4)$

41. $(-8, -4)$ and $(3, 5)$ **42.** $(6, -1)$ and $(-4, -10)$

△ *Determine whether the lines through each pair of points are parallel, perpendicular, or neither. See Examples 5 and 6.*

43. $(0, 6)$ and $(-2, 0)$ **44.** $(1, -1)$ and $(-1, -11)$
    $(0, 5)$ and $(1, 8)$ $(-2, -8)$ and $(0, 2)$

45. $(2, 6)$ and $(-2, 8)$ **46.** $(-1, 7)$ and $(1, 10)$
    $(0, 3)$ and $(1, 5)$ $(0, 3)$ and $(1, 5)$

47. $(3, 6)$ and $(7, 8)$ **48.** $(4, 2)$ and $(6, 6)$
    $(0, 6)$ and $(2, 7)$ $(0, -2)$ and $(1, 0)$

49. $(2, -3)$ and $(6, -5)$ **50.** $(-1, -5)$ and $(4, 4)$
    $(5, -2)$ and $(-3, -4)$ $(10, 8)$ and $(-7, 4)$

51. $(-4, -3)$ and $(-1, 0)$ **52.** $(2, -2)$ and $(4, -8)$
    $(4, -4)$ and $(0, 0)$ $(0, 7)$ and $(3, 8)$

△ *Solve. See Examples 5 and 6.*

53. Find the slope of a line parallel to the line passing through $(-7, -5)$ and $(-2, -6)$.

54. Find the slope of a line parallel to the line passing through the origin and $(-2, 10)$.

55. Find the slope of a line perpendicular to the line passing through the origin and $(1, -3)$.

56. Find the slope of the line perpendicular to the line passing through $(-1, 3)$ and $(2, -8)$.

57. Find the slope of a line parallel to the line passing through $(3, 3)$ and $(-3, -3)$.

58. Find the slope of a line parallel to the line passing through $(0, 3)$ and $(6, 0)$.

The pitch of a roof is its slope. Find the pitch of each roof shown. See Example 7.

59.

60.

The grade of a road is its slope written as a percent. Find the grade of each road shown. See Example 6.

61.

62.

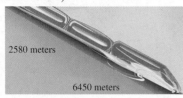

63. One of Japan's superconducting "bullet" trains is researched and tested at the Yamanashi Maglev Test Line near Otsuki City. The steepest section of the track has a rise of 2580 for a horizontal distance of 6450 meters. What is the grade for this section of track? (*Source:* Japan Railways Central Co.)

64. The steepest street is Baldwin Street in Dunedin, New Zealand. It has a maximum rise of 10 meters for a horizontal distance of 12.66 meters. Find the grade for this section of road. Round to the nearest whole percent. (*Source: The Guinness Book of Records*)

65. Professional plumbers suggest that a sewer pipe should rise 0.25 inch for every horizontal foot. Find the recommended slope for a sewer pipe. Round to the nearest hundredth.

66. According to federal regulations, a wheel chair ramp should rise no more than 1 foot for a horizontal distance of 12 feet. Write the slope as a grade. Round to the nearest tenth of a percent.

Find the slope of each line and write the slope as a rate of change. Don't forget to attach the proper units. See Example 8.

67. This graph approximates the number of U.S. internet users y (in millions) for year x.

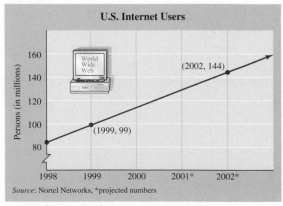

68. This graph approximates the total number of cosmetic surgeons for year x.

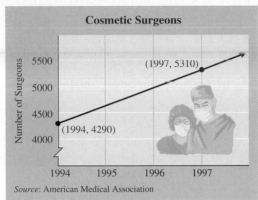

Cosmetic Surgeons

Number of Surgeons

(1997, 5310)

(1994, 4290)

1994 1995 1996 1997

Source: American Medical Association

69. The graph below shows the total cost y (in dollars) of owning and operating a compact car where x is the number of miles driven.

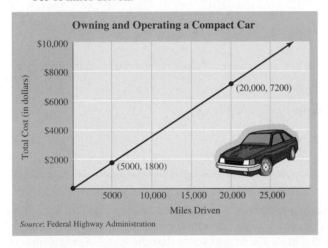

Owning and Operating a Compact Car

Total Cost (in dollars)

(20,000, 7200)

(5000, 1800)

5000 10,000 15,000 20,000 25,000

Miles Driven

Source: Federal Highway Administration

70. The graph below shows the total cost y (in dollars) of owning and operating a full-size pickup truck, where x is the number of miles driven.

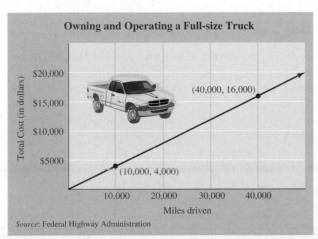

Owning and Operating a Full-size Truck

Total Cost (in dollars)

(40,000, 16,000)

(10,000, 4,000)

10,000 20,000 30,000 40,000

Miles driven

Source: Federal Highway Administration

The following line graph shows average miles per gallon of gasoline for cars, including taxis, during the years shown. Use the graph to answer Exercises 71–78.

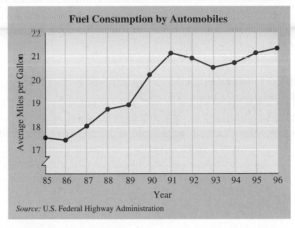

Fuel Consumption by Automobiles

Average Miles per Gallon

85 86 87 88 89 90 91 92 93 94 95 96

Year

Source: U.S. Federal Highway Administration

71. In 1988, cars averaged how many miles per gallon of fuel?

72. In 1996, cars averaged how many miles per gallon of fuel?

73. Find the increase of average miles per gallon for automobiles for the years 1985 to 1996.

74. Do you notice any trends from this graph?

75. What line segment has the greatest slope?

76. What line segment has the least slope?

77. What year showed the greatest increase in miles per gallon?

78. What year showed the least increase in miles per gallon?

79. Find x so that the pitch of the roof is $\frac{1}{3}$.

x

←18 feet→

80. Find x so that the pitch of the roof is $\frac{2}{5}$.

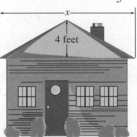

x

4 feet

△ **81.** Show that a triangle with vertices at the points $(2, 5)$, $(-4, 4)$, and $(-3, 0)$ is a right triangle.

△ **82.** Show that the quadrilateral with vertices $(1, 3)$, $(2, 1)$, $(-4, 0)$, and $(-3, -2)$ is a parallelogram.

Find the slope of the line through the given points.

83. $(2.1, 6.7)$ and $(-8.3, 9.3)$

84. $(-3.8, 1.2)$ and $(-2.2, 4.5)$

85. $(2.3, 0.2)$ and $(7.9, 5.1)$

86. $(14.3, -10.1)$ and $(9.8, -2.9)$

87. The graph of $y = -\frac{1}{3}x + 2$ has a slope of $-\frac{1}{3}$. The graph of $y = -2x + 2$ has a slope of -2. The graph of $y = -4x + 2$ has a slope of -4. Graph all three equations on a single coordinate system. As the absolute value of the slope becomes larger, how does the steepness of the line change?

REVIEW EXERCISES

Solve each linear inequality. See Section 2.9.

88. $-3x \le -9$

89. $-x > -16$

90. $\dfrac{x - 6}{2} < 3$

91. $\dfrac{2x + 1}{3} \ge -1$

Graph each linear equation in two variables. See Sections 3.2 and 3.3.

92. $x - y = 6$

93. $x = -2y$

94. $y = 3x$

95. $5x + 3y = 15$

96. $x = -2$

97. $y = 5$

3.5 GRAPHING LINEAR INEQUALITIES

CD-ROM   SSM

SSG   Video

▶ **OBJECTIVES**

1. Graph a linear inequality in two variables.

1 Recall that a linear equation in two variables is an equation that can be written in the form $Ax + By = C$ where A, B, and C are real numbers and A and B are not both 0. The definition of a linear inequality is the same except that the equal sign is replaced with an inequality sign.

A **linear inequality in two variables** is an inequality that can be written in one of the forms:

$$Ax + By < C \qquad Ax + By \le C$$
$$Ax + By > C \qquad Ax + By \ge C$$

where A, B, and C are real numbers and A and B are not both 0. Just as for linear equations in x and y, an ordered pair is a **solution** of an inequality in x and y if replacing the variables by coordinates of the ordered pair results in a true statement.

To graph a linear inequality in two variables, we will begin by graphing a related equation. For example, to graph $x - y \le 1$, we begin by graphing $x - y = 1$.

The linear equation $x - y = 1$ is graphed next. Recall that all points on the line correspond to ordered pairs that satisfy the equation $x - y = 1$. It can be shown that all the points above the line $x - y = 1$ have coordinates that satisfy the inequality $x - y < 1$. Similarly, all points below the line have coordinates that satisfy the inequality $x - y > 1$. To see this, a few points have been selected on one side of the line. These points all satisfy $x - y < 1$ as shown in the table on the next page.

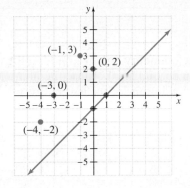

Point	$x - y < 1$	
Check $(0, 2)$	$0 - 2 < 1$	
	$-2 < 1$	Truc
Check $(-3, 0)$	$-3 - 0 < 1$	
	$-3 < 1$	True
Check $(-4, -2)$	$-4 - (-2) < 1$	
	$-2 < 1$	True
Check $(-1, 3)$	$-1 - 3 < 1$	
	$-4 < 1$	True

The graph of $x - y < 1$ is in blue below, and the graph of $x - y > 1$ is in red.

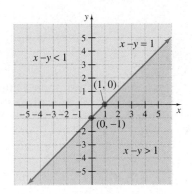

The region above the line and the region below the line are called **half-planes**. Every line divides the plane (similar to a sheet of paper extending indefinitely in all directions) into two half-planes; the line is called the **boundary**.

How do we graph $x - y \leq 1$? Recall that the inequality $x - y \leq 1$ means

$$x - y = 1 \quad \text{or} \quad x - y < 1$$

Thus, the graph of $x - y \leq 1$ is the half-plane $x - y < 1$ along with the boundary line $x - y = 1$.

GRAPHING A LINEAR INEQUALITY IN TWO VARIABLES

Step 1. Graph the boundary line found by replacing the inequality sign with an equal sign. If the inequality sign is > or <, graph a dashed boundary line (indicating that the points on the line are not solutions of the inequality). If the inequality sign is ≥ or ≤, graph a solid boundary line (indicating that the points on the line are solutions of the inequality).

Step 2. Choose a point, not on the boundary line, as a test point. Substitute the coordinates of this test point into the original inequality.

Step 3. If a true statement is obtained in Step 2, shade the half-plane that contains the test point. If a false statement is obtained, shade the half-plane that does not contain the test point.

Example 1   Graph: $x + y < 7$

Solution   First we graph the boundary line by graphing the equation $x + y = 7$. We graph this boundary as a dashed line because the inequality sign is $<$, and thus the points on the line are not solutions of the inequality $x + y < 7$.

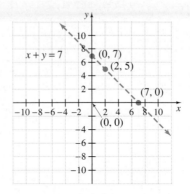

Next, choose a test point, being careful not to choose a point on the boundary line. We choose $(0, 0)$. Substitute the coordinates of $(0, 0)$ into $x + y < 7$.

$$x + y < 7 \qquad \text{Original inequality.}$$
$$0 + 0 \overset{?}{<} 7 \qquad \text{Replace } x \text{ with } 0 \text{ and } y \text{ with } 0.$$
$$0 < 7 \qquad \text{True.}$$

Since the result is a true statement, $(0, 0)$ is a solution of $x + y < 7$, and every point in the same half-plane as $(0, 0)$ is also a solution. To indicate this, shade the entire half-plane containing $(0, 0)$, as shown.

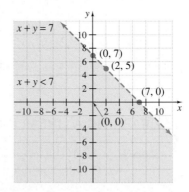

Example 2   Graph $2x - y \geq 3$.

Solution   Graph the boundary line by graphing $2x - y = 3$. Draw this line as a solid line since the inequality sign is $\geq$, and thus the points on the line are solutions of $2x - y \geq 3$. Once again, $(0, 0)$ is a convenient test point since it is not on the boundary line.

Substitute 0 for x and 0 for y into the **original inequality**.

$$2x - y \geq 3$$
$$2(0) - 0 \overset{?}{\geq} 3 \qquad \text{Let } x = 0 \text{ and } y = 0.$$
$$0 \geq 3 \qquad \text{False.}$$

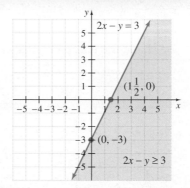

Since the statement is false, no point in the half-plane containing $(0, 0)$ is a solution. Shade the half-plane that does not contain $(0, 0)$. Every point in the shaded half-plane and every point on the boundary line satisfies $2x - y \geq 3$.   ■

> ▼
> **HELPFUL HINT**
> When graphing an inequality, make sure the test point is substituted into the **original inequality**. For Example 2, we substituted the test point $(0, 0)$ into the **original inequality** $2x - y \geq 3$, *not* $2x - y = 3$.

Example 3   Graph: $x > 2y$

Solution   We find the boundary line by graphing $x = 2y$. The boundary line is a dashed line since the inequality symbol is $>$. We cannot use $(0, 0)$ as a test point because it is a point on the boundary line. We choose instead $(0, 2)$.

$$x > 2y$$
$$0 \overset{?}{>} 2(2) \qquad \text{Let } x = 0 \text{ and } y = 2.$$
$$0 > 4 \qquad \text{False.}$$

Since the statement is false, we shade the half-plane that does not contain the test point $(0, 2)$, as shown.

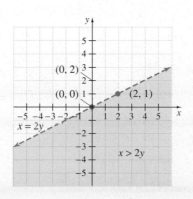

Example 4   Graph: $5x + 4y \leq 20$

Solution   We graph the solid boundary line $5x + 4y = 20$ and choose $(0, 0)$ as the test point.

$$5x + 4y \leq 20$$
$$5(0) + 4(0) \overset{?}{\leq} 20 \qquad \text{Let } x = 0 \text{ and } y = 0.$$
$$0 \leq 20 \qquad \text{True.}$$

We shade the half-plane that contains $(0, 0)$, as shown.

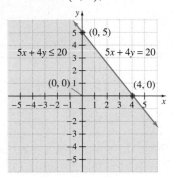

Example 5   Graph: $y > 3$

Solution   We graph the dashed boundary line $y = 3$ and choose $(0, 0)$ as the test point. (Recall that the graph of $y = 3$ is a horizontal line with y-intercept 3.)

$$y > 3$$
$$0 \overset{?}{>} 3 \qquad \text{Let } y = 0.$$
$$0 > 3 \qquad \text{False.}$$

We shade the half-plane that does not contain $(0, 0)$, as shown.

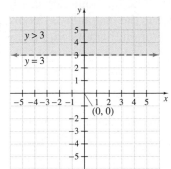

MENTAL MATH

State whether the graph of each inequality includes its corresponding boundary line.

1. $y \geq x + 4$ **2.** $x - y > -7$ **3.** $y \geq x$ **4.** $x > 0$

Decide whether $(0, 0)$ is a solution of each given inequality.

5. $x + y > -5$ **6.** $2x + 3y < 10$ **7.** $x - y \leq -1$ **8.** $\dfrac{2}{3}x + \dfrac{5}{6}y > 4$

Exercise Set 3.5

Determine which ordered pairs given are solutions of the linear inequality in two variables. See Example 1.

1. $x - y > 3$; $(2, -1)$, $(5, 1)$
2. $y - x < -2$; $(2, 1)$, $(5, -1)$
3. $3x - 5y \leq -4$; $(-1, -1)$, $(4, 0)$
4. $2x + y \geq 10$; $(-1, -4)$, $(5, 0)$
5. $x < -y$; $(0, 2)$, $(-5, 1)$
6. $y > 3x$; $(0, 0)$, $(-1, -4)$

Graph each inequality. See Examples 2 through 6.

7. $x + y \leq 1$
8. $x + y \geq -2$
9. $2x + y > -4$
10. $x + 3y \leq 3$
11. $x + 6y \leq -6$
12. $7x + y > -14$
13. $2x + 5y > -10$
14. $5x + 2y \leq 10$
15. $x + 2y \leq 3$
16. $2x + 3y > -5$
17. $2x + 7y > 5$
18. $3x + 5y \leq -2$
19. $x - 2y \geq 3$
20. $4x + y \leq 2$
21. $5x + y < 3$
22. $x + 2y > -7$
23. $4x + y < 8$
24. $9x + 2y \geq -9$
25. $y \geq 2x$
26. $x < 5y$
27. $x \geq 0$
28. $y \leq 0$
29. $y \leq -3$
30. $x > -\dfrac{2}{3}$
31. $2x - 7y > 0$
32. $5x + 2y \leq 0$
33. $3x - 7y \geq 0$
34. $-2x - 9y > 0$
35. $x > y$
36. $x \leq -y$
37. $x - y \leq 6$
38. $x - y > 10$
39. $-\dfrac{1}{4}y + \dfrac{1}{3}x > 1$
40. $\dfrac{1}{2}x - \dfrac{1}{3}y \leq -1$
41. $-x < 0.4y$
42. $0.3x \geq 0.1y$

Match each inequality with its graph.

a. $x > 2$ **b.** $y < 2$ **c.** $y < 2x$
d. $y \leq -3x$ **e.** $2x + 3y < 6$ **f.** $3x + 2y > 6$

43.

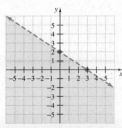

44.

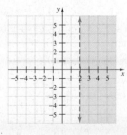

45.

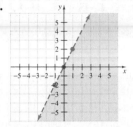

46.

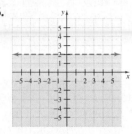

47.

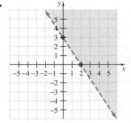

48.

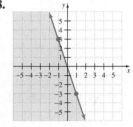

49. Write an inequality whose solutions are all pairs of numbers x and y whose sum is at least 13. Graph the inequality.

50. Write an inequality whose solutions are all the pairs of numbers x and y whose sum is at most -4. Graph the inequality.

51. Explain why a point on the boundary line should not be chosen as the test point.

52. Describe the graph of a linear inequality.

REVIEW EXERCISES

Evaluate. See Section 1.3.

53. 2^3
54. 3^4
55. $(-2)^5$
56. -2^5
57. $3 \cdot 4^2$
58. $4 \cdot 3^3$

Evaluate each expression for the given replacement value. See Section 1.7.

59. x^2 if x is -5
60. x^3 if x is -5
61. $2x^3$ if x is -1
62. $3x^2$ if x is -1

3

For additional Chapter Projects, visit the Real World Activities
Website by going to http://www.prenhall.com/martin-gay.

CHAPTER PROJECT

Financial Analysis

Investment analysts investigate a company's sales, net profit, debt, and assets to decide whether investing in it is a wise choice. One way to analyze this data is to graph it and look for trends over time. Another way is to find algebraically the rate at which the data change over time.

The table below gives the net profits in millions of dollars for the leading U.S. businesses in the pharmaceutical industry for the years 1997 and 1998. In this project, you will analyze the performances of these companies and, based on this information alone, make an investment recommendation. This project may be completed by working in groups or individually.

PHARMACEUTICAL INDUSTRY NET PROFITS (IN MILLIONS OF DOLLARS)

Company	1997	1998
Abbott Laboratories	2094.5	2333.2
American Home Products Corp.	2043.1	2474.3
Bristol Myers Squibb Co.	3205.0	3141.0
Eli Lilly & Co.	−385.1	2097.9
Johnson & Johnson	3303.0	3059.0
Merck & Co. Inc.	4614.1	5248.2
Pfizer Inc.	2213.0	3351.0
Pharmacia & Upjohn Inc.	323.0	691.0
Schering Plough Corp.	1444.0	1756.0
Warner Lambert Co.	869.5	1254.0

(*Source:* Disclosure Incorporated)

1. Scan the table. Did any of the companies have a loss during the years shown? If so, which company and when? What does this mean?

2. Write the data for each company as two ordered pairs of the form (year, net profit). Assuming that the trends in net profit are linear, use graph paper to graph the line represented by the ordered pairs for each company. Describe the trend shown by each graph.

3. Find the slope of the line for each company.

4. Which of the lines, if any, have positive slopes? What does that mean in this context? Which of the lines, if any, have negative slopes? What does that mean in this context?

5. Of these pharmaceutical companies, which one(s) would you recommend as an investment choice? Why?

6. Do you think it is wise to make a decision after looking at only 2 years of net profits? What other factors do you think should be taken into consideration when making an investment choice?

7. (Optional) Use financial magazines, company annual reports, or online investing information to find net profit information for two different years for two to four companies in the same industry. Analyze the net profits and make an investment recommendation.

CHAPTER 3 VOCABULARY CHECK

Fill in each blank with one of the words listed below.

y-axis	x-axis	solution	linear	standard
x-intercept	y-intercept	y	x	slope

1. An ordered pair is a _____ of an equation in two variables if replacing the variables by the coordinates of the ordered pair results in a true statement.

2. The vertical number line in the rectangular coordinate system is called the _____.

3. A _____ equation can be written in the form $Ax + By = C$.

4. A(n) _____ is a point of the graph where the graph crosses the x-axis.

5. The form $Ax + By = C$ is called _____ form.

6. A(n) _____ is a point of the graph where the graph crosses the y-axis.

7. To find an x-intercept of a graph, let ___ = 0.
8. The horizontal number line in the rectangular coordinate system is called the _____.
9. To find a y-intercept of a graph, let ___ = 0.
10. The _____ of a line measures the steepness or tilt of a line.

CHAPTER 3 HIGHLIGHTS

DEFINITIONS AND CONCEPTS	EXAMPLES

Section 3.1 The Rectangular Coordinate System

The **rectangular coordinate system** consists of a plane and a vertical and a horizontal number line intersecting at their 0 coordinate. The vertical number line is called the **y-axis** and the horizontal number line is called the **x-axis**. The point of intersection of the axes is called the **origin**.

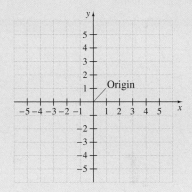

To **plot** or **graph** an ordered pair means to find its corresponding point on a rectangular coordinate system.

To plot or graph an ordered pair such as $(3, -2)$, start at the origin. Move 3 units to the right and from there, 2 units down.

To plot or graph $(-3, 4)$ start at the origin. Move 3 units to the left and from there, 4 units up.

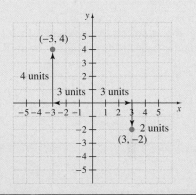

An ordered pair is a **solution** of an equation in two variables if replacing the variables by the coordinates of the ordered pair results in a true statement.

Determine whether $(-1, 5)$ is a solution of $2x + 3y = 13$.

$$2x + 3y = 13$$
$$2(-1) + 3 \cdot 5 = 13 \quad \text{Let } x = -1, y = 5$$
$$-2 + 15 = 13$$
$$13 = 13 \quad \text{True.}$$

If one coordinate of an ordered pair solution is known, the other value can be determined by substitution.

Complete the ordered pair solution $(0, \ )$ for the equation $x - 6y = 12$.

$$x - 6y = 12$$
$$0 - 6y = 12 \quad \text{Let } x = 0.$$
$$\frac{-6y}{-6} = \frac{12}{-6} \quad \text{Divide by } -6.$$
$$y = -2$$

The ordered pair solution is $(0, -2)$.

(continued)

DEFINITIONS AND CONCEPTS	EXAMPLES

Section 3.2 Graphing Linear Equations

A **linear equation in two variables** is an equation that can be written in the form $Ax + By = C$ where A and B are not both 0. The form $Ax + By = C$ is called **standard form**.

Linear Equations

$$3x + 2y = -6 \qquad x = -5$$
$$y = 3 \qquad y = -x + 10$$

$x + y = 10$ is in standard form.

To graph a linear equation in two variables, find three ordered pair solutions. Plot the solution points and draw the line connecting the points.

Graph $x - 2y = 5$.

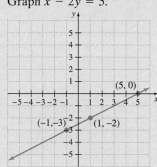

x	y
5	0
1	-2
-1	-3

Section 3.3 Intercepts

An **intercept** of a graph is a point where the graph intersects an axis. If a graph intersects the x-axis at a, then $(a, 0)$ is the **x-intercept**. If a graph intersects the y-axis at b, then $(0, b)$ is the **y-intercept**.

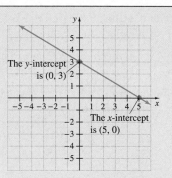

To find the x-intercept, let $y = 0$ and solve for x.
To find the y-intercept, let $x = 0$ and solve for y.

Graph $2x - 5y = -10$ by finding intercepts.

If $y = 0$, then

$$2x - 5 \cdot 0 = -10$$
$$2x = -10$$
$$\frac{2x}{2} = \frac{-10}{2}$$
$$x = -5$$

If $x = 0$, then

$$2 \cdot 0 - 5y = -10$$
$$-5y = -10$$
$$\frac{-5y}{-5} = \frac{-10}{-5}$$
$$y = 2$$

The x-intercept is $(-5, 0)$. The y-intercept is $(0, 2)$.

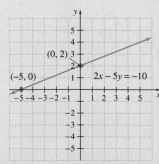

(continued)

DEFINITIONS AND CONCEPTS	EXAMPLES

Section 3.3 Intercepts

The graph of $x = c$ is a vertical line with x-intercept $(c, 0)$.

The graph of $y = c$ is a horizontal line with y-intercept $(0, c)$.

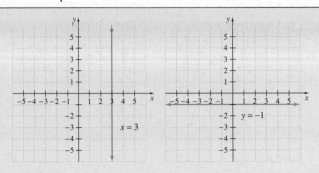

Section 3.4 Slope

The **slope m** of the line through points (x_1, y_1) and (x_2, y_2) is given by

$$m = \frac{y_2 - y_1}{x_2 - x_1} \quad \text{as long as } x_2 \neq x_1$$

A horizontal line has slope 0.

The slope of a vertical line is undefined.

Two nonvertical lines are perpendicular if the slope of one is the negative reciprocal of the slope of the other.

The slope of the line through points $(-1, 6)$ and $(-5, 8)$ is

$$m = \frac{y_2 - y_1}{x_2 - x_1} = \frac{8 - 6}{-5 - (-1)} = \frac{2}{-4} = -\frac{1}{2}$$

The slope of the line $y = -5$ is 0.

The line $x = 3$ has undefined slope.

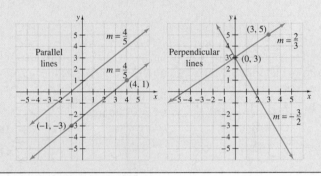

Section 3.5 Graphing Linear Inequalities

A **linear inequality in two variables** is an inequality that can be written in one of the forms:

$$Ax + By < C \qquad Ax + By \leq C$$

$$Ax + By > C \qquad Ax + By \geq C$$

To graph a linear inequality

1. Graph the boundary line by graphing the related equation. Draw the line solid if the inequality symbol is $\leq$ or $\geq$. Draw the line dashed if the inequality symbol is $<$ or $>$.

2. Choose a test point not on the line. Substitute its coordinates into the original inequality.

Linear Inequalities

$$2x - 5y < 6 \qquad\qquad x \geq -5$$

$$y > -8x \qquad\qquad y \leq 2$$

Graph $2x - y \leq 4$.

1. Graph $2x - y = 4$. Draw a solid line because the inequality symbol is $\leq$.

2. Check the test point $(0, 0)$ in the inequality $2x - y \leq 4$.

$$2 \cdot 0 - 0 \leq 4 \qquad \text{Let } x = 0 \text{ and } y = 0.$$

$$0 \leq 4 \qquad \text{True.}$$

(continued)

DEFINITIONS AND CONCEPTS	EXAMPLES

Section 3.5 Graphing Linear Inequalities

3. If the resulting inequality is true, shade the half-plane that contains the test point. If the inequality is not true, shade the half-plane that does not contain the test point.

3. The inequality is true so we shade the half-plane containing $(0, 0)$.

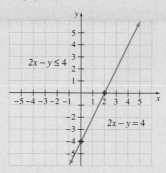

$2x - y \leq 4$

$2x - y = 4$

CHAPTER 3 REVIEW

(3.1) *Plot the following ordered pairs on a Cartesian coordinate system.*

1. $(-7, 0)$

2. $\left(0, 4\frac{4}{5}\right)$

3. $(-2, -5)$

4. $(1, -3)$

5. $(0.7, 0.7)$

6. $(-6, 4)$

7. A local lumberyard uses quantity pricing. The table shows the price per board for different amounts of lumber purchased.

Price per Board (in dollars)	Number of Boards Purchased
8.00	1
7.50	10
6.50	25
5.00	50
2.00	100

a. Write each paired data as an ordered pair of the form (price per board, number of boards purchased).

b. Create a scatter diagram of the paired data. Be sure to label the axes appropriately.

8. The table shows the average cost in cents per mile of owning and operating a vehicle for the years shown.

Year	Average Cost of Operating a Vehicle (cents per mile)
1992	46
1993	45
1994	47
1995	49
1996	51
1997	53

(*Source:* American Automobile Manufacturers Association, Inc.)

a. Write each paired data as an ordered pair of the form (year, number of cents per mile).

b. Create a scatter diagram of the paired data. Be sure to label the axes properly.

Determine whether each ordered pair is a solution of the given equation.

9. $7x - 8y = 56$; $(0, 56)$, $(8, 0)$

10. $-2x + 5y = 10$; $(-5, 0)$, $(1, 1)$

11. $x = 13$; $(13, 5)$, $(13, 13)$ **12.** $y = 2$; $(7, 2)$, $(2, 7)$

Complete the ordered pairs so that each is a solution of the given equation.

13. $-2 + y = 6x$; $(7,)$ **14.** $y = 3x + 5$; $(, -8)$

Complete the table of values for each given equation; then plot the ordered pairs. Use a single coordinate system for each exercise.

15. $9 = -3x + 4y$

x	y
	0
	3
9	

16. $y = 5$

x	y
7	
-7	
0	

17. $x = 2y$

x	y
	0
	5
	-5

18. The cost in dollars of producing x compact disk holders is given by $y = 5x + 2000$.

    a. Complete the following table.

x	1	100	1000
y			

    b. Find the number of compact disk holders that can be produced for $6430.

(3.2) *Graph each linear equation.*

19. $x - y = 1$ **20.** $x + y = 6$

21. $x - 3y = 12$ **22.** $5x - y = -8$

23. $x = 3y$ **24.** $y = -2x$

25. $2x - 3y = 6$ **26.** $4x - 3y = 12$

27. The total amount y paid out each year for Social Security payments (in billions of dollars) from 1990 through 1998 is given by $y = 16x + 250$. In the equation, x represents the number of years after 1990. Graph this equation and use the graph to predict the amount paid out for Social Security payments in the year 2006. (*Source:* Based on data from U.S. Bureau of the Census, *Statistical Abstract of the United States: 1998*)

(3.3) *Identify the intercepts and intercept points.*

28. 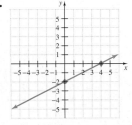   **29.**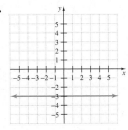

30. 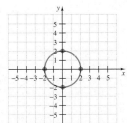   **31.**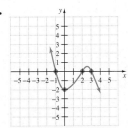

Graph each linear equation by finding its intercepts.

32. $x - 3y = 12$ **33.** $-4x + y = 8$

34. $y = -3$ **35.** $x = 5$

36. $y = -3x$ **37.** $x = 5y$

38. $x - 2 = 0$ **39.** $y + 6 = 0$

(3.4) *Find the slope of each line.*

40. 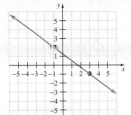   **41.**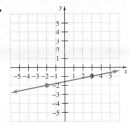

Match each line with its slope.

a. 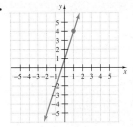   **b.**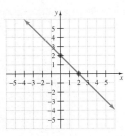

c. 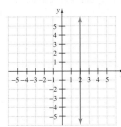   **d.**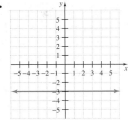

e.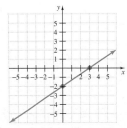

42. $m = 0$ **43.** $m = -1$

44. no slope   **45.** $m = 3$

46. $m = \dfrac{2}{3}$

Find the slope of the line that goes through the given points.

47. $(2, 5)$ and $(6, 8)$ **48.** $(4, 7)$ and $(1, 2)$

49. $(1, 3)$ and $(-2, -9)$ **50.** $(-4, 1)$ and $(3, -6)$

Find the slope of each line.

51. $x = 5$ **52.** $y = -1$

53. $y = -2$ **54.** $x = 0$

Determine whether the lines through the pairs of points are parallel, perpendicular, or neither.

△ **55.** $(-3, 1)$ and $(1, -2)$ △ **56.** $(-7, 6)$ and $(0, 4)$
       $(2, 4)$ and $(6, 1)$ $(-9, -3)$ and $(1, 5)$

△ **57.** (9, 10) and (8, −7)
(−1, −3) and (2, −8)

△ **58.** (−1, 3) and (3, −2)
(−2, −2) and (3, 2)

60. The graph below approximates the number of U.S. travelers y (in millions) that are vacationing per year x.

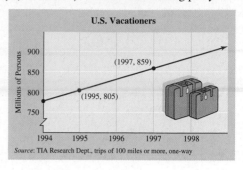

U.S. Vacationers

(1997, 859)

(1995, 805)

Millions of Persons: 900, 850, 800, 750

1994 1995 1996 1997 1998

Source: TIA Research Dept., trips of 100 miles or more, one-way

Find the slope of each line and write the slope as a rate of change. Don't forget to attach the proper units.

59. The graph below approximates the number of U.S. persons y (in millions) who have a bachelors degree or higher per year x.

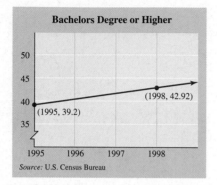

Bachelors Degree or Higher

50

45

40

35

(1998, 42.92)

(1995, 39.2)

1995 1996 1997 1998

Source: U.S. Census Bureau

(3.5) *Graph the following inequalities.*

61. $3x - 4y \leq 0$ **62.** $3x - 4y \geq 0$
63. $x + 6y < 6$ **64.** $x + y > -2$
65. $y \geq -7$ **66.** $y \leq -4$
67. $-x \leq y$

CHAPTER 3 TEST

Determine whether the ordered pairs are solutions of the equations.

1. $x - 2y = 3$; (1, 1)   **2.** $2x + 3y = 6$; (0, −2)

Complete the ordered pair solution for the following equations.

3. $12y - 7x = 5$; (1,)   **4.** $y = 17$; (−4,)

Find the slopes of the following lines.

5.
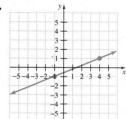

6.
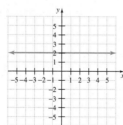

7. Through (6, −5) and (−1, 2)
8. Through (0, −8) and (−1, −1)
9. $-3x + y = 5$ **10.** $x = 6$

Determine whether the lines through the pairs of points are parallel, perpendicular, or neither.

△ **11.** (−1, 3) and (1, −3)
(2, −1) and (4, −7)

△ **12.** (−6, −6) and (−1, −2)
(−4, 3) and (3, −3)

Graph the following.

13. $2x + y = 8$ **14.** $-x + 4y = 5$
15. $x - y \geq -2$ **16.** $y \geq -4x$

17. $5x - 7y = 10$ **18.** $2x - 3y > -6$
19. $6x + y > -1$ **20.** $y = -1$
21. $x - 3 = 0$ **22.** $5x - 3y = 15$

Write each statement as an equation or inequality in two variables. Then graph the equation.

23. The y-value is 1 more than twice the x-value.
24. The x-value added to four times the y-value is less than −4.

△ **25.** The perimeter of the parallelogram below is 42 meters. Write a linear equation in two variables for the perimeter. Use this equation to find x when y is 8.

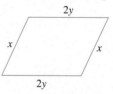

2y

x   x

2y

26. The table gives the number of cable TV subscribers (in millions) for the years shown. (*Source: Television and Cable Factbook,* Warren Publishing, Inc., Washington, D.C.)

Year	Cable TV Subscribers (in millions)
1986	38
1988	44
1990	50
1992	53
1994	57
1996	62
1997	64

a. Write this data as a set of ordered pairs of the form (year, number of cable TV subscribers in millions).

b. Create a scatter diagram of the data. Be sure to label the axes properly.

27. This graph approximates the movie ticket sales y (in millions) for the year x. Find the slope of the line and write the slope as a rate of change. Don't forget to attach the proper units.

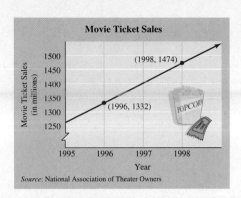

CHAPTER 3 CUMULATIVE REVIEW

1. Insert $<, >,$ or $=$ in the space between the paired numbers to make each statement true.

  a. 2   3   **b.** 7   4

  c. 72   27

2. Find the product of $\dfrac{2}{15}$ and $\dfrac{5}{13}$. Write the product in lowest terms.

3. Simplify: $\dfrac{3 + |4 - 3| + 2^2}{6 - 3}$

4. Add.

  a. $-8 + (-11)$ **b.** $-5 + 35$

  c. $0.6 + (-1.1)$ **d.** $-\dfrac{7}{10} + \left(-\dfrac{1}{10}\right)$

  e. $11.4 + (-4.7)$ **f.** $-\dfrac{3}{8} + \dfrac{2}{5}$

5. Simplify: $-14 - 8 + 10 - (-6)$

6. If $x = -2$ and $y = -4$, evaluate each expression.

  a. $5x - y$ **b.** $x^3 - y^2$

  c. $\dfrac{3x}{2y}$

7. Simplify each expression.

  a. $10 + (x + 12)$ **b.** $-3(7x)$

8. Identify the numerical coefficient in each term.

  a. $-3y$ **b.** $22z^4$

  c. y **d.** $-x$

  e. $\dfrac{x}{7}$

9. Solve $y + 0.6 = -1.0$ for y.

10. Solve: $-\dfrac{2}{3}x = -\dfrac{5}{2}$

11. If x is the first of three consecutive integers, express the sum of the three integers in terms of x. Simplify if possible.

12. Solve: $\dfrac{2(a + 3)}{3} = 6a + 2$

13. In a recent year, Congress had 8 more Republican senators than Democratic. If the total number of senators is 100, how many senators of each party were there?

14. Charles Pecot can afford enough fencing to enclose a rectangular garden with a perimeter of 140 feet. If the width of his garden is to be 30 feet, find the length.

15. Solve $y = mx + b$ for x.

16. Write each number as a percent.

  a. 0.73   **b.** 1.39

  c. $\dfrac{1}{4}$

17. The number 120 is 15% of what number?

18. A chemist working on his doctoral degree at Massachusetts Institute of Technology needs 12 liters of a 50% acid solution for a lab experiment. The stockroom has only 40% and 70% solutions. How much of each solution should be mixed together to form 12 liters of a 50% solution?

19. Graph $x \geq -1$.

20. Solve $-1 \leq 2x - 3 < 5$, and graph the solution set.

21. Determine whether each ordered pair is a solution of the equation $x - 2y = 6$.

  a. $(6, 0)$ **b.** $(0, 3)$

  c. $\left(1, -\dfrac{5}{2}\right)$

22. Identify the linear equations in two variables.

  a. $x - 1.5y = -1.6$ **b.** $y = -2x$

  c. $x + y^2 = 9$ **d.** $x = 5$

23. Find the slope of the line $y = -1$.

24. Find the slope of a line perpendicular to the line passing through the points $(-1, 7)$ and $(2, 2)$.

25. Graph $y > 3$

A Fast Growing Occupation

According to the U.S. Bureau of Labor Statistics, the paralegal profession is expected to be one of the fastest growing occupations for the beginning of the 21st century. Paralegals assist lawyers by

- doing legal research
- organizing and tracking case files
- helping draw up legal agreements
- preparing reports to assist attorneys.

Although paralegals provide lawyers with invaluable support in many areas, they are prohibited from arguing cases in court and giving legal advice.

Paralegals should have good communication, research, and problem-solving skills. A well-rounded education is also an asset. Because paralegals collect, organize, and analyze all types of information, including numerical data, a firm grasp of mathematics is helpful.

 For more information about a career as a paralegal, visit the National Federation of Paralegal Association's Website (www.paralegals.org).

In the Spotlight on Decision Making feature on page 253, you will have the opportunity to make a decision concerning experts' estimates in a legal case as a paralegal.

EXPONENTS AND POLYNOMIALS

4.1 EXPONENTS

4.2 ADDING AND SUBTRACTING POLYNOMIALS

4.3 MULTIPLYING POLYNOMIALS

4.4 SPECIAL PRODUCTS

4.5 NEGATIVE EXPONENTS AND SCIENTIFIC NOTATION

4.6 DIVISION OF POLYNOMIALS

Recall from Chapter 1 that an exponent is a shorthand notation for repeated factors. This chapter explores additional concepts about exponents and exponential expressions. An especially useful type of exponential expression is a polynomial. Polynomials model many real-world phenomena. This chapter will focus on operations on polynomials.

4.1 EXPONENTS

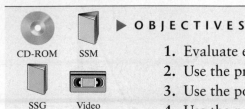

▶ O B J E C T I V E S

1. Evaluate exponential expressions.
2. Use the product rule for exponents.
3. Use the power rule for exponents.
4. Use the power rules for products and quotients.
5. Use the quotient rule for exponents, and define a number raised to the 0 power.

1

As we reviewed in Section 1.3, an exponent is a shorthand notation for repeated factors. For example, $2 \cdot 2 \cdot 2 \cdot 2 \cdot 2$ can be written as 2^5. The expression 2^5 is called an **exponential expression**. It is also called the fifth **power** of 2, or we say that 2 is **raised** to the fifth power.

$$5^6 = \underbrace{5 \cdot 5 \cdot 5 \cdot 5 \cdot 5 \cdot 5}_{6 \text{ factors; each factor is } 5} \qquad \text{and} \qquad (-3)^4 = \underbrace{(-3) \cdot (-3) \cdot (-3) \cdot (-3)}_{4 \text{ factors; each factor is } -3}$$

The **base** of an exponential expression is the repeated factor. The **exponent** is the number of times that the base is used as a factor.

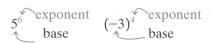

Example 1   Evaluate each expression.

    a. 2^3 **b.** 3^1 **c.** $(-4)^2$ **d.** -4^2 **e.** $\left(\dfrac{1}{2}\right)^4$ **f.** $4 \cdot 3^2$

Solution
    a. $2^3 = 2 \cdot 2 \cdot 2 = 8$
    b. To raise 3 to the first power means to use 3 as a factor only once. Therefore, $3^1 = 3$. Also, when no exponent is shown, the exponent is assumed to be 1.
    c. $(-4)^2 = (-4)(-4) = 16$
    d. $-4^2 = -(4 \cdot 4) = -16$
    e. $\left(\dfrac{1}{2}\right)^4 = \dfrac{1}{2} \cdot \dfrac{1}{2} \cdot \dfrac{1}{2} \cdot \dfrac{1}{2} = \dfrac{1}{16}$
    f. $4 \cdot 3^2 = 4 \cdot 9 = 36$

Notice how similar -4^2 is to $(-4)^2$ in the example above. The difference between the two is the parentheses. In $(-4)^2$, the parentheses tell us that the base, or repeated factor, is -4. In -4^2, only 4 is the base.

> **HELPFUL HINT**
> Be careful when identifying the base of an exponential expression. Pay close attention to the use of parentheses.
>
> $(-3)^2$ -3^2 $2 \cdot 3^2$
> The base is -3.   The base is 3.   The base is 3.
> $(-3)^2 = (-3)(-3) = 9$ $-3^2 = -(3 \cdot 3) = -9$ $2 \cdot 3^2 = 2 \cdot 3 \cdot 3 = 18$

2   An exponent has the same meaning whether the base is a number or a variable. If x is a real number and n is a positive integer, then x^n is the product of n factors, each of which is x.

$$x^n = \underbrace{x \cdot x \cdot x \cdot x \cdot x \ldots x}_{n \text{ factors of } x}$$

Example 2   Evaluate for the given value of x.

a. $2x^3$; x is 5

b. $\dfrac{9}{x^2}$; x is -3

Solution   **a.** If x is 5, $2x^3 = 2 \cdot (5)^3$

$\qquad\qquad\qquad = 2 \cdot (5 \cdot 5 \cdot 5)$

$\qquad\qquad\qquad = 2 \cdot 125$

$\qquad\qquad\qquad = 250$

b. If x is -3, $\dfrac{9}{x^2} = \dfrac{9}{(-3)^2}$

$\qquad\qquad\qquad\qquad = \dfrac{9}{(-3)(-3)}$

$\qquad\qquad\qquad\qquad = \dfrac{9}{9} = 1$

Exponential expressions can be multiplied, divided, added, subtracted, and themselves raised to powers. By our definition of an exponent,

$$5^4 \cdot 5^3 = \underbrace{(5 \cdot 5 \cdot 5 \cdot 5)}_{4 \text{ factors of } 5} \cdot \underbrace{(5 \cdot 5 \cdot 5)}_{3 \text{ factors of } 5}$$

$$= \underbrace{5 \cdot 5 \cdot 5 \cdot 5 \cdot 5 \cdot 5 \cdot 5}_{7 \text{ factors of } 5}$$

$$= 5^7$$

Also,

$$x^2 \cdot x^3 = (x \cdot x) \cdot (x \cdot x \cdot x)$$
$$= x \cdot x \cdot x \cdot x \cdot x$$
$$= x^5$$

In both cases, notice that the result is exactly the same if the exponents are added.

$$5^4 \cdot 5^3 = 5^{4+3} = 5^7 \qquad \text{and} \qquad x^2 \cdot x^3 = x^{2+3} = x^5$$

This suggests the following rule.

PRODUCT RULE FOR EXPONENTS

If m and n are positive integers and a is a real number, then

$$a^m \cdot a^n = a^{m+n}$$

In other words, to multiply two exponential expressions with a **common base**, keep the base and add the exponents.

Example 3   Use the product rule to simplify.

        a. $4^2 \cdot 4^5$ **b.** $x^2 \cdot x^5$ **c.** $y^3 \cdot y$ **d.** $y^3 \cdot y^2 \cdot y^7$ **e.** $(-5)^7 \cdot (-5)^8$

Solution   **a.** $4^2 \cdot 4^5 = 4^{2+5} = 4^7$
          b. $x^2 \cdot x^5 = x^{2+5} = x^7$
          c. $y^3 \cdot y = y^3 \cdot y^1$
                   $= y^{3+1}$
                   $= y^4$

> **HELPFUL HINT**
> Don't forget that if no exponent is written, it is assumed to be 1.

          d. $y^3 \cdot y^2 \cdot y^7 = y^{3+2+7} = y^{12}$
          e. $(-5)^7 \cdot (-5)^8 = (-5)^{7+8} = (-5)^{15}$

Example 4   Use the product rule to simplify $(2x^2)(-3x^5)$.

Solution   Recall that $2x^2$ means $2 \cdot x^2$ and $-3x^5$ means $-3 \cdot x^5$.

$$
\begin{aligned}
(2x^2)(-3x^5) &= 2 \cdot x^2 \cdot -3 \cdot x^5 &&\text{Remove parentheses.}\\
&= 2 \cdot -3 \cdot x^2 \cdot x^5 &&\text{Group factors with common bases.}\\
&= -6x^7 &&\text{Simplify.}
\end{aligned}
$$

> **HELPFUL HINT**
> These examples will remind you of the difference between adding and multiplying terms.
>
> ***Addition***
> $$5x^3 + 3x^3 = (5 + 3)x^3 = 8x^3$$
> $$7x + 4x^2 = 7x + 4x^2$$
>
> ***Multiplication***
> $$(5x^3)(3x^3) = 5 \cdot 3 \cdot x^3 \cdot x^3 = 15x^{3+3} = 15x^6$$
> $$(7x)(4x^2) = 7 \cdot 4 \cdot x \cdot x^2 = 28x^{1+2} = 28x^3$$

3   Exponential expressions can themselves be raised to powers. Let's try to discover a rule that simplifies an expression like $(x^2)^3$. By definition,

$$(x^2)^3 = \underbrace{(x^2)(x^2)(x^2)}_{3\text{ factors of } x^2}$$

which can be simplified by the product rule for exponents.

$$(x^2)^3 = (x^2)(x^2)(x^2) = x^{2+2+2} = x^6$$

Notice that the result is exactly the same if we multiply the exponents.

$$(x^2)^3 = x^{2 \cdot 3} = x^6$$

The following property states this result.

POWER RULE FOR EXPONENTS

If m and n are positive integers and a is a real number, then

$$(a^m)^n = a^{mn}$$

To raise a power to a power, keep the base and multiply the exponents.

Example 5 Simplify each of the following expressions.

    a. $(x^2)^5$ **b.** $(y^8)^2$ **c.** $[(-5)^3]^7$

Solution   **a.** $(x^2)^5 = x^{2 \cdot 5} = x^{10}$
        b. $(y^8)^2 = y^{8 \cdot 2} = y^{16}$
        c. $[(-5)^3]^7 = (-5)^{21}$

HELPFUL HINT
Take a moment to make sure that you understand when to apply the product rule
and when to apply the power rule.

Product Rule $\rightarrow$ *Add Exponents*	*Power Rule* $\rightarrow$ *Multiply Exponents*
$x^5 \cdot x^7 = x^{5+7} = x^{12}$ $y^6 \cdot y^2 = y^{6+2} = y^8$	$(x^5)^7 = x^{5 \cdot 7} = x^{35}$ $(y^6)^2 = y^{6 \cdot 2} = y^{12}$

---

4   When the base of an exponential expression is a product, the definition of x^n still applies. To simplify $(xy)^3$, for example,

$$(xy)^3 = (xy)(xy)(xy) \qquad \text{$(xy)^3$ means 3 factors of (xy).}$$
$$= x \cdot x \cdot x \cdot y \cdot y \cdot y \qquad \text{Group factors with common bases.}$$
$$= x^3 y^3 \qquad \text{Simplify.}$$

Notice that to simplify the expression $(xy)^3$, we raise each factor within the parentheses to a power of 3.

$$(xy)^3 = x^3 y^3$$

In general, we have the following rule.

POWER OF A PRODUCT RULE

If n is a positive integer and a and b are real numbers, then

$$(ab)^n = a^n b^n$$

In other words, to raise a product to a power, we raise each factor to the power.

Example 6 Simplify each expression.

a. $(st)^4$ b. $(2a)^3$ c. $(-5x^2y^3z)^2$

Solution
a. $(st)^4 = s^4 \cdot t^4 = s^4t^4$ *Use the power of a product rule.*

b. $(2a)^3 = 2^3 \cdot a^3 = 8a^3$ *Use the power of a product rule.*

c. $(-5x^2y^3z)^2 = (-5)^2 \cdot (x^2)^2 \cdot (y^3)^2 \cdot (z^1)^2$ *Use the power of a product rule.*

             $= 25x^4y^6z^2$ *Use the power rule for exponents.* ∎

Let's see what happens when we raise a quotient to a power. To simplify $\left(\dfrac{x}{y}\right)^3$, for example,

$$\left(\frac{x}{y}\right)^3 = \left(\frac{x}{y}\right)\left(\frac{x}{y}\right)\left(\frac{x}{y}\right) \qquad \left(\frac{x}{y}\right)^3 \text{ means 3 factors of } \left(\frac{x}{y}\right).$$

$$= \frac{x \cdot x \cdot x}{y \cdot y \cdot y} \qquad \text{Multiply fractions.}$$

$$= \frac{x^3}{y^3} \qquad \text{Simplify.}$$

Notice that to simplify the expression $\left(\dfrac{x}{y}\right)^3$, we raise both the numerator and the denominator to a power of 3.

$$\left(\frac{x}{y}\right)^3 = \frac{x^3}{y^3}$$

In general, we have the following.

POWER OF A QUOTIENT RULE

If n is a positive integer and a and c are real numbers, then

$$\left(\frac{a}{c}\right)^n = \frac{a^n}{c^n}, \quad c \neq 0$$

In other words, to raise a quotient to a power, we raise both the numerator and the denominator to the power.

Example 7 Simplify each expression.

a. $\left(\dfrac{m}{n}\right)^7$ b. $\left(\dfrac{x^3}{3y^5}\right)^4$

Solution
a. $\left(\dfrac{m}{n}\right)^7 = \dfrac{m^7}{n^7}, n \neq 0$ *Use the power of a quotient rule.*

b. $\left(\dfrac{x^3}{3y^5}\right)^4 = \dfrac{(x^3)^4}{3^4 \cdot (y^5)^4}$ *Use the power of a product or quotient rule.*

            $= \dfrac{x^{12}}{81y^{20}}, y \neq 0$ *Use the power rule for exponents.* ∎

5 Another pattern for simplifying exponential expressions involves quotients.

To simplify an expression like $\dfrac{x^5}{x^3}$, in which the numerator and the denominator have a common base, we can apply the fundamental principle of fractions and divide the numerator and the denominator by the common base factors. Assume for the remainder of this section that denominators are not 0.

$$\frac{x^5}{x^3} = \frac{x \cdot x \cdot x \cdot x \cdot x}{x \cdot x \cdot x}$$

$$= \frac{x \cdot x \cdot x \cdot x \cdot x}{x \cdot x \cdot x}$$

$$= x \cdot x$$

$$= x^2$$

Notice that the result is exactly the same if we subtract exponents of the common bases.

$$\frac{x^5}{x^3} = x^{5-3} = x^2$$

The quotient rule for exponents states this result in a general way.

QUOTIENT RULE FOR EXPONENTS

If m and n are positive integers and a is a real number, then

$$\frac{a^m}{a^n} = a^{m-n}$$

as long as a is not 0.

In other words, to divide one exponential expression by another with a common base, keep the base and subtract exponents.

Example 8 Simplify each quotient.

a. $\dfrac{x^5}{x^2}$ b. $\dfrac{4^7}{4^3}$ c. $\dfrac{(-3)^5}{(-3)^2}$ d. $\dfrac{2x^5y^2}{xy}$

Solution   a. $\dfrac{x^5}{x^2} = x^{5-2} = x^3$ *Use the quotient rule.*

b. $\dfrac{4^7}{4^3} = 4^{7-3} = 4^4 = 256$ *Use the quotient rule.*

c. $\dfrac{(-3)^5}{(-3)^2} = (-3)^3 = -27$

d. Begin by grouping common bases.

$$\frac{2x^5y^2}{xy} = 2 \cdot \frac{x^5}{x^1} \cdot \frac{y^2}{y^1}$$

$$= 2 \cdot \left(x^{5-1}\right) \cdot \left(y^{2-1}\right) \qquad \textit{Use the quotient rule.}$$

$$= 2x^4y^1 \quad \text{or} \quad 2x^4y$$

Let's now give meaning to an expression such as x^0. To do so, we will simplify $\dfrac{x^3}{x^3}$ in two ways and compare the results.

$$\frac{x^3}{x^3} = x^{3-3} = x^0 \qquad \text{\small{\textit{Apply the quotient rule.}}}$$

$$\frac{x^3}{x^3} = \frac{x \cdot x \cdot x}{x \cdot x \cdot x} = 1 \qquad \text{\small{\textit{Apply the fundamental principle for fractions.}}}$$

Since $\dfrac{x^3}{x^3} = x^0$ and $\dfrac{x^3}{x^3} = 1$, we define that $x^0 = 1$ as long as x is not 0.

ZERO EXPONENT

$a^0 = 1$, as long as a is not 0.

In other words, a base raised to the 0 power is 1, as long as the base is not 0.

Example 9 Simplify the following expressions.

    a. 3^0 **b.** $(ab)^0$ **c.** $(-5)^0$ **d.** -5^0

Solution **a.** $3^0 = 1$

    b. Assume that neither a nor b is zero.

$$(ab)^0 = a^0 \cdot b^0 = 1 \cdot 1 = 1$$

    c. $(-5)^0 = 1$

    d. $-5^0 = -1 \cdot 5^0 = -1 \cdot 1 = -1$

In the next example, exponential expressions are simplified using two or more of the exponent rules presented in this section.

Example 10 Simplify the following.

    a. $\left(\dfrac{-5x^2}{y^3}\right)^2$ **b.** $\dfrac{(x^3)^4 x}{x^7}$ **c.** $\dfrac{(2x)^5}{x^3}$ **d.** $\dfrac{(a^2b)^3}{a^3b^2}$

Solution **a.** Use the power of a product or quotient rule; then use the power rule for exponents.

$$\left(\frac{-5x^2}{y^3}\right)^2 = \frac{(-5)^2(x^2)^2}{(y^3)^2} = \frac{25x^4}{y^6}$$

    b. $\dfrac{(x^3)^4 x}{x^7} = \dfrac{x^{12} \cdot x}{x^7} = \dfrac{x^{12+1}}{x^7} = \dfrac{x^{13}}{x^7} = x^{13-7} = x^6$

    c. Use the power of a product rule; then use the quotient rule.

$$\frac{(2x)^5}{x^3} = \frac{2^5 \cdot x^5}{x^3} = 2^5 \cdot x^{5-3} = 32x^2$$

d. Begin by applying the power of a product rule to the numerator.

$$\frac{(a^2b)^3}{a^3b^2} = \frac{(a^2)^3 \cdot b^3}{a^3 \cdot b^2}$$

$$= \frac{a^6b^3}{a^3b^2} \qquad \textit{Use the power rule for exponents.}$$

$$= a^{6-3}b^{3-2} \qquad \textit{Use the quotient rule.}$$

$$= a^3b^1 \quad \text{or} \quad a^3b$$

SPOTLIGHT ON DECISION MAKING

Suppose you are an auto mechanic and amateur racing enthusiast. You have been modifying the engine in your car and would like to enter a local amateur race. The racing classes depend on the size, or displacement, of the engine.

Engine displacement can be calculated using the formula

$d = \dfrac{\pi}{4} b^2 sc.$ In the formula,

$d =$ the engine displacement in cubic centimeters (cc)

$b =$ the bore or engine cylinder diameter in centimeters

$s =$ the stroke or distance the piston travels up or down within the cylinder in centimeters

$c =$ the number of cylinders the engine has. You have made the following measurements on your modified engine: 8.4 cm bore, 7.6 cm stroke, and 6 cylinders. In which racing class would you enter your car? Explain.

BRENTWOOD AMATEUR RACING CLUB

Racing Class	Displacement Limit
A	up to 2000 cc
B	up to 2400 cc
C	up to 2650 cc
D	up to 3000 cc

MENTAL MATH

State the bases and the exponents for each of the following expressions.

1. 3^2
2. 5^4
3. $(-3)^6$
4. -3^7
5. -4^2
6. $(-4)^3$
7. $5 \cdot 3^4$
8. $9 \cdot 7^6$
9. $5x^2$
10. $(5x)^2$

Exercise Set 4.1

Evaluate each expression. See Example 1.

1. 7^2
2. -3^2
3. $(-5)^1$
4. $(-3)^2$
5. -2^4
6. -4^3
7. $(-2)^4$
8. $(-4)^3$
9. $\left(\dfrac{1}{3}\right)^3$
10. $\left(-\dfrac{1}{9}\right)^2$
11. $7 \cdot 2^4$
12. $9 \cdot 1^2$
13. Explain why $(-5)^4 = 625$, while $-5^4 = -625$.
14. Explain why $5 \cdot 4^2 = 80$, while $(5 \cdot 4)^2 = 400$.

Evaluate each expression given the replacement values for x. See Example 2.

15. x^2; $x = -2$ **16.** x^3; $x = -2$

17. $5x^3$; $x = 3$ **18.** $4x^2$; $x = -1$

19. $2xy^2$; $x = 3$ and $y = 5$

20. $-4x^2y^3$; $x = 2$ and $y = -1$

21. $\dfrac{2z^4}{5}$; $z = -2$ **22.** $\dfrac{10}{3y^3}$; $y = 5$

△ **23.** The formula $V = x^3$ can be used to find the volume V of a cube with side length x. Find the volume of a cube with side length 7 meters. (Volume is measured in cubic units.)

△ **24.** The formula $S = 6x^2$ can be used to find the surface area S of a cube with side length x. Find the surface area of the cube with side length 5 meters. (Surface area is measured in square units.)

△ **25.** To find the amount of water that a swimming pool in the shape of a cube can hold, do we use the formula for volume of the cube or surface area of the cube? (See Exercises 23 and 24.)

△ **26.** To find the amount of material needed to cover an ottoman in the shape of a cube, do we use the formula for volume of the cube or surface area of the cube? (See Exercises 23 and 24.)

Use the product rule to simplify each expression. Write the results using exponents. See Examples 3 and 4.

27. $x^2 \cdot x^5$ **28.** $y^2 \cdot y$

29. $(-3)^3 \cdot (-3)^9$ **30.** $(-5)^7 \cdot (-5)^6$

31. $(5y^4)(3y)$ **32.** $(-2z^3)(-2z^2)$

33. $(4z^{10})(-6z^7)(z^3)$ **34.** $(12x^5)(-x^6)(x^4)$

Use the power rule and the power of a product or quotient rule to simplify each expression. See Examples 5 through 7.

35. $(pq)^7$ **36.** $(4s)^3$

37. $\left(\dfrac{m}{n}\right)^9$ **38.** $\left(\dfrac{xy}{7}\right)^2$

39. $(x^2y^3)^5$ **40.** $(a^4b)^7$

41. $\left(\dfrac{-2xz}{y^5}\right)^2$ **42.** $\left(\dfrac{y^4}{-3z^3}\right)^3$

Use the quotient rule and simplify each expression. See Example 8.

43. $\dfrac{x^3}{x}$ **44.** $\dfrac{y^{10}}{y^9}$

45. $\dfrac{(-2)^5}{(-2)^3}$ **46.** $\dfrac{(-5)^{14}}{(-5)^{11}}$

47. $\dfrac{p^7q^{20}}{pq^{15}}$ **48.** $\dfrac{x^8y^6}{y^5}$

49. $\dfrac{7x^2y^6}{14x^2y^3}$ **50.** $\dfrac{9a^4b^7}{3ab^2}$

Simplify the following. See Example 9.

51. $(2x)^0$ **52.** $-4x^0$

53. $-2x^0$ **54.** $(4y)^0$

55. $5^0 + y^0$ **56.** $-3^0 + 4^0$

Simplify the following. See Example 10.

57. $\left(\dfrac{-3a^2}{b^3}\right)^3$ **58.** $\left(\dfrac{q^7}{-2p^5}\right)^5$

59. $\dfrac{(x^5)^7 \cdot x^8}{x^4}$ **60.** $\dfrac{y^{20}}{(y^2)^3 \cdot y^9}$

61. $\dfrac{(z^3)^6}{(5z)^4}$ **62.** $\dfrac{(3x)^4}{(x^2)^2}$

63. $\dfrac{(6mn)^5}{mn^2}$ **64.** $\dfrac{(6xy)^2}{9x^2y^2}$

Simplify the following.

65. -5^2 **66.** $(-5)^2$

67. $\left(\dfrac{1}{4}\right)^3$ **68.** $\left(\dfrac{2}{3}\right)^3$

69. $(9xy)^2$ **70.** $(2ab)^5$

71. $(6b)^0$ **72.** $(5ab)^0$

73. $2^3 + 2^5$ **74.** $7^2 - 7^0$

75. b^4b^2 **76.** y^4y^1

77. $a^2a^3a^4$ **78.** $x^2x^{15}x^9$

79. $(2x^3)(-8x^4)$ **80.** $(3y^4)(-5y)$

81. $(4a)^3$ **82.** $(2ab)^4$

83. $(-6xyz^3)^2$ **84.** $(-3xy^2a^3b)^3$

85. $\left(\dfrac{3y^5}{6x^4}\right)^3$ **86.** $\left(\dfrac{2ab}{6yz}\right)^4$

87. $\dfrac{x^5}{x^4}$ **88.** $\dfrac{5x^9}{x^3}$

89. $\dfrac{2x^3y^2z}{xyz}$ **90.** $\dfrac{x^{12}y^{13}}{x^5y^7}$

91. $\dfrac{(3x^2y^5)^5}{x^3y}$ **92.** $\dfrac{(4a^2)^4}{a^4b}$

93. In your own words, explain why $5^0 = 1$.

94. In your own words, explain when $(-3)^n$ is positive and when it is negative.

Find the area of each figure.

△ **95.** The following rectangle has width $4x^2$ feet and length $5x^3$ feet. Find its area. (Recall that Area = $l \cdot w$)

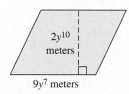

$4x^2$ feet

$5x^3$ feet

△ **96.** The following parallelogram has base length $9y^7$ meters and height $2y^{10}$ meters. Find its area. (Recall that Area = $b \cdot h$)

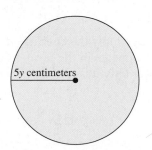

$2y^{10}$ meters

$9y^7$ meters

△ **97.** The following circle has a radius $5y$ centimeters, find its area. Do not approximate π.

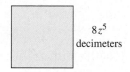

$5y$ centimeters

△ **98.** The square shown has sides of length $8z^5$ decimeters. Find its area.

$8z^5$ decimeters

Find the volume of each figure.

△ **99.** The following safe is in the shape of a cube. Each side is $3y^4$ feet, find its volume.

$3y^4$ feet

$3y^4$ feet

$3y^4$ feet

△ **100.** The silo shown is in the shape of a cylinder. Its radius is $4x$ meters and its height is $5x^3$ meters, find its volume. Do not approximate π.

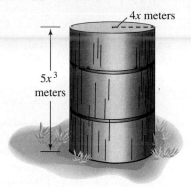

$4x$ meters

$5x^3$ meters

REVIEW EXERCISES

Simplify each expression by combining any like terms. Use the distributive property to remove any parentheses. See Section 2.1.

101. $3x - 5x + 7$ **102.** $7w + w - 2w$

103. $y - 10 + y$ **104.** $-6z + 20 - 3z$

105. $7x + 2 - 8x - 6$ **106.** $10y - 14 - y - 14$

107. $2(x - 5) + 3(5 - x)$

108. $-3(w + 7) + 5(w + 1)$

A Look Ahead

Example
Simplify $x^a \cdot x^{3a}$.

Solution:
Like bases, so add exponents.

$$x^a \cdot x^{3a} = x^{a+3a} = x^{4a}$$

Simplify each expression. Assume that variables represent positive integers. See the example.

109. $x^{5a}x^{4a}$ **110.** $b^{9a}b^{4a}$

111. $(a^b)^5$ **112.** $(2a^{4b})^4$

113. $\dfrac{x^{9a}}{x^{4a}}$ **114.** $\dfrac{y^{15b}}{y^{6b}}$

115. $(x^a y^b z^c)^{5a}$

116. $(9a^2 b^3 c^4 d^5)^{ab}$

4.2 ADDING AND SUBTRACTING POLYNOMIALS

CD-ROM   SSM

SSG   Video

▶ **OBJECTIVES**

1. Define monomial, binomial, trinomial, polynomial, and degree.
2. Find the value of a polynomial given replacement values for the variables.
3. Combine like terms.
4. Add and subtract polynomials.

1

In this section, we introduce a special algebraic expression called a polynomial. Let's first review some definitions presented in Section 2.1.

Recall that a term is a number or the product of a number and variables raised to powers. The terms of the expression $4x^2 + 3x$ are $4x^2$ and $3x$. The terms of the expression $9x^4 - 7x - 1$ are $9x^4, -7x$, and -1.

Expression	*Terms*
$4x^2 + 3x$	$4x^2, 3x$
$9x^4 - 7x - 1$	$9x^4, -7x, -1$
$7y^3$	$7y^3$

The **numerical coefficient** of a term, or simply the **coefficient**, is the numerical factor of each term. If no numerical factor appears in the term, then the coefficient is understood to be 1. If the term is a number only, it is called a **constant** term or simply a constant.

Term	*Coefficient*
x^5	1
$3x^2$	3
$-4x$	-4
$-x^2y$	-1
3 (constant)	3

A **polynomial in** x is a finite sum of terms of the form ax^n, where a is a real number and n is a whole number. For example,

$$x^5 - 3x^3 + 2x^2 - 5x + 1$$

is a polynomial. Notice that this polynomial is written in **descending powers** of x because the powers of x decrease from left to right. (Recall that the term 1 can be thought of as $1x^0$.)

On the other hand,

$$x^{-5} + 2x - 3$$

is **not** a polynomial because it contains an exponent, -5, that is not a whole number. (We study negative exponents in Section 5 of this chapter.)

Some polynomials are given special names.

A **monomial** is a polynomial with exactly one term.
A **binomial** is a polynomial with exactly two terms.
A **trinomial** is a polynomial with exactly three terms.

The following are examples of monomials, binomials, and trinomials. Each of these examples is also a polynomial.

POLYNOMIALS

Monomials	Binomials	Trinomials	None of These
ax^2	$x + y$	$x^2 + 4xy + y^2$	$5x^3 - 6x^2 + 3x - 6$
$-3z$	$3p + 2$	$x^5 + 7x^2 - x$	$-y^5 + y^4 - 3y^3 - y^2 + y$
4	$4x^2 - 7$	$-q^4 + q^3 - 2q$	$x^6 + x^4 - x^3 + 1$

Each term of a polynomial has a **degree**.

DEGREE OF A TERM

The degree of a term is the sum of the exponents on the variables contained in the term.

Example 1   Find the degree of each term.

a. $-3x^2$ **b.** $5x^3yz$ **c.** 2

Solution   **a.** The exponent on x is 2, so the degree of the term is 2.
b. $5x^3yz$ can be written as $5x^3y^1z^1$. The degree of the term is the sum of its exponents, so the degree is $3 + 1 + 1$ or 5.
c. The constant, 2, can be written as $2x^0$ (since $x^0 = 1$). The degree of 2 or $2x^0$ is 0. ■

From the preceding, we can say that **the degree of a constant is 0**.
Each polynomial also has a degree.

DEGREE OF A POLYNOMIAL

The degree of a polynomial is the greatest degree of any term of the polynomial.

Example 2   Find the degree of each polynomial and tell whether the polynomial is a monomial, binomial, trinomial, or none of these.

a. $-2t^2 + 3t + 6$ **b.** $15x - 10$ **c.** $7x + 3x^3 + 2x^2 - 1$

Solution   **a.** The degree of the trinomial $-2t^2 + 3t + 6$ is 2, the greatest degree of any of its terms.
b. The degree of the binomial $15x - 10$ or $15x^1 - 10$ is 1.

c. The degree of the polynomial $7x + 3x^3 + 2x^2 - 1$ is 3. ■

Example 3   Complete the table for the polynomial

$$7x^2y - 6xy + x^2 - 3y + 7$$

Use the table to give the degree of the polynomial.

Solution

Term	Numerical Coefficient	Degree of Term
$7x^2y$	7	3
$-6xy$	-6	2
x^2	1	2
$-3y$	-3	1
7	7	0

The degree of the polynomial is 3. ←

2   Polynomials have different values depending on replacement values for the variables.

Example 4   Find the value of the polynomial $3x^2 - 2x + 1$ when $x = -2$.

Solution   Replace x with -2 and simplify.

$$
\begin{aligned}
3x^2 - 2x + 1 &= 3(-2)^2 - 2(-2) + 1 \\
&= 3(4) + 4 + 1 \\
&= 12 + 4 + 1 \\
&= 17
\end{aligned}
$$

Many physical phenomena can be modeled by polynomials.

Example 5   **FINDING THE HEIGHT OF A DROPPED OBJECT**

The CN Tower in Toronto, Ontario, is 1821 feet tall and is the world's tallest self-supporting structure. An object is dropped from the top of this building. Neglecting air resistance, the height of the object at time t seconds is given by the polynomial $-16t^2 + 1821$. Find the height of the object when $t = 1$ second and when $t = 10$ seconds.

Solution   To find each height, we evaluate the polynomial when $t = 1$ and when $t = 10$.

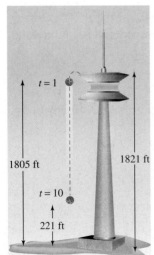

$$
\begin{aligned}
-16t^2 + 1821 &= -16(1)^2 + 1821 \qquad \text{Replace } t \text{ with 1.} \\
&= -16(1) + 1821 \\
&= -16 + 1821 \\
&= 1805
\end{aligned}
$$

The height of the object at 1 second is 1805 feet.

$$
\begin{aligned}
-16t^2 + 1821 &= -16(10)^2 + 1821 \qquad \text{Replace } t \text{ with 10.} \\
&= -16(100) + 1821 \\
&= -1600 + 1821 \\
&= 221
\end{aligned}
$$

The height of the object at 10 seconds is 221 feet.

3 Polynomials with like terms can be simplified by combining like terms. Recall that like terms are terms that contain exactly the same variables raised to exactly the same powers.

Like Terms	Unlike Terms
$5x^2, -7x^2$	$3x, 3y$
$y, 2y$	$-2x^2, -5x$
$\dfrac{1}{2}a^2b, -a^2b$	$6st^2, 4s^2t$

Only like terms can be combined. We combine like terms by applying the distributive property.

Example 6 Combine like terms.

    a. $-3x + 7x$ **b.** $11x^2 + 5 + 2x^2 - 7$ **c.** $\dfrac{2}{5}x^4 + \dfrac{2}{3}x^3 - x^2 + \dfrac{1}{10}x^4 - \dfrac{1}{6}x^3$

Solution   **a.** $-3x + 7x = (-3 + 7)x = 4x$

    b. $11x^2 + 5 + 2x^2 - 7 = 11x^2 + 2x^2 + 5 - 7$
                              $= 13x^2 - 2$ Combine like terms.

    c. $\dfrac{2}{5}x^4 + \dfrac{2}{3}x^3 - x^2 + \dfrac{1}{10}x^4 - \dfrac{1}{6}x^3$

$$= \left(\frac{2}{5} + \frac{1}{10}\right)x^4 + \left(\frac{2}{3} - \frac{1}{6}\right)x^3 - x^2$$

$$= \left(\frac{4}{10} + \frac{1}{10}\right)x^4 + \left(\frac{4}{6} - \frac{1}{6}\right)x^3 - x^2$$

$$= \frac{5}{10}x^4 + \frac{3}{6}x^3 - x^2$$

$$= \frac{1}{2}x^4 + \frac{1}{2}x^3 - x^2$$

Example 7 Combine like terms to simplify.

$$3xy - 5y^2 + 7xy - 9x^2$$

Solution $3xy - 5y^2 + 7xy - 9x^2 = (3 + 7)xy - 5y^2 - 9x^2$
                                    $= 10xy - 5y^2 - 9x^2$

> **HELPFUL HINT**
> This term can be written as $10xy$ or $10yx$.

4 We now practice adding and subtracting polynomials.

ADDING POLYNOMIALS

To add polynomials, combine all like terms.

Example 8 Add $\left(-2x^2 + 5x - 1\right)$ and $\left(-2x^2 + x + 3\right)$.

Solution
$$\begin{aligned}
\left(-2x^2 + 5x - 1\right) + \left(-2x^2 + x + 3\right) &= -2x^2 + 5x - 1 - 2x^2 + x + 3 \\
&= \left(-2x^2 - 2x^2\right) + \left(5x + 1x\right) + \left(-1 + 3\right) \\
&= -4x^2 + 6x + 2
\end{aligned}$$

Example 9 Add: $\left(4x^3 - 6x^2 + 2x + 7\right) + \left(5x^2 - 2x\right)$.

Solution
$$\begin{aligned}
\left(4x^3 - 6x^2 + 2x + 7\right) + \left(5x^2 - 2x\right) &= 4x^3 - 6x^2 + 2x + 7 + 5x^2 - 2x \\
&= 4x^3 + \left(-6x^2 + 5x^2\right) + \left(2x - 2x\right) + 7 \\
&= 4x^3 - x^2 + 7
\end{aligned}$$

Polynomials can be added vertically if we line up like terms underneath one another.

Example 10 Add $\left(7y^3 - 2y^2 + 7\right)$ and $\left(6y^2 + 1\right)$ using the vertical format.

Solution Vertically line up like terms and add.

$$\begin{array}{r}
7y^3 - 2y^2 + 7 \\
6y^2 + 1 \\
\hline
7y^3 + 4y^2 + 8
\end{array}$$

To subtract one polynomial from another, recall the definition of subtraction. To subtract a number, we add its opposite: $a - b = a + (-b)$. To subtract a polynomial, we also add its opposite. Just as $-b$ is the opposite of b, $-\left(x^2 + 5\right)$ is the opposite of $\left(x^2 + 5\right)$.

Example 11 Subtract: $(5x - 3) - (2x - 11)$.

Solution From the definition of subtraction, we have

$$\begin{aligned}
(5x - 3) - (2x - 11) &= (5x - 3) + \left[-(2x - 11)\right] && \text{Add the opposite.} \\
&= (5x - 3) + (-2x + 11) && \text{Apply the distributive property.} \\
&= 3x + 8
\end{aligned}$$

SUBTRACTING POLYNOMIALS

To subtract two polynomials, change the signs of the terms of the polynomial being subtracted and then add.

Example 12   Subtract: $(2x^3 + 8x^2 - 6x) - (2x^3 - x^2 + 1)$.

Solution   First, change the sign of each term of the second polynomial and then add.

> **HELPFUL HINT**
> Notice the sign of
> each term is changed.

$$(2x^3 + 8x^2 - 6x) - (2x^3 - x^2 + 1) = (2x^3 + 8x^2 - 6x) + (-2x^3 + x^2 - 1)$$
$$= 2x^3 - 2x^3 + 8x^2 + x^2 - 6x - 1$$
$$= 9x^2 - 6x - 1 \qquad \text{Combine like terms.} \blacksquare$$

Example 13   Subtract $(5y^2 + 2y - 6)$ from $(-3y^2 - 2y + 11)$ using the vertical format.

Solution   Arrange the polynomials in vertical format, lining up like terms.

$$
\begin{array}{r}
-3y^2 - 2y + 11 \\
-(5y^2 + 2y - 6) \\
\hline
\end{array}
\qquad
\begin{array}{r}
-3y^2 - 2y + 11 \\
-5y^2 - 2y + \ 6 \\
\hline
-8y^2 - 4y + 17
\end{array}
$$

$\blacksquare$

Example 14   Subtract $(5z - 7)$ from the sum of $(8z + 11)$ and $(9z - 2)$.

Solution   Notice that $(5z - 7)$ is to be subtracted **from** a sum. The translation is

$$[(8z + 11) + (9z - 2)] - (5z - 7)$$
$$= 8z + 11 + 9z - 2 - 5z + 7 \qquad \text{Remove grouping symbols.}$$
$$= 8z + 9z - 5z + 11 - 2 + 7 \qquad \text{Group like terms.}$$
$$= 12z + 16 \qquad \text{Combine like terms.} \quad \blacksquare$$

Example 15   Add or subtract as indicated.

   a. $(3x^2 - 6xy + 5y^2) + (-2x^2 + 8xy - y^2)$
   b. $(9a^2b^2 + 6ab - 3ab^2) - (5b^2a + 2ab - 3 - 9b^2)$

Solution   **a.** $(3x^2 - 6xy + 5y^2) + (-2x^2 + 8xy - y^2)$
$$= 3x^2 - 6xy + 5y^2 - 2x^2 + 8xy - y^2$$
$$= x^2 + 2xy + 4y^2 \qquad \text{Combine like terms.}$$

   b. $(9a^2b^2 + 6ab - 3ab^2) - (5b^2a + 2ab - 3 - 9b^2)$
$$= 9a^2b^2 + 6ab - 3ab^2 - 5b^2a - 2ab + 3 + 9b^2$$

Change the sign of
each term of the
polynomial being
subtracted.

$$= 9a^2b^2 + 4ab - 8ab^2 + 3 + 9b^2 \qquad \text{Combine like terms.} \quad \blacksquare$$

MENTAL MATH

Combine like terms.

1. $-9y - 5y$

2. $6m^5 + 7m^5$

3. $4y^3 + 3y^3$

4. $21y^5 - 19y^5$

5. $x + 6x$

6. $7z - z$

Exercise Set 4.2

Find the degree of each of the following polynomials and determine whether it is a monomial, binomial, trinomial, or none of these. See Examples 1 through 3.

1. $x + 2$

2. $-6y + y^2 + 4$

3. $9m^3 - 5m^2 + 4m - 8$

4. $5a^2 + 3a^3 - 4a^4$

5. $12x^4y - x^2y^2 - 12x^2y^4$

6. $7r^2s^2 + 2r - 3s^5$

7. $3zx - 5x^2$

8. $5y + 2$

Match each polynomial in the first column with its degree in the second column. See Examples 1 through 3.

Polynomial	Degree
9. $3xy^2 - 4$	4
10. $8x^2y^2$	2
11. $5a^2 - 2a + 1$	6
12. $4z^6 + 3z^2$	3

13. Describe how to find the degree of a term.

14. Describe how to find the degree of a polynomial.

15. Explain why xyz is a monomial while $x + y + z$ is a trinomial.

16. Explain why the degree of the term $5y^3$ is 3 and the degree of the polynomial $2y + y + 2y$ is 1.

Find the value of each polynomial when **(a)** $x = 0$ *and* **(b)** $x = -1$. *See Examples 4 and 5.*

17. $x + 6$

18. $2x - 10$

19. $x^2 - 5x - 2$

20. $x^2 - 4$

21. $x^3 - 15$

22. $-2x^3 + 3x^2 - 6$

23. Find the height of the object in Example 5 when t is 10.8 seconds. Explain your result.

24. Approximate how long (to the nearest tenth of a second) before the object in Example 5 hits the ground.

Simplify each of the following by combining like terms. See Examples 6 and 7.

25. $14x^2 + 9x^2$

26. $18x^3 - 4x^3$

27. $15x^2 - 3x^2 - y$

28. $12k^3 - 9k^3 + 11$

29. $8s - 5s + 4s$

30. $5y + 7y - 6y$

31. $0.1y^2 - 1.2y^2 + 6.7 - 1.9$

32. $7.6y + 3.2y^2 - 8y - 2.5y^2$

33. $\frac{2}{5}x^2 - \frac{1}{3}x^3 + x^2 - \frac{1}{4}x^3 + 6$

34. $\frac{1}{6}x^4 - \frac{1}{7}x^2 + 5 - \frac{1}{2}x^4 - \frac{3}{7}x^2 + \frac{1}{3}$

35. $6a^2 - 4ab + 7b^2 - a^2 - 5ab + 9b^2$

36. $x^2y + xy - y + 10x^2y - 2y + xy$

Perform the indicated operations. See Examples 8 through 12.

37. $(3x + 7) + (9x + 5)$

38. $(3x^2 + 7) + (3x^2 + 9)$

39. $(-7x + 5) + (-3x^2 + 7x + 5)$

40. $(3x - 8) + (4x^2 - 3x + 3)$

41. $(2x^2 + 5) - (3x^2 - 9)$

42. $(5x^2 + 4) - (-2y^2 + 4)$

43. $3x - (5x - 9)$

44. $4 - (-y - 4)$

45. $(2x^2 + 3x - 9) - (-4x + 7)$

46. $(-7x^2 + 4x + 7) - (-8x + 2)$

47. Given the following triangle, find its perimeter.

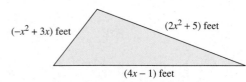

$(-x^2 + 3x)$ feet $(2x^2 + 5)$ feet

$(4x - 1)$ feet

48. Given the following quadrilateral, find its perimeter.

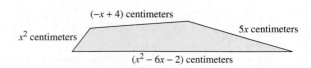

$(-x + 4)$ centimeters

$5x$ centimeters

x^2 centimeters

$(x^2 - 6x - 2)$ centimeters

△ **49.** A wooden beam is $(4y^2 + 4y + 1)$ meters long. If a piece $(y^2 - 10)$ meters is cut, express the length of the remaining piece of beam as a polynomial in y.

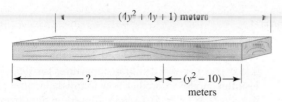

△ **50.** A piece of quarter-round molding is $(13x - 7)$ inches long. If a piece $(2x + 2)$ inches is removed, express the length of the remaining piece of molding as a polynomial in x.

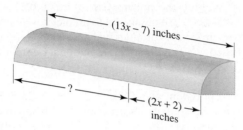

Perform the indicated operations. See Examples 10 and 13.

51.
$$3t^2 + 4$$
$$\underline{+5t^2 - 8}$$

52.
$$7x^3 + 3$$
$$\underline{+2x^3 + 1}$$

53.
$$4z^2 - 8z + 3$$
$$\underline{-(6z^2 + 8z - 3)}$$

54.
$$5u^5 - 4u^2 + 3u - 7$$
$$\underline{-(3u^5 + 6u^2 - 8u + 2)}$$

55.
$$5x^3 - 4x^2 + 6x - 2$$
$$\underline{-(3x^3 - 2x^2 - x - 4)}$$

56.
$$7a^2 - 9a + 6$$
$$\underline{-(11a^2 - 4a + 2)}$$

Perform the indicated operations. See Examples 13 and 14.

57. Subtract $(19x^2 + 5))$ from $(81x^2 + 10)$.

58. Subtract $(2x + xy)$ from $(3x - 9xy)$.

59. Subtract $(2x + 2)$ from the sum of $(8x + 1)$ and $(6x + 3)$.

60. Subtract $(-12x - 3)$ from the sum of $(-5x - 7)$ and $(12x + 3)$.

Perform the indicated operations.

61. $-15x - (-4x)$

62. $16y - (-4y)$

63. $2x - 5 + 5x - 8$

64. $x - 3 + 8x + 10$

65. $(-3y^2 - 4y) + (2y^2 + y - 1)$

66. $(7x^2 + 2x - 9) + (-3^2 + 5)$

67. $(5x + 8) - (-2x^2 - 6x + 8)$

68. $(-6y^2 + 3y - 4) - (9y^2 - 3y)$

69. $(-8x^4 + 7x) + (-8x^4 + x + 9)$

70. $(6y^5 - 6y^3 + 4) + (-2y^5 - 8y^3 - 7)$

71. $(3x^2 + 5x - 8) + (5x^2 + 9x + 12) - (x^2 - 14)$

72. $(-a^2 + 1) - (a^2 - 3) + (5a^2 - 6a + 7)$

73. Subtract $4x$ from $7x - 3$.

74. Subtract y from $y^2 - 4y + 1$.

75. Subtract $(5x + 7)$ from $(7x^2 + 3x + 9)$.

76. Subtract $(5y^2 + 8y + 2)$ from $(7y^2 + 9y - 8)$.

77. Subtract $(4y^2 - 6y - 3)$ from the sum of $(8y^2 + 7)$ and $(6y + 9)$.

78. Subtract $(5y + 7x^2)$ from the sum of $(8y - x)$ and $(3 + 8x^2)$.

79. Subtract $(-2x^2 + 4x - 12)$ from the sum of $(-x^2 - 2x)$ and $(5x^2 + x + 9)$.

80. Subtract $(4x^2 - 2x + 2)$ from the sum of $(x^2 + 7x + 1)$ and $(7x + 5)$.

81. $[(1.2x^2 - 3x + 9.1) - (7.8x^2 - 3.1 + 8)] + (1.2x - 6)$

82. $[(7.9y^4 - 6.8y^3 + 3.3y) + (6.1y^3 - 5)] - (4.2y^4 + 1.1y - 1)$

Find the area of each figure. Write a polynomial that describes the total area of the rectangles and squares shown in Exercises 83–84. Then simplify the polynomial.

△ **83.**

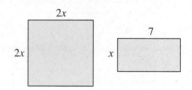

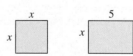

△ **84.**

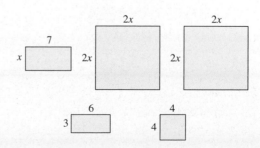

85. A rocket is fired upward from the ground with an initial velocity of 200 feet per second. Neglecting air resistance, the height of the rocket at any time t can be described by the polynomial $-16t^2 + 200t$. Find the height of the rocket at the given times.

    a. $t = 1$ second   **b.** $t = 5$ seconds

    c. $t = 7.6$ seconds   **d.** $t = 10.3$ seconds

86. Explain why the height of the rocket in Exercise 85 increases and then decreases as time passes.

87. Approximate (to the nearest tenth of a second) how long before the rocket in Exercise 85 hits the ground.

88. Write a polynomial that describes the surface area of the given figure in terms of its sides x and y. (Recall that the surface area of a solid is the sum of the areas of the faces or sides of the solid.)

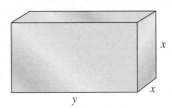

Add or subtract as indicated. See Example 15.

89. $(9a + 6b - 5) + (-11a - 7b + 6)$

90. $(3x - 2 + 6y) + (7x - 2 - y)$

91. $(4x^2 + y^2 + 3) - (x^2 + y^2 - 2)$

92. $(7a^2 - 3b^2 + 10) - (-2a^2 + b^2 - 12)$

93. $(x^2 + 2xy - y^2) + (5x^2 - 4xy + 20y^2)$

94. $(a^2 - ab + 4b^2) + (6a^2 + 8ab - b^2)$

95. $(11r^2s + 16rs - 3 - 2r^2s^2) - (3sr^2 + 5 - 9r^2s^2)$

96. $(3x^2y - 6xy + x^2y^2 - 5) - (11x^2y^2 - 1 + 5yx^2)$

Solve.

97. The number of U.S. airline departures (in millions) x years after 1993 is given by the polynomial $0.15x^2 + 0.15x + 7.2$ million departures for 1993–1995. Use this model to predict how many departures there were in 1998 ($x = 5$). (*Source:* Based on data from the National Transportation Safety Board)

98. The average number of physician visits per person age 75 years or older in the United States x years after 1995 is given by the polynomial $-0.1x^2 + 0.5x + 5.9$ for 1995–1997. Use this model to predict how many visits a person 75 years or older had in 2000 ($x = 5$). (*Source:* Based on data from the U.S. National Center of Health Statistics)

99. The polynomial $152.5x^2 - 187.5x + 25{,}525$ represents the annual U.S. production of beef (in millions of pounds) during 1996–1998. The polynomial $775.5x^2 - 670.5x + 17{,}763$ represents the combined annual U.S. production of veal, lamb, and pork (in millions of pounds) during 1996–1998. In both polynomials, x represents the number of years after 1996. Find a polynomial for the *total* U.S. meat production in millions of pounds during this period. (*Source:* Based on data from the U.S. Department of Agriculture)

100. The polynomial $61x^2 - 58x + 2719$ represents the total number of traffic fatalities in Florida for 1993–1995. The polynomial $944 + 48x - 31x^2$ represents the total number of alcohol-related traffic fatalities in Florida for 1993–1995. In both polynomials, x represents the number of years after 1993. Find a polynomial for the number of nonalcohol-related traffic fatalities in Florida during this period by subtracting the polynomials. (*Source:* Based on data from the Florida Highway Patrol)

REVIEW EXERCISES

Multiply. See Section 4.1.

101. $3x(2x)$ **102.** $-7x(x)$

103. $(12x^3)(-x^5)$ **104.** $6r^3(7r^{10})$

105. $10x^2(20xy^2)$ **106.** $-z^2y(11zy)$

4.3   MULTIPLYING POLYNOMIALS

CD-ROM

SSM

SSG   Video

► **OBJECTIVES**

1. Use the distributive property to multiply polynomials.
2. Multiply polynomials vertically.

1   To multiply polynomials, we apply our knowledge of the rules and definitions of exponents.

To multiply two monomials such as $(-5x^3)$ and $(-2x^4)$, use the associative and commutative properties and regroup. Remember that to multiply exponential expressions with a common base we add exponents.

$$(-5x^3)(-2x^4) = (-5)(-2)(x^3)(x^4) = 10x^7$$

To multiply polynomials that are not monomials, use the distributive property.

Example 1   Use the distributive property to find each product.

    a. $5x(2x^3 + 6)$ **b.** $-3x^2(5x^2 + 6x - 1)$

Solution **a.** $5x(2x^3 + 6) = 5x(2x^3) + 5x(6)$ *Use the distributive property.*

               $= 10x^4 + 30x$ *Multiply.*

  b. $-3x^2(5x^2 + 6x - 1)$
      $= (-3x^2)(5x^2) + (-3x^2)(6x) + (-3x^2)(-1)$ *Use the distributive property.*
      $= -15x^4 - 18x^3 + 3x^2$ *Multiply.*

We also use the distributive property to multiply two binomials. To multiply $(x + 3)$ by $(x + 1)$, distribute the factor $(x + 1)$ first.

$$(x + 3)(x + 1) = x(x + 1) + 3(x + 1) \qquad \text{Distribute } (x + 1).$$

$$= x(x) + x(1) + 3(x) + 3(1) \qquad \text{Apply distributive property a second time.}$$

$$= x^2 + x + 3x + 3 \qquad \text{Multiply.}$$

$$= x^2 + 4x + 3 \qquad \text{Combine like terms.}$$

This idea can be expanded so that we can multiply any two polynomials.

MULTIPLYING POLYNOMIALS

Multiply each term of the first polynomial by each term of the second polynomial, and then combine like terms.

Example 2   Find the product: $(3x + 2)(2x - 5)$.

Solution   Multiply each term of the first binomial by each term of the second.

$$(3x + 2)(2x - 5) = 3x(2x) + 3x(-5) + 2(2x) + 2(-5)$$
$$= 6x^2 - 15x + 4x - 10 \qquad \text{Multiply.}$$
$$= 6x^2 - 11x - 10 \qquad \text{Combine like terms.} \quad \blacksquare$$

Example 3   Multiply: $(2x - y)^2$.

Solution   Recall that $a^2 = a \cdot a$, so $(2x - y)^2 = (2x - y)(2x - y)$. Multiply each term of the first polynomial by each term of the second.

$$(2x - y)(2x - y) = 2x(2x) + 2x(-y) + (-y)(2x) + (-y)(-y)$$
$$= 4x^2 - 2xy - 2xy + y^2 \qquad \text{Multiply.}$$
$$= 4x^2 - 4xy + y^2 \qquad \text{Combine like terms.} \quad \blacksquare$$

Example 4   Multiply: $(t + 2)$ by $(3t^2 - 4t + 2)$.

Solution   Multiply each term of the first polynomial by each term of the second.

$$(t + 2)(3t^2 - 4t + 2) = t(3t^2) + t(-4t) + t(2) + 2(3t^2) + 2(-4t) + 2(2)$$
$$= 3t^3 - 4t^2 + 2t + 6t^2 - 8t + 4$$
$$= 3t^3 + 2t^2 - 6t + 4 \qquad \text{Combine like terms.} \quad \blacksquare$$

Example 5   Multiply: $(3a + b)^3$.

Solution   Write $(3a + b)^3$ as $(3a + b)(3a + b)(3a + b)$.

$$(3a + b)(3a + b)(3a + b) = (9a^2 + 3ab + 3ab + b^2)(3a + b)$$
$$= (9a^2 + 6ab + b^2)(3a + b)$$
$$= (9a^2 + 6ab + b^2)3a + (9a^2 + 6ab + b^2)b$$
$$= 27a^3 + 18a^2b + 3ab^2 + 9a^2b + 6ab^2 + b^3$$
$$= 27a^3 + 27a^2b + 9ab^2 + b^3 \quad \blacksquare$$

2   Another convenient method for multiplying polynomials is to use a vertical format similar to the format used to multiply real numbers. We demonstrate this method by multiplying $(3y^2 - 4y + 1)$ by $(y + 2)$.

Example 6   Multiply: $(3y^2 - 4y + 1)(y + 2)$. Use a vertical format.

Solution   **Step 1.**   Multiply 2 by each term of the top polynomial. Write the first **partial product** below the line.

$$\begin{array}{r} 3y^2 - 4y + 1 \\ \times \qquad y + 2 \\ \hline 6y^2 - 8y + 2 \end{array} \qquad \text{Partial product}$$

Step 2. Multiply y by each term of the top polynomial. Write this partial product underneath the previous one, being careful to line up like terms.

$$
\begin{array}{r}
3y^2 - 4y + 1 \\
\times \qquad y + 2 \\
\hline
6y^2 - 8y + 2 \\
3y^3 - 4y^2 + \ y
\end{array}
$$

Partial product.

Partial product.

Step 3. Combine like terms of the partial products.

$$
\begin{array}{r}
3y^2 - 4y + 1 \\
\times \qquad y + 2 \\
\hline
6y^2 - 8y + 2 \\
3y^3 - 4y^2 + \ y \\
\hline
3y^3 + 2y^2 - 7y + 2
\end{array}
$$

Combine like terms.

Thus, $(y + 2)(3y^2 - 4y + 1) = 3y^3 + 2y^2 - 7y + 2$. ▬

When multiplying vertically, be careful if a power is missing, you may want to leave space in the partial products and take care that like terms are lined up.

Example 7 Multiply: $(2x^3 - 3x + 4)(x^2 + 1)$. Use a vertical format.

Solution

$$
\begin{array}{r}
2x^3 - 3x + 4 \\
\times \qquad x^2 + 1 \\
\hline
2x^3 \qquad - 3x + 4 \\
2x^5 - 3x^3 + 4x^2 \\
\hline
2x^5 - \ x^3 + 4x^2 - 3x + 4
\end{array}
$$

Leave space for missing powers of x.

Combine like terms.

▬

MENTAL MATH

Find the following products mentally.

1. $5x(2y)$
2. $7a(4b)$
3. $x^2 \cdot x^5$
4. $z \cdot z^4$
5. $6x(3x^2)$
6. $5a^2(3a^2)$

Exercise Set 4.3

Find the following products. See Example 1.

1. $2a(2a - 4)$
2. $3a(2a + 7)$
3. $7x(x^2 + 2x - 1)$
4. $-5y(y^2 + y - 10)$
5. $3x^2(2x^2 - x)$
6. $-4y^2(5y - 6y^2)$

△ 7. The area of the larger rectangle below is $x(x + 3)$. Find another expression for this area by finding the sum of the areas of the smaller rectangles.

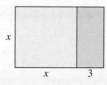

△ **8.** Write an expression for the area of the larger rectangle below in two different ways.

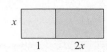

Find the following products. See Examples 2 and 3.

9. $(a + 7)(a - 2)$ **10.** $(y + 5)(y + 7)$
11. $(2y - 4)^2$ **12.** $(6x - 7)^2$
13. $(5x - 9y)(6x - 5y)$ **14.** $(3x - 7y)(7x + 2y)$
15. $(2x^2 - 5)^2$ **16.** $(x^2 - 4)^2$

△ **17.** The area of the figure below is $(x + 2)(x + 3)$. Find another expression for this area by finding the sum of the areas of the smaller rectangles.

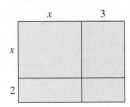

△ **18.** Write an expression for the area of the figure below in two different ways.

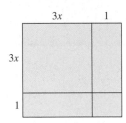

Find the following products. See Example 4.

19. $(x - 2)(x^2 - 3x + 7)$ **20.** $(x + 3)(x^2 + 5x - 8)$
21. $(x + 5)(x^3 - 3x + 4)$ **22.** $(a + 2)(a^3 - 3a^2 + 7)$
23. $(2a - 3)(5a^2 - 6a + 4)$
24. $(3 + b)(2 - 5b - 3b^2)$

Find the following products. See Example 5.

25. $(x + 2)^3$ **26.** $(y - 1)^3$
27. $(2y - 3)^3$ **28.** $(3x + 4)^3$

Find the following products. Use the vertical multiplication method. See Examples 6 and 7.

29. $(x + 3)(2x^2 + 4x - 1)$
30. $(2x - 5)(3x^2 - 4x + 7)$
31. $(x^3 + 5x - 7)(x^2 - 9)$
32. $(3x^4 - x^2 + 2)(2x^3 + 1)$

33. Evaluate each of the following.
  a. $(2 + 3)^2; 2^2 + 3^2$
  b. $(8 + 10)^2; 8^2 + 10^2$
  c. Does $(a + b)^2 = a^2 + b^2$ no matter what the values of a and b are? Why or why not?

34. Perform the indicated operations. Explain the difference between the two expressions.
  a. $(3x + 5) + (3x + 7)$
  b. $(3x + 5)(3x + 7)$

Find the following products.

35. $2a(a + 4)$ **36.** $-3a(2a + 7)$
37. $3x(2x^2 - 3x + 4)$ **38.** $-4x(5x^2 - 6x - 10)$
39. $(5x + 9y)(3x + 2y)$ **40.** $(5x - 5y)(2x - y)$
41. $(x + 2)(x^2 + 5x + 6)$
42. $(x - 7)(x^2 - 15x + 56)$
43. $(7x + 4)^2$ **44.** $(3x - 2)^2$
45. $-2a^2(3a^2 - 2a + 3)$
46. $-4b^2(3b^3 - 12b^2 - 6)$
47. $(x + 3)(x^2 + 7x + 12)$
48. $(n + 1)(n^2 - 7n - 9)$
49. $(a + 1)^3$ **50.** $(x - y)^3$
51. $(x + y)(x + y)$ **52.** $(x + 3)(7x + 1)$
53. $(x - 7)(x - 6)$ **54.** $(4x + 5)(-3x + 2)$
55. $3a(a^2 + 2)$ **56.** $x^3(x + 12)$
57. $-4y(y^2 + 3y - 11)$ **58.** $-2x(5x^2 - 6x + 1)$
59. $(5x + 1)(5x - 1)$ **60.** $(2x + y)(3x - y)$
61. $(5x + 4)(x^2 - x + 4)$ **62.** $(x - 2)(x^2 - x + 3)$
63. $(2x - 5)^3$ **64.** $(3y - 1)^3$
65. $(4x + 5)(8x^2 + 2x - 4)$ **66.** $(x + 7)(x^2 - 7x - 8)$
67. $(7xy - y)^2$ **68.** $(x + y)^2$
69. $(5y^2 - y + 3)(y^2 - 3y - 2)$
70. $(2x^2 + x - 1)(x^2 + 3x + 4)$
71. $(3x^2 + 2x - 4)(2x^2 - 4x + 3)$
72. $(a^2 + 3a - 2)(2a^2 - 5a - 1)$

Express each of the following as polynomials.

△ **73.** Find the area of the rectangle.

△ **74.** Find the area of the square field.

←——— $(x + 4)$ feet ———→

△ **75.** Find the area of the triangle.

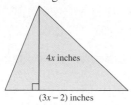

$4x$ inches

$(3x - 2)$ inches

△ **76.** Find the volume of the cube-shaped glass block.

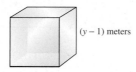

$(y - 1)$ meters

△ **77.** Write a polynomial that describes the area of the shaded region.

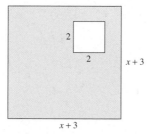

2

2

$x + 3$

$x + 3$

△ **78.** Write a polynomial that describes the area of the shaded region.

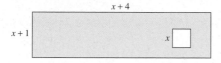

$x + 4$

$x + 1$

x

79. Multiply the following polynomials.
a. $(a + b)(a - b)$
b. $(2x + 3y)(2x - 3y)$
c. $(4x + 7)(4x - 7)$
d. Can you make a general statement about all products of the form $(x + y)(x - y)$?

REVIEW EXERCISES

Perform the indicated operation. See Section 4.1.

80. $(5x)^2$ **81.** $(4p)^2$
82. $(-3y^3)^2$ **83.** $(-7m^2)^2$

*For income tax purposes, Rob Calcutta, the owner of Copy Services, uses a method called **straight-line depreciation** to show the depreciated (or decreased) value of a copy machine he recently purchased. Rob assumes that he can use the machine for 7 years. The graph below shows the depreciated value of the machine over the years. Use this graph to answer Exercises 84–89. See Sections 1.9 and 3.1.*

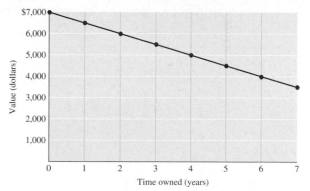

84. What was the purchase price of the copy machine?
85. What is the depreciated value of the machine in 7 years?
86. What loss in value occurred during the first year?
87. What loss in value occurred during the second year?
88. Why do you think this method of depreciating is called straight-line depreciation?
89. Why is the line tilted downward?

4.4 SPECIAL PRODUCTS

CD-ROM   SSM

SSG   Video

▶ **OBJECTIVES**

1. Multiply two binomials using the FOIL method.
2. Square a binomial.
3. Multiply the sum and difference of two terms.

1   In this section, we multiply binomials using special products. First, a special order for multiplying binomials called the FOIL order or method is introduced. This method is demonstrated by multiplying $(3x + 1)$ by $(2x + 5)$.

F stands for the product of the **First** terms. $(3x + 1)(2x + 5)$

$$(3x)(2x) = 6x^2 \quad \textbf{F}$$

O stands for the product of the **Outer** terms. $(3x + 1)(2x + 5)$

$$(3x)(5) = 15x \quad \textbf{O}$$

I stands for the product of the **Inner** terms. $(3x + 1)(2x + 5)$

$$(1)(2x) = 2x \quad \textbf{I}$$

L stands for the product of the **Last** terms. $(3x + 1)(3x + 5)$

$$(1)(5) = 5 \quad \textbf{L}$$

$$
\begin{array}{cccc}
\text{F} & \text{O} & \text{I} & \text{L}
\end{array}
$$
$$(3x + 1)(2x + 5) = 6x^2 + 15x + 2x + 5$$
$$= 6x^2 + 17x + 5 \qquad \textit{Combine like terms.}$$

Example 1 Find $(x - 3)(x + 4)$ by the FOIL method.

Solution
$$
\begin{array}{cccc}
 & \text{F} & \text{O} & \text{I} & \text{L}
\end{array}
$$
$$(x - 3)(x + 4) = (x)(x) + (x)(4) + (-3)(x) + (-3)(4)$$

$$= x^2 + 4x - 3x - 12$$
$$= x^2 + x - 12 \qquad \textit{Combine like terms.}$$

Example 2 Find $(5x - 7)(x - 2)$ by the FOIL method.

Solution
$$
\begin{array}{cccc}
\text{F} & \text{O} & \text{I} & \text{L}
\end{array}
$$
$$(5x - 7)(x - 2) = 5x(x) + 5x(-2) + (-7)(x) + (-7)(-2)$$

$$= 5x^2 - 10x - 7x + 14$$
$$= 5x^2 - 17x + 14 \qquad \textit{Combine like terms.}$$

Example 3 Multiply: $(y + 6)(2y - 1)$.

Solution
$$
\begin{array}{cccc}
\text{F} & \text{O} & \text{I} & \text{L}
\end{array}
$$
$$(y + 6)(2y - 1) = 2y^2 - 1y + 12y - 6$$
$$= 2y^2 + 11y - 6$$

2   Now, try squaring a binomial using the FOIL method.

Example 4   Multiply: $(3y + 1)^2$.

Solution $(3y + 1)^2 = (3y + 1)(3y + 1)$

$$\begin{array}{cccc} \text{F} & \text{O} & \text{I} & \text{L} \end{array}$$
$$= (3y)(3y) + (3y)(1) + 1(3y) + 1(1)$$
$$= 9y^2 + 3y + 3y + 1$$
$$= 9y^2 + 6y + 1$$

Notice the pattern that appears in Example 4.

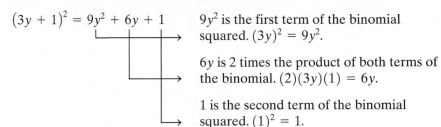

$$(3y + 1)^2 = 9y^2 + 6y + 1$$

$9y^2$ is the first term of the binomial squared. $(3y)^2 = 9y^2$.

$6y$ is 2 times the product of both terms of the binomial. $(2)(3y)(1) = 6y$.

1 is the second term of the binomial squared. $(1)^2 = 1$.

This pattern leads to the following, which can be used when squaring a binomial. We call these **special products**.

SQUARING A BINOMIAL

A binomial squared is equal to the square of the first term plus or minus twice the product of both terms plus the square of the second term.

$$(a + b)^2 = a^2 + 2ab + b^2$$
$$(a - b)^2 = a^2 - 2ab + b^2$$

This product can be visualized geometrically.

The area of the large square is side · side.

Area $= (a + b)(a + b) = (a + b)^2$

The area of the large square is also the sum of the areas of the smaller rectangles.

Area $= a^2 + ab + ab + b^2 = a^2 + 2ab + b^2$

Thus, $(a + b)^2 = a^2 + 2ab + b^2$.

Example 5 Use a special product to square each binomial.

a. $(t + 2)^2$ **b.** $(p - q)^2$ **c.** $(2x + 5)^2$ **d.** $(x^2 - 7y)^2$

Solution

first term squared	plus or minus	twice the product of the terms	plus	second term squared

a. $(t + 2)^2 = t^2 + 2(t)(2) + 2^2 = t^2 + 4t + 4$
b. $(p - q)^2 = p^2 - 2(p)(q) + q^2 = p^2 - 2pq + q^2$
c. $(2x + 5)^2 = (2x)^2 + 2(2x)(5) + 5^2 = 4x^2 + 20x + 25$
d. $(x^2 - 7y)^2 = (x^2)^2 - 2(x^2)(7y) + (7y^2) = x^4 - 14x^2y + 49y^2$ ■

HELPFUL HINT
Notice that

$$(a + b)^2 \neq a^2 + b^2 \qquad \text{The middle term } 2ab \text{ is missing.}$$

$$(a + b)^2 = (a + b)(a + b) = a^2 + 2ab + b^2$$

Likewise,
$$(a - b)^2 \neq a^2 - b^2$$

$$(a - b)^2 = (a - b)(a - b) = a^2 - 2ab + b^2$$

Another special product is the product of the sum and difference of the same two terms, such as $(x + y)(x - y)$. Finding this product by the FOIL method, we see a pattern emerge.

$$(x + y)(x - y) = x^2 - xy + xy - y^2$$

$$= x^2 - y^2$$

Notice that the middle two terms subtract out. This is because the **O**uter product is the opposite of the **I**nner product. Only the **difference of squares** remains.

---

3 | **MULTIPLYING THE SUM AND DIFFERENCE OF TWO TERMS**

The product of the sum and difference of two terms is the square of the first term minus the square of the second term.

$$(a + b)(a - b) = a^2 - b^2$$

Example 6 Use a special product to multiply.

    a. $(x + 4)(x - 4)$ **b.** $(6t + 7)(6t - 7)$ **c.** $\left(x - \dfrac{1}{4}\right)\left(x + \dfrac{1}{4}\right)$

    d. $(2p - q)(2p + q)$ **e.** $(3x^2 - 5y)(3x^2 + 5y)$

Solution

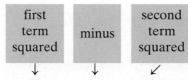

first term squared	minus	second term squared

    a. $(x + 4)(x - 4) = x^2 \qquad - \qquad 4^2 = x^2 - 16$

    b. $(6t + 7)(6t - 7) = (6t)^2 \qquad - \qquad 7^2 = 36t^2 - 49$

    c. $\left(x - \dfrac{1}{4}\right)\left(x + \dfrac{1}{4}\right) = x^2 - \left(\dfrac{1}{4}\right)^2 = x^2 - \dfrac{1}{16}$

    d. $(2p - q)(2p + q) = (2p)^2 - q^2 = 4p^2 - q^2$

    e. $(3x^2 - 5y)(3x^2 + 5y) = (3x^2)^2 - (5y)^2 = 9x^4 - 25y^2$

    Let's now practice multiplying polynomials in general. If possible, use a special product.

Example 7 Multiply.

    a. $(x - 5)(3x + 4)$ **b.** $(7x + 4)^2$ **c.** $(y - 0.6)(y + 0.6)$

    d. $(a - 3)(a^2 + 2a - 1)$

Solution   **a.** $(x - 5)(3x + 4) = 3x^2 + 4x - 15x - 20$ FOIL.
                      $= 3x^2 - 11x - 20$

    b. $(7x + 4)^2 = (7x)^2 + 2(7x)(4) + 4^2$ Squaring a binomial.
                   $= 49x^2 + 56x + 16$

    c. $(y - 0.6)(y + 0.6) = y^2 - (0.6)^2 = y^2 - 0.36$ Multiplying the sum
                                       and difference of 2 terms.

    d. $(a - 3)(a^2 + 2a - 1) = a(a^2 + 2a - 1) - 3(a^2 + 2a - 1)$ Multiplying each
                        $= a^3 + 2a^2 - a - 3a^2 - 6a + 3$ term of the binomial
                        $= a^3 - a^2 - 7a + 3$ by each term of the
                                               trinomial.

MENTAL MATH

Answer each exercise true or false.

  1. $(x + 4)^2 = x^2 + 16$ **2.** For $(x + 6)(2x - 1)$ the product of the first terms is $2x^2$.

  3. $(x + 4)(x - 4) = x^2 + 16$ **4.** The product $(x - 1)(x^3 + 3x - 1)$ is a polynomial of degree 5.

Exercise Set 4.4

Find each product using the FOIL method. See Examples 1 through 3.

1. $(x + 3)(x + 4)$ **2.** $(x + 5)(x - 1)$
3. $(x - 5)(x + 10)$ **4.** $(y - 12)(y + 4)$
5. $(5x - 6)(x + 2)$ **6.** $(3y - 5)(2y - 7)$
7. $(y - 6)(4y - 1)$ **8.** $(2x - 9)(x - 11)$
9. $(2x + 5)(3x - 1)$ **10.** $(6x + 2)(x - 2)$

Find each product. See Examples 4 and 5.

11. $(x - 2)^2$ **12.** $(x + 7)^2$
13. $(2x - 1)^2$ **14.** $(7x - 3)^2$
15. $(3a - 5)^2$ **16.** $(5a + 2)^2$
17. $(5x + 9)^2$ **18.** $(6s - 2)^2$

19. Using your own words, explain how to square a binomial such as $(a + b)^2$.

20. Explain how to find the product of two binomials using the FOIL method.

Find each product. See Example 6.

21. $(a - 7)(a + 7)$ **22.** $(b + 3)(b - 3)$
23. $(3x - 1)(3x + 1)$ **24.** $(4x - 5)(4x + 5)$
25. $\left(3x - \dfrac{1}{2}\right)\left(3x + \dfrac{1}{2}\right)$ **26.** $\left(10x + \dfrac{2}{7}\right)\left(10x - \dfrac{2}{7}\right)$
27 $(9x + y)(9x - y)$ **28.** $(2x - y)(2x + y)$

Express each of the following as a polynomial in x.

29. Find the area of the square rug shown if its side is $(2x + 1)$ feet.

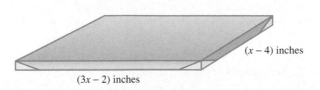

(2x + 1) feet

(2x + 1) feet

30. Find the area of the rectangular canvas if its length is $(3x - 2)$ inches and its width is $(x - 4)$ inches.

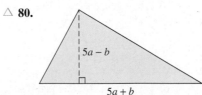

(x – 4) inches

(3x – 2) inches

Find each product. See Example 7.

31. $(a + 5)(a + 4)$ **32.** $(a - 5)(a - 7)$
33. $(a + 7)^2$ **34.** $(b - 2)^2$
35. $(4a + 1)(3a - 1)$ **36.** $(6a + 7)(6a + 5)$
37. $(x + 2)(x - 2)$ **38.** $(x - 10)(x + 10)$
39. $(3a + 1)^2$ **40.** $(4a - 2)^2$
41. $(x + y)(4x - y)$ **42.** $(3x + 2)(4x - 2)$
43. $(x + 3)(x^2 - 6x + 1)$ **44.** $(x - 2)(x^2 - 4x + 2)$
45. $(2a - 3)^2$ **46.** $(5b - 4x)^2$
47. $(5x - 6z)(5x + 6z)$ **48.** $(11x - 7y)(11x + 7y)$
49. $(x - 3)(x - 5)$ **50.** $(a + 5b)(a + 6b)$
51. $\left(x - \dfrac{1}{3}\right)\left(x + \dfrac{1}{3}\right)$ **52.** $\left(3x + \dfrac{1}{5}\right)\left(3x - \dfrac{1}{5}\right)$
53. $(a + 11)(a - 3)$ **54.** $(2x + 5)(x - 8)$
55. $(x - 2)^2$ **56.** $(3b + 7)^2$
57. $(3b + 7)(2b - 5)$ **58.** $(3y - 13)(y - 3)$
59. $(7p - 8)(7p + 8)$ **60.** $(3s - 4)(3s + 4)$
61. $\left(\dfrac{1}{3}a^2 - 7\right)\left(\dfrac{1}{3}a^2 + 7\right)$ **62.** $\left(\dfrac{2}{3}a - b^2\right)\left(\dfrac{2}{3}a - b^2\right)$
63. $5x^2(3x^2 - x + 2)$ **64.** $4x^3(2x^2 + 5x - 1)$
65. $(2r - 3s)(2r + 3s)$ **66.** $(6r - 2x)(6r + 2x)$
67. $(3x - 7y)^2$ **68.** $(4s - 2y)^2$
69. $(4x + 5)(4x - 5)$ **70.** $(3x + 5)(3x - 5)$
71. $(x + 4)(x + 4)$ **72.** $(3x + 2)(3x + 2)$
73. $\left(a - \dfrac{1}{2}y\right)\left(a + \dfrac{1}{2}y\right)$ **74.** $\left(\dfrac{a}{2} + 4y\right)\left(\dfrac{a}{2} - 4y\right)$
75. $\left(\dfrac{1}{5}x - y\right)\left(\dfrac{1}{5}x + y\right)$ **76.** $\left(\dfrac{y}{6} - 8\right)\left(\dfrac{y}{6} + 8\right)$
77. $(a + 1)(3a^2 - a + 1)$ **78.** $(b + 3)(2b^2 + b - 3)$

Find the area of each shaded region.

79.

2x + 3

2x – 3

x
x

80.

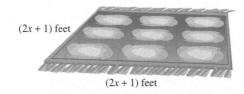

5a – b

5a + b

△ **81.**

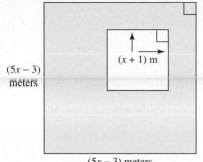

(5x − 3) meters

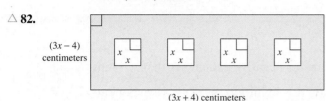

(x + 1) m

(5x − 3) meters

△ **82.**

(3x − 4) centimeters

x x

x x

x x

x x

(3x + 4) centimeters

83. When is the product of two binomials a binomial? To help you answer this question, find the product (x + 2)(x + 2) and then find the product (x + 2)(x − 2).

REVIEW EXERCISES

Simplify each expression. See Section 4.1.

84. $\dfrac{50b^{10}}{70b^5}$

85. $\dfrac{x^3y^6}{xy^2}$

86. $\dfrac{8a^{17}b^5}{-4a^7b^{10}}$

87. $\dfrac{-6a^8y}{3a^4y}$

88. $\dfrac{2x^4y^{12}}{3x^4y^4}$

89. $\dfrac{-48ab^6}{32ab^3}$

Find the slope of each line. See Section 3.4.

90.

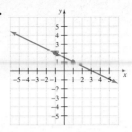

91.

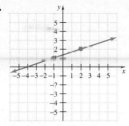

92.

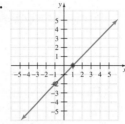

93.

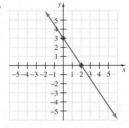

A Look Ahead

Example

Find the product $[(a + b) - 2][(a + b) + 2]$.

Solution:

If we think of $(a + b)$ as one term, we can think of $[(a + b) - 2][(a + b) + 2]$ as the product of the sum and difference of two terms.

$$[(a + b) - 2][(a + b) + 2] = (a + b)^2 - 2^2$$

Next, square $(a + b)$. $= a^2 + 2ab + b^2 - 4$

Find each product.

94. $[(x + y) - 3][(x + y) + 3]$

95. $[(a + c) - 5][(a + c) + 5]$

96. $[(a - 3) + b][(a - 3) - b]$

97. $[(x - 2) + y][(x - 2) - y]$

4.5 NEGATIVE EXPONENTS AND SCIENTIFIC NOTATION

CD-ROM

SSM

SSG   Video

▶ **OBJECTIVES**

1. Evaluate numbers raised to negative integer powers.
2. Use all the rules and definitions for exponents to simplify exponential expressions.
3. Write numbers in scientific notation.
4. Convert numbers from scientific notation to standard form.

1 Our work with exponential expressions so far has been limited to exponents that are positive integers or 0. Here we expand to give meaning to an expression like x^{-3}.

Suppose that we wish to simplify the expression $\dfrac{x^2}{x^5}$. If we use the quotient rule for exponents, we subtract exponents:

$$\frac{x^2}{x^5} = x^{2-5} = x^{-3}, \quad x \neq 0$$

But what does x^{-3} mean? Let's simplify $\frac{x^2}{x^5}$ using the definition of x^n.

$$\frac{x^2}{x^5} = \frac{x \cdot x}{x \cdot x \cdot x \cdot x \cdot x}$$

$$= \frac{x \cdot x}{x \cdot x \cdot x \cdot x \cdot x}$$
Divide numerator and denominator by common factors by applying the fundamental principle for fractions.

$$= \frac{1}{x^3}$$

If the quotient rule is to hold true for negative exponents, then x^{-3} must equal $\frac{1}{x^3}$.

From this example, we state the definition for negative exponents.

NEGATIVE EXPONENTS

If a is a real number other than 0 and n is an integer, then

$$a^{-n} = \frac{1}{a^n}$$

In other words, another way to write a^{-n} is to take its reciprocal and change the sign of its exponent.

◆ **Example 1**   Simplify by writing each expression with positive exponents only.

a. 3^{-2}

b. $2x^{-3}$

c. $2^{-1} + 4^{-1}$

d. $(-2)^{-4}$

Solution   a. $3^{-2} = \frac{1}{3^2} = \frac{1}{9}$ Use the definition of negative exponents.

b. $2x^{-3} = 2 \cdot \frac{1}{x^3} = \frac{2}{x^3}$ Use the definition of negative exponents.

> **HELPFUL HINT**
> Don't forget that since there are no parentheses, only x is the base for the exponent -3.

c. $2^{-1} + 4^{-1} = \frac{1}{2} + \frac{1}{4} = \frac{2}{4} + \frac{1}{4} = \frac{3}{4}$

d. $(-2)^{-4} = \frac{1}{(-2)^4} = \frac{1}{(-2)(-2)(-2)(-2)} = \frac{1}{16}$

> ▼
> **HELPFUL HINT**
> A negative exponent *does not affect* the sign of its base.
> Remember: Another way to write a^{-n} is to take its reciprocal and change the sign of its exponent, $a^{-n} = \dfrac{1}{a^n}$. For example,
>
> $$x^{-2} = \frac{1}{x^2}, \qquad 2^{-3} = \frac{1}{2^3} \quad \text{or} \quad \frac{1}{8}$$
>
> $$\frac{1}{y^{-4}} = \frac{1}{\frac{1}{y^4}} = y^4, \quad \frac{1}{5^{-2}} = 5^2 \quad \text{or} \quad 25$$

Example 2   Simplify each expression. Write results using positive exponents only.

   a. $\left(\dfrac{2}{3}\right)^{-3}$ 　　　　　　　**b.** $\dfrac{1}{x^{-3}}$ 　　　　　　　**c.** $\dfrac{p^{-4}}{q^{-9}}$

Solution   **a.** $\left(\dfrac{2}{3}\right)^{-3} = \dfrac{2^{-3}}{3^{-3}} = \dfrac{3^3}{2^3} = \dfrac{27}{8}$ 　　Use the negative exponent rule.

   b. $\dfrac{1}{x^{-3}} = x^3$

   c. $\dfrac{p^{-4}}{q^{-9}} = \dfrac{q^9}{p^4}$

Example 3   Simplify each expression. Write answers with positive exponents.

   a. $\dfrac{y}{y^{-2}}$ 　　　　　　　**b.** $\dfrac{3}{x^{-4}}$ 　　　　　　　**c.** $\dfrac{x^{-5}}{x^7}$

Solution   **a.** $\dfrac{y}{y^{-2}} = \dfrac{y^1}{y^{-2}} = y^{1-(-2)} = y^3$

   b. $\dfrac{3}{x^{-4}} = 3 \cdot \dfrac{1}{x^{-4}} = 3 \cdot x^4 \quad \text{or} \quad 3x^4$

   c. $\dfrac{x^{-5}}{x^7} = x^{-5-7} = x^{-12} = \dfrac{1}{x^{12}}$

<u>2</u>   All the previously stated rules for exponents apply for negative exponents also. Here is a summary of the rules and definitions for exponents.

SUMMARY OF EXPONENT RULES

If m and n are integers and a, b, and c are real numbers, then:

Product rule for exponents: $a^m \cdot a^n = a^{m+n}$

Power rule for exponents: $(a^m)^n = a^{m \cdot n}$

Power of a product: $(ab)^n = a^n b^n$

Power of a quotient: $\left(\dfrac{a}{c}\right)^n = \dfrac{a^n}{c^n}, \quad c \neq 0$

Quotient rule for exponents: $\dfrac{a^m}{a^n} = a^{m-n}, \quad a \neq 0$

Zero exponent: $a^0 = 1, \quad a \neq 0$

Negative exponent: $a^{-n} = \dfrac{1}{a^n}, \quad a \neq 0$

Example 4 Simplify the following expressions. Write each result using positive exponents only.

a. $\dfrac{(x^3)^4 x}{x^7}$

b. $\left(\dfrac{3a^2}{b}\right)^{-3}$

c. $\dfrac{4^{-1}x^{-3}y}{4^{-3}x^2 y^{-6}}$

d. $(y^{-3}z^6)^{-6}$

e. $\left(\dfrac{-2x^3 y}{xy^{-1}}\right)^3$

Solution

a. $\dfrac{(x^3)^4 x}{x^7} = \dfrac{x^{12} \cdot x}{x^7} = \dfrac{x^{12+1}}{x^7} = \dfrac{x^{13}}{x^7} = x^{13-7} = x^6$ Use the power rule.

b. $\left(\dfrac{3a^2}{b}\right)^{-3} = \dfrac{3^{-3}(a^2)^{-3}}{b^{-3}}$ Raise each factor in the numerator and the denominator to the -3 power.

$= \dfrac{3^{-3}a^{-6}}{b^{-3}}$ Use the power rule.

$= \dfrac{b^3}{3^3 a^6}$ Use the negative exponent rule.

$= \dfrac{b^3}{27a^6}$ Write 3^3 as 27.

c. $\dfrac{4^{-1}x^{-3}y}{4^{-3}x^2 y^{-6}} = 4^{-1-(-3)}x^{-3-2}y^{1-(-6)} = 4^2 x^{-5} y^7 = \dfrac{4^2 y^7}{x^5} = \dfrac{16 y^7}{x^5}$

d. $(y^{-3}z^6)^{-6} = y^{18} \cdot z^{-36} = \dfrac{y^{18}}{z^{36}}$

e. $\left(\dfrac{-2x^3 y}{xy^{-1}}\right)^3 = \dfrac{(-2)^3 x^9 y^3}{x^3 y^{-3}} = \dfrac{-8x^9 y^3}{x^3 y^{-3}} = -8x^{9-3}y^{3-(-3)} = -8x^6 y^6$

3 Both very large and very small numbers frequently occur in many fields of science. For example, the distance between the sun and the planet Pluto is approximately 5,906,000,000 kilometers, and the mass of a proton is approximately 0.00000000000000000000000165 gram. It can be tedious to write these numbers in this standard decimal notation, so **scientific notation** is used as a convenient shorthand for expressing very large and very small numbers.

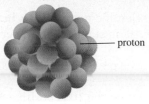

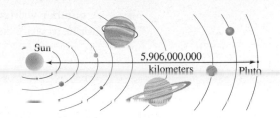

5,906,000,000 kilometers

Mass of proton is approximately
0.000 000 000 000 000 000 000 001 65 gram

SCIENTIFIC NOTATION

A positive number is written in scientific notation if it is written as the product of a number a, where $1 \leq a < 10$, and an integer power r of 10:
$$a \times 10^r$$

The numbers below are written in scientific notation. The $\times$ sign for multiplication is used as part of the notation.

2.03×10^2 $\quad$ 7.362×10^7 $\quad$ 5.906×10^9 $\quad$ (Distance between the sun and Pluto)

1×10^{-3} $\quad$ 8.1×10^{-5} $\quad$ 1.65×10^{-24} $\quad$ (Mass of a proton)

The following steps are useful when writing numbers in scientific notation.

WRITING A NUMBER IN SCIENTIFIC NOTATION

Step 1. Move the decimal point in the original number so that the new number has a value between 1 and 10.

Step 2. Count the number of decimal places the decimal point is moved in Step 1. If the decimal point is moved to the left, the count is positive. If the decimal point is moved to the right, the count is negative.

Step 3. Multiply the new number in Step 1 by 10 raised to an exponent equal to the count found in Step 2.

Example 5 Write each number in scientific notation.

a. 367,000,000 $\qquad$ **b.** 0.000003 $\qquad$ **c.** 20,520,000,000 $\qquad$ **d.** 0.00085

Solution **a. Step 1.** Move the decimal point until the number is between 1 and 10.

367,000,000

8 places

Step 2. The decimal point is moved to the left 8 places, so the count is positive 8.

Step 3. $367,000,000 = 3.67 \times 10^8$.

b. Step 1. Move the decimal point until the number is between 1 and 10.

0.000003

6 places

Step 2. The decimal point is moved 6 places to the right, so the count is -6.

Step 3. $0.000003 = 3.0 \times 10^{-6}$

c. $20,520,000,000 = 2.052 \times 10^{10}$

d. $0.00085 = 8.5 \times 10^{-4}$

4 A number written in scientific notation can be rewritten in standard form. For example, to write 8.63×10^3 in standard form, recall that $10^3 = 1000$.

$$8.63 \times 10^3 = 8.63(1000) = 8630$$

Notice that the exponent on the 10 is positive 3, and we moved the decimal point 3 places to the right.

To write 7.29×10^{-3} in standard form, recall that $10^{-3} = \dfrac{1}{10^3} = \dfrac{1}{1000}$.

$$7.29 \times 10^{-3} = 7.29\left(\frac{1}{1000}\right) = \frac{7.29}{1000} = 0.00729$$

The exponent on the 10 is negative 3, and we moved the decimal to the left 3 places.

In general, **to write a scientific notation number in standard form**, move the decimal point the same number of places as the exponent on 10. If the exponent is positive, move the decimal point to the right; if the exponent is negative, move the decimal point to the left.

Example 6   Write each number in standard notation, without exponents.

a. 1.02×10^5 **b.** 7.358×10^{-3} **c.** 8.4×10^7 **d.** 3.007×10^{-5}

Solution   **a.** Move the decimal point 5 places to the right.

$$1.02 \times 10^5 = 102,000.$$

b. Move the decimal point 3 places to the left.

$$7.358 \times 10^{-3} = 0.007358$$

c. $8.4 \times 10^7 = 84,000,000.$ *7 places to the right*

d. $3.007 \times 10^{-5} = 0.00003007$ *5 places to the left*

Performing operations on numbers written in scientific notation makes use of the rules and definitions for exponents.

Example 7   Perform each indicated operation. Write each result in standard decimal notation.

a. $(8 \times 10^{-6})(7 \times 10^3)$

b. $\dfrac{12 \times 10^2}{6 \times 10^{-3}}$

Solution   **a.** $(8 \times 10^{-6})(7 \times 10^3) = 8 \cdot 7 \cdot 10^{-6} \cdot 10^3$
$$= 56 \times 10^{-3}$$
$$= 0.056$$

b. $\dfrac{12 \times 10^2}{6 \times 10^{-3}} = \dfrac{12}{6} \times 10^{2-(-3)} = 2 \times 10^5 = 200{,}000$

CALCULATOR EXPLORATIONS

Scientific Notation

To enter a number written in scientific notation on a scientific calculator, locate the scientific notation key, which may be marked $\boxed{\text{EE}}$ or $\boxed{\text{EXP}}$. To enter 3.1×10^7, press $\boxed{3.1}$ $\boxed{\text{EE}}$ $\boxed{7}$. The display should read $\boxed{3.1 \quad 07}$.

Enter each number written in scientific notation on your calculator.

  1. 5.31×10^3 **2.** -4.8×10^{14}
  3. 6.6×10^{-9} **4.** -9.9811×10^{-2}

Multiply each of the following on your calculator. Notice the form of the result.

  5. $3{,}000{,}000 \times 5{,}000{,}000$ **6.** $230{,}000 \times 1000$

Multiply each of the following on your calculator. Write the product in scientific notation.

  7. $(3.26 \times 10^6)(2.5 \times 10^{13})$ **8.** $(8.76 \times 10^{-4})(1.237 \times 10^9)$

SPOTLIGHT ON DECISION MAKING

Suppose you are a paralegal for a law firm. You are investigating the facts of a mineral rights case. Drilling on property adjacent to the client's has struck a natural gas reserve. The client believes that a portion of this reserve lies within her own property boundaries and she is, therefore, entitled to a portion of the proceeds from selling the natural gas. As part of your investigation, you contact two different experts, who fax you the following estimates for the size of the natural gas reserve:

Expert A	*Expert B*
Estimate of entire reserve:	Estimate of entire reserve:
4.6×10^7 cubic feet	6.7×10^6 cubic feet
Estimate of size of reserve on client's property:	Estimate of size of reserve on client's property:
1.84×10^7 cubic feet	2.68×10^6 cubic feet

How, if at all, would you use these estimates in the case? Explain.

MENTAL MATH

State each expression using positive exponents only.

1. $5x^{-2}$

2. $3x^{-3}$

3. $\dfrac{1}{y^{-6}}$

4. $\dfrac{1}{x^{-3}}$

5. $\dfrac{4}{y^{-3}}$

6. $\dfrac{16}{y^{-7}}$

Exercise Set 4.5

Simplify each expression. Write each result using positive exponents only. See Examples 1 through 3.

1. 4^{-3}

2. 6^{-2}

3. $7x^{-3}$

4. $(7x)^{-3}$

5. $\left(-\dfrac{1}{4}\right)^{-3}$

6. $\left(-\dfrac{1}{8}\right)^{-2}$

7. $3^{-1} + 2^{-1}$

8. $4^{-1} + 4^{-2}$

9. $\dfrac{1}{p^{-3}}$

10. $\dfrac{1}{q^{-5}}$

11. $\dfrac{p^{-5}}{q^{-4}}$

12. $\dfrac{r^{-5}}{s^{-2}}$

13. $\dfrac{x^{-2}}{x}$

14. $\dfrac{y}{y^{-3}}$

15. $\dfrac{z^{-4}}{z^{-7}}$

16. $\dfrac{x^{-4}}{x^{-1}}$

17. $2^0 + 3^{-1}$

18. $4^{-2} - 4^{-3}$

19. $(-3)^{-2}$

20. $(-2)^{-6}$

21. $\dfrac{-1}{p^{-4}}$

22. $\dfrac{-1}{y^{-6}}$

23. $-2^0 - 3^0$

24. $5^0 + (-5)^0$

Simplify each expression. Write each result using positive exponents only. See Example 4.

25. $\dfrac{x^2 x^5}{x^3}$

26. $\dfrac{y^4 y^5}{y^6}$

27. $\dfrac{p^2 p}{p^{-1}}$

28. $\dfrac{y^3 y}{y^{-2}}$

29. $\dfrac{(m^5)^4 m}{m^{10}}$

30. $\dfrac{(x^2)^8 x}{x^9}$

31. $\dfrac{r}{r^{-3} r^{-2}}$

32. $\dfrac{p}{p^{-3} q^{-5}}$

33. $(x^5 y^3)^{-3}$

34. $(z^5 x^5)^{-3}$

35. $\dfrac{(x^2)^3}{x^{10}}$

36. $\dfrac{(y^4)^2}{y^{12}}$

37. $\dfrac{(a^5)^2}{(a^3)^4}$

38. $\dfrac{(x^2)^5}{(x^4)^3}$

39. $\dfrac{8k^4}{2k}$

40. $\dfrac{27r^4}{3r^6}$

41. $\dfrac{-6m^4}{-2m^3}$

42. $\dfrac{15a^4}{-15a^5}$

43. $\dfrac{-24a^6 b}{6ab^2}$

44. $\dfrac{-5x^4 y^5}{15x^4 y^2}$

45. $\dfrac{6x^2 y^3}{-7xy^5}$

46. $\dfrac{-8xa^2 b}{-5xa^5 b}$

47. $(a^{-5} b^2)^{-6}$

48. $(4^{-1} x^5)^{-2}$

49. $\left(\dfrac{x^{-2} y^4}{x^3 y^7}\right)^2$

50. $\left(\dfrac{a^5 b}{a^7 b^{-2}}\right)^{-3}$

51. $\dfrac{4^2 z^{-3}}{4^3 z^{-5}}$

52. $\dfrac{3^{-1} x^4}{3^3 x^{-7}}$

53. $\dfrac{2^{-3} x^{-4}}{2^2 x}$

54. $\dfrac{5^{-1} z^7}{5^{-2} z^9}$

55. $\dfrac{7ab^{-4}}{7^{-1} a^{-3} b^2}$

56. $\dfrac{6^{-5} x^{-1} y^2}{6^{-2} x^{-4} y^4}$

57. $\left(\dfrac{a^{-5} b}{ab^3}\right)^{-4}$

58. $\left(\dfrac{r^{-2} s^{-3}}{r^{-4} s^{-3}}\right)^{-3}$

59. $\dfrac{(xy^3)^5}{(xy)^{-4}}$

60. $\dfrac{(rs)^{-3}}{(r^2 s^3)^2}$

61. $\dfrac{(-2xy^{-3})^{-3}}{(xy^{-1})^{-1}}$

62. $\dfrac{(-3x^2 y^2)^{-2}}{(xyz)^{-2}}$

△ **63.** Find the volume of the cube.

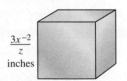

$\dfrac{3x^{-2}}{z}$ inches

△ **64.** Find the area of the triangle.

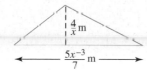

Write each number in scientific notation. See Example 5.

65. 78,000   **66.** 9,300,000,000

67. 0.00000167   **68.** 0.00000017

69. 0.00635   **70.** 0.00194

71. 1,160,000   **72.** 700,000

73. The temperature at the interior of the Earth is 20,000,000 degrees Celsius. Write 20,000,000 in scientific notation.

74. The half-life of a carbon isotope is 5000 years. Write 5000 in scientific notation.

75. The distance between the Earth and the sun is 93,000,000 miles. Write 93,000,000 in scientific notation.

76. The population of the world is 6,067,000,000. Write 6,067,000,000 in scientific notation. (*Source*: U.S. Bureau of the Census)

77. On March 23, 1997, Comet Hale-Bopp passed its closest to Earth. It was 120,000,000 miles away. (*Source: World Almanac and Book of Facts,* 1997) Write this number in scientific notation.

78. The highest price paid for a diamond is $16,550,000. The diamond is 100.10 carets and pear-shaped and was sold in 1995 in Switzerland. Write this money amount in scientific notation. (*Source: The Guinness Book of Records,* 1999)

Write each number in standard notation. See Example 6.

79. 8.673×10^{-10}

80. 9.056×10^{-4}

81. 3.3×10^{-2}

82. 4.8×10^{-6}

83. 2.032×10^{4}

84. 9.07×10^{10}

85. One coulomb of electricity is 6.25×10^{18} electrons. Write this number in standard notation.

86. The mass of a hydrogen atom is 1.7×10^{-24} grams. Write this number in standard notation.

87. The distance light travels in 1 year is 9.460×10^{12} kilometers. Write this number in standard notation.

88. The population of the United States is 2.68×10^{8}. Write this number in standard notation. (*Source:* U.S. Bureau of the Census)

Evaluate each expression using exponential rules. Write each result in standard notation. See Example 7.

89. $(1.2 \times 10^{-3})(3 \times 10^{-2})$

90. $(2.5 \times 10^{6})(2 \times 10^{-6})$

91. $(4 \times 10^{-10})(7 \times 10^{-9})$

92. $(5 \times 10^{6})(4 \times 10^{-8})$

93. $\dfrac{8 \times 10^{-1}}{16 \times 10^{5}}$ **94.** $\dfrac{25 \times 10^{-4}}{5 \times 10^{-9}}$

95. $\dfrac{1.4 \times 10^{-2}}{7 \times 10^{-8}}$ **96.** $\dfrac{0.4 \times 10^{5}}{0.2 \times 10^{11}}$

97. The average amount of water flowing past the mouth of the Amazon River is 4.2×10^{6} cubic feet per second. How much water flows past in an hour? (*Hint:* 1 hour equals 3600 seconds.) Write the result in scientific notation.

98. A beam of light travels 9.460×10^{12} kilometers per year. How far does light travel in 10,000 years? Write the result in scientific notation.

99. Explain why $(a^{-1})^{3}$ has the same value as $(a^{3})^{-1}$.

100. Determine whether each statement is true or false.

   a. $5^{-1} < 5^{-2}$

   b. $\left(\dfrac{1}{5}\right)^{-1} < \left(\dfrac{1}{5}\right)^{-2}$

   c. $a^{-1} < a^{-2}$ for all nonzero numbers.

101. If $a = \dfrac{1}{10}$, then find the value of a^{-2}.

102. It was stated earlier that, for an integer n,

$$x^{-n} = \dfrac{1}{x^n}, \quad x \neq 0$$

Explain why x may not equal 0.

Simplify. Write results in standard notation.

103. $(2.63 \times 10^{12})(-1.5 \times 10^{-10})$

104. $(6.785 \times 10^{-4})(4.68 \times 10^{10})$

Light travels at a rate of 1.86×10^5 miles per second. Use this information and the distance formula $d = r \cdot t$ to answer Exercises 105 and 106.

105. If the distance from the moon to the Earth is 238,857 miles, find how long it takes the reflected light of the moon to reach the Earth. (Round to the nearest tenth of a second.)

106. If the distance from the sun to the Earth is 93,000,000 miles, find how long it takes the light of the sun to reach the Earth. (Round to the nearest second.)

REVIEW EXERCISES

Simplify the following. See Section 4.1.

107. $\dfrac{5x^7}{3x^4}$

108. $\dfrac{27y^{14}}{3y^7}$

109. $\dfrac{15z^4 y^3}{21zy}$

110. $\dfrac{18a^7 b^{17}}{30a^7 b}$

Use the distributive property and multiply. See Section 4.3.

111. $\dfrac{1}{y}(5y^2 - 6y + 5)$

112. $\dfrac{2}{x}(3x^5 + x^4 - 2)$

113. $2x^2\left(10x - 6 + \dfrac{1}{x}\right)$

114. $-5y^3\left(2y^2 - 4y + 2 - \dfrac{3}{y}\right)$

A Look Ahead

Example

Simplify the following expressions. Assume that the variable in the exponent represents an integer value.

   a. $x^{m+1} \cdot x^m$ **b.** $(z^{2x+1})^x$ **c.** $\dfrac{y^{6a}}{y^{4a}}$

Solution:

   a. $x^{m+1} \cdot x^m = x^{(m+1)+m} = x^{2m+1}$

   b. $(z^{2x+1})^x = z^{(2x+1)x} = z^{2x^2+x}$

   c. $\dfrac{y^{6a}}{y^{4a}} = y^{6a-4a} = y^{2a}$

Simplify each expression. Assume that variables represent positive integers. See the example.

115. $a^{-4m} \cdot a^{5m}$

116. $(x^{-3s})^3$

117. $(3y^{2z})^3$

118. $a^{4m+1} \cdot a^4$

119. $\dfrac{y^{4a}}{y^{-a}}$

120. $\dfrac{y^{-6a}}{zy^{6a}}$

121. $(z^{3a+2})^{-2}$

122. $(a^{4x-1})^{-1}$

4.6 DIVISION OF POLYNOMIALS

CD-ROM   SSM

SSG   Video

▶ **OBJECTIVES**

1. Divide a polynomial by a monomial.
2. Use long division to divide a polynomial by another polynomial.

1   Now that we know how to add, subtract, and multiply polynomials, we practice dividing polynomials.

To divide a polynomial by a monomial, recall addition of fractions. Fractions that have a common denominator are added by adding the numerators:

$$\frac{a}{c} + \frac{b}{c} = \frac{a+b}{c}$$

If we read this equation from right to left and let a, b, and c be monomials, $c \neq 0$, we have the following:

DIVIDING A POLYNOMIAL BY A MONOMIAL

Divide each term of the polynomial by the monomial.

$$\frac{a + b}{c} = \frac{a}{c} + \frac{b}{c}, \quad c \neq 0$$

Throughout this section, we assume that denominators are not 0.

Example 1   Divide: $6m^2 + 2m$ by $2m$

Solution   We begin by writing the quotient in fraction form. Then we divide each term of the polynomial $6m^2 + 2m$ by the monomial $2m$.

$$\frac{6m^2 + 2m}{2m} = \frac{6m^2}{2m} + \frac{2m}{2m}$$

$$= 3m + 1 \qquad \text{Simplify.}$$

Check   To check, we multiply.

$$2m(3m + 1) = 2m(3m) + 2m(1) = 6m^2 + 2m$$

The quotient $3m + 1$ checks.

Example 2   Divide: $\dfrac{9x^5 - 12x^2 + 3x}{3x^2}$

Solution $\dfrac{9x^5 - 12x^2 + 3x}{3x^2} = \dfrac{9x^5}{3x^2} - \dfrac{12x^2}{3x^2} + \dfrac{3x}{3x^2}$ Divide each term by $3x^2$.

$$= 3x^3 - 4 + \frac{1}{x} \qquad \text{Simplify.}$$

Notice that the quotient is not a polynomial because of the term $\dfrac{1}{x}$. This expression is called a rational expression—we will study rational expressions further in Chapter 5. Although the quotient of two polynomials is not always a polynomial, we may still check by multiplying.

Check $3x^2\left(3x^3 - 4 + \dfrac{1}{x}\right) = 3x^2(3x^3) - 3x^2(4) + 3x^2\left(\dfrac{1}{x}\right)$

$$= 9x^5 - 12x^2 + 3x$$

Example 3   Divide: $\dfrac{8x^2y^2 - 16xy + 2x}{4xy}$

Solution $\dfrac{8x^2y^2 - 16xy + 2x}{4xy} = \dfrac{8x^2y^2}{4xy} - \dfrac{16xy}{4xy} + \dfrac{2x}{4xy}$ Divide each term by $4xy$.

$$= 2xy - 4 + \dfrac{1}{2y}$$ Simplify.

Check $4xy\left(2xy - 4 + \dfrac{1}{2y}\right) = 4xy(2xy) - 4xy(4) + 4xy\left(\dfrac{1}{2y}\right)$

$$= 8x^2y^2 - 16xy + 2x$$

2   To divide a polynomial by a polynomial other than a monomial, we use a process known as long division. Polynomial long division is similar to number long division, so we review long division by dividing 13 into 3660.

$$
\begin{array}{r}
281 \\
13\overline{)3660} \\
26 \\
\hline
106 \\
104 \\
\hline
20 \\
13 \\
\hline
7
\end{array}
$$

$2 \cdot 13 = 26$

Subtract and bring down the next digit in the dividend.

$8 \cdot 13 = 104$

Subtract and bring down the next digit in the dividend.

$1 \cdot 13 = 13$

Subtract. There are no more digits to bring down, so the remainder is 7.

The quotient is 281   R   7, which can be written as $281 \dfrac{7}{13}$ $\begin{array}{l}\leftarrow \text{remainder}\\ \leftarrow \text{divisor}\end{array}$.

Recall that division can be checked by multiplication. To check a division problem such as this one, we see that

$$13 \cdot 281 + 7 = 3660$$

Now we demonstrate long division of polynomials.

Example 4   Divide $x^2 + 7x + 12$ by $x + 3$ using long division.

Solution

To subtract, change the signs of these terms and add.

$$
\begin{array}{r}
x \\
x + 3\overline{)x^2 + 7x + 12} \\
x^2 + 3x \\
\hline
4x + 12
\end{array}
$$

How many times does x divide x^2? $\dfrac{x^2}{x} = x$.

Multiply: $x(x + 3)$.

Subtract and bring down the next term.

Now we repeat this process.

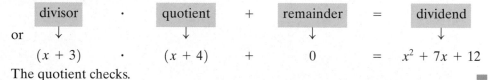

$$\begin{array}{r} x + 4 \\ x + 3 \overline{)x^2 + 7x + 12} \\ x^2 + 3x \\ \hline 4x + 12 \\ \end{array}$$

How many times does x divide $4x$? $\dfrac{4x}{x} = 4.$

To subtract, change the signs of these terms and add. $\longrightarrow$ $\overline{}4x \not{+} 12$

0

Multiply: $4(x + 3)$.

Subtract. The remainder is 0.

The quotient is $x + 4$.

Check We check by multiplying.

divisor	·	quotient	+	remainder	=	dividend
↓		↓		↓		↓

or

$(x + 3)$ · $(x + 4)$ + 0 = $x^2 + 7x + 12$

The quotient checks.

Example 5 Divide $6x^2 + 10x - 5$ by $3x - 1$ using long division.

Solution

$$\begin{array}{r} 2x + 4 \\ 3x - 1 \overline{)6x^2 + 10x - 5} \\ \overline{}6x^2 \not{} 2x \downarrow \\ \hline 12x - 5 \\ \overline{}12x \not{} 4 \\ \hline -1 \end{array}$$

$\dfrac{6x^2}{3x} = 2x$, so $2x$ is a term of the quotient.

Multiply $2x(3x - 1)$

Subtract and bring down the next term.

$\dfrac{12x}{3x} = 4, 4(3x - 1)$

Subtract. The remainder is -1.

Thus $(6x^2 + 10x - 5)$ divided by $(3x - 1)$ is $(2x + 4)$ with a remainder of -1. This can be written as

$$\frac{6x^2 + 10x - 5}{3x - 1} = 2x + 4 + \frac{-1}{3x - 1} \quad \begin{array}{l}\leftarrow \text{ remainder} \\ \leftarrow \text{ divisor}\end{array}$$

Check To check, we multiply $(3x - 1)(2x + 4)$. Then we add the remainder, -1, to this product.

$$(3x - 1)(2x + 4) + (-1) = (6x^2 + 12x - 2x - 4) - 1$$
$$= 6x^2 + 10x - 5$$

The quotient checks.

Notice that the division process is continued until the degree of the remainder polynomial is less than the degree of the divisor polynomial.

Example 6 Divide: $\dfrac{4x^2 + 7 + 8x^3}{2x + 3}$

Solution Before we begin the division process, we rewrite $4x^2 + 7 + 8x^3$ as $8x^3 + 4x^2 + 0x + 7$. Notice that we have written the polynomial in descending order and have represented the missing x term by $0x$.

$$
\begin{array}{r}
4x^2 - \ 4x + \ 6 \\
2x + 3 \overline{)8x^3 + \ 4x^2 + \ 0x + \ 7} \\
\underline{8x^3 \mp 12x^2 } \\
-8x^2 + \ 0x \\
\underline{\mp \ 8x^2 \mp 12x } \\
12x + \ 7 \\
\underline{^{-}12x \mp 18} \\
-11 \qquad \text{Remainder.}
\end{array}
$$

Thus, $\dfrac{4x^2 + 7 + 8x^3}{2x + 3} = 4x^2 - 4x + 6 + \dfrac{-11}{2x + 3}.$ ▪

Example 7 Divide: $\dfrac{2x^4 - x^3 + 3x^2 + x - 1}{x^2 + 1}$

Solution Before dividing, rewrite the divisor polynomial $(x^2 + 1)$ as $(x^2 + 0x + 1)$. The $0x$ term represents the missing x^1 term in the divisor.

$$
\begin{array}{r}
2x^2 - \ x + 1 \\
x^2 + 0x + 1 \overline{)2x^4 - \ x^3 + 3x^2 + \ x - 1} \\
\underline{^{-}2x^4 \mp 0x^3 \mp 2x^2 } \\
-x^3 + \ x^2 + \ x \\
\underline{\mp \ x^3 \mp 0x^2 \mp \ x } \\
x^2 + 2x - 1 \\
\underline{^{-}x^2 \mp 0x \mp 1} \\
2x - 2 \qquad \text{Remainder polynomial.}
\end{array}
$$

Thus, $\dfrac{2x^4 - x^3 + 3x^2 + x - 1}{x^2 + 1} = 2x^2 - x + 1 + \dfrac{2x - 2}{x^2 + 1}.$ ▪

M E N T A L   M A T H

Simplify each expression mentally.

1. $\dfrac{a^6}{a^4}$ **2.** $\dfrac{y^2}{y}$ **3.** $\dfrac{a^3}{a}$ **4.** $\dfrac{p^8}{p^3}$ **5.** $\dfrac{k^5}{k^2}$

6. $\dfrac{k^7}{k^5}$ **7.** $\dfrac{p^8}{p^3}$ **8.** $\dfrac{k^5}{k^2}$ **9.** $\dfrac{k^7}{k^5}$

Exercise Set 4.6

Perform each division. See Examples 1 through 3.

1. $\dfrac{15p^3 + 18p^7}{3p}$

2. $\dfrac{14m^2 - 27m^3}{7m}$

3. $\dfrac{9x^4 + 18x^5}{6x^5}$

4. $\dfrac{6x^5 + 3x^4}{3x^4}$

5. $\dfrac{-9x^5 + 3x^4 - 12}{3x^3}$

6. $\dfrac{6a^2 - 4a + 12}{2a^2}$

7. $\dfrac{4x^4 - 6x^3 + 7}{-4x^4}$

8. $\dfrac{-12a^3 + 36a - 15}{3a}$

9. $\dfrac{25x^5 - 15x^3 + 5}{5x^2}$

10. $\dfrac{-4y^2 + 4y + 6}{2y}$

△ **11.** The perimeter of a square is $(12x^3 + 4x - 16)$ feet. Find the length of its side.

Perimeter is
$(12x^3 + 4x - 16)$
feet

△ **12.** The volume of the swimming pool shown is $(36x^5 - 12x^3 + 6x^2)$ cubic feet. If its height is $2x$ feet and its width is $3x$ feet, find its length.

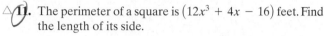

3x feet

2x feet

Perform each division. See Examples 4 through 7.

13. $\dfrac{x^2 + 4x + 3}{x + 3}$

14. $\dfrac{x^2 + 7x + 10}{x + 5}$

15. $\dfrac{2x^2 + 13x + 15}{x + 5}$

16. $\dfrac{3x^2 + 8x + 4}{x + 2}$

17. $\dfrac{2x^2 - 7x + 3}{x - 4}$

18. $\dfrac{3x^2 - x - 4}{x - 1}$

19. $\dfrac{8x^2 + 6x - 27}{2x - 3}$

20. $\dfrac{18w^2 + 18w - 8}{3w + 4}$

21. $\dfrac{9a^3 - 3a^2 - 3a + 4}{3a + 2}$

22. $\dfrac{-x^3 - 6x^2 + 2x - 3}{x - 1}$

23. $\dfrac{2b^3 + 9b^2 + 6b - 4}{b + 4}$

24. $\dfrac{2x^3 + 3x^2 - 3x + 4}{x + 2}$

25. Explain how to check a polynomial long division result when the remainder is 0.

26. Explain how to check a polynomial long division result when the remainder is 0.

△ **27.** The area of the following parallelogram is $(10x^2 + 31x + 15)$ square meters. If its base is $(5x + 3)$ meters, find its height.

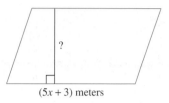

$(5x + 3)$ meters

△ **28.** The area of the top of the Ping-Pong table is $(49x^2 + 70x - 200)$ square inches. If its length is $(7x + 20)$ inches, find its width.

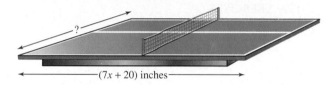

?

$(7x + 20)$ inches

Perform each division.

29. $\dfrac{20x^2 + 5x + 9}{5x^3}$

30. $\dfrac{8x^3 - 4x^2 + 6x + 2}{2x^2}$

31. $\dfrac{5x^2 + 28x - 10}{x + 6}$

32. $\dfrac{2x^2 + x - 15}{x + 3}$

33. $\dfrac{10x^3 - 24x^2 - 10x}{10x}$

34. $\dfrac{2x^3 + 12x^2 + 16}{4x^2}$

35. $\dfrac{6x^2 + 17x - 4}{x + 3}$

36. $\dfrac{2x^2 - 9x + 15}{x - 6}$

37. $\dfrac{12x^4 + 3x^2}{3x^2}$

38. $\dfrac{15x^2 - 9x^5}{9x^5}$

39. $\dfrac{2x^3 + 2x^2 - 17x + 8}{x - 2}$

40. $\dfrac{4x^3 + 11x^2 - 8x - 10}{x + 3}$

41. $\dfrac{30x^2 - 17x + 2}{5x - 2}$

42. $\dfrac{4x^2 - 13x - 12}{4x + 3}$

43. $\dfrac{3x^4 - 9x^3 + 12}{-3x}$

44. $\dfrac{8y^6 - 3y^2 - 4y}{4y}$

45. $\dfrac{8x^2 + 10x + 1}{2x + 1}$

46. $\dfrac{3x^2 + 17x + 7}{3x + 2}$

47. $\dfrac{4x^2 - 81}{2x - 9}$

48. $\dfrac{16x^2 - 36}{4x + 6}$

49. $\dfrac{4x^3 + 12x^2 + x - 12}{2x + 3}$

50. $\dfrac{6x^2 + 11x - 10}{3x - 2}$

51. $\dfrac{x^3 - 27}{x - 3}$

52. $\dfrac{x^3 + 64}{x + 4}$

53. $\dfrac{x^3 + 1}{x + 1}$

54. $\dfrac{x^5 + x^2}{x^2 + x}$

55. $\dfrac{1 - 3x^2}{x + 2}$

56. $\dfrac{7 - 5x^2}{x + 3}$

57. $\dfrac{-4b + 4b^2 - 5}{2b - 1}$

58. $\dfrac{-3y + 2y^2 - 15}{2y + 5}$

65. $9ab(ab^2c + 4bc - 8)$

66. $-7sr(6s^2r + 9sr^2 + 9rs + 8)$

Use the bar graph below to answer Exercises 67–70. See Section 1.9.

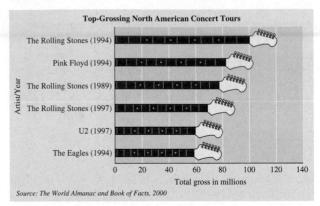

Source: The World Almanac and Book of Facts, 2000

REVIEW EXERCISES

Multiply each expression. See Section 4.3.

59. $2a(a^2 + 1)$

60. $-4a(3a^2 - 4)$

61. $2x(x^2 + 7x - 5)$

62. $4y(y^2 - 8y - 4)$

63. $-3xy(xy^2 + 7x^2y + 8)$

64. $-9xy(4xyz + 7xy^2z + 2)$

67. Which artist has grossed the most money on tour?

68. Estimate the amount of money made by the concert tour of Pink Floyd.

69. Estimate the amount of money made by the 1997 concert tour of U2.

70. Which artist shown has grossed the least amount of money on a tour?

For additional Chapter Projects, visit the Real World Activities Website by going to http://www.prenhall.com/martin-gay.

CHAPTER PROJECT

Modeling with Polynomials

The polynomial model $1.65x^2 + 14.77x + 364.27$ dollars represents consumer spending per person per year on all U.S. media from 1990 to 1996. This includes spending on subscription TV services, recorded music, newspapers, magazines, books, home video, theater movies, video games, and educational software. The polynomial model $0.81x^2 + 3.41x + 88.98$ dollars represents consumer spending per person per year on subscription TV services alone during this same period. In both models, x is the number of years after 1990. (*Source:* Based on data from Veronis, Suhler & Associates, New York, NY, *Communications Industry Report*, annual)

In this project, you will have the opportunity to investigate these polynomial models numerically, algebraically, and graphically. This project may be completed by working in groups or individually.

1. Use the polynomials to complete the following table showing the annual consumer spending per person over the period 1990–1996 by evaluating each polynomial at the given values of x. Then subtract each value in the fourth column from the corresponding value in the third column. Record the result in the last column, "Difference." What do you think these values represent? What trends do you notice in the data?

Year	x	Consumer Spending per Person per Year on ALL U.S. Media	Consumer spending per Person per Year on Subscription TV	Difference
1990	0			
1992	2			
1994	4			
1996	6			

2. Use the polynomial models to find a new polynomial model representing the amount of consumer spending per person on U.S. media other than subscription TV services (such as recorded music, newspapers, magazines, books, home video, theater movies, video games, and educational software). Then use this new polynomial model to complete the following table.

Year	x	Consumer Spending per Person per Year on Media Other Than Subscription TV
1990	0	
1992	2	
1994	4	
1996	6	

3. Compare the values in the last column of the table in Question 1 to the values in the last column of the table in Question 2. What do you notice? What can you conclude?

4. Use the polynomial models to estimate consumer spending on (a) all U.S. media, (b) subscription TV, and (c) media other than subscription TV for the year 1998.

5. Use the polynomial models to estimate consumer spending on (a) all U.S. media, (b) subscription TV, and (c) media other than subscription TV for the year 2000.

6. Create a bar graph that represents the data for consumer spending on all U.S. media in the years 1990, 1992, 1994, and 1996 along with your estimates for 1998 and 2000. Study your bar graph. Discuss what the graph implies about the future.

CHAPTER 4 VOCABULARY CHECK

Fill in each blank with one of the words or phrases listed below.

term   coefficient   monomial   binomial   trinomial
polynomials   degree of a term   degree of a polynomial   FOIL

1. A _____ is a number or the product of numbers and variables raised to powers.
2. The _____ method may be used when multiplying two binomials.
3. A polynomial with exactly 3 terms is called a _____.
4. The _____ is the greatest degree of any term of the polynomial.
5. A polynomial with exactly 2 terms is called a _____.
6. The _____ of a term is its numerical factor.
7. The _____ is the sum of the exponents on the variables in the term.
8. A polynomial with exactly 1 term is called a _____.
9. Monomials, binomials, and trinomials are all examples of _____.

CHAPTER 4 HIGHLIGHTS

DEFINITIONS AND CONCEPTS	EXAMPLES

Section 4.1 Exponents

a^n means the product of n factors, each of which is a.

$3^2 = 3 \cdot 3 = 9$

$(-5)^3 = (-5)(-5)(-5) = -125$

$\left(\dfrac{1}{2}\right)^4 = \dfrac{1}{2} \cdot \dfrac{1}{2} \cdot \dfrac{1}{2} \cdot \dfrac{1}{2} = \dfrac{1}{16}$

If m and n are integers and no denominators are 0,

Product Rule: $a^m \cdot a^n = a^{m+n}$

$x^2 \cdot x^7 = x^{2+7} = x^9$

Power Rule: $(a^m)^n = a^{mn}$

$(5^3)^8 = 5^{3 \cdot 8} = 5^{24}$

Power of a Product Rule: $(ab)^n = a^n b^n$

$(7y)^4 = 7^4 y^4$

Power of a Quotient Rule: $\left(\dfrac{a}{b}\right)^n = \dfrac{a^n}{b^n}$

$\left(\dfrac{x}{8}\right)^3 = \dfrac{x^3}{8^3}$

Quotient Rule: $\dfrac{a^m}{a^n} = a^{m-n}$

$\dfrac{x^9}{x^4} = x^{9-4} = x^5$

Zero Exponent: $a^0 = 1, a \neq 0$.

$5^0 = 1, x^0 = 1, x \neq 0$

Section 4.2 Adding and Subtracting Polynomials

A **term** is a number or the product of numbers and variables raised to powers.

Terms

$-5x, 7a^2b, \dfrac{1}{4}y^4, 0.2$

The **numerical coefficient** or **coefficient** of a term is its numerical factor.

Term	*Coefficient*
$7x^2$	7
y	1
$-a^2b$	-1

A **polynomial** is a term or a finite sum of terms in which variables may appear in the numerator raised to whole number powers only.
A **monomial** is a polynomial with exactly 1 term.
A **binomial** is a polynomial with exactly 2 terms.
A **trinomial** is a polynomial with exactly 3 terms.

Polynomials

$3x^2 - 2x + 1$	(Trinomial)
$-0.2a^2b - 5b^2$	(Binomial)
$\dfrac{5}{6}y^3$	(Monomial)

The **degree of a term** is the sum of the exponents on the variables in the term.

Term	*Degree*
$-5x^3$	3
3 (or $3x^0$)	0
$2a^2b^2c$	5

The **degree of a polynomial** is the greatest degree of any term of the polynomial.

Polynomial	*Degree*
$5x^2 - 3x + 2$	2
$7y + 8y^2z^3 - 12$	$2 + 3 = 5$

To add polynomials, add like terms.

Add:
$(7x^2 - 3x + 2) + (-5x - 6) = 7x^2 - 3x + 2 - 5x - 6$
$= 7x^2 - 8x - 4$

To subtract two polynomials, change the signs of the terms of the second polynomial, then add.

Subtract:
$(17y^2 - 2y + 1) - (-3y^3 + 5y - 6)$
$= (17y^2 - 2y + 1) + (3y^3 - 5y + 6)$
$= 17y^2 - 2y + 1 + 3y^3 - 5y + 6$
$= 3y^3 + 17y^2 - 7y + 7$ *(continued)*

DEFINITIONS AND CONCEPTS	EXAMPLES

Section 4.3 Multiplying Polynomials

To multiply two polynomials, multiply each term of one polynomial by each term of the other polynomial, and then combine like terms.

Multiply:
$(2x + 1)(5x^2 - 6x + 2)$

$$= 2x(5x^2 - 6x + 2) + 1(5x^2 - 6x + 2)$$
$$= 10x^3 - 12x^2 + 4x + 5x^2 - 6x + 2$$
$$= 10x^3 - 7x^2 - 2x + 2$$

Section 4.4 Special Products

The **FOIL method** may be used when multiplying two binomials.

Multiply: $(5x - 3)(2x + 3)$

$$(5x - 3)(2x + 3) = (5x)(2x) + (5x)(3) + (-3)(2x) + (-3)(3)$$
$$= 10x^2 + 15x - 6x - 9$$
$$= 10x^2 + 9x - 9$$

Squaring a Binomial

$$(a + b)^2 = a^2 + 2ab + b^2$$

$$(a - b)^2 = a^2 - 2ab + b^2$$

Square each binomial.

$$(x + 5)^2 = x^2 + 2(x)(5) + 5^2$$
$$= x^2 + 10x + 25$$

$$(3x - 2y)^2 = (3x)^2 - 2(3x)(2y) + (2y)^2$$
$$= 9x^2 - 12xy + 4y^2$$

Multiplying the Sum and Difference of Two Terms

$$(a + b)(a - b) = a^2 - b^2$$

Multiply.

$$(6y + 5)(6y - 5) = (6y)^2 - 5^2$$
$$= 36y^2 - 25$$

Section 4.5 Negative Exponents and Scientific Notation

If $a \neq 0$ and n is an integer,

$$a^{-n} = \frac{1}{a^n}$$

Rules for exponents are true for positive and negative integers.

$$3^{-2} = \frac{1}{3^2} = \frac{1}{9}; 5x^{-2} = \frac{5}{x^2}$$

Simplify: $\left(\dfrac{x^{-2}y}{x^5}\right)^{-2} = \dfrac{x^4y^{-2}}{x^{-10}}$

$$= x^{4-(-10)}y^{-2}$$
$$= \frac{x^{14}}{y^2}$$

A positive number is written in scientific notation if it is as the product of a number a, $1 \leq a < 10$, and an integer power r of 10.

$$a \times 10^r$$

Write each number in scientific notation.

$$12,000 = 1.2 \times 10^4$$

$$0.00000568 = 5.68 \times 10^{-6}$$

(*continued*)

DEFINITIONS AND CONCEPTS	EXAMPLES

Section 4.6 Division of Polynomials

To divide a polynomial by a monomial:

$$\frac{a+b}{c} = \frac{a}{c} + \frac{b}{c}$$

To divide a polynomial by a polynomial other than a monomial, use long division.

Divide:

$$\frac{15x^5 - 10x^3 + 5x^2 - 2x}{5x^2} = \frac{15x^5}{5x^2} - \frac{10x^3}{5x^2} + \frac{5x^2}{5x^2} - \frac{2x}{5x^2}$$

$$= 3x^3 - 2x + 1 - \frac{2}{5x}$$

$$5x - 1 + \frac{-4}{2x+3}$$

$$2x+3 \overline{)10x^2 + 13x - 7}$$
$$\underline{10x^2 + 15x}$$
$$-2x - 7$$
$$\underline{-2x - 3}$$
$$-4$$

CHAPTER 4 REVIEW

(4.1) *State the base and the exponent for each expression.*

1. 3^2

2. $(-5)^4$

3. -5^4

Evaluate each expression.

4. 8^3

5. $(-6)^2$

6. -6^2

7. $-4^3 - 4^0$

8. $(3b)^0$

9. $\dfrac{8b}{8b}$

Simplify each expression.

10. $5b^3b^5a^6$

11. $2^3 \cdot x^0$

12. $\left[(-3)^2\right]^3$

13. $(2x^3)(-5x^2)$

14. $\left(\dfrac{mn}{q}\right)^2 \cdot \left(\dfrac{mn}{q}\right)$

15. $\left(\dfrac{3ab^2}{6ab}\right)^4$

16. $\dfrac{x^9}{x^4}$

17. $\dfrac{2x^7y^8}{8xy^2}$

18. $\dfrac{12xy^6}{3x^4y^{10}}$

19. $5a^7(2a^4)^3$

20. $(2x)^2(9x)$

21. $\dfrac{(-4)^2(3^3)}{(4^5)(3^2)}$

22. $\dfrac{(-7)^2(3^5)}{(-7)^3(3^4)}$

23. $\dfrac{(2x)^0(-4)^2}{16x}$

24. $\dfrac{(8xy)(3xy)}{18x^2y^2}$

25. $m^0 + p^0 + 3q^0$

26. $(-5a)^0 + 7^0 + 8^0$

27. $\left(3xy^2 + 8x + 9\right)^0$

28. $8x^0 + 9^0$

29. $6\left(a^2b^3\right)^3$

30. $\dfrac{\left(x^3z\right)^a}{x^2z^2}$

(4.2) *Find the degree of each term.*

31. $-5x^4y^3$

32. $10x^3y^2z$

33. $35a^5bc^2$

34. $95xyz$

Find the degree of each polynomial.

35. $y^5 + 7x - 8x^4$

36. $9y^2 + 30y + 25$

37. $-14x^2y - 28x^2y^3 - 42x^2y^2$

38. $6x^2y^2z^2 + 5x^2y^3 - 12xyz$

39. a. Complete the table for the polynomial $3a^2b - 2a^2 + ab - b^2 - 6$.

Term	Numerical Coefficient	Degree of Term
$3a^2b$		
$-2a^2$		
ab		
$-b^2$		
-6		

b. What is the degree of the polynomial?

40. a. Complete the table for the polynomial
$x^2y^2 + 5x^2 - 7y^2 + 11xy - 1$.

Term	Numerical Coefficient	Degree of Term
x^2y^2		
$5x^2$		
$-7y^2$		
$11xy$		
-1		

b. What is the degree of the polynomial?

△ **41.** The surface area of a box with a square base and a height of 5 units is given by the polynomial $2x^2 + 20x$. Fill in the table below by evaluating $2x^2 + 20x$ for the given values of x.

x	1	3	5.1	10
$2x^2 + 20x$				

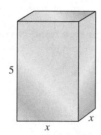

Combine like terms.

42. $6a^2b^2 + 4ab + 9a^2b^2$

43. $21x^2y^3 + 3xy + x^2y^3 + 6$

44. $4a^2b - 3b^2 - 8q^2 - 10a^2b + 7q^2$

45. $2s^{14} + 3s^{13} + 12s^{12} - s^{10}$

Add or subtract as indicated.

46. $(3k^2 + 2k + 6) + (5k^2 + k)$

47. $(2s^5 + 3s^4 + 4s^3 + 5s^2) - (4s^2 + 7s + 6)$

48. $(2m^7 + 3x^4 + 7m^6) - (8m^7 + 4m^2 + 6x^4)$

49. $(11r^2 + 16rs - 2s^2) - (3r^2 + 5rs - 9s^2)$

50. $(3x^2 - 6xy + y^2) - (11x^2 - xy + 5y^2)$

51. Subtract $(3x - y)$ from $(7x - 14y)$.

52. Subtract $(4x^2 + 8x - 7)$ from the sum of $(x^2 + 7x + 9)$ and $(x^2 + 4)$.

53. The average room rate per night for hotels and motels in the United States x years after 1990 is given by the polynomial $0.365x^2 - 0.244x + 57.958$ dollars for 1990 through 1996. Use this model to predict the room rate in 2001. (*Source:* Based on data from the American Hotel & Motel Association, Washington, D.C.)

(4.3) *Multiply each expression.*

54. $9x(x^2y)$

55. $-7(8xz^2)$

56. $(6xa^2)(xya^3)$

57. $(4xy)(-3xa^2y^3)$

58. $6(x + 5)$

59. $9(x - 7)$

60. $4(2a + 7)$

61. $9(6a - 3)$

62. $-7x(x^2 + 5)$

63. $-8y(4y^2 - 6)$

64. $-2(x^3 - 9x^2 + x)$

65. $-3a(a^2b + ab + b^2)$

66. $(3a^3 - 4a + 1)(-2a)$

67. $(6b^3 - 4b + 2)(7b)$

68. $(2x + 2)(x - 7)$

69. $(2x - 5)(3x + 2)$

70. $(4a - 1)(a + 7)$

71. $(6a - 1)(7a + 3)$

72. $(x + 7)(x^3 + 4x - 5)$

73. $(x + 2)(x^5 + x + 1)$

74. $(x^2 + 2x + 4)(x^2 + 2x - 4)$

75. $(x^3 + 4x + 4)(x^3 + 4x - 4)$

76. $(x + 7)^3$

77. $(2x - 5)^3$

(4.4) *Multiply.*

78. $2x(3x^2 - 7x + 1)$

79. $3y(5y^2 - y + 2)$

80. $(6x - 1)(4x + 3)$

81. $(4a - 1)(3a + 7)$

82. $(x + 7)^2$

83. $(x - 5)^2$

84. $(3x - 7)^2$

85. $(4x + 2)^2$

86. $(y + 1)(y^2 - 6y - 5)$

87. $(x - 2)(x^2 - x - 2)$

88. $(5x - 9)^2$

89. $(5x + 1)(5x - 1)$

90. $(7x + 4)(7x - 4)$

91. $(a + 2b)(a - 2b)$

92. $(2x - 6)(2x + 6)$

93. $(4a^2 - 2b)(4a^2 + 2b)$

(4.5) *Simplify each expression.*

94. 7^{-2}

95. -7^{-2}

96. $2x^{-4}$

97. $(2x)^{-4}$

98. $\left(\dfrac{1}{5}\right)^{-3}$

99. $\left(\dfrac{-2}{3}\right)^{-2}$

100. $2^0 + 2^{-4}$

101. $6^{-1} - 7^{-1}$

Simplify each expression. Assume that variables in an exponent represent positive integers only. Write each answer using positive exponents.

102. $\dfrac{1}{(2q)^{-3}}$

103. $\dfrac{-1}{(qr)^{-3}}$

104. $\dfrac{r^{-3}}{s^{-4}}$

105. $\dfrac{rs^{-3}}{r^{-4}}$

106. $\dfrac{-6}{8x^{-3}r^4}$

107. $\dfrac{-4s}{16s^{-3}}$

108. $\left(2x^{-5}\right)^{-3}$

109. $\left(3y^{-6}\right)^{-1}$

110. $\left(3a^{-1}b^{-1}c^{-2}\right)^{-2}$

111. $\left(4x^{-2}y^{-3}z\right)^{-3}$

112. $\dfrac{5^{-2}x^8}{5^{-3}x^{11}}$

113. $\dfrac{7^5y^{-2}}{7^7y^{-10}}$

114. $\left(\dfrac{bc^{-2}}{bc^{-3}}\right)^4$

115. $\left(\dfrac{x^{-3}y^{-4}}{x^{-2}y^{-5}}\right)^{-3}$

116. $\dfrac{x^{-4}y^{-6}}{x^2y^7}$

117. $\dfrac{a^5b^{-5}}{a^{-5}b^5}$

118. $-2^0 + 2^{-4}$

119. $-3^{-2} - 3^{-3}$

120. $a^{6m}a^{5m}$

121. $\dfrac{\left(x^{5+h}\right)^3}{x^5}$

122. $\left(3xy^{2z}\right)^3$

123. $a^{m+2}a^{m+3}$

Write each number in scientific notation.

124. 0.00027

125. 0.8868

126. 80,800,000

127. −868,000

128. The population of California is 32,667,000. Write this number in scientific notation. (*Source:* Federal–State Cooperative Program for Population Estimates)

△ **129.** The radius of the Earth is approximately 4000 miles. Write 4000 in scientific notation.

4000 miles

Write each number in standard form.

130. 8.67×10^5

131. 3.86×10^{-3}

132. 8.6×10^{-4}

133. 8.936×10^5

134. The number of photons of light emitted by a 100-watt bulb every second is 1×10^{20}. Write 1×10^{20} in standard notation.

135. The real mass of all the galaxies in the constellation of Virgo is 3×10^{-25}. Write 3×10^{-25} in standard notation.

Simplify. Express each result in standard form.

136. $\left(8 \times 10^4\right)\left(2 \times 10^{-7}\right)$

137. $\dfrac{8 \times 10^4}{2 \times 10^{-7}}$

(4.6) *Perform each division.*

138. $\dfrac{x^2 + 21x + 49}{7x^2}$

139. $\dfrac{5a^3b - 15ab^2 + 20ab}{-5ab}$

140. $\left(a^2 - a + 4\right) \div (a - 2)$

141. $\left(4x^2 + 20x + 7\right) \div (x + 5)$

142. $\dfrac{a^3 + a^2 + 2a + 6}{a - 2}$

143. $\dfrac{9b^3 - 18b^2 + 8b - 1}{3b - 2}$

144. $\dfrac{4x^4 - 4x^3 + x^2 + 4x - 3}{2x - 1}$

145. $\dfrac{-10x^2 - x^3 - 21x + 18}{x - 6}$

CHAPTER 4 TEST

Evaluate each expression.

1. 2^5

2. $(-3)^4$

3. -3^4

4. 4^{-3}

Simplify each exponential expression.

5. $(3x^2)(-5x^9)$

6. $\dfrac{y^7}{y^2}$

7. $\dfrac{r^{-8}}{r^{-3}}$

Simplify each expression. Write the result using only positive exponents.

8. $\left(\dfrac{x^2y^3}{x^3y^{-4}}\right)^2$

9. $\dfrac{6^2x^{-4}y^{-1}}{6^3x^{-3}y^7}$

Express each number in scientific notation.

10. 563,000

11. 0.0000863

Write each number in standard form.

12. 1.5×10^{-3}

13. 6.23×10^4

14. Simplify. Write the answer in standard form.
$(1.2 \times 10^5)(3 \times 10^{-7})$

15. a. Complete the table for the polynomial $4xy^2 + 7xyz + x^3y - 2$.

Term	Numerical Coefficient	Degree of Term
$4xy^2$		
$7xyz$		
x^3y		
-2		

b. What is the degree of the polynomial?

16. Simplify by combining like terms.
$5x^2 + 4xy - 7x^2 + 11 + 8xy$

Perform each indicated operation.

17. $(8x^3 + 7x^2 + 4x - 7) + (8x^3 - 7x - 6)$

18. $\begin{aligned}5x^3 + x^2 + 5x - 2 \\ -(8x^3 - 4x^2 + x - 7)\end{aligned}$

19. Subtract $(4x + 2)$ from the sum of $(8x^2 + 7x + 5)$ and $(x^3 - 8)$.

Multiply.

20. $(3x + 7)(x^2 + 5x + 2)$

21. $3x^2(2x^2 - 3x + 7)$

22. $(x + 7)(3x - 5)$

23. $(3x - 7)(3x + 7)$

24. $(4x - 2)^2$

25. $(8x + 3)^2$

26. $(x^2 - 9b)(x^2 + 9b)$

27. The height of the Bank of China in Hong Kong is 1001 feet. Neglecting air resistance, the height of an object dropped from this building at time t seconds is given by the polynomial $-16t^2 + 1001$. Find the height of the object at the given times below.

t	0 seconds	1 second	3 seconds	5 seconds
$-16t^2 + 1001$				

Divide.

28. $\dfrac{8xy^2}{4x^3y^3z}$

29. $\dfrac{4x^2 + 2xy - 7x}{8xy}$

30. $(x^2 + 7x + 10) \div (x + 5)$

31. $\dfrac{27x^3 - 8}{3x + 2}$

32. The number of bankruptcy cases (in thousands) filed in the United States x years after 1993 is given by the polynomial $62x^2 - 149x + 922$ thousand cases for 1993 through 1997. Use this model to predict the number of bankruptcy cases in 2000. (*Source:* Based on data from the Administrative Office of the U.S. Courts)

CHAPTER 4 CUMULATIVE REVIEW

1. Tell whether each statement is true or false.

  a. $8 \geq 8$ **b.** $8 \leq 8$

  c. $23 \leq 0$ **d.** $23 \geq 0$

2. Find each quotient. Write all answers in lowest terms.

  a. $\dfrac{4}{5} \div \dfrac{5}{16}$ **b.** $\dfrac{7}{10} \div 14$

  c. $\dfrac{3}{8} \div \dfrac{3}{10}$

3. Evaluate the following:

  a. 3^2 **b.** 5^3

  c. 2^4 **d.** 7^1

  e. $\left(\dfrac{3}{7}\right)^2$

4. Add.

  a. $-3 + (-7)$ **b.** $-1 + (-20)$

  c. $-2 + (-10)$

5. Subtract 8 from -4.

6. Find the reciprocal of each number.

  a. 22 **b.** $\dfrac{3}{16}$

  c. -10 **d.** $-\dfrac{9}{13}$

7. Use an associative property to complete each statement.

  a. $5 + (4 + 6) = $ _____

  b. $(-1 \cdot 2) \cdot 5 = $ _____

8. The following bar graph shows the cents charged per kilowatt-hour for selected electricity companies.

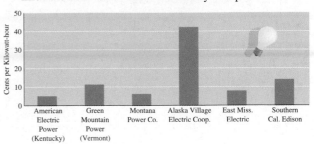

Source: Electric Company Listed

  a. Which company charges the highest rate?

  b. Which company charges the lowest rate?

  c. Approximate the electricity rate charged by the first four companies listed.

  d. Approximate the difference in the rates charged by the companies in parts (a) and (b).

9. Find each product by using the distributive property to remove parentheses.

  a. $5(x + 2)$

  b. $-2(y + 0.3z - 1)$

  c. $-(x + y - 2z + 6)$

10. Solve $x - 7 = 10$ for x.

11. Solve for x: $\dfrac{5}{2}x = 15$

12. Twice a number, added to seven, is the same as three subtracted from the number. Find the number.

13. Solve $F = \dfrac{9}{5}C + 32$ for C.

14. Find 72% of 200.

15. The length of a rectangular road sign is 2 feet less than three times its width. Find the dimensions if the perimeter is 28 feet.

16. Graph $2 < x \leq 4$.

17. Complete the following ordered pair solutions for the equation $3x + y = 12$.

    a. $(0, \quad)$ **b.** $(\quad, 6)$

    c. $(-1, \quad)$

18. Graph the linear equation $2x + y = 5$

19. Graph $x = 2$.

20. Find the slope of the line $x = 5$.

21. Graph: $x + y < 7$

22. Use the product rule to simplify $(2x^2)(-3x^3)$.

23. Add $(-2x^3 + 3x - 1)$ and $(-2x^2 + x + 3)$.

24. Multiply: $(2x - y)^2$

25. Divide: $6m^2 + 2m$ by $2m$

Outdoor Opportunities

Professional landscapers, landscape technicians, and landscape architects work together to design, install, and maintain attractive outdoor planting schemes in parks, around buildings, and in homeowners' yards. There are over 70,000 professional lawn care and landscape companies in the United States, employing more than 623,000 people. In recent years, the demand for lawn care and landscape services has skyrocketed.

People working in the landscape industry enjoy working outdoors. They also are knowledgeable about plants and how best to care for them. Landscapers find math and geometry useful in situations such as estimating and pricing landscaping jobs, mixing fertilizers or pesticides, and designing or laying out areas to be landscaped.

For more information about a career in the landscaping industry, visit the Associated Landscape Contractors of America Website by first going to www.prenhall.com/martin-gay.

In the Spotlight on Decision Making feature on page 318, you will have the opportunity as a landscaper to make a decision about a flower bed design.

FACTORING POLYNOMIALS

5.1 THE GREATEST COMMON FACTOR AND FACTORING BY GROUPING

5.2 FACTORING TRINOMIALS OF THE FORM $x^2 + bx + c$

5.3 FACTORING TRINOMIALS OF THE FORM $ax^2 + bx + c$

5.4 FACTORING BINOMIALS

5.5 CHOOSING A FACTORING STRATEGY

5.6 SOLVING QUADRATIC EQUATIONS BY FACTORING

5.7 QUADRATIC EQUATIONS AND PROBLEM SOLVING

In Chapter 4, you learned how to multiply polynomials. This chapter deals with an operation that is the reverse process of multiplying, called *factoring*. Factoring is an important algebraic skill because this process allows us to write a sum as a product.

At the end of this chapter, we use factoring to help us solve equations other than linear equations, and in Chapter 6 we use factoring to simplify and perform arithmetic operations on rational expressions.

5.1 THE GREATEST COMMON FACTOR AND FACTORING BY GROUPING

CD-ROM   SSM

SSG   Video

▶ **O B J E C T I V E S**

1. Find the greatest common factor of a list of integers.
2. Find the greatest common factor of a list of terms.
3. Factor out the greatest common factor from a polynomial.
4. Factor a polynomial by grouping.

When an integer is written as the product of two or more other integers, each of these integers is called a **factor** of the product. This is true for polynomials, also. When a polynomial is written as the product of two or more other polynomials, each of these polynomials is called a factor of the product.

The process of writing a polynomial as a product is called **factoring** the polynomial.

$$2 \cdot 3 = 6 \qquad x^2 \cdot x^3 = x^5 \qquad (x + 2)(x + 3) = x^2 + 5x + 6$$

factor   factor   product   factor   factor   product   factor   factor   product

Notice that factoring is the reverse process of multiplying.

$$\overset{\text{factoring}}{x^2 + 5x + 6 = (x + 2)(x + 3)} \\ \underset{\text{multiplying}}{}$$

The first step in factoring a polynomial is to see whether the terms of the polynomial have a common factor. If there is one, we can write the polynomial as a product by **factoring out** the common factor. We will usually factor out the **greatest common factor (GCF)**.

1   The GCF of a list of integers is the largest integer that is a factor of all the integers in the list. For example, the GCF of 12 and 20 is 4 because 4 is the largest integer that is a factor of both 12 and 20. With large integers, the GCF may not be easily found by inspection. When this happens, use the following steps.

FINDING THE GCF OF A LIST OF INTEGERS

Step 1. Write each number as a product of prime numbers.
Step 2. Identify the common prime factors.
Step 3. The product of all common prime factors found in step 2 is the greatest common factor. If there are no common prime factors, the greatest common factor is 1.

Recall from Section 1.2 that a prime number is a whole number other than 1, whose only factors are 1 and itself.

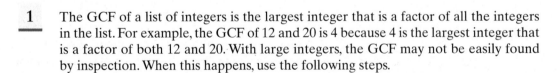

Example 1   Find the GCF of each list of numbers.

a. 28 and 40   **b.** 55 and 21   **c.** 15, 18, and 66

Solution   **a.** Write each number as a product of primes.

$$28 = 2 \cdot 2 \cdot 7 = 2^2 \cdot 7$$
$$40 = 2 \cdot 2 \cdot 2 \cdot 5 = 2^3 \cdot 5$$

There are two common factors, each of which is 2, so the GCF is

$$\text{GCF} = 2 \cdot 2 = 4$$

b. $55 = 5 \cdot 11$
$21 = 3 \cdot 7$
There are no common prime factors; thus, the GCF is 1.

c. $15 = 3 \cdot 5$
$18 = 2 \cdot 3 \cdot 3 = 2 \cdot 3^2$
$66 = 2 \cdot 3 \cdot 11$
The only prime factor common to all three numbers is 3, so the GCF is

$$\text{GCF} = 3$$

---

2   The greatest common factor of a list of variables raised to powers is found in a similar way. For example, the GCF of x^2, x^3, and x^5 is x^2 because each term contains a factor of x^2 and no higher power of x is a factor of each term.

$$x^2 = x \cdot x$$
$$x^3 = x \cdot x \cdot x$$
$$x^5 = x \cdot x \cdot x \cdot x \cdot x$$

There are two common factors of x, so the GCF $= x \cdot x$ or x^2. From this example, we see that **the GCF of a list of common variables raised to powers is the variable raised to the smallest exponent in the list.**

◆ **Example 2**   Find the GCF of each list of terms.

a. x^3, x^7, and x^5

b. y, y^4, and y^7

Solution   **a.** The GCF is x^3, since 3 is the smallest exponent to which x is raised.

b. The GCF is y^1 or y, since 1 is the smallest exponent on y.

In general, the **greatest common factor (GCF) of a list of terms** is the product of the GCF of the numerical coefficients and the GCF of the variable factors.

Example 3   Find the greatest common factor of each list of terms.

a. $6x^2$, $10x^3$, and $-8x$

b. $8y^2$, y^3, and y^5

c. a^3b^2, a^5b, and a^6b^2

Solution   **a.** The GCF of the numerical coefficients 6, 10, and -8 is 2.
The GCF of variable factors x^2, x^3, and x is x.
Thus, the GCF of the terms $6x^2$, $10x^3$, and $-8x$ is $2x$.

b. The GCF of the numerical coefficients 8, 1, and 1 is 1.
The GCF of variable factors y^2, y^3, and y^5 is y^2.
Thus, the GCF of the terms $8y^2$, y^3, and y^5 is $1y^2$ or y^2.

c. The GCF of a^3, a^5, and a^6 is a^3.
The GCF of b^2, b, and b^2 is b. Thus, the GCF of the terms is a^3b.

3 The first step in factoring a polynomial is to find the GCF of its terms. Once we do so, we can write the polynomial as a product by **factoring out** the GCF.

The polynomial $8x + 14$, for example, contains two terms: $8x$ and 14. The GCF of these terms is 2. We factor out 2 from each term by writing each term as a product of 2 and the term's remaining factors.

$$8x + 14 = 2 \cdot 4x + 2 \cdot 7$$

Using the distributive property, we can write

$$8x + 14 = 2 \cdot 4x + 2 \cdot 7$$
$$= 2(4x + 7)$$

Thus, a factored form of $8x + 14$ is $2(4x + 7)$.

> **HELPFUL HINT**
> A factored form of $8x + 14$ is *not*
>
> $$2 \cdot 4x + 2 \cdot 7$$
>
> Although the *terms* have been factored (written as a product), the *polynomial* $8x + 14$ has not been factored. A factored form of $8x + 14$ is the *product* $2(4x + 7)$.

Example 4 Factor each polynomial by factoring out the GCF.

a. $6t + 18$ **b.** $y^5 - y^7$

Solution **a.** The GCF of terms $6t$ and 18 is 6.

$$6t + 18 = 6 \cdot t + 6 \cdot 3$$
$$= 6(t + 3) \qquad \text{Apply the distributive property.}$$

Our work can be checked by multiplying 6 and $(t + 3)$.

$$6(t + 3) = 6 \cdot t + 6 \cdot 3 = 6t + 18, \text{ the original polynomial.}$$

b. The GCF of y^5 and y^7 is y^5. Thus,

$$y^5 - y^7 = (y^5)1 - (y^5)y^2$$
$$= y^5(1 - y^2)$$

> **HELPFUL HINT**
> Don't forget the 1.

Example 5 Factor $-9a^5 + 18a^2 - 3a$.

Solution **a.** $-9a^5 + 18a^2 - 3a = (3a)(-3a^4) + (3a)(6a) + (3a)(-1)$
$$= 3a(-3a^4 + 6a - 1)$$

In Example 5 we could have chosen to factor out a $-3a$ instead of $3a$. If we factor out a $-3a$, we have

$$-9a^5 + 18a^2 - 3a = (-3a)(3a^4) + (-3a)(-6a) + (-3a)(1)$$
$$= -3a(3a^4 - 6a + 1)$$

Example 6   Factor $25x^4z + 15x^3z + 5x^2z$.

Solution   The greatest common factor is $5x^2z$.

$$25x^4z + 15x^3z + 5x^2z = 5x^2z(5x^2 + 3x + 1)$$

> **HELPFUL HINT**
> Be careful when the GCF of the terms is the same as one of the terms in the poly-nomial. The greatest common factor of the terms of $8x^2 - 6x^3 + 2x$ is $2x$. When factoring out $2x$ from the terms of $8x^2 - 6x^3 + 2x$, don't forget a term of 1.
>
> $$8x^2 - 6x^3 + 2x = 2x(4x) - 2x(3x^2) + 2x(1)$$
> $$= 2x(4x - 3x^2 + 1)$$
>
> Check by multiplying.
>
> $$2x(4x - 3x^2 + 1) = 8x^2 - 6x^3 + 2x$$

Example 7   Factor $5(x + 3) + y(x + 3)$.

Solution   The binomial $(x + 3)$ is the greatest common factor. Use the distributive property to factor out $(x + 3)$.

$$5(x + 3) + y(x + 3) = (x + 3)(5 + y)$$

Example 8   Factor $3m^2n(a + b) - (a + b)$.

Solution   The greatest common factor is $(a + b)$.

$$3m^2n(a + b) - 1(a + b) = (a + b)(3m^2n - 1)$$

4   Once the GCF is factored out, we can often continue to factor the polynomial, using a variety of techniques. We discuss here a technique for factoring polynomials called **grouping**.

Example 9   Factor $xy + 2x + 3y + 6$ by grouping. Check by multiplying.

Solution   The GCF of the first two terms is x, and the GCF of the last two terms is 3.

$$xy + 2x + 3y + 6 = \underbrace{x(y + 2) + 3(y + 2)}$$

> **HELPFUL HINT**
> Notice that this is *not* a factored form of the original polynomial. It is a sum, not a product.

Next, factor out the common binomial factor of $(y + 2)$.

$$x(y + 2) + 3(y + 2) = (y + 2)(x + 3)$$

To check, multiply $(y + 2)$ by $(x + 3)$.

$$(y + 2)(x + 3) = xy + 2x + 3y + 6, \text{ the original polynomial.}$$

Thus, the factored form of $xy + 2x + 3y + 6$ is $(y + 2)(x + 3)$. ▬

FACTORING A FOUR-TERM POLYNOMIAL BY GROUPING

Step 1. Arrange the terms so that the first two terms have a common factor and the last two terms have a common factor.
Step 2. For each pair of terms, use the distributive property to factor out the pair's greatest common factor.
Step 3. If there is now a common binomial factor, factor it out.
Step 4. If there is no common binomial factor in step 3, begin again, rearranging the terms differently. If no rearrangement leads to a common binomial factor, the polynomial cannot be factored.

Example 10   Factor $3x^2 + 4xy - 3x - 4y$ by grouping.

Solution   The first two terms have a common factor x. Factor -1 from the last two terms so that the common binomial factor of $(3x + 4y)$ appears.

$$3x^2 + 4xy - 3x - 4y = x(3x + 4y) - 1(3x + 4y)$$

Next, factor out the common factor $(3x + 4y)$.

$$= (3x + 4y)(x - 1)$$ ▬

> **HELPFUL HINT**
> One more reminder: When **factoring** a polynomial, make sure the polynomial is written as a **product**. For example, it is true that
>
> $$3x^2 + 4xy - 3x - 4y = x(3x + 4y) - 1(3x + 4y)$$
>
> but $x(3x + 4y) - 1(3x + 4y)$ is not a **factored form** of the original polynomial since it is a **sum**, not a **product**. The factored form of $3x^2 + 4xy - 3x - 4y$ is $(3x + 4y)(x - 1)$.

Factoring out a greatest common factor first makes factoring by any method easier, as we see in the next example.

Example 11   Factor $4ax - 4ab - 2bx + 2b^2$.

Solution   First, factor out the common factor 2 from all four terms.

$$4ax - 4ab - 2bx + 2b^2$$
$$= 2(2ax - 2ab - bx + b^2) \qquad \text{Factor out 2 from all four terms.}$$
$$= 2[2a(x - b) - b(x - b)] \qquad \text{Factor out common factors from each pair of terms.}$$
$$= 2(x - b)(2a - b) \qquad \text{Factor out the common binomial.}$$

Notice that we factored out $-b$ instead of b from the second pair of terms so that the binomial factor of each pair is the same. ▬

MENTAL MATH

Find the prime factorization of the following integers.

1. 14   **2.** 15   **3.** 10   **4.** 70

Find the GCF of the following pairs of integers.

5. 6, 15   **6.** 20, 15   **7.** 3, 18   **8.** 14, 35

Exercise Set 5.1

Find the GCF for each list. See Examples 1 through 3.

1. 32, 36   **2.** 36, 90

3. 12, 18, 36   **4.** 24, 14, 21

5. y^2, y^4, y^7 **6.** x^3, x^2, x^3

7. $x^{10}y^2, xy^2, x^3y^3$ **8.** p^7q, p^8q^2, p^9q^3

9. $8x, 4$ **10.** $9y, y$

11. $12y^4, 20y^3$ **12.** $32x, 18x^2$

13. $12x^3, 6x^4, 3x^5$ **14.** $15y^2, 5y^7, 20y^3$

15. $18x^2y, 9x^3y^3, 36x^3y$ **16.** $7x, 21x^2y^2, 14xy$

Factor out the GCF from each polynomial. See Examples 4 through 6.

17. $30x - 15$ **18.** $42x - 7$

19. $24cd^3 - 18c^2d$ **20.** $25x^4y^3 - 15x^2y^2$

21. $-24a^4x + 18a^3x$ **22.** $-15a^2x + 9ax$

23. $12x^3 + 16x^2 - 8x$ **24.** $6x^3 - 9x^2 + 12x$

25. $5x^3y - 15x^2y + 10xy$ **26.** $14x^3y + 7x^2y - 7xy$

27. Construct a binomial whose greatest common factor is $5a^3$.

28. Construct a trinomial whose greatest common factor is $2x^2$.

Factor out the GCF from each polynomial. See Examples 7 and 8.

29. $y(x + 2) + 3(x + 2)$ **30.** $z(y + 4) + 3(y + 4)$

31. $x(y - 3) - 4(y - 3)$ **32.** $6(x + 2) - y(x + 2)$

33. $2x(x + y) - (x + y)$ **34.** $xy(y + 1) - (y + 1)$

Factor the following four-term polynomials by grouping. See Examples 9 through 11.

35. $5x + 15 + xy + 3y$ **36.** $xy + y + 2x + 2$

37. $2y - 8 + xy - 4x$ **38.** $6x - 42 + xy - 7y$

39. $3xy - 6x + 8y - 16$ **40.** $xy - 2yz + 5x - 10z$

41. $y^3 + 3y^2 + y + 3$ **42.** $x^3 + 4x + x^2 + 4$

Write an expression for the area of each shaded region. Then write the expression as a factored polynomial.

△ **43.**

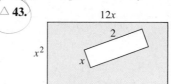

△ **44.**

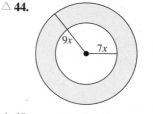

△ **45.**

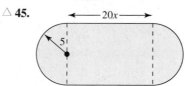

△ **46.**

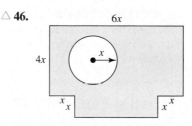

Factor the following polynomials.

47. $3x - 6$ **48.** $4x - 16$

49. $32xy - 18x^2$ **50.** $10xy - 15x^2$

51. $4x - 8y + 4$ **52.** $7x + 21y - 7$

53. $8(x + 2) - y(x + 2)$

54. $x(y^2 + 1)) - 3(y^2 + 1))$

55. $-40x^8y^6 - 16x^9y^5$ **56.** $-21x^3y - 49x^2y^2$

57. $-3x + 12$ **58.** $-10x + 20$

59. $18x^3y^3 - 12x^3y^2 + 6x^5y^2$

60. $32x^3y^3 - 24x^2y^3 + 8x^2y^4$

61. $y^2(x - 2) + (x - 2)$ **62.** $x(y + 4) + (y + 4)$
63. $5xy + 15x + 6y + 18$ **64.** $2x^3 + x^2 + 8x + 4$
65. $4x^2 - 8xy - 3x + 6y$ **66.** $2x^3 - x^2 - 10x + 5$
67. $126x^3yz + 210y^4z^3$ **68.** $231x^3y^2z - 143yz^2$
69. $3y - 5x + 15 - xy$ **70.** $2x - 9y + 18 - xy$
71. $12x^2y - 42x^2 - 4y + 14$
72. $90 + 15y^2 - 18x - 3xy^2$
73. Explain how you can tell whether a polynomial is written in factored form.
74. Construct a 4-term polynomial that can be factored by grouping.

Which of the following expressions is factored?

75. $(a + 6)(b - 2)$ **76.** $(x + 5)(x + y)$
77. $5(2y + z) - b(2y + z)$
78. $3x(a + 2b) + 2(a + 2b)$

Write an expression for the length of each rectangle.

△ **79.**

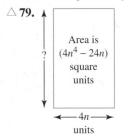

Area is $(4n^4 - 24n)$ square units
? (height)
$4n$ units (width)

△ **80.**

Area is $(5x^5 - 5x^2)$ square units
$5x^2$ units (height)
? (width)

81. The number (in millions) of CDs sold annually in the United States each year during 1996–1998 can be modeled by the polynomial $60x^2 - 85x + 780$, where x is the number of years since 1996. (*Source:* Recording Industry Association of America)

a. Find the number of CDs sold in 1998. To do so, let $x = 2$ and evaluate $60x^2 - 85x + 780$.

b. Use this expression to predict the number of CDs sold in 2001.

c. Factor the polynomial $60x^2 - 85x + 780$.

82. The number (in thousands) of students who graduated from U.S. high schools each year during 1993–1997 can be modeled by $8x^2 + 20x + 2488$, where x is the number of years since 1993. (*Source:* U.S. Bureau of the Census)

a. Find the number of students who graduated from U.S. high schools in 1995. To do so, let $x = 2$ and evaluate $8x^2 + 20x + 2488$.

b. Use this expression to predict the number of students who will graduate from U.S. high schools in 2002.

c. Factor the polynomial $8x^2 + 20x + 2488$.

REVIEW EXERCISES

Multiply. See Section 4.4.

83. $(x + 2)(x + 5)$ **84.** $(y + 3)(y + 6)$
85. $(a - 7)(a - 8)$ **86.** $(z - 4)(z - 4)$

Fill in the chart by finding two numbers that have the given product and sum. The first row is filled in for you.

	Two Numbers	Their Product	Their Sum
	4, 7	28	11
87.		12	8
88.		20	9
89.		8	−9
90.		16	−10
91.		−10	3
92.		−9	0
93.		−24	−5
94.		−36	−5

5.2 FACTORING TRINOMIALS OF THE FORM $x^2 + bx + c$

CD-ROM   SSM

SSG   Video

▶ **OBJECTIVES**

1. Factor trinomials of the form $x^2 + bx + c$.
2. Factor out the greatest common factor and then factor a trinomial of the form $x^2 + bx + c$.

1   In this section, we factor trinomials of the form $x^2 + bx + c$, such as

$$x^2 + 4x + 3, \quad x^2 - 8x + 15, \quad x^2 + 4x - 12, \quad r^2 - r - 42$$

Notice that for these trinomials, the coefficient of the squared variable is 1.

Recall that factoring means to write as a product and that factoring and multiplying are reverse processes. Using the FOIL method of multiplying binomials, we have that

$$\overset{\text{F} \quad \text{O} \quad \text{I} \quad \text{L}}{(x + 3)(x + 1) = x^2 + 1x + 3x + 3}$$

$$= x^2 + 4x + 3$$

Thus, a factored form of $x^2 + 4x + 3$ is $(x + 3)(x + 1)$.

Notice that the product of the first terms of the binomials is $x \cdot x = x^2$, the first term of the trinomial. Also, the product of the last two terms of the binomials is $3 \cdot 1 = 3$, the third term of the trinomial. The sum of these same terms is $3 + 1 = 4$, the coefficient of the middle term, x, of the trinomial.

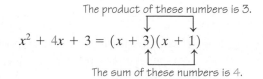

Many trinomials, such as the one above, factor into two binomials. To factor $x^2 + 7x + 10$, let's assume that it factors into two binomials and begin by writing two pairs of parentheses. The first term of the trinomial is x^2, so we use x and x as the first terms of the binomial factors.

$$x^2 + 7x + 10 = (x + \quad)(x + \quad)$$

To determine the last term of each binomial factor, we look for two integers whose product is 10 and whose sum is 7. Since our numbers must have a positive product and a positive sum, we list pairs of positive integer factors of 10 only.

Positive Factors of 10	*Sum of Factors*
1, 10	$1 + 10 = 11$
2, 5	$2 + 5 = 7$

The correct pair of numbers is 2 and 5 because their product is 10 and their sum is 7. Now we can fill in the last terms of the binomial factors.

$$x^2 + 7x + 10 = (x + 2)(x + 5)$$

To see if we have factored correctly, multiply.

$$(x + 2)(x + 5) = x^2 + 2x + 5x + 10$$

$$= x^2 + 7x + 10 \qquad \text{Combine like terms.}$$

▼
HELPFUL HINT
Since multiplication is commutative, the factored form of $x^2 + 7x + 10$ can be written as either $(x + 2)(x + 5)$ or $(x + 5)(x + 2)$.

> **FACTORING A TRINOMIAL OF THE FORM $x^2 + bx + c$**
>
> To factor a trinomial of the form $x^2 + bx + c$, look for two numbers whose product is c and whose sum is b. The factored form of $x^2 + bx + c$ is
>
> $$\overbrace{(x + \text{one number})(x + \text{other number})}^{\text{product is } c}$$
> $$\underbrace{\phantom{(x + \text{one number})(x + \text{other number})}}_{\text{sum is } b}$$

Example 1 Factor $x^2 + 7x + 12$.

Solution Begin by writing the first terms of the binomial factors.

$$(x + \quad)(x + \quad)$$

Next, look for two numbers whose product is 12 and whose sum is 7. Since our numbers must have a positive product and a positive sum, we look at positive pairs of factors of 12 only.

Positive Factors of 12	Sum of Factors
1, 12	$1 + 12 = 13$
2, 6	$2 + 6 = 8$
3, 4	$3 + 4 = 7$

The correct pair of numbers is 3 and 4 because their product is 12 and their sum is 7. Use these numbers as the last terms of the binomial factors.

$$x^2 + 7x + 12 = (x + 3)(x + 4)$$

To check, multiply $(x + 3)$ by $(x + 4)$.

Example 2 Factor $x^2 - 8x + 15$.

Solution Begin by writing the first terms of the binomials.

$$(x + \quad)(x + \quad)$$

Now look for two numbers whose product is 15 and whose sum is -8. Since our numbers must have a positive product and a negative sum, we look at negative factors of 15 only.

Negative Factors of 15	Sum of Factors
$-1, -15$	$-1 + (-15) = -16$
$-3, -5$	$-3 + (-5) = -8$

The correct pair of numbers is -3 and -5 because their product is 15 and their sum is -8. Then

$$x^2 - 8x + 15 = (x - 3)(x - 5)$$

Example 3   Factor $x^2 + 4x - 12$.

Solution $x^2 + 4x - 12 = (x + \quad)(x + \quad)$
Look for two numbers whose product is -12 and whose sum is 4. Since our numbers have a negative product, their signs must be different.

Factors of -12	*Sum of Factors*
$-1, 12$	$-1 + 12 = 11$
$1, -12$	$1 + (-12) = -11$
$-2, 6$	$-2 + 6 = 4$
$2, -6$	$2 + (-6) = -4$
$-3, 4$	$-3 + 4 = 1$
$3, -4$	$3 + (-4) = -1$

The correct pair of numbers is -2 and 6 since their product is -12 and their sum is 4. Hence

$$x^2 + 4x - 12 = (x - 2)(x + 6)$$

Example 4   Factor $r^2 - r - 42$.

Solution   Because the variable in this trinomial is r, the first term of each binomial factor is r.

$$r^2 - r - 42 = (r + \quad)(r + \quad)$$

Find two numbers whose product is -42 and whose sum is -1, the numerical coefficient of r. The numbers are 6 and -7. Therefore,

$$r^2 - r - 42 = (r + 6)(r - 7)$$

Example 5   Factor $a^2 + 2a + 10$.

Solution   Look for two numbers whose product is 10 and whose sum is 2. Neither 1 and 10 nor 2 and 5 give the required sum, 2. We conclude that $a^2 + 2a + 10$ is not factorable with integers. The polynomial $a^2 + 2a + 10$ is called a **prime polynomial**.

Example 6   Factor $x^2 + 5yx + 6y^2$.

Solution $x^2 + 5yx + 6y^2 = (x + \quad)(x + \quad)$
Look for two terms whose product is $6y^2$ and whose sum is $5y$, the coefficient of x in the middle term of the trinomial. The terms are $2y$ and $3y$ because $2y \cdot 3y = 6y^2$ and $2y + 3y = 5y$. Therefore,

$$x^2 + 5yx + 6y^2 = (x + 2y)(x + 3y)$$

The following sign patterns may be useful when factoring trinomials.

HELPFUL HINT—SIGN PATTERNS

A positive constant in a trinomial tells us to look for two numbers with the same sign. The sign of the coefficient of the middle term tells us whether the signs are both positive or both negative.

both   same
positive sign
↓   ↓

$$x^2 + 10x + 16 = (x + 2)(x + 8)$$

both   same
negative sign
↓   ↓

$$x^2 - 10x + 16 = (x - 2)(x - 8)$$

A negative constant in a trinomial tells us to look for two numbers with opposite signs.

opposite   opposite
signs   signs
↓   ↓

$$x^2 + 6x - 16 = (x + 8)(x - 2) \quad x^2 - 6x - 16 = (x - 8)(x + 2)$$

---

2   Remember that the first step in factoring any polynomial is to factor out the greatest common factor (if there is one other than 1 or −1).

Example 7   Factor $3m^2 - 24m - 60$.

Solution   First factor out the greatest common factor, 3, from each term.

$$3m^2 - 24m - 60 = 3(m^2 - 8m - 20)$$

Next, factor $m^2 - 8m - 20$ by looking for two factors of −20 whose sum is −8. The factors are −10 and 2.

$$3m^2 - 24m - 60 = 3(m + 2)(m - 10)$$

Remember to write the common factor 3 as part of the answer.
Check by multiplying.

$$3(m + 2)(m - 10) = 3(m^2 - 8m - 20)$$
$$= 3m^2 - 24m - 60$$

HELPFUL HINT

When factoring a polynomial, remember that factored out common factors are part of the final factored form. For example,

$$5x^2 - 15x - 50 = 5(x^2 - 3x - 10)$$

$$= 5(x + 2)(x - 5)$$

Thus, $5x^2 - 15x - 50$ **factored completely** is $5(x + 2)(x - 5)$.

MENTAL MATH

Complete the following.

1. $x^2 + 9x + 20 = (x + 4)(x \quad)$ 2. $x^2 + 12x + 35 = (x + 5)(x \quad)$ 3. $x^2 + 7x + 12 = (x \quad 4)(x \quad)$

4. $x^2 - 13x + 22 = (x - 2)(x \quad)$ 5. $x^2 + 4x + 4 = (x + 2)(x \quad)$ 6. $x^2 + 10x + 24 = (x + 6)(x \quad)$

Exercise Set 5.2

Factor each trinomial completely. If the polynomial can't be factored, write prime. See Examples 1 through 5.

1. $x^2 + 7x + 6$
2. $x^2 + 6x + 8$
3. $x^2 + 9x + 20$
4. $x^2 + 13x + 30$
5. $x^2 - 8x + 15$
6. $x^2 - 9x + 14$
7. $x^2 - 10x + 9$
8. $x^2 - 6x + 9$
9. $x^2 - 15x + 5$
10. $x^2 - 13x + 30$
11. $x^2 - 3x - 18$
12. $x^2 - x - 30$
13. $x^2 + 5x + 2$
14. $x^2 - 7x + 5$

Factor each trinomial completely. See Example 6.

15. $x^2 + 8xy + 15y^2$
16. $x^2 + 6xy + 8y^2$
17. $x^2 - 2xy + y^2$
18. $x^2 - 11xy + 30y^2$
19. $x^2 - 3xy - 4y^2$
20. $x^2 - 4xy - 77y^2$

Factor each trinomial completely. See Example 7.

21. $2z^2 + 20z + 32$
22. $3x^2 + 30x + 63$
23. $2x^3 - 18x^2 + 40x$
24. $x^3 - x^2 - 56x$
25. $7x^2 + 14xy - 21y^2$
26. $6r^2 - 3rs - 3s^2$

27. To factor $x^2 + 13x + 42$, think of two numbers whose _____ is 42 and whose _____ is 13.

28. Write a polynomial that factors as $(x - 3)(x + 8)$.

Factor each trinomial completely.

29. $x^2 + 15x + 36$
30. $x^2 + 19x + 60$
31. $x^2 - x - 2$
32. $x^2 - 5x - 14$
33. $r^2 - 16r + 48$
34. $r^2 - 10r + 21$
35. $x^2 - 4x - 21$
36. $x^2 - 4x - 32$
37. $x^2 + 7xy + 10y^2$
38. $x^2 - 3xy - 4y^2$
39. $r^2 - 3r + 6$
40. $x^2 + 4x - 10$
41. $2t^2 + 24t + 64$
42. $2t^2 + 20t + 50$
43. $x^3 - 2x^2 - 24x$
44. $x^3 - 3x^2 - 28x$
45. $x^2 - 16x + 63$
46. $x^2 - 19x + 88$

47. $x^2 + xy - 2y^2$
48. $x^2 - xy - 6y^2$
49. $3x^2 - 60x + 108$
50. $2x^2 - 24x + 70$
51. $x^2 - 18x - 144$
52. $x^2 + x - 42$
53. $6x^3 + 54x^2 + 120x$
54. $3x^3 + 3x^2 - 126x$
55. $2t^5 - 14t^4 + 24t^3$
56. $3x^6 + 30x^5 + 72x^4$
57. $5x^3y - 25x^2y^2 - 120xy^3$
58. $3x^2 - 6xy - 72y^2$
59. $4x^2y + 4xy - 12y$
60. $3x^2y - 9xy + 45y$
61. $2a^2b - 20ab^2 + 42b^3$
62. $-1x^2z + 14xz^2 - 28z^3$

Find a positive value of b so that each trinomial is factorable.

63. $x^2 + bx + 15$
64. $y^2 + by + 20$
65. $m^2 + bm - 27$
66. $x^2 + bx - 14$

Find a positive value of c so that each trinomial is factorable.

67. $x^2 + 6x + c$
68. $t^2 + 8t + c$
69. $y^2 - 4y + c$
70. $n^2 - 16n + c$

Complete the following sentences in your own words.

71. If $x^2 + bx + c$ is factorable and c is negative, then the signs of the last term factors of the binomial are opposite because . . . .

72. If $x^2 + bx + c$ is factorable and c is positive, then the signs of the last term factors of the binomials are the same because . . . .

REVIEW EXERCISES

Multiply. See Section 4.4.

73. $(2x + 1)(x + 5)$
74. $(3x + 2)(x + 4)$
75. $(5y - 4)(3y - 1)$
76. $(4z - 7)(7z - 1)$
77. $(a + 3)(9a - 4)$
78. $(y - 5)(6y + 5)$

Graph each linear equation. See Section 3.2.

79. $y = -3x$ **80.** $y = 5x$

81. $y = 2x - 7$ **82.** $y = -x + 4$

A Look Ahead

Example

Factor $2t^5y^2 - 22t^4y^2 + 56t^3y^2$.

Solution:
First, factor out the greatest common factor of $2t^3y^2$.

$$2t^5y^2 - 22t^4y^2 + 56t^3y^2 = 2t^3y^2(t^2 - 11t + 28)$$
$$= 2t^3y^2(t - 4)(t - 7)$$

Factor each trinomial completely.

83. $2x^2y + 30xy + 100y$

84. $3x^2z^2 + 9xz^2 + 6z^2$

85. $-12x^2y^3 - 24xy^3 - 36y^3$

86. $-4x^2t^4 + 4xt^4 + 24t^4$

87. $y^2(x + 1) - 2y(x + 1) - 15(x + 1)$

88. $z^2(x + 1) - 3z(x + 1) - 70(x + 1)$

5.3 FACTORING TRINOMIALS OF THE FORM $ax^2 + bx + c$

CD-ROM   SSM

SSG   Video

▶ **OBJECTIVES**

1. Factor trinomials of the form $ax^2 + bx + c$.
2. Factor out a GCF before factoring a trinomial of the form $ax^2 + bx + c$.
3. Factor perfect square trinomials.
4. Factor trinomials of the form $ax^2 + bx + c$ by an alternate method.

1   In this section, we factor trinomials of the form $ax^2 + bx + c$, such as

$$3x^2 + 11x + 6, \qquad 8x^2 - 22x + 5, \qquad 2x^2 + 13x - 7$$

Notice that the coefficient of the squared variable in these trinomials is a number other than 1. We will factor these trinomials using a trial-and-check method based on our work in the last section.

To begin, let's review the relationship between the numerical coefficients of the trinomial and the numerical coefficients of its factored form. For example, since $(2x + 1)(x + 6) = 2x^2 + 13x + 6$, the factored form of $2x^2 + 13x + 6$ is

$$2x^2 + 13x + 6 = (2x + 1)(x + 6)$$

Notice that $2x$ and x are factors of $2x^2$, the first term of the trinomial. Also, 6 and 1 are factors of 6, the last term of the trinomial, as shown:

$$2x^2 + 13x + 6 = (2x + 1)(x + 6)$$

with $2x \cdot x$ indicated over the first terms and $1 \cdot 6$ indicated under the last terms.

Also notice that $13x$, the middle term, is the sum of the following products:

$$2x^2 + 13x + 6 = (2x + 1)(x + 6)$$

$$\begin{array}{r} 1x \\ + \ 12x \\ \hline 13x \end{array} \quad \text{Middle term}$$

Let's use this pattern to factor $5x^2 + 7x + 2$. First, we find factors of $5x^2$. Since all numerical coefficients in this trinomial are positive, we will use factors with positive numerical coefficients only. Thus, the factors of $5x^2$ are $5x$ and x. Let's try these factors as first terms of the binomials. Thus far, we have

$$5x^2 + 7x + 2 = (5x + \quad)(x + \quad)$$

Next, we need to find positive factors of 2. Positive factors of 2 are 1 and 2. Now we try possible combinations of these factors as second terms of the binomials until we obtain a middle term of $7x$.

$$(5x + 1)(x + 2) = 5x^2 + 11x + 2$$

$$1x$$
$$+ 10x$$
$$11x \longrightarrow \textbf{Incorrect} \text{ middle term}$$

Let's try switching factors 2 and 1.

$$(5x + 2)(x + 1) = 5x^2 + 7x + 2$$

$$2x$$
$$+ 5x$$
$$7x \longrightarrow \textbf{Correct} \text{ middle term}$$

Thus the factored form of $5x^2 + 7x + 2$ is $(5x + 2)(x + 1)$. To check, we multiply $(5x + 2)$ and $(x + 1)$. The product is $5x^2 + 7x + 2$.

Example 1   Factor: $3x^2 + 11x + 6$

Solution   Since all numerical coefficients are positive, we use factors with positive numerical coefficients. We first find factors of $3x^2$.

Factors of $3x^2$: $3x^2 = 3x \cdot x$

If factorable, the trinomial will be of the form

$$3x^2 + 11x + 6 = (3x + \quad)(x + \quad)$$

Next we factor 6.

Factors of 6: $6 = 1 \cdot 6, \qquad 6 = 2 \cdot 3$

Now we try combinations of factors of 6 until a middle term of $11x$ is obtained. Let's try 1 and 6 first.

$$(3x + 1)(x + 6) = 3x^2 + 19x + 6$$

$$1x$$
$$+ 18x$$
$$19x \longrightarrow \textbf{Incorrect} \text{ middle term}$$

Now let's next try 6 and 1.

$$(3x + 6)(x + 1)$$

Before multiplying, notice that the terms of the factor $3x + 6$ have a common factor of 3. The terms of the original trinomial $3x^2 + 11x + 6$ have no common factor other than 1, so the terms of the factored form of $3x^2 + 11x + 6$ can contain no common factor other than 1. This means that $(3x + 6)(x + 1)$ is not a factored form.

Next let's try 2 and 3 as last terms.

$$(3x + 2)(x + 3) = 3x^2 + 11x + 6$$

$$\underbrace{\begin{array}{c} 2x \\ +\ \ 9x \\ \hline 11x \end{array}}\longrightarrow \textbf{Correct } \text{middle term}$$

Thus the factored form of $3x^2 + 11x + 6$ is $(3x + 2)(x + 3)$.

> **HELPFUL HINT**
> If the terms of a trinomial have no common factor (other than 1), then the terms of neither of its binomial factors will contain a common factor (other than 1).

Example 2   Factor: $8x^2 - 22x + 5$

Solution   Factors of $8x^2$: $8x^2 = 8x \cdot x$, $8x^2 = 4x \cdot 2x$
We'll try $8x$ and x.

$$8x^2 - 22x + 5 = (8x +\ \ )(x +\ \ )$$

Since the middle term, $-22x$, has a negative numerical coefficient, we factor 5 into negative factors.

$$\text{Factors of 5:}\quad 5 = -1 \cdot -5$$

Let's try -1 and -5.

$$(8x - 1)(x - 5) = 8x^2 - 41x + 5$$

$$\underbrace{\begin{array}{c} -1x \\ +\ (-40x) \\ \hline -41x \end{array}}\longrightarrow \textbf{Incorrect } \text{middle term}$$

Now let's try -5 and -1.

$$(8x - 5)(x - 1) = 8x^2 - 13x + 5$$

$$\underbrace{\begin{array}{c} -5x \\ +\ (-8x) \\ \hline -13x \end{array}}\longrightarrow \textbf{Incorrect } \text{middle term}$$

Don't give up yet! We can still try other factors of $8x^2$. Let's try $4x$ and $2x$ with -1 and -5.

$$(4x - 1)(2x - 5) = 8x^2 - 22x + 5$$

$$\underbrace{\begin{array}{c} -2x \\ +\ (-20x) \\ \hline -22x \end{array}}\longrightarrow \textbf{Correct } \text{middle term}$$

The factored form of $8x^2 - 22x + 5$ is $(4x - 1)(2x - 5)$.

Example 3 Factor: $2x^2 + 13x - 7$

Solution Factors of $2x^2$: $2x^2 = 2x \cdot x$

Factors of -7: $-7 = -1 \cdot 7,$ $-7 = 1 \cdot -7$

We try possible combinations of these factors:

$$(2x + 1)(x - 7) = 2x^2 - 13x - 7 \qquad \text{Incorrect middle term}$$
$$(2x - 1)(x + 7) = 2x^2 + 13x - 7 \qquad \text{Correct middle term}$$

The factored form of $2x^2 + 13x - 7$ is $(2x - 1)(x + 7)$.

Example 4 Factor: $10x^2 - 13xy - 3y^2$

Solution Factors of $10x^2$: $10x^2 = 10x \cdot x,$ $10x^2 = 2x \cdot 5x$

Factors of $-3y^2$: $-3y^2 = -3y \cdot y,$ $-3y^2 = 3y \cdot -y$

We try some combinations of these factors:

$$(10x - 3y)(x + y) = 10x^2 + 7xy - 3y^2$$
$$(x + 3y)(10x - y) = 10x^2 + 29xy - 3y^2$$
$$(5x + 3y)(2x - y) = 10x^2 + xy - 3y^2$$
$$(2x - 3y)(5x + y) = 10x^2 - 13xy - 3y^2 \qquad \text{Correct middle term}$$

The factored form of $10x^2 - 13xy - 3y^2$ is $(2x - 3y)(5x + y)$.

2 Don't forget that the best first step in factoring any polynomial is to look for a common factor to factor out.

Example 5 Factor $24x^4 + 40x^3 + 6x^2$.

Solution Notice that all three terms have a common factor of $2x^2$. First, factor out $2x^2$.

$$24x^4 + 40x^3 + 6x^2 = 2x^2(12x^2 + 20x + 3)$$

Next, factor $12x^2 + 20x + 3$.

Factors of $12x^2$: $12x^2 = 6x \cdot 2x,$ $12x^2 = 4x \cdot 3x,$ $12x^2 = 12x \cdot x$

Since all terms in the trinomial have positive numerical coefficients, factor 3 using positive factors only.

Factors of 3: $3 = 1 \cdot 3$

We try some combinations of the factors.

$$2x^2(4x + 3)(3x + 1) = 2x^2(12x^2 + 13x + 3)$$
$$2x^2(12x + 1)(x + 3) = 2x^2(12x^2 + 37x + 3)$$
$$2x^2(2x + 3)(6x + 1) = 2x^2(12x^2 + 20x + 3) \qquad \textbf{Correct} \text{ middle term}$$

The factored form of $24x^4 + 40x^3 + 6x^2$ is $2x^2(2x + 3)(6x + 1)$. ▬

> ▼
> **HELPFUL HINT**
> Don't forget to include the common factor in the factored form.

Example 6 Factor $4x^2 - 12x + 9$.

Solution Factors of $4x^2$: $\quad 4x^2 = 2x \cdot 2x, \qquad 4x^2 = 4x \cdot x$
Since the middle term $-12x$ has a negative numerical coefficient, factor 9 into negative factors only.

Factors of 9: $\quad 9 = -3 \cdot -3, \qquad 9 = -1 \cdot -9$

The correct combination is

$$(2x - 3)(2x - 3) = 4x^2 - 12x + 9$$

$$\underbrace{-6x}$$
$$+ (-6x)$$
$$-12x \longrightarrow \textbf{Correct} \text{ middle term}$$

Thus, $4x^2 - 12x + 9 = (2x - 3)(2x - 3)$, which can also be written as $(2x - 3)^2$. ▬

3 Notice in Example 6 that $4x^2 - 12x + 9 = (2x - 3)^2$. The trinomial $4x^2 - 12x + 9$ is called a **perfect square trinomial** since it is the square of the binomial $2x - 3$.

In the last chapter, we learned a shortcut special product for squaring a binomial, recognizing that

$$(a + b)^2 = a^2 + 2ab + b^2$$

The trinomial $a^2 + 2ab + b^2$ is a perfect square trinomial, since it is the square of the binomial $a + b$. We can use this pattern to help us factor perfect square trinomials. To use this pattern, we must first be able to recognize a perfect square trinomial. A trinomial is a perfect square when its first term is the square of some expression a, its last term is the square of some expression b, and its middle term is twice the product of the expressions a and b. When a trinomial fits this description, its factored form is $(a + b)^2$.

PERFECT SQUARE TRINOMIALS

$$a^2 + 2ab + b^2 = (a + b)^2$$
$$a^2 - 2ab + b^2 = (a - b)^2$$

Example 7   Factor $x^2 + 12x + 36$.

Solution   This trinomial is a perfect square trinomial since:
1. The first term is the square of x: $x^2 = (x)^2$.
2. The last term is the square of 6: $36 = (6)^2$.
3. The middle term is twice the product of x and 6: $12x = 2 \cdot x \cdot 6$.
Thus, $x^2 + 12x + 36 = (x + 6)^2$.

Example 8   Factor $25x^2 + 25xy + 4y^2$.

Solution   Determine whether or not this trinomial is a perfect square by considering the same three questions.
1. Is the first term a square? Yes, $25x^2 = (5x)^2$.
2. Is the last term a square? Yes, $4y^2 = (2y)^2$.
3. Is the middle term twice the product of $5x$ and $2y$? **No**. $2 \cdot 5x \cdot 2y = 20xy$, not $25xy$.

Therefore, $25x^2 + 25xy + 4y^2$ is not a perfect square trinomial. It is factorable, though. Using earlier techniques, we find that $25x^2 + 25xy + 4y^2 = (5x + 4y)(5x + y)$.

> **HELPFUL HINT**
> A perfect square trinomial that is not recognized as such can be factored by other methods.

Example 9   Factor $4m^2 - 4m + 1$.

Solution   This is a perfect square trinomial since $4m^2 = (2m)^2, 1 = (1)^2$, and $4m = 2 \cdot 2m \cdot 1$.

$$4m^2 - 4m + 1 = (2m - 1)^2$$

4   Grouping can also be used to factor trinomials of the form $ax^2 + bx + c$. To use this method, write the trinomial as a four-term polynomial. For example, to factor $2x^2 + 11x + 12$ using grouping, find two numbers whose product is $2 \cdot 12 = 24$ and whose sum is 11. Since we want a positive product and a positive sum, we consider positive pairs of factors of 24 only.

Factors of 24	*Sum of Factors*
1, 24	$1 + 24 = 25$
2, 12	$2 + 12 = 14$
3, 8	$3 + 8 = 11$

The factors are 3 and 8. Use these factors to write the middle term $11x$ as $3x + 8x$. Replace $11x$ with $3x + 8x$ in the original trinomial and factor by grouping.

$$2x^2 + 11x + 12 = 2x^2 + 3x + 8x + 12$$
$$= (2x^2 + 3x) + (8x + 12)$$
$$= x(2x + 3) + 4(2x + 3)$$
$$= (2x + 3)(x + 4)$$

In general, we have the following:

FACTORING TRINOMIALS OF THE FORM $ax^2 + bx + c$ BY GROUPING

Step 1. Find two numbers whose product is $a \cdot c$ and whose sum is b.
Step 2. Write the middle term, bx, using the factors found in Step 1.
Step 3. Factor by grouping.

◈ Example 10 Factor $8x^2 - 14x + 5$ by grouping.

Solution This trinomial is of the form $ax^2 + bx + c$ with $a = 8$, $b = -14$, and $c = 5$.

Step 1. Find two numbers whose product is $a \cdot c$ or $8 \cdot 5 = 40$, and whose sum is b or -14. The numbers are -4 and -10.

Step 2. Write $-14x$ as $-4x - 10x$ so that

$$8x^2 - 14x + 5 = 8x^2 - 4x - 10x + 5$$

Step 3. Factor by grouping.

$$8x^2 - 4x - 10x + 5 = 4x(2x - 1) - 5(2x - 1)$$
$$= (2x - 1)(4x - 5)$$

Example 11 Factor $3x^2 - x - 10$ by grouping.

Solution In $3x^2 - x - 10$, $a = 3$, $b = -1$, and $c = -10$.

Step 1. Find two numbers whose product is $a \cdot c$ or $3(-10) = -30$ and whose sum is b or -1. The numbers are -6 and 5.

Step 2. $3x^2 - x - 10 = 3x^2 - 6x + 5x - 10$

Step 3. $\qquad\qquad = 3x(x - 2) + 5(x - 2)$
$$= (x - 2)(3x + 5)$$

MENTAL MATH

State whether or not each trinomial is a perfect trinomial square.

1. $x^2 + 14x + 49$

2. $9x^2 - 12x + 4$

3. $y^2 + 2y + 4$

4. $x^2 - 4x + 2$

5. $9y^2 + 6y + 1$

6. $y^2 - 16y + 64$

Exercise Set 5.3

Factor completely. See Examples 1 through 5. (See Examples 10 and 11 for alternate method.)

1. $2x^2 + 13x + 15$

2. $3x^2 + 8x + 4$

3. $2x^2 - 9x - 5$

4. $3x^2 + 20x - 63$

5. $2y^2 - y - 6$

6. $8y^2 - 17y + 9$

7. $16a^2 - 24a + 9$

8. $25x^2 + 20x + 4$

9. $36r^2 - 5r - 24$

10. $20r^2 + 27r - 8$

11. $10x^2 + 17x + 3$

12. $21x^2 - 41x + 10$

13. $21x^2 - 48x - 45$

14. $12x^2 - 14x - 10$

15. $12x^2 - 14x - 6$

16. $20x^2 - 2x + 6$

17. $4x^3 - 9x^2 - 9x$

18. $6x^3 - 31x^2 + 5x$

Factor the following perfect square trinomials. See Examples 6 through 9.

19. $x^2 + 22x + 121$

20. $x^2 + 18x + 81$

21. $x^2 - 16x + 64$

22. $x^2 - 12x + 36$

23. $16y^2 - 40y + 25$

24. $9y^2 + 48y + 64$

25. $x^2y^2 - 10xy + 25$

26. $4x^2y^2 - 28xy + 49$

27. Describe a perfect square trinomial.

28. Write a perfect square trinomial that factors as $(x + 3y)^2$.

Factor the following completely.

29. $2x^2 - 7x - 99$

30. $2x^2 + 7x - 72$

31. $4x^2 - 8x - 21$

32. $6x^2 - 11x - 10$

33. $30x^2 - 53x + 21$

34. $21x^3 - 6x - 30$

35. $24x^2 - 58x + 9$

36. $36x^2 + 55x - 14$

37. $9x^2 - 24xy + 16y^2$

38. $25x^2 + 60xy + 36y^2$

39. $x^2 - 14xy + 49y^2$

40. $x^2 + 10xy + 25y^2$

41. $2x^2 + 7x + 5$

42. $2x^2 + 7x + 3$

43. $3x^2 - 5x + 1$

44. $3x^2 - 7x + 6$

45. $-2y^2 + y + 10$

46. $-4x^2 - 23x + 6$

47. $16x^2 + 24xy + 9y^2$

48. $4x^2 - 36xy + 81y^2$

49. $8x^2y + 34xy - 84y$

50. $6x^2y^2 - 2xy^2 - 60y^2$

51. $3x^2 + x - 2$

52. $8y^2 + y - 9$

53. $x^2y^2 + 4xy + 4$

54. $x^2y^2 - 6xy + 9$

55. $49y^2 + 42xy + 9x^2$

56. $16x^2 - 8xy + y^2$

57. $3x^2 - 42x + 63$

58. $5x^2 - 75x + 60$

59. $42a^2 - 43a + 6$

60. $54a^2 + 39ab - 8b^2$

61. $18x^2 - 9x - 14$

62. $8x^2 + 6x - 27$

63. $25p^2 - 70pq + 49q^2$

64. $36p^2 - 18pq + 9q^2$

65. $15x^2 - 16x - 15$

66. $12x^2 + 7x - 12$

67. $-27t + 7t^2 - 4$

68. $4t^2 - 7 - 3t$

The area of the largest square in the figure is $(a + b)^2$. Use this figure to answer Exercises 69–70.

△ **69.** Write the area of the largest square as the sum of the areas of the smaller squares and rectangles.

△ **70.** What factoring formula from this section is visually represented by this square?

Find a positive value of b so that each trinomial is factorable.

71. $3x^2 + bx - 5$

72. $2y^2 + by + 3$

73. $2z^2 + bz - 7$

74. $5x^2 + bx - 1$

Find a positive value of c so that each trinomial is factorable.

75. $5x^2 + 7x + c$

76. $7x^2 + 22x + c$

77. $3x^2 - 8x + c$

78. $11y^2 - 40y + c$

REVIEW EXERCISES

Multiply the following. See Section 4.4.

79. $(x - 2)(x + 2)$

80. $(y^2 + 3)(y^2 - 3)$

81. $(a + 3)(a^2 - 3a + 9)$

82. $(z - 2)(z^2 + 2z + 4)$

83. $(y - 5)(y^2 + 5y + 25)$

84. $(m + 7)(m^2 - 7m + 49)$

As of 1998, approximately 42% of U.S. households had a personal computer. The following graph shows the percent of selected households having a computer grouped according to household income. See Section 1.9.

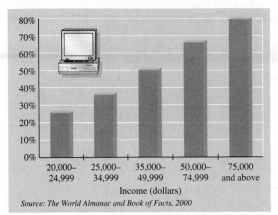

Income (dollars)

Source: The World Almanac and Book of Facts, 2000

85. Which range of household income corresponds to the greatest percent of households having a personal computer?

86. Which range of household income corresponds to the greatest increase in percent of households having a personal computer?

87. Describe any trend you notice from this graph.

88. Why don't the percents shown in the graph add to 42%?

A Look Ahead

Example

Factor $10x^4y - 5x^3y^2 - 15x^2y^3$.

Solution:

Start by factoring out the greatest common factor of $5x^2y$.

$$10x^4y - 5x^3y^2 - 15x^2y^3 = 5x^2y(2x^2 - xy - 3y^2)$$

Next, factor $2x^2 - xy - 3y^2$. Try $2x$ and x as factors of $2x^2$ and $-3y$ and y as factors of $-3y^2$.

$$\underbrace{(2x - \underset{-3xy}{3y})(x + y)}_{\underset{-xy}{+2xy}}$$
    The middle term is correct.

Hence $10x^4y - 5x^3y^2 - 15x^2y^3 = 5x^2y(2x - 3y)(x + y)$.

Factor completely. See the example.

89. $-12x^3y^2 + 3x^2y^2 + 15xy^2$

90. $-12r^3x^2 + 38r^2x^2 + 14rx^2$

91. $-30p^3q + 88p^2q^2 + 6pq^3$

92. $3x^3y^2 + 3x^2y^3 - 18xy^4$

93. $4x^2(y - 1)^2 + 10x(y - 1)^2 + 25(y - 1)^2$

94. $3x^2(a + 3)^3 - 28x(a + 3)^3 + 25(a + 3)^3$

5.4 FACTORING BINOMIALS

CD-ROM   SSM

SSG   Video

▶ **OBJECTIVES**

1. Factor the difference of two squares.
2. Factor the sum or difference of two cubes.

1   When learning to multiply binomials in Chapter 4, we studied a special product, the product of the sum and difference of two terms, *a* and *b*:

$$(a + b)(a - b) = a^2 - b^2$$

For example, the product of $x + 3$ and $x - 3$ is

$$(x + 3)(x - 3) = x^2 - 9$$

The binomial $x^2 - 9$ is called a **difference of squares**. In this section, we use the pattern for the product of a sum and difference to factor the binomial difference of squares.

To use this pattern to help us factor, we must be able to recognize a difference of squares. A binomial is a difference of squares when it is the difference of the square of some expression *a* and the square of some expression *b*.

DIFFERENCE OF TWO SQUARES

$$a^2 - b^2 = (a + b)(a - b)$$

Example 1   Factor $x^2 - 25$.

Solution $x^2 - 25$ is the difference of two squares since $x^2 - 25 = x^2 - 5^2$. Therefore,

$$x^2 - 25 = x^2 - 5^2 = (x + 5)(x - 5)$$

Multiply to check.

Example 2   Factor each difference of squares.

    a. $4x^2 - 1$ **b.** $25a^2 - 9b^2$

Solution   **a.** $4x^2 - 1 = (2x)^2 - 1^2 = (2x + 1)(2x - 1)$
              b. $25a^2 - 9b^2 = (5a)^2 - (3b)^2 = (5a + 3b)(5a - 3b)$

Example 3   Factor $x^4 - y^6$.

Solution   Write x^4 as $\left(x^2\right)^2$ and y^6 as $\left(y^3\right)^2$.

$$x^4 - y^6 = \left(x^2\right)^2 - \left(y^3\right)^2 = \left(x^2 + y^3\right)\left(x^2 - y^3\right)$$

Example 4   Factor $9x^2 - 36$.

Solution   Remember when factoring always to check first for common factors. If there are common factors, factor out the GCF and then factor the resulting polynomial.

$$\begin{aligned} 9x^2 - 36 &= 9\left(x^2 - 4\right) \qquad \text{\textit{Factor out the GCF 9.}} \\ &= 9\left(x^2 - 2^2\right) \\ &= 9(x + 2)(x - 2) \end{aligned}$$

In this example, if we forget to factor out the GCF first, we still have the difference of two squares.

$$9x^2 - 36 = (3x)^2 - (6)^2 = (3x + 6)(3x - 6)$$

This binomial has not been factored completely since both terms of both binomial factors have a common factor of 3.

$$3x + 6 = 3(x + 2) \quad \text{and} \quad 3x - 6 = 3(x - 2)$$

Then

$$9x^2 - 36 = (3x + 6)(3x - 6) = 3(x + 2)3(x - 2) = 9(x + 2)(x - 2)$$

Factoring is easier if the GCF is factored out first before using other methods.

◆ Example 5   Factor $x^2 + 4$.

Solution   The binomial $x^2 + 4$ is the **sum** of squares since we can write $x^2 + 4$ as $x^2 + 2^2$. We might try to factor using $(x + 2)(x + 2)$ or $(x - 2)(x - 2)$. But when multiplying to check, neither factoring is correct.

$$(x + 2)(x + 2) = x^2 + 4x + 4$$
$$(x - 2)(x - 2) = x^2 - 4x + 4$$

In both cases, the product is a trinomial, not the required binomial. In fact, $x^2 + 4$ is a prime polynomial.

> **HELPFUL HINT**
> After the greatest common factor has been removed, the *sum* of two squares cannot be factored further using real numbers.

---

2   Although the sum of two squares usually does not factor, the sum or difference of two cubes can be factored and reveals factoring patterns. The pattern for the sum of cubes is illustrated by multiplying the binomial $x + y$ and the trinomial $x^2 - xy + y^2$.

$$
\begin{array}{r}
x^2 - xy + y^2 \\
x + y \\
\hline
x^2y - xy^2 + y^3 \\
x^3 - x^2y + xy^2 \quad\quad \\
\hline
x^3 \quad\quad\quad\quad\quad + y^3
\end{array}
$$

$$(x + y)(x^2 - xy + y^2) = x^3 + y^3 \qquad \text{Sum of cubes.}$$

The pattern for the difference of two cubes is illustrated by multiplying the binomial $x - y$ by the trinomial $x^2 + xy + y^2$. The result is

$$(x - y)(x^2 + xy + y^2) = x^3 - y^3 \qquad \text{Difference of cubes.}$$

SUM OR DIFFERENCE OF TWO CUBES

$$a^3 + b^3 = (a + b)(a^2 - ab + b^2)$$

$$a^3 - b^3 = (a - b)(a^2 + ab + b^2)$$

Example 6   Factor $x^3 + 8$.

Solution   First, write the binomial in the form $a^3 + b^3$.

$$x^3 + 8 = x^3 + 2^3 \qquad \text{Write in the form } a^3 + b^3.$$

If we replace a with x and b with 2 in the formula above, we have

$$x^3 + 2^3 = (x + 2)\left[x^2 - (x)(2) + 2^2\right]$$
$$= (x + 2)(x^2 - 2x + 4)$$

> **HELPFUL HINT**
> When factoring sums or differences of cubes, notice the sign patterns.
>
> same sign
> $$x^3 + y^3 = (x + y)(x^2 - xy + y^2)$$
> opposite sign   always positive
>
> same sign
> $$x^3 - y^3 = (x - y)(x^2 + xy + y^2)$$
> opposite sign   always positive

Example 7 Factor $y^3 - 27$.

Solution
$$y^3 - 27 = y^3 - 3^3 \qquad \text{Write in the form } a^3 - b^3.$$
$$= (y - 3)\left[y^2 + (y)(3) + 3^2\right]$$
$$= (y - 3)(y^2 + 3y + 9)$$

Example 8 Factor $64x^3 + 1$.

Solution
$$64x^3 + 1 = (4x)^3 + 1^3$$
$$= (4x + 1)\left[(4x)^2 - (4x)(1) + 1^2\right]$$
$$= (4x + 1)(16x^2 - 4x + 1)$$

◆ **Example 9** Factor $54a^3 - 16b^3$.

Solution Remember to factor out common factors first before using other factoring methods.

$$54a^3 - 16b^3 = 2(27a^3 - 8b^3) \qquad \text{Factor out the GCF 2.}$$
$$= 2\left[(3a)^3 - (2b)^3\right] \qquad \text{Difference of two cubes.}$$
$$= 2(3a - 2b)\left[(3a)^2 + (3a)(2b) + (2b)^2\right]$$
$$= 2(3a - 2b)(9a^2 + 6ab + 4b^2)$$

GRAPHING CALCULATOR EXPLORATIONS

A graphing calculator is a convenient tool for evaluating an expression at a given replacement value. For example, let's evaluate $x^2 - 6x$ when $x = 2$. To do so, store the value 2 in the variable x and then enter and evaluate the algebraic expression.

```
2 → X
                2
X² − 6X
               −8
```

The value of $x^2 - 6x$ when $x = 2$ is -8. You may want to use this method for evaluating expressions as you explore the following.

We can use a graphing calculator to explore factoring patterns numerically. Use your calculator to evaluate $x^2 - 2x + 1$, $x^2 - 2x - 1$, and $(x - 1)^2$ for each value of x given in the table. What do you observe?

	$x^2 - 2x + 1$	$x^2 - 2x - 1$	$(x - 1)^2$
$x = 5$			
$x = -3$			
$x = 2.7$			
$x = -12.1$			
$x = 0$			

Notice in each case that $x^2 - 2x - 1 \neq (x - 1)^2$. Because for each x in the table the value of $x^2 - 2x + 1$ and the value of $(x - 1)^2$ are the same, we might guess that $x^2 - 2x + 1 = (x - 1)^2$. We can verify our guess algebraically with multiplication:

$$(x - 1)(x - 1) = x^2 - x - x + 1 = x^2 - 2x + 1$$

MENTAL MATH

State each number as a square.

1. 1   **2.** 25   **3.** 81
4. 64   **5.** 9   **6.** 100

State each number as a cube.

7. 1   **8.** 64   **9.** 8   **10.** 27

Exercise Set 5.4

Factor the difference of two squares. See Examples 1 through 4.

1. $x^2 - 4$ **2.** $y^2 - 81$

3. $y^2 - 49$ **4.** $x^2 - 100$

5. $25y^2 - 9$ **6.** $49a^2 - 16$

7. $121 - 100x^2$ **8.** $144 - 81x^2$

9. $12x^2 - 27$ **10.** $36x^2 - 64$

11. $169a^2 - 49b^2$ **12.** $225a^2 - 81b^2$

13. $x^2y^2 - 1$ **14.** $16 - a^2b^2$

15. $x^4 - 9$ **16.** $y^4 - 25$

17. $49a^4 - 16$ **18.** $49b^4 - 1$

19. $x^4 - y^{10}$ **20.** $x^{14} - y^4$

21. What binomial multiplied by $(x - 6)$ gives the difference of two squares?

22. What binomial multiplied by $(5 + y)$ gives the difference of two squares?

Factor the sum or difference of two cubes. See Examples 6 through 9.

23. $a^3 + 27$ **24.** $b^3 - 8$

25. $8a^3 + 1$ **26.** $64x^3 - 1$

27. $5k^3 + 40$ **28.** $6r^3 - 162$

29. $x^3y^3 - 64$ **30.** $8x^3 - y^3$

31. $x^3 + 125$ **32.** $a^3 - 216$

33. $24x^4 - 81xy^3$ **34.** $375y^6 - 24y^3$

35. What binomial multiplied by $\left(4x^2 - 2xy + y^2\right)$ gives the sum or difference of two cubes?

36. What binomial multiplied by $\left(1 + 4y + 16y^2\right)$ gives the sum or difference of two cubes?

Factor the binomials completely.

37. $x^2 - 4$ **38.** $x^2 - 36$

39. $81 - p^2$ **40.** $100 - t^2$

41. $4r^2 - 1$ **42.** $9t^2 - 1$

43. $9x^2 - 16$ **44.** $36y^2 - 25$

45. $16r^2 + 1$ **46.** $49y^2 + 1$

47. $27 - t^3$ **48.** $125 + r^3$

49. $8r^3 - 64$ **50.** $54r^3 + 2$

51. $t^3 - 343$ **52.** $s^3 + 216$

53. $x^2 - 169y^2$ **54.** $x^2 - 225y^2$

55. $x^2y^2 - z^2$ **56.** $x^3y^3 - z^3$

57. $x^3y^3 + 1$ **58.** $x^2y^2 + z^2$

59. $s^3 - 64t^3$ **60.** $8t^3 + s^3$

61. $18r^2 - 8$ **62.** $32t^2 - 50$

63. $9xy^2 - 4x$ **64.** $16xy^2 - 64x$

65. $25y^4 - 100y^2$ **66.** $xy^3 - 9xyz^2$

67. $x^3y - 4xy^3$ **68.** $12s^3t^3 + 192s^5t$

69. $8s^6t^3 + 100s^3t^6$ **70.** $25x^5y + 121x^3y$

71. $27x^2y^3 - xy^2$ **72.** $8x^3y^3 + x^3y$

73. An object is dropped from the top of Pittsburgh's USX Towers, which is 841 feet tall. (*Source: World Almanac research*) The height of the object after t seconds is given by the expression $841 - 16t^2$.

     a. Find the height of the object after 2 seconds.

     b. Find the height of the object after 5 seconds.

     c. To the nearest whole second, estimate when the object hits the ground.

     d. Factor $841 - 16t^2$.

841 feet

74. A worker on the top of the Aetna Life Building in San Francisco accidentally drops a bolt. The Aetna Life Building is 529 feet tall. (*Source: World Almanac research*) The height of the bolt after t seconds is given by the expression $529 - 16t^2$.

     a. Find the height of the bolt after 1 second.

     b. Find the height of the bolt after 4 seconds.

     c. To the nearest whole second, estimate when the bolt hits the ground.

     d. Factor $529 - 16t^2$.

75. In your own words, explain how to tell whether a binomial is a difference of squares. Then explain how to factor a difference of squares.

76. In your own words, explain how to tell whether a binomial is a sum of cubes. Then explain how to factor a sum of cubes.

REVIEW EXERCISES

Divide the following. See Section 4.6.

77. $\dfrac{8x^4 + 4x^3 - 2x + 6}{2x}$

78. $\dfrac{3y^4 + 9y^2 - 6y + 1}{3y^2}$

Use long division to divide the following. See Section 4.6.

79. $\dfrac{2x^2 - 3x - 2}{x - 2}$

80. $\dfrac{4x^2 - 21x + 21}{x - 3}$

81. $\dfrac{3x^2 + 13x + 10}{x + 3}$

82. $\dfrac{5x^2 + 14x + 12}{x + 2}$

A Look Ahead

Example

Factor $(x + y)^2 - (x - y)^2$.

Solution:

Use the method for factoring the difference of squares.

$$(x + y)^2 - (x - y)^2 = \big[(x + y) + (x - y)\big]$$
$$\big[(x + y) - (x - y)\big]$$
$$= (x + y + x - y)(x + y - x + y)$$
$$= (2x)(2y)$$
$$= 4xy$$

Factor each difference of squares. See the example.

83. $a^2 - (2 + b)^2$

84. $(x + 3)^2 - y^2$

85. $\left(x^2 - 4\right)^2 - (x - 2)^2$

86. $\left(x^2 - 9\right) - (3 - x)$

5.5 CHOOSING A FACTORING STRATEGY

CD-ROM   SSM

SSG   Video

▶ **OBJECTIVE**

1. Factor polynomials completely.

1   A polynomial is factored completely when it is written as the product of prime polynomials. This section uses the various methods of factoring polynomials that have been discussed in earlier sections. Since these methods are applied throughout the remainder of this text, as well as in later courses, it is important to master the skills of factoring. The following is a set of guidelines for factoring polynomials.

FACTORING A POLYNOMIAL

Step 1. Are there any common factors? If so, factor out the GCF.

Step 2. How many terms are in the polynomial?

    a. If there are **two** terms, decide if one of the following can be applied.

        i. Difference of two squares: $a^2 - b^2 = (a - b)(a + b)$.

        ii. Difference of two cubes: $a^3 - b^3 = (a - b)\left(a^2 + ab + b^2\right)$.

        iii. Sum of two cubes: $a^3 + b^3 = (a + b)\left(a^2 - ab + b^2\right)$.

    b. If there are **three** terms, try one of the following.

        i. Perfect square trinomial: $a^2 + 2ab + b^2 = (a + b)^2$.

        ii. If not a perfect square trinomial, factor using the methods presented in Sections 5.2 and 5.3.

    c. If there are **four** or more terms, try factoring by grouping.

Step 3. See if any factors in the factored polynomial can be factored further.

Step 4. Check by multiplying.

Example 1 Factor $10t^2 - 17t + 3$.

Solution **Step 1.** The terms of this polynomial have no common factor (other than 1).

Step 2. There are three terms, so this polynomial is a trinomial. This trinomial is not a perfect square trinomial, so factor using methods from earlier sections.

Factors of $10t^2$: $\quad 10t^2 = 2t \cdot 5t, \qquad 10t^2 = t \cdot 10t$

Since the middle term, $-17t$, has a negative numerical coefficient, find negative factors of 3.

Factors of 3: $\quad 3 = -1 \cdot -3$

Try different combinations of these factors. The correct combination is

$$\underbrace{(2t \; \underline{-3})(\underline{5t} \; - 1)}_{\substack{-15t \\ -2t \\ \overline{-17t}}} = 10t^2 - 17t + 3$$

Correct middle term.

Step 3. No factor can be factored further, so we have factored completely.

Step 4. To check, multiply $2t - 3$ and $5t - 1$.

$$(2t - 3)(5t - 1) = 10t^2 - 2t - 15t + 3 = 10t^2 - 17t + 3$$

The factored form of $10t^2 - 17t + 3$ is $(2t - 3)(5t - 1)$. ∎

Example 2 Factor $2x^3 + 3x^2 - 2x - 3$.

Solution **Step 1.** There are no factors common to all terms.

Step 2. Try factoring by grouping since this polynomial has four terms.

$$2x^3 + 3x^2 - 2x - 3 = x^2(2x + 3) - 1(2x + 3) \quad \text{\small Factor out the greatest common factor for each pair of terms.}$$

$$= (2x + 3)\left(x^2 - 1\right) \quad \text{\small Factor out } 2x + 3.$$

Step 3. The binomial $x^2 - 1$ can be factored further. It is the difference of two squares.

$$= (2x + 3)(x + 1)(x - 1) \quad \text{\small Factor } x^2 - 1 \text{ as a difference of squares.}$$

Step 4. Check by finding the product of the three binomials. The polynomial factored completely is $(2x + 3)(x + 1)(x - 1)$. ∎

Example 3 Factor $12m^2 - 3n^2$.

Solution **Step 1.** The terms of this binomial contain a greatest common factor of 3.

$$12m^2 - 3n^2 = 3\left(4m^2 - n^2\right) \quad \text{\small Factor out the greatest common factor.}$$

Step 2.   The binomial $4m^2 - n^2$ is a difference of squares.

$$= 3(2m + n)(2m - n) \quad \text{Factor the difference of squares.}$$

Step 3.   No factor can be factored further.

Step 4.   We check by multiplying.

$$3(2m + n)(2m - n) = 3(4m^2 - n^2) = 12m^2 - 3n^2$$

The factored form of $12m^2 - 3n^2$ is $3(2m + n)(2m - n)$.   ▬

Example 4   Factor $x^3 + 27y^3$.

Solution   **Step 1.**   The terms of this binomial contain no common factor (other than 1).

Step 2.   This binomial is the sum of two cubes.

$$
\begin{aligned}
x^3 + 27y^3 &= (x)^3 + (3y)^3 \\
&= (x + 3y)\left[x^2 - 3xy + (3y)^2\right] \\
&= (x + 3y)(x^2 - 3xy + 9y^2)
\end{aligned}
$$

Step 3.   No factor can be factored further.

Step 4.   We check by multiplying.

$$
\begin{aligned}
(x + 3y)(x^2 - 3xy + 9y^2) &= x(x^2 - 3xy + 9y^2) \\
&\quad + 3y(x^2 - 3xy + 9y^2) \\
&= x^3 - 3x^2y + 9xy^2 + 3x^2y \\
&\quad - 9xy^2 + 27y^3 \\
&= x^3 + 27y^3
\end{aligned}
$$

Thus, $x^3 + 27y^3$ factored completely is $(x + 3y)(x^2 - 3xy + 9y^2)$.   ▬

Example 5   Factor $30a^2b^3 + 55a^2b^2 - 35a^2b$.

Solution   **Step 1.** $30a^2b^3 + 55a^2b^2 - 35a^2b = 5a^2b(6b^2 + 11b - 7)$ Factor out the GCF.

Step 2. $= 5a^2b(2b - 1)(3b + 7)$ Factor the resulting trinomial.

Step 3.   No factor can be factored further.

Step 4.   Check by multiplying.

The trinomial factored completely is $5a^2b(2b - 1)(3b + 7)$.   ▬

SPOTLIGHT ON DECISION MAKING

Suppose you are shopping for a credit card and have received the following credit card offers in the mail.

Credit Card Offer #1	Credit Card Offer #2	Credit Card Offer #3
We offer a low 9.8% APR interest rate, coupled with a generous 20-day grace period and low $35 annual fee. We can offer you a $2000 credit limit. In addition, you'll earn one frequent flier mile for every dollar charged on your card. And every time you use your card to pay for airline tickets, we'll even give you $250,000 in flight insurance– absolutely free!	With our card, you'll get a 14.5% APR interest rate, absolutely *no* annual fee, a 25-day grace period, and a $2500 credit limit. And, only with our card, you can receive a year-end cash bonus of up to 1% of the total value of purchases made to your card during the year! That's our way of saying "Thank you" for choosing our card.	*Just say "yes" to this offer, and you can get our lowest 17% APR interest rate, along with a 30-day grace period and easy-to-swallow $20 annual fee. You also qualify for a $3000 credit limit.* *That's not all: you earn gift certificates for a local mall with every purchase! Plus, you'll receive $100,000 of flight insurance whenever you charge airline tickets.*

You construct a decision grid to help make your choice. In the decision grid, give each of the decision criteria a rank reflecting its importance to you, with 1 being not important to 10 being very important. Then for each card offer, decide how well the criteria are supported, assigning a rating of 1 for poor support to a rating of 10 for excellent support. For Credit Card Offer #1, fill in the Score column by multiplying rank by rating for each criteria. Repeat for each credit card offer. Finally, total the scores for each credit card offer. The offer with the highest score is likely to be the best choice for you.

Based on your decision-grid analysis, which credit card would you choose? Explain.

Criteria	Rank	CREDIT CARD OFFER #1		CREDIT CARD OFFER #2		CREDIT CARD OFFER #3	
		Rating	Score	Rating	Score	Rating	Score
Interest rate							
Annual fee							
Grace period							
Credit limit							
Automatic flight insurance							
Rebates/ incentives/bonuses							
TOTAL							

Exercise Set 5.5

Factor the following completely. See Examples 1 through 5.

1. $a^2 + 2ab + b^2$

2. $a^2 - 2ab + b^2$

3. $a^2 + a - 12$

4. $a^2 - 7a + 10$

5. $a^2 - a - 6$

6. $a^2 + 2a + 1$

7. $x^2 + 2x + 1$

8. $x^2 + x - 2$

9. $x^2 + 4x + 3$

10. $x^2 + x - 6$

11. $x^2 + 7x + 12$

12. $x^2 + x - 12$

13. $x^2 + 3x - 4$

14. $x^2 - 7x + 10$

15. $x^2 + 2x - 15$

16. $x^2 + 11x + 30$

17. $x^2 - x - 30$

18. $x^2 + 11x + 24$

19. $2x^2 - 98$

20. $3x^2 - 75$

21. $x^2 + x + xy + 3y$

22. $3y - 21 + xy - 7x$

23. $x^2 + 6x - 16$

24. $x^2 - 3x - 28$

25. $4x^3 + 20x^2 - 56x$

26. $6x^3 - 6x^2 - 120x$

27. $12x^2 + 34x + 24$

28. $8a^2 + 6ab - 5b^2$

29. $4a^2 - b^2$

30. $28 - 13x - 6x^2$

31. $20 - 3x - 2x^2$

32. $x^2 - 2x + 4$

33. $a^2 + a - 3$

34. $6y^2 + y - 15$

35. $4x^2 - x - 5$

36. $x^2y - y^3$

37. $4t^2 + 36$

38. $x^2 + x + xy + y$

39. $ax + 2x + a + 2$

40. $18x^3 - 63x^2 + 9x$

41. $12a^3 - 24a^2 + 4a$

42. $x^2 + 14x - 32$

43. $x^2 - 14x - 48$

44. $16a^2 - 56ab + 49b^2$

45. $25p^2 - 70pq + 49q^2$

46. $7x^2 + 24xy + 9y^2$

47. $125 - 8y^3$

48. $64x^3 + 27$

49. $-x^2 - x + 30$

50. $-x^2 + 6x - 8$

51. $14 + 5x - x^2$

52. $3 - 2x - x^2$

53. $3x^4y + 6x^3y - 72x^2y$

54. $2x^3y + 8x^2y^2 - 10xy^3$

55. $5x^3y^2 - 40x^2y^3 + 35xy^4$

56. $4x^4y - 8x^3y - 60x^2y$

57. $12x^3y + 243xy$

58. $6x^3y^2 + 8xy^2$

59. $(x - y)^2 - z^2$

60. $(x + 2y)^2 - 9$

61. $3rs - s + 12r - 4$

62. $x^3 - 2x^2 + 3x - 6$

63. $4x^2 - 8xy - 3x + 6y$

64. $4x^2 - 2xy - 7yz + 14xz$

65. $6x^2 + 18xy + 12y^2$

66. $12x^2 + 46xy - 8y^2$

67. $xy^2 - 4x + 3y^2 - 12$

68. $x^2y^2 - 9x^2 + 3y^2 - 27$

69. $5(x + y) + x(x + y)$

70. $7(x - y) + y(x - y)$

71. $14t^2 - 9t + 1$

72. $3t^2 - 5t + 1$

73. $3x^2 + 2x - 5$

74. $7x^2 + 19x - 6$

75. $x^2 + 9xy - 36y^2$

76. $3x^2 + 10xy - 8y^2$

77. $1 - 8ab - 20a^2b^2$

78. $1 - 7ab - 60a^2b^2$

79. $x^4 - 10x^2 + 9$

80. $x^4 - 13x^2 + 36$

81. $x^4 - 14x^2 - 32$

82. $x^4 - 22x^2 - 75$

83. $x^2 - 23x + 120$

84. $y^2 + 22y + 96$

85. $6x^3 - 28x^2 + 16x$

86. $6y^3 - 8y^2 - 30y$

87. $27x^3 - 125y^3$

88. $216y^3 - z^3$

89. $x^3y^3 + 8z^3$

90. $27a^3b^3 + 8$

91. $2xy - 72x^3y$

92. $2x^3 - 18x$

93. $x^3 + 6x^2 - 4x - 24$

94. $x^3 - 2x^2 - 36x + 72$

95. $6a^3 + 10a^2$

96. $4n^2 - 6n$

97. $a^2(a + 2) + 2(a + 2)$

98. $a - b + x(a - b)$

99. $x^3 - 28 + 7x^2 - 4x$

100. $a^3 - 45 - 9a + 5a^2$

101. Explain why it makes good sense to factor out the GCF first, before using other methods of factoring.

102. The sum of two squares usually does not factor. Is the sum of two squares $9x^2 + 81y^2$ factorable?

REVIEW EXERCISES

Solve each equation. See Section 2.4.

103. $x - 6 = 0$

104. $y + 5 = 0$

105. $2m + 4 = 0$

106. $3x - 9 = 0$

107. $5z - 1 = 0$

108. $4a + 2 = 0$

Solve the following. See Section 2.6.

△ **109.** A suitcase has a volume of 960 cubic inches. Find x.

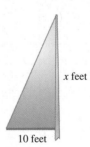

10 inches

12 inches

x inches

△ **110.** The sail shown has an area of 25 square feet. Find its height, x.

x feet

10 feet

List the x- and y-intercepts for each graph. See Section 3.3.

111.

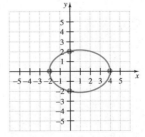

112.

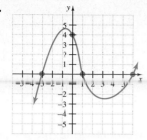

114.

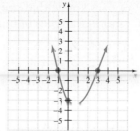

113.

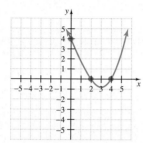

5.6   SOLVING QUADRATIC EQUATIONS BY FACTORING

CD-ROM   SSM

SSG   Video

▶ **OBJECTIVES**

1. Define quadratic equation.
2. Solve quadratic equations by factoring.
3. Solve equations with degree greater than 2 by factoring.

1   Linear equations, while versatile, are not versatile enough to model many real-life phenomena. For example, let's suppose an object is dropped from the top of a 256-foot cliff and we want to know how long before the object strikes the ground. The answer to this question is found by solving the equation $-16t^2 + 256 = 0$. (See Example 1 in the next section.) This equation is called a **quadratic equation** because it contains a variable with an exponent of 2 and no other variable in the equation contains an exponent greater than 2. In this section, we solve quadratic equations by factoring.

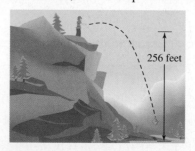

256 feet

QUADRATIC EQUATION

A quadratic equation is one that can be written in the form
$$ax^2 + bx + c = 0,$$
where a, b, and c are real numbers, and $a \neq 0$.

Notice that the degree of the polynomial $ax^2 + bx + c$ is 2. Here are a few more examples of quadratic equations.

Quadratic Equations

$$3x^2 + 5x + 6 = 0 \qquad x^2 = 9 \qquad y^2 + y = 1$$

The form $ax^2 + bx + c = 0$ is called the **standard form** of a quadratic equation. The quadratic equations $3x^2 + 5x + 6 = 0$ and $-16t^2 + 256 = 0$ are in standard form. One side of the equation is 0 and the other side is a polynomial of degree 2 written in descending powers of the variable.

2   Some quadratic equations can be solved by making use of factoring and the **zero factor theorem**.

ZERO FACTOR THEOREM

If a and b are real numbers and if $ab = 0$, then $a = 0$ or $b = 0$.

This theorem states that if the product of two numbers is 0 then at least one of the numbers must be 0.

Example 1   Solve $(x - 3)(x + 1) = 0$.

Solution   If this equation is to be a true statement, then either the factor $x - 3$ must be 0 or the factor $x + 1$ must be 0. In other words, either

$$x - 3 = 0 \qquad \text{or} \qquad x + 1 = 0$$

If we solve these two linear equations, we have

$$x = 3 \qquad \text{or} \qquad x = -1$$

Thus, 3 and -1 are both solutions of the equation $(x - 3)(x + 1) = 0$. To check, we replace x with 3 in the original equation. Then we replace x with -1 in the original equation.

Check
$$(x - 3)(x + 1) = 0 \qquad\qquad\qquad (x - 3)(x + 1) = 0$$
$$(3 - 3)(3 + 1) \stackrel{?}{=} 0 \quad \text{Replace x with 3.} \qquad (-1 - 3)(-1 + 1) \stackrel{?}{=} 0 \quad \text{Replace x with -1.}$$
$$0(4) = 0 \quad \text{True.} \qquad\qquad\qquad\qquad (-4)(0) = 0 \quad \text{True.}$$

The solutions are 3 and -1.

HELPFUL HINT

The zero factor property says that *if a product is 0, then a factor is 0.*

If $a \cdot b = 0$, then $a = 0$ or $b = 0$.

If $x(x + 5) = 0$, then $x = 0$ or $x + 5 = 0$.

If $(x + 7)(2x - 3) = 0$, then $x + 7 = 0$ or $2x - 3 = 0$.

Use this property only when the product is 0. For example, if $a \cdot b = 8$, we do not know the value of a or b. The values may be $a = 2, b = 4$ or $a = 8, b = 1$, or any other two numbers whose product is 8.

Example 2   Solve $x^2 - 9x = -20$.

Solution   First, write the equation in standard form; then factor.

$$x^2 - 9x = -20$$
$$x^2 - 9x + 20 = 0 \qquad \text{Write in standard form by adding 20 to both sides.}$$
$$(x - 4)(x - 5) = 0 \qquad \text{Factor.}$$

Next, use the zero factor theorem and set each factor equal to 0.

$$x - 4 = 0 \quad \text{or} \quad x - 5 = 0 \qquad \text{Set each factor equal to 0.}$$
$$x = 4 \quad \text{or} \quad x = 5 \qquad \text{Solve.}$$

Check the solutions by replacing x with each value in the original equation. The solutions are 4 and 5.

The following steps may be used to solve a quadratic equation by factoring.

SOLVING QUADRATIC EQUATIONS BY FACTORING

Step 1. Write the equation in standard form: $ax^2 + bx + c = 0$.
Step 2. Factor the quadratic completely.
Step 3. Set each factor containing a variable equal to 0.
Step 4. Solve the resulting equations.
Step 5. Check each solution in the original equation.

Since it is not always possible to factor a quadratic polynomial, not all quadratic equations can be solved by factoring. Other methods of solving quadratic equations are presented in Chapter 10.

Example 3   Solve $x(2x - 7) = 4$.

Solution   First, write the equation in standard form; then factor.

$$x(2x - 7) = 4$$
$$2x^2 - 7x = 4 \qquad \text{Multiply.}$$
$$2x^2 - 7x - 4 = 0 \qquad \text{Write in standard form.}$$
$$(2x + 1)(x - 4) = 0 \qquad \text{Factor.}$$
$$2x + 1 = 0 \quad \text{or} \quad x - 4 = 0 \qquad \text{Set each factor equal to zero.}$$
$$2x = -1 \quad \text{or} \quad x = 4 \qquad \text{Solve.}$$
$$x = -\frac{1}{2}$$

Check both solutions $-\dfrac{1}{2}$ and 4.

> **HELPFUL HINT**
> One more reminder: To apply the zero factor theorem, one side of the equation must be 0 and the other side of the equation must be factored. To solve the equation $x(2x - 7) = 4$, for example, you may **not** set each factor equal to 4.

Example 4 Solve $-2x^2 - 4x + 30 = 0$.

Solution The equation is in standard form so we begin by factoring out a common factor of -2.

$$-2x^2 - 4x + 30 = 0$$
$$-2(x^2 + 2x - 15) = 0 \qquad \text{Factor out } -2.$$
$$-2(x + 5)(x - 3) = 0 \qquad \text{Factor the quadratic.}$$

Next, set each factor **containing a variable** equal to 0.

$$\begin{array}{ccc} x + 5 = 0 & \text{or} & x - 3 = 0 \\ x = -5 & \text{or} & x = 3 \end{array}$$

Set each factor containing a variable equal to 0.

Solve.

Note that the factor -2 is a constant term containing no variables and can never equal 0. The solutions are -5 and 3. ■

3 Some equations involving polynomials of degree higher than 2 may also be solved by factoring and then applying the zero factor theorem.

Example 5 Solve $3x^3 - 12x = 0$.

Solution Factor the left side of the equation. Begin by factoring out the common factor of $3x$.

$$3x^3 - 12x = 0$$
$$3x(x^2 - 4) = 0 \qquad \text{Factor out the GCF } 3x.$$
$$3x(x + 2)(x - 2) = 0 \qquad \text{Factor } x^2 - 4, \text{ a difference of squares.}$$

$$\begin{array}{ccccc} 3x = 0 & \text{or} & x + 2 = 0 & \text{or} & x - 2 = 0 \\ x = 0 & \text{or} & x = -2 & \text{or} & x = 2 \end{array}$$

Set each factor equal to 0.

Solve.

Thus, the equation $3x^3 - 12x = 0$ has three solutions: $0, -2$, and 2. To check, replace x with each solution in the original equation.

Let $x = 0$.   **Let $x = -2$.**   **Let $x = 2$.**

$$3(0)^3 - 12(0) \overset{?}{=} 0 \qquad 3(-2)^3 - 12(-2) \overset{?}{=} 0 \qquad 3(2)^3 - 12(2) \overset{?}{=} 0$$
$$0 = 0 \qquad\qquad 3(-8) + 24 \overset{?}{=} 0 \qquad 3(8) - 24 \overset{?}{=} 0$$
$$0 = 0 \qquad\qquad\qquad 0 = 0$$

Substituting $0, -2$, or 2 into the original equation results each time in a true equation. The solutions are $0, -2$, and 2. ■

Example 6 Solve $(5x - 1)(2x^2 + 15x + 18) = 0$.

Solution

$$(5x - 1)(2x^2 + 15x + 18) = 0$$
$$(5x - 1)(2x + 3)(x + 6) = 0 \qquad \text{Factor the trinomial.}$$

$$\begin{array}{ccccc} 5x - 1 = 0 & \text{or} & 2x + 3 = 0 & \text{or} & x + 6 = 0 \\ 5x = 1 & \text{or} & 2x = -3 & \text{or} & x = -6 \end{array}$$

Set each factor equal to 0.

Solve.

$$x = \frac{1}{5} \qquad \text{or} \qquad x = -\frac{3}{2}$$

The solutions are $\frac{1}{5}, -\frac{3}{2}$, and -6. Check by replacing x with each solution in the original equation. The solutions are $-6, -\frac{3}{2}$, and $\frac{1}{5}$. ■

Example 7   Solve $2x^3 - 4x^2 - 30x = 0$.

Solution   Begin by factoring out the GCF $2x$.

$$2x^3 - 4x^2 - 30x = 0$$

$$2x(x^2 - 2x - 15) = 0 \qquad \text{Factor out the GCF } 2x.$$

$$2x(x - 5)(x + 3) = 0 \qquad \text{Factor the quadratic.}$$

$$2x = 0 \quad \text{or} \quad x - 5 = 0 \quad \text{or} \quad x + 3 = 0 \qquad \begin{array}{l}\text{Set each factor containing}\\ \text{a variable equal to 0.}\end{array}$$

$$x = 0 \quad \text{or} \qquad x = 5 \quad \text{or} \qquad x = -3 \qquad \text{Solve.}$$

Check by replacing x with each solution in the cubic equation. The solutions are -3, 0, and 5. ■

In Chapter 3, we graphed linear equations in two variables, such as $y = 5x - 6$. Recall that to find the x-intercept of the graph of a linear equation, let $y = 0$ and solve for x. This is also how to find the x-intercepts of the graph of a **quadratic equation in two variables**, such as $y = x^2 - 5x + 4$.

Example 8   Find the x-intercepts of the graph of $y = x^2 - 5x + 4$.

Solution   Let $y = 0$ and solve for x.

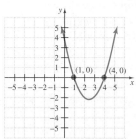

$$y = x^2 - 5x + 4$$

$$0 = x^2 - 5x + 4 \qquad \text{Let } y = 0.$$

$$0 = (x - 1)(x - 4) \qquad \text{Factor.}$$

$$x - 1 = 0 \quad \text{or} \quad x - 4 = 0 \qquad \text{Set each factor equal to 0.}$$

$$x = 1 \quad \text{or} \qquad x = 4 \qquad \text{Solve.}$$

The x-intercepts of the graph of $y = x^2 - 5x + 4$ are $(1, 0)$ and $(4, 0)$.
The graph of $y = x^2 - 5x + 4$ is shown in the margin. ■

In general, a quadratic equation in two variables is one that can be written in the form $y = ax^2 + bx + c$ where $a \neq 0$. The graph of such an equation is called a **parabola** and will open up or down depending on the value of a.

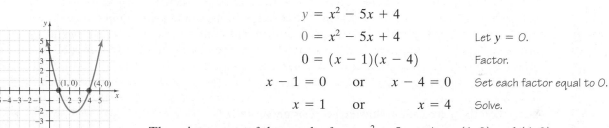

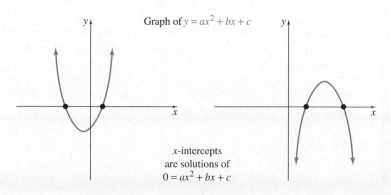

Notice that the x-intercepts of the graph of $y = ax^2 + bx + c$ are the real number solutions of $0 = ax^2 + bx + c$. Also, the real number solutions of $0 = ax^2 + bx + c$ are the x-intercepts of the graph of $y = ax^2 + bx + c$. We study more about graphs of quadratic equations in two variables in Chapter 10.

GRAPHING CALCULATOR EXPLORATIONS

A grapher may be used to find solutions of a quadratic equation whether the related quadratic polynomial is factorable or not. For example, let's use a grapher to approximate the solutions of $0 = x^2 + 4x - 3$. To do so, graph $y_1 = x^2 + 4x - 3$. Recall that the x-intercepts of this graph are the solutions of $0 = x^2 + 4x - 3$.

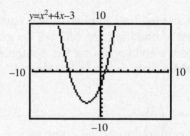

Notice that the graph appears to have an x-intercept between -5 and -4 and one between 0 and 1. Many graphers contain a TRACE feature. This feature activates a graph cursor that can be used to *trace* along a graph while the corresponding x- and y-coordinates are shown on the screen. Use the TRACE feature to confirm that x-intercepts lie between -5 and -4 and also 0 and 1. To approximate the x-intercepts to the nearest tenth, use a ROOT or a ZOOM feature on your grapher or redefine the viewing window. (A ROOT feature calculates the x-intercept. A ZOOM feature magnifies the viewing window around a specific location such as the graph cursor.) If we redefine the window to $[0, 1]$ on the x-axis and $[-1, 1]$ on the y-axis, the following graph is generated.

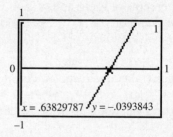

By using the TRACE feature, we can conclude that one x-intercept is approximately 0.6 to the nearest tenth. By repeating these steps for the other x-intercept, we find that it is approximately -4.6.

Use a grapher to approximate to the nearest tenth the real number solutions. If an equation has no real number solution, state so.

1. $3x^2 - 4x - 6 = 0$ **2.** $x^2 - x - 9 = 0$

3. $2x^2 + x + 2 = 0$ **4.** $-4x^2 - 5x - 4 = 0$

5. $-x^2 + x + 5 = 0$ **6.** $10x^2 + 6x - 3 = 0$

MENTAL MATH

Solve each equation by inspection.

1. $(a - 3)(a - 7) = 0$ **2.** $(a - 5)(a - 2) = 0$

3. $(x + 8)(x + 6) = 0$ **4.** $(x + 2)(x + 3) = 0$

5. $(x + 1)(x - 3) = 0$ **6.** $(x - 1)(x + 2) = 0$

Exercise Set 5.6

Solve each equation. See Example 1.

1. $(x - 2)(x + 1) = 0$ **2.** $(x + 3)(x + 2) = 0$

3. $x(x + 6) = 0$ **4.** $2x(x - 7) = 0$

5. $(2x + 3)(4x - 5) = 0$ **6.** $(3x - 2)(5x + 1) = 0$

7. $(2x - 7)(7x + 2) = 0$ **8.** $(9x + 1)(4x - 3) = 0$

9. Write a quadratic equation that has two solutions, 6 and −1. Leave the polynomial in the equation in factored form.

10. Write a quadratic equation that has two solutions, 0 and −2. Leave the polynomial in the equation in factored form.

Solve each equation. See Examples 2 through 4.

11. $x^2 - 13x + 36 = 0$ **12.** $x^2 + 2x - 63 = 0$

13. $x^2 + 2x - 8 = 0$ **14.** $x^2 - 5x + 6 = 0$

15. $x^2 - 4x = 32$ **16.** $x^2 - 5x = 24$

17. $x(3x - 1) = 14$ **18.** $x(4x - 11) = 3$

19. $3x^2 + 19x - 72 = 0$ **20.** $36x^2 + x - 21 = 0$

21. Write a quadratic equation in standard form that has two solutions, 5 and 7.

22. Write an equation that has three solutions, 0, 1, and 2.

Solve each equation. See Examples 5 through 7.

23. $x^3 - 12x^2 + 32x = 0$ **24.** $x^3 - 14x^2 + 49x = 0$

25. $(4x - 3)(16x^2 - 24x + 9) = 0$

26. $(2x + 5)(4x^2 - 10x + 25) = 0$

27. $4x^3 - x = 0$ **28.** $4y^3 - 36y = 0$

29. $32x^3 - 4x^2 - 6x = 0$ **30.** $15x^3 + 24x^2 - 63x = 0$

Solve each equation. Be careful. Some of the equations are quadratic and higher degree and some are linear.

31. $x(x + 7) = 0$ **32.** $y(6 - y) = 0$

33. $(x + 5)(x - 4) = 0$ **34.** $(x - 8)(x - 1) = 0$

35. $x^2 - x = 30$ **36.** $x^2 + 13x = -36$

37. $6y^2 - 22y - 40 = 0$ **38.** $3x^2 - 6x - 9 = 0$

39. $(2x + 3)(2x^2 - 5x - 3) = 0$

40. $(2x - 9)(x^2 + 5x - 36) = 0$

41. $x^2 - 15 = -2x$ **42.** $x^2 - 26 = -11x$

43. $x^2 - 16x = 0$ **44.** $x^2 + 5x = 0$

45. $-18y^2 - 33y + 216 = 0$

46. $-20y^2 + 145y - 35 = 0$

47. $12x^2 - 59x + 55 = 0$

48. $30x^2 - 97x + 60 = 0$

49. $18x^2 + 9x - 2 = 0$ **50.** $28x^2 - 27x - 10 = 0$

51. $x(6x + 7) = 5$ **52.** $4x(8x + 9) = 5$

53. $4(x - 7) = 6$ **54.** $5(3 - 4x) = 9$

55. $5x^2 - 6x - 8 = 0$ **56.** $9x^2 + 6x + 2 = 0$

57. $(y - 2)(y + 3) = 6$ **58.** $(y - 5)(y - 2) = 28$

59. $4y^2 - 1 = 0$ **60.** $4y^2 - 81 = 0$

61. $t^2 + 13t + 22 = 0$ **62.** $x^2 - 9x + 18 = 0$

63. $5t - 3 = 12$ **64.** $9 - t = -1$

65. $x^2 + 6x - 17 = -26$ **66.** $x^2 - 8x - 4 = -20$

67. $12x^2 + 7x - 12 = 0$ **68.** $30x^2 - 11x - 30 = 0$

69. $10t^3 - 25t - 15t^2 = 0$ **70.** $36t^3 - 48t - 12t^2 = 0$

Find the x-intercepts of the graph of each equation. See Example 8.

71. $y = (3x + 4)(x - 1)$

72. $y = (5x - 3)(x - 4)$

73. $y = x^2 - 3x - 10$

74. $y = x^2 + 7x + 6$

75. $y = 2x^2 + 11x - 6$

76. $y = 4x^2 + 11x + 6$

For Exercises 77 through 82, match each equation with its graph.

A
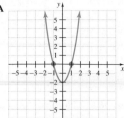

B
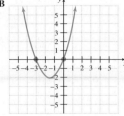

C
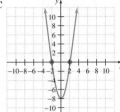

D
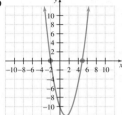

E
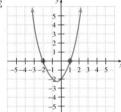

F
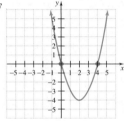

77. $y = (x + 2)(x - 1)$

78. $y = (x - 5)(x + 2)$

79. $y = x(x + 3)$

80. $y = x(x - 4)$

81. $y = 2x^2 - 8$

82. $y = 2x^2 - 2$

 83. A compass is accidentally thrown upward and out of an air balloon at a height of 300 feet. The height, y, of the compass at time x in seconds is given by the equation
$$y = -16x^2 + 20x + 300$$

300 feet

a. Find the height of the compass at the given times by filling in the table below.

time x	0	1	2	3	4	5	6
height y							

b. Use the table to determine when the compass strikes the ground.

c. Use the table to approximate the maximum height of the compass.

d. Plot the points (x, y) on a rectangular coordinate system and connect them with a smooth curve. Explain your results.

 84. A rocket is fired upward from the ground with an initial velocity of 100 feet per second. The height, y, of the rocket at any time x is given by the equation
$$y = -16x^2 + 100x$$

a. Find the height of the rocket at the given times by filling in the table below.

time x	0	1	2	3	4	5	6	7
height y								

b. Use the table to approximate when the rocket strikes the ground to the nearest second.

c. Use the table to approximate the maximum height of the rocket.

d. Plot the points (x, y) on a rectangular coordinate system and connect them with a smooth curve. Explain your results.

REVIEW EXERCISES

Perform the following operations. Write all results in lowest terms. See Section 1.2.

85. $\dfrac{3}{5} + \dfrac{4}{9}$

86. $\dfrac{2}{3} + \dfrac{3}{7}$

87. $\dfrac{7}{10} - \dfrac{5}{12}$

88. $\dfrac{5}{9} - \dfrac{5}{12}$

89. $\dfrac{7}{8} \div \dfrac{7}{15}$

90. $\dfrac{5}{12} - \dfrac{3}{10}$

91. $\dfrac{4}{5} \cdot \dfrac{7}{8}$

92. $\dfrac{3}{7} \cdot \dfrac{12}{17}$

A Look Ahead

Example
Solve $(x - 6)(2x - 3) = (x + 2)(x + 9)$.

Solution:
$$(x - 6)(2x - 3) = (x + 2)(x + 9)$$
$$2x^2 - 15x + 18 = x^2 + 11x + 18$$
$$x^2 - 26x = 0$$
$$x(x - 26) = 0$$
$$x = 0 \quad \text{or} \quad x - 26 = 0$$
$$x = 26$$

Solve each equation. See the example.

93. $(x - 3)(3x + 4) = (x + 2)(x - 6)$

94. $(2x - 3)(x + 6) = (x - 9)(x + 2)$

95. $(2x - 3)(x + 8) - (x - 6)(x + 4)$

96. $(x + 6)(x - 6) = (2x - 9)(x + 4)$

97. $(4x - 1)(x - 8) = (x + 2)(x + 4)$

98. $(5x - 2)(x + 3) = (2x - 3)(x + 2)$

5.7 QUADRATIC EQUATIONS AND PROBLEM SOLVING

CD-ROM SSM

SSG Video

▶ **OBJECTIVE**

1. Solve problems that can be modeled by quadratic equations.

1 Some problems may be modeled by quadratic equations. To solve these problems, we use the same problem-solving steps that were introduced in Section 2.5. When solving these problems, keep in mind that a solution of an equation that models a problem may not be a solution to the problem. For example, a person's age or the length of a rectangle is always a positive number. Discard solutions that do not make sense as solutions of the problem.

Example 1 **FINDING THE LENGTH OF TIME**

For a TV commercial, a piece of luggage is dropped from a cliff 256 feet above the ground to show the durability of the luggage. Neglecting air resistance, the height h in feet of the luggage above the ground after t seconds is given by the quadratic equation

$$h = -16t^2 + 256$$

Find how long it takes for the luggage to hit the ground.

Solution **1. UNDERSTAND.** Read and reread the problem. Then draw a picture of the problem.

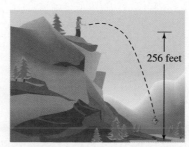

256 feet

The equation $h = -16t^2 + 256$ models the height of the falling luggage at time t. Familiarize yourself with this equation by finding the height of the luggage at $t = 1$ second and $t = 2$ seconds.

When $t = 1$ second, the height of the suitcase is
$h = -16(1)^2 + 256 = 240$ feet.

When $t = 2$ seconds, the height of the suitcase is
$h = -16(2)^2 + 256 = 192$ feet.

2. TRANSLATE. To find how long it takes the luggage to hit the ground, we want to know the value of t for which the height $h = 0$.
$$0 = -16t^2 + 256$$

3. SOLVE. We solve the quadratic equation by factoring.

$$0 = -16t^2 + 256$$
$$0 = -16(t^2 - 16)$$
$$0 = -16(t - 4)(t + 4)$$
$$t - 4 = 0 \qquad \text{or} \qquad t + 4 = 0$$
$$t = 4 \qquad\qquad\qquad t = -4$$

4. INTERPRET. Since the time t cannot be negative, the proposed solution is 4 seconds.

Check: Verify that the height of the luggage when t is 4 seconds is 0.
When $t = 4$ seconds, $h = -16(4)^2 + 256 = -256 + 256 = 0$ feet.

State: The solution checks and the luggage hits the ground 4 seconds after it is dropped.

Example 2 FINDING THE LENGTH AND WIDTH OF A DEN

In May 1995, 24 inches of rain fell on Slidell, Louisiana, in just two days, causing approximately one-third of the houses in that community to flood. When a home floods, all flooring in the home must be removed and replaced. The Callacs' home flooded and the soiled rug in their den was removed and now needs to be replaced. The length of their den is 4 feet more than the width. If the area of the floor is 117 square feet, find its length and width.

Solution

1. UNDERSTAND. Read and reread the problem. Propose and check a solution. Let

$$x = \text{the width of the floor; then}$$

$$x + 4 = \text{the length of the floor since it is 4 feet longer.}$$

2. TRANSLATE. Here, we use the formula for the area of a rectangle.

In words: | width | · | length | = | area |

Translate: x · $(x + 4)$ = 117

3. SOLVE. $x(x + 4) = 117$

$$x^2 + 4x = 117 \qquad \text{Multiply.}$$
$$x^2 + 4x - 117 = 0 \qquad \text{Write in standard form.}$$
$$(x + 13)(x - 9) = 0 \qquad \text{Factor.}$$

Next, set each factor equal to 0.

$$x + 13 = 0 \qquad \text{or} \qquad x - 9 = 0$$
$$x = -13 \qquad \text{or} \qquad x = 9 \qquad \text{Solve.}$$

4. INTERPRET. The solutions are -13 and 9. Since x represents the width of the room, the solution -13 must be discarded. The proposed width is 9 feet and the proposed length is $x + 4$ or $9 + 4$ or 13 feet.

   Check: The area of a 9-foot by 13-foot room is (9 feet)(13 feet) = 117 square feet. The proposed solution checks.

   State: The floor is 9 feet by 13 feet.

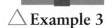

△ Example 3 FINDING THE BASE AND HEIGHT OF A SAIL

The height of a triangular sail is 2 meters less than twice the length of the base. If the sail has an area of 30 square meters, find the length of its base and the height.

Solution

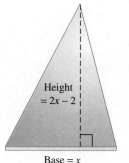

Height
$= 2x - 2$

Base $= x$

1. UNDERSTAND. Read and reread the problem. Since we are finding the length of the base and the height, we let

$$x = \text{the length of the base}$$

and since the height is 2 meters less than twice the base,

$$2x - 2 = \text{the height}$$

An illustration is shown to the left.

2. TRANSLATE. We are given that the area of the triangle is 30 square meters, so we use the formula for area of a triangle.

area of triangle	=	$\frac{1}{2}$	·	base	·	height
↓		↓		↓		↓
30	=	$\frac{1}{2}$	·	x	·	$(2x - 2)$

3. SOLVE. Now we solve the quadratic equation.

$$30 = \frac{1}{2}x(2x - 2)$$

$$30 = x^2 - x \qquad \text{Multiply.}$$
$$x^2 - x - 30 = 0 \qquad \text{Write in standard form.}$$
$$(x - 6)(x + 5) = 0 \qquad \text{Factor.}$$
$$x - 6 = 0 \qquad \text{or} \qquad x + 5 = 0 \qquad \text{Set each factor equal to 0.}$$
$$x = 6 \qquad\qquad x = -5$$

4. INTERPRET. Since x represents the length of the base, we discard the solution -5. The base of a triangle cannot be negative. The base is then 6 feet and the height is $2(6) - 2 = 10$ feet.

Check: To check this problem, we recall that $\frac{1}{2}$ base $\cdot$ height $=$ area, or

$$\frac{1}{2}(6)(10) = 30 \quad \text{The required area}$$

State: The base of the triangular sail is 6 meters and the height is 10 meters.

The next example makes use of the **Pythagorean theorem** and consecutive integers. Before we review this theorem, recall that a **right triangle** is a triangle that contains a 90° or right angle. The **hypotenuse** of a right triangle is the side opposite the right angle and is the longest side of the triangle. The **legs** of a right triangle are the other sides of the triangle.

PYTHAGOREAN THEOREM

HELPFUL HINT
If you use this formula, don't forget that c represents the length of the hypotenuse.

In a right triangle, the sum of the squares of the lengths of the two legs is equal to the square of the length of the hypotenuse.

$$(\text{leg})^2 + (\text{leg})^2 = (\text{hypotenuse})^2 \quad \text{or} \quad a^2 + b^2 = c^2$$

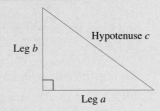

Study the following diagrams for a review of consecutive integers.

Consecutive integers: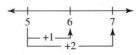

If x is the first integer: $x, x + 1, x + 2$

Consecutive even integers: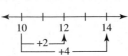

If x is the first even integer: $x, x + 2, x + 4$

Consecutive odd integers: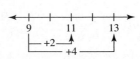

If x is the first odd integer: $x, x + 2, x + 4$

△ **Example 4**   **FINDING THE DIMENSIONS OF A TRIANGLE**

Find the lengths of the sides of a right triangle if the lengths can be expressed as three consecutive even integers.

Solution   1. UNDERSTAND. Read and reread the problem. Let's suppose that the length of one leg of the right triangle is 4 units. Then the other leg is the next even integer, or 6 units, and the hypotenuse of the triangle is the next even integer, or 8 units. Remember that the hypotenuse is the longest side. Let's see if a triangle with sides of these lengths forms a right triangle. To do this, we check to see whether the Pythagorean theorem holds true.

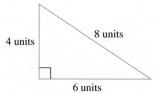

$$4^2 + 6^2 \overset{?}{=} 8^2$$
$$16 + 36 \overset{?}{=} 64$$
$$52 = 64 \qquad \text{False.}$$

Our proposed numbers do not check, but we now have a better understanding of the problem.

We let x, $x + 2$, and $x + 4$ be three consecutive even integers. Since these integers represent lengths of the sides of a right triangle, we have

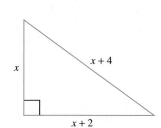

$$x = \text{one leg}$$
$$x + 2 = \text{other leg}$$
$$x + 4 = \text{hypotenuse (longest side)}$$

2. TRANSLATE. By the Pythagorean theorem, we have that

$$(\text{hypotenuse})^2 = (\text{leg})^2 + (\text{leg})^2$$
$$(x + 4)^2 = (x)^2 + (x + 2)^2$$

3. SOLVE. Now we solve the equation.

$$(x + 4)^2 = x^2 + (x + 2)^2$$

$x^2 + 8x + 16 = x^2 + x^2 + 4x + 4$	Multiply.
$x^2 + 8x + 16 = 2x^2 + 4x + 4$	Combine like terms.
$x^2 - 4x - 12 = 0$	Write in standard form.
$(x - 6)(x + 2) = 0$	Factor.
$x - 6 = 0 \quad \text{or} \quad x + 2 = 0$	Set each factor equal to 0.
$x = 6 \qquad\qquad x = -2$	

4. INTERPRET. We discard $x = -2$ since length cannot be negative. If $x = 6$, then $x + 2 = 8$ and $x + 4 = 10$.

Check: Verify that $(\text{hypotenuse})^2 = (\text{leg})^2 + (\text{leg})^2$, or $10^2 = 6^2 + 8^2$, or $100 = 36 + 64$.

State: The sides of the right triangle have lengths 6 units, 8 units, and 10 units.

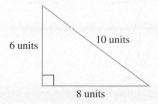

△SPOTLIGHT ON DECISION MAKING

Suppose you are a landscaper. You are landscaping a public park and have just put in a flower bed measuring 8 feet by 12 feet. You would also like to surround the bed with a decorative floral border consisting of low-growing, spreading plants. Each plant will cover approximately 1 square foot when mature, and you have 224 plants to use. How wide of a strip of ground should you prepare around the flower bed for the border? Explain.

1 foot

Grows to cover 1 sq ft

1 foot

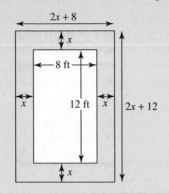

2x + 8

x

8 ft

x   12 ft   x   2x + 12

x

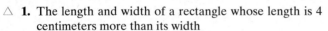

Exercise Set 5.7

See Examples 1 through 4 for all exercises. Represent each given condition using a single variable, x.

△ **1.** The length and width of a rectangle whose length is 4 centimeters more than its width

x

△ **2.** The length and width of a rectangle whose length is twice its width

x

3. Two consecutive odd integers

4. Two consecutive even integers

△ **5.** The base and height of a triangle whose height is one more than four times its base

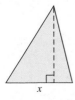
x

△ **6.** The base and height of a trapezoid whose base is three less than five times its height

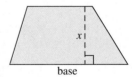
x

base

Use the information given to find the dimensions of each figure.

△ **7.**

x

The *area* of the square is 121 square units. Find the length of its sides.

△ **8.**

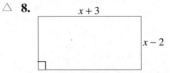
x + 3

x − 2

The *area* of the rectangle is 84 square inches. Find its length and width.

△ **9.**

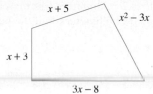

The *perimeter* of the quadrilateral is 120 centimeters. Find the lengths of the sides.

△ **10.**

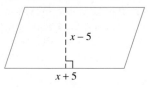

The *perimeter* of the triangle is 85 feet. Find the lengths of its sides.

△ **11.**

The *area* of the parallelogram is 96 square miles. Find its base and height.

△ **12.**

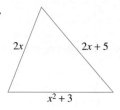

The *area* of the circle is 25π square kilometers. Find its radius.

Solve.

🔷 **13.** An object is thrown upward from the top of an 80-foot building with an initial velocity of 64 feet per second. The height h of the object after t seconds is given by the quadratic equation $h = -16t^2 + 64t + 80$. When will the object hit the ground?

14. A hang glider pilot accidentally drops her compass from the top of a 400-foot cliff. The height h of the compass after t seconds is given by the quadratic equation $h = -16t^2 + 400$. When will the compass hit the ground?

△ **15.** The length of a rectangle is 7 centimeters less than twice its width. Its area is 30 square centimeters. Find the dimensions of the rectangle.

△ **16.** The length of a rectangle is 9 inches more than its width. Its area is 112 square inches. Find the dimensions of the rectangle.

The equation $D = \frac{1}{2}n(n - 3)$ gives the number of diagonals D for a polygon with n sides. For example, a polygon with 6 sides has $D = \frac{1}{2} \cdot 6(6 - 3)$ or $D = 9$ diagonals. (See if you can count all 9 diagonals. Some are shown in the figure.) Use this equation, $D = \frac{1}{2}n(n - 3)$, for Exercises 17–20.

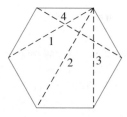

△ **17.** Find the number of diagonals for a polygon that has 12 sides.

△ **18.** Find the number of diagonals for a polygon that has 15 sides.

△ **19.** Find the number of sides n for a polygon that has 35 diagonals.

△ **20.** Find the number of sides n for a polygon that has 14 diagonals.

Solve.

21. The sum of a number and its square is 132. Find the number.

22. The sum of a number and its square is 182. Find the number.

△ **23.** Two boats travel at a right angle to each other after leaving the same dock at the same time. One hour later the boats are 17 miles apart. If one boat travels 7 miles per hour faster than the other boat, find the rate of each boat.

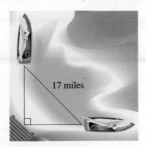

17 miles

△ **24.** The side of a square equals the width of a rectangle. The length of the rectangle is 6 meters longer than its width. The sum of the areas of the square and the rectangle is 176 square meters. Find the side of the square.

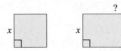

?

x x

25. The sum of two numbers is 20, and the sum of their squares is 218. Find the numbers.

26. The sum of two numbers is 25, and the sum of their squares is 325. Find the numbers.

△ **27.** If the sides of a square are increased by 3 inches, the area becomes 64 square inches. Find the length of the sides of the original square.

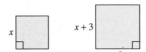

x $x + 3$

△ **28.** If the sides of a square are increased by 5 meters, the area becomes 100 square meters. Find the length of the sides of the original square.

x $x + 5$

◆ **29.** One leg of a right triangle is 4 millimeters longer than
△ the smaller leg and the hypotenuse is 8 millimeters longer than the smaller leg. Find the lengths of the sides of the triangle.

△ **30.** One leg of a right triangle is 9 centimeters longer than the other leg and the hypotenuse is 45 centimeters. Find the lengths of the legs of the triangle.

△ **31.** The length of the base of a triangle is twice its height. If the area of the triangle is 100 square kilometers, find the height.

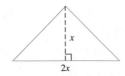

x

$2x$

△ **32.** The height of a triangle is 2 millimeters less than the base. If the area is 60 square millimeters, find the base.

$x - 2$

x

△ **33.** Find the length of the shorter leg of a right triangle if the longer leg is 12 feet more than the shorter leg and the hypotenuse is 12 feet less than twice the shorter leg.

△ **34.** Find the length of the shorter leg of a right triangle if the longer leg is 10 miles more than the shorter leg and the hypotenuse is 10 miles less than twice the shorter leg.

35. An object is dropped from the top of the 625-foot-tall Waldorf-Astoria Hotel on Park Avenue in New York City. (*Source: World Almanac* research) The height h of the object after t seconds is given by the equation $h = -16t^2 + 625$. Find how many seconds pass before the object reaches the ground.

625 feet

36. A 6-foot-tall person drops an object from the top of the Westin Peachtree Plaza in Atlanta, Georgia. The Westin building is 723 feet tall. (*Source: World Almanac* research) The height h of the object after t seconds is given by the equation $h = -16t^2 + 729$. Find how many seconds pass before the object reaches the ground.

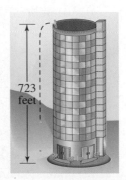

723 feet

37. At the end of 2 years, P dollars invested at an interest rate r compounded annually increases to an amount, A dollars, given by

$$A = P(1 + r)^2$$

Find the interest rate if $100 increased to $144 in 2 years.

38. At the end of 2 years, P dollars invested at an interest rate r compounded annually increases to an amount, A dollars, given by

$$A = P(1 + r)^2$$

Find the interest rate if $2000 increased to $2420 in 2 years.

△ **39.** Find the dimensions of a rectangle whose width is 7 miles less than its length and whose area is 120 square miles.

△ **40.** Find the dimensions of a rectangle whose width is 2 inches less than half its length and whose area is 160 square inches.

41. If the cost, C, for manufacturing x units of a certain product is given by $C = x^2 - 15x + 50$, find the number of units manufactured at a cost of $9500.

42. If a switchboard handles n telephones, the number C of telephone connections it can make simultaneously is given by the equation $C = \dfrac{n(n - 1)}{2}$. Find how many telephones are handled by a switchboard making 120 telephone connections simultaneously.

△ **43.** According to the International America's Cup Class (IACC) rule, a sailboat competing in the America's Cup match must have a 110-foot-tall mast and a combined mainsail and jib sail area of 3000 square feet. *(Source: America's Cup Organizing Committee)* A design for an IACC-class sailboat calls for the mainsail to be 60% of the combined sail area. If the height of the triangular mainsail is 28 feet more than twice the length of the boom, find the length of the boom and the height of the mainsail.

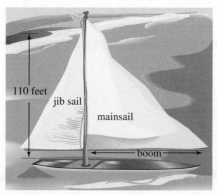

REVIEW EXERCISES

The following double line graph shows a comparison of the number of farms in the United States and the size of the average farm. Use this graph to answer Exercises 44–50. See Section 1.8.

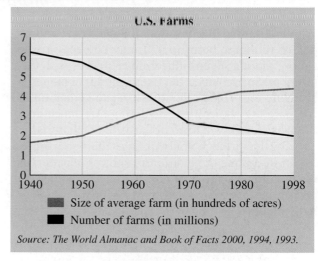

Source: The World Almanac and Book of Facts 2000, 1994, 1993.

△ **44.** Approximate the size of the average farm in 1940.

△ **45.** Approximate the size of the average farm in 1998.

46. Approximate the number of farms in 1940.

47. Approximate the number of farms in 1998.

48. Approximate the year that the colored lines in this graph intersect.

49. In your own words, explain the meaning of the point of intersection in the graph.

50. Describe the trends shown in this graph and speculate as to why these trends have occurred.

Write each fraction in simplest form. See Section 1.2.

51. $\dfrac{20}{35}$

52. $\dfrac{24}{32}$

53. $\dfrac{27}{18}$

54. $\dfrac{15}{27}$

55. $\dfrac{14}{42}$

56. $\dfrac{45}{50}$

5

 For additional Chapter Projects, visit the Real World Activities Website by going to http://www.prenhall.com/martin-gay.

CHAPTER PROJECT

Choosing Among Building Options

Whether putting in a new floor, hanging new wallpaper, or retiling a bathroom, it may be necessary to choose among several different materials with different pricing schemes. If a fixed amount of money is available for projects like these, it can be helpful to compare the choices by calculating how much area can be covered by a fixed dollar-value of material.

In this project, you will have the opportunity to choose among three different choices of materials for building a patio around a swimming pool. This project may be completed by working in groups or individually.

Option	Material	Price
A	Poured cement	$5 per square foot
B	Brick	$7.50 per square foot plus a $30 flat fee for delivering the bricks
C	Outdoor carpeting	$4.50 per square foot plus $10.86 per foot of the pool's perimeter to install an edging

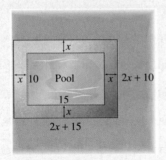

Situation: Suppose you have just had a 10-foot-by-15-foot inground swimming pool installed in your backyard. You have $3000 left from the building project that you would like to spend on surrounding the pool with a patio, equally wide on all sides (see figure). You have talked to several local suppliers about options for building this patio and must choose among the following.

△ 1. Find the area of the swimming pool.

△ 2. Write an algebraic expression for the total area of the region containing both the pool and the patio.

△ 3. Use subtraction to find an algebraic expression for the area of just the patio (not including the area of the pool).

△ 4. Find the perimeter of the swimming pool alone.

5. For each patio material option, write an algebraic expression for the total cost of installing the patio based on its area and the given price information.

6. If you plan to spend the entire $3000 on the patio, how wide would the patio in option A be?

7. If you plan to spend the entire $3000 on the patio, how wide would the patio in option B be?

8. If you plan to spend the entire $3000 on the patio, how wide would the patio in option C be?

9. Which option would you choose? Why? Discuss the pros and cons of each option.

CHAPTER 5 VOCABULARY CHECK

Fill in each blank with one of the words or phrases listed below.

factoring   quadratic equation   perfect square trinomial
greatest common factor

1. An equation that can be written in the form $ax^2 + bx + c = 0$ (with a not 0) is called a

    _____ .

2. _____ is the process of writing an expression as a product.

3. The _____ of a list of terms is the product of all common factors.

4. A trinomial that is the square of some binomial is called a _____ .

CHAPTER 5 HIGHLIGHTS

DEFINITIONS AND CONCEPTS	EXAMPLES

Section 5.1 The Greatest Common Factor and Factoring by Grouping

Factoring is the process of writing an expression as a product.

Factor: $6 = 2 \cdot 3$

$$x^2 + 5x + 6 = (x + 2)(x + 3)$$

To find the GCF of a list of integers,

Step 1. Write each number as a product of primes.

Step 2. Identify the common prime factors.

Step 3. The product of all common factors is the greatest common factor. If there are no common prime factors, the GCF is 1.

Find the GCF of 12, 36, and 48.

$12 = 2 \cdot 2 \cdot 3$

$36 = 2 \cdot 2 \cdot 3 \cdot 3$

$48 = 2 \cdot 2 \cdot 2 \cdot 2 \cdot 3$

$GCF = 2 \cdot 2 \cdot 3 = 12.$

The GCF of a list of common variables raised to powers is the variable raised to the smallest exponent in the list.

The GCF of z^5, z^3, and z^{10} is z^3.

The GCF of a list of terms is the product of all common factors.

Find the GCF of $8x^2y$, $10x^3y^2$, and $25x^2y^3$.

The GCF of 8, 10, and 25 *is* 2.

The GCF of x^2, x^3, and x^2 is x^2.

The GCF of y, y^2, and y^3 is y.

The GCF of the terms is $2x^2y$.

To Factor by Grouping

Step 1. Arrange the terms so that the first two terms have a common factor and the last two have a common factor.

Step 2. For each pair of terms, factor out the pair's GCF.

Step 3. If there is now a common binomial factor, factor it out.

Step 4. If there is no common binomial factor, begin again, rearranging the terms differently. If no rearrangement leads to a common binomial factor, the polynomial cannot be factored.

Factor $10ax + 15a - 6xy - 9y$.

Step 1. $10ax + 15a - 6xy - 9y$

Step 2. $5a(2x + 3) - 3y(2x + 3)$

Step 3. $(2x + 3)(5a - 3y)$

Section 5.2 Factoring Trinomials of the Form $x^2 + bx + c$

To factor a trinomial of the form $x^2 + bx + c$, look for two numbers whose product is c and whose sum is b. The factored form is

$(x + \text{one number})(x + \text{other number})$

Factor: $x^2 + 7x + 12$

$3 + 4 = 7 \qquad 3 \cdot 4 = 12$

$(x + 3)(x + 4)$

(continued)

DEFINITIONS AND CONCEPTS	EXAMPLES

Section 5.3 Factoring Trinomials of the Form $ax^2 + bx + c$

Method 1: To factor $ax^2 + bx + c$, try various combinations of factors of ax^2 and c until a middle term of bx is obtained when checking.	Factor: $3x^2 + 14x - 5$ Factors of $3x^2$: $3x, x$ Factors of -5: $-1, 5$ and $1, -5$. $(3x - \underline{1})(x + 5)$ $\underbrace{-1x}$ $\dfrac{15x}{14x}$ **Correct** middle term

Method 2: Factor $ax^2 + bx + c$ by grouping. **Step 1.** Find two numbers whose product is $a \cdot c$ and whose sum is b. **Step 2.** Rewrite bx, using the factors found in step 1. **Step 3.** Factor by grouping.	Factor: $3x^2 + 14x - 5$ **Step 1.** Find two numbers whose product is $3 \cdot (-5)$ or -15 and whose sum is 14. They are 15 and -1. **Step 2.** $3x^2 + 14x - 5$ $= 3x^2 + 15x - 1x - 5$ **Step 3.** $= 3x(x + 5) - 1(x + 5)$ $= (x + 5)(3x - 1)$

A **perfect square trinomial** is a trinomial that is the square of some binomial.	Perfect square trinomial = square of binomial $x^2 + 4x + 4 \qquad = (x + 2)^2$ $25x^2 - 10x + 1 \qquad = (5x - 1)^2$

Factoring perfect square trinomials: $a^2 + 2ab + b^2 = (a + b)^2$ $a^2 - 2ab + b^2 = (a - b)^2$	Factor: $x^2 + 6x + 9 = x^2 + 2 \cdot x \cdot 3 + 3^2 = (x + 3)^2$ $4x^2 - 12x + 9 = (2x)^2 - 2 \cdot 2x \cdot 3 + 3^2 = (2x - 3)^2$

Section 5.4 Factoring Binomials

Difference of Squares $a^2 - b^2 = (a + b)(a - b)$ **Sum or Difference of Cubes** $a^3 + b^3 = (a + b)(a^2 - ab + b^2)$ $a^3 - b^3 = (a - b)(a^2 + ab + b^2)$	Factor: $x^2 - 9 = x^2 - 3^2 = (x + 3)(x - 3)$ $y^3 + 8 = y^3 + 2^3 = (y + 2)(y^2 - 2y + 4)$ $125z^3 - 1 = (5z)^3 - 1^3 = (5z - 1)(25z^2 + 5z + 1)$

Section 5.5 Choosing a Factoring Strategy

To factor a polynomial, **Step 1.** Factor out the GCF. **Step 2.** **a.** If two terms, **i.** $a^2 - b^2 = (a - b)(a + b)$ **ii.** $a^3 - b^3 = (a - b)(a^2 + ab + b^2)$ **iii.** $a^3 + b^3 = (a + b)(a^2 - ab + b^2)$	Factor: $2x^4 - 6x^2 - 8$ **Step 1.** $2x^4 - 6x^2 - 8 = 2(x^4 - 3x^2 - 4)$ *(continued)*

DEFINITIONS AND CONCEPTS	EXAMPLES

Section 5.5 Choosing a Factoring Strategy

b. If three terms,

    i. $a^2 + 2ab + b^2 = (a + b)^2$

    ii. Methods in Sections 5.2 and 5.3

c. If four or more terms, try factoring by grouping.

Step 3. See if any factors can be factored further.

Step 4. Check by multiplying.

Step 2. b, ii. $= 2(x^2 + 1)(x^2 - 1)$

Step 3. $= 2(x^2 + 1)(x + 2)(x - 2)$

Step 4. Check by multiplying.

$$2(x^2 + 1)(x + 2)(x - 2) = 2(x^2 + 1)(x^2 - 4)$$
$$= 2(x^4 - 3x^2 - 4)$$
$$= 2x^4 - 6x^2 - 8$$

Section 5.6 Solving Quadratic Equations by Factoring

A **quadratic equation** is an equation that can be written in the form $ax^2 + bx + c = 0$ with a not 0.

The form $ax^2 + bx + c = 0$ is called the **standard form** of a quadratic equation.

Quadratic Equation	Standard Form
$x^2 = 16$	$x^2 - 16 = 0$
$y = -2y^2 + 5$	$2y^2 + y - 5 = 0$

Zero Factor Theorem

If a and b are real numbers and if $ab = 0$, then $a = 0$ or $b = 0$.

If $(x + 3)(x - 1) = 0$, then $x + 3 = 0$ or $x - 1 = 0$

To solve quadratic equations by factoring,

Solve: $3x^2 = 13x - 4$

Step 1. Write the equation in standard form: $ax^2 + bx + c = 0$.

Step 1. $3x^2 - 13x + 4 = 0$

Step 2. Factor the quadratic.

Step 2. $(3x - 1)(x - 4) = 0$

Step 3. Set each factor containing a variable equal to 0.

Step 3. $3x - 1 = 0$ or $x - 4 = 0$

Step 4. Solve the equations.

Step 4. $3x = 1$ or $x = 4$

$$x = \frac{1}{3}$$

Step 5. Check in the original equation.

Step 5. Check both $\frac{1}{3}$ and 4 in the original equation.

Section 5.7 Quadratic Equations and Problem Solving

Problem-Solving Steps

1. UNDERSTAND the problem.

A garden is in the shape of a rectangle whose length is two feet more than its width. If the area of the garden is 35 square feet, find its dimensions.

1. Read and reread the problem. Guess a solution and check your guess.

Let x be the width of the rectangular garden. Then $x + 2$ is the length.

(continued)

DEFINITIONS AND CONCEPTS	EXAMPLES
Section 5.7 Quadratic Equations and Problem Solving	

2. TRANSLATE.	**2.** In words: $\boxed{\text{length}} \cdot \boxed{\text{width}} = \boxed{\text{area}}$
	Translate: $(x + 2) \cdot x = 35$
3. SOLVE.	**3.** $\quad (x + 2)x = 35$
	$\quad x^2 + 2x - 35 = 0$
	$\quad (x - 5)(x + 7) = 0$
	$x - 5 = 0 \quad \text{or} \quad x + 7 = 0$
	$x = 5 \quad \text{or} \quad x = -7$
4. INTERPRET.	**4.** Discard the solution of -7 since x represents width. *Check:* If x is 5 feet then $x + 2 = 5 + 2 = 7$ feet. The area of a rectangle whose width is 5 feet and whose length is 7 feet is (5 feet)(7 feet) or 35 square feet. *State:* The garden is 5 feet by 7 feet.

CHAPTER 5 REVIEW

(5.1) *Complete the factoring.*

1. $6x^2 - 15x = 3x(\quad)$

2. $2x^3y - 6x^2y^2 - 8xy^3 = 2xy(\quad)$

Factor the GCF from each polynomial.

3. $20x^2 + 12x$ 　　　　**4.** $6x^2y^2 - 3xy^3$

5. $-8x^3y + 6x^2y^2$

6. $3x(2x + 3) - 5(2x + 3)$

7. $5x(x + 1) - (x + 1)$

Factor.

8. $3x^2 - 3x + 2x - 2$

9. $6x^2 + 10x - 3x - 5$ 　　**10.** $3a^2 + 9ab + 3b^2 + ab$

(5.2) *Factor each trinomial.*

11. $x^2 + 6x + 8$ 　　　　**12.** $x^2 - 11x + 24$

13. $x^2 + x + 2$ 　　　　**14.** $x^2 - 5x - 6$

15. $x^2 + 2x - 8$ 　　　　**16.** $x^2 + 4xy - 12y^2$

17. $x^2 + 8xy + 15y^2$ 　　**18.** $3x^2y + 6xy^2 + 3y^3$

19. $72 - 18x - 2x^2$ 　　　**20.** $32 + 12x - 4x^2$

(5.3) *Factor each trinomial.*

21. $2x^2 + 11x - 6$ 　　　**22.** $4x^2 - 7x + 4$

23. $4x^2 + 4x - 3$ 　　　**24.** $6x^2 + 5xy - 4y^2$

25. $6x^2 - 25xy + 4y^2$ 　　**26.** $18x^2 - 60x + 50$

27. $2x^2 - 23xy - 39y^2$ 　**28.** $4x^2 - 28xy + 49y^2$

29. $18x^2 - 9xy - 20y^2$ 　**30.** $36x^3y + 24x^2y^2 - 45xy^3$

(5.4) *Factor each binomial.*

31. $4x^2 - 9$ 　　　　　**32.** $9t^2 - 25s^2$

33. $16x^2 + y^2$ 　　　　**34.** $x^3 - 8y^3$

35. $8x^3 + 27$ 　　　　**36.** $2x^3 + 8x$

37. $54 - 2x^3y^3$ 　　　　**38.** $9x^2 - 4y^2$

39. $16x^4 - 1$ 　　　　　**40.** $x^4 + 16$

(5.5) *Factor*

41. $2x^2 + 5x - 12$ 　　　**42.** $3x^2 - 12$

43. $x(x - 1) + 3(x - 1)$ 　**44.** $x^2 + xy - 3x - 3y$

45. $4x^2y - 6xy^2$ 　　　　**46.** $8x^2 - 15x - x^3$

47. $125x^3 + 27$ 　　　　**48.** $24x^2 - 3x - 18$

49. $(x + 7)^2 - y^2$ 　　　**50.** $x^2(x + 3) - 4(x + 3)$

(5.6) *Solve the following equations.*

51. $(x + 6)(x - 2) = 0$

52. $3x(x + 1)(7x - 2) = 0$

53. $4(5x + 1)(x + 3) = 0$

54. $x^2 + 8x + 7 = 0$ **55.** $x^2 - 2x - 24 = 0$

56. $x^2 + 10x = -25$

57. $x(x - 10) = -16$

58. $(3x - 1)(9x^2 + 3x + 1) = 0$

59. $56x^2 - 5x - 6 = 0$

60. $20x^2 - 7x - 6 = 0$

61. $5(3x + 2) = 4$

62. $6x^2 - 3x + 8 = 0$

63. $12 - 5t = -3$ **64.** $5x^3 + 20x^2 + 20x = 0$

65. $4t^3 - 5t^2 - 21t = 0$

(5.7) *Solve the following problems.*

△ **66.** A flag for a local organization is in the shape of a rectangle whose length is 15 inches less than twice its width. If the area of the flag is 500 square inches, find its dimensions.

△ **67.** The base of a triangular sail is four times its height. If the area of the triangle is 162 square yards, find the base.

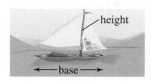

68. Find two consecutive positive integers whose product is 380.

69. A rocket is fired from the ground with an initial velocity of 440 feet per second. Its height h after t seconds is given by the equation

$$h = -16t^2 + 440t$$

a. Find how many seconds pass before the rocket reaches a height of 2800 feet. Explain why two answers are obtained.

b. Find how many seconds pass before the rocket reaches the ground again.

△ **70.** An architect's squaring instrument is in the shape of a right triangle. Find the length of the long leg of the right triangle if the hypotenuse is 8 centimeters longer than the long leg and the short leg is 8 centimeters shorter than the long leg.

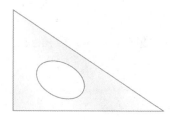

CHAPTER 5 TEST

Factor each polynomial completely. If a polynomial cannot be factored, write "prime."

1. $9x^3 + 39x^2 + 12x$ **2.** $x^2 + x - 10$

3. $x^2 + 4$ **4.** $y^2 - 8y - 48$

5. $3a^2 + 3ab - 7a - 7b$ **6.** $3x^2 - 5x + 2$

7. $x^2 + 20x + 90$ **8.** $x^2 + 14xy + 24y^2$

9. $26x^6 - x^4$ **10.** $50x^3 + 10x^2 - 35x$

11. $180 - 5x^2$ **12.** $64x^3 - 1$

13. $6t^2 - t - 5$

14. $xy^2 - 7y^2 - 4x + 28$

15. $x - x^5$ **16.** $-xy^3 - x^3y$

Solve each equation.

17. $x^2 + 5x = 14$ **18.** $(x + 3)^2 = 16$

19. $3x(2x - 3)(3x + 4) = 0$

20. $5t^3 - 45t = 0$

21. $3x^2 = -12x$

22. $t^2 - 2t - 15 = 0$

23. $7x^2 = 168 + 35x$

24. $6x^2 = 15x$

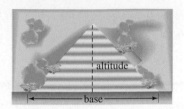

Solve each problem.

△ **25.** Find the dimensions of a rectangular garden whose length is 5 feet longer than its width and whose area is 66 square feet.

△ **26.** A deck for a home is in the shape of a triangle. The length of the base of the triangle is 9 feet longer than its altitude. If the area of the triangle is 68 square feet, find the length of the base.

27. The sum of two numbers is 17, and the sum of their squares is 145. Find the numbers.

28. An object is dropped from the top of the Woolworth Building on Broadway in New York City. The height h of the object after t seconds is given by the equation

$$h = -16t^2 + 784$$

Find how many seconds pass before the object reaches the ground.

CHAPTER 5 CUMULATIVE REVIEW

1. Translate each sentence into a mathematical statement.
  a. Nine is less than or equal to eleven.
  b. Eight is greater than one.
  c. Three is not equal to four.

2. Write each fraction in lowest terms.
  a. $\dfrac{42}{49}$
  b. $\dfrac{11}{27}$
  c. $\dfrac{88}{20}$

3. Simplify: $\dfrac{8 + 2 \cdot 3}{2^2 - 1}$

4. Add.
  a. $3 + (-7) + (-8)$
  b. $[7 + (-10)] + [-2 + (-4)]$

5. Find the product.
  a. $(-6)(4)$
  b. $2(-1)$
  c. $(-5)(-10)$

6. Simplify each expression by combining like terms.
  a. $7x - 3x$
  b. $10y^2 + y^2$
  c. $8x^2 + 2x - 3x$

Solve.

7. $3 - x = 7$

8. $-3x = 33$

9. $8(2 - t) = -5t$

10. A 10-foot board is to be cut into two pieces so that the longer piece is 4 times the shorter. Find the length of each piece.

11. Graph the linear equation $y = -\dfrac{1}{3}x$.

12. Is the line passing through the points $(-6, 0)$ and $(-2, 3)$ parallel to the line passing through the points $(5, 4)$ and $(9, 7)$?

13. Evaluate for the given value of x.
  a. $2x^3$; x is 5
  b. $\dfrac{9}{x^2}$; x is -3

14. Find the degree of each term.
  a. $-3x^2$
  b. $5x^3yz$
  c. 2

15. Subtract: $(2x^3 + 8x^2 - 6x) - (2x^3 - x^2 + 1)$

16. Find the product: $(3x + 2)(2x - 5)$

17. Multiply: $(3y + 1)^2$

18. Simplify by writing each expression with positive exponents only.
  a. 3^{-2}
  b. $2x^{-3}$ —
  c. $2^{-1} + 4^{-1}$
  d. $(-2)^{-4}$

19. Write each number in scientific notation.
  a. 367,000,000
  b. 0.000003
  c. 20,520,000,000
  d. 0.00085

20. Divide $x^2 + 7x + 12$ by $x + 3$ using long division.

21. Find the GCF of each list of terms.

    a. x^3, x^7, x^5

    b. y, y^4, and y^7

Factor.

22. $x^2 + 7x + 12$

23. $8x^2 - 22x + 5$

24. $25a^2 - 9b^2$

25. Solve: $(x - 3)(x + 1) = 0$

A Widely Known Aspect of Forensic Science

Forensic science is the application of science and technology to the resolution of criminal and civil issues. Areas of forensic science include forensic engineering, forensic pathology, forensic dentistry, forensic psychiatry, physical anthropology, and forensic toxicology. One of the most widely known aspects of forensic science is criminalistics—the analysis of crime scenes and the examination of physical evidence. Criminalistic investigations are often carried out in what is popularly known as a crime laboratory.

Crime lab workers might be expected to analyze and identify chemical compounds, human hair and tissues, or materials such as metals, glass, or wood. They might also make measurements on items of physical evidence and compare footprints, firearms, or bullets. Anyone wishing to work in criminalistics, or the forensic sciences in general, should have a firm grasp of scientific principles. They also need excellent problem-solving and communication skills. Because mathematics goes hand-in-hand with the sciences, forensic scientists should also have a good mathematical background.

 For more information about a career in the forensic sciences, visit the American Academy of Forensic Sciences Website by first going to www.prenhall.com/martin-gay.

In the Spotlight on Decision Making feature on page 336, you will have the opportunity, as a forensic lab technician, to decide what material a piece of physical evidence is made of.

RATIONAL EXPRESSIONS

6.1 SIMPLIFYING RATIONAL EXPRESSIONS

6.2 MULTIPLYING AND DIVIDING RATIONAL EXPRESSIONS

6.3 ADDING AND SUBTRACTING RATIONAL EXPRESSIONS WITH COMMON DENOMINATORS AND LEAST COMMON DENOMINATOR

6.4 ADDING AND SUBTRACTING RATIONAL EXPRESSIONS WITH UNLIKE DENOMINATORS

6.5 SIMPLIFYING COMPLEX FRACTIONS

6.6 SOLVING EQUATIONS CONTAINING RATIONAL EXPRESSIONS

6.7 RATIO AND PROPORTION

6.8 RATIONAL EQUATIONS AND PROBLEM SOLVING

In this chapter, we expand our knowledge of algebraic expressions to include another category called rational expressions, such as $\frac{x+1}{x}$. We explore the operations of addition, subtraction, multiplication, and division for these algebraic fractions, using principles similar to the principles for number fractions. Thus, the material in this chapter will make full use of your knowledge of number fractions.

6.1 SIMPLIFYING RATIONAL EXPRESSIONS

▶ **OBJECTIVES**

CD-ROM
SSM
SSG
Video

1. Find the value of a rational expression given a replacement number.
2. Identify values for which a rational expression is undefined.
3. Write rational expressions in lowest terms.

1 As we reviewed in Chapter 1, a rational number is a number that can be written as a quotient of integers. A **rational expression** is also a quotient; it is a quotient of polynomials.

RATIONAL EXPRESSION

A rational expression is an expression that can be written in the form $\dfrac{P}{Q}$, where P and Q are polynomials and Q does not equal 0.

Rational Expressions
$$\frac{3y^3}{8} \qquad \frac{-4p}{p^3 + 2p + 1} \qquad \frac{5x^2 - 3x + 2}{3x + 7}$$

Rational expressions have different values depending on what value replaces the variable. Next, we review the standard order of operations by finding values of rational expressions at given replacement values of the variable.

Example 1 Find the value of $\dfrac{x + 4}{2x - 3}$ for the given replacement values.

a. $x = 5$ **b.** $x = -2$

Solution **a.** Replace each x in the expression with 5 and then simplify.
$$\frac{x + 4}{2x - 3} = \frac{5 + 4}{2(5) - 3} = \frac{9}{10 - 3} = \frac{9}{7}$$

b. Replace each x in the expression with -2 and then simplify.
$$\frac{x + 4}{2x - 3} = \frac{-2 + 4}{2(-2) - 3} = \frac{2}{-7} \quad \text{or} \quad -\frac{2}{7}$$

For a negative fraction such as $\dfrac{2}{-7}$, recall from Chapter 1 that
$$\frac{2}{-7} = \frac{-2}{7} = -\frac{2}{7}$$

In general, for any fraction

$$\frac{-a}{b} = \frac{a}{-b} = -\frac{a}{b}, \qquad b \neq 0$$

This is also true for rational expressions. For example,

Notice the parentheses.

$$\frac{-(x + 2)}{x} = \frac{x + 2}{-x} = -\frac{x + 2}{x}$$

2   In the box on the previous page, notice that we wrote $b \neq 0$ for the denominator b. This is because the denominator of a rational expression must not equal 0 since division by 0 is not defined. This means we must be careful when replacing the variable in a rational expression by a number. For example, suppose we replace x with 5 in the rational expression $\dfrac{2 + x}{x - 5}$. The expression becomes

$$\frac{2 + x}{x - 5} = \frac{2 + 5}{5 - 5} = \frac{7}{0}$$

But division by 0 is undefined. Therefore, in this expression we can allow x to be any real number *except* 5. A rational expression is undefined for values that make the denominator 0.

Example 2   Are there any values for x for which each expression is undefined?

a. $\dfrac{x}{x - 3}$ **b.** $\dfrac{x^2 + 2}{x^2 - 3x + 2}$ **c.** $\dfrac{x^3 - 6x^2 - 10x}{3}$ **d.** $\dfrac{2}{x^2 + 1}$

Solution   To find values for which a rational expression is undefined, find values that make the denominator 0.

a. The denominator of $\dfrac{x}{x - 3}$ is 0 when $x - 3 = 0$ or when $x = 3$. Thus, when $x = 3$, the expression $\dfrac{x}{x - 3}$ is undefined.

b. Set the denominator equal to zero.

$$x^2 - 3x + 2 = 0$$
$$(x - 2)(x - 1) = 0 \qquad \text{Factor.}$$
$$x - 2 = 0 \quad \text{or} \quad x - 1 = 0 \qquad \text{Set each factor equal to zero.}$$
$$x = 2 \quad \text{or} \quad x = 1 \qquad \text{Solve.}$$

Thus, when $x = 2$ or $x = 1$, the denominator $x^2 - 3x + 2$ is 0. So the rational expression $\dfrac{x^2 + 2}{x^2 - 3x + 2}$ is undefined when $x = 2$ or when $x = 1$.

c. The denominator of $\dfrac{x^3 - 6x^2 - 10x}{3}$ is never zero, so there are no values of x for which this expression is undefined.

d. No matter which real number x is replaced by, the denominator $x^2 + 1$ does not equal 0, so there are no real numbers for which this expression is undefined.

3 A fraction is said to be written in lowest terms or simplest form when the numerator and denominator have no common factors other than 1 (or -1). For example, the fraction $\frac{7}{10}$ is in lowest terms since the numerator and denominator have no common factors other than 1 (or -1).

The process of writing a rational expression in lowest terms or simplest form is called **simplifying** a rational expression. The following fundamental principle of rational expressions is used to simplify a rational expression.

FUNDAMENTAL PRINCIPLE OF RATIONAL EXPRESSIONS

If P, Q, and R are polynomials, and Q and R are not 0,

$$\frac{PR}{QR} = \frac{P}{Q}$$

Simplifying a rational expression is similar to simplifying a fraction. To simplify the fraction $\frac{15}{20}$, we factor the numerator and the denominator, look for common factors in both, and then use the fundamental principle.

$$\frac{15}{20} = \frac{3 \cdot 5}{2 \cdot 2 \cdot 5} = \frac{3}{2 \cdot 2} = \frac{3}{4}$$

To simplify the rational expression $\frac{x^2 - 9}{x^2 + x - 6}$, we also factor the numerator and denominator, look for common factors in both, and then use the fundamental principle of rational expressions.

$$\frac{x^2 - 9}{x^2 + x - 6} = \frac{(x - 3)(x + 3)}{(x - 2)(x + 3)} = \frac{x - 3}{x - 2}$$

This means that the rational expression $\frac{x^2 - 9}{x^2 + x - 6}$ has the same value as the rational expression $\frac{x - 3}{x - 2}$ for all values of x except 2 and -3. (Remember that when x is 2, the denominator of both rational expressions is 0 and when x is -3, the original rational expression has a denominator of 0.)

As we simplify rational expressions, we will assume that the simplified rational expression is equal to the original rational expression for all real numbers except those for which either denominator is 0. The following steps may be used to simplify rational expressions.

SIMPLIFYING A RATIONAL EXPRESSION

Step 1. Completely factor the numerator and denominator.
Step 2. Apply the fundamental principle of rational expressions to divide out common factors.

Example 3 Write $\frac{21a^2b}{3a^5b}$ in simplest form.

Solution   Factor the numerator and denominator. Then apply the fundamental principle.

$$\frac{21a^2b}{3a^5b} = \frac{7 \cdot 3 \cdot a^2 \cdot b}{3 \cdot a^3 \cdot a^2 \cdot b} = \frac{7}{a^3}$$

Example 4   Simplify: $\dfrac{6a - 33}{15}$.

Solution   Factor the numerator and denominator. Then apply the fundamental principle.

$$\frac{6a - 33}{15} = \frac{3(2a - 11)}{3 \cdot 5} = \frac{2a - 11}{5}$$

Example 5   Simplify: $\dfrac{5x - 5}{x^3 - x^2}$.

Solution   Factor the numerator and denominator, if possible, and then apply the fundamental principle.

$$\frac{5x - 5}{x^3 - x^2} = \frac{5(x - 1)}{x^2(x - 1)} = \frac{5}{x^2}$$

Example 6   Write $\dfrac{x^2 + 8x + 7}{x^2 - 4x - 5}$ in simplest form.

Solution   Factor the numerator and denominator and apply the fundamental principle.

$$\frac{x^2 + 8x + 7}{x^2 - 4x - 5} = \frac{(x + 7)(x + 1)}{(x - 5)(x + 1)} = \frac{x + 7}{x - 5}$$

HELPFUL HINT

When simplifying a rational expression, the fundamental principle applies to **common *factors*, not common *terms***.

$$\frac{x \cdot (x + 2)}{x \cdot x} = \frac{x + 2}{x}$$

Common factors. These can be divided out.

$$\frac{x + 2}{x}$$

Common terms. Fundamental principle does not apply. This is in simplest form.

Example 7 Simplify each rational expression.

a. $\dfrac{x+y}{y+x}$ b. $\dfrac{x-y}{y-x}$

Solution **a.** The expression $\dfrac{x+y}{y+x}$ can be simplified by using the commutative property of addition to rewrite the denominator $y+x$ as $x+y$.

$$\frac{x+y}{y+x} = \frac{x+y}{x+y} = 1$$

b. The expression $\dfrac{x-y}{y-x}$ can be simplified by recognizing that $y-x$ and $x-y$ are opposites. In other words, $y-x = -1(x-y)$. Proceed as follows:

$$\frac{x-y}{y-x} = \frac{1 \cdot (x-y)}{(-1)(x-y)} = \frac{1}{-1} = -1$$

Example 8 Simplify: $\dfrac{4-x^2}{3x^2-5x-2}$.

Solution
$$\frac{4-x^2}{3x^2-5x-2} = \frac{(2-x)(2+x)}{(x-2)(3x+1)} \qquad \text{Factor.}$$

$$= \frac{(-1)(x-2)(2+x)}{(x-2)(3x+1)} \qquad \text{Write } 2-x \text{ as } -1(x-2).$$

$$= \frac{(-1)(2+x)}{3x+1} \quad \text{or} \quad \frac{-2-x}{3x+1} \qquad \text{Simplify.}$$

SPOTLIGHT ON DECISION MAKING

Suppose you are a forensic lab technician. You have been asked to try to identify a piece of metal found at a crime scene. You know that one way to analyze the piece of metal is to find its density (mass per unit volume), using the formula $density = \dfrac{mass}{volume}$. After weighing the metal, you find its mass as 36.2 grams. You have also determined the volume of the metal to be 4.5 milliliters. Use this information to decide which type of metal this piece is most likely made of, and explain your reasoning. What other characteristics might help the identification?

Densities	
Metal	*Density (g/ml)*
Aluminum	2.7
Iron	7.8
Lead	11.5
Silver	10.5

M E N T A L   M A T H

Find any real numbers for which each rational expression is undefined. See Example 2.

1. $\dfrac{x + 5}{x}$

2. $\dfrac{x^2 + 5x}{x - 3}$

3. $\dfrac{x^2 + 4x - 2}{x(x - 1)}$

4. $\dfrac{x + 2}{(x - 5)(x - 6)}$

Exercise Set 6.1

Find the value of the following expressions when $x = 2$, $y = -2$, and $z = -5$. See Example 1.

1. $\dfrac{x + 5}{x + 2}$

2. $\dfrac{x + 8}{2x + 5}$

3. $\dfrac{z - 8}{z + 2}$

4. $\dfrac{y - 2}{-5 + y}$

5. $\dfrac{x^2 + 8x + 2}{x^2 - x - 6}$

6. $\dfrac{z^2 + 8}{z^3 - 25z}$

7. $\dfrac{x + 5}{x^2 + 4x - 8}$

8. $\dfrac{z^3 + 1}{z^2 + 1}$

9. $\dfrac{y^3}{y^2 - 1}$

10. $\dfrac{z}{z^2 - 5}$

Find any real numbers for which each rational expression is undefined. See Example 2.

11. $\dfrac{x + 3}{x + 2}$

12. $\dfrac{5x + 1}{x - 3}$

13. $\dfrac{4x^2 + 9}{2x - 8}$

14. $\dfrac{9x^3 + 4x}{15x + 45}$

15. $\dfrac{9x^3 + 4}{15x + 30}$

16. $\dfrac{19x^3 + 2}{x^3 - x}$

17. $\dfrac{x^2 - 5x - 2}{x^2 + 4}$

18. $\dfrac{9y^5 + y^3}{x^2 + 9}$

19. Explain why the denominator of a fraction or a rational expression must not equal zero.

20. Does $\dfrac{(x - 3)(x + 3)}{x - 3}$ have the same value as $x + 3$ for all real numbers? Explain why or why not.

Simplify each expression. See Examples 3 through 6.

21. $\dfrac{8x^5}{4x^9}$

22. $\dfrac{12y^7}{-2y^6}$

23. $\dfrac{5(x - 2)}{(x - 2)(x + 1)}$

24. $\dfrac{9(x - 7)(x + 7)}{3(x - 7)}$

25. $\dfrac{-5a - 5b}{a + b}$

26. $\dfrac{7x + 35}{x^2 + 5x}$

27. $\dfrac{x + 5}{x^2 - 4x - 45}$

28. $\dfrac{x - 3}{x^2 - 6x + 9}$

29. $\dfrac{5x^2 + 11x + 2}{x + 2}$

30. $\dfrac{12x^2 + 4x - 1}{2x + 1}$

31. $\dfrac{x^2 + x - 12}{2x^2 - 5x - 3}$

32. $\dfrac{x^2 + 3x - 4}{x^2 - x - 20}$

33. Explain how to write a fraction in lowest terms.

34. Explain how to write a rational expression in lowest terms.

Simplify each expression. See Examples 7 and 8.

35. $\dfrac{x - 7}{7 - x}$

36. $\dfrac{y - z}{z - y}$

37. $\dfrac{y^2 - 2y}{4 - 2y}$

38. $\dfrac{x^2 + 5x}{20 + 4x}$

39. $\dfrac{x^2 - 4x + 4}{4 - x^2}$

40. $\dfrac{x^2 + 10x + 21}{-2x - 14}$

Simplify each expression.

41. $\dfrac{15x^4y^8}{-5x^8y^3}$

42. $\dfrac{24a^3b^3}{6a^2b^4}$

43. $\dfrac{(x - 2)(x + 3)}{5(x + 3)}$

44. $\dfrac{-2(y - 9)}{(y - 9)^2}$

45. $\dfrac{-6a - 6b}{a + b}$

46. $\dfrac{4a - 4y}{4y - 4a}$

47. $\dfrac{2x^2 - 8}{4x - 8}$

48. $\dfrac{5x^2 - 500}{35x + 350}$

49. $\dfrac{11x^2 - 22x^3}{6x - 12x^2}$

50. $\dfrac{16r^2 - 4s^2}{4r - 2s}$

51. $\dfrac{x + 7}{x^2 + 5x - 14}$

52. $\dfrac{x - 10}{x^2 - 17x + 70}$

53. $\dfrac{2x^2 + 3x - 2}{2x - 1}$

54. $\dfrac{4x^2 + 24x}{x + 6}$

55. $\dfrac{x^2 - 1}{x^2 - 2x + 1}$

56. $\dfrac{x^2 - 16}{x^2 - 8x + 16}$

57. $\dfrac{m^2 - 6m + 9}{m^2 - 9}$

58. $\dfrac{m^2 - 4m + 4}{m^2 + m - 6}$

59. $\dfrac{-2a^2 + 12a - 18}{9 - a^2}$

60. $\dfrac{-4a^2 + 8a - 4}{2a^2 - 2}$

61. $\dfrac{2 - x}{x - 2}$

62. $\dfrac{7 - y}{y - 7}$

63. $\dfrac{x^2 - 1}{1 - x}$

64. $\dfrac{x^2 - xy}{2y - 2x}$

65. $\dfrac{x^2 + 7x + 10}{x^2 - 3x - 10}$

66. $\dfrac{2x^2 + 7x - 4}{x^2 + 3x - 4}$

67. $\dfrac{3x^2 + 7x + 2}{3x^2 + 13x + 4}$

68. $\dfrac{4x^2 - 4x + 1}{2x^2 + 9x - 5}$

Solve. See Example 1.

69. The dose of medicine prescribed for a child depends on the child's age A in years and the adult dose D for the medication. Young's Rule is a formula used by pediatricians that gives a child's dose C as

$$C = \frac{DA}{A + 12}$$

Suppose that an 8-year-old child needs medication, and the normal adult dose is 1000 mg. What size dose should the child receive?

70. During a storm, water treatment engineers monitor how quickly rain is falling. If too much rain comes too fast, there is a danger of sewers backing up. A formula that gives the rainfall intensity i in millimeters per hour for a certain strength storm in eastern Virginia is

$$i = \frac{5840}{t + 29}$$

where t is the duration of the storm in minutes. What rainfall intensity should engineers expect for a storm of this strength in eastern Virginia that lasts for 80 minutes? Round your answer to one decimal place.

71. Calculating body-mass index is a way to gauge whether a person should lose weight. Doctors recommend that body-mass index values fall between 19 and 25. The formula for body-mass index B is

$$B = \frac{705w}{h^2}$$

where w is weight in pounds and h is height in inches. Should a 148-pound person who is 5 feet 6 inches tall lose weight?

72. Anthropologists and forensic scientists use a measure called the cephalic index to help classify skulls. The cephalic index of a skull with width W and length L from front to back is given by the formula

$$C = \frac{100W}{L}$$

A long skull has an index value less than 75, a medium skull has an index value between 75 and 85, and a broad skull has an index value over 85. Find the cephalic index of a skull that is 5 inches wide and 6.4 inches long. Classify the skull.

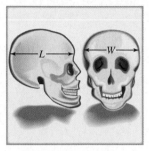

73. The total revenue R from the sale of a popular music compact disc is approximately given by the equation

$$R = \frac{150x^2}{x^2 + 3}$$

where x is the number of years since the CD has been released and revenue R is in millions of dollars.

a. Find the total revenue generated by the end of the first year.

b. Find the total revenue generated by the end of the second year.

c. Find the total revenue generated in the second year only.

74. For a certain model fax machine, the manufacturing cost C per machine is given by the equation

$$C = \frac{250x + 10,000}{x}$$

where x is the number of fax machines manufactured and cost C is in dollars per machine.

a. Find the cost per fax machine when manufacturing 100 fax machines.

b. Find the cost per fax machine when manufacturing 1000 fax machines.

c. Does the cost per machine decrease or increase when more machines are manufactured? Explain why this is so.

75. Recall that the fundamental principle applies to common *factors* only. Which of the following are *not* true? Explain why.

a. $\dfrac{3 - 1}{3 + 5} = -\dfrac{1}{5}$

b. $\dfrac{2x + 10}{2} = x + 5$

c. $\dfrac{37}{72} = \dfrac{3}{2}$

d. $\dfrac{2x + 6}{2} = x + 3$

Simplify. These expressions contain 4-term polynomials and sums and differences of cubes.

76. $\dfrac{x^2 + xy + 2x + 2y}{x + 2}$

77. $\dfrac{ab + ac + b^2 + bc}{b + c}$

78. $\dfrac{5x + 15 - xy - 3y}{2x + 6}$

79. $\dfrac{xy - 6x + 2y - 12}{y^2 - 6y}$

80. $\dfrac{x^3 + 8}{x + 2}$

81. $\dfrac{x^3 + 64}{x + 4}$

82. $\dfrac{x^3 - 1}{1 - x}$

83. $\dfrac{3 - x}{x^3 - 27}$

How does the graph of $y = \frac{x^2 - 9}{x - 3}$ compare to the graph of $y = x + 3$?

Recall that $\frac{x^2 - 9}{x - 3} = \frac{(x + 3)(x - 3)}{x - 3} = x + 3$ as long as x is not 3. This means that the graph of $y = \frac{x^2 - 9}{x - 3}$ is the same as the graph of $y = x + 3$ with $x \neq 3$. To graph $y = \frac{x^2 - 9}{x - 3}$, then, graph the linear equation $y = x + 3$ and place an open dot on the graph at 3. This open dot or interruption of the line at 3 means $x \neq 3$.

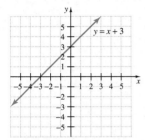

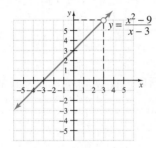

84. Graph $y = \dfrac{x^2 - 25}{x + 5}$.

85. Graph $y = \dfrac{x^2 - 16}{x - 4}$.

86. Graph $y = \dfrac{x^2 + x - 12}{x + 4}$.

87. Graph $y = \dfrac{x^2 - 6x + 8}{x - 2}$.

REVIEW EXERCISES

Perform the indicated operations. See Section 1.2.

88. $\dfrac{1}{3} \cdot \dfrac{9}{11}$

89. $\dfrac{5}{27} \cdot \dfrac{2}{5}$

90. $\dfrac{1}{3} \div \dfrac{1}{4}$

91. $\dfrac{7}{8} \div \dfrac{1}{2}$

92. $\dfrac{5}{6} \cdot \dfrac{10}{11} \cdot \dfrac{2}{3}$

93. $\dfrac{4}{3} \cdot \dfrac{1}{7} \cdot \dfrac{10}{13}$

94. $\dfrac{13}{20} \div \dfrac{2}{9}$

95. $\dfrac{8}{15} \div \dfrac{5}{8}$

6.2 MULTIPLYING AND DIVIDING RATIONAL EXPRESSIONS

CD-ROM   SSM

SSG   Video

▶ **OBJECTIVES**

1. Multiply rational expressions.
2. Divide rational expressions.

1 Just as simplifying rational expressions is similar to simplifying number fractions, multiplying and dividing rational expressions is similar to multiplying and dividing number fractions. To find the product of fractions and rational expressions, multiply the numerators and multiply the denominators.

$$\frac{3}{5} \cdot \frac{1}{4} = \frac{3 \cdot 1}{5 \cdot 4} = \frac{3}{20} \quad \text{and} \quad \frac{x}{y + 1} \cdot \frac{x + 3}{y - 1} = \frac{x(x + 3)}{(y + 1)(y - 1)}$$

MULTIPLYING RATIONAL EXPRESSIONS

Let P, Q, R, and S be polynomials. Then

$$\frac{P}{Q} \cdot \frac{R}{S} = \frac{PR}{QS}$$

as long as $Q \neq 0$ and $S \neq 0$.

Example 1 Multiply.

   a. $\dfrac{25x}{2} \cdot \dfrac{1}{y^3}$ **b.** $\dfrac{-7x^2}{5y} \cdot \dfrac{3y^5}{14x^2}$

Solution   To multiply rational expressions, multiply the numerators and then multiply the denominators of both expressions. Then simplify if possible.

   a. $\dfrac{25x}{2} \cdot \dfrac{1}{y^3} = \dfrac{25x \cdot 1}{2 \cdot y^3} = \dfrac{25x}{2y^3}$

   The expression $\dfrac{25x}{2y^3}$ is in simplest form.

   b. $\dfrac{-7x^2}{5y} \cdot \dfrac{3y^5}{14x^2} = \dfrac{-7x^2 \cdot 3y^5}{5y \cdot 14x^2}$ Multiply.

   The expression $\dfrac{-7x^2 \cdot 3y^5}{5y \cdot 14x^2}$ is not in simplest form, so we factor the numerator and the denominator and apply the fundamental principle.

$$= \frac{-1 \cdot 7 \cdot 3 \cdot x^2 \cdot y \cdot y^4}{5 \cdot 2 \cdot 7 \cdot x^2 \cdot y}$$

$$= -\frac{3y^4}{10}$$

When multiplying rational expressions, it is usually best to factor each numerator and denominator. This will help us when we apply the fundamental principle to write the product in lowest terms.

Example 2 Multiply: $\dfrac{x^2 + x}{3x} \cdot \dfrac{6}{5x + 5}$.

Solution $\dfrac{x^2 + x}{3x} \cdot \dfrac{6}{5x + 5} = \dfrac{x(x + 1)}{3x} \cdot \dfrac{2 \cdot 3}{5(x + 1)}$ Factor numerators and denominators.

$$= \frac{x(x + 1) \cdot 2 \cdot 3}{3x \cdot 5(x + 1)}$$ Multiply.

$$= \frac{2}{5}$$ Simplify.

The following steps may be used to multiply rational expressions.

MULTIPLYING RATIONAL EXPRESSIONS

Step 1. Completely factor numerators and denominators.
Step 2. Multiply numerators and multiply denominators.
Step 3. Simplify or write the product in lowest terms by applying the fundamental principle to all common factors.

Example 3   Multiply: $\dfrac{3x + 3}{5x - 5x^2} \cdot \dfrac{2x^2 + x - 3}{4x^2 - 9}$.

Solution

$$\frac{3x + 3}{5x - 5x^2} \cdot \frac{2x^2 + x - 3}{4x^2 - 9} = \frac{3(x + 1)}{5x(1 - x)} \cdot \frac{(2x + 3)(x - 1)}{(2x - 3)(2x + 3)} \qquad \text{Factor.}$$

$$= \frac{3(x + 1)(2x + 3)(x - 1)}{5x(1 - x)(2x - 3)(2x + 3)} \qquad \text{Multiply.}$$

$$= \frac{3(x + 1)(x - 1)}{5x(1 - x)(2x - 3)} \qquad \begin{array}{l}\text{Apply the fundamental}\\ \text{principle.}\end{array}$$

Next, recall that $x - 1$ and $1 - x$ are opposites so that $x - 1 = -1(1 - x)$.

$$= \frac{3(x + 1)(-1)(1 - x)}{5x(1 - x)(2x - 3)} \qquad \text{Write } x - 1 \text{ as } -1(1 - x).$$

$$= \frac{-3(x + 1)}{5x(2x - 3)} \quad \text{or} \quad -\frac{3(x + 1)}{5x(2x - 3)} \qquad \text{Apply the fundamental principle.}$$

2   We can divide by a rational expression in the same way we divide by a fraction. To divide by a fraction, multiply by its reciprocal.

> **HELPFUL HINT**
>
> Don't forget how to find reciprocals. The reciprocal of $\dfrac{a}{b}$ is $\dfrac{b}{a}$, $a \neq 0, b \neq 0$.

For example, to divide $\frac{3}{2}$ by $\frac{7}{8}$, multiply $\frac{3}{2}$ by $\frac{8}{7}$.

$$\frac{3}{2} \div \frac{7}{8} = \frac{3}{2} \cdot \frac{8}{7} = \frac{3 \cdot 4 \cdot 2}{2 \cdot 7} = \frac{12}{7}$$

DIVIDING RATIONAL EXPRESSIONS

Let P, Q, R, and S be polynomials. Then,

$$\frac{P}{Q} \div \frac{R}{S} = \frac{P}{Q} \cdot \frac{S}{R} = \frac{PS}{QR}$$

as long as $Q \neq 0$, $S \neq 0$, and $R \neq 0$.

Example 4   Divide: $\dfrac{3x^3y^7}{40} \div \dfrac{4x^3}{y^2}$.

Solution $\dfrac{3x^3y^7}{40} \div \dfrac{4x^3}{y^2} = \dfrac{3x^3y^7}{40} \cdot \dfrac{y^2}{4x^3}$ Multiply by the reciprocal of $\dfrac{4x^3}{y^2}$.

$= \dfrac{3x^3y^9}{160x^3}$

$= \dfrac{3y^9}{160}$ Simplify.

Example 5   Divide $\dfrac{(x-1)(x+2)}{10}$ by $\dfrac{2x+4}{5}$.

Solution $\dfrac{(x-1)(x+2)}{10} \div \dfrac{2x+4}{5} = \dfrac{(x-1)(x+2)}{10} \cdot \dfrac{5}{2x+4}$ Multiply by the reciprocal of $\dfrac{2x+4}{5}$.

$= \dfrac{(x-1)(x+2) \cdot 5}{5 \cdot 2 \cdot 2 \cdot (x+2)}$ Factor and multiply.

$= \dfrac{x-1}{4}$ Simplify.

The following may be used to divide by a rational expression.

DIVIDING BY A RATIONAL EXPRESSION

Multiply by its reciprocal.

Example 6   Divide: $\dfrac{6x+2}{x^2-1} \div \dfrac{3x^2+x}{x-1}$.

Solution $\dfrac{6x+2}{x^2-1} \div \dfrac{3x^2+x}{x-1} = \dfrac{6x+2}{x^2-1} \cdot \dfrac{x-1}{3x^2+x}$ Multiply by the reciprocal.

$= \dfrac{2(3x+1)(x-1)}{(x+1)(x-1) \cdot x(3x+1)}$ Factor and multiply.

$= \dfrac{2}{x(x+1)}$ Simplify.

Example 7   Divide: $\dfrac{2x^2-11x+5}{5x-25} \div \dfrac{4x-2}{10}$.

Solution

$$\frac{2x^2 - 11x + 5}{5x - 25} \div \frac{4x - 2}{10} = \frac{2x^2 - 11x + 5}{5x - 25} \cdot \frac{10}{4x - 2}$$ Multiply by the reciprocal.

$$= \frac{(2x - 1)(x - 5) \cdot 2 \cdot 5}{5(x - 5) \cdot 2(2x - 1)}$$ Factor and multiply.

$$= \frac{1}{1} \quad \text{or} \quad 1$$ Simplify.

Now that we know how to multiply fractions and rational expressions, we can use this knowledge to help us convert between units of measure. To do so, we will use **unit fractions**. A unit fraction is a fraction that equals 1. For example, since 12 in. = 1 ft, we have the unit fractions

$$\frac{12 \text{ in.}}{1 \text{ ft}} = 1 \quad \text{and} \quad \frac{1 \text{ ft}}{12 \text{ in.}} = 1$$

Example 8   **CONVERTING FROM SQUARE YARDS TO SQUARE FEET**

The largest casino in the world is the Foxwoods Resort Casino in Ledyard, CT. The gaming area for this casino is 21,444 *square yards*. Find the size of the gaming area in *square feet*. (*Source: The Guinness Book of Records*, 1996)

Solution   There are 9 square feet in 1 square yard.

Unit fraction

$$21,444 \text{ square yards} = 21,444 \, \cancel{\text{sq. yd}} \cdot \frac{9 \text{ sq. ft}}{1 \, \cancel{\text{sq. yd}}}$$

$$= 192,996 \text{ square feet}$$

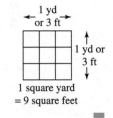

1 yd or 3 ft

1 yd or 3 ft

1 square yard = 9 square feet

HELPFUL HINT
When converting a unit of measurement, if possible, write the unit fraction so that **the numerator contains units converting to** and **the denominator contains original units.**

Unit fraction

$$48 \text{ in.} = \frac{48 \, \cancel{\text{in.}}}{1} \cdot \frac{1 \text{ ft}}{12 \, \cancel{\text{in.}}} \quad \leftarrow \text{Units converting to} \\ \leftarrow \text{Original units}$$

$$= \frac{48}{12} \text{ ft} = 4 \text{ ft}$$

MENTAL MATH

Find the following products. See Example 1.

1. $\dfrac{2}{y} \cdot \dfrac{x}{3}$

2. $\dfrac{3x}{4} \cdot \dfrac{1}{y}$

3. $\dfrac{5}{7} \cdot \dfrac{y^2}{x^2}$

4. $\dfrac{x^5}{11} \cdot \dfrac{4}{z^3}$

5. $\dfrac{9}{x} \cdot \dfrac{x}{5}$

6. $\dfrac{y}{7} \cdot \dfrac{3}{y}$

Exercise Set 6.2

Multiply. Simplify if possible. See Examples 1 through 3.

1. $\dfrac{3x}{y^2} \cdot \dfrac{7y}{4x}$

2. $\dfrac{9x^2}{y} \cdot \dfrac{4y}{3x^2}$

3. $\dfrac{8x}{2} \cdot \dfrac{x^5}{4x^2}$

4. $\dfrac{6x^2}{10x^3} \cdot \dfrac{5x}{12}$

5. $-\dfrac{5a^2b}{30a^2b^2} \cdot b^3$

6. $-\dfrac{9x^3y^2}{18xy^5} \cdot y^3$

7. $\dfrac{x}{2x - 14} \cdot \dfrac{x^2 - 7x}{5}$

8. $\dfrac{4x - 24}{20x} \cdot \dfrac{5}{x - 6}$

9. $\dfrac{6x + 6}{5} \cdot \dfrac{10}{36x + 36}$

10. $\dfrac{x^2 + x}{8} \cdot \dfrac{16}{x + 1}$

11. $\dfrac{m^2 - n^2}{m + n} \cdot \dfrac{m}{m^2 - mn}$

12. $\dfrac{(m - n)^2}{m + n} \cdot \dfrac{m}{m^2 - mn}$

13. $\dfrac{x^2 - 25}{x^2 - 3x - 10} \cdot \dfrac{x + 2}{x}$

14. $\dfrac{a^2 + 6a + 9}{a^2 - 4} \cdot \dfrac{a + 3}{a - 2}$

15. Find the area of the following rectangle.

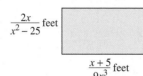

$\dfrac{2x}{x^2 - 25}$ feet

$\dfrac{x + 5}{9x^3}$ feet

16. Find the area of the following square.

$\dfrac{2x}{5x^2 + 3x}$ meters

Divide. Simplify if possible. See Examples 4 through 7.

17. $\dfrac{5x^7}{2x^5} \div \dfrac{10x}{4x^3}$

18. $\dfrac{9y^4}{6y} \div \dfrac{y^2}{3}$

19. $\dfrac{8x^2}{y^3} \div \dfrac{4x^2y^3}{6}$

20. $\dfrac{7a^2b}{3ab^2} \div \dfrac{21a^2b^2}{14ab}$

21. $\dfrac{(x - 6)(x + 4)}{4x} \div \dfrac{2x - 12}{8x^2}$

22. $\dfrac{(x + 3)^2}{5} \div \dfrac{5x + 15}{25}$

23. $\dfrac{3x^2}{x^2 - 1} \div \dfrac{x^5}{(x + 1)^2}$

24. $\dfrac{(x + 1)}{(x + 1)(2x + 3)} \div \dfrac{20}{2x + 3}$

25. $\dfrac{m^2 - n^2}{m + n} \div \dfrac{m}{m^2 + nm}$

26. $\dfrac{(m - n)^2}{m + n} \div \dfrac{m^2 - mn}{m}$

27. $\dfrac{x + 2}{7 - x} \div \dfrac{x^2 - 5x + 6}{x^2 - 9x + 14}$

28. $(x - 3) \div \dfrac{x^2 + 3x - 18}{x}$

29. $\dfrac{x^2 + 7x + 10}{1 - x} \div \dfrac{x^2 + 2x - 15}{x - 1}$

30. $\dfrac{a^2 - b^2}{9} \div \dfrac{3b - 3a}{27x^2}$

31. Explain how to multiply rational expressions.

32. Explain how to divide rational expressions.

Perform the indicated operations.

33. $\dfrac{5a^2b}{30a^2b^2} \cdot \dfrac{1}{b^3}$

34. $\dfrac{9x^2y^2}{42xy^5} \cdot \dfrac{6}{x^5}$

35. $\dfrac{12x^3y}{8xy^7} \div \dfrac{7x^5y}{6x}$

36. $\dfrac{4y^2z}{3y^7z^7} \div \dfrac{12y}{6z}$

37. $\dfrac{5x - 10}{12} \div \dfrac{4x - 8}{8}$

38. $\dfrac{6x + 6}{5} \div \dfrac{3x + 3}{10}$

39. $\dfrac{x^2 + 5x}{8} \cdot \dfrac{9}{3x + 15}$

40. $\dfrac{3x^2 + 12x}{6} \cdot \dfrac{9}{2x + 8}$

41. $\dfrac{7}{6p^2 + q} \div \dfrac{14}{18p^2 + 3q}$

42. $\dfrac{5x - 10}{12} \div \dfrac{4x - 8}{8}$

43. $\dfrac{3x + 4y}{x^2 + 4xy + 4y^2} \cdot \dfrac{x + 2y}{2}$ **44.** $\dfrac{2a + 2b}{3} \div \dfrac{a^2 - b^2}{a - b}$

45. $\dfrac{x^2 - 9}{x^2 + 8} \div \dfrac{3 - x}{2x^2 + 16}$

46. $\dfrac{x^2 - y^2}{3x^2 + 3xy} \cdot \dfrac{3x^2 + 6x}{3x^2 - 2xy - y^2}$

47. $\dfrac{(x + 2)^2}{x - 2} \div \dfrac{x^2 - 4}{2x - 4}$ **48.** $\dfrac{x^2 - 4}{2y} \div \dfrac{2 - x}{6xy}$

49. $\dfrac{a^2 + 7a + 12}{a^2 + 5a + 6} \cdot \dfrac{a^2 + 8a + 15}{a^2 + 5a + 4}$

50. $\dfrac{b^2 + 2b - 3}{b^2 + b - 2} \cdot \dfrac{b^2 - 4}{b^2 + 6b + 8}$

51. $\dfrac{1}{-x - 4} \div \dfrac{x^2 - 7x}{x^2 - 3x - 28}$

52. $\dfrac{x^2 - 10x + 21}{7 - x} \div (x + 3)$

53. $\dfrac{x^2 - 5x - 24}{2x^2 - 2x - 24} \cdot \dfrac{4x^2 + 4x - 24}{x^2 - 10x + 16}$

54. $\dfrac{a^2 - b^2}{a} \cdot \dfrac{a + b}{a^3 + ab}$ **55.** $(x - 5) \div \dfrac{5 - x}{x^2 + 2}$

56. $\dfrac{2x^2 + 3xy + y^2}{x^2 - y^2} \div \dfrac{1}{2x + 2y}$

57. $\dfrac{x^2 - y^2}{x^2 - 2xy + y^2} \cdot \dfrac{y - x}{x + y}$ **58.** $\dfrac{x + 3}{x^2 - 9} \cdot \dfrac{x^2 - 8x + 15}{5x}$

59. Find the quotient of $\dfrac{x^2 - 9}{2x}$ and $\dfrac{x + 3}{8x^4}$.

60. Find the quotient of $\dfrac{4x^2 + 4x + 1}{4x + 2}$ and $\dfrac{4x + 2}{16}$.

Convert as indicated. See Example 8.

61. 10 square feet = _____ square inches.

62. 1008 square inches = _____ square feet.

63. The Pentagon, headquarters for the Department of Defense, contains 3,707,745 square feet of office space. Convert this to square yards. (Round to the nearest square yard.) (*Source: World Almanac, 2000*)

64. The Empire State building in Manhattan contains approximately 137,300 square yards of office space. Convert this to square feet.

65. 50 miles per hour = _____ feet per second (round to the nearest whole).

66. 10 feet per second = _____ miles per hour (round to the nearest tenth).

67. The speed of sound is 5023 feet per second in ocean water whose temperature is 77°F. Convert this speed of sound to miles per hour. Round to the nearest tenth. (*Source: CRC Handbook of Chemistry and Physics*, 65th edition)

68. On October 28, 1996, Craig Breedlove tried unsuccessfully to break the world land speed record. He reached an unofficial speed of 675 mph before losing control of his car, the Spirit of America. Find this speed in feet per second. (*Source: The World Almanac and Book of Facts*, 1997)

Simplify. These expressions contain 4-term polynomials and sums and differences of cubes.

69. $\dfrac{a^2 + ac + ba + bc}{a - b} \div \dfrac{a + c}{a + b}$

70. $\dfrac{x^2 + 2x - xy - 2y}{x^2 - y^2} \div \dfrac{2x + 4}{x + y}$

71. $\dfrac{3x^2 + 8x + 5}{x^2 + 8x + 7} \cdot \dfrac{x + 7}{x^2 + 4}$

72. $\dfrac{16x^2 + 2x}{16x^2 + 10x + 1} \cdot \dfrac{1}{4x^2 + 2x}$

73. $\dfrac{x^3 + 8}{x^2 - 2x + 4} \cdot \dfrac{4}{x^2 - 4}$

74. $\dfrac{9y}{3y - 3} \cdot \dfrac{y^3 - 1}{y^3 + y^2 + y}$

75. $\dfrac{a^2 - ab}{6a^2 + 6ab} \div \dfrac{a^3 - b^3}{a^2 - b^2}$

76. $\dfrac{x^3 + 27y^3}{6x} \div \dfrac{x^2 - 9y^2}{x^2 - 3xy}$

REVIEW EXERCISES

Perform each operation. See Section 1.3.

77. $\dfrac{1}{5} + \dfrac{4}{5}$　　　　　**78.** $\dfrac{3}{15} + \dfrac{6}{15}$

79. $\dfrac{9}{9} - \dfrac{19}{9}$　　　　　**80.** $\dfrac{4}{3} - \dfrac{8}{3}$

81. $\dfrac{6}{5} + \left(\dfrac{1}{5} - \dfrac{8}{5}\right)$　　　　**82.** $-\dfrac{3}{2} + \left(\dfrac{1}{2} - \dfrac{3}{2}\right)$

See Section 3.2.

83. Graph the linear equation $x - 2y = 6$.
84. Graph the linear equation $5x + y = 10$.

A Look Ahead

Example
Perform the indicated operations.

$$\frac{15x^2 - x - 6}{12x^3} \cdot \frac{4x}{9 - 25x^2} \div \frac{x}{3x - 2}$$

Solution:

$$\frac{15x^2 - x - 6}{12x^3} \cdot \frac{4x}{9 - 25x^2} \div \frac{x}{3x - 2}$$

$$= \left(\frac{15x^2 - x - 6}{12x^3} \cdot \frac{4x}{9 - 25x^2}\right) \cdot \frac{3x - 2}{x}$$

$$= \frac{(3x - 2)(5x + 3) \cdot 4x(3x - 2)}{12x^3(3 - 5x)(3 + 5x) \cdot x}$$

$$= \frac{(3x - 2)^2}{3x^3(3 - 5x)}$$

Perform the following operations.

85. $\left(\dfrac{x^2 - y^2}{x^2 + y^2} \div \dfrac{x^2 - y^2}{3x}\right) \cdot \dfrac{x^2 + y^2}{6}$

86. $\left(\dfrac{x^2 - 9}{x^2 - 1} \cdot \dfrac{x^2 + 2x + 1}{2x^2 + 9x + 9}\right) \div \dfrac{2x + 3}{1 - x}$

87. $\left(\dfrac{2a + b}{b^2} \cdot \dfrac{3a^2 - 2ab}{ab + 2b^2}\right) \div \dfrac{a^2 - 3ab + 2b^2}{5ab - 10b^2}$

88. $\left(\dfrac{x^2y^2 - xy}{4x - 4y} \div \dfrac{3y - 3x}{8x - 8y}\right) \cdot \dfrac{y - x}{8}$

6.3 ADDING AND SUBTRACTING RATIONAL EXPRESSIONS WITH COMMON DENOMINATORS AND LEAST COMMON DENOMINATOR

CD-ROM

SSM

SSG

Video

▶ **OBJECTIVES**

1. Add and subtract rational expressions with the same denominator.
2. Find the least common denominator of a list of rational expressions.
3. Write a rational expression as an equivalent expression whose denominator is given.

1 Like multiplication and division, addition and subtraction of rational expressions is similar to addition and subtraction of rational numbers. In this section, we add and subtract rational expressions with a common (or the same) denominator.

Add: $\dfrac{6}{5} + \dfrac{2}{5}$　　　　　　　Add: $\dfrac{9}{x + 2} + \dfrac{3}{x + 2}$

Add the numerators and place the sum over the common denominator.

$\dfrac{6}{5} + \dfrac{2}{5} = \dfrac{6 + 2}{5}$　　　　　　$\dfrac{9}{x + 2} + \dfrac{3}{x + 2} = \dfrac{9 + 3}{x + 2}$

$\quad\quad = \dfrac{8}{5}$　Simplify.　　　　　$\quad\quad = \dfrac{12}{x + 2}$　Simplify.

ADDING AND SUBTRACTING RATIONAL EXPRESSIONS WITH COMMON DENOMINATORS

If $\dfrac{P}{R}$ and $\dfrac{Q}{R}$ are rational expressions, then

$$\frac{P}{R} + \frac{Q}{R} = \frac{P + Q}{R} \quad \text{and} \quad \frac{P}{R} - \frac{Q}{R} = \frac{P - Q}{R}$$

To add or subtract rational expressions, add or subtract numerators and place the sum or difference over the common denominator.

Example 1   Add: $\dfrac{5m}{2n} + \dfrac{m}{2n}$

Solution $\dfrac{5m}{2n} + \dfrac{m}{2n} = \dfrac{5m + m}{2n}$ Add the numerators.

$= \dfrac{6m}{2n}$ Simplify the numerator by combining like terms.

$= \dfrac{3m}{n}$ Simplify by applying the fundamental principle.

Example 2   Subtract: $\dfrac{2y}{2y - 7} - \dfrac{7}{2y - 7}$

Solution $\dfrac{2y}{2y - 7} - \dfrac{7}{2y - 7} = \dfrac{2y - 7}{2y - 7}$ Subtract the numerators.

$= \dfrac{1}{1}$ or 1   Simplify.

Example 3   Subtract: $\dfrac{3x^2 + 2x}{x - 1} - \dfrac{10x - 5}{x - 1}$

Solution $\dfrac{3x^2 + 2x}{x - 1} - \dfrac{10x - 5}{x - 1} = \dfrac{3x^2 + 2x - (10x - 5)}{x - 1}$ Subtract the numerators. Notice the parentheses.

$= \dfrac{3x^2 + 2x - 10x + 5}{x - 1}$ Use the distributive property.

$= \dfrac{3x^2 - 8x + 5}{x - 1}$ Combine like terms.

$= \dfrac{(x - 1)(3x - 5)}{x - 1}$ Factor.

$= 3x - 5$ Simplify.

▼
HELPFUL HINT
Notice how the numerator $10x - 5$ has been subtracted in Example 3.

This $-$ sign applies to the entire numerator of $10x - 5$.   So parentheses are inserted here to indicate this.

$$\frac{3x^2 + 2x}{x - 1} - \frac{10x - 5}{x - 1} = \frac{3x^2 + 2x - (10x - 5)}{x - 1}$$

---

2 To add and subtract fractions with **unlike** denominators, first find a least common denominator (LCD), and then write all fractions as equivalent fractions with the LCD.

For example, suppose we add $\frac{8}{3}$ and $\frac{2}{5}$. The LCD of denominators 3 and 5 is 15, since 15 is the smallest number that both 3 and 5 divide into evenly. Rewrite each fraction so that its denominator is 15. (Notice how we apply the fundamental principle.)

$$\frac{8}{3} + \frac{2}{5} = \frac{8(5)}{3(5)} + \frac{2(3)}{5(3)} = \frac{40}{15} + \frac{6}{15} = \frac{40 + 6}{15} = \frac{46}{15}$$

To add or subtract rational expressions with unlike denominators, we also first find an LCD and then write all rational expressions as equivalent expressions with the LCD. The **least common denominator (LCD) of a list of rational expressions** is a polynomial of least degree whose factors include all the factors of the denominators in the list.

FINDING THE LEAST COMMON DENOMINATOR (LCD)

Step 1.   Factor each denominator completely.
Step 2.   The least common denominator (LCD) is the product of all unique factors found in step 1, each raised to a power equal to the greatest number of times that the factor appears in any one factored denominator.

Example 4   Find the LCD for each pair.

a. $\dfrac{1}{8}, \dfrac{3}{22}$ **b.** $\dfrac{7}{5x}, \dfrac{6}{15x^2}$

Solution   **a.** Start by finding the prime factorization of each denominator.

$$8 = 2 \cdot 2 \cdot 2 = 2^3 \qquad \text{and} \qquad 22 = 2 \cdot 11$$

Next, write the product of all the unique factors, each raised to a power equal to the greatest number of times that the factor appears.

The greatest number of times that the factor 2 appears is 3.
The greatest number of times that the factor 11 appears is 1.

$$\text{LCD} = 2^3 \cdot 11^1 = 8 \cdot 11 = 88$$

b. Factor each denominator.

$$5x = 5 \cdot x \qquad \text{and} \qquad 15x^2 = 3 \cdot 5 \cdot x^2$$

The greatest number of times that the factor 5 appears is 1.
The greatest number of times that the factor 3 appears is 1.
The greatest number of times that the factor x appears is 2.

$$LCD = 3^1 \cdot 5^1 \cdot x^2 = 15x^2$$

Example 5 Find the LCD of $\dfrac{7x}{x + 2}$ and $\dfrac{5x^2}{x - 2}$.

Solution The denominators $x + 2$ and $x - 2$ are completely factored already. The factor $x + 2$ appears once and the factor $x - 2$ appears once.

$$LCD = (x + 2)(x - 2)$$

Example 6 Find the LCD of $\dfrac{6m^2}{3m + 15}$ and $\dfrac{2}{(m + 5)^2}$.

Solution Factor each denominator.

$$3m + 15 = 3(m + 5)$$
$$(m + 5)^2 \text{ is already factored.}$$

The greatest number of times that the factor 3 appears is 1.
The greatest number of times that the factor $m + 5$ appears *in any one denominator* is 2.

$$LCD = 3(m + 5)^2$$

Example 7 Find the LCD of $\dfrac{t - 10}{t^2 - t - 6}$ and $\dfrac{t + 5}{t^2 + 3t + 2}$.

Solution Start by factoring each denominator.

$$t^2 - t - 6 = (t - 3)(t + 2)$$
$$t^2 + 3t + 2 = (t + 1)(t + 2)$$
$$LCD = (t - 3)(t + 2)(t + 1)$$

Example 8 Find the LCD of $\dfrac{2}{x - 2}$ and $\dfrac{10}{2 - x}$.

Solution The denominators $x - 2$ and $2 - x$ are opposites. That is, $2 - x = -1(x - 2)$. Use $x - 2$ or $2 - x$ as the LCD.

$$LCD = x - 2 \qquad \text{or} \qquad LCD = 2 - x$$

3   Next we practice writing a rational expression as an equivalent rational expression with a given denominator. To do this, we apply the fundamental principle, which says that $\frac{PR}{QR} = \frac{P}{Q}$, or equivalently that $\frac{P}{Q} = \frac{PR}{QR}$. This can be seen by recalling that multiplying an expression by 1 produces an equivalent expression. In other words,

$$\frac{P}{Q} = \frac{P}{Q} \cdot 1 = \frac{P}{Q} \cdot \frac{R}{R} = \frac{PR}{QR}.$$

Example 9   Write $\dfrac{4b}{9a}$ as an equivalent fraction with the given denominator.

$$\frac{4b}{9a} = \frac{}{27a^2b}$$

Solution   Ask yourself: "What do we multiply $9a$ by to get $27a^2b$?" The answer is $3ab$, since $9a(3ab) = 27a^2b$. Multiply the numerator and denominator by $3ab$.

$$\frac{4b}{9a} = \frac{4b(3ab)}{9a(3ab)} = \frac{12ab^2}{27a^2b}$$

Example 10   Write the rational expression as an equivalent rational expression with the given denominator.

$$\frac{5}{x^2 - 4} = \frac{}{(x - 2)(x + 2)(x - 4)}$$

Solution   First, factor the denominator $x^2 - 4$ as $(x - 2)(x + 2)$.
If we multiply the original denominator $(x - 2)(x + 2)$ by $x - 4$, the result is the new denominator $(x + 2)(x - 2)(x - 4)$. Thus, multiply the numerator and the denominator by $x - 4$.

$$\frac{5}{\underbrace{x^2 - 4}} = \frac{5}{\underbrace{(x - 2)(x + 2)}} = \frac{5(x - 4)}{(x - 2)(x + 2)(x - 4)}$$

Factored denominator

$$= \frac{5x - 20}{(x - 2)(x + 2)(x - 4)}$$

MENTAL MATH

Perform the indicated operations.

1. $\dfrac{2}{3} + \dfrac{1}{3}$

2. $\dfrac{5}{11} + \dfrac{1}{11}$

3. $\dfrac{3x}{9} + \dfrac{4x}{9}$

4. $\dfrac{3y}{8} + \dfrac{2y}{8}$

5. $\dfrac{8}{9} - \dfrac{7}{9}$

6. $-\dfrac{4}{12} - \dfrac{3}{12}$

7. $\dfrac{7}{5} - \dfrac{10y}{5}$

8. $\dfrac{12x}{7} - \dfrac{4x}{7}$

Exercise Set 6.3

Add or subtract as indicated. Simplify the result if possible. See Examples 1 through 3.

1. $\dfrac{a}{13} + \dfrac{9}{13}$

2. $\dfrac{x+1}{7} + \dfrac{6}{7}$

3. $\dfrac{9}{3+y} + \dfrac{y+1}{3+y}$

4. $\dfrac{9}{y+9} + \dfrac{y}{y+9}$

5. $\dfrac{4m}{3n} + \dfrac{5m}{3n}$

6. $\dfrac{3p}{2} + \dfrac{11p}{2}$

7. $\dfrac{2x+1}{x-3} + \dfrac{3x+6}{x-3}$

8. $\dfrac{4p-3}{2p+7} + \dfrac{3p+8}{2p+7}$

9. $\dfrac{7}{8} - \dfrac{3}{8}$

10. $\dfrac{4}{5} - \dfrac{13}{5}$

11. $\dfrac{4m}{m-6} - \dfrac{24}{m-6}$

12. $\dfrac{8y}{y-2} - \dfrac{16}{y-2}$

13. $\dfrac{2x^2}{x-5} - \dfrac{25+x^2}{x-5}$

14. $\dfrac{6x^2}{2x-5} - \dfrac{25+2x^2}{2x-5}$

15. $\dfrac{-3x^2-4}{x-4} - \dfrac{12-4x^2}{x-4}$

16. $\dfrac{7x^2-9}{2x-5} - \dfrac{16+3x^2}{2x-5}$

17. $\dfrac{2x+3}{x+1} - \dfrac{x+2}{x+1}$

18. $\dfrac{1}{x^2-2x-15} - \dfrac{4-x}{x^2-2x-15}$

19. $\dfrac{3}{x^3} + \dfrac{9}{x^3}$

20. $\dfrac{5}{xy} + \dfrac{8}{xy}$

21. $\dfrac{5}{x+4} - \dfrac{10}{x+4}$

22. $\dfrac{4}{2x+1} - \dfrac{8}{2x+1}$

23. $\dfrac{x}{x+y} - \dfrac{2}{x+y}$

24. $\dfrac{y+1}{y+2} - \dfrac{3}{y+2}$

25. $\dfrac{8x}{2x+5} + \dfrac{20}{2x+5}$

26. $\dfrac{12y-5}{3y-1} + \dfrac{1}{3y-1}$

27. $\dfrac{5x+4}{x-1} - \dfrac{2x+7}{x-1}$

28. $\dfrac{x^2+9x}{x+7} - \dfrac{4x+14}{x+7}$

29. $\dfrac{a}{a^2+2a-15} - \dfrac{3}{a^2+2a-15}$

30. $\dfrac{3y}{y^2+3y-10} - \dfrac{6}{y^2+3y-10}$

31. $\dfrac{2x+3}{x^2-x-30} - \dfrac{x-2}{x^2-x-30}$

32. $\dfrac{3x-1}{x^2+5x-6} - \dfrac{2x-7}{x^2+5x-6}$

△ **33.** A square-shaped pasture has a side of length $\dfrac{5}{x-2}$ meters. Express its perimeter as a rational expression.

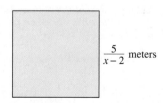

$\dfrac{5}{x-2}$ meters

△ **34.** The following trapezoid has sides of indicated length. Find its perimeter.

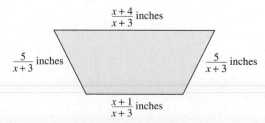

$\dfrac{x+4}{x+3}$ inches

$\dfrac{5}{x+3}$ inches $\dfrac{5}{x+3}$ inches

$\dfrac{x+1}{x+3}$ inches

35. Describe the process for adding and subtracting two rational expressions with the same denominators.

36. Explain the similarities between subtracting $\dfrac{3}{8}$ from $\dfrac{7}{8}$ and subtracting $\dfrac{6}{x+3}$ from $\dfrac{9}{x+3}$.

Find the LCD for the following lists of rational expressions. See Examples 4 through 8.

37. $\dfrac{2}{3}, \dfrac{4}{33}$

38. $\dfrac{8}{20}, \dfrac{4}{15}$

39. $\dfrac{19}{2x}, \dfrac{5}{4x^3}$

40. $\dfrac{17x}{4y^5}, \dfrac{2}{8y}$

41. $\dfrac{9}{8x}, \dfrac{3}{2x + 4}$

42. $\dfrac{1}{6y}, \dfrac{3x}{4y + 12}$

43. $\dfrac{1}{3x + 3}, \dfrac{8}{2x^2 + 4x + 2}$

44. $\dfrac{19x + 5}{4x - 12}, \dfrac{3}{2x^2 - 12x + 18}$

45. $\dfrac{5}{x - 8}, \dfrac{3}{8 - x}$

46. $\dfrac{2x + 5}{3x - 7}, \dfrac{5}{7 - 3x}$

47. $\dfrac{4 + x}{8x^2(x - 1)^2}, \dfrac{17}{10x^3(x - 1)}$

48. $\dfrac{2x + 3}{9x(x + 2)}, \dfrac{9x + 5}{12(x + 2)^2}$

49. $\dfrac{9x + 1}{2x + 1}, \dfrac{3x - 5}{2x - 1}$

50. $\dfrac{5}{4x - 2}, \dfrac{7}{4x + 2}$

51. $\dfrac{5x + 1}{2x^2 + 7x - 4}, \dfrac{3x}{2x^2 + 5x - 3}$

52. $\dfrac{4}{x^2 + 4x + 3}, \dfrac{4x - 2}{x^2 + 10x + 21}$

53. Write some instructions to help a friend who is having difficulty finding the LCD of two rational expressions.

54. Explain why the LCD of rational expressions $\frac{7}{x + 1}$ and $\frac{9x}{(x + 1)^2}$ is $(x + 1)^2$ and not $(x + 1)^3$.

Rewrite each rational expression as an equivalent rational expression whose denominator is the given polynomial. See Examples 9 and 10.

55. $\dfrac{3}{2x} = \dfrac{}{4x^2}$

56. $\dfrac{3}{9y^5} = \dfrac{}{72y^9}$

57. $\dfrac{6}{3a} = \dfrac{}{12ab^2}$

58. $\dfrac{17a}{4y^2x} = \dfrac{}{32y^3x^2z}$

59. $\dfrac{9}{x + 3} = \dfrac{}{2(x + 3)}$

60. $\dfrac{4x + 1}{3x + 6} = \dfrac{}{3y(x + 2)}$

61. $\dfrac{9a + 2}{5a + 10} = \dfrac{}{5b(a + 2)}$

62. $\dfrac{5 + y}{2x^2 + 10} = \dfrac{}{4(x^2 + 5)}$

63. $\dfrac{x}{x^2 + 6x + 8} = \dfrac{}{(x + 4)(x + 2)(x + 1)}$

64. $\dfrac{5x}{x^2 + 2x - 3} = \dfrac{}{(x - 1)(x - 5)(x + 3)}$

65. $\dfrac{9y - 1}{15x^2 - 30} = \dfrac{}{30x^2 - 60}$

66. $\dfrac{8x + 3}{7y^2 - 21} = \dfrac{}{21y^2 - 63}$

67. $\dfrac{5}{2x^2 - 9x - 5} = \dfrac{}{3x(2x + 1)(x - 7)(x - 5)}$

68. $\dfrac{x - 9}{3x^2 + 10x + 3} = \dfrac{}{x(x + 3)(x + 5)(3x + 1)}$

Write each rational expression as an equivalent expression with a denominator of $x - 2$.

69. $\dfrac{5}{2 - x}$

70. $\dfrac{8y}{2 - x}$

71. $-\dfrac{7 + x}{2 - x}$

72. $\dfrac{x - 3}{-(x - 2)}$

73. The planet Mercury revolves around the sun in 88 Earth days. It takes Jupiter 4332 Earth days to make one revolution around the sun. (*Source:* National Space Science Data Center) If the two planets are aligned as shown in the figure, how long will it take for them to align again?

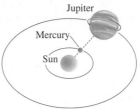

74. You are throwing a barbecue and you want to make sure that you purchase the same number of hot dogs as hot dog buns. Hot dogs come 8 to a package and hot dog buns come 12 to a package. What is the least number of each type of package you should buy?

75. An algebra student approaches you with a problem. He's tried to subtract two rational expressions, but his result does not match the book's. Check to see if the student has made an error. If so, correct his work shown below.

$$\dfrac{2x - 6}{x - 5} - \dfrac{x + 4}{x - 5}$$
$$= \dfrac{2x - 6 - x + 4}{x - 5}$$
$$= \dfrac{x - 2}{x - 5}$$

REVIEW EXERCISES

Solve the following quadratic equations by factoring. See Section 5.6.

76. $x(x - 3) = 0$

77. $2x(x + 5) = 0$

78. $x^2 + 6x + 5 = 0$

79. $x^2 - 6x + 5 = 0$

Perform each operation. See Section 1.2.

80. $\dfrac{2}{3} + \dfrac{5}{7}$

81. $\dfrac{9}{10} - \dfrac{3}{5}$

82. $\dfrac{2}{6} - \dfrac{3}{4}$

83. $\dfrac{11}{15} + \dfrac{5}{9}$

6.4 ADDING AND SUBTRACTING RATIONAL EXPRESSIONS WITH UNLIKE DENOMINATORS

CD-ROM   SSM

SSG   Video

▶ OBJECTIVE

1. Add and subtract rational expressions with unlike denominators.

1   In the previous section, we practiced all the skills we need to add and subtract rational expressions with unlike or different denominators. The steps are as follows:

ADDING OR SUBTRACTING RATIONAL EXPRESSIONS WITH UNLIKE DENOMINATORS

Step 1. Find the LCD of the rational expressions.
Step 2. Rewrite each rational expression as an equivalent expression whose denominator is the LCD found in Step 1.
Step 3. Add or subtract numerators and write the sum or difference over the common denominator.
Step 4. Simplify or write the rational expression in simplest form.

Example 1   Perform each indicated operation.

a. $\dfrac{a}{4} - \dfrac{2a}{8}$

b. $\dfrac{3}{10x^2} + \dfrac{7}{25x}$

Solution   **a.** First, we must find the LCD. Since $4 = 2^2$ and $8 = 2^3$, the LCD $= 2^3 = 8$. Next we write each fraction as an equivalent fraction with the denominator 8, then we subtract.

$$\frac{a}{4} - \frac{2a}{8} = \frac{a(2)}{4(2)} - \frac{2a}{8} = \frac{2a}{8} - \frac{2a}{8} = \frac{2a - 2a}{8} = \frac{0}{8} = 0$$

b. Since $10x^2 = 2 \cdot 5 \cdot x \cdot x$ and $25x = 5 \cdot 5 \cdot x$, the LCD $= 2 \cdot 5^2 \cdot x^2 = 50x^2$. We write each fraction as an equivalent fraction with a denominator of $50x^2$.

$$\frac{3}{10x^2} + \frac{7}{25x} = \frac{3(5)}{10x^2(5)} + \frac{7(2x)}{25x(2x)}$$

$$= \frac{15}{50x^2} + \frac{14x}{50x^2}$$

$$= \frac{15 + 14x}{50x^2} \qquad \text{Add numerators. Write the sum over the common denominator.}$$

Example 2 Subtract: $\dfrac{6x}{x^2 - 4} - \dfrac{3}{x + 2}$

Solution Since $x^2 - 4 = (x + 2)(x - 2)$, the LCD $= (x - 2)(x + 2)$. We write equivalent expressions with the LCD as denominators.

$$\frac{6x}{x^2 - 4} - \frac{3}{x + 2} = \frac{6x}{(x - 2)(x + 2)} - \frac{3(x - 2)}{(x + 2)(x - 2)}$$

$$= \frac{6x - 3(x - 2)}{(x + 2)(x - 2)} \quad \text{Subtract numerators. Write the difference over the common denominator.}$$

$$= \frac{6x - 3x + 6}{(x + 2)(x - 2)} \quad \text{Apply the distributive property in the numerator.}$$

$$= \frac{3x + 6}{(x + 2)(x - 2)} \quad \text{Combine like terms in the numerator.}$$

Next we factor the numerator to see if this rational expression can be simplified.

$$= \frac{3(x + 2)}{(x + 2)(x - 2)} \quad \text{Factor.}$$

$$= \frac{3}{x - 2} \quad \text{Apply the fundamental principle to simplify.}$$

Example 3 Add: $\dfrac{2}{3t} + \dfrac{5}{t + 1}$

Solution The LCD is $3t(t + 1)$. We write each rational expression as an equivalent rational expression with a denominator of $3t(t + 1)$.

$$\frac{2}{3t} + \frac{5}{t + 1} = \frac{2(t + 1)}{3t(t + 1)} + \frac{5(3t)}{(t + 1)(3t)}$$

$$= \frac{2(t + 1) + 5(3t)}{3t(t + 1)} \quad \text{Add numerators. Write the sum over the common denominator.}$$

$$= \frac{2t + 2 + 15t}{3t(t + 1)} \quad \text{Apply the distributive property in the numerator.}$$

$$= \frac{17t + 2}{3t(t + 1)} \quad \text{Combine like terms in the numerator.}$$

Example 4 Subtract: $\dfrac{7}{x - 3} - \dfrac{9}{3 - x}$

Solution   To find a common denominator, we notice that $x - 3$ and $3 - x$ are opposites. That is, $3 - x = -(x - 3)$. We write the denominator $3 - x$ as $-(x - 3)$ and simplify.

$$\frac{7}{x - 3} - \frac{9}{3 - x} - \frac{7}{x - 3} - \frac{9}{-(x - 3)}$$

$$= \frac{7}{x - 3} - \frac{-9}{x - 3} \qquad \text{Apply } \frac{a}{-b} = \frac{-a}{b}.$$

$$= \frac{7 - (-9)}{x - 3} \qquad \text{Subtract numerators. Write the difference over the common denominator.}$$

$$= \frac{16}{x - 3}$$

■

Example 5   Add: $1 + \dfrac{m}{m + 1}$

Solution   Recall that 1 is the same as $\dfrac{1}{1}$. The LCD of and $\dfrac{1}{1}$ and $\dfrac{m}{m + 1}$ is $m + 1$.

$$1 + \frac{m}{m + 1} = \frac{1}{1} + \frac{m}{m + 1} \qquad \text{Write 1 as } \tfrac{1}{1}.$$

$$= \frac{1(m + 1)}{1(m + 1)} + \frac{m}{m + 1} \qquad \begin{array}{l}\text{Multiply both the numerator and the}\\ \text{denominator of } \tfrac{1}{1} \text{ by } m + 1.\end{array}$$

$$= \frac{m + 1 + m}{m + 1} \qquad \begin{array}{l}\text{Add numerators. Write the sum over the}\\ \text{common denominator.}\end{array}$$

$$= \frac{2m + 1}{m + 1} \qquad \text{Combine like terms in the numerator.}$$

■

Example 6   Subtract: $\dfrac{3}{2x^2 + x} - \dfrac{2x}{6x + 3}$

Solution   First, we factor the denominators.

$$\frac{3}{2x^2 + x} - \frac{2x}{6x + 3} = \frac{3}{x(2x + 1)} - \frac{2x}{3(2x + 1)}$$

The LCD is $3x(2x + 1)$. We write equivalent expressions with denominators of $3x(2x + 1)$.

$$= \frac{3(3)}{x(2x + 1)(3)} - \frac{2x(x)}{3(2x + 1)(x)}$$

$$= \frac{9 - 2x^2}{3x(2x + 1)} \qquad \begin{array}{l}\text{Subtract numerators. Write the difference}\\ \text{over the common denominator.}\end{array}$$

■

Example 7   Add: $\dfrac{2x}{x^2 + 2x + 1} + \dfrac{x}{x^2 - 1}$

Solution   First we factor the denominators.

$$\frac{2x}{x^2 + 2x + 1} + \frac{x}{x^2 - 1} = \frac{2x}{(x + 1)(x + 1)} + \frac{x}{(x + 1)(x - 1)}$$

Now we write the rational expressions as equivalent expressions with denominators of $(x + 1)(x + 1)(x - 1)$, the LCD.

$$= \frac{2x(x - 1)}{(x + 1)(x + 1)(x - 1)} + \frac{x(x + 1)}{(x + 1)(x - 1)(x + 1)}$$

$$= \frac{2x(x - 1) + x(x + 1)}{(x + 1)^2(x - 1)} \quad \text{Add numerators. Write the sum over the common denominator.}$$

$$= \frac{2x^2 - 2x + x^2 + x}{(x + 1)^2(x - 1)} \quad \text{Apply the distributive property in the numerator.}$$

$$= \frac{3x^2 - x}{(x + 1)^2(x - 1)} \quad \text{or} \quad \frac{x(3x - 1)}{(x + 1)^2(x - 1)}$$

The numerator was factored as a last step to see if the rational expression could be simplified further. Since there are no factors common to the numerator and the denominator, we can't simplify further.

MENTAL MATH

Match each exercise with the first *step needed to perform the operation. Do not actually perform the operation.*

1. $\dfrac{3}{4} - \dfrac{y}{4}$ **2.** $\dfrac{2}{a} \cdot \dfrac{3}{(a + 6)}$ **3.** $\dfrac{x + 1}{x} \div \dfrac{x - 1}{x}$ **4.** $\dfrac{9}{x - 2} - \dfrac{x}{x + 2}$

A. Multiply the first rational expression by the reciprocal of the second rational expression.

B. Find the LCD. Write each expression as an equivalent expression with the LCD as denominator.

C. Multiply numerators, then multiply denominators.

D. Subtract numerators. Place the difference over a common denominator.

Exercise Set 6.4

Perform the indicated operations. See Example 1.

1. $\dfrac{4}{2x} + \dfrac{9}{3x}$ **2.** $\dfrac{15}{7a} + \dfrac{8}{6a}$

3. $\dfrac{15a}{b} + \dfrac{6b}{5}$ **4.** $\dfrac{4c}{d} + \dfrac{8x}{5} - \dfrac{20c + 8dx}{5d}$

5. $\dfrac{3}{x} + \dfrac{5}{2x^2}$ **6.** $\dfrac{14}{3x^2} + \dfrac{6}{x}$

Perform the indicated operations. See Examples 2 and 3.

7. $\dfrac{6}{x + 1} + \dfrac{9}{2x + 2}$ **8.** $\dfrac{8}{x + 4} - \dfrac{3}{3x + 12}$

9. $\dfrac{15}{2x - 4} + \dfrac{x}{x^2 - 4}$ **10.** $\dfrac{3}{x + 2} - \dfrac{1}{x^2 - 4}$

11. $\dfrac{3}{4x} + \dfrac{8}{x - 2}$ **12.** $\dfrac{x}{x + 1} + \dfrac{3}{x - 1}$

13. $\dfrac{5}{y^2} - \dfrac{y}{2y + 1}$ **14.** $\dfrac{x}{4x - 3} - \dfrac{3}{8x - 6}$

15. In your own words, explain how to add two rational expressions with unlike denominators.

16. In your own words, explain how to subtract two rational expressions with unlike denominators.

Add or subtract as indicated. See Example 4.

17. $\dfrac{6}{x - 3} + \dfrac{8}{3 - x}$ **18.** $\dfrac{9}{x - 3} + \dfrac{9}{3 - x}$

19. $\dfrac{-8}{x^2 - 1} - \dfrac{7}{1 - x^2}$ **20.** $\dfrac{-9}{25x^2 - 1} + \dfrac{7}{1 - 25x^2}$

21. $\dfrac{x}{x^2 - 4} - \dfrac{2}{4 - x^2}$ **22.** $\dfrac{5}{2x - 6} - \dfrac{3}{6 - 2x}$

Add or subtract as indicated. See Example 5.

23. $\dfrac{5}{x} + 2$ **24.** $\dfrac{7}{x^2} - 5x$

25. $\dfrac{5}{x - 2} + 6$ **26.** $\dfrac{6y}{y + 5} + 1$

27. $\dfrac{y + 2}{y + 3} - 2$ **28.** $\dfrac{7}{2x - 3} - 3$

△ **29.** Two angles are said to be complementary if their sum is 90°. If one angle measures $\frac{40}{x}$ degrees, find the measure of its complement.

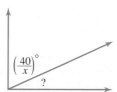

△ **30.** Two angles are said to be supplementary if their sum is 180°. If one angle measures $\frac{x + 2}{x}$ degrees, find the measure of its supplement.

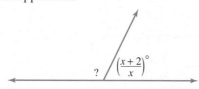

Perform the indicated operations. See Examples 1 through 7.

31. $\dfrac{5x}{x + 2} - \dfrac{3x - 4}{x + 2}$ **32.** $\dfrac{7x}{x - 3} - \dfrac{4x + 9}{x - 3}$

33. $\dfrac{3x^4}{x} - \dfrac{4x^2}{x^2}$ **34.** $\dfrac{5x}{6} + \dfrac{15x^2}{2}$

35. $\dfrac{1}{x + 3} - \dfrac{1}{(x + 3)^2}$ **36.** $\dfrac{5x}{(x - 2)^2} - \dfrac{3}{x - 2}$

37. $\dfrac{4}{5b} + \dfrac{1}{b - 1}$ **38.** $\dfrac{1}{y + 5} + \dfrac{2}{3y}$

39. $\dfrac{2}{m} + 1$ **40.** $\dfrac{6}{x} - 1$

41. $\dfrac{6}{1 - 2x} - \dfrac{4}{2x - 1}$ **42.** $\dfrac{10}{3n - 4} - \dfrac{5}{4 - 3n}$

43. $\dfrac{7}{(x + 1)(x - 1)} + \dfrac{8}{(x + 1)^2}$

44. $\dfrac{5x + 2}{(x + 1)(x + 5)} - \dfrac{2}{x + 5}$

45. $\dfrac{x}{x^2 - 1} - \dfrac{2}{x^2 - 2x + 1}$

46. $\dfrac{x}{x^2 - 4} - \dfrac{5}{x^2 - 4x + 4}$

47. $\dfrac{3a}{2a + 6} - \dfrac{a - 1}{a + 3}$ **48.** $\dfrac{1}{x + y} - \dfrac{y}{x^2 - y^2}$

49. $\dfrac{5}{2 - x} + \dfrac{x}{2x - 4}$ **50.** $\dfrac{-1}{a - 2} + \dfrac{4}{4 - 2a}$

51. $\dfrac{-7}{y^2 - 3y + 2} - \dfrac{2}{y - 1}$ **52.** $\dfrac{2}{x^2 + 4x + 4} + \dfrac{1}{x + 2}$

53. $\dfrac{13}{x^2 - 5x + 6} - \dfrac{5}{x - 3}$ **54.** $\dfrac{27}{y^2 - 81} + \dfrac{3}{2(y + 9)}$

55. $\dfrac{8}{(x + 2)(x - 2)} + \dfrac{4}{(x + 2)(x - 3)}$

56. $\dfrac{5}{6x^2(x + 2)} + \dfrac{4x}{x(x + 2)^2}$ **57.** $\dfrac{5}{9x^2 - 4} + \dfrac{2}{3x - 2}$

58. $\dfrac{4}{x^2 - x - 6} + \dfrac{x}{x^2 + 5x + 6}$

59. $\dfrac{x + 8}{x^2 - 5x - 6} + \dfrac{x + 1}{x^2 - 4x - 5}$

60. $\dfrac{x}{x^2 + 12x + 20} - \dfrac{1}{x^2 + 8x - 20}$

61. A board of length $\frac{3}{x + 4}$ inches was cut into two pieces. If one piece is $\frac{1}{x - 4}$ inches, express the length of the other board as a rational expression.

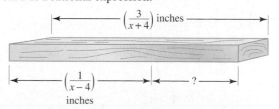

△ **62.** The length of the rectangle is $\frac{3}{y - 5}$ feet, while its width is $\frac{2}{y}$ feet. Find its perimeter and then find its area.

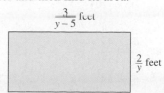

Multiple choice. Select the correct result.

63. $\dfrac{3}{x} + \dfrac{y}{x} =$ **A.** $\dfrac{3+y}{x^2}$ **B.** $\dfrac{3+y}{2x}$ **C.** $\dfrac{3+y}{x}$

64. $\dfrac{3}{x} - \dfrac{y}{x} =$ **A.** $\dfrac{3-y}{x^2}$ **B.** $\dfrac{3-y}{2x}$ **C.** $\dfrac{3-y}{x}$

65. $\dfrac{3}{x} \cdot \dfrac{y}{x} =$ **A.** $\dfrac{3y}{x}$ **B.** $\dfrac{3y}{x^2}$ **C.** $3y$

66. $\dfrac{3}{x} \div \dfrac{y}{x} =$ **A.** $\dfrac{3}{y}$ **B.** $\dfrac{y}{3}$ **C.** $\dfrac{3}{x^2 y}$

Perform the indicated operations. Addition, subtraction, multiplication, and division of rational expressions are included here.

67. $\dfrac{15x}{x+8} \cdot \dfrac{2x+16}{3x}$

68. $\dfrac{9z+5}{15} \cdot \dfrac{5z}{81z^2-25}$

69. $\dfrac{8x+7}{3x+5} - \dfrac{2x-3}{3x+5}$

70. $\dfrac{2z^2}{4z-1} - \dfrac{z-2z^2}{4z-1}$

71. $\dfrac{5a+10}{18} \div \dfrac{a^2-4}{10a}$

72. $\dfrac{9}{x^2-1} \div \dfrac{12}{3x+3}$

73. $\dfrac{5}{x^2-3x+2} + \dfrac{1}{x-2}$

74. $\dfrac{4}{2x^2+5x-3} + \dfrac{2}{x+3}$

75. Explain when the LCD is the product of the denominators.

76. Explain when the LCD is the same as one of the denominators of a rational expression to be added or subtracted.

REVIEW EXERCISES

Simplify. Follow the order shown.

77. $\dfrac{\left.\dfrac{3}{4} + \dfrac{1}{4}\right\}}{\left.\dfrac{3}{8} + \dfrac{13}{8}\right\}}$ ① Add.
       ← ③ Divide.
       ② Add.

78. $\dfrac{\left.\dfrac{9}{5} + \dfrac{6}{5}\right\}}{\left.\dfrac{17}{6} + \dfrac{7}{6}\right\}}$ ① Add.
       ← ③ Divide.
       ② Add.

79. $\dfrac{\left.\dfrac{2}{5} + \dfrac{1}{5}\right\}}{\left.\dfrac{7}{10} + \dfrac{7}{10}\right\}}$ ① Add.
       ← ③ Divide.
       ② Add.

80. $\dfrac{\left.\dfrac{1}{4} + \dfrac{5}{4}\right\}}{\left.\dfrac{3}{8} + \dfrac{7}{8}\right\}}$ ① Add.
       ← ③ Divide.
       ② Add.

Find the slope of each line. See Section 3.4.

81.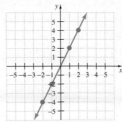

82.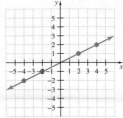

83.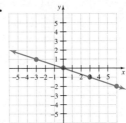

84.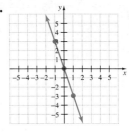

A Look Ahead

Example

Perform the indicated operations:

$$\frac{3}{x^2-16} + \frac{2}{x^2-9x+20} - \frac{4}{x^2-x-20}$$

Solution:

Factor the denominators.

$$= \frac{3}{(x+4)(x-4)} + \frac{2}{(x-4)(x-5)} - \frac{4}{(x-5)(x+4)}$$

Write each expression with an LCD of $(x-4)(x+4)(x-5)$.

$$= \frac{3(x-5)}{(x-4)(x+4)(x-5)} + \frac{2(x+4)}{(x-4)(x+4)(x-5)}$$

$$- \frac{4(x-4)}{(x-4)(x+4)(x-5)}$$

$$= \frac{3(x-5) + 2(x+4) - 4(x-4)}{(x-4)(x+4)(x-5)}$$

$$= \frac{3x - 15 + 2x + 8 - 4x + 16}{(x-4)(x+4)(x-5)}$$

$$= \frac{x+9}{(x-4)(x+4)(x-5)}$$

Add or subtract as indicated.

85. $\dfrac{5}{x^2-4} + \dfrac{2}{x^2-4x+4} - \dfrac{3}{x^2-x-6}$

86. $\dfrac{8}{x^2+6x+5} - \dfrac{3x}{x^2+4x-5} + \dfrac{2}{x^2-1}$

87. $\dfrac{5+x}{x^3-27} + \dfrac{x}{x^3+3x^2+9x}$

88. $\dfrac{x+5}{x^3+1} - \dfrac{3}{2x^2-2x+2}$

6.5   SIMPLIFYING COMPLEX FRACTIONS

CD-ROM   SSM

SSG   Video

▶ **OBJECTIVES**

1. Simplify complex fractions using method 1.
2. Simplify complex fractions using method 2.

1

A rational expression whose numerator or denominator or both numerator and denominator contain fractions is called a **complex rational expression** or a **complex fraction**. Some examples are

$$\frac{4}{2 - \dfrac{1}{2}}, \quad \frac{\dfrac{3}{2}}{\dfrac{4}{7} - x}, \quad \frac{\dfrac{1}{x + 2}}{x + 2 - \dfrac{1}{x}}$$

← Numerator of complex fraction
← Main fraction bar
← Denominator of complex fraction

Our goal in this section is to write complex fractions in simplest form. A complex fraction is in simplest form when it is in the form $\dfrac{P}{Q}$, where P and Q are polynomials that have no common factors.

In this section, two methods of simplifying complex fractions are represented. The first method presented uses the fact that the main fraction bar indicates division.

METHOD 1: SIMPLIFYING A COMPLEX FRACTION

Step 1. Add or subtract fractions in the numerator or denominator so that the numerator is a single fraction and the denominator is a single fraction.

Step 2. Perform the indicated division by multiplying the numerator of the complex fraction by the reciprocal of the denominator of the complex fraction.

Step 3. Write the rational expression in simplest form.

Example 1   Simplify the complex fraction $\dfrac{\dfrac{5}{8}}{\dfrac{2}{3}}$.

Solution   Since the numerator and denominator of the complex fraction are already single fractions, we proceed to step 2: perform the indicated division by multiplying the numerator $\dfrac{5}{8}$ by the reciprocal of the denominator $\dfrac{2}{3}$.

$$\frac{\dfrac{5}{8}}{\dfrac{2}{3}} = \frac{5}{8} \cdot \frac{3}{2} = \frac{15}{16}$$

The reciprocal of $\frac{2}{3}$ is $\frac{3}{2}$.

Example 2 Simplify: $\dfrac{\dfrac{2}{3} + \dfrac{1}{5}}{\dfrac{2}{3} - \dfrac{2}{9}}$.

Solution Simplify above and below the main fraction bar separately. First, add $\dfrac{2}{3}$ and $\dfrac{1}{5}$ to obtain a single fraction in the numerator; then subtract $\dfrac{2}{9}$ from $\dfrac{2}{3}$ to obtain a single fraction in the denominator.

$$\dfrac{\dfrac{2}{3} + \dfrac{1}{5}}{\dfrac{2}{3} - \dfrac{2}{9}} = \dfrac{\dfrac{2(5)}{3(5)} + \dfrac{1(3)}{5(3)}}{\dfrac{2(3)}{3(3)} - \dfrac{2}{9}}$$
The LCD of the numerator's fractions is 15.

The LCD of the denominator's fractions is 9.

$$= \dfrac{\dfrac{10}{15} + \dfrac{3}{15}}{\dfrac{6}{9} - \dfrac{2}{9}}$$
Simplify.

$$= \dfrac{\dfrac{13}{15}}{\dfrac{4}{9}}$$
Add the numerator's fractions.

Subtract the denominator's fractions.

Next, perform the indicated division by multiplying the numerator of the complex fraction by the reciprocal of the denominator of the complex fraction.

$$\dfrac{\dfrac{13}{15}}{\dfrac{4}{9}} = \dfrac{13}{15} \cdot \dfrac{9}{4}$$
The reciprocal of $\dfrac{4}{9}$ is $\dfrac{9}{4}$.

$$= \dfrac{13 \cdot 3 \cdot \boxed{3}}{\boxed{3} \cdot 5 \cdot 4} = \dfrac{39}{20}$$

Example 3 Simplify: $\dfrac{\dfrac{1}{z} - \dfrac{1}{2}}{\dfrac{1}{3} - \dfrac{z}{6}}$.

Solution Subtract to get a single fraction in the numerator and a single fraction in the denominator of the complex fraction.

$$\dfrac{\dfrac{1}{z} - \dfrac{1}{2}}{\dfrac{1}{3} - \dfrac{z}{6}} = \dfrac{\dfrac{2}{2z} - \dfrac{z}{2z}}{\dfrac{2}{6} - \dfrac{z}{6}}$$
The LCD of the numerator's fractions is $2z$.

The LCD of the denominator's fractions is 6.

$$= \frac{\dfrac{2-z}{2z}}{\dfrac{2-z}{6}}$$

$$= \frac{2-z}{2z} \cdot \frac{6}{2-z} \qquad \text{Multiply by the reciprocal of } \frac{2-z}{6}.$$

$$= \frac{2 \cdot 3 \cdot (2-z)}{2 \cdot z \cdot (2-z)} \qquad \text{Factor.}$$

$$= \frac{3}{z} \qquad \text{Write in simplest form.} \quad \blacksquare$$

2 Next we study a second method for simplifying complex fractions. In this method, we multiply the numerator and the denominator of the complex fraction by the LCD of all fractions in the complex fraction.

> ### METHOD 2: SIMPLIFYING A COMPLEX FRACTION
>
> **Step 1.** Find the LCD of all the fractions in the complex fraction.
> **Step 2.** Multiply both the numerator and the denominator of the complex fraction by the LCD from Step 1.
> **Step 3.** Perform the indicated operations and write the result in simplest form.

We use method 2 to rework Example 2.

Example 4 Simplify: $\dfrac{\dfrac{2}{3} + \dfrac{1}{5}}{\dfrac{2}{3} - \dfrac{2}{9}}$

Solution The LCD of $\dfrac{2}{3}, \dfrac{1}{5}, \dfrac{2}{3}$ and $\dfrac{2}{9}$ is 45, so we multiply the numerator and the denominator of the complex fraction by 45. Then we perform the indicated operations, and write in simplest form.

$$\frac{\dfrac{2}{3} + \dfrac{1}{5}}{\dfrac{2}{3} - \dfrac{2}{9}} = \frac{45\left(\dfrac{2}{3} + \dfrac{1}{5}\right)}{45\left(\dfrac{2}{3} - \dfrac{2}{9}\right)}$$

$$= \frac{45\left(\dfrac{2}{3}\right) + 45\left(\dfrac{1}{5}\right)}{45\left(\dfrac{2}{3}\right) - 45\left(\dfrac{2}{9}\right)} \qquad \text{Apply the distributive property.}$$

$$= \frac{30 + 9}{30 - 10} = \frac{39}{20} \qquad \text{Simplify.} \quad \blacksquare$$

▼
HELPFUL HINT
The same complex fraction was simplified using two different methods in
Examples 2 and 4. Notice that each time the simplified result is the same.

Example 5 Simplify: $\dfrac{\dfrac{x+1}{y}}{\dfrac{x}{y}+2}$

Solution The LCD of $\dfrac{x+1}{y}, \dfrac{x}{y}$, and $\dfrac{2}{1}$ is y, so we multiply the numerator and the denominator of the complex fraction by y.

$$\dfrac{\dfrac{x+1}{y}}{\dfrac{x}{y}+2} = \dfrac{y\left(\dfrac{x+1}{y}\right)}{y\left(\dfrac{x}{y}+2\right)}$$

$$= \dfrac{y\left(\dfrac{x+1}{y}\right)}{y\left(\dfrac{x}{y}\right)+y\cdot 2} \qquad \text{Apply the distributive property in the denominator.}$$

$$= \dfrac{x+1}{x+2y} \qquad \text{Simplify.}$$

■

Example 6 Simplify: $\dfrac{\dfrac{x}{y}+\dfrac{3}{2x}}{\dfrac{x}{2}+y}$

Solution The LCD of $\dfrac{x}{y}, \dfrac{3}{2x}, \dfrac{x}{2}$, and $\dfrac{y}{1}$ is $2xy$, so we multiply both the numerator and the denominator of the complex fraction by $2xy$.

$$\dfrac{\dfrac{x}{y}+\dfrac{3}{2x}}{\dfrac{x}{2}+y} = \dfrac{2xy\left(\dfrac{x}{y}+\dfrac{3}{2x}\right)}{2xy\left(\dfrac{x}{2}+y\right)}$$

$$= \dfrac{2xy\left(\dfrac{x}{y}\right)+2xy\left(\dfrac{3}{2x}\right)}{2xy\left(\dfrac{x}{2}\right)+2xy(y)} \qquad \text{Apply the distributive property.}$$

$$= \dfrac{2x^2+3y}{x^2y+2xy^2}$$

$$\text{or } \dfrac{2x^2+3y}{xy(x+2y)}$$

■

MENTAL MATH

Complete the steps by stating the simplified complex fraction.

1. $\dfrac{\dfrac{y}{2}}{\dfrac{5x}{2}} = \dfrac{2\left(\dfrac{y}{2}\right)}{2\left(\dfrac{5x}{2}\right)} = \dfrac{?}{?}$

2. $\dfrac{\dfrac{10}{x}}{\dfrac{z}{x}} = \dfrac{x\left(\dfrac{10}{x}\right)}{x\left(\dfrac{z}{x}\right)} = \dfrac{?}{?}$

3. $\dfrac{\dfrac{3}{x}}{\dfrac{5}{x^2}} = \dfrac{x^2\left(\dfrac{3}{x}\right)}{x^2\left(\dfrac{5}{x^2}\right)} = \dfrac{?}{?}$

4. $\dfrac{\dfrac{a}{10}}{\dfrac{b}{20}} = \dfrac{20\left(\dfrac{a}{10}\right)}{20\left(\dfrac{b}{20}\right)} = \dfrac{?}{?}$

Exercise Set 6.5

Simplify each complex fraction. See Examples 1 through 6.

1. $\dfrac{\dfrac{1}{2}}{\dfrac{3}{4}}$

2. $\dfrac{\dfrac{1}{8}}{-\dfrac{5}{12}}$

3. $\dfrac{-\dfrac{4x}{9}}{-\dfrac{2x}{3}}$

4. $\dfrac{-\dfrac{6y}{11}}{\dfrac{4y}{9}}$

5. $\dfrac{\dfrac{1+x}{6}}{\dfrac{1+x}{3}}$

6. $\dfrac{\dfrac{6x-3}{5x^2}}{\dfrac{2x-1}{10x}}$

7. $\dfrac{\dfrac{1}{2}+\dfrac{2}{3}}{\dfrac{5}{9}-\dfrac{5}{6}}$

8. $\dfrac{\dfrac{3}{4}-\dfrac{1}{2}}{\dfrac{3}{8}+\dfrac{1}{6}}$

9. $\dfrac{2+\dfrac{7}{10}}{1+\dfrac{3}{5}}$

10. $\dfrac{4-\dfrac{11}{12}}{5+\dfrac{1}{4}}$

11. $\dfrac{\dfrac{1}{3}}{\dfrac{1}{2}-\dfrac{1}{4}}$

12. $\dfrac{\dfrac{7}{10}-\dfrac{3}{5}}{\dfrac{1}{2}}$

13. $\dfrac{-\dfrac{2}{9}}{-\dfrac{14}{3}}$

14. $\dfrac{\dfrac{3}{8}}{\dfrac{4}{15}}$

15. $\dfrac{-\dfrac{5}{12x^2}}{\dfrac{25}{16x^3}}$

16. $\dfrac{-\dfrac{7}{8y}}{\dfrac{21}{4y}}$

17. $\dfrac{\dfrac{m}{n}-1}{\dfrac{m}{n}+1}$

18. $\dfrac{\dfrac{x}{2}+2}{\dfrac{x}{2}-2}$

19. $\dfrac{\dfrac{1}{5}-\dfrac{1}{x}}{\dfrac{7}{10}+\dfrac{1}{x^2}}$

20. $\dfrac{\dfrac{1}{y^2}+\dfrac{2}{3}}{\dfrac{1}{y}-\dfrac{5}{6}}$

21. $\dfrac{1+\dfrac{1}{y-2}}{y+\dfrac{1}{y-2}}$

22. $\dfrac{x-\dfrac{1}{2x+1}}{1-\dfrac{x}{2x+1}}$

23. $\dfrac{\dfrac{4y-8}{16}}{\dfrac{6y-12}{4}}$

24. $\dfrac{\dfrac{7y+21}{3}}{\dfrac{3y+9}{8}}$

25. $\dfrac{\dfrac{x}{y}+1}{\dfrac{x}{y}-1}$

26. $\dfrac{\dfrac{3}{5y}+8}{\dfrac{3}{5y}-8}$

27. $\dfrac{1}{2+\dfrac{1}{3}}$

28. $\dfrac{3}{1-\dfrac{4}{3}}$

29. $\dfrac{\dfrac{ax+ab}{x^2-b^2}}{\dfrac{x+b}{x-b}}$

30. $\dfrac{\dfrac{m+2}{m-2}}{\dfrac{2m+4}{m^2-4}}$

31. $\dfrac{\dfrac{-3 + y}{4}}{\dfrac{8 + y}{28}}$

32. $\dfrac{\dfrac{-x + 2}{18}}{\dfrac{8}{9}}$

33. $\dfrac{3 + \dfrac{12}{x}}{1 - \dfrac{16}{x^2}}$

34. $\dfrac{2 + \dfrac{6}{x}}{1 - \dfrac{9}{x^2}}$

35. $\dfrac{\dfrac{8}{x + 4} + 2}{\dfrac{12}{x + 4} - 2}$

36. $\dfrac{\dfrac{25}{x + 5} + 5}{\dfrac{3}{x + 5} - 5}$

37. $\dfrac{\dfrac{s}{r} + \dfrac{r}{s}}{\dfrac{s}{r} - \dfrac{r}{s}}$

38. $\dfrac{\dfrac{2}{x} + \dfrac{x}{2}}{\dfrac{2}{x} - \dfrac{x}{2}}$

39. Explain how to simplify a complex fraction using method 1.

40. Explain how to simplify a complex fraction using method 2.

To find the average of two numbers, we find their sum and divide by 2. For example, the average of 65 and 81 is found by simplifying $\frac{65 + 81}{2}$. This simplifies to $\frac{146}{2} = 73$.

41. Find the average of $\frac{1}{3}$ and $\frac{3}{4}$.

42. Write the average of $\dfrac{3}{n}$ and $\dfrac{5}{n^2}$ as a simplified rational expression.

Solve.

43. In electronics, when two resistors R_1 (read R sub 1) and R_2 (read R sub 2) are connected in parallel, the total resistance is given by the complex fraction $\dfrac{1}{\dfrac{1}{R_1} + \dfrac{1}{R_2}}$. Simplify this expression.

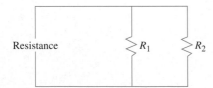

Resistance R_1 R_2

44. Astronomers occasionally need to know the day of the week a particular date fell on. The complex fraction

$$\dfrac{J + \dfrac{3}{2}}{7},$$

where J is the *Julian day number*, is used to make this calculation. Simplify this expression.

45. If the distance formula $d = r \cdot t$ is solved for t, then $t = \frac{d}{r}$. Use this formula to find t if distance d is $\frac{20x}{3}$ miles and rate r is $\frac{5x}{9}$ miles per hour. Write t in simplified form.

△ **46.** If the formula for area of a rectangle, $A = l \cdot w$, is solved for w, then $w = \frac{A}{l}$. Use this formula to find w if area A is $\frac{4x - 2}{3}$ square meters and length l is $\frac{6x - 3}{5}$ meters. Write w in simplified form.

REVIEW EXERCISES

Solve the following linear and quadratic equations. See Sections 2.4 and 5.6.

47. $3x + 5 = 7$

48. $5x - 1 = 8$

49. $2x^2 - x - 1 = 0$

50. $4x^2 - 9 = 0$

Simplify the following rational expressions. See Section 6.1.

51. $\dfrac{2 + x}{x + 2}$

52. $\dfrac{y^2 + y}{y + y^2}$

53. $\dfrac{2 - x}{x - 2}$

54. $\dfrac{z - 4}{4 - z}$

A Look Ahead

Example

Simplify $\dfrac{1 + x^{-1}}{x^{-1}}$.

Solution:

$$\dfrac{1 + x^{-1}}{x^{-1}} = \dfrac{1 + \dfrac{1}{x}}{\dfrac{1}{x}}$$

$$= \dfrac{x\left(1 + \dfrac{1}{x}\right)}{x\left(\dfrac{1}{x}\right)}$$

$$= \dfrac{x + 1}{1} \text{ or } x + 1$$

Simplify the following.

55. $\dfrac{x^{-1} + 2^{-1}}{x^{-2} - 4^{-1}}$

56. $\dfrac{3^{-1} - x^{-1}}{9^{-1} - x^{-2}}$

57. $\dfrac{x + y^{-1}}{\dfrac{x}{y}}$

58. $\dfrac{x - xy^{-1}}{\dfrac{1 + x}{y}}$

59. $\dfrac{y^{-2}}{1 - y^{-2}}$

60. $\dfrac{4 + x^{-1}}{3 + x^{-1}}$

6.6 SOLVING EQUATIONS CONTAINING RATIONAL EXPRESSIONS

▶ **OBJECTIVES**

CD-ROM SSM

SSG Video

1. Solve equations containing rational expressions.
2. Understand the difference between solving an equation and performing arithmetic operations on rational expressions.
3. Solve equations containing rational expressions for a specified variable.

1 In Chapter 2, we solved equations containing fractions. In this section, we continue the work we began in Chapter 2 by solving equations containing rational expressions.

Examples of Equations Containing Rational Expressions

$$\frac{x}{5} + \frac{x+2}{9} = 8 \quad \text{and} \quad \frac{x+1}{9x-5} = \frac{2}{3x}$$

To solve equations such as these, use the multiplication property of equality to clear the equation of fractions by multiplying both sides of the equation by the LCD.

Example 1 Solve: $\dfrac{x}{2} + \dfrac{8}{3} = \dfrac{1}{6}$

Solution The LCD of denominators 2, 3, and 6 is 6, so we multiply both sides of the equation by 6.

$$6\left(\frac{x}{2} + \frac{8}{3}\right) = 6\left(\frac{1}{6}\right)$$

HELPFUL HINT
Make sure that *each* term is multiplied by the LCD.

$$6\left(\frac{x}{2}\right) + 6\left(\frac{8}{3}\right) = 6\left(\frac{1}{6}\right) \quad \text{Use the distributive property.}$$

$$3 \cdot x + 16 = 1 \quad \text{Multiply and simplify.}$$

$$3x = -15 \quad \text{Subtract 16 from both sides.}$$

$$x = -5 \quad \text{Divide both sides by 3.}$$

Check To check, we replace x with -5 in the original equation.

$$\frac{-5}{2} + \frac{8}{3} \stackrel{?}{=} \frac{1}{6} \quad \text{Replace } x \text{ with } -5.$$

$$\frac{1}{6} = \frac{1}{6} \quad \text{True.}$$

This number checks, so the solution is -5.

Example 2 Solve: $\dfrac{t-4}{2} - \dfrac{t-3}{9} = \dfrac{5}{18}$

Solution The LCD of denominators 2, 9, and 18 is 18, so we multiply both sides of the equation by 18.

$$18\left(\frac{t-4}{2} - \frac{t-3}{9}\right) = 18\left(\frac{5}{18}\right)$$

$$18\left(\frac{t-4}{2}\right) - 18\left(\frac{t-3}{9}\right) = 18\left(\frac{5}{18}\right) \qquad \text{Use the distributive property.}$$

> **HELPFUL HINT**
> Multiply *each* term by 18.

$$9(t-4) - 2(t-3) = 5 \qquad \text{Simplify.}$$

$$9t - 36 - 2t + 6 = 5 \qquad \text{Use the distributive property.}$$

$$7t - 30 = 5 \qquad \text{Combine like terms.}$$

$$7t = 35$$

$$t = 5 \qquad \text{Solve for } t.$$

Check

$$\frac{t-4}{2} - \frac{t-3}{9} = \frac{5}{18}$$

$$\frac{5-4}{2} - \frac{5-3}{9} \stackrel{?}{=} \frac{5}{18} \qquad \text{Replace } t \text{ with 5.}$$

$$\frac{1}{2} - \frac{2}{9} \stackrel{?}{=} \frac{5}{18} \qquad \text{Simplify.}$$

$$\frac{5}{18} = \frac{5}{18} \qquad \text{True.}$$

The solution is 5.

Recall from Section 5.1 that a rational expression is defined for all real numbers except those that make the denominator of the expression 0. This means that if an equation contains *rational expressions with variables in the denominator*, we must be certain that the proposed solution does not make the denominator 0. If replacing the variable with the proposed solution makes the denominator 0, the rational expression is undefined and this proposed solution must be rejected.

Example 3 Solve: $3 - \dfrac{6}{x} = x + 8$

Solution In this equation, 0 cannot be a solution because if x is 0, the rational expression $\dfrac{6}{x}$ is undefined. The LCD is x, so we multiply both sides of the equation by x.

$$x\left(3 - \frac{6}{r}\right) = x(x + 8)$$

> **HELPFUL HINT**
> Multiply *each* term by x.

$$x(3) - x\left(\frac{6}{x}\right) = x \cdot x + x \cdot 8 \qquad \text{Use the distributive property.}$$

$$3x - 6 = x^2 + 8x \qquad \text{Simplify.}$$

Now we write the quadratic equation in standard form and solve for x.

$$0 = x^2 + 5x + 6$$

$$0 = (x + 3)(x + 2) \qquad \text{Factor.}$$

$$x + 3 = 0 \qquad \text{or} \qquad x + 2 = 0 \qquad \text{Set each factor equal to 0 and solve.}$$

$$x = -3 \qquad\qquad\qquad x = -2$$

Notice that neither -3 nor -2 makes the denominator in the original equation equal to 0.

Check To check these solutions, we replace x in the original equation by -3, and then by -2.

If $x = -3$:

$$3 - \frac{6}{x} = x + 8$$

$$3 - \frac{6}{-3} \overset{?}{=} -3 + 8$$

$$3 - (-2) \overset{?}{=} 5$$

$$5 = 5 \qquad \text{True.}$$

If $x = -2$:

$$3 - \frac{6}{x} = x + 8$$

$$3 - \frac{6}{-2} \overset{?}{=} -2 + 8$$

$$3 - (-3) \overset{?}{=} 6$$

$$6 = 6 \qquad \text{True.}$$

Both -3 and -2 are solutions.

The following steps may be used to solve an equation containing rational expressions.

SOLVING AN EQUATION CONTAINING RATIONAL EXPRESSIONS

Step 1. Multiply both sides of the equation by the LCD of all rational expressions in the equation.

Step 2. Remove any grouping symbols and solve the resulting equation.

Step 3. Check the solution in the original equation.

Example 4 Solve: $\dfrac{4x}{x^2 - 25} + \dfrac{2}{x - 5} = \dfrac{1}{x + 5}$

Solution The denominator $x^2 - 25$ factors as $(x + 5)(x - 5)$. The LCD is then $(x + 5)(x - 5)$, so we multiply both sides of the equation by this LCD.

$$(x + 5)(x - 5)\left(\frac{4x}{(x + 5)(x - 5)} + \frac{2}{x - 5}\right) = (x + 5)(x - 5)\left(\frac{1}{x + 5}\right)$$

Multiply by the LCD. Notice that -5 and 5 cannot be solutions.

$$(x + 5)(x - 5) \cdot \frac{4x}{x^2 - 25} + (x + 5)(x - 5) \cdot \frac{2}{x - 5}$$

Use the distributive property.

$$= (x + 5)(x - 5) \cdot \frac{1}{x + 5}$$

$$4x + 2(x + 5) = x - 5 \qquad \text{Simplify.}$$
$$4x + 2x + 10 = x - 5 \qquad \text{Use the distributive property.}$$
$$6x + 10 = x - 5 \qquad \text{Combine like terms.}$$
$$5x = -15$$
$$x = -3 \qquad \text{Divide both sides by 5.}$$

Check Check by replacing x with -3 in the original equation. The solution is -3. ▬

Example 5 Solve: $\dfrac{2x}{x - 4} = \dfrac{8}{x - 4} + 1$

Solution Multiply both sides by the LCD, $x - 4$.

$$(x - 4)\left(\frac{2x}{x - 4}\right) = (x - 4)\left(\frac{8}{x - 4} + 1\right)$$

Multiply by the LCD. Notice that 4 cannot be a solution.

$$(x - 4) \cdot \frac{2x}{x - 4} = (x - 4) \cdot \frac{8}{x - 4} + (x - 4) \cdot 1$$

Use the distributive property.

$$2x = 8 + (x - 4) \qquad \text{Simplify.}$$
$$2x = 4 + x$$
$$x = 4$$

Notice that 4 makes the denominator 0 in the original equation. Therefore, 4 is *not* a solution and this equation has *no solution*. ▬

> ▼
> **HELPFUL HINT**
> As we can see from Example 5, it is important to check the proposed solution(s) in the *original* equation.

Example 6   Solve: $x + \dfrac{14}{x-2} = \dfrac{7x}{x-2} + 1$

Solution   Notice the denominators in this equation. We can see that 2 can't be a solution. The LCD is $x - 2$, so we multiply both sides of the equation by $x - 2$.

$$(x-2)\left(x + \frac{14}{x-2}\right) = (x-2)\left(\frac{7x}{x-2} + 1\right)$$

$$(x-2)(x) + (x-2)\left(\frac{14}{x-2}\right) = (x-2)\left(\frac{7x}{x-2}\right) + (x-2)(1)$$

$$x^2 - 2x + 14 = 7x + x - 2 \qquad \text{Simplify.}$$

$$x^2 - 2x + 14 = 8x - 2 \qquad \text{Combine like terms.}$$

$$x^2 - 10x + 16 = 0 \qquad \begin{array}{l}\text{Write the quadratic equa-}\\ \text{tion in standard form.}\end{array}$$

$$(x-8)(x-2) = 0 \qquad \text{Factor.}$$

$$x - 8 = 0 \quad \text{or} \quad x - 2 = 0 \qquad \text{Set each factor equal to 0.}$$

$$x = 8 \qquad\qquad\quad x = 2 \qquad \text{Solve.}$$

As we have already noted, 2 can't be a solution of the original equation. So we need only replace x with 8 in the original equation. We find that 8 is a solution; the only solution is 8.   ■

2   At this point, let's make sure you understand the difference between solving an equation containing rational expressions and performing operations on rational expressions.

Example 7   **a.** Solve for x: $\dfrac{x}{4} + 2x = 9$.   **b.** Add: $\dfrac{x}{4} + 2x$.

Solution   **a.** This is an equation to solve for x. Begin by multiplying both sides by the LCD, 4.

$$4\left(\frac{x}{4} + 2x\right) = 4(9)$$

$$4\left(\frac{x}{4}\right) + 4(2x) = 4(9) \qquad \text{Apply the distributive property.}$$

$$x + 8x = 36 \qquad \text{Simplify.}$$

$$9x = 36 \qquad \text{Combine like terms.}$$

$$x = 4 \qquad \text{Solve.}$$

Check to see that 4 is the solution.

b. This example is **not an equation** to solve; it is an addition to perform. To add these rational expressions, find the LCD and write each rational expression as an equivalent expression whose denominator is the LCD. The LCD is 4.

$$\frac{x}{4} + 2x = \frac{x}{4} + \frac{2x(4)}{4}$$

$$= \frac{x + 8x}{4} \qquad \text{Add.}$$

$$= \frac{9x}{4} \qquad \text{Combine like terms in the numerator.} \quad \blacksquare$$

3   The last example in this section is an equation containing several variables. We are directed to solve for one of them. The steps used in the preceeding examples can be applied to solve equations for a specified variable as well.

Example 8   Solve $\dfrac{1}{a} + \dfrac{1}{b} = \dfrac{1}{x}$ for x.

Solution   (This type of equation often models a work problem, as we shall see in Section 6.8.) The LCD is abx, so we multiply both sides by abx.

$$abx\left(\frac{1}{a} + \frac{1}{b}\right) = abx\left(\frac{1}{x}\right)$$

$$abx\left(\frac{1}{a}\right) + abx\left(\frac{1}{b}\right) = abx \cdot \frac{1}{x}$$

$$bx + ax = ab \qquad \text{Simplify.}$$

$$x(b + a) = ab \qquad \text{Factor out } x \text{ from each term on the left side.}$$

$$\frac{x(b + a)}{b + a} = \frac{ab}{b + a} \qquad \text{Divide both sides by } b + a.$$

$$x = \frac{ab}{b + a} \qquad \text{Simplify.}$$

This equation is now solved for x. $\blacksquare$

GRAPHING CALCULATOR EXPLORATIONS

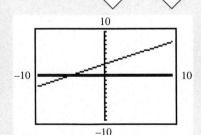

A grapher may be used to check solutions of equations containing rational expressions. For example, to check the solution of Example 1, $\dfrac{x}{2} + \dfrac{8}{3} = \dfrac{1}{6}$, graph $y_1 = x/2 + 8/3$ and $y_2 = 1/6$.

Use TRACE and ZOOM, or use INTERSECT, to find the point of intersection. The point of intersection has an x-value of -5, so the solution of the equation is -5.

Use a grapher to check the examples of this section.

  1. Example 2   **2.** Example 7a.   **3.** Example 3

  4. Example 4   **5.** Example 5   **6.** Example 6

MENTAL MATH

Solve each equation for the variable.

1. $\dfrac{x}{5} = 2$ **2.** $\dfrac{x}{8} = 4$ **3.** $\dfrac{z}{6} = 0$ **4.** $\dfrac{y}{7} = 8$

Exercise Set 6.6

Solve each equation. See Examples 1 and 2.

1. $\dfrac{x}{5} + 3 = 9$

2. $\dfrac{x}{5} - 2 = 9$

3. $\dfrac{x}{2} + \dfrac{5x}{4} = \dfrac{x}{12}$

4. $\dfrac{x}{6} + \dfrac{4x}{3} = \dfrac{x}{18}$

5. $2 + \dfrac{10}{x} = x + 5$

6. $6 + \dfrac{5}{y} = y - \dfrac{2}{y}$

7. $\dfrac{a}{5} = \dfrac{a - 3}{2}$

8. $\dfrac{2b}{5} = \dfrac{b + 2}{6}$

9. $\dfrac{x - 3}{5} + \dfrac{x - 2}{2} = \dfrac{1}{2}$

10. $\dfrac{a + 5}{4} + \dfrac{a + 5}{2} = \dfrac{a}{8}$

Recall that two angles are supplementary if the sum of their measures is 180°. Find the measures of the following supplementary angles.

△ **11.**

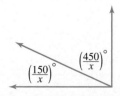
$\left(\dfrac{20x}{3}\right)°$ $\left(\dfrac{32x}{6}\right)°$

△ **12.**

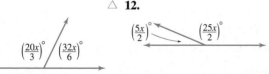
$\left(\dfrac{5x}{2}\right)°$ $\left(\dfrac{25x}{2}\right)°$

Recall that two angles are complementary if the sum of their measures is 90°. Find the measures of the following complementary angles.

△ **13.**

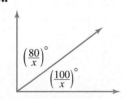
$\left(\dfrac{450}{x}\right)°$ $\left(\dfrac{150}{x}\right)°$

△ **14.**

$\left(\dfrac{80}{x}\right)°$ $\left(\dfrac{100}{x}\right)°$

Solve each equation. See Examples 3 through 6.

15. $\dfrac{9}{2a - 5} = -2$

16. $\dfrac{6}{4 - 3x} = 3$

17. $\dfrac{y}{y + 4} + \dfrac{4}{y + 4} = 3$

18. $\dfrac{5y}{y + 1} - \dfrac{3}{y + 1} = 4$

19. $\dfrac{2x}{x + 2} - 2 = \dfrac{x - 8}{x - 2}$

20. $\dfrac{4y}{y - 3} - 3 = \dfrac{3y - 1}{y + 3}$

21. $\dfrac{4y}{y - 4} + 5 = \dfrac{5y}{y - 4}$

22. $\dfrac{2a}{a + 2} - 5 = \dfrac{7a}{a + 2}$

23. $\dfrac{7}{x - 2} + 1 = \dfrac{x}{x + 2}$

24. $1 + \dfrac{3}{x + 1} = \dfrac{x}{x - 1}$

25. $\dfrac{x + 1}{x + 3} = \dfrac{2x^2 - 15x}{x^2 + x - 6} - \dfrac{x - 3}{x - 2}$

26. $\dfrac{3}{x + 3} = \dfrac{12x + 19}{x^2 + 7x + 12} - \dfrac{5}{x + 4}$

27. $\dfrac{y}{2y + 2} + \dfrac{2y - 16}{4y + 4} = \dfrac{2y - 3}{y + 1}$

28. $\dfrac{1}{x + 2} = \dfrac{4}{x^2 - 4} - \dfrac{1}{x - 2}$

Determine whether each of the following is an equation or an expression. If it is an equation, then solve it for its variable. If it is an expression, perform the indicated operation. See Example 7.

29. $\dfrac{1}{x} + \dfrac{2}{3}$

30. $\dfrac{3}{a} + \dfrac{5}{6}$

31. $\dfrac{1}{x} + \dfrac{2}{3} = \dfrac{3}{x}$

32. $\dfrac{3}{a} + \dfrac{5}{6} = 1$

33. $\dfrac{2}{x + 1} - \dfrac{1}{x}$

34. $\dfrac{4}{x - 3} - \dfrac{1}{x}$

35. $\dfrac{2}{x + 1} - \dfrac{1}{x} = 1$

36. $\dfrac{4}{x - 3} - \dfrac{1}{x} = \dfrac{6}{x(x - 3)}$

37. Explain the difference between solving an equation such as $\dfrac{x}{2} + \dfrac{3}{4} = \dfrac{x}{4}$ for x and performing an operation such as adding $\dfrac{x}{2} + \dfrac{3}{4}$.

38. When solving an equation such as $\dfrac{y}{4} = \dfrac{y}{2} - \dfrac{1}{4}$, we may multiply all terms by 4. When subtracting two rational expressions such as $\dfrac{y}{2} - \dfrac{1}{4}$, we may not. Explain why.

Solve each equation.

39. $\dfrac{2x}{7} - 5x = 9$

40. $\dfrac{4x}{8} - 5x = 10$

 41. $\dfrac{2}{y} + \dfrac{1}{2} = \dfrac{5}{2y}$

42. $\dfrac{6}{3y} + \dfrac{3}{y} = 1$

43. $\dfrac{4x + 10}{7} = \dfrac{8}{2}$

44. $\dfrac{1}{2} = \dfrac{x + 1}{8}$

45. $2 + \dfrac{3}{a - 3} = \dfrac{a}{a - 3}$

46. $\dfrac{2y}{y - 2} - \dfrac{4}{y - 2} = 4$

47. $\dfrac{5}{x} + \dfrac{2}{3} = \dfrac{7}{2x}$

48. $\dfrac{5}{3} - \dfrac{3}{2x} = \dfrac{5}{4}$

49. $\dfrac{2a}{a + 4} = \dfrac{3}{a - 1}$

50. $\dfrac{5}{3x - 8} = \dfrac{x}{x - 2}$

51. $\dfrac{x + 1}{3} - \dfrac{x - 1}{6} = \dfrac{1}{6}$

52. $\dfrac{3x}{5} - \dfrac{x - 6}{3} = \dfrac{1}{5}$

53. $\dfrac{4r - 1}{r^2 + 5r - 14} + \dfrac{2}{r + 7} = \dfrac{1}{r - 2}$

54. $\dfrac{2t + 3}{t - 1} - \dfrac{2}{t + 3} = \dfrac{5 - 6t}{t^2 + 2t - 3}$

55. $\dfrac{t}{t - 4} = \dfrac{t + 4}{6}$

56. $\dfrac{15}{x + 4} = \dfrac{x - 4}{x}$

57. $\dfrac{x}{2x + 6} + \dfrac{x + 1}{3x + 9} = \dfrac{2}{4x + 12}$

58. $\dfrac{a}{5a - 5} - \dfrac{a - 2}{2a - 2} = \dfrac{5}{4a - 4}$

Solve each equation for the indicated variable. See Example 8.

59. $\dfrac{D}{R} = T$; for R

△ **60.** $\dfrac{A}{W} = L$; for W

61. $\dfrac{3}{x} = \dfrac{5y}{x + 2}$; for y

62. $\dfrac{7x - 1}{2x} = \dfrac{5}{y}$; for y

63. $\dfrac{3a + 2}{3b - 2} = -\dfrac{4}{2a}$; for b

64. $\dfrac{6x + y}{7x} = \dfrac{3x}{h}$; for h

△ **65.** $\dfrac{A}{BH} = \dfrac{1}{2}$; for B

△ **66.** $\dfrac{V}{\pi r^2 h} = 1$; for h

△ **67.** $\dfrac{C}{\pi r} = 2$; for r

△ **68.** $\dfrac{3V}{A} = H$; for V

69. $\dfrac{1}{a} = \dfrac{1}{b} + \dfrac{1}{c}$; for a

70. $\dfrac{1}{2} - \dfrac{1}{x} = \dfrac{1}{y}$; for x

71. $\dfrac{m^2}{6} - \dfrac{n}{3} = \dfrac{p}{2}$; for n

72. $\dfrac{x^2}{r} + \dfrac{y^2}{t} = 1$; for r

Solve each equation.

73. $\dfrac{5}{a^2 + 4a + 3} + \dfrac{2}{a^2 + a - 6} - \dfrac{3}{a^2 - a - 2} = 0$

74. $-\dfrac{2}{a^2 + 2a - 8} + \dfrac{1}{a^2 + 9a + 20} = \dfrac{-4}{a^2 + 3a - 10}$

REVIEW EXERCISES

Identify the x- and y-intercepts. See Section 3.3.

75.

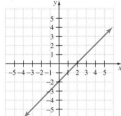

76.

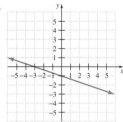

77.

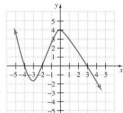

78.

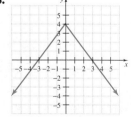

6.7 RATIO AND PROPORTION

CD-ROM SSM

SSG Video

▶ **OBJECTIVES**

1. Use fractional notation to express ratios.
2. Identify and solve proportions.
3. Use proportions to solve problems.
4. Determine unit pricing.

1   A **ratio** is the quotient of two numbers or two quantities.

> **RATIO**
>
> If a and b are two numbers and $b \neq 0$, the **ratio of a to b** is the quotient of a and b. The ratio of a to b can also be written as
>
> $$a:b \text{ or as } \frac{a}{b}$$

Example 1   Write a ratio for each phrase. Use fractional notation.

     a. The ratio of 2 parts salt to 5 parts water
     b. The ratio of 18 inches to 2 feet

Solution   **a.** The ratio of 2 parts salt to 5 parts water is $\dfrac{2}{5}$.

     b. When comparing measurements, use the same unit of measurement in the numerator as in the denominator. Here, we write 2 feet as 2 · 12 inches, or 24 inches. The ratio of 18 inches to 2 feet is then $\dfrac{18}{24} = \dfrac{3}{4}$ in simplest form. ▬

2   If two ratios are equal, we say the ratios are **in proportion** to each other. A **proportion** is a mathematical statement that two ratios are equal.

     For example, the equation $\dfrac{1}{2} = \dfrac{4}{8}$ is a proportion, as is $\dfrac{x}{5} = \dfrac{8}{10}$, because both sides of the equations are ratios. When we want to emphasize the equation as a proportion, we

     read the proportion $\dfrac{1}{2} = \dfrac{4}{8}$ as "one is to two as four is to eight"

     In a proportion, cross products are equal. To understand cross products, let's start with the proportion

$$\frac{a}{b} = \frac{c}{d}$$

and multiply both sides by the LCD, bd.

$$bd\left(\frac{a}{b}\right) = bd\left(\frac{c}{d}\right) \qquad \text{Multiply both sides by the LCD, } bd.$$

$$\underbrace{ad}_{\text{Cross product}} = \underbrace{bc}_{\text{Cross product}} \qquad \text{Simplify.}$$

Notice why ad and bc are called cross products.

$$\frac{a}{b} \;\; \begin{matrix} bc \\ = \\ \end{matrix} \;\; \frac{c}{d} \\ ad$$

CROSS PRODUCTS

If $\dfrac{a}{b} = \dfrac{c}{d}$, then $ad = bc$.

For example, since

$$\frac{1}{2} = \frac{4}{8}, \quad \text{then} \quad \begin{array}{c} 1 \cdot 8 = 2 \cdot 4 \quad \text{or} \\ 8 = 8 \end{array}$$

Example 2 Solve for x: $\dfrac{45}{x} = \dfrac{5}{7}$

Solution To solve, we set cross products equal.

$$\frac{45}{x} \;\;=\;\; \frac{5}{7}$$

$$45 \cdot 7 = x \cdot 5 \qquad \text{Set cross products equal.}$$
$$315 = 5x \qquad \text{Multiply.}$$
$$\frac{315}{5} = \frac{5x}{5} \qquad \text{Divide both sides by 5.}$$
$$63 = x \qquad \text{Simplify.}$$

Check To check, substitute 63 for x in the original proportion. The solution is 63.

Example 3 Solve for x: $\dfrac{x-5}{3} = \dfrac{x+2}{5}$

Solution

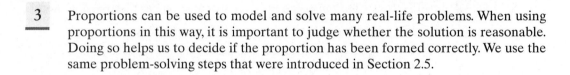

$$\frac{x-5}{3} = \frac{x+2}{5}$$

$5(x - 5) = 3(x + 2)$ Set cross products equal.

$5x - 25 = 3x + 6$ Multiply.

$5x = 3x + 31$ Add 25 to both sides.

$2x = 31$ Subtract $3x$ from both sides.

$\dfrac{2x}{2} = \dfrac{31}{2}$ Divide both sides by 2.

$x = \dfrac{31}{2}$

Check   Verify that $\dfrac{31}{2}$ is the solution.

3   Proportions can be used to model and solve many real-life problems. When using proportions in this way, it is important to judge whether the solution is reasonable. Doing so helps us to decide if the proportion has been formed correctly. We use the same problem-solving steps that were introduced in Section 2.5.

Example 4   **CALCULATING THE COST OF RECORDABLE COMPACT DISCS**

Three boxes of CD-Rs (recordable compact discs) cost $37.47. How much should 5 boxes cost?

Solution   1. UNDERSTAND. Read and reread the problem. We know that the cost of 5 boxes is more than the cost of 3 boxes, or $37.47, and less than the cost of 6 boxes, which is double the cost of 3 boxes, or 2($37.47) = $74.94. Let's suppose that 5 boxes cost $60.00. To check, we see if 3 boxes is to 5 boxes as the *price* of 3 boxes is to the *price* of 5 boxes. In other words, we see if

$$\frac{3 \text{ boxes}}{5 \text{ boxes}} = \frac{\text{price of 3 boxes}}{\text{price of 5 boxes}}$$

or

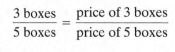

$$\frac{3}{5} = \frac{37.47}{60.00}$$

$3(60.00) = 5(37.47)$ Set cross products equal.

or

$180.00 = 187.35$ Not a true statement.

Thus, $60 is not correct, but we now have a better understanding of the problem.

Let x = price of 5 boxes of CD-Rs.

2. TRANSLATE.

$$\frac{3 \text{ boxes}}{5 \text{ boxes}} = \frac{\text{price of 3 boxes}}{\text{price of 5 boxes}}$$

$$\frac{3}{5} = \frac{37.47}{x}$$

3. SOLVE.

$$\frac{3}{5} = \frac{37.47}{x}$$

$$3x = 5(37.47) \qquad \text{Set cross products equal.}$$

$$3x = 187.35$$

$$x = 62.45 \qquad \text{Divide both sides by 3.}$$

4. INTERPRET.

Check: Verify that 3 boxes is to 5 boxes as $37.47 is to $62.45. Also, notice that our solution is a reasonable one as discussed in Step 1.

State: Five boxes of CD-Rs cost $62.45.

HELPFUL HINT

The proportion $\dfrac{5 \text{ boxes}}{3 \text{ boxes}} = \dfrac{\text{price of 5 boxes}}{\text{price of 3 boxes}}$ could also have been used to solve the problem above. Notice that the cross products are the same.

---

4 When shopping for an item offered in many different sizes, it is important to be able to determine the best buy, or the best price per unit. To find the unit price of an item, divide the total price of the item by the total number of units.

$$\text{unit price} = \frac{\text{total price}}{\text{number of units}}$$

For example, if a 16-ounce can of green beans is priced at $0.88, its unit price is

$$\text{unit price} = \frac{\$0.88}{16} = \$0.055$$

Example 5   **COMPARING CEREAL COSTS**

A supermarket offers a 14-ounce box of cereal for $3.79 and an 18-ounce box of the same brand of cereal for $4.99. Which is the better buy?

Solution   To find the better buy, we compare unit prices. The following unit prices were rounded to three decimal places.

Size	Price	Unit Price
14-ounce	$3.79	$\frac{\$3.79}{14} \approx \$0.271$
18-ounce	$4.99	$\frac{\$4.99}{18} \approx \$0.277$

The 14-ounce box of cereal has the lower unit price so it is the better buy.

SPOTLIGHT ON DECISION MAKING

Suppose you must select a child care center for your two-year-old daughter. You have compiled information on two possible choices in the table shown at the right. Which child care center would you choose? Why?

	Center A	Center B
Weekly cost	$130	$110
Number of 2-year-olds	7	13
Number of adults for 2-year-olds	2	3
Distance from home	6 miles	11 miles

Exercise Set 6.7

Write each ratio in fractional notation in simplest form. See Example 1.

1. 2 megabytes to 15 megabytes
2. 18 disks to 41 disks
△ 3. 10 inches to 12 inches
4. 15 miles to 40 miles
5. 5 quarts to 3 gallons
6. 8 inches to 3 feet
 7. 4 nickels to 2 dollars
8. 12 quarters to 2 dollars
9. 175 centimeters to 5 meters
10. 90 centimeters to 4 meters
11. 190 minutes to 3 hours
12. 60 hours to 2 days
13. Suppose someone tells you that the ratio of 11 inches to 2 feet is $\frac{11}{2}$. How do you correct that person and explain the error?
14. Write a ratio that can be written in fractional notation as $\frac{3}{2}$.

Solve each proportion. See Examples 2 and 3.

15. $\frac{2}{3} = \frac{x}{6}$

16. $\frac{x}{2} = \frac{16}{6}$

17. $\frac{x}{10} = \frac{5}{9}$

18. $\frac{9}{4x} = \frac{6}{2}$

19. $\frac{4x}{6} = \frac{7}{2}$

20. $\frac{a}{5} = \frac{3}{2}$

21. $\frac{a}{25} = \frac{12}{10}$

22. $\frac{n}{10} = 9$

23. $\frac{x-3}{x} = \frac{4}{7}$

24. $\frac{y}{y-16} = \frac{5}{3}$

25. $\frac{5x+1}{x} = \frac{6}{3}$

26. $\frac{3x-2}{5} = \frac{4x}{1}$

27. $\frac{x+1}{2x+3} = \frac{2}{3}$

28. $\frac{x+1}{x+2} = \frac{5}{3}$

29. $\frac{9}{5} = \frac{12}{3x+2}$

30. $\frac{6}{11} = \frac{27}{3x-2}$

31. $\frac{3}{x+1} = \frac{5}{2x}$

32. $\frac{7}{x-3} = \frac{8}{2x}$

Solve. See Example 4.

33. The ratio of the weight of an object on Earth to the weight of the same object on Pluto is 100 to 3. If an elephant weighs 4100 pounds on Earth, find the elephant's weight on Pluto.

34. If a 170-pound person weighs approximately 65 pounds on Mars, how much does a 9000-pound satellite weigh?

35. There are 110 calories per 28.4 grams of Crispy Rice cereal. Find how many calories are in 42.6 grams of this cereal.

36. On an architect's blueprint, 1 inch corresponds to 4 feet. Find the length of a wall represented by a line that is $3\frac{7}{8}$ inches long on the blueprint.

37. A recent headline read, "Women Earn Bigger Checks in 1 of Every 6 Couples." If there are 23,000 couples in a nearby metropolitan area, how many women would you expect to earn bigger paychecks?

38. A human factors expert recommends that there be at least 9 square feet of floor space in a college classroom for every student in the class. Find the minimum floor space that 40 students need.

39. To mix weed killer with water correctly, it is necessary to mix 8 teaspoons of weed killer with 2 gallons of water. Find how many gallons of water are needed to mix with the entire box if it contains 36 teaspoons of weed killer.

40. Ken Hall, a tailback, holds the high school sports record for total yards rushed in a season. In 1953, he rushed for 4045 total yards in 12 games. Find his average rushing yards per game.

41. To estimate the number of people in Jackson, population 50,000, who have no health insurance, 250 people were polled. Of those polled, 39 had no insurance. How many people in the city might we expect to be uninsured?

42. The manufacturers of cans of salted mixed nuts state that the ratio of peanuts to other nuts is 3 to 2. If 324 peanuts are in a can, find how many other nuts should also be in the can.

43. There are 1280 calories in a 14-ounce portion of Eagle Brand Milk. Find how many calories are in 2 ounces of Eagle Brand Milk.

44. Due to space problems at a local university, a 20-foot by 12-foot conference room is converted into a classroom. Find the maximum number of students the room can accommodate. (See Exercise 38.)

Given the following prices charged for various sizes of an item, find the best buy. See Example 5.

45. Laundry detergent
110 ounces for $5.79
240 ounces for $13.99

46. Jelly
10 ounces for $1.14
15 ounces for $1.69

47. Tuna (in cans)
6 ounces for $0.69
8 ounces for $0.90
16 ounces for $1.89

48. Picante sauce
10 ounces for $0.99
16 ounces for $1.69
30 ounces for $3.29

49. Blank video cassettes
4-pack for $8.99
6-pack for $13.99

50. Recordable compact disc
10-pack for $14.99
100-pack for $99.99

51. Milk
1 quart for $1.57
half gallon (2 qt) $2.10
gallon (4 qt) for $3.99

52. Pens
3-pack for $1.99
6-pack for $3.69
dozen for $6.99

The following bar graph shows the capacity of the United States to generate electricity from the wind in the years shown. Use this graph for Exercises 53 and 54.

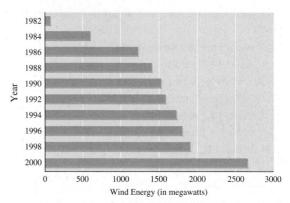

53. Find the approximate increase in megawatt capacity during the 2-year period from 1982 to 1984.

54. Find the approximate increase in megawatt capacity during the 2-year period from 1998 to 2000.

In general, 1000 megawatts will serve the average electricity needs of 560,000 people. Use this fact and the preceding graph to answer Exercises 55 and 56.

55. In 2000, the number of megawatts that can be generated from wind will serve the electricity needs of how many people?

56. How many megawatts of electricity are needed to serve the city or town in which you live?

57. If x is 10, is $\frac{2}{x}$ in proportion to $\frac{x}{50}$? Explain why or why not.

58. For what value of x is $\frac{x}{x-1}$ in proportion to $\frac{x+1}{x}$? Explain your result.

59. For which of the following equations can we immediately use cross products to solve for x?

a. $\dfrac{2-x}{5} - \dfrac{1+x}{3}$ b. $\dfrac{2}{5} = x - \dfrac{1+x}{3}$

REVIEW EXERCISES

Find the slope of the line through each pair of points. Use the slope to determine whether the line is vertical, horizontal, or moves upward or downward from left to right. See Section 3.4.

60. $(-2, 3), (4, -3)$ **61.** $(0, 4) (2, 10)$

62. $(-3, -6)(1, 5)$ **63.** $(-2, 7)(3, -2)$

64. $(3, 7) (3, -2)$ **65.** $(0, -4) (2, -4)$

6.8 RATIONAL EQUATIONS AND PROBLEM SOLVING

CD-ROM   SSM

SSG   Video

▶ **OBJECTIVES**

1. Translate sentences to equations containing rational expressions.
2. Solve problems involving work.
3. Solve problems involving distance.
4. Solve problems involving similar triangles.

1   In this section, we solve problems that can be modeled by equations containing rational expressions. To solve these problems, we use the same problem-solving steps that were first introduced in Section 2.5. In our first example, our goal is to find an unknown number.

Example 1   **FINDING AN UNKNOWN NUMBER**

The quotient of a number and 6 minus $\dfrac{5}{3}$ is the quotient of the number and 2. Find the number.

Solution   1. UNDERSTAND. Read and reread the problem. Suppose that the unknown number is 2, then we see if the quotient of 2 and 6, or $\dfrac{2}{6}$, minus $\dfrac{5}{3}$ is equal to the quotient of 2 and 2, or $\dfrac{2}{2}$.

$$\frac{2}{6} - \frac{5}{3} = \frac{1}{3} - \frac{5}{3} = -\frac{4}{3}, \text{ not } \frac{2}{2}$$

Don't forget that the purpose of a proposed solution is to better understand the problem.

Let x = the unknown number.

2. TRANSLATE.

In words:	the quotient of x and 6	minus	$\dfrac{5}{3}$	is	the quotient of x and 2
	↓	↓	↓	↓	↓
Translate:	$\dfrac{x}{6}$	$-$	$\dfrac{5}{3}$	$=$	$\dfrac{x}{2}$

3. SOLVE. Here, we solve the equation $\frac{x}{6} - \frac{5}{3} = \frac{x}{2}$. We begin by multiplying both sides of the equation by the LCD, 6.

$$6\left(\frac{x}{6} - \frac{5}{3}\right) = 6\left(\frac{x}{2}\right)$$

$$6\left(\frac{x}{6}\right) - 6\left(\frac{5}{3}\right) = 6\left(\frac{x}{2}\right) \quad \text{Apply the distributive property.}$$

$$x - 10 = 3x \quad \text{Simplify.}$$

$$-10 = 2x \quad \text{Subtract } x \text{ from both sides.}$$

$$-\frac{10}{2} = \frac{2x}{2} \quad \text{Divide both sides by 2.}$$

$$-5 = x \quad \text{Simplify.}$$

4. INTERPRET.

Check: To check, we verify that "the quotient of -5 and 6 minus $\frac{5}{3}$ is the quotient of -5 and 2," or $-\frac{5}{6} - \frac{5}{3} = -\frac{5}{2}$.

State: The unknown number is -5.

---

2   The next example is often called a work problem. Work problems usually involve people or machines doing a certain task.

Example 2   **FINDING WORK RATES**

Sam Waterton and Frank Schaffer work in a plant that manufactures automobiles. Sam can complete a quality control tour of the plant in 3 hours while his assistant, Frank, needs 7 hours to complete the same job. The regional manager is coming to inspect the plant facilities, so both Sam and Frank are directed to complete a quality control tour together. How long will this take?

Solution   1. UNDERSTAND. Read and reread the problem. The key idea here is the relationship between the **time** (hours) it takes to complete the job and the **part of the job** completed in 1 unit of time (hour). For example, if the **time** it takes Sam to complete the job is 3 hours, the **part of the job** he can complete in 1 hour is $\frac{1}{3}$. Similarly, Frank can complete $\frac{1}{7}$ of the job in 1 hour.

Let $x =$ the **time** in hours it takes Sam and Frank to complete the job together.

Then $\frac{1}{x} =$ the **part of the job** they complete in 1 hour.

	Hours to Complete Total Job	Part of Job Completed in 1 Hour
Sam	3	$\frac{1}{3}$
Frank	7	$\frac{1}{7}$
Together	x	$\frac{1}{x}$

2. TRANSLATE.

In words:

part of job Sam completed in 1 hour	added to	part of job Frank completed in 1 hour	is equal to	part of job they completed together in 1 hour
↓	↓	↓	↓	↓

Translate:

$$\frac{1}{3} \quad + \quad \frac{1}{7} \quad = \quad \frac{1}{x}$$

3. SOLVE. Here, we solve the equation $\frac{1}{3} + \frac{1}{7} = \frac{1}{x}$. We begin by multiplying both sides of the equation by the LCD, $21x$.

$$21x\left(\frac{1}{3}\right) + 21x\left(\frac{1}{7}\right) = 21x\left(\frac{1}{x}\right)$$

$$7x + 3x = 21 \qquad\qquad \text{Simplify.}$$

$$10x = 21$$

$$x = \frac{21}{10} \quad \text{or} \quad 2\frac{1}{10} \text{ hours}$$

4. INTERPRET.

Check: Our proposed solution is $2\frac{1}{10}$ hours. This proposed solution is reasonable since $2\frac{1}{10}$ hours is more than half of Sam's time and less than half of Frank's time. Check this solution in the originally *stated* problem.

State: Sam and Frank can complete the quality control tour in $2\frac{1}{10}$ hours.   ▬

---

3 Next we look at a problem solved by the distance formula.

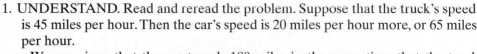

Example 3 **FINDING SPEEDS OF VEHICLES**

A car travels 180 miles in the same time that a truck travels 120 miles. If the car's speed is 20 miles per hour faster than the truck's, find the car's speed and the truck's speed.

Solution 1. UNDERSTAND. Read and reread the problem. Suppose that the truck's speed is 45 miles per hour. Then the car's speed is 20 miles per hour more, or 65 miles per hour.

We are given that the car travels 180 miles in the same time that the truck travels 120 miles. To find the time it takes the car to travel 180 miles, remember that since $d = rt$, we know that $\frac{d}{r} = t$.

Car's Time	*Truck's Time*
$t = \dfrac{d}{r} = \dfrac{180}{65} = 2\dfrac{50}{65} = 2\dfrac{10}{13}$ hours	$t = \dfrac{d}{r} = \dfrac{120}{45} = 2\dfrac{30}{45} = 2\dfrac{2}{3}$ hours

Since the times are not the same, our proposed solution is not correct. But we have a better understanding of the problem.

Let $x =$ the speed of the truck.

Since the car's speed is 20 miles per hour faster than the truck's, then

$$x + 20 = \text{the speed of the car}$$

Use the formula $d = r \cdot t$ or **d**istance = **r**ate · **t**ime. Prepare a chart to organize the information in the problem.

	Distance	=	Rate	·	Time
TRUCK	120		x		$\dfrac{120}{x}$ $\leftarrow$ distance $\leftarrow$ rate
CAR	180		$x + 20$		$\dfrac{180}{x + 20}$ $\leftarrow$ distance $\leftarrow$ rate

HELPFUL HINT

If $d = r \cdot t$,

then $t = \dfrac{d}{r}$

or $time = \dfrac{distance}{rate}$.

2. TRANSLATE. Since the car and the truck traveled the same amount of time, we have that

In words:

car's time	=	truck's time
$\downarrow$		$\downarrow$

Translate: $\dfrac{180}{x + 20} = \dfrac{120}{x}$

3. SOLVE. We begin by multiplying both sides of the equation by the LCD, $x(x + 20)$, or cross multiplying.

$$\frac{180}{x + 20} = \frac{120}{x}$$

$$180x = 120(x + 20)$$

$$180x = 120x + 2400 \qquad \text{Use the distributive property.}$$

$$60x = 2400 \qquad \text{Subtract } 120x \text{ from both sides.}$$

$$x = 40 \qquad \text{Divide both sides by 60.}$$

4. INTERPRET. The speed of the truck is 40 miles per hour. The speed of the car must then be $x + 20$ or 60 miles per hour.

Check: Find the time it takes the car to travel 180 miles and the time it takes the truck to travel 120 miles.

Car's Time	*Truck's Time*
$t = \dfrac{d}{r} = \dfrac{180}{60} = 3 \text{ hours}$	$t = \dfrac{d}{r} = \dfrac{120}{40} = 3 \text{ hours}$

Since both travel the same amount of time, the proposed solution is correct.

State: The car's speed is 60 miles per hour and the truck's speed is 40 miles per hour.

---

4 **Similar triangles** have the same shape but not necessarily the same size. In similar triangles, the measures of corresponding angles are equal, and corresponding sides are in proportion.

If triangle ABC and triangle XYZ shown are similar, then we know that the measure of angle A = the measure of angle X, the measure of angle B = the measure of angle Y, and the measure of angle C = the measure of angle Z. We also know that corresponding sides are in proportion: $\dfrac{a}{x} = \dfrac{b}{y} = \dfrac{c}{z}$.

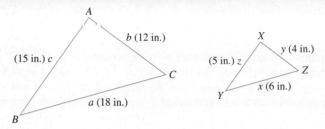

In this section, we will position similar triangles so that they have the same orientation.

To show that corresponding sides are in proportion for the triangles above, we write the ratios of the corresponding sides.

$$\frac{a}{x} = \frac{18}{6} = 3 \qquad \frac{b}{y} = \frac{12}{4} = 3 \qquad \frac{c}{z} = \frac{15}{5} = 3$$

△ **Example 4**   **FINDING THE LENGTH OF A SIDE OF A TRIANGLE**

If the following two triangles are similar, find the missing length x.

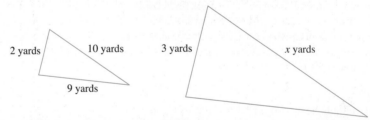

Solution   Since the triangles are similar, their corresponding sides are in proportion and we have

$$\frac{2}{3} = \frac{10}{x}$$

To solve, we multiply both sides by the LCD, $3x$, or cross multiply.

$$2x = 30$$
$$x = 15 \qquad \text{Divide both sides by 2.}$$

The missing length is 15 yards.

SPOTLIGHT ON DECISION MAKING

Suppose you coach your company's softball team. Two new employees are interested in playing on the company team, and you have only one open position. Employee A reports that last season he had 32 hits in 122 times at bat. Employee B reports that last season she had 19 hits in 56 times at bat. Which would you try to recruit first? Why? What other factors would you want to consider?

Exercise Set 6.8

Solve the following. See Example 1.

1. Three times the reciprocal of a number equals 9 times the reciprocal of 6. Find the number.

2. Twelve divided by the sum of x and 2 equals the quotient of 4 and the difference of x and 2. Find x.

3. If twice a number added to 3 is divided by the number plus 1, the result is three halves. Find the number.

4. A number added to the product of 6 and the reciprocal of the number equals -5. Find the number.

See Example 2.

5. Smith Engineering found that an experienced surveyor surveys a roadbed in 4 hours. An apprentice surveyor needs 5 hours to survey the same stretch of road. If the two work together, find how long it takes them to complete the job.

6. An experienced bricklayer constructs a small wall in 3 hours. The apprentice completes the job in 6 hours. Find how long it takes if they work together.

7. In 2 minutes, a conveyor belt moves 300 pounds of recyclable aluminum from the delivery truck to a storage area. A smaller belt moves the same quantity of cans the same distance in 6 minutes. If both belts are used, find how long it takes to move the cans to the storage area.

8. Find how long it takes the conveyor belts described in Exercise 7 to move 1200 pounds of cans. (*Hint:* Think of 1200 pounds as four 300-pound jobs.)

See Example 3.

9. A jogger begins her workout by jogging to the park, a distance of 12 miles. She then jogs home at the same speed but along a different route. This return trip is 18 miles and her time is one hour longer. Find her jogging speed. Complete the accompanying chart and use it to find her jogging speed.

	distance	=	rate	·	time
Trip to park	12				x
Return trip	18				$x + 1$

10. A boat can travel 9 miles upstream in the same amount of time it takes to travel 11 miles downstream. If the current of the river is 3 miles per hour, complete the chart below and use it to find the speed of the boat in still water.

	distance	=	rate	·	time
Upstream	9		$r - 3$		
Downstream	11		$r + 3$		

11. A cyclist rode the first 20-mile portion of his workout at a constant speed. For the 16-mile cooldown portion of his workout, he reduced his speed by 2 miles per hour. Each portion of the workout took the same time. Find the cyclist's speed during the first portion and find his speed during the cooldown portion.

12. A semi truck travels 300 miles through the flatland in the same amount of time that it travels 180 miles through mountains. The rate of the truck is 20 miles per hour slower in the mountains than in the flatland. Find both the flatland rate and mountain rate.

Given that the following pairs of triangles are similar, find the missing length, x. See Example 4.

13.
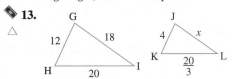

14.
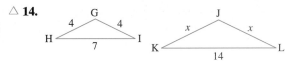

15.
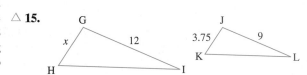

16.
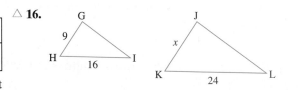

Solve the following.

17. One-fourth equals the quotient of a number and 8. Find the number.

18. Four times a number added to 5 is divided by 6. The result is $\frac{7}{2}$. Find the number.

19. Marcus and Tony work for Lombardo's Pipe and Concrete. Mr. Lombardo is preparing an estimate for a customer. He knows that Marcus lays a slab of concrete in

6 hours. Tony lays the same size slab in 4 hours. If both work on the job and the cost of labor is $45.00 per hour, decide what the labor estimate should be.

20. Mr. Dodson can paint his house by himself in 4 days. His son needs an additional day to complete the job if he works by himself. If they work together, find how long it takes to paint the house.

21. While road testing a new make of car, the editor of a consumer magazine finds that he can go 10 miles into a 3-mile-per-hour wind in the same amount of time he can go 11 miles with a 3-mile-per-hour wind behind him. Find the speed of the car in still air.

22. A fisherman on Pearl River rows 9 miles downstream in the same amount of time he rows 3 miles upstream. If the current is 6 miles per hour, find how long it takes him to cover the 12 miles.

Find the unknown length y in the following pairs of similar triangles.

△ **23.**

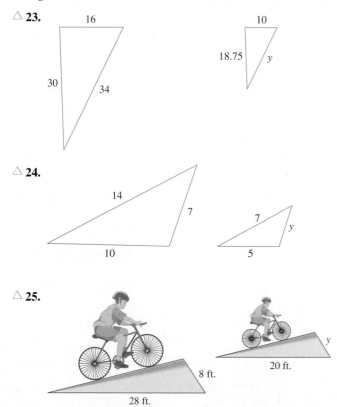

△ **24.**

△ **25.**

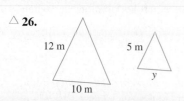

△ **26.**

△ **27.**

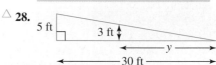

△ **28.**

Solve the following.

29. Two divided by the difference of a number and 3 minus 4 divided by a number plus 3, equals 8 times the reciprocal of the difference of the number squared and 9. What is the number?

30. If 15 times the reciprocal of a number is added to the ratio of 9 times a number minus 7 and the number plus 2, the result is 9. What is the number?

31. A pilot flies 630 miles with a tail wind of 35 miles per hour. Against the wind, he flies only 455 miles in the same amount of time. Find the rate of the plane in still air.

32. A marketing manager travels 1080 miles in a corporate jet and then an additional 240 miles by car. If the car ride takes one hour longer than the jet ride takes, and if the rate of the jet is 6 times the rate of the car, find the time the manager travels by jet and find the time the manager travels by car.

33. A cyclist rides 16 miles per hour on level ground on a still day. He finds that he rides 48 miles with the wind behind him in the same amount of time that he rides 16 miles into the wind. Find the rate of the wind.

34. The current on a portion of the Mississippi River is 3 miles per hour. A barge can go 6 miles upstream in the same amount of time it takes to go 10 miles downstream. Find the speed of the boat in still water.

35. One custodian cleans a suite of offices in 3 hours. When a second worker is asked to join the regular custodian, the job takes only $1\frac{1}{2}$ hours. How long does it take the second worker to do the same job alone?

36. One person proofreads a copy for a small newspaper in 4 hours. If a second proofreader is also employed, the job can be done in $2\frac{1}{2}$ hours. How long does it take for the second proofreader to do the same job alone?

37. One pipe fills a storage pool in 20 hours. A second pipe fills the same pool in 15 hours. When a third pipe is added and all three are used to fill the pool, it takes only 6 hours. Find how long it takes the third pipe to do the job.

38. One pump fills a tank 2 times as fast as another pump. If the pumps work together, they fill the tank in 18 minutes. How long does it take for each pump to fill the tank?

△ **39.** An architech is completing the plans for a triangular deck. Use the diagram below to find the missing dimension.

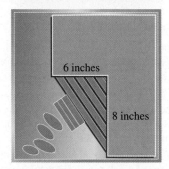

△ **40.** A student wishes to make a small model of a triangular mainsail in order to study the effects of wind on the sail. The smaller model will be the same shape as a regular-size sailboat's mainsail. Use the following diagram to find the missing dimensions.

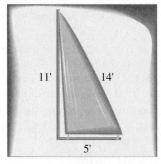

41. Andrew and Timothy Larson volunteer at a local recycling plant. Andrew can sort a batch of recyclables in 2 hours alone while his brother Timothy needs 3 hours to complete the same job. If they work together, how long will it take them to sort one batch?

42. A car travels 280 miles in the same time that a motorcycle travels 240 miles. If the car's speed is 10 miles per hour more than the motorcycle's, find the speed of the car and the speed of the motorcycle.

43. In 6 hours, an experienced cook prepares enough pies to supply a local restaurant's daily order. Another cook prepares the same number of pies in 7 hours. Together with a third cook, they prepare the pies in 2 hours. Find the work rate of the third cook.

44. Mrs. Smith balances the company books in 8 hours. It takes her assistant half again as long to do the same job. If they work together, find how long it takes them to balance the books.

45. One pump fills a tank 3 times as fast as another pump. If the pumps work together, they fill the tank in 21 minutes. How long does it take for each pump to fill the tank?

One of the great algebraists of ancient times was a man named Diophantus. Little is known of his life other than that he lived and worked in Alexandria. Some historians believe he lived during the first century of the Christian era, about the time of Nero. The only clue to his personal life is the following epigram found in a collection called the Palatine Anthology.

God granted him youth for a sixth of his life and added a twelfth part to this. He clothed his cheeks in down. He lit him the light of wedlock after a seventh part and five years after his marriage, He granted him a son. Alas, lateborn wretched child. After attaining the measure of half his father's life, cruel fate overtook him, thus leaving Diophantus during the last four years of his life only such consolation as the science of numbers. How old was Diophantus at his death?*

We are looking for Diophantus' age when he died, so let x represent that age. If we sum the parts of his life, we should get the total age.

Parts of his life
$$\begin{cases} \dfrac{1}{6} \cdot x + \dfrac{1}{12} \cdot x \text{ is the time of his youth.} \\[2mm] \dfrac{1}{7} \cdot x \text{ is the time between his youth and when he married.} \\[2mm] 5 \text{ years is the time between his marriage and the birth of his son.} \\[2mm] \dfrac{1}{2} \cdot x \text{ is the time Diophantus had with his son.} \\[2mm] 4 \text{ years is the time between his son's death and his own.} \end{cases}$$

The sum of these parts should equal Diophantus' age when he died.

$$\frac{1}{6} \cdot x + \frac{1}{12} \cdot x + \frac{1}{7} \cdot x + 5 + \frac{1}{2} \cdot x + 4 = x$$

*From *The Nature and Growth of Modern Mathematics*, Edna Kramer, 1970, Fawcett Premier Books, Vol. 1, pages 107–108.

46. Solve the epigram.

47. How old was Diophantus when his son was born? How old was the son when he died?

48. Solve the following epigram:

I was four when my mother packed my lunch and sent me off to school. Half my life was spent in school and another sixth was spent on a farm. Alas, hard times befell me. My crops and cattle fared poorly and my land was sold. I returned to school for 3 years and have spent one tenth of my life teaching. How old am I?

49. Write an epigram describing your life. Be sure that none of the time periods in your epigram overlap.

50. During the 2000 Grand Prix of Miami auto race, Juan Montoya posted the fastest lap speed but Max Papis won the race. The track is 1.5 miles long. When traveling at their fastest lap speeds, Papis drove 1.48 miles in the same time that it took Montoya to complete an entire lap. Montoya's fastest lap speed was 2.61 mph faster than Papis's fastest lap speed. Find each driver's fastest lap speed. (*Source:* Based on data from Championship Auto Racing Teams, Inc.)

51. A hyena spots a giraffe 0.5 mile away and begins running toward it. The giraffe starts running away from the hyena just as the hyena begins running toward it. A hyena can run at a speed of 40 mph and a giraffe can run at 32 mph. How long will it take for the hyena to overtake the giraffe? (*Source:* Based on data from *Natural History*, March 1974)

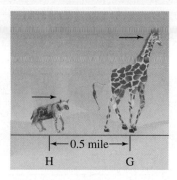

←— 0.5 mile —→

H   G

REVIEW EXERCISES

Graph each linear equation by finding intercepts. See Section 3.3.

52. $5x + y = 10$

53. $-x + 3y = 6$

54. $x = -3y$

55. $y = 2x$

56. $x - y = -2$

57. $y - x = -5$

For additional Chapter Projects, visit the Real World Activities Website by going to http://www.prenhall.com/martin-gay.

CHAPTER PROJECT

Comparing Dosage Formulas

In this project, you will have the opportunity to investigate two well-known formulas for predicting the correct doses of medication for children. This project may be completed by working in groups or individually.

Young's Rule and Cowling's Rule are dose formulas for prescribing medicines to children. Unlike formulas for, say area or distance, these dose formulas describe only an approximate relationship. The formulas relate a child's age A in years and an adult dose D of medication to the proper child's dose C. The formulas are most accurate when applied to children between the ages of 2 and 13.

$$\text{Young's Rule:} \quad C = \frac{DA}{A + 12}$$

$$\text{Cowling's Rule:} \quad C = \frac{D(A + 1)}{24}$$

1. Let the adult dose $D = 1000$ mg. Complete the Young's Rule and Cowling's Rule columns of the following table comparing the doses predicted by both formulas for ages 2 through 13.

Age A	Young's Rule	Cowling's Rule	Difference
2			
3			
4			
5			
6			
7			
8			
9			
10			
11			
12			
13			

2. Use the data from the table in Question 1 to form sets of ordered pairs of the form (age, child's dose) for each formula. Graph the ordered pairs for each formula on the same graph. Describe the shapes of the graphed data.

3. Use your table, graph, or both, to decide whether either formula will consistently predict a larger dose than the other. If so, which one? If not, is there an age at which the doses predicted by one becomes greater than the doses predicted by the other? If so, estimate that age.

4. Use your graph to estimate for what age the difference in the two predicted doses is greatest.

5. Return to the table in Question 1 and complete the last column, titled "Difference," by finding the absolute value of the difference between the Young's dose and the Cowling's dose for each age. Use this column in the table to verify your graphical estimate found in Question 4.

6. Does Cowling's Rule ever predict exactly the adult dose? If so, at what age? Explain. Does Young's Rule ever predict exactly the adult dose? If so, at what age? Explain.

7. Many doctors prefer to use formulas that relate doses to factors other than a child's age. Why is age not necessarily the most important factor when predicting a child's dose? What other factors might be used?

CHAPTER 6 VOCABULARY CHECK

Fill in each blank with one of the words or phrases listed below.

rational expression   complex fraction   ratio   proportion
cross products

1. A _____ is the quotient of two numbers. It can be written as a fraction, using a colon, or using the word *to*.

2. $\dfrac{x}{2} = \dfrac{7}{16}$ is an example of a _____.

3. If $a/b = c/d$, then ad and bc are called _____.

4. A _____ is an expression that can be written in the form P/Q, where P and Q are polynomials and Q is not 0.

5. In a _____, the numerator or denominator or both may contain fractions.

CHAPTER 6 HIGHLIGHTS

DEFINITIONS AND CONCEPTS	EXAMPLES

Section 6.1 Simplifying Rational Expressions

A **rational expression** is an expression that can be written in the form $\frac{P}{Q}$, where P and Q are polynomials and Q does not equal 0.

Rational Expressions

$$\frac{7y^9}{4}, \frac{x^2 + 6x + 1}{x - 3}, \frac{-5}{s^3 + 8}$$

To find values for which a rational expression is undefined, find values for which the denominator is 0.

Find any values for which the expression $\dfrac{5y}{y^2 - 4y + 3}$ is undefined.

$$y^2 - 4y + 3 = 0 \quad \text{Set denominator equal to 0.}$$

$$(y - 3)(y - 1) = 0 \quad \text{Factor.}$$

$$y - 3 = 0 \text{ or } y - 1 = 0 \quad \text{Set each factor equal to 0.}$$

$$y = 3 \text{ or } y = 1 \quad \text{Solve.}$$

The expression is undefined when y is 3 and when y is 1.

Fundamental Principle of Rational Expressions
If P and Q are polynomials, and Q and R are not 0, then

$$\frac{PR}{QR} = \frac{P}{Q}$$

By the fundamental principle,

$$\frac{(x - 3)(x + 1)}{x(x + 1)} = \frac{x - 3}{x}$$

as long as $x \neq 0$ and $x \neq -1$.

To simplify a rational expression,
Step 1. Factor the numerator and denominator.
Step 2. Apply the fundamental principle to divide out common factors.

Simplify: $\dfrac{4x + 20}{x^2 - 25}$

$$\frac{4x + 20}{x^2 - 25} = \frac{4(x + 5)}{(x + 5)(x - 5)} = \frac{4}{x - 5}$$

Section 6.2 Multiplying and Dividing Rational Expressions

To multiply rational expressions,
Step 1. Factor numerators and denominators.
Step 2. Multiply numerators and multiply denominators.
Step 3. Write the product in simplest form.

$$\frac{P}{Q} \cdot \frac{R}{S} = \frac{PR}{QS}$$

Multiply: $\dfrac{4x + 4}{2x - 3} \cdot \dfrac{2x^2 + x - 6}{x^2 - 1}$

$$\frac{4x + 4}{2x - 3} \cdot \frac{2x^2 + x - 6}{x^2 - 1} = \frac{4(x + 1)}{2x - 3} \cdot \frac{(2x - 3)(x + 2)}{(x + 1)(x - 1)}$$

$$= \frac{4(x + 1)(2x - 3)(x + 2)}{(2x - 3)(x + 1)(x - 1)}$$

$$= \frac{4(x + 2)}{x - 1}$$

To divide by a rational expression, multiply by the reciprocal.

$$\frac{P}{Q} \div \frac{R}{S} = \frac{P}{Q} \cdot \frac{S}{R} = \frac{PS}{QR}$$

Divide: $\dfrac{15x + 5}{3x^2 - 14x - 5} \div \dfrac{15}{3x - 12}$

$$\frac{15x + 5}{3x^2 - 14x - 5} \div \frac{15}{3x - 12} = \frac{5(3x + 1)}{(3x + 1)(x - 5)} \cdot \frac{3(x - 4)}{3 \cdot 5}$$

$$= \frac{x - 4}{x - 5}$$

(continued)

DEFINITIONS AND CONCEPTS	EXAMPLES

Section 6.3 Adding and Subtracting Rational Expressions with Common Denominators and Least Common Denominator

To add or subtract rational expressions with the same denominator, add or subtract numerators, and place the sum or difference over a common denominator.

$$\frac{P}{R} + \frac{Q}{R} = \frac{P + Q}{R}$$

$$\frac{P}{R} - \frac{Q}{R} = \frac{P - Q}{R}$$

Perform indicated operations.

$$\frac{5}{x + 1} + \frac{x}{x + 1} = \frac{5 + x}{x + 1}$$

$$\frac{2y + 7}{y^2 - 9} - \frac{y + 4}{y^2 - 9} = \frac{(2y + 7) - (y + 4)}{y^2 - 9}$$

$$= \frac{2y + 7 - y - 4}{y^2 - 9}$$

$$= \frac{y + 3}{(y + 3)(y - 3)}$$

$$= \frac{1}{y - 3}$$

To find the least common denominator (LCD),
Step 1. Factor the denominators.
Step 2. The LCD is the product of all unique factors, each raised to a power equal to the greatest number of times that it appears in any one factored denominator.

Find the LCD for

$$\frac{7x}{x^2 + 10x + 25} \text{ and } \frac{11}{3x^2 + 15x}$$

$$x^2 + 10x + 25 = (x + 5)(x + 5)$$

$$3x^2 + 15x = 3x(x + 5)$$

LCD is $3x(x + 5)(x + 5)$ or $3x(x + 5)^2$

Section 6.4 Adding and Subtracting Rational Expressions with Unlike Denominators

To add or subtract rational expressions with unlike denominators,
Step 1. Find the LCD.
Step 2. Rewrite each rational expression as an equivalent expression whose denominator is the LCD.
Step 3. Add or subtract numerators and place the sum or difference over the common denominator.
Step 4. Write the result in simplest form.

Perform the indicated operation.

$$\frac{9x + 3}{x^2 - 9} - \frac{5}{x - 3}$$

$$= \frac{9x + 3}{(x + 3)(x - 3)} - \frac{5}{x - 3}$$

LCD is $(x + 3)(x - 3)$.

$$= \frac{9x + 3}{(x + 3)(x - 3)} - \frac{5(x + 3)}{(x - 3)(x + 3)}$$

$$= \frac{9x + 3 - 5(x + 3)}{(x + 3)(x - 3)}$$

$$= \frac{9x + 3 - 5x - 15}{(x + 3)(x - 3)}$$

$$= \frac{4x - 12}{(x + 3)(x - 3)}$$

$$= \frac{4(x - 3)}{(x + 3)(x - 3)} = \frac{4}{x + 3}$$

(continued)

DEFINITIONS AND CONCEPTS	EXAMPLES

Section 6.5 Simplifying Complex Fractions

Method 1: To Simplify a Complex Fraction

Step 1. Add or subtract fractions in the numerator and the denominator of the complex fraction.

Step 2. Perform the indicated division.

Step 3. Write the result in simplest form.

Simplify

$$\frac{\frac{1}{x} + 2}{\frac{1}{x} - \frac{1}{y}} = \frac{\frac{1}{x} + \frac{2x}{x}}{\frac{y}{xy} - \frac{x}{xy}}$$

$$= \frac{\frac{1 + 2x}{x}}{\frac{y - x}{xy}}$$

$$= \frac{1 + 2x}{x} \cdot \frac{x\,y}{y - x}$$

$$= \frac{y(1 + 2x)}{y - x}$$

Method 2: To Simplify a Complex Fraction

Step 1. Find the LCD of all fractions in the complex fraction.

Step 2. Multiply the numerator and the denominator of the complex fraction by the LCD.

Step 3. Perform indicated operations and write in simplest form.

$$\frac{\frac{1}{x} + 2}{\frac{1}{x} - \frac{1}{y}} = \frac{xy\left(\frac{1}{x} + 2\right)}{xy\left(\frac{1}{x} - \frac{1}{y}\right)}$$

$$= \frac{xy\left(\frac{1}{x}\right) + xy(2)}{xy\left(\frac{1}{x}\right) - xy\left(\frac{1}{y}\right)}$$

$$= \frac{y + 2xy}{y - x} \quad \text{or} \quad \frac{y(1 + 2x)}{y - x}$$

Section 6.6 Solving Equations Containing Rational Expressions

To solve an equation containing rational expressions,

Step 1. Multiply both sides of the equation by the LCD of all rational expressions in the equation.

Step 2. Remove any grouping symbols and solve the resulting equation.

Step 3. Check the solution in the original equation.

Solve: $\dfrac{5x}{x + 2} + 3 = \dfrac{4x - 6}{x + 2}$

$$(x + 2)\left(\frac{5x}{x + 2} + 3\right) = (x + 2)\left(\frac{4x - 6}{x + 2}\right)$$

$$(x + 2)\left(\frac{5x}{x + 2}\right) + (x + 2)(3) = (x + 2)\left(\frac{4x - 6}{x + 2}\right)$$

$$5x + 3x + 6 = 4x - 6$$

$$4x = -12$$

$$x = -3$$

The solution checks and the solution is -3.

(continued)

DEFINITIONS AND CONCEPTS	EXAMPLES

Section 6.7 Ratio and Proportion

A **ratio** is the quotient of two numbers or two quantities. **The ratio of *a* to *b*** can be written as

Fractional Notation	*Colon Notation*
$\dfrac{a}{b}$	$a:b$

Write the ratio of 5 hours to 1 day using fractional notation.

$$\frac{5 \text{ hours}}{1 \text{ day}} = \frac{5 \text{ hours}}{24 \text{ hours}} = \frac{5}{24}$$

A **proportion** is a mathematical statement that two ratios are equal.

Proportions

$$\frac{2}{3} = \frac{8}{12} \qquad\qquad \frac{x}{7} = \frac{15}{35}$$

In the proportion $\frac{a}{b} = \frac{c}{d}$, the products ad and bc are called **cross products.**

Cross Products

$$\frac{2}{3} \diagdown\!\!\!\!= \diagup\!\!\!\! \frac{8}{12} \quad \begin{matrix} \rightarrow & 3 \cdot 8 \text{ or } 24 \\ \rightarrow & 2 \cdot 12 \text{ or } 24 \end{matrix}$$

Cross products:

If $\dfrac{a}{b} = \dfrac{c}{d}$, then $ad = bc$.

Solve: $\dfrac{3}{4} = \dfrac{x}{x-1}$

$$\frac{3}{4} \diagdown\!\!\!\!\nearrow\!\! \frac{x}{x-1}$$

$3(x - 1) = 4x$ Set cross products equal.

$3x - 3 = 4x$

$-3 = x$

Section 6.8 Rational Equations and Problem Solving

Problem-Solving Steps

1. UNDERSTAND. Read and reread the problem.

A small plane and a car leave Kansas City, Missouri, and head for Minneapolis, Minnesota, a distance of 450 miles. The speed of the plane is 3 times the speed of the car, and the plane arrives 6 hours ahead of the car. Find the speed of the car.

Let x = the speed of the car.
Then $3x$ = the speed of the plane.

	Distance =	*Rate* ·	*Time*
Car	450	x	$\dfrac{450}{x} \left(\dfrac{\text{distance}}{\text{rate}}\right)$
Plane	450	$3x$	$\dfrac{450}{3x} \left(\dfrac{\text{distance}}{\text{rate}}\right)$

In words:

plane's time	+	6 hours	=	car's time
↓	↓	↓		
$\dfrac{450}{3x}$	+	6	=	$\dfrac{450}{x}$

2. TRANSLATE.

Translate:

$$\frac{450}{3x} + 6 = \frac{450}{x}$$

3. SOLVE.

$$3x\left(\frac{450}{3x}\right) + 3x(6) = 3x\left(\frac{450}{x}\right)$$
$$450 + 18x = 1350$$
$$18x = 900$$
$$x = 50$$

4. INTERPRET.

Check the solution by replacing x with 50 in the original equation. *State* the conclusion: The speed of the car is 50 miles per hour.

CHAPTER 6 REVIEW

(6.1) *Find any real number for which each rational expression is undefined.*

1. $\dfrac{x + 5}{x^2 - 4}$

2. $\dfrac{5x + 9}{4x^2 - 4x - 15}$

Find the value of each rational expression when $x = 5$, $y = 7$, and $z = -2$.

3. $\dfrac{z^2 - z}{z + xy}$

4. $\dfrac{x^2 + xy - z^2}{x + y + z}$

Simplify each rational expression.

5. $\dfrac{x + 2}{x^2 - 3x - 10}$

6. $\dfrac{x + 4}{x^2 + 5x + 4}$

7. $\dfrac{x^3 - 4x}{x^2 + 3x + 2}$

8. $\dfrac{5x^2 - 125}{x^2 + 2x - 15}$

9. $\dfrac{x^2 - x - 6}{x^2 - 3x - 10}$

10. $\dfrac{x^2 - 2x}{x^2 + 2x - 8}$

11. $\dfrac{x^2 + 6x + 5}{2x^2 + 11x + 5}$

12. $\dfrac{x^2 + xa + xb + ab}{x^2 - xc + bx - bc}$

13. $\dfrac{x^2 + 5x - 2x - 10}{x^2 - 3x - 2x + 6}$

14. $\dfrac{x^2 - 9}{9 - x^2}$

15. $\dfrac{4 - x}{x^3 - 64}$

(6.2) *Perform the indicated operations and simplify.*

16. $\dfrac{15x^3y^2}{z} \cdot \dfrac{z}{5xy^3}$

17. $\dfrac{-y^3}{8} \cdot \dfrac{9x^2}{y^3}$

18. $\dfrac{x^2 - 9}{x^2 - 4} \cdot \dfrac{x - 2}{x + 3}$

19. $\dfrac{2x + 5}{x - 6} \cdot \dfrac{2x}{-x + 6}$

20. $\dfrac{x^2 - 5x - 24}{x^2 - x - 12} \div \dfrac{x^2 - 10x + 16}{x^2 + x - 6}$

21. $\dfrac{4x + 4y}{xy^2} \div \dfrac{3x + 3y}{x^2y}$

22. $\dfrac{x^2 + x - 42}{x - 3} \cdot \dfrac{(x - 3)^2}{x + 7}$

23. $\dfrac{2a + 2b}{3} \cdot \dfrac{a - b}{a^2 - b^2}$

24. $\dfrac{x^2 - 9x + 14}{x^2 - 5x + 6} \cdot \dfrac{x + 2}{x^2 - 5x - 14}$

25. $(x - 3) \cdot \dfrac{x}{x^2 + 3x - 18}$

26. $\dfrac{2x^2 - 9x + 9}{8x - 12} \div \dfrac{x^2 - 3x}{2x}$

27. $\dfrac{x^2 - y^2}{x^2 + xy} \div \dfrac{3x^2 - 2xy - y^2}{3x^2 + 6x}$

28. $\dfrac{x^2 - y^2}{8x^2 - 16xy + 8y^2} \div \dfrac{x + y}{4x - y}$

29. $\dfrac{x - y}{4} \cdot \dfrac{y^2 - 2y - xy + 2x}{16x + 24}$

30. $\dfrac{y - 3}{4x + 3} \div \dfrac{9 - y^2}{4x^2 - x - 3}$

(6.3) *Perform the indicated operations and simplify.*

31. $\dfrac{5x - 4}{3x - 1} + \dfrac{6}{3x - 1}$

32. $\dfrac{4x - 5}{3x^2} - \dfrac{2x + 5}{3x^2}$

33. $\dfrac{9x + 7}{6x^2} - \dfrac{3x + 4}{6x^2}$

Find the LCD of each pair of rational expressions.

34. $\dfrac{x + 4}{2x}, \dfrac{3}{7x}$

35. $\dfrac{x - 2}{x^2 - 5x - 24}, \dfrac{3}{x^2 + 11x + 24}$

Rewrite the following rational expressions as equivalent expressions whose denominator is the given polynomial.

36. $\dfrac{x + 2}{x^2 + 11x + 18}, (x + 2)(x - 5)(x + 9)$

37. $\dfrac{3x - 5}{x^2 + 4x + 4}, (x + 2)^2(x + 3)$

(6.4) *Perform the indicated operations and simplify.*

38. $\dfrac{4}{5x^2} - \dfrac{6}{y}$

39. $\dfrac{2}{x - 3} - \dfrac{4}{x - 1}$

40. $\dfrac{x + 7}{x + 3} - \dfrac{x - 3}{x + 7}$

41. $\dfrac{4}{x + 3} - 2$

42. $\dfrac{3}{x^2 + 2x - 8} + \dfrac{2}{x^2 - 3x + 2}$

43. $\dfrac{2x - 5}{6x + 9} - \dfrac{4}{2x^2 + 3x}$

44. $\dfrac{x - 1}{x^2 - 2x + 1} - \dfrac{x + 1}{x - 1}$

45. $\dfrac{x - 1}{x^2 + 4x + 4} + \dfrac{x - 1}{x + 2}$

Find the perimeter and the area of each figure.

△ **46.**

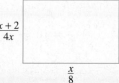

$\dfrac{x + 2}{4x}$

$\dfrac{x}{8}$

△ 47.

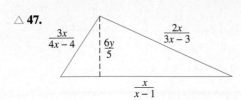

(6.5) *Simplify each complex fraction.*

48. $\dfrac{\dfrac{5x}{27}}{-\dfrac{10xy}{21}}$

49. $\dfrac{\dfrac{8x}{x^2 - 9}}{\dfrac{4}{x + 3}}$

50. $\dfrac{\dfrac{3}{5} + \dfrac{2}{7}}{\dfrac{1}{5} + \dfrac{5}{6}}$

51. $\dfrac{\dfrac{2}{a} + \dfrac{1}{2a}}{a + \dfrac{a}{2}}$

52. $\dfrac{3 - \dfrac{1}{y}}{2 - \dfrac{1}{y}}$

53. $\dfrac{2 + \dfrac{1}{x^2}}{\dfrac{1}{x} + \dfrac{2}{x^2}}$

54. $\dfrac{\dfrac{1}{a} + \dfrac{1}{b}}{\dfrac{1}{ab}}$

55. $\dfrac{\dfrac{6}{x + 2} + 4}{\dfrac{8}{x + 2} - 4}$

(6.6) *Solve each equation for the variable or perform the indicated operation.*

56. $\dfrac{x + 4}{9} = \dfrac{5}{9}$

57. $\dfrac{n}{10} = 9 - \dfrac{n}{5}$

58. $\dfrac{5y - 3}{7} = \dfrac{15y - 2}{28}$

59. $\dfrac{2}{x + 1} - \dfrac{1}{x - 2} = -\dfrac{1}{2}$

60. $\dfrac{1}{a + 3} + \dfrac{1}{a - 3} = -\dfrac{5}{a^2 - 9}$

61. $\dfrac{y}{2y + 2} + \dfrac{2y - 16}{4y + 4} = \dfrac{y - 3}{y + 1}$

62. $\dfrac{4}{x + 3} + \dfrac{8}{x^2 - 9} = 0$

63. $\dfrac{2}{x - 3} - \dfrac{4}{x + 3} = \dfrac{8}{x^2 - 9}$

64. $\dfrac{x - 3}{x + 1} - \dfrac{x - 6}{x + 5} = 0$ 65. $x + 5 = \dfrac{6}{x}$

Solve the equation for the indicated variable.

66. $\dfrac{4A}{5b} = x^2$, for b 67. $\dfrac{x}{7} + \dfrac{y}{8} = 10$, for y

(6.7) *Write each phrase as a ratio in fractional notation.*

68. 20 cents to 1 dollar

69. four parts red to six parts white

Solve each proportion.

70. $\dfrac{x}{2} = \dfrac{12}{4}$ 71. $\dfrac{20}{1} = \dfrac{x}{25}$

72. $\dfrac{32}{100} = \dfrac{100}{x}$ 73. $\dfrac{20}{2} = \dfrac{c}{5}$

74. $\dfrac{2}{x - 1} = \dfrac{3}{x + 3}$ 75. $\dfrac{4}{y - 3} = \dfrac{2}{y - 3}$

76. $\dfrac{y + 2}{y} = \dfrac{5}{3}$ 77. $\dfrac{x - 3}{3x + 2} = \dfrac{2}{6}$

Given the following prices charged for various sizes of an item, find the best buy.

78. Shampoo

   10 ounces for $1.29

   16 ounces for $2.15

79. Frozen green beans

   8 ounces for $0.89

   15 ounces for $1.63

   20 ounces for $2.36

Solve.

80. A machine can process 300 parts in 20 minutes. Find how many parts can be processed in 45 minutes.

81. As his consulting fee, Mr. Visconti charges $90.00 per day. Find how much he charges for 3 hours of consulting. Assume an 8-hour work day.

82. One fund raiser can address 100 letters in 35 minutes. Find how many he can address in 55 minutes.

(6.8) *Solve each problem.*

83. Five times the reciprocal of a number equals the sum of $\frac{3}{2}$ times the reciprocal of the number and $\frac{7}{6}$. What is the number?

84. The reciprocal of a number equals the reciprocal of the difference of 4 and the number. Find the number.

85. A car travels 90 miles in the same time that a car traveling 10 miles per hour slower travels 60 miles. Find the speed of each car.

86. The speed of a bayou near Lafayette, Louisiana, is 4 miles per hour. A paddle boat travels 48 miles upstream in the same amount of time it takes to travel 72 miles downstream. Find the speed of the boat in still water.

87. When Mark and Maria manicure Mr. Stergeon's lawn, it takes them 5 hours. If Mark works alone, it takes 7 hours. Find how long it takes Maria alone.

88. It takes pipe A 20 days to fill a fish pond. Pipe B takes 15 days. Find how long it takes both pipes together to fill the pond.

Given that the pairs of triangles are similar, find each missing length x.

△ **89.**

△ **90.**

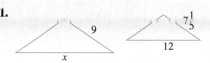

△ **91.**

△ **92.**

CHAPTER 6 TEST

1. Find any real numbers for which the following expression is undefined.

$$\frac{x + 5}{x^2 + 4x + 3}$$

2. For a certain computer desk, the manufacturing cost C per desk (in dollars) is

$$C = \frac{100x + 3000}{x}$$

where x is the number of desks manufactured.

a. Find the average cost per desk when manufacturing 200 computer desks.

b. Find the average cost per desk when manufacturing 1000 computer desks.

Simplify each rational expression.

3. $\dfrac{3x - 6}{5x - 10}$

4. $\dfrac{x + 10}{x^2 - 100}$

5. $\dfrac{x + 6}{x^2 + 12x + 36}$

6. $\dfrac{x + 3}{x^3 + 27}$

7. $\dfrac{2m^3 - 2m^2 - 12m}{m^2 - 5m + 6}$

8. $\dfrac{ay + 3a + 2y + 6}{ay + 3a + 5y + 15}$

9. $\dfrac{y - x}{x^2 - y^2}$

Perform the indicated operation and simplify if possible.

10. $\dfrac{x^2 - 13x + 42}{x^2 + 10x + 21} \div \dfrac{x^2 - 4}{x^2 + x - 6}$

11. $\dfrac{3}{x - 1} \cdot (5x - 5)$

12. $\dfrac{y^2 - 5y + 6}{2y + 4} \cdot \dfrac{y + 2}{2y - 6}$

13. $\dfrac{5}{2x + 5} - \dfrac{6}{2x + 5}$

14. $\dfrac{5a}{a^2 - a - 6} - \dfrac{2}{a - 3}$

15. $\dfrac{6}{x^2 - 1} + \dfrac{3}{x + 1}$

16. $\dfrac{x^2 - 9}{x^2 - 3x} \div \dfrac{xy + 5x + 3y + 15}{2x + 10}$

17. $\dfrac{x + 2}{x^2 + 11x + 18} + \dfrac{5}{x^2 - 3x - 10}$

18. $\dfrac{4y}{y^2 + 6y + 5} - \dfrac{3}{y^2 + 5y + 4}$

Solve each equation.

19. $\dfrac{4}{y} - \dfrac{5}{3} = \dfrac{-1}{5}$

20. $\dfrac{5}{y + 1} = \dfrac{4}{y + 2}$

21. $\dfrac{a}{a - 3} = \dfrac{3}{a - 3} - \dfrac{3}{2}$

22. $\dfrac{10}{x^2 - 25} = \dfrac{3}{x + 5} + \dfrac{1}{x - 5}$

Simplify each complex fraction.

23. $\dfrac{\dfrac{5x^2}{yz^2}}{\dfrac{10x}{z^3}}$

24. $\dfrac{\dfrac{b}{a} - \dfrac{a}{b}}{\dfrac{b}{a} + \dfrac{b}{a}}$

25. $\dfrac{5 - \dfrac{1}{y^2}}{\dfrac{1}{y} + \dfrac{2}{y^2}}$

26. In a sample of 85 fluorescent bulbs, 3 were found to be defective. At this rate, how many defective bulbs should be found in 510 bulbs?

27. One number plus five times its reciprocal is equal to six. Find the number.

28. A pleasure boat traveling down the Red River takes the same time to go 14 miles upstream as it takes to go 16 miles downstream. If the current of the river is 2 miles per hour, find the speed of the boat in still water.

29. An inlet pipe can fill a tank in 12 hours. A second pipe can fill the tank in 15 hours. If both pipes are used, find how long it takes to fill the tank.

▦ 30. Decide which is the best buy in crackers.

6 ounces for $1.19

10 ounces for $2.15

16 ounces for $3.25

△ 31. Given that the two triangles are similar, find x.

CHAPTER 6 CUMULATIVE REVIEW

1. Write each sentence as an equation. Let x represent the unknown number.
   a. The quotient of 15 and a number is 4.
   b. Three subtracted from 12 is a number.
   c. Four times a number added to 17 is 21.

2. Rajiv Puri invested part of his $20,000 inheritance in a mutual funds account that pays 7% simple interest yearly and the rest in a certificate of deposit that pays 9% simple interest yearly. At the end of one year, Rajiv's investments earned $1550. Find the amount he invested at each rate.

3. Graph $x - 3y = 6$ by finding and plotting intercepts.

4. Use the product rule to simplify.
   a. $4^2 \cdot 4^5$
   b. $x^2 \cdot x^5$
   c. $y^3 \cdot y$
   d. $y^3 \cdot y^2 \cdot y^7$
   e. $(-5)^7 \cdot (-5)^8$

5. Subtract $(5z - 7)$ from the sum of $(8z + 11)$ and $(9z - 2)$.

6. Multiply: $(3a + b)^3$

7. Use a special product to square each binomial.
   a. $(t + 2)^2$
   b. $(p - q)^2$
   c. $(2x + 5)^2$
   d. $(x^2 - 7y)^2$

8. Simplify each expression. Write results using positive exponents only.
   a. $\left(\dfrac{2}{3}\right)^{-3}$
   b. $\dfrac{1}{x^{-3}}$
   c. $\dfrac{p^{-4}}{q^{-9}}$

9. Divide: $\dfrac{4x^2 + 7 + 8x^3}{2x + 3}$

10. Find the GCF of each list of numbers.
    a. 28 and 40
    b. 55 and 21
    c. 15, 18, and 66

Factor.

11. $-9a^5 + 18a^2 - 3a$

12. $3m^2 - 24m - 60$

13. $3x^3 + 11x + 6$

14. $x^2 + 12x + 36$

15. $x^2 + 4$

16. $x^3 + 8$

17. $2x^3 + 3x^2 - 2x - 3$

18. $12m^2 - 3n^2$

19. Solve: $x(2x - 7) = 4$

20. Find the x-intercepts of the graph of $y = x^2 - 5x + 4$.

21. The height of a triangular sail is 2 meters less than twice the length of the base. If the sail has an area of 30 square meters, find the length of its base and the height.

22. Simplify: $\dfrac{5x - 5}{x^3 - x^2}$

23. Divide: $\dfrac{6x + 2}{x^2 - 1} \div \dfrac{3x^2 + x}{x - 1}$

24. Simplify: $\dfrac{\dfrac{x + 1}{y}}{\dfrac{x}{y} + 2}$

25. A supermarket offers a 14-ounce box of cereal for $3.79 and an 18-ounce box of the same brand of cereal for $4.99. Which is the better buy?

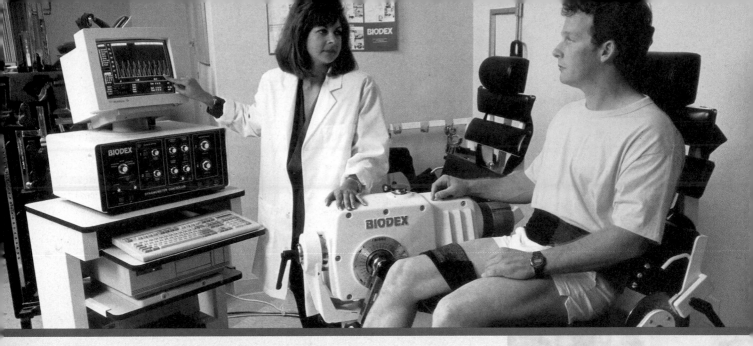

Helping Nearly One Million People Every Day

Do you know someone who has needed the services of a physical therapist? Chances are good that you do, because physical therapists and their staffs treat nearly one million people every day in this country. Working with patients with physical difficulties stemming from disease or injury, physical therapists evaluate mobility and endurance and then plan individualized treatment programs designed to rebuild strength, relieve pain, and recover mobility. Once a treatment strategy is in place, physical therapist assistants often administer treatment procedures to patients. Assistants also track patients' progress and report treatment outcomes to the physical therapist.

Physical therapists typically have a bachelor's or higher degree, and physical therapist assistants typically have an associate's degree. In either degree program, courses include first aid and CPR, biology, anatomy, physiology, mathematics, chemistry, and psychology. Strong math skills are helpful in situations such as tracking patients' treatment progress, measuring strength and range of motion, and properly setting up equipment used in treating patients.

 For more information about a career in physical therapy, visit the American Physical Therapy Association Website by first going to www.prenhall.com/martin-gay.

In the Spotlight on Decision Making feature on page 411, you will have the opportunity, as a physical therapist assistant, to make a decision concerning a patient's treatment plan.

FURTHER GRAPHING

7.1 THE SLOPE-INTERCEPT FORM

7.2 THE POINT-SLOPE FORM

7.3 GRAPHING NONLINEAR EQUATIONS

7.4 FUNCTIONS

In Chapter 3, linear equations and inequalities in two variables were graphed on a rectangular coordinate system. We found that the graph of a linear equation in two variables is simply a picture of its ordered pair solutions and this picture each time is a line. Chapter 3 also included a study of the tilt of a line, called its slope. Equations in two variables and the slope of their graphs have many applications in the fields of science and business, just to name a few. For example, every business would like a graph of its profits to show a positive slope over time.

  In this chapter, our study of slope and two variable equations is continued. Two-variable equations lead directly to the concept of function, perhaps the most important concept in all mathematics. Functions are introduced in the concluding section of this chapter.

7.1 THE SLOPE-INTERCEPT FORM

CD-ROM   SSM   ▶ **O B J E C T I V E S**

SSG   Video

△ **1.** Use the slope-intercept form to find the slope and the y-intercept of a line.

△ **2.** Use the slope-intercept form to determine whether two lines are parallel, perpendicular, or neither.

3. Use the slope-intercept form to write an equation of a line.

4. Use the slope-intercept form to graph a linear equation.

---

1   In Section 3.4, we learned that the slant of a line is called its slope. Recall that the slope of a line m is the ratio of vertical change (change in y) to the corresponding horizontal change (change in x) as we move along the line.

> ▼
> **H E L P F U L   H I N T**
> Don't forget that the slope of a line is the same no matter what ordered pairs of the line are used to calculate slope.

Example 1   Find the slope of the line whose equation is $y = \dfrac{3}{4}x + 6$.

Solution   To find the slope of this line, find any two ordered pair solutions of the equation. Let's find and use intercept points as our two ordered pair solutions. Recall from Section 3.2 that the graph of an equation of the form $y = mx + b$ has y-intercept b. This means that the graph of $y = \dfrac{3}{4}x + 6$ has a y-intercept of 6 as shown below.

To find the y-intercept, let $x = 0$. | To find the x-intercept, let $y = 0$.

Then $y = \dfrac{3}{4}x + 6$ becomes | Then $y = \dfrac{3}{4}x + 6$ becomes

$y = \dfrac{3}{4} \cdot 0 + 6$ or | $0 = \dfrac{3}{4}x + 6$ or

$y = 6$ | $0 = 3x + 24$ Multiply both sides by 4.

  | $-24 = 3x$ Subtract 24 from both sides.

  | $-8 = x$ Divide both sides by 3.

The y-intercept is $(0, 6)$, as expected. | The x-intercept is $(-8, 0)$.

Use the points $(0, 6)$ and $(-8, 0)$ to find the slope. Then

$$m = \frac{\text{change in } y}{\text{change in } x} = \frac{0 - 6}{-8 - 0} = \frac{-6}{-8} = \frac{3}{4}$$

The slope of the line is $\dfrac{3}{4}$.

Analyzing the results of Example 1, you may notice a striking pattern:

The slope of $y = \dfrac{3}{4} x + 6$ is $\dfrac{3}{4}$, the same as the coefficient of x.

Also, as mentioned earlier, the y-intercept is 6, the same as the constant term.

When a linear equation is written in the form $y = mx + b$, not only is b the y-intercept of the line, but m is its slope. The form $y = mx + b$ is appropriately called the **slope-intercept form**.

$\uparrow \quad \uparrow$
Slope y-intercept

SLOPE-INTERCEPT FORM

When a linear equation in two variables is written in slope-intercept form,

$$y = mx + b$$

m is the slope of the line and b is the y-intercept of the line.

Example 2 Find the slope and the y-intercept of the line whose equation is $5x + y = 2$.

Solution Write the equation in slope-intercept form by solving the equation for y.

$$5x + y = 2$$
$$y = -5x + 2 \qquad \text{Subtract } 5x \text{ from both sides.}$$

The coefficient of x, -5, is the slope and the constant term, 2, is the y-intercept. ■

Example 3 Find the slope and the y-intercept of the line whose equation is $3x - 4y = 4$.

Solution Write the equation in slope-intercept form by solving for y.

$$3x - 4y = 4$$
$$-4y = -3x + 4 \qquad \text{Subtract } 3x \text{ from both sides.}$$
$$\frac{-4y}{-4} = \frac{-3x}{-4} + \frac{4}{-4} \qquad \text{Divide both sides by } -4.$$
$$y = \frac{3}{4} x - 1 \qquad \text{Simplify.}$$

The coefficient of x, $\dfrac{3}{4}$, is the slope, and the constant term, -1, is the y-intercept. ■

---

2 The slope-intercept form can be used to determine whether two lines are parallel or perpendicular. Recall that nonvertical parallel lines have the same slope and different y-intercepts. Also, nonvertical perpendicular lines have slopes whose product is -1.

△ **Example 4**   Determine whether the graphs of $y = -\frac{1}{5}x + 1$ and $2x + 10y = 30$ are parallel lines, perpendicular lines, or neither.

    Solution   The graph of $y = -\frac{1}{5}x + 1$ is a line with slope $-\frac{1}{5}$ and with y-intercept 1. To find the slope and the y-intercept of the graph of $2x + 10y = 30$, write this equation in slope-intercept form. To do this, solve the equation for y.

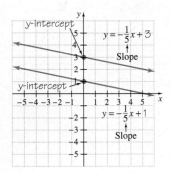

$$2x + 10y = 30$$

$$10y = -2x + 30 \qquad \text{Subtract } 2x \text{ from both sides.}$$

$$y = -\frac{1}{5}x + 3 \qquad \text{Divide both sides by 10.}$$

The graph of this equation is a line with slope $-\frac{1}{5}$ and y-intercept 3. Because both lines have the same slopes but different y-intercepts, these lines are parallel. ▬

3   The slope-intercept form can also be used to write the equation of a line given its slope and y-intercept.

Example 5   Find an equation of the line with y-intercept $(0, -3)$ and slope of $\frac{1}{4}$.

    Solution   We are given the slope and the y-intercept. Let $m = \frac{1}{4}$ and $b = -3$, and write the equation in slope-intercept form, $y = mx + b$.

$$y = mx + b$$

$$y = \frac{1}{4}x + (-3) \qquad \text{Let } m = \frac{1}{4} \text{ and } b = -3.$$

$$y = \frac{1}{4}x - 3 \qquad \text{Simplify.}$$ ▬

4   We now use the slope-intercept form of the equation of a line to graph a linear equation.

Example 6   Graph: $y = \frac{3}{5}x - 2$.

    Solution   Since the equation $y = \frac{3}{5}x - 2$ is written in slope-intercept form $y = mx + b$, the slope of its graph is $\frac{3}{5}$ and the y-intercept is $(0, -2)$. To graph, begin by plotting the intercept $(0, -2)$. From this point, find another point of the graph by using the slope $\frac{3}{5}$ and recalling that slope is $\frac{\text{rise}}{\text{run}}$. Start at the intercept point and move 3 units up since the numerator of the slope is 3; then move 5 units to the right since the denominator of the slope is 5. Stop at the point $(5, 1)$. The line through $(0, -2)$ and $(5, 1)$ is the graph of $y = \frac{3}{5}x - 2$.

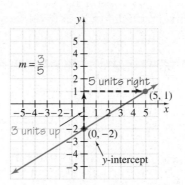

Example 7   Use the slope-intercept form to graph the equation $4x + y = 1$.

Solution   First, write the given equation in slope-intercept form.

$$4x + y = 1$$

$$y = -4x + 1$$

The graph of this equation will have slope -4 and y-intercept 1. To graph this line, first plot the intercept point $(0, 1)$. To find another point of the graph, use the slope -4, which can be written as $\frac{-4}{1}$. Start at the point $(0, 1)$ and move 4 units down (since the numerator of the slope is -4); then move 1 unit to the right (since the denominator of the slope is 1). We arrive at the point $(1, -3)$. The line through $(0, 1)$ and $(1, -3)$ is the graph of $4x + y = 1$.

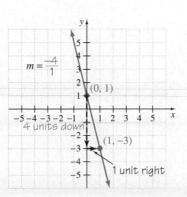

GRAPHING CALCULATOR EXPLORATIONS

A grapher is a very useful tool for discovering patterns. To discover the change in the graph of a linear equation caused by a change in slope, try the following. Use a standard window and graph a linear equation in the form $y = mx + b$. Recall that the graph of such an equation will have slope m and y-intercept b.

First graph $y = x + 3$. To do so, press the $\boxed{Y=}$ key and enter $Y_1 = x + 3$. Notice that this graph has slope 1 and that the y-intercept is 3. Next, on the same set of axes, graph $y = 2x + 3$ and $y = 3x + 3$ by pressing $\boxed{Y=}$ and entering $Y_2 = 2x + 3$ and $Y_3 = 3x + 3$.

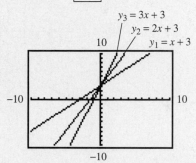

Notice the difference in the graph of each equation as the slope changes from 1 to 2 to 3. How would the graph of $y = 5x + 3$ appear? To see the change in the graph caused by a change in negative slope, try graphing $y = -x + 3$, $y = -2x + 3$, and $y = -3x + 3$ on the same set of axes.

Use a grapher to graph the following equations. For each exercise, graph the first equation and use its graph to predict the appearance of the other equations. Then graph the other equations on the same set of axes and check your prediction.

1. $y = x$; $y = 6x$, $y = -6x$

2. $y = -x$; $y = -5x$, $y = -10x$

3. $y = \frac{1}{2}x + 2$; $y = \frac{3}{4}x + 2$, $y = x + 2$

4. $y = x + 1$; $y = \frac{5}{4}x + 1$, $y = \frac{5}{2}x + 1$

5. $y = -7x + 5$; $y = 7x + 5$

6. $y = 3x - 1$; $y = -3x - 1$

MENTAL MATH

Identify the slope and the y-intercept of the graph of each equation.

1. $y = 2x - 1$

2. $y = -7x + 3$

3. $y = x + \frac{1}{3}$

4. $y = -x - \frac{2}{9}$

5. $y = \frac{5}{7}x - 4$

6. $y = -\frac{1}{4}x + \frac{3}{5}$

Exercise Set 7.1

Determine the slope and the y-intercept of the graph of each equation. See Examples 1 through 3.

1. $2x + y = 4$

2. $-3x + y = -5$

3. $x + 9y = 1$

4. $x + 7y = 6$

5. $4x - 3y = 12$

6. $2x - 6y = 12$

7. $x + y = 0$

8. $x - y = 0$

9. $y = -3$

10. $y + 2 = 0$

11. $-x + 5y = 20$

12. $-8x + 3y = 11$

Match each linear equation with its graph.

13. $y = 2x + 1$

14. $y = -x + 1$

15. $y = -3x - 2$

16. $y = \dfrac{5}{3}x - 2$

A.

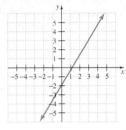

B.

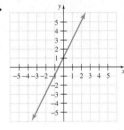

C.

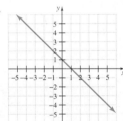

D.

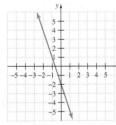

Determine whether the lines are parallel, perpendicular, or neither. See Example 4.

△ **17.** $x - 3y = -6$
    $3x - y = 0$

△ **18.** $-5x + y = -6$
    $x + 5y = 5$

△ **19.** $2x - 7y = 1$
    $2y = 7x - 2$

△ **20.** $y = 4x - 2$
    $4x + y = 5$

21. $10 + 3x = 5y$
△ $5x + 3y = 1$

△ **22.** $-x + 2y = -2$
    $2x = 4y + 3$

△ **23.** $6x = 5y + 1$
    $-12x + 10y = 1$

△ **24.** $11x + 10y = 3$
    $11y = 10x + 5$

25. Explain how to write the equation of a line if you are given the slope of the line and its y-intercept.

26. Explain why the graphs of $y = x$ and $y = -x$ are per-
△   pendicular lines.

Use the slope-intercept form of the linear equation to write an equation of each line with given slope and y-intercept. See Example 5.

27. Slope -1; y-intercept $(0, 1)$

28. Slope $\dfrac{1}{2}$; y-intercept $(0, -6)$

29. Slope 2; y-intercept $\left(0, \dfrac{3}{4}\right)$

30. Slope -3; y-intercept $\left(0, -\dfrac{1}{5}\right)$

31. Slope $\dfrac{2}{7}$; y-intercept $(0, 0)$

32. Slope $-\dfrac{4}{5}$; y-intercept $(0, 0)$

Use the slope-intercept form to graph each equation. See Examples 6 and 7.

33. $y = \dfrac{2}{3}x + 5$

34. $y = \dfrac{1}{4}x - 3$

35. $y = -\dfrac{3}{5}x - 2$

36. $y = -\dfrac{5}{7}x + 5$

37. $y = 2x + 1$

38. $y = -4x - 1$

39. $y = -5x$

40. $y = 6x$

41. $4x + y = 6$

42. $-3x + y = 2$

43. $x - y = -2$

44. $x + y = 3$

45. $3x + 5y = 10$

46. $2x + 3y = 5$

47. $4x - 7y = -14$

48. $3x - 4y = 4$

49. In 1992, durum yield in North Dakota was about 28 heads per square foot. In 1997, the yield had dropped to 22 heads per square foot. (*Source:* United States Department of Agriculture)

a. Write two ordered pairs of the form (years after 1992, durum yield) for this situation.

b. The relationship between years after 1992 and durum yield is approximately linear over this period. Use the ordered pairs from part (a) to write an equation of the line relating year to durum yield.

c. Use the linear equation from part (b) to estimate the durum yield in 2001.

Durum is a type of hard wheat. Flour made from it is used in spaghetti, macaroni, and other pastas.

50. In 1992, crude oil production by OPEC countries was 24.4 million barrels per day. In 1997, OPEC crude oil production had risen to 28.3 million barrels per day. (*Source:* Energy Information Administration)

a. Write two ordered pairs of the form (years after 1992, crude oil production) for this situation.

b. The relationship between years after 1992 and crude oil production is approximately linear over this period. Use the ordered pairs from part (a) to write an equation of the line relating year to crude oil production.

c. Use the linear equation from part (b) to estimate the crude oil production by OPEC countries in 2002.

51. The equation relating Celsius temperature and Fahrenheit temperature is linear and its graph is shown to the right. This line contains the point $(0, 32)$, which means that $0°$ Celsius is equivalent to $32°$ Fahrenheit. This line also passes through the point $(100, 212)$.

a. Write the meaning of the ordered pair $(100, 212)$ in words.

b. Use the graph and approximate the Fahrenheit temperature equivalent to $20°$ Celsius.

c. Use the graph and approximate the Celsius temperature equivalent to $80°$ Fahrenheit.

d. Use the ordered pairs $(0, 32)$ and $(100, 212)$ and the slope-intercept form and write an equation of the line relating Celsius temperatures and Fahrenheit temperatures. (*Hint:* Use both ordered pairs to find slope. Then substitute the slope and y-intercept in slope-intercept form.)

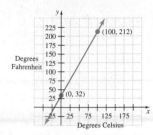

e. Use the linear equation found in part (d) to check your approximations from (b) and (c).

52. The equation relating temperatures on the Kelvin scale and the Celsius scale is linear. If $0°$ Celsius is equivalent to 273 Kelvin and $100°$ Celsius is equivalent to 373 Kelvin, find the linear equation that models the relationship between these two temperature scales. (*Hint:* See Exercise 51 and use the ordered pairs $(0, 273)$ and $(100, 373)$.)

53. Show that the slope of a line whose equation is $y = mx + b$ is m. To do so, find two ordered pair solutions of this equation and use the slope formula. For example, if $x = 0$, then $y = 0 \cdot x + b$ or $y = b$. Thus, one ordered pair solution is $(0, b)$. Find one more ordered pair solution and find the slope.

REVIEW EXERCISES

Graph each equation. See Section 3.3.

54. $y = 5$

55. $y = -1$

56. $x = -3$

57. $x = 2$

Write each equation in standard form: $Ax + By = C$. See Section 3.2.

58. $y - 5 = 2(x - 3)$

59. $y - 2 = 3(x - 6)$

60. $y + 2 = -1(x + 5)$

61. $y + 4 = -6(x + 1)$

7.2 THE POINT-SLOPE FORM

CD-ROM

SSM

SSG

Video

▶ **OBJECTIVES**

1. Use the point-slope form to find an equation of a line given its slope and a point of the line.
2. Use the point-slope form to find an equation of a line given two points of the line.
3. Find equations of vertical and horizontal lines.
4. Use the point-slope form to solve problems.

1 From the last section, we know that if the y-intercept of a line and its slope are given, then the line can be graphed and an equation of this line can be found. Our goal in this section is to answer the following: If any two points of a line are given, can an equation of the line be found? To answer this question, first let's see if we can find an equation of a line given one point (not necessarily the *y*-intercept) and the slope of the line.

Suppose that the slope of a line is −3 and the line contains the point (2, 4). For any other point with ordered pair (x, y) to be on the line, its coordinates must satisfy the slope equation

$$\frac{y - 4}{x - 2} = -3$$

Now multiply both sides of this equation by $x - 2$.

$$y - 4 = -3(x - 2)$$

This equation is a linear equation whose graph is a line that contains the point $(2, 4)$ and has slope -3. This form of a linear equation is called the **point-slope form.**

In general, when the slope of a line and any point on the line are known, the equation of the line can be found. To do this, use the slope formula to write the slope of a line that passes through points (x, y) and (x_1, y_1). We have

$$\frac{y - y_1}{x - x_1} = m$$

Multiply both sides of this equation by $x - x_1$ to obtain

$$y - y_1 = m(x - x_1)$$

$$\text{Slope}$$

POINT-SLOPE FORM OF THE EQUATION OF A LINE

The point-slope form of the equation of a line is $y - y_1 = m(x - x_1)$, where m is the slope of the line and (x_1, y_1) is a point on the line.

Example 1   Find an equation of the line passing through $(-1, 5)$ with slope -2. Write the equation in standard form: $Ax + By = C$.

Solution   Since the slope and a point on the line are given, use point-slope form $y - y_1 = m(x - x_1)$ to write the equation. Let $m = -2$ and $(-1, 5) = (x_1, y_1)$.

$$y - y_1 = m(x - x_1)$$
$$y - 5 = -2[x - (-1)] \qquad \text{Let } m = -2 \text{ and } (x_1, y_1) = (-1, 5).$$
$$y - 5 = -2(x + 1) \qquad \text{Simplify.}$$
$$y - 5 = -2x - 2 \qquad \text{Use the distributive property.}$$
$$y = -2x + 3 \qquad \text{Add 5 to both sides.}$$
$$2x + y = 3 \qquad \text{Add } 2x \text{ to both sides.}$$

In standard form, the equation is $2x + y = 3$.

2   We may also find the equation of a line given any two points of the line, as seen in the next example.

Example 2   Find an equation of the line through $(2, 5)$ and $(-3, 4)$. Write the equation in standard form.

Solution   First, use the two given points to find the slope of the line.

$$m = \frac{4 - 5}{-3 - 2} = \frac{-1}{-5} = \frac{1}{5}$$

Next, use the slope $\frac{1}{5}$ and either one of the given points to write the equation in point-slope form. We use $(2, 5)$. Let $x_1 = 2$, $y_1 = 5$, and $m = \frac{1}{5}$.

$y - y_1 = m(x - x_1)$	Use point-slope form.
$y - 5 = \frac{1}{5}(x - 2)$	Let $x_1 = 2$, $y_1 = 5$, and $m = \frac{1}{5}$.
$5(y - 5) = 5 \cdot \frac{1}{5}(x - 2)$	Multiply both sides by 5 to clear fractions.
$5y - 25 = x - 2$	Use the distributive property and simplify.
$-x + 5y - 25 = -2$	Subtract x from both sides.
$-x + 5y = 23$	Add 25 to both sides.

HELPFUL HINT
Multiply both sides of the equation $-x + 5y = 23$ by -1, and it becomes $x - 5y = -23$. Both $-x + 5y = 23$ and $x - 5y = -23$ are in standard form, and they are equations of the same line.

---

3   Recall from Section 3.3 that:

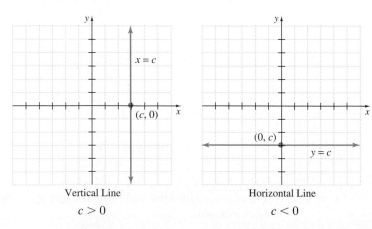

Vertical Line	Horizontal Line
$c > 0$	$c < 0$

Example 3   Find an equation of the vertical line through $(-1, 5)$.

Solution   The equation of a vertical line can be written in the form $x = c$, so an equation for a vertical line passing through $(-1, 5)$ is $x = -1$.

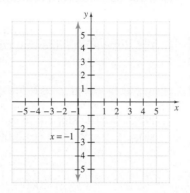

△ **Example 4**   Find an equation of the line parallel to the line $y = 5$ and passing through $(-2, -3)$.

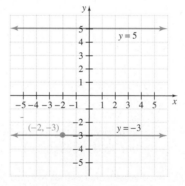

Solution   Since the graph of $y = 5$ is a horizontal line, any line parallel to it is also horizontal. The equation of a horizontal line can be written in the form $y = c$. An equation for the horizontal line passing through $(-2, -3)$ is $y = -3$.

4   Problems occurring in many fields can be modeled by linear equations in two variables. The next example is from the field of marketing and shows how consumer demand of a product depends on the price of the product.

Example 5   **PREDICTING THE SALES OF FRISBEES**

The Whammo Company has learned that by pricing a newly released Frisbee at $6, sales will reach 2000 Frisbees per day. Raising the price to $8 will cause the sales to fall to 1500 Frisbees per day.

a. Assume that the relationship between sales price and number of Frisbees sold is linear and write an equation describing this relationship. Write the equation in slope-intercept form.

b. Predict the daily sales of Frisbees if the price is $7.50.

Solution

a. First, use the given information and write two ordered pairs. Ordered pairs will be in the form (sales price, number sold) so that our ordered pairs are (6, 2000) and (8, 1500). Use the point-slope form to write an equation. To do so, we find the slope of the line that contains these points.

$$m = \frac{2000 - 1500}{6 - 8} = \frac{500}{-2} = -250$$

Next, use the slope and either one of the points to write the equation in point-slope form. We use (6, 2000).

$$y - y_1 = m(x - x_1) \qquad \text{Use point-slope form.}$$
$$y - 2000 = -250(x - 6) \qquad \text{Let } x_1 = 6, y_1 = 2000, \text{ and } m = -250.$$
$$y - 2000 = -250x + 1500 \qquad \text{Use the distributive property.}$$
$$y = -250x + 3500 \qquad \text{Write in slope-intercept form.}$$

b. To predict the sales if the price is $7.50, we find y when $x = 7.50$.

$$y = -250x + 3500$$
$$y = -250(7.50) + 3500 \qquad \text{Let } x = 7.50.$$
$$y = -1875 + 3500$$
$$y = 1625$$

If the price is $7.50, sales will reach 1625 Frisbees per day.

The preceding example may also be solved by using ordered pairs of the form (number sold, sales price).

FORMS OF LINEAR EQUATIONS

$Ax + By = C$	**Standard form** of a linear equation. A and B are not both 0.
$y = mx + b$	**Slope-intercept form** of a linear equation. The slope is m and the y-intercept is b.
$y - y_1 = m(x - x_1)$	**Point-slope form** of a linear equation. The slope is m and (x_1, y_1) is a point on the line.
$y = c$	**Horizontal line** The slope is 0 and the y-intercept is c.
$x = c$	**Vertical line** The slope is undefined and the x-intercept is c.

PARALLEL AND PERPENDICULAR LINES

Nonvertical parallel lines have the same slope.
The product of the slopes of two nonvertical perpendicular lines is −1.

SPOTLIGHT ON DECISION MAKING

Suppose you are a physical therapist. At weekly sessions, you are administering treadmill treatment to a patient. In the first phase of the treatment plan, the patient is to walk on a treadmill at a speed of 2 mph until fatigue sets in. Once the patient is able to walk 10 minutes at this speed, he will be ready for the next phase of treatment. However, you may change the treatment plan if it looks like it will take longer than 10 weeks to build up to 10 minutes of walking at 2 mph.

You record and plot the patient's progress on his chart for 5 weeks. Do you think you may need to change the first phase of treatment? Explain your reasoning. If not, when do you think the patient will be ready for the next phase of treatment?

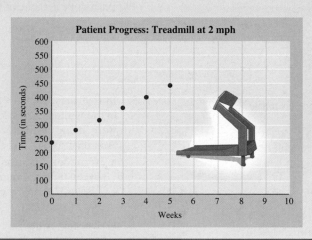

MENTAL MATH

The graph of each equation below is a line. Use the equation to identify the slope and a point of the line.

1. $y - 8 = 3(x - 4)$

2. $y - 1 = 5(x - 2)$

3. $y + 3 = -2(x - 10)$

4. $y + 6 = -7(x - 2)$

5. $y = \dfrac{2}{5}(x + 1)$

6. $y = \dfrac{3}{7}(x + 4)$

Exercise Set 7.2

Use the point-slope form of the linear equation to find an equation of each line with the given slope and passing through the given point. Then write the equation in standard form. See Example 1.

1. Slope 6; through $(2, 2)$

2. Slope 4; through $(1, 3)$

3. Slope -8; through $(-1, -5)$

4. Slope -2; through $(-11, -12)$

5. Slope $\dfrac{1}{2}$; through $(5, -6)$

6. Slope $\dfrac{2}{3}$; through $(-8, 9)$

Find an equation of the line through the given points. Write the equation in standard form. See Example 2.

7. Through $(3, 2)$ and $(5, 6)$

8. Through $(6, 2)$ and $(8, 8)$

9. Through $(-1, 3)$ and $(-2, -5)$

10. Through $(-4, 0)$ and $(6, -1)$

11. Through $(2, 3)$ and $(-1, -1)$

12. Through $(0, 0)$ and $\left(\dfrac{1}{2}, \dfrac{1}{3}\right)$

Find an equation of each line. See Example 3.

13. Vertical line through $(0, 2)$

14. Horizontal line through $(1, 4)$

15. Horizontal line through $(-1, 3)$

16. Vertical line through $(-1, 3)$

17. Vertical line through $\left(\dfrac{-7}{3}, \dfrac{-2}{5}\right)$

18. Horizontal line through $\left(\dfrac{2}{7}, 0\right)$

Find an equation of each line. See Example 4.

19. Parallel to $y = 5$, through $(1, 2)$

△ **20.** Perpendicular to $y = 5$, through $(1, 2)$

△ **21.** Perpendicular to $x = -3$, through $(-2, 5)$

△ **22.** Parallel to $y = -4$, through $(0, -3)$

△ **23.** Parallel to $x = 0$, through $(6, -8)$

△ **24.** Perpendicular to $x = 7$, through $(-5, 0)$

Find an equation of each line described. Write each equation in standard form.

25. With slope $-\dfrac{1}{2}$, through $\left(0, \dfrac{5}{3}\right)$

26. With slope $\dfrac{5}{7}$, through $(0, -3)$

27. Slope 1, through $(-7, 9)$

28. Slope 5, through $(6, -8)$

29. Through $(10, 7)$ and $(7, 10)$

30. Through $(5, -6)$ and $(-6, 5)$

△ **31.** Through $(6, 7)$, parallel to the x-axis

△ **32.** Through $(0, -5)$, parallel to the y-axis

33. Slope $-\dfrac{4}{7}$, through $(-1, -2)$

34. Slope $-\dfrac{3}{5}$, through $(4, 4)$

35. Through $(-8, 1)$ and $(0, 0)$

36. Through $(2, 3)$ and $(0, 0)$

37. Through $(0, 0)$ with slope 3

38. Through $(0, -2)$ with slope -1

39. Through $(-6, -6)$ and $(0, 0)$

40. Through $(0, 0)$ and $(4, 4)$

41. Slope -5, y-intercept 7

42. Slope -2; y-intercept -4

43. Through $(-1, 5)$ and $(0, -6)$

44. Through $(4, 0)$ and $(0, -5)$

45. With undefined slope, through $\left(-\dfrac{3}{4}, 1\right)$

46. With slope 0, through $(6.7, 12.1)$

△ **47.** Through $(-2, -3)$, perpendicular to the y-axis

△ **48.** Through $(0, 12)$, perpendicular to the x-axis

49. With slope 7, through $(1, 3)$

50. With slope -10, through $(5, -1)$

Solve. Assume that each exercise describes a linear relationship. See Example 5.

51. In 1998 there were 4760 electric-powered vehicles in use in the United States. In 1996 only 3280 electric vehicles were being used. (*Source:* U.S. Energy Information Administration)

   a. Write an equation describing the relationship between time and number of electric-powered vehicles. Use ordered pairs of the form (years past 1996, number of vehicles).

   b. Use this equation to predict the number of electric-powered vehicles in use in 2005.

52. In 1990 there were 403 thousand eating establishments in the United States. In 1996, there were 457 thousand eating establishments.

   a. Write an equation describing the relationship between time and number of eating establishments. Use ordered pairs of the form (years past 1990, number of eating establishments in thousands).

   b. Use this equation to predict the number of eating establishments in 2006.

53. A rock is dropped from the top of a 400-foot cliff. After 1 second, the rock is traveling 32 feet per second. After 3 seconds, the rock is traveling 96 feet per second.

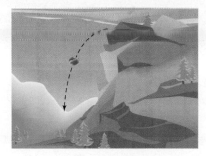

a. Assume that the relationship between time and speed is linear and write an equation describing this relationship. Use ordered pairs of the form (time, speed).

b. Use this equation to determine the speed of the rock 4 seconds after it was dropped.

54. In 1990 it cost a family of four (on average) $128.30 per week for food. In 1997 it cost an average of $127.30 per week to feed a family of four. (*Source:* U.S. Department of Agriculture)

a. Write an equation describing the relationship between time and cost to feed a family of four. Use ordered pairs of the form (years past 1990, cost).

b. Use this equation to predict the average cost of feeding a family of four in 2004.

55. In 1990 there were 150 thousand apparel and accessory stores in the United States. In 1996 there were a total of 135 thousand apparel and accessory stores. (*Source:* U.S. Bureau of the Census, *County Business Patterns*, annual)

a. Write an equation describing this relationship. Use ordered pairs of the form (years past 1990, number of stores in thousands).

b. Use this equation to predict the number of apparel and accessory stores in 2008.

56. The value of a building bought in 1985 may be depreciated (or decreased) as time passes for income tax purposes. Seven years after the building was bought, this value was $165,000 and 12 years after it was bought, this value was $140,000.

a. If the relationship between number of years past 1985 and the depreciated value of the building is linear, write an equation describing this relationship. Use or-

dered pairs of the form (years past 1985, value of building).

b. Use this equation to estimate the depreciated value of the building in 2005.

△ **57.** A linear equation can be written that relates the radius of a circle to its circumference. A circle with a radius of 10 centimeters has a circumference of approximately 63 centimeters. A circle with a radius of 15 centimeters has a circumference of approximately 94 centimeters. Use the ordered pairs $(10, 63)$ and $(15, 94)$ to write a linear equation in standard form that approximates the relationship between the radius of the circle and its circumference.

58. Del Monte Fruit Company is studying the sales of a pineapple sauce to see if this product is to be continued. At the end of its first year, profits on this product amounted to $30,000. At the end of the fourth year, profits are $66,000.

a. Assume that the relationship between years on the market and profit is linear and write an equation describing this relationship. Use ordered pairs of the form (years on the market, profit).

b. Use this equation to predict the profit at the end of 7 years.

59. The value of a computer bought in 1995 depreciates or decreases as time passes. Two years after the computer was bought, it was worth $2600 and 5 years after it was bought, it was worth $2000.

a. If the relationship between number of years past 1995 and value of the computer is linear, write an equation describing this relationship. Use ordered pairs of the form (years past 1995, value of computer).

b. Use this equation to estimate the value of the computer in the year 2005.

60. The Pool Fun Company has learned that, by pricing a newly released Fun Noodle at $3, sales will reach 10,000 Fun Noodles per day during the summer. Raising the price to $5 will cause the sales to fall to 8000 Fun Noodles per day.

a. Assume that the relationship between sales price and number of Fun Noodles sold is linear and write an equation describing this relationship. Use ordered pairs of the form (sales price, number sold).

b. Predict the daily sales of Fun Noodles if the price is $3.50.

61. Given the equation of a nonvertical line, explain how to find the slope without finding two points on the line.

62. Given two points on a nonvertical line, explain how to use the point-slope form to find the equation of the line.

△ **63.** Write an equation in standard form of the line that contains the point $(-1, 2)$ and is
a. parallel to the line $y = 3x - 1$.
b. perpendicular to the line $y = 3x - 1$.

△ **64.** Write an equation in standard form of the line that contains the point $(4, 0)$ and is
a. parallel to the line $y = -2x + 3$.
b. perpendicular to the line $y = -2x + 3$.

△ **65.** Write an equation in standard form of the line that contains the point $(3, -5)$ and is

a. parallel to the line $3x + 2y = 7$.
b. perpendicular to the line $3x + 2y = 7$.

△ **66.** Write an equation in standard form of the line that contains the point $(-2, 4)$ and is
a. parallel to the line $x + 3y = 6$.
b. perpendicular to the line $x + 3y = 6$.

REVIEW EXERCISES

Graph each linear equation. See Section 3.2.

67. $y = 2x - 6$

68. $y = -5x$

69. $x + 3y = 5$

70. $5x - y = 4$

71. $y = -2$

72. $x = 4$

7.3 GRAPHING NONLINEAR EQUATIONS

CD-ROM   SSM

SSG   Video

▶ **OBJECTIVES**

1. Identify and graph nonlinear equations.
2. Write ordered pairs from the graphs of nonlinear equations.

$y = x^2$ x

Recall from Section 3.2 that a linear equation in two variables is an equation that can be written in the form $Ax + By = C$ where A and B are not both 0. We also know that the graph of a linear equation is a line. Linear equations are important in their own right, but they are not sufficient for describing all relationships between two quantities. For example, the area, y, of a square is related to the length of its side x, by the nonlinear equation $y = x^2$.

1   In this section, we practice identifying and graphing nonlinear equations.

◆ **Example 1**   Graph the equation $y = x^2$.

Solution   This equation is not linear and its graph is not a line. Recall that the graph of an equation is a picture of its ordered pair solutions. To graph this equation, begin by finding ordered pair solutions. Because this equation is solved for y, we choose x-values and find corresponding y-values.

x	y
-3	9
-2	4
-1	1
0	0
1	1
2	4
3	9

If $x = -3$, then $y = (-3)^2$, or 9
If $x = -2$, then $y = (-2)^2$, or 4
If $x = -1$, then $y = (-1)^2$, or 1
If $x = 0$, then $y = 0^2$, or 0
If $x = 1$, then $y = 1^2$, or 1
If $x = 2$, then $y = 2^2$, or 4
If $x = 3$, then $y = 3^2$, or 9

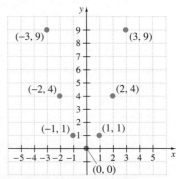

To complete the graph, connect these plotted points with a smooth curve as shown below.

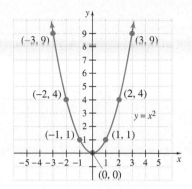

This curve is called a **parabola**. The equation $y = x^2$ is called a **quadratic equation** and the graph of a quadratic equation is a parabola. We will study more about parabolas in Chapter 10.

QUADRATIC EQUATION IN TWO VARIABLES

A quadratic equation in two variables is an equation that can be written in the form

$$y = ax^2 + bx + c$$

where a, b, and c are real numbers and a is not 0. The graph of a quadratic equation is a parabola.

◆**Example 2**   Graph the equation $y = |x|$.

Solution   This is not a linear equation and its graph is not a line. Because we do not know the shape of the graph of this equation, we find many ordered pair solutions. Choose x-values and substitute to find corresponding y-values.

x	y
-4	4
-2	2
0	0
$\frac{1}{2}$	$\frac{1}{2}$
1	1
3	3

If $x = -4$, then $y = |-4|$, or 4
If $x = -2$, then $y = |-2|$, or 2
If $x = 0$, then $y = |0|$, or 0
If $x = \frac{1}{2}$, then $y = \left|\frac{1}{2}\right|$, or $\frac{1}{2}$
If $x = 1$, then $y = |1|$, or 1
If $x = 3$, then $y = |3|$, or 3

From the plotted ordered pairs, we see that the graph of this absolute value equation is V-shaped. The completed graph is shown below.

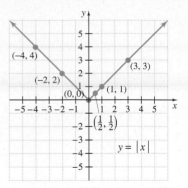

Example 3 Graph the equation $y = |x| - 3$.

Solution The graph of $y = |x| - 3$ is simply the graph of $y = |x|$ moved down 3 units. To see this, choose x-values and substitute to find corresponding y-values.

x	y
-4	1
-2	-1
0	-3
2	-1
4	1

If $x = -4$, then $y = |-4| - 3$, or 1
If $x = -2$, then $y = |-2| - 3$, or -1
If $x = 0$, then $y = |0| - 3$, or -3
If $x = 2$, then $y = |2| - 3$, or -1
If $x = 4$, then $y = |4| - 3$, or 1

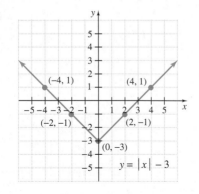

Notice that the graph of $y = |x| - 3$ is the graph of $y = |x|$ moved down 3 units.

In general, the shape of the graph of $y = |x| + k$ where k is a constant is V-shaped.

2 The concept of a relation and a function is introduced in the next section. To prepare for this section, we review reading ordered pairs from a graph.

Example 4 Find the missing coordinate so that each ordered pair is a solution of the equation graphed below.

a. $(-1, \ )$ **b.** $(-2, \ )$ **c.** $(\ , -1)$

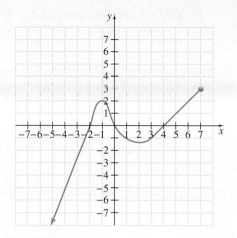

Solution **a.** To find the *y*-coordinate that corresponds to an *x*-value of −1, find −1 on the *x*-axis and move vertically until the graph is reached. From the point on the graph, move horizontally until the *y*-axis is reached and read the *y*-value. The *y*-value is 2, so the ordered pair is (−1, 2).

b. The completed ordered pair is (−2, 0).

c. The *y*-value of −1 corresponds to three *x*-values as shown below. The completed three ordered pairs are (−2.5, −1), (1, −1), and (3, −1).

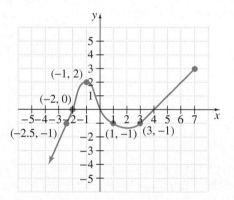

Example 5 Use the graph of Example 4 to

a. find the coordinates of the point on the graph with the greatest *y*-value;

b. find the coordinates of the point on the graph with the least *y*-value;

c. list the *x*-intercept(s) and the *y*-intercept(s).

Solution **a.** The point on the graph with the greatest *y*-value is the "highest" point. This point has coordinates (7, 3).

b. The point on the graph with the least *y*-value is the "lowest" point. The arrow shown on the graph means that the graph continues in the same manner. This means that this graph contains no lowest point, that is, no point with the least *y*-value.

c. The x-intercepts are $(-2, 0)$, $(0, 0)$, and $(4, 0)$. The y-intercept is $(0, 0)$.

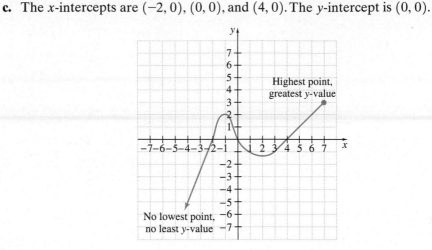

GRAPHING CALCULATOR EXPLORATIONS

We can use a grapher to approximate the x-intercepts of the graph of an equation such as the quadratic equation $y = x^2 - 2x - 4$. When we use a standard window, the graph of the equation looks like this:

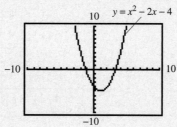

The graph appears to have one x-intercept between -2 and -1 and one between 3 and 4. To find the x-intercept between 3 and 4 to the nearest hundredth, we can use a ZERO (or ROOT) feature, a ZOOM feature (which magnifies a portion of the graph around the cursor), or we can redefine our window. If we redefine our window to

$$\text{Xmin} = 2 \qquad \text{Ymin} = -1$$

$$\text{Xmax} = 5 \qquad \text{Ymax} = 1$$

$$\text{Xscl} = 1 \qquad \text{Yscl} = 1$$

and trace along the graph, the resulting screen is

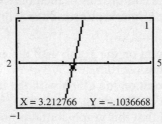

(continued)

By using the TRACE feature, we can now see that one of the intercepts is between 3.21 and 3.25. To approximate to the nearest hundredth, ZOOM again or redefine the window to

Xmin = 3.2   Ymin = −0.1

Xmax = 3.3   Ymax = 0.1

Xscl = 1   Yscl = 1

If we use the TRACE feature again, we see that, to the nearest hundredth, the x-intercept is 3.24. By repeating this process, we can approximate the other x-intercept to be −1.24. To check, find y when $x = 3.24$ and when $x = -1.24$. Both of these y-values should be close to 0. (They will not be exactly 0 since we approximated these solutions.)

Approximate to two decimal places the x-intercepts of the graph of each equation.

1. $y = x^2 + 3x - 2$

2. $y = 5x^2 - 7x + 1$

3. $y = 2.3x^2 - 4.4x - 5.6$

4. $y = 0.2x^2 + 6.2x + 2.1$

5. $y = 2|x| - 1.3$

6. $y = |-3x| + 2.7$

7. $y = 1.7x^3 + 5.9$

8. $y = -3.6x^3 - 1.3$

Exercise Set 7.3

Graph each nonlinear equation. See Examples 1 through 3.

1. $y = x^2 + 2$

2. $y = x^2 - 2$

3. $y = |x| - 1$

4. $y = |x| + 3$

5. $y = -x^2$

6. $y = -|x|$

7. $y = |x + 5|$

8. $y = (x - 5)^2$

Match each equation with its graph.

9. $y = |x| + 1$

10. $y = |x| - 4$

11. $y = x^2 - 6$

12. $y = x^2 + 2$

13. $y = 3x - 2$

14. $y = x + 1$

A.

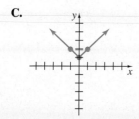

B.

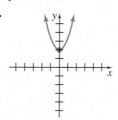

C.

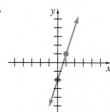

D.

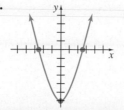

E.

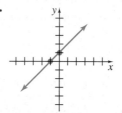

F.

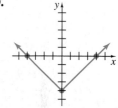

Use the graph to complete each ordered pair. See Example 4.

15. $(0, \ ) \ (\ , 0)$

16. $(3, \ ) \ (\ , -2)$

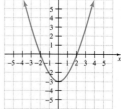

17. $(2, \ ) \ (\ , -2)$

18. $(2, \ ) \ (1, \ )$

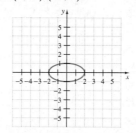

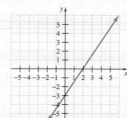

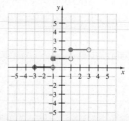

19. (2,)

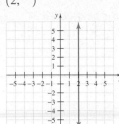

20. (, −3)

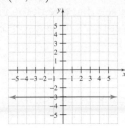

Answer the following questions about the given graph. See Example 5.

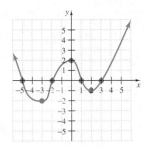

21. Find the coordinates of the point with the greatest *y*-value.

22. Find the coordinates of the point with the least *y*-value.

23. Between *x* = 1 and *x* = 3, find the coordinates of the point with the least *y*-value.

24. List the *x*- and *y*-intercept points for the graph.

Determine whether each equation is linear or not. Then graph each equation.

25. $x + y = 3$ **26.** $y - x = 8$

27. $y = 4x$ **28.** $y = 6x$

29. $y = 4x - 2$ **30.** $y = 6x - 5$

31. $y = |x| + 3$ **32.** $y = |x| + 2$

33. $2x - y = 5$ **34.** $4x - y = 7$

35. $y = 2x^2$ **36.** $y = 3x^2$

37. $y = (x - 3)^2$ **38.** $y = (x + 3)^2$

39. $y = -2x$ **40.** $y = -3x$

41. $y = -2x + 3$ **42.** $y = -3x + 2$

43. $y = x^2 + 4x$ **44.** $y = x^2 - 2x$

45. Graph the equation $y = x^3$ by completing the table of values, plotting the ordered pair solutions on a rectangular coordinate system, and then connecting the points with a smooth curve.

x	−3	−2	−1	0	1	2	3
y							

46. Graph the equation $y = x^4$ by completing the table of values, plotting the ordered pair solutions on a rectangular coordinate system, and then connecting the points with a smooth curve.

x	−3	−2	−1	0	1	2	3
y							

47. Draw a curve that has four *x*-intercepts and one *y*-intercept.

48. What is the greatest number of *x*-intercepts that a graph will have? Explain your answer.

49. What is the greatest number of *y*-intercepts that a graph will have? Explain your answer.

REVIEW EXERCISES

Find the value of $x^2 - 3x + 1$ for each given value of x. See Section 2.1.

50. 2   **51.** 5

52. −1   **53.** −3

For each graph, determine whether any x-values correspond to two or more y-values. See Section 7.3.

54.

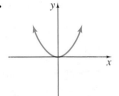

55.

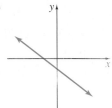

56.

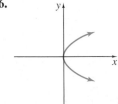

57.

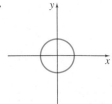

7.4 FUNCTIONS

CD-ROM SSM

SSG Video

▶ **OBJECTIVES**

1. Identify relations, domains, and ranges.
2. Identify functions.
3. Use the vertical line test.
4. Use function notation.

1 In previous sections, we have discussed the relationships between two quantities. For example, the relationship between the length of the side of a square x and its area y is described by the equation $y = x^2$. These variables x and y are related in the following way: for any given value of x, we can find the corresponding value of y by squaring the x-value. Ordered pairs can be used to write down solutions of this equation. For example, $(2, 4)$ is a solution of $y = x^2$, and this notation tells us that the x-value 2 is related to the y-value 4 for this equation. In other words, when the length of the side of a square is 2 units, its area is 4 units square.

A set of ordered pairs is called a **relation**. The set of all x-coordinates is called the **domain** of a relation, and the set of all y-coordinates is called the **range** of a relation. Equations such as $y = x^2$ are also called relations since equations in two variables define a set of ordered pair solutions.

Example 1 Find the domain and the range of the relation $\{(0, 2), (3, 3), (-1, 0), (3, -2)\}$.

Solution The domain is the set of all x-values or $\{-1, 0, 3\}$, and the range is the set of all y-values, or $\{-2, 0, 2, 3\}$.

2 Some relations are also functions.

> **FUNCTION**
>
> A function is a set of ordered pairs that assigns to each x-value exactly one y-value.

Example 2 Which of the following relations are also functions?

a. $\{(-1, 1), (2, 3), (7, 3), (8, 6)\}$ **b.** $\{(0, -2), (1, 5), (0, 3), (7, 7)\}$

Solution **a.** Although the ordered pairs $(2, 3)$ and $(7, 3)$ have the same y-value, each x-value is assigned to only one y-value so this set of ordered pairs is a function.

b. The x-value 0 is assigned to two y-values, -2 and 3, so this set of ordered pairs is not a function.

Relations and functions can be described by a graph of their ordered pairs.

Example 3 Which graph is the graph of a function?

a.
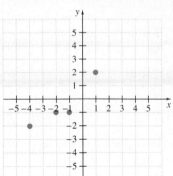

b.
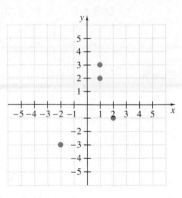

Solution **a.** This is the graph of the relation $\{(-4, -2), (-2, -1)\ (-1, -1), (1, 2)\}$. Each x-coordinate has exactly one y-coordinate, so this is the graph of a function.

b. This is the graph of the relation $\{(-2, -3), (1, 2), (1, 3), (2, -1)\}$. The x-coordinate 1 is paired with two y-coordinates, 2 and 3, so this is not the graph of a function.

3 The graph in Example 3(b) was not the graph of a function because the x-coordinate 1 was paired with two y-coordinates, 2 and 3. Notice that when an x-coordinate is paired with more than one y-coordinate, a vertical line can be drawn that will intersect the graph at more than one point. We can use this fact to determine whether a relation is also a function. We call this the **vertical line test**.

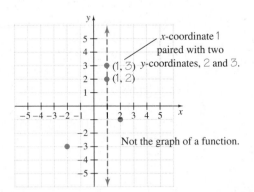

VERTICAL LINE TEST

If a vertical line can be drawn so that it intersects a graph more than once, the graph is not the graph of a function.

This vertical line test works for all types of graphs on the rectangular coordinate system.

Example 4   Use the vertical line test to determine whether each graph is the graph of a function.

a.

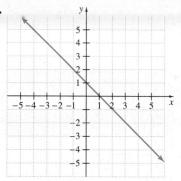

b.

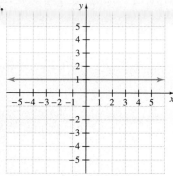

c.

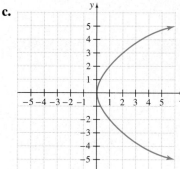

d.

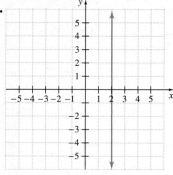

Solution   **a.** This graph is the graph of a function since no vertical line will intersect this graph more than once.

b. This graph is also the graph of a function; no vertical line will intersect it more than once.

c. This graph is not the graph of a function. Vertical lines can be drawn that intersect the graph in two points. An example of one is shown.

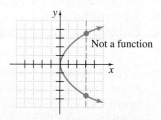

Not a function

d. This graph is not the graph of a function. A vertical line can be drawn that intersects this line at every point.

Recall that the graph of a linear equation is a line, and a line that is not vertical will pass the vertical line test. **Thus, all linear equations are functions except those of the form $x = c$, which are vertical lines.**

Example 5 Which of the following linear equations are functions?

a. $y = x$ **b.** $y = 2x + 1$

c. $y = 5$ **d.** $x = -1$

Solution **a**, **b**, and **c** are functions because their graphs are nonvertical lines.
d is not a function because its graph is a vertical line.

Examples of functions can often be found in magazines, newspapers, books, and other printed material in the form of tables or graphs such as that in Example 6.

Example 6 The graph shows the sunrise time for Indianapolis, Indiana, for the year. Use this graph to answer the questions.

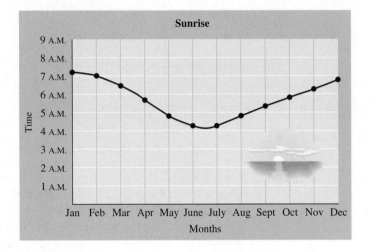

a. Approximate the time of sunrise on February 1.
b. Approximately when does the sun rise at 5 A.M.?
c. Is this the graph of a function?

Solution **a.** To approximate the time of sunrise on February 1, we find the mark on the horizontal axis that corresponds to February 1. From this mark, we move vertically upward until the graph is reached. From that point on the graph, we move horizontally to the left until the vertical axis is reached. The vertical axis there reads 7 A.M.

b. To approximate when the sun rises at 5 A.M., we find 5 A.M. on the time axis and move horizontally to the right. Notice that we will reach the graph twice, corresponding to two dates for which the sun rises at 5 A.M. We follow both points on the graph vertically downward until the horizontal axis is reached. The sun rises at 5 A.M. at approximately the end of the month of April and the middle of the month of August.

Sunrise

[graph showing sunrise times with Time (A.M.) on vertical axis from 1 to 9, and Months on horizontal axis from Jan to Dec]

c. The graph is the graph of a function since it passes the vertical line test. In other words, for every day of the year in Indianapolis, there is exactly one sunrise time.

4 The graph of the linear equation $y = 2x + 1$ passes the vertical line test, so we say that $y = 2x + 1$ is a function. In other words, $y = 2x + 1$ gives us a rule for writing ordered pairs where every x-coordinate is paired with one y-coordinate.

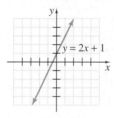

We often use letters such as f, g, and h to name functions. For example, the symbol $f(x)$ means *function of x* and is read "*f* of *x*." This notation is called **function notation**. The equation $y = 2x + 1$ can be written as $f(x) = 2x + 1$ using function notation, and these equations mean the same thing. In other words, $y = f(x)$.

The notation $f(1)$ means to replace x with 1 and find the resulting y or function value. Since

$$f(x) = 2x + 1$$

then

$$f(1) = 2(1) + 1 = 3$$

This means that, when $x = 1$, y or $f(x) = 3$, and we have the ordered pair $(1, 3)$. Now let's find $f(2)$, $f(0)$, and $f(-1)$.

$f(x) = 2x + 1$	$f(x) = 2x + 1$	$f(x) = 2x + 1$
$f(2) = 2(2) + 1$	$f(0) = 2(0) + 1$	$f(-1) = 2(-1) + 1$
$= 4 + 1$	$= 0 + 1$	$= -2 + 1$
$= 5$	$= 1$	$= -1$

Ordered
Pair: $(2, 5)$ $(0, 1)$ $(-1, -1)$

HELPFUL HINT
Note that $f(x)$ is a special symbol in mathematics used to denote a function. The symbol $f(x)$ is read "*f* of *x*." It does **not** mean $f \cdot x$ (*f* times *x*).

Example 7 Given $g(x) = x^2 - 3$, find the following.

    a. $g(2)$ **b.** $g(-2)$ **c.** $g(0)$

Solution   **a.** $g(x) = x^2 - 3$ **b.** $g(x) = x^2 - 3$ **c.** $g(x) = x^2 - 3$

               $g(2) = 2^2 - 3$ $g(-2) = (-2)^2 - 3$ $g(0) = 0^2 - 3$

                    $= 4 - 3$ $= 4 - 3$ $= 0 - 3$

                    $= 1$ $= 1$ $= -3$

We now practice finding the domain and the range of a function. The domain of our functions will be the set of all possible real numbers that x can be replaced by. The range is the set of corresponding y-values.

Example 8 Find the domain of each function.

    a. $g(x) = \dfrac{1}{x}$ **b.** $f(x) = 2x + 1$

Solution   **a.** Recall that we cannot divide by 0 so that the domain of $g(x)$ is the set of all real numbers except 0.

        b. In this function, x can be any real number. The domain of $f(x)$ is the set of all real numbers.

Example 9 Find the domain and the range of each function graphed.

    a.

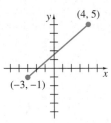

    b.

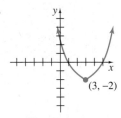

Solution   **a.**

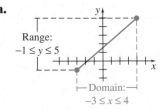

    b.

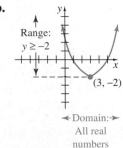

SPOTLIGHT ON DECISION MAKING

Suppose you are the property and grounds manager for a small company in Portland, Oregon. A recent heat wave has caused employees to ask for air-conditioning in the building. As you begin to research air-conditioning installation, you discover this graph of average high temperatures for Portland. Because the building is surrounded by trees, the current ventilation system (without air-conditioning) is able to keep the temperature 10°F cooler than the outside temperature on a warm day. A comfortable temperature range for working is 68°F to 74°F. Would you recommend installing air-conditioning in the building? Explain your reasoning. What other factors would you want to consider?

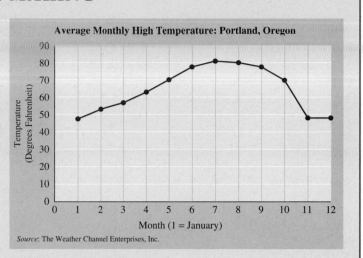

Source: The Weather Channel Enterprises, Inc.

Exercise Set 7.4

Find the domain and the range of each relation. See Example 1.

1. $\{(2, 4), (0, 0), (-7, 10), (10, -7)\}$

2. $\{(3, -6), (1, 4), (-2, -2)\}$

3. $\{(0, -2), (1, -2) (5, -2)\}$

4. $\{(5, 0), (5, -3), (5, 4), (5, 3)\}$

Determine which relations are also functions. See Example 2.

5. $\{(1, 1), (2, 2), (-3, -3), (0, 0)\}$

6. $\{(1, 2), (3, 2), (4, 2)\}$

7. $\{(-1, 0), (-1, 6), (-1, 8)\}$

8. $\{(11, 6), (-1, -2), (0, 0), (3, -2)\}$

Use the vertical line test to determine whether each graph is the graph of a function. See Examples 3 and 4.

9.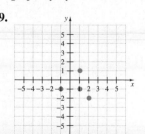

10.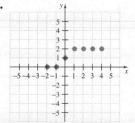

11.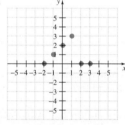

12.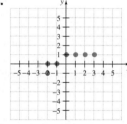

13.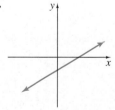

14.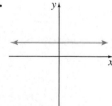

15.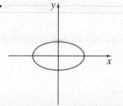

16.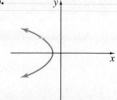

17.

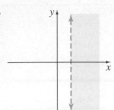

18.

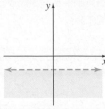

19.

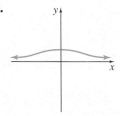

20.

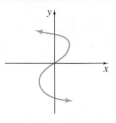

Decide whether the equation describes a function. See Example 5.

21. $y = x + 1$

22. $y = x - 1$

23. $y - x = 7$

24. $2x - 3y = 9$

25. $y = 6$

26. $x = 3$

27. $x = -2$

28. $y = -9$

29. $y < 2x + 1$

30. $y \leq x - 4$

Use the graph in Example 6 to answer Exercises 17–20.

31. Approximate the time of sunrise on September 1 in Indianapolis.

32. Approximate the date(s) when the sun rises in Indianapolis at 7 A.M.

33. Describe the change in sunrise over the year for Indianapolis.

34. When, in Indianapolis, is the earliest sunrise? What point on the graph does this correspond to?

This graph shows the U.S. hourly minimum wage at the beginning of each year shown. Use this graph to answer Exercises 35 through 40.

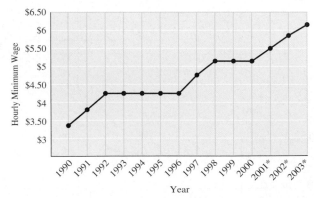

*Already passed by Congress

35. Approximate the minimum wage at the beginning of 1997.

36. Approximate the minimum wage at the beginning of 1999.

37. Approximate the year when the minimum wage will increase to over $6.00 per hour.

38. Approximate the year when the minimum wage increased to over $5.00 per hour.

39. Is this graph the graph of a function? Why or why not?

40. Do you think that a similar graph of your hourly wage on January 1 of every year (whether you are working or not) will be the graph of a function? Why or why not?

Given the following functions, find $f(-2)$, $f(0)$, and $f(3)$. See Example 7.

41. $f(x) = 2x - 5$

42. $f(x) = 3 - 7x$

43. $f(x) = x^2 + 2$

44. $f(x) = x^2 - 4$

45. $f(x) = x^3$

46. $f(x) = -x^3$

47. $f(x) = |x|$

48. $f(x) = |2 - x|$

Given the following functions, find $h(-1)$, $h(0)$, and $h(4)$. See Example 7.

49. $h(x) = 5x$

50. $h(x) = -3x$

51. $h(x) = 2x^2 + 3$

52. $h(x) = -x^2$

53. $h(x) = -x^2 - 2x + 3$

54. $h(x) = -x^2 + 4x - 3$

55. $h(x) = 6$

56. $h(x) = -12$

Find the domain of each function. See Example 8.

57. $f(x) = 3x - 7$

58. $g(x) = 5 - 2x$

59. $h(x) = \dfrac{1}{x + 5}$

60. $f(x) = \dfrac{1}{x - 6}$

61. $g(x) = |x + 1|$

62. $h(x) = |2x|$

Find the domain and the range of each function graphed. See Example 9.

63.

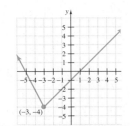

64.

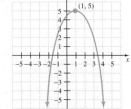

65.

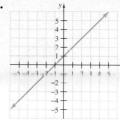

66.

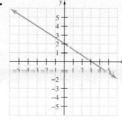

67.

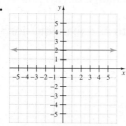

68.

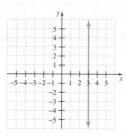

69. Forensic scientists use the function
$$H(x) = 2.59x + 47.24$$
to estimate the height of a woman given the length x of her femur bone.

a. Estimate the height of a woman whose femur measures 46 centimeters.

b. Estimate the height of a woman whose femur measures 39 centimeters.

70. The dosage in milligrams D of Ivermectin, a heartworm preventive for a dog who weighs x pounds, is given by the function
$$D(x) = \frac{136}{25} x$$

a. Find the proper dosage for a dog that weighs 35 pounds.

b. Find the proper dosage for a dog that weighs 70 pounds.

71. In your own words define **(a)** function; **(b)** domain; **(c)** range.

72. Explain the vertical line test and how it is used.

73. Since $y - x + 7$ is a function, rewrite the equation using function notation.

REVIEW EXERCISES

Find the coordinates of the point of intersection. See Section 3.1.

74.

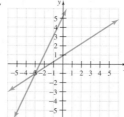

75.

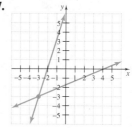

76.

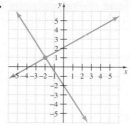

77.

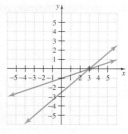

A Look Ahead

Example

If $f(x) = x^2 + 2x + 1$, find $f(\pi)$.

Solution:

$$f(x) = x^2 + 2x + 1$$
$$f(\pi) = \pi^2 + 2\pi + 1$$

Given the following functions, find the indicated values.

78. $f(x) = 2x + 7$;

    a. $f(2)$ **b.** $f(a)$

79. $g(x) = -3x + 12$;

    a. $g(s)$ **b.** $g(r)$

80. $h(x) = x^2 + 7$;

    a. $h(3)$ **b.** $h(a)$

81. $f(x) = x^2 - 12$;

    a. $f(12)$ **b.** $f(a)$

For additional Chapter Projects, visit the Real World Activities Website by going to http://www.prenhall.com/martin-gay.

CHAPTER PROJECT

Matching Descriptions of Linear Data to Their Equations and Graphs

Many situations can be described using a linear equation. With the data's annual rate of change of two points on the line, it is possible to find an equation that summarizes the situation.

In this project, you will have the opportunity to match a description of a situation to its linear equation and its graph. This project may be completed by working in groups or individually.

1. Each of the following equations **a** through **j** is represented in graphs **(i)** through **(x)** below. Match each equation to its graph.

   a. $y = -2.5x + 165$

   b. $y = -8x + 289$

   c. $y = 25$

   d. $y = 2x - 25$

   e. $y = 8x + 289$

   f. $y = -38.5x + 1019$

   g. $y = 2x + 25$

   h. $y = -2.5x + 350$

   i. $y = 38.5x + 1019$

   j. $y = 12$

Graphs

(i)

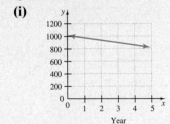

(ii)

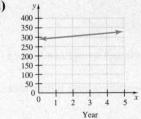

(iii)

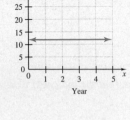

(iv)

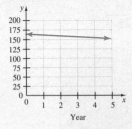

(v)

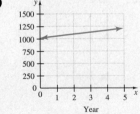

(vi)

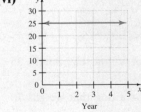

(vii)

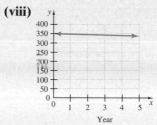

(viii)

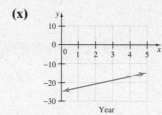

(ix)

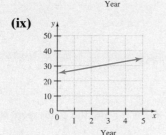

(x)

2. For each of the following five descriptions, match the situation to one of the ten graphs/equation pairs from Question 1 that models the situation. Assume that the relationship described in each situation is linear, and assume that $x = 0$ represents 1995. Explain how you made your choice in each case.

Situations
   a. The average time U.S. citizens spent watching movies in theaters was 12 hours in 1993. In 1998 the average time was still 12 hours.
   b. In 1995 the average time U.S. citizens spent reading newspapers was 165 hours and has since declined at a rate of 2.5 hours per year.
   c. The average time U.S. citizens spent playing home video games was 29 hours in 1997 and 33 hours in 1999.
   d. The average time U.S. citizens spent watching broadcast television was 1019 hours in 1995 and 942 hours in 1997.
   e. The average time U.S. citizens spent listening to recorded music was 289 hours in 1995 and has been increasing at a rate of 8 hours per year.

(*Source:* Based on data from Veronis, Suhler, & Associates Inc., New York, NY)

CHAPTER 7 VOCABULARY CHECK

Fill in each blank with one of the words or phrases listed below.

slope-intercept	point-slope	function	domain
quadratic equation	relation	range	

   1. A set of ordered pairs that assigns to each x-value exactly one y-value is called a

   _____ .
   2. A _____ is one that can be written in the form $y = ax^2 + bx + c$.
   3. The equation $y = 7x - 5$ is written in _____ form.
   4. The set of all x-coordinates of a relation is called the _____ of the relation.
   5. The set of all y-coordinates of a relation is called the _____ of the relation.
   6. A set of ordered pairs is called a _____ .
   7. The equation $y + 1 = 7(x - 2)$ is written in _____ form.

CHAPTER 7 HIGHLIGHTS

DEFINITIONS AND CONCEPTS	EXAMPLES

Section 7.1 The Slope-Intercept Form

Slope-Intercept Form

$$y = mx + b$$

m is the slope of the line.
b is the y-intercept.

Find the slope and the y-intercept of the line whose equation is $2x + 3y = 6$.

Solve for y:

$$2x + 3y = 6$$

$$3y = -2x + 6 \qquad \text{Subtract } 2x.$$

$$y = -\frac{2}{3}x + 2 \qquad \text{Divide by 3.}$$

The slope of the line is $-\frac{2}{3}$ and the y-intercept is 2.

Find the equation of the line with slope 3 and y-intercept -1.
The equation is $y = 3x - 1$.

Section 7.2 The Point-Slope Form

Point-Slope Form

$$y - y_1 = m(x - x_1)$$

m is the slope.
(x_1, y_1) is a point on the line.

Find an equation of the line with slope $\frac{3}{4}$ that contains the point $(-1, 5)$.

$$y - 5 = \frac{3}{4}\left[x - (-1)\right]$$

$$4(y - 5) = 3(x + 1) \qquad \text{Multiply by 4.}$$

$$4y - 20 = 3x + 3 \qquad \text{Distribute.}$$

$$-3x + 4y = 23 \qquad \begin{array}{l}\text{Subtract } 3x \\ \text{and add 20.}\end{array}$$

Section 7.3 Graphing Nonlinear Equations

A **quadratic equation in two variables** is an equation that can be written in the form $y = ax^2 + bx + c$ where a, b, and c are real numbers and a is not 0.
The graph of a quadratic equation is a parabola.

Quadratic Equations
$y = x^2 + 3$, $y = 7x^2 - 2x + 3$
Graph $y = x^2 + 2x$.

x	y
-3	3
-2	0
-1	-1
0	0
1	3

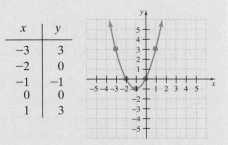

DEFINITIONS AND CONCEPTS	EXAMPLES

Section 7.4 Functions

A set of ordered pairs is a **relation**. The set of all x coordinates is called the **domain** of the relation and the set of all y-coordinates is called the **range** of the relation.

The domain of the relation $\{(0, 5), (2, 5), (4, 5), (5, -2)\}$ is $\{0, 2, 4, 5\}$. The range is $\{-2, 5\}$.

A **function** is a set of ordered pairs that assigns to each x-value exactly one y-value.

Which are graphs of functions?

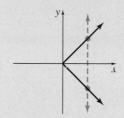 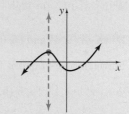

Vertical Line Test
If a vertical line can be drawn so that it intersects a graph more than once, the graph is not the graph of a function.

This graph is not the graph of a function.   This graph is the graph of a function.

The symbol $f(x)$ means **function of x**. This notation is called **function** notation.

If $f(x) = 2x^2 + 6x - 1$, find $f(3)$.

$$f(3) = 2(3)^2 + 6 \cdot 3 - 1$$
$$= 2 \cdot 9 + 18 - 1$$
$$= 18 + 18 - 1$$
$$= 35$$

CHAPTER 7 REVIEW

(7.1) *Determine the slope and the y-intercept of the graph of each equation.*

1. $3x + y = 7$

2. $x - 6y = -1$

3. $y = 2$

4. $x = -5$

Determine whether the lines are parallel, perpendicular, or neither.

△ **5.** $x - y = -6$
$x + y = 3$

△ **6.** $3x + y = 7$
$-3x - y = 10$

△ **7.** $y = 4x + \dfrac{1}{2}$
$4x + 2y = 1$

Write an equation of each line in slope-intercept form.

8. slope -5; y-intercept $\dfrac{1}{2}$

9. slope $\dfrac{2}{3}$; y-intercept 6

Use the slope-intercept form to graph each equation.

10. $y = -3x$

11. $y = 3x - 1$

12. $-x + 2y = 8$

13. $5x - 3y = 15$

Match each equation with its graph.

14. $y = -2x + 1$

15. $y = -4x$

16. $y = 2x$

17. $y = 2x - 1$

A.

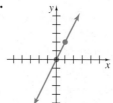

B.

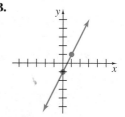

C.

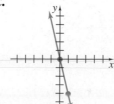

D.

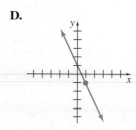

(7.2) *Write an equation of each line in standard form.*

18. With slope 4, through $(2, 0)$

19. With slope -3, through $(0, -5)$

20. With slope $\dfrac{1}{2}$, through $\left(0, -\dfrac{7}{2}\right)$

21. With slope 0, through $(-2, -3)$

22. With 0 slope, through the origin

23. With slope -6, through $(2, -1)$

24. With slope 12, through $\left(\dfrac{1}{2}, 5\right)$

25. Through $(0, 6)$ and $(6, 0)$

26. Through $(0, -4)$ and $(-8, 0)$

27. Vertical line, through $(5, 7)$

28. Horizontal line, through $(-6, 8)$

△ **29.** Through $(6, 0)$, perpendicular to $y = 8$

△ **30.** Through $(10, 12)$, perpendicular to $x = -2$

△ **31.** Write an equation in standard form of the line that contains $(5, 0)$ and is
  a. parallel to the line $y = -3x + 7$.
  b. perpendicular to the line $y = -3x + 7$.

(7.3) Graph each equation.

32. $y = x + 5$ **33.** $y = x^2 + 5$

34. $y = |x| + 5$ **35.** $y = -x^2 + 1$

36. $y = |5x|$ **37.** $y = x^2 + 4x$

Use the given graph to answer Exercises 38–41.

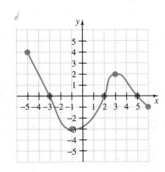

38. Find the coordinates of the point with the greatest y-value.

39. Find the coordinates of the point with the least y-value.

40. Find the coordinates of the point with the least x-value.

41. Find the coordinates of the point with the greatest x-value.

(7.4) Determine which of the following are functions.

42. $\{(7, 1), (7, 5), (2, 6)\}$

43. $\{(0, -1), (5, -1), (2, 2)\}$

44. $7x - 6y = 1$

45. $y = 7$ **46.** $x = 2$

47. $y = x^3$ **48.** $x + y < 6$

49. 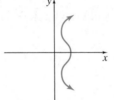   **50.**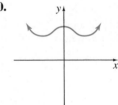

Given the following functions, find the indicated function value.

51. Given $f(x) = -2x + 6$, find
  a. $f(0)$ **b.** $f(-2)$ **c.** $f\left(\dfrac{1}{2}\right)$

52. Given $h(x) = -5 - 3x$, find
  a. $h(2)$ **b.** $h(-3)$ **c.** $h(0)$

53. Given $g(x) = x^2 + 12x$, find
  a. $g(3)$ **b.** $g(-5)$ **c.** $g(0)$

54. Given $h(x) = 6 - |x|$, find
  a. $h(-1)$ **b.** $h(1)$ **c.** $h(-4)$

Find the domain of each function.

55. $f(x) = 2x + 7$

56. $g(x) = \dfrac{7}{x - 2}$

Find the domain and the range of each function graphed.

57.

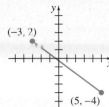

(−3, 2)

(5, −4)

58.

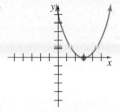

59.

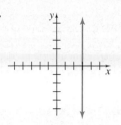

60.

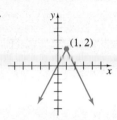

(1, 2)

CHAPTER 7 TEST

1. Determine the slope and the y-intercept of the graph of $7x - 3y = 2$.

△ 2. Determine whether the graphs of $y = 2x - 6$ and $-4x = 2y$ are parallel lines, perpendicular lines, or neither.

Find equations of the following lines. Write the equation in standard form.

3. With slope of $-\dfrac{1}{4}$, through $(2, 2)$

4. Through the origin and $(6, -7)$

5. Through $(2, -5)$ and $(1, 3)$

△ 6. Through $(-5, -1)$ and parallel to $x = 7$

7. With slope $\dfrac{1}{8}$ and y-intercept 12

Graph each equation.

8. $x - 4y = 4$ 9. $y = -3x$

10. $y = |x + 1|$ 11. $y = x^2 + 1$

Which of the following are functions?

12. $7x - y = 12$ 13. $y = -5$

14.

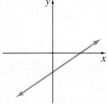

15.

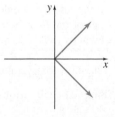

16.

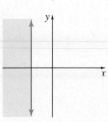

Given the following functions, find the indicated function values.

17. $f(x) = 2x - 4$
   a. $f(-2)$ **b.** $f(0.2)$ **c.** $f(0)$

18. $h(x) = x^3 - x$
   a. $h(-1)$ **b.** $h(0)$ **c.** $h(4)$

19. $g(x) = 6$
   a. $g(0)$ **b.** $g(a)$ **c.** $g(242)$

20. Find the domain of $y = \dfrac{1}{x + 1}$.

Find the domain and the range of each function graphed.

21.

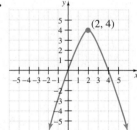

(2, 4)

22.

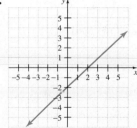

This graph shows the sunset times for Seward, Alaska. Use this graph to answer Exercises 23–28.

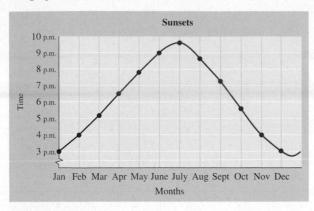

Sunsets

23. Approximate the time of sunset on June 1.

24. Approximate the time of sunset on November 1.

25. Approximate the date(s) when the sunset is 3 p.m.

26. Approximate the date(s) when the sunset is 9 p.m.

27. Is this graph the graph of a function? Why or why not?

28. Do you think a graph of sunset times for any location will always be a function? Why or why not?

CHAPTER 7 CUMULATIVE REVIEW

1. Name the property illustrated by each true statement.
   a. $3 \cdot y = y \cdot 3$
   b. $(x + 7) + 9 = x + (7 + 9)$
   c. $(b + 0) + 3 = b + 3$
   d. $2 \cdot (z \cdot 5) = 2 \cdot (5 \cdot z)$
   e. $-2 \cdot \left(-\dfrac{1}{2}\right) = 1$
   f. $-2 + 2 = 0$
   g. $-6 \cdot (y \cdot 2) = (-6 \cdot 2) \cdot y$

2. Solve: $\dfrac{x}{2} - 1 = \dfrac{2}{3}x - 3$

3. Graph: $2x - y \geq 3$

4. Simplify each expression.
   a. $\left(\dfrac{m}{n}\right)^7$ b. $\left(\dfrac{x^3}{3y^5}\right)^4$

5. Combine like terms.
   a. $-3x + 7x$
   b. $11x^2 + 5 + 2x^2 - 7$
   c. $\dfrac{2}{5}x^4 + \dfrac{2}{3}x^3 - x^2 + \dfrac{1}{10}x^4 - \dfrac{1}{6}x^3$

6. Multiply: $(t + 2)$ by $(3t^2 - 4t + 2)$

7. Find $(x - 3)(x + 4)$ by the FOIL method.

8. Perform each indicated operation. Write each result in standard decimal notation.
   a. $(8 \times 10^{-6})(7 \times 10^3)$
   b. $\dfrac{12 \times 10^2}{6 \times 10^{-3}}$

9. Divide: $\dfrac{9x^5 - 12x^2 + 3x}{3x^2}$

10. Factor: $5(x + 3) + y(x + 3)$

11. $x^2 + 4x - 12$

12. $10x^2 - 13xy - 3y^2$

13. $x^2 - 25$

14. $30a^2b^3 + 55a^2b^2 - 35a^2b$

15. Solve: $3x^3 - 12x = 0$

16. For a TV commercial, a piece of luggage is dropped from a cliff 256 feet above the ground to show the durability of the luggage. Neglecting air resistance, the height h in feet of the luggage above the ground after t seconds is given by the quadratic equation
$$h = -16t^2 + 256$$
Find how long it takes for the luggage to hit the ground.

17. Simplify each rational expression.
   a. $\dfrac{x + y}{y + x}$ b. $\dfrac{x - y}{y - x}$

18. Multiply: $\dfrac{3x + 3}{5x - 5x^2} \cdot \dfrac{2x^2 + x - 3}{4x^2 - 9}$

19. Subtract: $\dfrac{3x^2 + 2x}{x - 1} - \dfrac{10x - 5}{x - 1}$

20. Add: $\dfrac{2}{3t} + \dfrac{6}{t + 1}$

21. Simplify: $\dfrac{\dfrac{x}{y} + \dfrac{3}{2x}}{\dfrac{x}{2} + y}$

22. Solve: $\dfrac{t - 4}{2} - \dfrac{t - 3}{9} = \dfrac{5}{18}$

23. Find the slope of the line whose equation is $y - \dfrac{3}{4}x + 6$.

24. Find an equation of the vertical line through $(-1, 5)$.

25. Which of the following relations are also functions?
   a. $\{(-1, 1), (2, 3), (7, 3), (8, 6)\}$
   b. $\{(0, -2), (1, 5), (0, 3), (7, 7)\}$

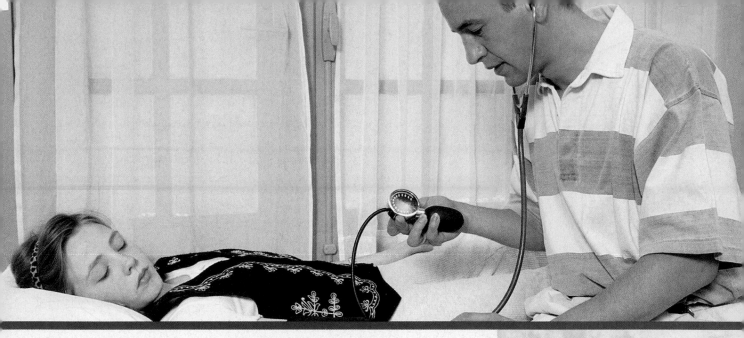

Largest Health Care Occupation in the U.S.

Did you know that nurses make up the largest health care occupation in the United States? Over 2.5 million people work as nurses in settings as varied as hospitals, private homes, corporate offices, nursing homes, overnight camps, doctors' offices, and community health centers. The U.S. Bureau of Labor Statistics predicts that the demand for nurses will continue to grow rapidly as the American population ages, requiring more long-term health care and home health care.

Registered nurses must be licensed in the state in which they work. About 25% of registered nurses hold a diploma from a hospital program, 35% hold an associate's degree, 30% hold a bachelor's degree, and 10% hold a higher degree. Although the focus of study in a nursing degree program is on areas such as anatomy, chemistry, microbiology, nutrition, and clinical experience, it isn't hard to see how good math skills would be useful. In fact, nurses use math skills nearly every day in taking and comparing vital signs, administering medications, and tracking fluid intake and output.

 For more information about a nursing career, visit the National League for Nursing Website by first going to www.prenhall.com/martin-gay.

In the Spotlight on Decision Making feature on page 471, you will have the opportunity, as a registered nurse, to make a decision concerning a patient's blood pressure.

SOLVING SYSTEMS OF LINEAR EQUATIONS

8.1   SOLVING SYSTEMS OF LINEAR EQUATIONS BY GRAPHING

8.2   SOLVING SYSTEMS OF LINEAR EQUATIONS BY SUBSTITUTION

8.3   SOLVING SYSTEMS OF LINEAR EQUATIONS BY ADDITION

8.4   SYSTEMS OF LINEAR EQUATIONS AND PROBLEM SOLVING

8.5   SYSTEMS OF LINEAR INEQUALITIES

In Chapters 3 and 7, we graphed equations containing two variables. Equations like these are often needed to represent relationships between two different values. For example, an economist attempts to predict what effects a price change will have on the sales prospects of calculators. There are many real-life opportunities to compare and contrast two such equations, called a system of equations. This chapter presents linear systems and ways we solve these systems and apply them to real-life situations.

8.1 SOLVING SYSTEMS OF LINEAR EQUATIONS BY GRAPHING

CD-ROM   SSM

SSG   Video

▶ **OBJECTIVES**

1. Determine if an ordered pair is a solution of a system of equations in two variables.
2. Solve a system of linear equations by graphing.
3. Without graphing, determine the number of solutions of a system.

1 A **system of linear equations** consists of two or more linear equations. In this section, we focus on solving systems of linear equations containing two equations in two variables. Examples of such linear systems are

$$\begin{cases} 3x - 3y = 0 \\ \quad x = 2y \end{cases} \qquad \begin{cases} x - y = 0 \\ 2x + y = 10 \end{cases} \qquad \begin{cases} y = 7x - 1 \\ y = 4 \end{cases}$$

A **solution** of a system of two equations in two variables is an ordered pair of numbers that is a solution of both equations in the system.

Example 1 Which of the following ordered pairs is a solution of the given system?

$$\begin{cases} 2x - 3y = 6 & \text{First equation} \\ \quad x = 2y & \text{Second equation} \end{cases}$$

a. $(12, 6)$ **b.** $(0, -2)$

Solution If an ordered pair is a solution of both equations, it is a solution of the system.
a. Replace x with 12 and y with 6 in both equations.

$2x - 3y = 6$	First equation	$x = 2y$	Second equation
$2(12) - 3(6) \stackrel{?}{=} 6$	Let $x = 12$ and $y = 6$.	$12 \stackrel{?}{=} 2(6)$	Let $x = 12$ and $y = 6$.
$24 - 18 \stackrel{?}{=} 6$	Simplify.	$12 = 12$	True
$6 = 6$	True		

Since $(12, 6)$ is a solution of both equations, it is a solution of the system.
b. Start by replacing x with 0 and y with -2 in both equations.

$2x - 3y = 6$	First equation	$x = 2y$	Second equation
$2(0) - 3(-2) \stackrel{?}{=} 6$	Let $x = 0$ and $y = -2$.	$0 \stackrel{?}{=} 2(-2)$	Let $x = 0$ and $y = -2$.
$0 + 6 \stackrel{?}{=} 6$	Simplify.	$0 = -4$	False
$6 = 6$	True		

While $(0, -2)$ is a solution of the first equation, it is not a solution of the second equation, so it is **not** a solution of the system. ▬

2 Since a solution of a system of two equations in two variables is a solution common to both equations, it is also a point common to the graphs of both equations. Let's practice finding solutions of both equations in a system—that is, solutions of a system—by graphing and identifying points of intersection.

Example 2   Solve the system of equations by graphing.

$$\begin{cases} -x + 3y = 10 \\ x + y = 2 \end{cases}$$

Solution   On a single set of axes, graph each linear equation.

$-x + 3y = 10$

x	y
0	$\dfrac{10}{3}$
-4	2
2	4

$x + y = 2$

x	y
0	2
2	0
1	1

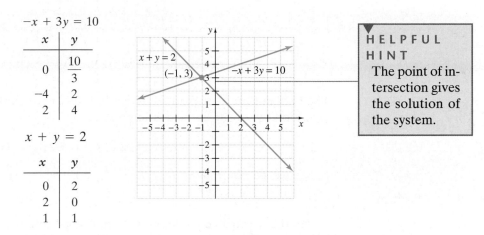

> **HELPFUL HINT**
> The point of intersection gives the solution of the system.

The two lines appear to intersect at the point $(-1, 3)$. To check, we replace x with -1 and y with 3 in both equations.

$-x + 3y = 10$	*First equation*	$x + y = 2$	*Second equation*
$-(-1) + 3(3) \stackrel{?}{=} 10$	*Let $x = -1$ and $y = 3$.*	$-1 + 3 \stackrel{?}{=} 2$	*Let $x = -1$ and $y = 3$.*
$1 + 9 \stackrel{?}{=} 10$	*Simplify.*	$2 = 2$	*True*
$10 = 10$	*True*		

$(-1, 3)$ checks, so it is the solution of the system.

> **HELPFUL HINT**
> Neatly drawn graphs can help when you are estimating the solution of a system of linear equations by graphing.

A system of equations that has at least one solution as in Example 2 is said to be a **consistent system**. A system that has no solution is said to be an **inconsistent system**.

◆ Example 3   Solve the following system of equations by graphing.

$$\begin{cases} 2x + y = 7 \\ 2y = -4x \end{cases}$$

Solution   Graph each of the two lines in the system.

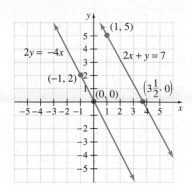

The lines **appear** to be parallel. To confirm this, write both equations in slope-intercept form by solving each equation for y.

$2x + y = 7$ First equation $2y = -4x$ Second equation

$\quad y = -2x + 7$ Subtract $2x$ from both sides. $\dfrac{2y}{2} = \dfrac{-4x}{2}$ Divide both sides by 2.

$\qquad\qquad\qquad\qquad\qquad\qquad\qquad\qquad\quad y = -2x$

Recall that when an equation is written in slope-intercept form, the coefficient of x is the slope. Since both equations have the same slope, -2, but different y-intercepts, the lines are parallel and have no points in common. Thus, there is no solution of the system and the system is inconsistent.

   In Examples 2 and 3, the graphs of the two linear equations of each system are different. When this happens, we call these equations **independent equations**. If the graphs of the two equations in a system are identical, we call the equations **dependent equations**.

◆ **Example 4**   Solve the system of equations by graphing.

$$\begin{cases} x - y = 3 \\ -x + y = -3 \end{cases}$$

Solution   Graph each line.

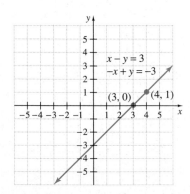

These graphs **appear** to be identical. To confirm this, write each equation in slope-intercept form.

$$x - y = 3 \qquad \text{First equation}$$

$$-y = -x + 3 \qquad \text{Subtract } x \text{ from both sides.}$$

$$\frac{-y}{-1} = \frac{-x}{-1} + \frac{3}{-1} \qquad \text{Divide both sides by } -1.$$

$$y = x - 3$$

$$-x + y = -3 \qquad \text{Second equation}$$

$$y = x - 3 \qquad \text{Add } x \text{ to both sides.}$$

The equations are identical and so must be their graphs. The lines have an infinite number of points in common. Thus, there is an infinite number of solutions of the system and this is a consistent system. The equations are dependent equations. ■

As we have seen, three different situations can occur when graphing the two lines associated with the equations in a linear system:

One point of intersection: one solution

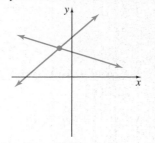

Parallel lines: no solution

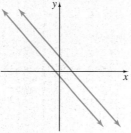

Same line: infinite number of solutions

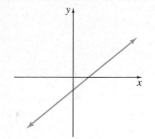

Consistent system
(at least one solution)
Independent equations
(graphs of equations differ)

Inconsistent system
(no solution)
Independent equations
(graphs of equations differ)

Consistent system
(at least one solution)
Dependent equations
(graphs of equations identical)

3 You may have suspected by now that graphing alone is not an accurate way to solve a system of linear equations. For example, a solution of $\left(\frac{1}{2}, \frac{2}{9}\right)$ is unlikely to be read correctly from a graph. The next two sections present two accurate methods of solving these systems. In the meantime, we can decide how many solutions a system has by writing each equation in the slope-intercept form.

Example 5 Without graphing, determine the number of solutions of the system.

$$\begin{cases} \dfrac{1}{2}x - y = 2 \\ x = 2y + 5 \end{cases}$$

Solution First write each equation in slope-intercept form.

$$\frac{1}{2}x - y = 2 \qquad \text{First equation}$$

$$\frac{1}{2}x = y + 2 \qquad \text{Add } y \text{ to both sides.}$$

$$\frac{1}{2}x - 2 = y \qquad \text{Subtract 2 from both sides.}$$

$$x = 2y + 5 \qquad \text{Second equation}$$

$$x - 5 = 2y \qquad \text{Subtract 5 from both sides.}$$

$$\frac{x}{2} - \frac{5}{2} = \frac{2y}{2} \qquad \text{Divide both sides by 2.}$$

$$\frac{1}{2}x - \frac{5}{2} = y \qquad \text{Simplify.}$$

The slope of each line is $\frac{1}{2}$, but they have different y-intercepts. This tells us that the lines representing these equations are parallel. Since the lines are parallel, the system has no solution and is inconsistent.

Example 6   Determine the number of solutions of the system.

$$\begin{cases} 3x - y = 4 \\ x + 2y = 8 \end{cases}$$

Solution   Once again, the slope-intercept form helps determine how many solutions this system has.

$3x - y = 4$	First equation		$x + 2y = 8$	Second equation
$3x = y + 4$	Add y to both sides.		$x = -2y + 8$	Subtract $2y$ from both sides.
$3x - 4 = y$	Subtract 4 from both sides.		$x - 8 = -2y$	Subtract 8 from both sides.
			$\dfrac{x}{-2} - \dfrac{8}{-2} = \dfrac{-2y}{-2}$	Divide both sides by -2.
			$-\dfrac{1}{2}x + 4 = y$	Simplify.

The slope of the second line is $-\frac{1}{2}$, whereas the slope of the first line is 3. Since the slopes are not equal, the two lines are neither parallel nor identical and must intersect. Therefore, this system has one solution and is consistent.

GRAPHING CALCULATOR EXPLORATIONS

A graphing calculator may be used to approximate solutions of systems of equations. For example, to approximate the solution of the system

$$\begin{cases} y = -3.14x - 1.35 \\ y = 4.88x + 5.25, \end{cases}$$

first graph each equation on the same set of axes. Then use the intersect feature of your calculator to approximate the point of intersection.

The approximate point of intersection is $(-0.82, 1.23)$.

Solve each system of equations. Approximate the solutions to two decimal places.

1. $\begin{cases} y = -2.68x + 1.21 \\ y = 5.22x - 1.68 \end{cases}$ **2.** $\begin{cases} y = 4.25x + 3.89 \\ y = -1.88x + 3.21 \end{cases}$

3. $\begin{cases} 4.3x - 2.9y = 5.6 \\ 8.1x + 7.6y = -14.1 \end{cases}$ **4.** $\begin{cases} -3.6x - 8.6y = 10 \\ -4.5x + 9.6y = -7.7 \end{cases}$

MENTAL MATH

Each rectangular coordinate system shows the graph of the equations in a system of equations. Use each graph to determine the number of solutions for each associated system. If the system has only one solution, give its coordinates.

1.
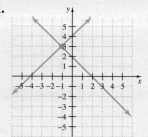

2.
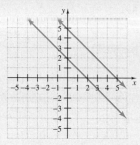

3.
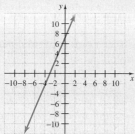

4.
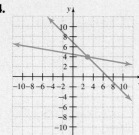

5.
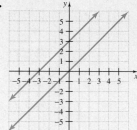

6.
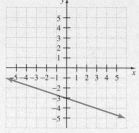

7.
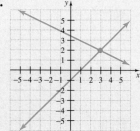

8.
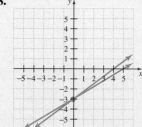

Exercise Set 8.1

Determine whether any ordered pairs satisfy the system of the linear equations. See Example 1.

1. $\begin{cases} x + y = 8 \\ 3x + 2y = 21 \end{cases}$
a. $(2, 4)$
b. $(5, 3)$
c. $(1, 9)$

2. $\begin{cases} 2x + y = 5 \\ x + 3y = 5 \end{cases}$
a. $(5, 0)$
b. $(1, 2)$
c. $(2, 1)$

3. $\begin{cases} 3x - y = 5 \\ x + 2y = 11 \end{cases}$
a. $(2, -1)$
b. $(3, 4)$
c. $(0, -5)$

4. $\begin{cases} 2x - 3y = 8 \\ x - 2y = 6 \end{cases}$
a. $(4, 0)$
b. $(-2, -4)$
c. $(7, 2)$

5. $\begin{cases} 2y = 4x \\ 2x - y = 0 \end{cases}$

  a. $(-3, -6)$
  b. $(0, 0)$
  c. $(1, 2)$

6. $\begin{cases} 4x = 1 - y \\ x - 3y = -8 \end{cases}$

  a. $(0, 1)$
  b. $(1, -3)$
  c. $(-2, 2)$

7. Construct a system of two linear equations that has $(2, 5)$ as a solution.

8. Construct a system of two linear equations that has $(0, 1)$ as a solution.

Solve each system of equations by graphing the equations on the same set of axes. Tell whether the system is consistent or inconsistent and whether the equations are dependent or independent. See Examples 2 through 4.

9. $\begin{cases} y = x + 1 \\ y = 2x - 1 \end{cases}$

10. $\begin{cases} y = 3x - 4 \\ y = x + 2 \end{cases}$

11. $\begin{cases} 2x + y = 0 \\ 3x + y = 1 \end{cases}$

12. $\begin{cases} 2x + y = 1 \\ 3x + y = 0 \end{cases}$

13. $\begin{cases} y = -x - 1 \\ y = 2x + 5 \end{cases}$

14. $\begin{cases} y = x - 1 \\ y = -3x - 5 \end{cases}$

15. $\begin{cases} 2x - y = 6 \\ y = 2 \end{cases}$

16. $\begin{cases} x + y = 5 \\ x = 4 \end{cases}$

17. $\begin{cases} x + y = 5 \\ x + y = 6 \end{cases}$

18. $\begin{cases} 2x + y = 4 \\ x + y = 2 \end{cases}$

19. $\begin{cases} y - 3x = -2 \\ 6x - 2y = 4 \end{cases}$

20. $\begin{cases} y + 2x = 3 \\ 4x = 2 - 2y \end{cases}$

21. $\begin{cases} x - 2y = 2 \\ 3x + 2y = -2 \end{cases}$

22. $\begin{cases} x + 3y = 7 \\ 2x - 3y = -4 \end{cases}$

23. $\begin{cases} \dfrac{1}{2}x + y = -1 \\ x = 4 \end{cases}$

24. $\begin{cases} x + \dfrac{3}{4}y = 2 \\ x = -1 \end{cases}$

25. $\begin{cases} y = x - 2 \\ y = 2x + 3 \end{cases}$

26. $\begin{cases} y = x + 5 \\ y = -2x - 4 \end{cases}$

27. $\begin{cases} x + y = 7 \\ x - y = 3 \end{cases}$

28. $\begin{cases} x + y = -4 \\ x - y = 2 \end{cases}$

29. Explain how to use a graph to determine the number of solutions of a system.

30. The ordered pair $(-2, 3)$ is a solution of all three independent equations:

$$x + y = 1$$
$$2x - y = -7$$
$$x + 3y = 7$$

Describe the graph of all three equations on the same axes.

Without graphing, decide.

  a. *Are the graphs of the equations identical lines, parallel lines, or lines intersecting at a single point?*

  b. *How many solutions does the system have? See Examples 5 and 6.*

31. $\begin{cases} 4x + y = 24 \\ x + 2y = 2 \end{cases}$

32. $\begin{cases} 3x + y = 1 \\ 3x + 2y = 6 \end{cases}$

33. $\begin{cases} 2x + y = 0 \\ 2y = 6 - 4x \end{cases}$

34. $\begin{cases} 3x + y = 0 \\ 2y = -6x \end{cases}$

35. $\begin{cases} 6x - y = 4 \\ \dfrac{1}{2}y = -2 + 3x \end{cases}$

36. $\begin{cases} 3x - y = 2 \\ \dfrac{1}{3}y = -2 + 3x \end{cases}$

37. $\begin{cases} x = 5 \\ y = -2 \end{cases}$

38. $\begin{cases} y = 3 \\ x = -4 \end{cases}$

39. $\begin{cases} 3y - 2x = 3 \\ x + 2y = 9 \end{cases}$

40. $\begin{cases} 2y = x + 2 \\ y + 2x = 3 \end{cases}$

41. $\begin{cases} 6y + 4x = 6 \\ 3y - 3 = -2x \end{cases}$

42. $\begin{cases} 8y + 6x = 4 \\ 4y - 2 = 3x \end{cases}$

43. $\begin{cases} x + y = 4 \\ x + y = 3 \end{cases}$

44. $\begin{cases} 2x + y = 0 \\ y = -2x + 1 \end{cases}$

45. Explain how writing each equation in a linear system in the point-slope form helps determine the number of solutions of a system.

46. Is it possible for a system of two linear equations in two variables to be inconsistent, but with dependent equations? Why or why not?

The double line graph below shows the number of pounds of fishery products from U.S. domestic catch and from imports. Use this graph for Exercises 47 and 48.

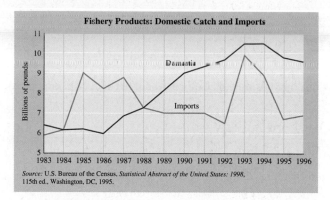

Fishery Products: Domestic Catch and Imports

Source: U.S. Bureau of the Census, *Statistical Abstract of the United States: 1998,* 115th ed., Washington, DC, 1995.

47. In what year(s) is the number of pounds of fishery products imported equal to the number of pounds of domestic catch?

48. In what year(s) is the number of pounds of fishery products imported greater than the number of pounds of domestic catch?

49. Below are tables of values for two linear equations. Using the tables,

   a. find a solution of the corresponding system.

   b. graph several ordered pairs from each table and sketch the two lines.

Does your graph confirm the solution from part (a)?

x	y		x	y
1	3		1	6
2	5		2	7
3	7		3	8
4	9		4	9
5	11		5	10

REVIEW EXERCISES

Solve each equation. See Section 2.4.

50. $5(x - 3) + 3x = 1$

51. $-2x + 3(x + 6) = 17$

52. $4\left(\dfrac{y + 1}{2}\right) + 3y = 0$

53. $-y + 12\left(\dfrac{y - 1}{4}\right) = 3$

54. $8a - 2(3a - 1) = 6$

55. $3z - (4z - 2) = 9$

8.2 SOLVING SYSTEMS OF LINEAR EQUATIONS BY SUBSTITUTION

CD-ROM   SSM

SSG   Video

▶ **OBJECTIVE**

1. Use the substitution method to solve a system of linear equations.

1   As we stated in the preceding section, graphing alone is not an accurate way to solve a system of linear equations. In this section, we discuss a second, more accurate method for solving systems of equations. This method is called the **substitution method** and is introduced in the next example.

Example 1   Solve the system:

$$\begin{cases} 2x + y = 10 & \text{First equation} \\ x = y + 2 & \text{Second equation} \end{cases}$$

Solution   The second equation in this system is $x = y + 2$. This tells us that x and $y + 2$ have the same value. This means that we may substitute $y + 2$ for x in the first equation.

$$2x + y = 10 \quad \text{First equation}$$

$$2(y + 2) + y = 10 \quad \text{Substitute } y + 2 \text{ for } x \text{ since } x = y + 2.$$

Notice that this equation now has one variable, y. Let's now solve this equation for y.

$$2(y + 2) + y = 10$$

	$2y + 4 + y = 10$	Use the distributive property.
	$3y + 4 = 10$	Combine like terms.
	$3y = 6$	Subtract 4 from both sides.
	$y = 2$	Divide both sides by 3.

HELPFUL HINT
Don't forget the distributive property.

Now we know that the y-value of the ordered pair solution of the system is 2. To find the corresponding x-value, we replace y with 2 in the equation $x = y + 2$ and solve for x.

$$x = y + 2$$
$$x = 2 + 2 \quad \text{Let } y = 2.$$
$$x = 4$$

The solution of the system is the ordered pair $(4, 2)$. Since an ordered pair solution must satisfy both linear equations in the system, we could have chosen the equation $2x + y = 10$ to find the corresponding x-value. The resulting x-value is the same.

Check   We check to see that $(4, 2)$ satisfies both equations of the original system.

First Equation	***Second Equation***	
$2x + y = 10$	$x = y + 2$	
$2(4) + 2 \stackrel{?}{=} 10$	$4 \stackrel{?}{=} 2 + 2$	Let $x = 4$ and $y = 2$.
$10 = 10$ True	$4 = 4$ True	

The solution of the system is $(4, 2)$.

A graph of the two equations shows the two lines intersecting at the point $(4, 2)$.

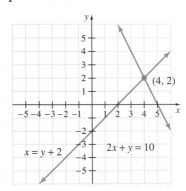

To solve a system of equations by substitution, we first need an equation solved for one of its variables.

Example 2 Solve the system:

$$\begin{cases} x + 2y = 7 \\ 2x + 2y = 13 \end{cases}$$

Solution We choose one of the equations and solve for x or y. We will solve the first equation for x by subtracting $2y$ from both sides.

$$x + 2y = 7 \qquad \text{First equation}$$
$$x = 7 - 2y \qquad \text{Subtract } 2y \text{ from both sides.}$$

Since $x = 7 - 2y$, we now substitute $7 - 2y$ for x in the second equation and solve for y.

$$2x + 2y = 13 \qquad \text{Second equation}$$
$$2(7 - 2y) + 2y = 13 \qquad \text{Let } x = 7 - 2y.$$
$$14 - 4y + 2y = 13 \qquad \text{Use the distributive property.}$$
$$14 - 2y = 13 \qquad \text{Simplify.}$$
$$-2y = -1 \qquad \text{Subtract 14 from both sides.}$$
$$y = \frac{1}{2} \qquad \text{Divide both sides by } -2.$$

> **HELPFUL HINT**
> Don't forget to insert parentheses when substituting $7 - 2y$ for x.

To find x, we let $y = \dfrac{1}{2}$ in the equation $x = 7 - 2y$.

$$x = 7 - 2y$$
$$x = 7 - 2\left(\frac{1}{2}\right) \qquad \text{Let } y = \frac{1}{2}.$$
$$x = 7 - 1$$
$$x = 6$$

The solution is $\left(6, \dfrac{1}{2}\right)$. Check the solution in both equations of the original system.

The following steps may be used to solve a system of equations by the substitution method.

SOLVING A SYSTEM OF LINEAR EQUATIONS BY THE SUBSTITUTION METHOD

Step 1. Solve one of the equations for one of its variables.

Step 2. Substitute the expression for the variable found in step 1 into the other equation.

Step 3. Solve the equation from step 2 to find the value of one variable.

Step 4. Substitute the value found in step 3 in any equation containing both variables to find the value of the other variable.

Step 5. Check the proposed solution in the original system.

Example 3 Solve the system: $\begin{cases} 7x - 3y = -14 \\ -3x + y = 6 \end{cases}$

Solution To avoid introducing fractions, we will solve the second equation for y.

$$-3x + y = 6 \qquad \text{\textit{Second equation}}$$
$$y = 3x + 6$$

Next, substitute $3x + 6$ for y in the first equation.

$$7x - 3y = -14 \qquad \text{\textit{First equation}}$$

$$7x - 3(3x + 6) = -14$$

$$7x - 9x - 18 = -14$$

$$-2x - 18 = -14$$

$$-2x = 4$$

$$\frac{-2x}{-2} = \frac{4}{-2}$$

$$x = -2$$

To find the corresponding y-value, substitute -2 for x in the equation $y = 3x + 6$. Then $y = 3(-2) + 6$ or $y = 0$. The solution of the system is $(-2, 0)$. Check this solution in both equations of the system.

> **HELPFUL HINT**
> When solving a system of equations by the substitution method, begin by solving an equation for one of its variables. If possible, solve for a variable that has a coefficient of 1 or -1. This way, we avoid working with time-consuming fractions.

Example 4 Solve the system: $\begin{cases} \dfrac{1}{2}x - y = 3 \\ x = 6 + 2y \end{cases}$

Solution The second equation is already solved for x in terms of y. Thus we substitute $6 + 2y$ for x in the first equation and solve for y.

$$\frac{1}{2}x - y = 3 \qquad \text{\textit{First equation.}}$$

$$\frac{1}{2}(6 + 2y) - y = 3 \qquad \text{\textit{Let } } x = 6 + 2y.$$

$$3 + y - y = 3$$

$$3 = 3$$

Arriving at a true statement such as $3 = 3$ indicates that the two linear equations in the original system are equivalent. This means that their graphs are identical and there is an infinite number of solutions of the system. Any solution of one equation is also a solution of the other.

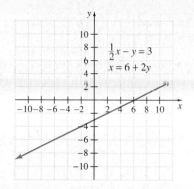

Example 5 Use substitution to solve the system.

$$\begin{cases} 6x + 12y = 5 \\ -4x - 8y = 0 \end{cases}$$

Solution Choose the second equation and solve for y.

$$-4x - 8y = 0 \qquad \text{Second equation}$$

$$-8y = 4x \qquad \text{Add } 4x \text{ to both sides.}$$

$$\frac{-8y}{-8} = \frac{4x}{-8} \qquad \text{Divide both sides by } -8.$$

$$y = -\frac{1}{2}x \qquad \text{Simplify.}$$

Now replace y with $-\dfrac{1}{2}x$ in the first equation.

$$6x + 12y = 5 \qquad \text{First equation}$$

$$6x + 12\left(-\frac{1}{2}x\right) = 5 \qquad \text{Let } y = -\frac{1}{2}x.$$

$$6x + (-6x) = 5 \qquad \text{Simplify.}$$

$$0 = 5 \qquad \text{Combine like terms.}$$

The false statement $0 = 5$ indicates that this system has no solution and is inconsistent. The graph of the linear equations in the system is a pair of parallel lines.

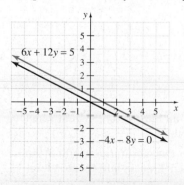

Exercise Set 8.2

Solve each system of equations by the substitution method. See Examples 1 through 5.

1. $\begin{cases} x + y = 3 \\ x = 2y \end{cases}$

2. $\begin{cases} x + y = 20 \\ x = 3y \end{cases}$

3. $\begin{cases} x + y = 6 \\ y = -3x \end{cases}$

4. $\begin{cases} x + y = 6 \\ y = -4x \end{cases}$

5. $\begin{cases} 3x + 2y = 16 \\ x = 3y - 2 \end{cases}$

6. $\begin{cases} 2x + 3y = 18 \\ x = 2y - 5 \end{cases}$

7. $\begin{cases} 3x - 4y = 10 \\ x = 2y \end{cases}$

8. $\begin{cases} 3x - 4y = 10 \\ y = 2x \end{cases}$

9. $\begin{cases} y = 3x + 1 \\ 4y - 8x = 12 \end{cases}$

10. $\begin{cases} y = 2x + 3 \\ 5y - 7x = 18 \end{cases}$

11. $\begin{cases} x + 2y = 6 \\ 2x + 3y = 8 \end{cases}$

12. $\begin{cases} x + 3y = -5 \\ 2x + 2y = 6 \end{cases}$

13. $\begin{cases} 2x - 5y = 1 \\ 3x + y = -7 \end{cases}$

14. $\begin{cases} 4x + 2y = 5 \\ 2x + y = -4 \end{cases}$

15. $\begin{cases} 2y = x + 2 \\ 6x - 12y = 0 \end{cases}$

16. $\begin{cases} 3y = x + 6 \\ 4x + 12y = 0 \end{cases}$

17. $\begin{cases} \frac{1}{3}x - y = 2 \\ x - 3y = 6 \end{cases}$

18. $\begin{cases} \frac{1}{4}x - 2y = 1 \\ x - 8y = 4 \end{cases}$

19. $\begin{cases} 4x + y = 11 \\ 2x + 5y = 1 \end{cases}$

20. $\begin{cases} 3x + y = -14 \\ 4x + 3y = -22 \end{cases}$

21. $\begin{cases} 2x - 3y = -9 \\ 3x = y + 4 \end{cases}$

22. $\begin{cases} 8x - 3y = -4 \\ 7x = y + 3 \end{cases}$

23. $\begin{cases} 6x - 3y = 5 \\ x + 2y = 0 \end{cases}$

24. $\begin{cases} 10x - 5y = -21 \\ x + 3y = 0 \end{cases}$

25. $\begin{cases} 3x - y = 1 \\ 2x - 3y = 10 \end{cases}$

26. $\begin{cases} 2x - y = -7 \\ 4x - 3y = -11 \end{cases}$

27. $\begin{cases} -x + 2y = 10 \\ -2x + 3y = 18 \end{cases}$

28. $\begin{cases} -x + 3y = 18 \\ -3x + 2y = 19 \end{cases}$

29. $\begin{cases} 5x + 10y = 20 \\ 2x + 6y = 10 \end{cases}$

30. $\begin{cases} 2x + 4y = 6 \\ 5x + 10y = 15 \end{cases}$

31. $\begin{cases} 3x + 6y = 9 \\ 4x + 8y = 16 \end{cases}$

32. $\begin{cases} 6x + 3y = 12 \\ 9x + 6y = 15 \end{cases}$

33. $\begin{cases} y = 2x + 9 \\ y = 7x + 10 \end{cases}$

34. $\begin{cases} y = 5x - 3 \\ y = 8x + 4 \end{cases}$

35. Explain how to identify an inconsistent system when using the substitution method.

36. Occasionally, when using the substitution method, the equation $0 = 0$ is obtained. Explain how this result indicates that the equations are dependent.

Solve each system by the substitution method. First simplify each equation by combining like terms.

37. $-5y + 6y = 3x + 2(x - 5) - 3x + 5$
$4(x + y) - x + y = -12$

38. $5x + 2y - 4x - 2y = 2(2y + 6) - 7$
$3(2x - y) - 4x = 1 + 9$

39. For the years 1960 through 1995, the annual percentage y of U.S. households that used fuel oil to heat their homes is given by the equation $y = -0.65x + 32.02$, where x is the number of years since 1960. For the same period, the annual percentage y of U.S. households that used electricity to heat their homes is given by the equation $y = 0.78x + 1.32$, where x is the number of years since 1960. (*Source:* Based on data from the U.S. Bureau of the Census)

  a. Use the substitution method to solve this system of equations. (Round your final results to the nearest whole numbers.)

  b. Explain the meaning of your answer to part (a).

  c. Sketch a graph of the system of equations. Write a sentence describing the use of fuel oil and electricity for heating homes between 1960 and 1995.

40. The number of music CDs (in millions) shipped to retailers in the United States from 1990 through 1998 is given by the equation $y = 74.5x + 289.3$, where x is the number of years since 1990. The number y of music cassettes (in millions) shipped to retailers in the United States from 1990 through 1998 is given by the equation $y = -34.1x + 434.5$, where x is the number of years since 1990. (*Source:* Based on data from the Recording Industry Association of America)

  a. Use the substitution method to solve this system of equations. (Round your final results to the nearest tenth.)

  b. Explain the meaning of your answer to part (a).

  c. Sketch a graph of the system of equations. Write a sentence describing the trends in the popularity of these two types of music formats.

▦ *Use a graphing calculator to solve each system.*

41. $\begin{cases} y = 5.1x + 14.56 \\ y = -2x - 3.9 \end{cases}$

42. $\begin{cases} y = 3.1x - 16.35 \\ y = -9.7x + 28.45 \end{cases}$

43. $\begin{cases} 3x + 2y = 14.05 \\ 5x + y = 18.5 \end{cases}$

44. $\begin{cases} x + y - -15.2 \\ -2x + 5y = -19.3 \end{cases}$

47. $-4x + y = 3; 3$

48. $5a - 7b = -4; -4$

Add the binomials. See Section 4.2.

$3n + 6m$

49. $\underline{2n - 6m}$

$-2x + 5y$

50. $\underline{2x + 11y}$

REVIEW EXERCISES

Write equivalent equations by multiplying both sides of the given equation by the given nonzero number. See Section 2.4.

45. $3x + 2y = 6; -2$

46. $-x + y = 10; 5$

$-5a - 7b$

51. $\underline{5a - 8b}$

$9q + p$

52. $\underline{-9q - p}$

8.3 SOLVING SYSTEMS OF LINEAR EQUATIONS BY ADDITION

CD-ROM SSM SSG Video

▶ **OBJECTIVE**

1. Use the addition method to solve a system of linear equations.

1 We have seen that substitution is an accurate way to solve a linear system. Another method for solving a system of equations accurately is the **addition** or **elimination method**. The addition method is based on the addition property of equality: adding equal quantities to both sides of an equation does not change the solution of the equation. In symbols,

$$\text{if } A = B \text{ and } C = D, \text{ then } A + C = B + D.$$

Example 1 Solve the system: $\begin{cases} x + y = 7 \\ x - y = 5 \end{cases}$

Solution Since the left side of each equation is equal to the right side, we add equal quantities by adding the left sides of the equations together and the right sides of the equations together. If we choose wisely, this adding gives us an equation in one variable, x, which we can solve for x.

$x + y = 7$	First equation
$\underline{x - y = 5}$	Second equation
$2x \quad\quad = 12$	Add the equations.
$x = 6$	Divide both sides by 2.

The x-value of the solution is 6. To find the corresponding y-value, let $x = 6$ in either equation of the system. We will use the first equation.

$x + y = 7$	First equation
$6 + y = 7$	Let $x = 6$.
$y = 7 - 6$	Solve for y.
$y = 1$	Simplify.

The solution is $(6, 1)$. Check this in both equations.

First Equation	**Second Equation**	
$x + y = 7$	$x - y = 5$	
$6 + 1 \stackrel{?}{=} 7$	$6 - 1 \stackrel{?}{=} 5$	Let $x = 6$ and $y = 1$.
$7 = 7$ True	$5 = 5$ True	

Thus, the solution of the system is $(6, 1)$ and the graphs of the two equations intersect at the point $(6, 1)$ as shown.

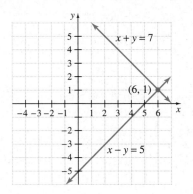

Example 2   Solve the system: $\begin{cases} -2x + y = 2 \\ -x + 3y = -4 \end{cases}$

Solution   If we simply add the two equations, the result is still an equation in two variables. However, our goal is to eliminate one of the variables. Notice what happens if we multiply *both sides* of the first equation by -3, which we are allowed to do by the multiplication property of equality. The system

$$\begin{cases} -3(-2x + y) = -3(2) \\ -x + 3y = -4 \end{cases} \quad \text{simplifies to} \quad \begin{cases} 6x - 3y = -6 \\ -x + 3y = -4 \end{cases}$$

Now add the resulting equations and the y variable is eliminated.

$$\begin{aligned} 6x - 3y &= -6 \\ \underline{-x + 3y} &= \underline{-4} \\ 5x &= -10 \qquad \text{Add.} \\ x &= -2 \qquad \text{Divide both sides by 5.} \end{aligned}$$

To find the corresponding y-value, let $x = -2$ in any of the preceding equations containing both variables. We use the first equation of the original system.

$$\begin{aligned} -2x + y &= 2 \qquad \text{First equation} \\ -2(-2) + y &= 2 \qquad \text{Let } x = -2. \\ 4 + y &= 2 \\ y &= -2 \end{aligned}$$

The solution is $(-2, -2)$. Check this ordered pair in both equations of the original system.

In Example 2, the decision to multiply the first equation by -3 was no accident. **To eliminate a variable** when adding two equations, **the coefficient of the variable in one equation must be the opposite of its coefficient in the other equation**.

> **HELPFUL HINT**
> Be sure to multiply *both sides* of an equation by a chosen number when solving by the addition method. A common mistake is to multiply only the side containing the variables.

Example 3   Solve the system: $\begin{cases} 2x - y = 7 \\ 8x - 4y = 1 \end{cases}$

Solution   Multiply both sides of the first equation by -4 and the resulting coefficient of x is -8, the opposite of 8, the coefficient of x in the second equation. The system

> **HELPFUL HINT**
> Don't forget to multiply both sides by -4.

$$\begin{cases} -4(2x - y) = -4(7) \\ 8x - 4y = 1 \end{cases} \quad \text{simplifies to} \quad \begin{cases} -8x + 4y = -28 \\ 8x - 4y = 1 \end{cases}$$

Now add the resulting equations.

$$\begin{array}{r} -8x + 4y = -28 \\ 8x - 4y = 1 \\ \hline 0 = -27 \quad \text{False} \end{array}$$

When we add the equations, both variables are eliminated and we have $0 = -27$, a false statement. This means that the system has no solution. The equations, if graphed, are parallel lines.

Example 4   Solve the system: $\begin{cases} 3x - 2y = 2 \\ -9x + 6y = -6 \end{cases}$

Solution   First we multiply both sides of the first equation by 3, then we add the resulting equations.

$$\begin{cases} 3(3x - 2y) = 3(2) \\ -9x + 6y = -6 \end{cases} \quad \text{simplifies to} \quad \begin{array}{r} \begin{cases} 9x - 6y = 6 \\ -9x + 6y = -6 \end{cases} \quad \text{Add the equations.} \\ \hline 0 = 0 \end{array}$$

Both variables are eliminated and we have $0 = 0$, a true statement. Whenever you eliminate a variable and get the equation $0 = 0$, the system has an infinite number of solutions.

Example 5   Solve the system: $\begin{cases} 3x + 4y = 13 \\ 5x - 9y = 6 \end{cases}$

Solution   We can eliminate the variable y by multiplying the first equation by 9 and the second equation by 4.

$$\begin{cases} 9(3x + 4y) = 9(13) \\ 4(5x - 9y) = 4(6) \end{cases} \quad \text{simplifies to} \quad \begin{array}{r} \begin{cases} 27x + 36y = 117 \\ 20x - 36y = 24 \end{cases} \\ \hline 47x = 141 \\ x = 3 \end{array} \quad \text{Add the equations.}$$

To find the corresponding y-value, we let $x = 3$ in any equation in this example containing two variables. Doing so in any of these equations will give $y = 1$. The solution to this system is $(3, 1)$. Check to see that $(3, 1)$ satisfies each equation in the original system. ▬

If we had decided to eliminate x instead of y in Example 5, the first equation could have been multiplied by 5 and the second by -3. Try solving the original system this way to check that the solution is $(3, 1)$.

The following steps summarize how to solve a system of linear equations by the addition method.

SOLVING A SYSTEM OF TWO LINEAR EQUATIONS BY THE ADDITION METHOD

Step 1. Rewrite each equation in standard form $Ax + By = C$.
Step 2. If necessary, multiply one or both equations by a nonzero number so that the coefficients of a chosen variable in the system are opposites.
Step 3. Add the equations.
Step 4. Find the value of one variable by solving the resulting equation from Step 3.
Step 5. Find the value of the second variable by substituting the value found in Step 4 into either of the original equations.
Step 6. Check the proposed solution in the original system.

Example 6   Solve the system: $\begin{cases} -x - \dfrac{y}{2} = \dfrac{5}{2} \\ -\dfrac{x}{2} + \dfrac{y}{4} = 0 \end{cases}$

Solution   We begin by clearing each equation of fractions. To do so, we multiply both sides of the first equation by the LCD 2 and both sides of the second equation by the LCD 4. Then the system

$$\begin{cases} 2\left(-x - \dfrac{y}{2}\right) = 2\left(\dfrac{5}{2}\right) \\ 4\left(-\dfrac{x}{2} + \dfrac{y}{4}\right) = 4(0) \end{cases} \quad \text{simplifies to} \quad \begin{cases} -2x - y = 5 \\ -2x + y = 0 \end{cases}$$

Now we add the resulting equations in the simplified system.

$$\begin{array}{r} -2x - y = 5 \\ \underline{-2x + y = 0} \\ -4x \qquad = 5 \end{array} \quad \text{Add.}$$

$$x = -\dfrac{5}{4}$$

To find y, we could replace x with $-\dfrac{5}{4}$ in one of the equations with two variables.

Instead, let's go back to the simplified system and multiply by appropriate factors to eliminate the variable x and solve for y. To do this, we multiply the first equation in the simplified system by -1. Then the system

$$\begin{cases} -1(-2x - y) = -1(5) \\ -2x + y = 0 \end{cases} \quad \text{simplifies to} \quad \begin{cases} 2x + y = -5 \\ \underline{-2x + y = 0} \\ 2y = -5 \quad \text{Add.} \\ y = -\dfrac{5}{2} \end{cases}$$

Check the ordered pair $\left(-\dfrac{5}{4}, -\dfrac{5}{2}\right)$ in both equations of the original system. The solution is $\left(-\dfrac{5}{4}, -\dfrac{5}{2}\right)$. ▬

SPOTLIGHT ON DECISION MAKING

Suppose you have been offered two similar positions as a sales associate. In one position, you would be paid a monthly salary of $1500 plus a 2% commission on all sales you make during the month. In the other position, you would be paid a monthly salary of $500 plus a 6% commission on all sales you make during the month. Which position would you choose? Explain your reasoning. Would knowing that the sales positions were at a car dealership affect your choice? What if the positions were at a shoe store?

Exercise Set 8.3

Solve each system of equations by the addition method. See Examples 1 through 5.

1. $\begin{cases} 3x + y = 5 \\ 6x - y = 4 \end{cases}$

2. $\begin{cases} 4x + y = 13 \\ 2x - y = 5 \end{cases}$

3. $\begin{cases} x - 2y = 8 \\ -x + 5y = -17 \end{cases}$

4. $\begin{cases} x - 2y = -11 \\ -x + 5y = 23 \end{cases}$

5. $\begin{cases} x + y = 6 \\ x - y = 6 \end{cases}$

6. $\begin{cases} x - y = 1 \\ -x + 2y = 0 \end{cases}$

7. $\begin{cases} 3x + y = 4 \\ 9x + 3y = 6 \end{cases}$

8. $\begin{cases} 2x + y = 6 \\ 4x + 2y = 12 \end{cases}$

9. $\begin{cases} 3x - 2y = 7 \\ 5x + 4y = 8 \end{cases}$

10. $\begin{cases} 6x - 5y = 25 \\ 4x + 15y = 13 \end{cases}$

11. $\begin{cases} \dfrac{2}{3}x + 4y = -4 \\ 5x + 6y = 18 \end{cases}$

12. $\begin{cases} \dfrac{3}{2}x + 4y = 1 \\ 9x + 24y = 5 \end{cases}$

13. $\begin{cases} 4x - 6y = 8 \\ 6x - 9y = 12 \end{cases}$

14. $\begin{cases} 9x - 3y = 12 \\ 12x - 4y = 18 \end{cases}$

15. $\begin{cases} 3x + y = -11 \\ 6x - 2y = -2 \end{cases}$

16. $\begin{cases} 4x + y = -13 \\ 6x - 3y = -15 \end{cases}$

17. $\begin{cases} 3x + 2y = 11 \\ 5x - 2y = 29 \end{cases}$

18. $\begin{cases} 4x + 2y = 2 \\ 3x - 2y = 12 \end{cases}$

19. $\begin{cases} x + 5y = 18 \\ 3x + 2y = -11 \end{cases}$

20. $\begin{cases} x + 4y = 14 \\ 5x + 3y = 2 \end{cases}$

21. $\begin{cases} 2x - 5y = 4 \\ 3x - 2y = 4 \end{cases}$

22. $\begin{cases} 6x - 5y = 7 \\ 4x - 6y = 7 \end{cases}$

23. $\begin{cases} 2x + 3y = 0 \\ 4x + 6y = 3 \end{cases}$

24. $\begin{cases} -x + 5y = -1 \\ 3x - 15y = 3 \end{cases}$

Solve each system of equations by the addition method. See Example 6.

25. $\begin{cases} \dfrac{x}{3} + \dfrac{y}{6} = 1 \\ \dfrac{x}{2} - \dfrac{y}{4} = 0 \end{cases}$ **26.** $\begin{cases} \dfrac{x}{2} + \dfrac{y}{8} = 3 \\ x - \dfrac{y}{4} = 0 \end{cases}$

27. $\begin{cases} x - \dfrac{y}{3} = -1 \\ -\dfrac{x}{2} + \dfrac{y}{8} = \dfrac{1}{4} \end{cases}$ **28.** $\begin{cases} 2x - \dfrac{3y}{4} = -3 \\ x + \dfrac{y}{9} = \dfrac{13}{3} \end{cases}$

29. $\begin{cases} \dfrac{x}{3} - y = 2 \\ -\dfrac{x}{2} + \dfrac{3y}{2} = -3 \end{cases}$ **30.** $\begin{cases} \dfrac{x}{2} + \dfrac{y}{4} = 1 \\ -\dfrac{x}{4} - \dfrac{y}{8} = 1 \end{cases}$

31. $\begin{cases} 8x = -11y - 16 \\ 2x + 3y = -4 \end{cases}$ **32.** $\begin{cases} 10x + 3y = -12 \\ 5x = -4y - 16 \end{cases}$

33. When solving a system of equations by the addition method, how do we know when the system has no solution?

34. To solve the system $\begin{cases} 2x - 3y = 5 \\ 5x + 2y = 6 \end{cases}$, explain why the addition method might be preferred rather than the substitution method.

Solve each system by either the addition method or the substitution method.

35. $\begin{cases} 2x - 3y = -11 \\ y = 4x - 3 \end{cases}$ **36.** $\begin{cases} 4x - 5y = 6 \\ y = 3x - 10 \end{cases}$

37. $\begin{cases} x + 2y = 1 \\ 3x + 4y = -1 \end{cases}$ **38.** $\begin{cases} x + 3y = 5 \\ 5x + 6y = -2 \end{cases}$

39. $\begin{cases} 2y = x + 6 \\ 3x - 2y = -6 \end{cases}$ **40.** $\begin{cases} 3y = x + 14 \\ 2x - 3y = -16 \end{cases}$

41. $\begin{cases} y = 2x - 3 \\ y = 5x - 18 \end{cases}$ **42.** $\begin{cases} y = 6x - 5 \\ y = 4x - 11 \end{cases}$

43. $\begin{cases} x + \dfrac{1}{6}y = \dfrac{1}{2} \\ 3x + 2y = 3 \end{cases}$ **44.** $\begin{cases} x + \dfrac{1}{3}y = \dfrac{5}{12} \\ 8x + 3y = 4 \end{cases}$

45. $\begin{cases} \dfrac{x+2}{2} = \dfrac{y+11}{3} \\ \dfrac{x}{2} = \dfrac{2y+16}{6} \end{cases}$ **46.** $\begin{cases} \dfrac{x+5}{2} = \dfrac{y+14}{4} \\ \dfrac{x}{3} = \dfrac{2y+2}{6} \end{cases}$

Solve each system by the addition method.

47. $\begin{cases} 2x + 3y = 14 \\ 3x - 4y = -69.1 \end{cases}$ **48.** $\begin{cases} 5x - 2y = -19.8 \\ -3x + 5y = -3.7 \end{cases}$

49. Commercial broadcast television stations can be divided into VHF stations (channels 2 through 13) and UHF stations (channels 14 through 83). The number y of VHF stations in the United States from 1985 through 1997 is given by the equation $10x - 2y = -986$, where x is the number of years after 1980. The number y of UHF stations in the United States from 1985 through 1997 is given by the equation $-21x + y = 295$, where x is the number of years after 1980. (*Source:* Based on data from the Television Bureau of Advertising, Inc.)

a. Use the addition method to solve this system of equations. (Round your final results to the nearest whole numbers.)

b. Interpret your solution from part (a).

c. During which years were there more UHF commercial television stations than VHF stations?

50. In recent years, the number of daily newspapers printed as morning editions has been increasing and the number of daily newspapers printed as evening editions has been decreasing. The number y of daily morning newspapers in existence from 1980 through 1998 is given by the equation $-665x + 36y = 13,800$, where x is the number of years after 1980. The number y of daily evening newspapers in existence from 1980 through 1998 is given by the equation $3239x + 96y = 134,013.6$, where x is the number of years after 1980. (*Source:* Based on data from the *Editor and Publisher International Year Book*, Editor and Publisher Co., New York, NY, annual)

a. Suppose these trends continue in the future. Use the addition method to predict the year in which the number of morning newspapers will equal the number of evening newspapers. (Round to the nearest whole number.)

b. How many of each type of newspaper will be in existence in that year?

51. Use the system of linear equations below to answer the questions.

$$\begin{cases} x + y = 5 \\ 3x + 3y = b \end{cases}$$

a. Find the value of b so that the equations are dependent and the system has an infinite number of solutions.

b. Find a value of b so that the system is inconsistent and there are no solutions to the system.

52. Use the system of linear equations below to answer the questions.

$$\begin{cases} x + y = 4 \\ 2x + by = 8 \end{cases}$$

a. Find the value of b so that the equations are dependent and the system has an infinite number of solutions.

b. Find a value of b so that the system is consistent and the system has a single solution.

53. Suppose you are solving the system

$$\begin{cases} -4x + 7y = 6 \\ x + 2y = 5 \end{cases}$$

by the addition method.

a. What step(s) should you take if you wish to eliminate x when adding the equations?

b. What step(s) should you take if you wish to eliminate y when adding the equations?

54. Suppose you are solving the system

$$\begin{cases} 3x + 8y = -5 \\ 2x - 4y = 3 \end{cases}$$

You decide to use the addition method by multiplying both sides of the second equation by 2. In which of the following was the multiplication performed correctly? Explain.

a. $4x - 8y = 3$ b. $4x - 8y = 6$

REVIEW EXERCISES

Rewrite the following sentences using mathematical symbols. Do not solve the equations. See Sections 2.4 and 2.5.

55. Twice a number added to 6 is 3 less than the number.

56. The sum of three consecutive integers is 66.

57. Three times a number subtracted from 20 is 2.

58. Twice the sum of 8 and a number is the difference of the number and 20.

59. The product of 4 and the sum of a number and 6 is twice a number.

60. The quotient of twice a number and 7 is subtracted from the reciprocal of the number.

8.4 SYSTEMS OF LINEAR EQUATIONS AND PROBLEM SOLVING

CD-ROM   SSM

SSG   Video

▶ **OBJECTIVE**

1. Use a system of equations to solve problems.

1   Many of the word problems solved earlier using one-variable equations can also be solved using two equations in **two** variables. We use the same problem-solving steps that have been used throughout this text. The only difference is that two variables are assigned to represent the two unknown quantities and that the problem is translated into **two** equations.

PROBLEM-SOLVING STEPS

1. UNDERSTAND the problem. During this step, become comfortable with the problem. Some ways of doing this are to

   Read and reread the problem.
   Choose two variables to represent the two unknowns.
   Construct a drawing.
   Propose a solution and check. Pay careful attention to how you check your proposed solution. This will help when writing equations to model the problem.

2. TRANSLATE the problem into two equations.

3. SOLVE the system of equations.

4. INTERPRET the results: **Check** the proposed solution in the stated problem and **state** your conclusion.

◆**Example 1**   **FINDING UNKNOWN NUMBERS**

Find two numbers whose sum is 37 and whose difference is 21.

Solution   1. UNDERSTAND. Read and reread the problem. Suppose that one number is 20. If their sum is 37, the other number is 17 because $20 + 17 = 37$. Is their difference 21? No; $20 - 17 = 3$. Our proposed solution is incorrect, but we now have a better understanding of the problem.

Since we are looking for two numbers, we let

x = first number
y = second number

2. TRANSLATE. Since we have assigned two variables to this problem, we translate our problem into two equations.

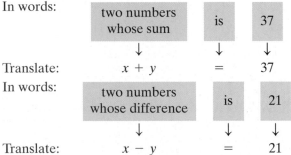

In words:

two numbers whose sum	is	37
↓	↓	↓

Translate: $x + y$ =   37

In words:

two numbers whose difference	is	21
↓	↓	↓

Translate: $x - y$ =   21

3. SOLVE. Now we solve the system

$$\begin{cases} x + y = 37 \\ x - y = 21 \end{cases}$$

Notice that the coefficients of the variable y are opposites. Let's then solve by the addition method and begin by adding the equations.

$$\begin{array}{r} x + y = 37 \\ \underline{x - y = 21} \\ 2x \quad\;\; = 58 \end{array}$$ Add the equations.

$$x = \frac{58}{2} = 29$$ Divide both sides by 2.

Now we let $x = 29$ in the first equation to find y.

$$x + y = 37 \qquad \text{First equation.}$$
$$29 + y = 37$$
$$y = 37 - 29 = 8$$

4. **INTERPRET.** The solution of the system is $(29, 8)$.

   Check: Notice that the sum of 29 and 8 is $29 + 8 = 37$, the required sum. Their difference is $29 - 8 = 21$, the required difference.

   State: The numbers are 29 and 8.

Example 2   **SOLVING A PROBLEM ABOUT PRICES**

A local high school is presenting the play "Grease." Admission for 4 adults and 2 children is $22, while admission for 2 adults and 3 children is $16.

a. What is the price of an adult's ticket?
b. What is the price of a child's ticket?
c. A special rate of $60 is charged for groups of 20 persons. Should a group of 4 adults and 16 children use the group rate? Why or why not?

Solution   1. **UNDERSTAND.** Read and reread the problem and guess a solution. Let's suppose that the price of an adult's ticket is $5 and the price of a child's ticket is $4. To check our proposed solution, let's see if admission for 4 adults and 2 children is $22. Admission for 4 adults is 4($5) or $20 and admission for 2 children is 2($4) or $8. This gives a total admission of $20 + $8 = $28, not the required $22. Again though, we have accomplished the purpose of this process: We have a better understanding of the problem. To continue, we let

A = the price of an adult's ticket

C = the price of a child's ticket

2. **TRANSLATE.** We translate the problem into two equations using both variables.

In words:

admission for 4 adults	and	admission for 2 children	is	$22
↓	↓	↓	↓	↓
$4A$	$+$	$2C$	$=$	22

Translate: shown above

In words:

admission for 2 adults	and	admission for 3 children	is	$16
↓	↓	↓	↓	↓
$2A$	$+$	$3C$	$=$	16

Translate: shown above

3. **SOLVE.** We solve the system

$$\begin{cases} 4A + 2C = 22 \\ 2A + 3C = 16 \end{cases}$$

Since both equations are written in standard form, we solve by the addition method. First we multiply the second equation by -2 to eliminate the variable A. Then the system

$$\begin{cases} 4A + 2C = 22 \\ -2(2A + 3C) = -2(16) \end{cases}$$ simplifies to $$\begin{cases} 4A + 2C = 22 \\ -4A - 6C = -32 \end{cases}$$ Add the equations.

$$\overline{\; -4C = -10}$$

$$C = \frac{-10}{-4} = \frac{5}{2}$$

$C = \dfrac{5}{2} = 2.5$ or \$2.50, the children's ticket price.

To find A, we replace C with 2.5 in the first equation.

$$4A + 2C = 22 \qquad \text{First equation}$$
$$4A + 2(2.5) = 22 \qquad \text{Let } C = 2.5.$$
$$4A + 5 = 22$$
$$4A = 17$$
$$A = \frac{17}{4} = 4.25 \text{ or \$4.25, the adult's ticket price.}$$

4. **INTERPRET.**

Check: Notice that 4 adults and 2 children will pay $4(\$4.25) + 2(\$2.50) = \$17 + \$5 = \$22$, the required amount. Also, the price for 2 adults and 3 children is $2(\$4.25) + 3(\$2.50) = \$8.50 + \$7.50 = \$16$, the required amount.

State: Answer the three original questions.

a. Since $A = 4.25$, the price of an adult's ticket is \$4.25.

b. Since $C = 2.5$, the price of a child's ticket is \$2.50.

c. The regular admission price for 4 adults and 16 children is

$$4(\$4.25) + 16(\$2.50) = \$17.00 + \$40.00$$
$$= \$57.00$$

This is \$3 less than the special group rate of \$60, so they should *not* request the group rate. ▬

Example 3 FINDING RATES

Albert and Louis live 15 miles away from each other. They decide to meet one day by walking toward one another. After 2 hours they meet. If Louis walks one mile per hour faster than Albert, find both walking speeds.

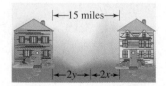

Solution 1. **UNDERSTAND.** Read and reread the problem. Let's propose a solution and use the formula $d = r \cdot t$ to check. Suppose that Louis's rate is 4 miles per hour. Since Louis's rate is 1 mile per hour faster, Albert's rate is 3 miles per hour. To check, see if they can walk a total of 15 miles in 2 hours. Louis's distance is rate · time $= 4(2) = 8$ miles and Albert's distance is rate time $= 3(2) = 6$ miles. Their total distance is 8 miles $+$ 6 miles $= 14$ miles, not the required 15 miles. Now that we have a better understanding of the problem, let's model it with a system of equations.

First, we let

$x =$ Albert's rate in miles per hour

$y =$ Louis's rate in miles per hour

Now we use the facts stated in the problem and the formula $d = rt$ to fill in the following chart.

	r	$\cdot$	t	$=$	d
ALBERT	x		2		$2x$
LOUIS	y		2		$2y$

2. TRANSLATE. We translate the problem into two equations using both variables.

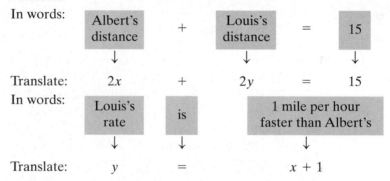

In words:

Albert's distance	$+$	Louis's distance	$=$	15
$\downarrow$		$\downarrow$		$\downarrow$

Translate: $2x$ $+$ $2y$ $=$ 15

In words:

Louis's rate	is	1 mile per hour faster than Albert's
$\downarrow$	$\downarrow$	$\downarrow$

Translate: y $=$ $x + 1$

3. SOLVE. The system of equations we are solving is

$$\begin{cases} 2x + 2y = 15 \\ y = x + 1 \end{cases}$$

Let's use substitution to solve the system since the second equation is solved for y.

$$2x + 2y = 15 \quad \text{First equation}$$

$$2x + 2(x + 1) = 15 \quad \text{Replace } y \text{ with } x + 1.$$

$$2x + 2x + 2 = 15$$

$$4x = 13$$

$$x = \frac{13}{4} = 3.25$$

$$y = x + 1 = 3.25 + 1 = 4.25$$

4. INTERPRET. Albert's proposed rate is 3.25 miles per hour and Louis's proposed rate is 4.25 miles per hour.

Check: Use the formula $d = rt$ and find that in 2 hours, Albert's distance is $(3.25)(2)$ miles or 6.5 miles. In 2 hours, Louis's distance is $(4.25)(2)$ miles or 8.5 miles. The total distance walked is 6.5 miles $+$ 8.5 miles or 15 miles, the given distance.

State: Albert walks at a rate of 3.25 miles per hour and Louis walks at a rate of 4.25 miles per hour. ∎

Example 4 **FINDING AMOUNTS OF SOLUTIONS**

Eric Daly, a chemistry teaching assistant, needs 10 liters of a 20% saline solution (salt water) for his 2 p.m. laboratory class. Unfortunately, the only mixtures on hand are a 5% saline solution and a 25% saline solution. How much of each solution should he mix to produce the 20% solution?

Solution   1. UNDERSTAND. Read and reread the problem. Suppose that we need 4 liters of the 5% solution. Then we need $10 - 4 = 6$ liters of the 25% solution. To see if this gives us 10 liters of a 20% saline solution, let's find the amount of pure salt in each solution.

	concentration rate	×	amount of solution	=	amount of pure salt
	↓		↓		↓
5% solution:	0.05	×	4 liters	=	0.2 liters
25% solution:	0.25	×	6 liters	=	1.5 liters
20% solution:	0.20	×	10 liters	=	2 liters

Since 0.2 liters $+ 1.5$ liters $= 1.7$ liters, not 2 liters, our proposed solution is incorrect. But we have gained some insight into how to model and check this problem.

We let

x = number of liters of 5% solution
y = number of liters of 25% solution

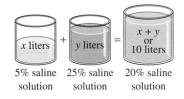

5% saline   25% saline   20% saline
solution   solution   solution

Now we use a table to organize the given data.

	Concentration Rate	Liters of Solution	Liters of Pure Salt
FIRST SOLUTION	5%	x	$0.05x$
SECOND SOLUTION	25%	y	$0.25y$
MIXTURE NEEDED	20%	10	$(0.20)(10)$

2. TRANSLATE. We translate into two equations using both variables.

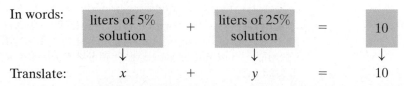

In words:

liters of 5% solution	+	liters of 25% solution	=	10
↓		↓		↓

Translate: x + y =   10

In words:

salt in 5% solution	$+$	salt in 25% solution	$=$	salt in mixture

$\downarrow$ $\downarrow$ $\downarrow$

Translate: $0.05x$ $+$ $0.25y$ $=$ $(0.20)(10)$

3. SOLVE. Here we solve the system

$$\begin{cases} x + y = 10 \\ 0.05x + 0.25y = 2 \end{cases}$$

To solve by the addition method, we first multiply the first equation by -25 and the second equation by 100. Then the system

$$\begin{cases} -25(x + y) = -25(10) \\ 100(0.05x + 0.25y) = 100(2) \end{cases} \text{ simplifies to } \begin{cases} -25x - 25y = -250 \\ \underline{5x + 25y = 200} \\ -20x \quad\quad = -50 \quad \text{Add.} \\ x = 2.5 \end{cases}$$

To find y, we let $x = 2.5$ in the first equation of the original system.

$$x + y = 10$$

$$2.5 + y = 10 \quad \text{Let } x = 2.5.$$

$$y = 7.5$$

4. INTERPRET. Thus, we propose that Eric needs to mix 2.5 liters of 5% saline solution with 7.5 liters of 25% saline solution.

Check: Notice that $2.5 + 7.5 = 10$, the required number of liters. Also, the sum of the liters of salt in the two solutions equals the liters of salt in the required mixture:

$$0.05(2.5) + 0.25(7.5) = 0.20(10)$$

$$0.125 + 1.875 = 2$$

State: Eric needs 2.5 liters of the 5% saline solution and 7.5 liters of the 25% solution. ■

Exercise Set 8.4

Without actually solving each problem, choose each correct solution by deciding which choice satisfies the given conditions.

△ **1.** The length of a rectangle is 3 feet longer than the width. The perimeter is 30 feet. Find the dimensions of the rectangle.
  a. length = 8 feet; width = 5 feet
  b. length = 8 feet; width = 7 feet
  c. length = 9 feet; width = 6 feet

△ **2.** An isosceles triangle, a triangle with two sides of equal length, has a perimeter of 20 inches. Each of the equal sides is one inch longer than the third side. Find the lengths of the three sides.
  a. 6 inches, 6 inches, and 7 inches

  b. 7 inches, 7 inches, and 6 inches
  c. 6 inches, 7 inches, and 8 inches

3. Two computer disks and three notebooks cost $17. However, five computer disks and four notebooks cost $32. Find the price of each.
  a. notebook = $4;   computer disk = $3
  b. notebook = $3;   computer disk = $4
  c. notebook = $5;   computer disk = $2

4. Two music CDs and four music cassette tapes cost a total of $40. However, three music CDs and five cassette tapes cost $55. Find the price of each.
  a. CD = $12; cassette = $4
  b. CD = $15; cassette = $2
  c. CD = $10; cassette = $5

5. Kesha has a total of 100 coins, all of which are either dimes or quarters. The total value of the coins is $13.00. Find the number of each type of coin.
   a. 80 dimes; 20 quarters **b.** 20 dimes; 44 quarters
   c. 60 dimes; 40 quarters

6. Yolanda has 28 gallons of saline solution available in two large containers at her pharmacy. One container holds three times as much as the other container. Find the capacity of each container.
   a. 15 gallons; 5 gallons **b.** 20 gallons; 8 gallons
   c. 21 gallons; 7 gallons

Write a system of equations describing each situation. Do not solve the system. See Example 1.

7. Two numbers add up to 15 and have a difference of 7.

8. The total of two numbers is 16. The first number plus 2 more than 3 times the second equals 18.

9. Keiko has a total of $6500, which she has invested in two accounts. The larger account is $800 greater than the smaller account.

10. Dominique has four times as much money in his savings account as in his checking account. The total amount is $2300.

Solve. See Examples 1 through 4.

11. Two numbers total 83 and have a difference of 17. Find the two numbers.

12. The sum of two numbers is 76 and their difference is 52. Find the two numbers.

13. A first number plus twice a second number is 8. Twice the first number plus the second totals 25. Find the numbers.

14. One number is 4 more than twice the second number. Their total is 25. Find the numbers.

15. The highest scorer during the WNBA 1999 regular season was Cynthia Cooper of the Houston Comets. Over the season, Cooper scored 101 more points than the second-highest scorer, her teammate Sheryl Swoopes. Together, Cooper and Swoopes scored 1271 points during the 1999 regular season. How many points did each player score over the course of the season? (*Source:* Women's National Basketball Association)

16. During the 1998–1999 regular NHL season, Teemu Selanne, of the Mighty Ducks of Anaheim, scored 20 fewer points than the Pittsburgh Penguins' Jaromir Jagr. Together, they scored 234 points during the 1998–1999 regular season. How many points each did Selanne and Jagr score? (*Source:* National Hockey League)

17. Ann Marie Jones has been pricing Amtrak train fares for a group trip to New York. Three adults and four children must pay $159. Two adults and three children must pay $112. Find the price of an adult's ticket, and find the price of a child's ticket.

18. Last month, Jerry Papa purchased five cassettes and two compact discs at Wall-to-Wall Sound for $65. This month he bought three cassettes and four compact discs for $81. Find the price of each cassette, and find the price of each compact disc.

19. Johnston and Betsy Waring have a jar containing 80 coins, all of which are either quarters or nickels. The total value of the coins is $14.60. How many of each type of coin do they have?

20. Art and Bette Meish purchased 40 stamps, a mixture of 32¢ and 19¢ stamps. Find the number of each type of stamp if they spent $12.15.

21. Fred and Staci Whittingham own 50 shares of IBM stock and 40 shares of GA Financial stock. At the close of the markets on March 24, 2000, their stock portfolio was worth $6485.90. The closing price of GA Financial stock was $64.25 more per share than the closing price of IBM stock on that day. What was the price of each stock on March 24, 2000? (*Source:* Based on data from Standard & Poor's ComStock)

22. Edie Hall has an investment in Kroger and General Motors stock. On March 24, 2000, Kroger stock closed at $17.875 per share and General Motors stock closed at $85.375 per share. Edie's portfolio was worth $11,317.50 at the end of the day. If Edie owns 60 more shares of General Motors stock than Kroger stock, how many of each type of stock does she own? (*Source:* Based on data from Standard & Poor's Com-Stock)

23. Pratap Puri rowed 18 miles down the Delaware River in 2 hours, but the return trip took him $4\frac{1}{2}$ hours. Find the rate Pratap could row in still water, and find the rate of the current.

	d	$=$ r	$\cdot$ t
Downstream	18	$x + y$	?
Upstream	18	$x - y$	$4\frac{1}{2}$

24. The Jonathan Schultz family took a canoe 10 miles down the Allegheny River in 1 hour and 15 minutes. After lunch it took them 4 hours to return. Find the rate of the current.

	d	$=$ r	$\cdot$ t
Downstream	10	$x + y$	$1\frac{1}{4}$
Upstream	10	$x - y$	4

25. Dave and Sandy Hartranft are frequent flyers with Delta Airlines. They often fly from Philadelphia to Chicago, a distance of 780 miles. On one particular trip they fly into the wind, and the flight takes 2 hours. The return trip, with the wind behind them, only takes $1\frac{1}{2}$ hours. Find the speed of the wind and find the speed of the plane in still air.

26. With a strong wind behind it, a United Airlines jet flies 2400 miles from Los Angeles to Orlando in 4 hours and 45 minutes. The return trip takes 6 hours, as the plane flies into the wind. Find the speed of the plane in still air, and find the wind speed to the nearest tenth mile per hour.

27. Dorren Schmidt is a chemist with Gemco Pharmaceutical. She needs to prepare 12 ounces of a 9% hydrochloric acid solution. Find the amount of 4% and the amount of 12% solution she should mix to get this solution.

28. Elise Everly is preparing 15 liters of a 25% saline solution. Elise has two other saline solutions with strengths of 40% and 10%. Find the amount of 40% solution and the amount of 10% solution she should mix to get 15 liters of a 25% solution.

29. Wayne Osby blends coffee for Maxwell House. He needs to prepare 200 pounds of blended coffee beans selling for $3.95 per pound. He intends to do this by blending together a high-quality bean costing $4.95 per pound and a cheaper bean costing $2.65 per pound. To the nearest pound, find how much high-quality coffee bean and how much cheaper coffee bean he should blend.

30. Macadamia nuts cost an astounding $16.50 per pound, but research by Planter's Peanuts says that mixed nuts sell better if macadamias are included. The standard mix costs $9.25 per pound. Find how many pounds of macadamias and how many pounds of the standard mix should be combined to produce 40 pounds that will cost $10 per pound. Find the amounts to the nearest tenth of a pound.

△ **31.** Find the measures of two complementary angles if one angle is twice the other. (Recall that two angles are complementary if their sum is 90°.)

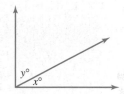

△ **32.** Find the measures of two supplementary angles if one angle is 20° more than four times the other. (Recall that two angles are supplementary if their sum is 180°.)

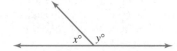

△ **33.** Find the measures of two complementary angles if one angle is 10° more than three times the other.

△ **34.** Find the measures of two supplementary angles if one angle is 18° more than twice the other.

35. Barb Hayes, a pharmacist, needs 50 liters of a 60% alcohol solution. She currently has available a 20% solution and a 70% solution. How many liters of each does she need to make the needed 50 liters of 60% alcohol solution?

36. Two cars are 440 miles apart and traveling toward each other. They meet in 3 hours. If one car's speed is 10 miles per hour faster than the other car's speed, find the speed of each car.

37. Carrie and Raymond McCormick had a pottery stand at the annual Skippack Craft Fair. They sold some of their pottery at the original price of $9.50 each, but later decreased the price of each by $2. If they sold all 90 pieces and took in $721, find how many they sold at the original price and how many they sold at the reduced price.

38. Trinity Church held its annual spaghetti supper and fed a total of 387 people. They charged $6.80 for adults and half-price for children. If they took in $2444.60, find how many adults and how many children attended the supper.

△ **39.** Dale Maxfield has decided to fence off a garden plot behind his house, using his house as the "fence" along one side of the garden. The length (which runs parallel to the house) is 3 feet less than twice the width. Find the dimensions if 33 feet of fencing is used along the three sides requiring it.

△ **40.** Judy McElroy plans to erect 152 feet of fencing around her rectangular horse pasture. A river bank serves as one side of the rectangle. If each width is 4 feet longer than half the length, find the dimensions.

41. Jim Williamson began a 186-mile bicycle trip to build up stamina for a triathlete competition. Unfortunately, his bicycle chain broke, so he finished the trip walking. The whole trip took 6 hours. If Jim walks at a rate of 4 miles per hour and rides at 40 mph, find the amount of time he spent on the bicycle.

42. Joan Gundersen rented a car from Hertz, which rents its cars for a daily fee plus an additional charge per mile driven. Joan recalls that a car rented for 5 days and driven for 300 miles cost her $178, while a car rented for 4 days and driven for 500 miles cost $197. Find the daily fee, and find the mileage charge.

43. In Canada, eastbound and westbound trains travel along the same track, with sidings to pull onto to avoid accidents. Two trains are now 150 miles apart, with the westbound train traveling twice as fast as the eastbound train. A warning must be issued to pull one train onto a siding or else the trains will crash in $1\frac{1}{4}$ hours. Find the speed of the eastbound train and the speed of the westbound train.

44. Cyril and Anoa Nantambu operate a small construction and supply company. In July they charged the Shaffers $1702.50 for 65 hours of labor and 3 tons of material. In August the Shaffers paid $1349 for 49 hours of labor and $2\frac{1}{2}$ tons of material. Find the cost per hour of labor and the cost per ton of material.

45. Suppose you mix an amount of 30% acid solution with an amount of 50% acid solution. Which of the following acid strengths would be possible for the resulting acid mixture? Explain why.

   A. 22%   **B.** 44%   **C.** 63%

REVIEW EXERCISES

Graph each linear inequality. See Section 6.5.

46. $y < 3 - x$ **47.** $y \geq 4 - 2x$

48. $2x - y \geq 6$ **49.** $3x + 5y < 15$

8.5 SYSTEMS OF LINEAR INEQUALITIES

CD-ROM   SSM

▶ **OBJECTIVE**

1. Solve a system of linear inequalities.

SSG   Video

1 Earlier we solved linear inequalities in two variables. Just as two linear equations make a system of linear equations, two linear inequalities make a **system of linear inequalities**. Systems of inequalities are very important in a process called linear programming. Many businesses use linear programming to find the most profitable way to use limited resources such as employees, machines, or buildings.

    A **solution of a system of linear inequalities** is an ordered pair that satisfies each inequality in the system. The set of all such ordered pairs is the solution set of the system. Graphing this set gives us a picture of the solution set. We can graph a system of inequalities by graphing each inequality in the system and identifying the region of overlap.

Example 1   Graph the solution of the system: $\begin{cases} 3x \geq y \\ x + 2y \leq 8 \end{cases}$

Solution   We begin by graphing each inequality on the same set of axes. The graph of the solution of the system is the region contained in the graphs of both inequalities. It is their intersection.

First, graph $3x \geq y$. The boundary line is the graph of $3x = y$. Sketch a solid boundary line since the inequality $3x \geq y$ means $3x > y$ or $3x = y$. The test point $(1, 0)$ satisfies the inequality, so shade the half-plane that includes $(1, 0)$.

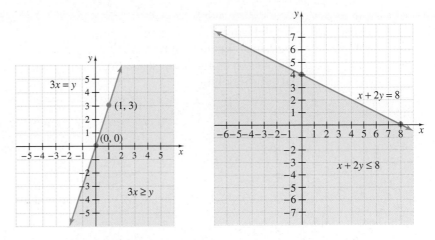

Next, sketch a solid boundary line $x + 2y = 8$ on the same set of axes. The test point $(0, 0)$ satisfies the inequality $x + 2y \leq 8$, so shade the half-plane that includes $(0, 0)$. (For clarity, the graph of $x + 2y \leq 8$ is shown on a separate set of axes.)

An ordered pair solution of the system must satisfy both inequalities. These solutions are points that lie in both shaded regions. The solution of the system is the darkest shaded region. This solution includes parts of both boundary lines.

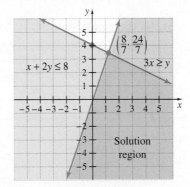

In linear programming, it is sometimes necessary to find the coordinates of the **corner point**: the point at which the two boundary lines intersect. To find the point of intersection, solve the related linear system

$$\begin{cases} 3x = y \\ x + 2y = 8 \end{cases}$$

by the substitution method or the addition method. The lines intersect at $\left(\frac{8}{7}, \frac{24}{7}\right)$, the corner point of the graph.

GRAPHING THE SOLUTION OF A SYSTEM OF LINEAR INEQUALITIES

Step 1. Graph each inequality in the system on the same set of axes.
Step 2. The solutions of the system are the points common to the graphs of all the inequalities in the system.

Example 2 Graph the solution of the system $\begin{cases} x - y < 2 \\ x + 2y > -1 \end{cases}$

Solution Graph both inequalities on the same set of axes. Both boundary lines are dashed lines since the inequality symbols are $<$ and $>$. The solution of the system is the region shown by the darkest shading. In this example, the boundary lines are not a part of the solution.

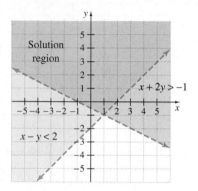

Example 3 Graph the solution of the system $\begin{cases} -3x + 4y < 12 \\ x \geq 2 \end{cases}$

Solution Graph both inequalities on the same set of axes.

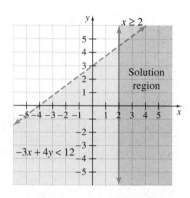

The solution of the system is the darkest shaded region, including a portion of the line $x = 2$.

SPOTLIGHT ON DECISION MAKING

Suppose you are a registered nurse. Today you are working at a health fair providing free blood pressure screenings. You measure an attendee's blood pressure as 168/82 (read as "168 over 82," where the systolic blood pressure is listed first and the diastolic blood pressure is listed second). What would you recommend that this health fair attendee do? Explain.

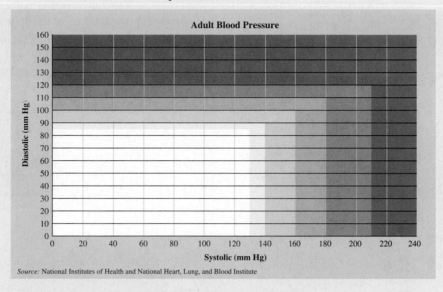

Adult Blood Pressure

Source: National Institutes of Health and National Heart, Lung, and Blood Institute

Blood Pressure Category and Recommended Follow-up:
- ☐ Normal: recheck in 2 years
- ☐ High normal: recheck in 1 year
- ☐ Mild hypertension: confirm within 2 months
- ☐ Moderate hypertension: see primary care physician within 1 month
- ☐ Severe hypertension: see primary care physician within 1 week
- ☐ Very severe hypertension: see primary care physician immediately

Exercise Set 8.5

Graph the solution of each system of linear inequalities. See Examples 1 through 3.

1. $\begin{cases} y \ge x + 1 \\ y \ge 3 - x \end{cases}$

2. $\begin{cases} y \ge x - 3 \\ y \ge -1 - x \end{cases}$

3. $\begin{cases} y < 3x - 4 \\ y \le x + 2 \end{cases}$

4. $\begin{cases} y \le 2x + 1 \\ y > x + 2 \end{cases}$

5. $\begin{cases} y \le -2x - 2 \\ y \ge x + 4 \end{cases}$

6. $\begin{cases} y \le 2x + 4 \\ y \ge -x - 5 \end{cases}$

7. $\begin{cases} y \ge -x + 2 \\ y \le 2x + 5 \end{cases}$

8. $\begin{cases} y \ge x - 5 \\ y \le -3x + 3 \end{cases}$

9. $\begin{cases} x \ge 3y \\ x + 3y \le 6 \end{cases}$

10. $\begin{cases} -2x < y \\ x + 2y < 3 \end{cases}$

11. $\begin{cases} y + 2x \ge 0 \\ 5x - 3y \le 12 \end{cases}$

12. $\begin{cases} y + 2x \le 0 \\ 5x + 3y \ge -2 \end{cases}$

13. $\begin{cases} 3x - 4y \ge -6 \\ 2x + y \le 7 \end{cases}$

14. $\begin{cases} 4x - y \ge -2 \\ 2x + 3y \le -8 \end{cases}$

15. $\begin{cases} x \le 2 \\ y \ge -3 \end{cases}$

16. $\begin{cases} x \ge -3 \\ y \ge -2 \end{cases}$

17. $\begin{cases} y \ge 1 \\ x < -3 \end{cases}$

18. $\begin{cases} y > 2 \\ x \ge -1 \end{cases}$

19. $\begin{cases} 2x + 3y < -8 \\ x \ge -4 \end{cases}$

20. $\begin{cases} 3x + 2y \le 6 \\ x < 2 \end{cases}$

21. $\begin{cases} 2x - 5y \le 9 \\ y \le -3 \end{cases}$

22. $\begin{cases} 2x + 5y \le -10 \\ y \ge 1 \end{cases}$

23. $\begin{cases} y \ge \frac{1}{2}x + 2 \\ y \le \frac{1}{2}x - 3 \end{cases}$

24. $\begin{cases} y \ge \frac{-3}{2}x + 3 \\ y < \frac{-3}{2}x + 6 \end{cases}$

For each system of inequalities, choose the corresponding graph.

25. $\begin{cases} y < 5 \\ x > 3 \end{cases}$ **26.** $\begin{cases} y > 5 \\ x < 3 \end{cases}$

27. $\begin{cases} y \le 5 \\ x < 3 \end{cases}$ **28.** $\begin{cases} y > 5 \\ x \ge 3 \end{cases}$

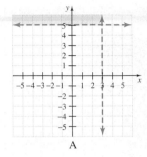
A

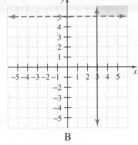
B

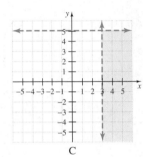
C

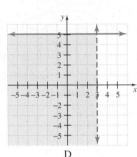
D

29. Tony Noellert budgets his time at work today. Part of the day he can write bills; the rest of the day he can use to write purchase orders. The total time available is at most 8 hours. Less than 3 hours is to be spent writing bills.
  a. Write a system of inequalities to describe the situation. (Let x = hours available for writing bills and y = hours available for writing purchase orders.)
  b. Graph the solution of the system.

30. Marty Thieme plans to invest money in two different accounts: a standard savings account and a riskier money market fund. The total of the two investments can be no more than $1000. At least $300 is to be put into the money market fund.
  a. Write a system of inequalities to describe the situation. Let x = amount invested in the savings account and y = amount invested in the money market fund.
  b. Graph the solutions of the system.

31. Explain how to decide which region to shade to show the solution region of the following system.
$$\begin{cases} x \ge 3 \\ y \ge -2 \end{cases}$$

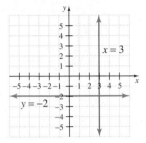

32. Describe the location of the solution region of the system
$$\begin{cases} x > 0 \\ y > 0. \end{cases}$$

REVIEW EXERCISES

Find the square of each expression. For example, the square of 7 is 7^2 or 49. The square of $5x$ is $(5x)^2$ or $25x^2$. See Section 3.1.

33. 4   **34.** 3

35. $6x$ **36.** $11y$

37. $10y^3$ **38.** $8x^5$

For additional Chapter Projects, visit the Real World Activities Website by going to http://www.prenhall.com/martin-gay.

CHAPTER PROJECT

Analyzing the Courses of Ships

From overhead photographs or satellite imagery of ships on the ocean, defense analysts can tell a lot about a ship's immediate course by looking at its wake. Assuming that two ships will maintain their present courses, it is possible to extend their paths, based on the wakes visible in the photograph. This can be used to find possible points of collision for the two ships.

In this project, you will investigate the courses and possibility of collision of the two ships shown in the figure. This project may be completed by working in groups or individually.

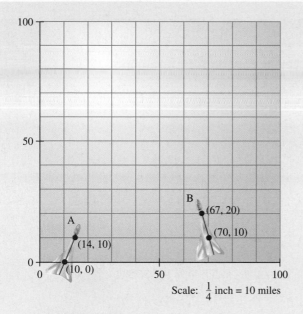

Scale: $\frac{1}{4}$ inch = 10 miles

1. Using each ship's wake as a guide, use a straight-edge to extend the paths of the ships on the figure. Estimate the coordinates of the point of intersection of the ships' courses from the grid. If the ships continue in these courses, they could possibly collide at the point of intersection of their paths.

2. What factor will govern whether or not the ships actually collide at the point found in Question 1?

3. Using the coordinates labeled on each ship's wake, find a linear equation that describes each path.

4. (Optional) Use a graphing calculator to graph both equations in the same window. Use the Intersect or Trace feature to estimate the point of intersection of the two paths. Compare this estimate to your estimate in Question 1.

5. Solve the system of two linear equations using one of the methods in this chapter. The solution is the point of intersection of the two paths. Compare your answer to your estimates from Questions 1 and 4.

6. Plot the point of intersection you found in Question 5 on the figure. Use the figure's scale to find each ship's distance from this point of collision by measuring from the bow (tip) of each ship with a ruler. Suppose that the speed of Ship A is r_1 and the speed of ship B is r_2. Given the present positions and courses of the two ships, find a relationship between their speeds that would ensure their collision.

CHAPTER 8 VOCABULARY CHECK

Fill in each blank with one of the words or phrases listed below.

| system of linear equations | solution | consistent | independent |
| dependent | inconsistent | substitution | addition |

1. In a system of linear equations in two variables, if the graphs of the equations are the same, the equations are _____ equations.

2. Two or more linear equations are called a _____.

3. A system of equations that has at least one solution is called a(n) _____ system.

4. A _____ of a system of two equations in two variables is an ordered pair of numbers that is a solution of both equations in the system.

5. Two algebraic methods for solving systems of equations are _____ and _____.

6. A system of equations that has no solution is called a(n) _____ system.

7. In a system of linear equations in two variables, if the graphs of the equations are different, the equations are _____ equations.

CHAPTER 8 HIGHLIGHTS

DEFINITIONS AND CONCEPTS	EXAMPLES

Section 8.1 Solving Systems of Linear Equations by Graphing

A **solution** of a system of two equations in two variables is an ordered pair of numbers that is a solution of both equations in the system.

Determine whether $(-1, 3)$ is a solution of the system:

$$\begin{cases} 2x - y = -5 \\ x = 3y - 10 \end{cases}$$

Replace x with -1 and y with 3 in both equations.

$2x - y = -5$ 　　　　　　 $x = 3y - 10$

$2(-1) - 3 \stackrel{?}{=} -5$ 　　　 $-1 \stackrel{?}{=} 3 \cdot 3 - 10$

$-5 = -5$ 　True 　　 $-1 = -1$ 　　　　True

$(-1, 3)$ is a solution of the system.

Graphically, a solution of a system is a point common to the graphs of both equations.

Solve by graphing: $\begin{cases} 3x - 2y = -3 \\ x + y = 4 \end{cases}$

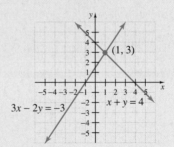

A system of equations with at least one solution is a **consistent system**. A system that has no solution is an **inconsistent system**.

If the graphs of two linear equations are identical, the equations are **dependent**. If their graphs are different, the equations are **independent**.

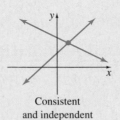

Consistent
and independent

Consistent
and dependent

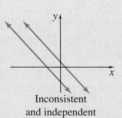

Inconsistent
and independent

(continued)

DEFINITIONS AND CONCEPTS	EXAMPLES

Section 8.2 Solving Systems of Linear Equations by Substitution

To solve a system of linear equations by the substitution method

Step 1. Solve one equation for a variable.

Step 2. Substitute the expression for the variable into the other equation.

Step 3. Solve the equation from Step 2 to find the value of one variable.

Step 4. Substitute the value from Step 3 in either original equation to find the value of the other variable.

Step 5. Check the solution in both equations.

Solve by substitution.

$$\begin{cases} 3x + 2y = 1 \\ x = y - 3 \end{cases}$$

Substitute $y - 3$ for x in the first equation.

$$3x + 2y = 1$$

$$3(y - 3) + 2y = 1$$

$$3y - 9 + 2y = 1$$

$$5y = 10$$

$$y = 2 \qquad \text{Divide by 5.}$$

To find x, substitute 2 for y in $x = y - 3$ so that $x = 2 - 3$ or -1. The solution $(-1, 2)$ checks.

Section 8.3 Solving Systems of Linear Equations by Addition

To solve a system of linear equations by the addition method

Step 1. Rewrite each equation in standard form $Ax + By = C$.

Step 2. Multiply one or both equations by a nonzero number so that the coefficients of a variable are opposites.

Step 3. Add the equations.

Step 4. Find the value of one variable by solving the resulting equation.

Step 5. Substitute the value from Step 4 into either original equation to find the value of the other variable.

Step 6. Check the solution in both equations.

If solving a system of linear equations by substitution or addition yields a true statement such as $-2 = -2$, then the graphs of the equations in the system are identical and there is an infinite number of solutions of the system.

Solve by addition.

$$\begin{cases} x - 2y = 8 \\ 3x + y = -4 \end{cases}$$

Multiply both sides of the first equation by -3.

$$\begin{cases} -3x + 6y = -24 \\ \underline{3x + y = -4} \end{cases}$$
$$7y = -28 \qquad \text{Add.}$$
$$y = -4 \qquad \text{Divide by 7.}$$

To find x, let $y = -4$ in an original equation.

$$x - 2(-4) = 8 \qquad \text{First equation}$$

$$x + 8 = 8$$

$$x = 0$$

The solution $(0, -4)$ checks.

Solve: $\begin{cases} 2x - 6y = -2 \\ x = 3y - 1 \end{cases}$

Substitute $3y - 1$ for x in the first equation.

$$2(3y - 1) - 6y = -2$$

$$6y - 2 - 6y = -2$$

$$-2 = -2 \qquad \text{True}$$

The system has an infinite number of solutions.

(continued)

DEFINITIONS AND CONCEPTS	EXAMPLES

Section 8.4 Systems of Linear Equations and Problem Solving

Problem-solving steps

  1. UNDERSTAND. Read and reread the problem.

Two angles are supplementary if their sum is 180°. The larger of two supplementary angles is three times the smaller, decreased by twelve. Find the measure of each angle. Let

$$x = \text{measure of smaller angle}$$
$$y = \text{measure of larger angle}$$

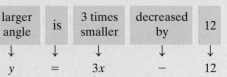

  2. TRANSLATE.

In words:

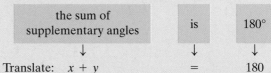

Translate: $x + y$ $=$ 180

In words:

larger angle	is	3 times smaller	decreased by	12
↓	↓	↓	↓	↓

Translate: y $=$ $3x$ $-$ 12

  3. SOLVE.

Solve the system:

$$\begin{cases} x + y = 180 \\ y = 3x - 12 \end{cases}$$

Use the substitution method and replace y with $3x - 12$ in the first equation.

$$x + y = 180$$
$$x + (3x - 12) = 180$$
$$4x = 192$$
$$x = 48$$

Since $y = 3x - 12$, then $y = 3 \cdot 48 - 12$ or 132.

  4. INTERPRET.

The solution checks. The smaller angle measures 48° and the larger angle measures 132°.

Section 8.5 Systems of Linear Inequalities

A system of linear inequalities consists of two or more linear inequalities.

To graph a system of inequalities, graph each inequality in the system. The overlapping region is the solution of the system.

System of Linear Inequalities

$$\begin{cases} x - y \geq 3 \\ y \leq -2x \end{cases}$$

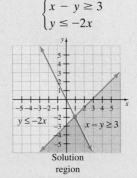

CHAPTER 8 REVIEW

(8.1) *Determine whether any of the following ordered pairs satisfy the system of linear equations.*

1. $\begin{cases} 2x - 3y = 12 \\ 3x + 4y = 1 \end{cases}$

   a. $(12, 4)$

   b. $(3, -2)$

   c. $(-3, 6)$

2. $\begin{cases} 4x + y = 0 \\ -8x - 5y = 9 \end{cases}$

   a. $\left(\dfrac{3}{4}, -3\right)$

   b. $(-2, 8)$

   c. $\left(\dfrac{1}{2}, -2\right)$

3. $\begin{cases} 5x - 6y = 18 \\ 2y - x = -4 \end{cases}$

   a. $(-6, -8)$

   b. $\left(3, \dfrac{5}{2}\right)$

   c. $\left(3, -\dfrac{1}{2}\right)$

4. $\begin{cases} 2x + 3y = 1 \\ 3y - x = 4 \end{cases}$

   a. $(2, 2)$

   b. $(-1, 1)$

   c. $(2, -1)$

Solve each system of equations by graphing.

5. $\begin{cases} 2x + y = 5 \\ 3y = -x \end{cases}$

6. $\begin{cases} 3x + y = -2 \\ 2x - y = -3 \end{cases}$

7. $\begin{cases} y - 2x = 4 \\ x + y = -5 \end{cases}$

8. $\begin{cases} y - 3x = 0 \\ 2y - 3 = 6x \end{cases}$

9. $\begin{cases} 3x + y = 2 \\ 3x - 6 = -9y \end{cases}$

10. $\begin{cases} 2y + x = 2 \\ x - y = 5 \end{cases}$

Without graphing, ***(a)*** *decide whether the graphs of the system are identical lines, parallel lines, or lines intersecting at a single point and* ***(b)*** *determine the number of solutions for each system.*

11. $\begin{cases} 2x - y = 3 \\ y = 3x + 1 \end{cases}$

12. $\begin{cases} 3x + y = 4 \\ y = -3x + 1 \end{cases}$

13. $\begin{cases} \dfrac{2}{3}x + \dfrac{1}{6}y = 0 \\ \phantom{\dfrac{2}{3}x +} y = -4x \end{cases}$

14. $\begin{cases} \dfrac{1}{4}x + \dfrac{1}{8}y = 0 \\ \phantom{\dfrac{1}{4}x +} y = -6x \end{cases}$

(8.2) *Solve the following systems of equations by the substitution method. If there is a single solution, give the ordered pair. If not, state whether the system is inconsistent or whether the equations are dependent.*

15. $\begin{cases} y = 2x + 6 \\ 3x - 2y = -11 \end{cases}$

16. $\begin{cases} y = 3x - 7 \\ 2x - 3y = 7 \end{cases}$

17. $\begin{cases} x + 3y = -3 \\ 2x + y = 4 \end{cases}$

18. $\begin{cases} 3x + y = 11 \\ x + 2y = 12 \end{cases}$

19. $\begin{cases} 4y = 2x - 3 \\ x - 2y = 4 \end{cases}$

20. $\begin{cases} 2x = 3y - 18 \\ x + 4y = 2 \end{cases}$

21. $\begin{cases} 2(3x - y) = 7x - 5 \\ 3(x - y) - 4x = 6 \end{cases}$

22. $\begin{cases} 4(x - 3y) = 3x - 1 \\ 3(4y - 3x) = 1 - 8x \end{cases}$

23. $\begin{cases} \dfrac{3}{4}x + \dfrac{2}{3}y = 2 \\ 3x + y = 18 \end{cases}$

24. $\begin{cases} \dfrac{2}{5}x + \dfrac{3}{4}y = 1 \\ x + 3y = -2 \end{cases}$

(8.3) *Solve the following systems of equations by the addition method. If there is a single solution, give the ordered pair. If not, state whether the system is inconsistent or whether the equations are dependent.*

25. $\begin{cases} 2x + 3y = -6 \\ x - 3y = -12 \end{cases}$

26. $\begin{cases} 4x + y = 15 \\ -4x + 3y = -19 \end{cases}$

27. $\begin{cases} 2x - 3y = -15 \\ x + 4y = 31 \end{cases}$

28. $\begin{cases} x - 5y = -22 \\ 4x + 3y = 4 \end{cases}$

29. $\begin{cases} 2x = 6y - 1 \\ \dfrac{1}{3}x - y = \dfrac{-1}{6} \end{cases}$

30. $\begin{cases} 8x = 3y - 2 \\ \dfrac{4}{7}x - y = \dfrac{-5}{2} \end{cases}$

31. $\begin{cases} 5x = 6y + 25 \\ -2y = 7x - 9 \end{cases}$

32. $\begin{cases} -4x = 8 + 6y \\ -3y = 2x - 3 \end{cases}$

33. $\begin{cases} 3(x - 4) = -2y \\ 2x = 3(y - 19) \end{cases}$

34. $\begin{cases} 4(x + 5) = -3y \\ 3x - 2(y + 18) = 0 \end{cases}$

35. $\begin{cases} \dfrac{2x + 9}{3} = \dfrac{y + 1}{2} \\ \dfrac{x}{3} = \dfrac{y - 7}{6} \end{cases}$

36. $\begin{cases} \dfrac{2 - 5x}{4} = \dfrac{2y - 4}{2} \\ \dfrac{x + 5}{3} = \dfrac{y}{5} \end{cases}$

(8.4) *Solve by writing and solving a system of linear equations.*

37. The sum of two numbers is 16. Three times the larger number decreased by the smaller number is 72. Find the two numbers.

38. The Forrest Theater can seat a total of 360 people. They take in $15,150 when every seat is sold. If orchestra section tickets cost $45 and balcony tickets cost $35, find the number of people that can be seated in the orchestra section.

39. A riverboat can head 340 miles upriver in 19 hours, but the return trip takes only 14 hours. Find the current of the river and find the speed of the ship in still water to the nearest tenth of a mile.

	D	=	R	·	T
Upriver	340		$x - y$		19
Downriver	340		$x + y$		14

40. Sam Abney invested $9000 one year ago. Part of the money was invested at 6%, the rest at 10%. If the total interest earned in one year was $652.80, find how much was invested at each rate.

△ **41.** Ancient Greeks thought that the most pleasing dimensions for a picture are those where the length is approximately 1.6 times longer than the width. This ratio is known as the Golden Ratio. If Sandreka Walker has 6 feet of framing material, find the dimensions of the largest frame she can make that satisfies the Golden Ratio. Find the dimensions to the nearest hundredth of a foot.

42. Find the amount of 6% acid solution and the amount of 14% acid solution Pat should combine to prepare 50 cc (cubic centimeters) of a 12% solution.

43. The Deli charges $3.80 for a breakfast of 3 eggs and 4 strips of bacon. The charge is $2.75 for 2 eggs and 3 strips of bacon. Find the cost of each egg and the cost of each strip of bacon.

44. An exercise enthusiast alternates between jogging and walking. He traveled 15 miles during the past 3 hours. He jogs at a rate of 7.5 miles per hour and walks at a rate of 4 miles per hour. Find how much time, to the nearest hundredth of an hour, he actually spent jogging.

(8.5) *Graph the solutions of the following systems of linear inequalities.*

45. $\begin{cases} y \geq 2x - 3 \\ y \leq -2x + 1 \end{cases}$

46. $\begin{cases} y \leq -3x - 3 \\ y \leq 2x + 7 \end{cases}$

47. $\begin{cases} x + 2y > 0 \\ x - y \leq 6 \end{cases}$

48. $\begin{cases} x - 2y \geq 7 \\ x + y \leq -5 \end{cases}$

49. $\begin{cases} 3x - 2y \leq 4 \\ 2x + y \geq 5 \end{cases}$

50. $\begin{cases} 4x - y \leq 0 \\ 3x - 2y \geq -5 \end{cases}$

51. $\begin{cases} -3x + 2y > -1 \\ y < -2 \end{cases}$

52. $\begin{cases} -2x + 3y > -7 \\ x \geq -2 \end{cases}$

CHAPTER 8 TEST

Answer each question true or false.

1. A system of two linear equations in two variables can have exactly two solutions.

2. Although (1, 4) is not a solution of $x + 2y = 6$, it can still be a solution of the system $\begin{cases} x + 2y = 6 \\ x + y = 5 \end{cases}$.

3. If the two equations in a system of linear equations are added and the result is $3 = 0$, the system has no solution.

4. If the two equations in a system of linear equations are added and the result is $3x = 0$, the system has no solution.

Is the ordered pair a solution of the given linear system?

5. $\begin{cases} 2x - 3y = 5 \\ 6x + y = 1 \end{cases}$; $(1, -1)$

6. $\begin{cases} 4x - 3y = 24 \\ 4x + 5y = -8 \end{cases}$; $(3, -4)$

7. Use graphing to find the solutions of the system
$\begin{cases} y - x = 6 \\ y + 2x = -6 \end{cases}$

8. Use the substitution method to solve the system
$\begin{cases} 3x - 2y = -14 \\ x + 3y = -1 \end{cases}$

9. Use the substitution method to solve the system
$\begin{cases} \dfrac{1}{2}x + 2y = -\dfrac{15}{4} \\ 4x = -y \end{cases}$

10. Use the addition method to solve the system
$\begin{cases} 3x + 5y = 2 \\ 2x - 3y = 14 \end{cases}$

11. Use the addition method to solve the system
$\begin{cases} 5x - 6y = 7 \\ 7x - 4y = 12 \end{cases}$

Solve each system using the substitution method or the addition method.

12. $\begin{cases} 3x + y = 7 \\ 4x + 3y = 1 \end{cases}$

13. $\begin{cases} 3(2x + y) = 4x + 20 \\ x - 2y = 3 \end{cases}$

14.
$$\begin{cases} \dfrac{x-3}{2} = \dfrac{2-y}{4} \\ \dfrac{7-2x}{3} = \dfrac{y}{2} \end{cases}$$

15. Lisa has a bundle of money consisting of $1 bills and $5 bills. There are 62 bills in the bundle. The total value of the bundle is $230. Find the number of $1 bills and the number of $5 bills.

16. Don has invested $4000, part at 5% simple annual interest and the rest at 9%. Find how much he invested at each rate if the total interest after 1 year is $311.

17. Although the number of farms in the U.S. is still decreasing, small farms are making a comeback. Texas and Missouri are the states with the most number of farms. Texas has 116 thousand more farms than Missouri and the total number of farms for these two states is 336 thousand. Find the number of farms for each state.

Graph the solutions of the following systems of linear inequalities.

18. $\begin{cases} y + 2x \le 4 \\ y \ge 2 \end{cases}$

19. $\begin{cases} 2y - x \ge 1 \\ x + y \ge -4 \end{cases}$

CHAPTER 8 CUMULATIVE REVIEW

1. Insert $<$, $>$, or $=$ in the space between the paired numbers to make each statement true.
   a. $-1 \quad 0$
   b. $7 \quad \frac{14}{2}$
   c. $-5 \quad -6$

2. Solve $5t - 5 = 6t + 2$ for t.

3. Graph $x = -2y$ by plotting intercepts.

4. Find $(5x - 7)(x - 2)$ by the FOIL method.

Factor.

5. $x^2 - 8x + 15$

6. $4m^2 - 4m + 1$

7. $54a^3 - 16b^3$

8. $10t^2 - 17t + 3$

9. Solve: $x^2 - 9x = -20$

10. Find the value of $\dfrac{x+4}{2x-3}$ for the given replacement values.
   a. $x = 5$
   b. $x = -2$

11. Multiply: $\dfrac{x^2+x}{3x} \cdot \dfrac{6}{5x+5}$

12. Find the LCD of $\dfrac{2}{x-2}$ and $\dfrac{10}{2-x}$.

13. Subtract: $\dfrac{6x}{x^2-4} - \dfrac{3}{x+2}$

14. Simplify the complex fraction $\dfrac{\frac{5}{8}}{\frac{2}{3}}$.

15. Solve: $\dfrac{2x}{x-4} = \dfrac{8}{x-4} + 1$

16. Solve for x: $\dfrac{45}{x} = \dfrac{5}{7}$

17. The quotient of a number and 6 minus $\dfrac{5}{3}$ is the quotient of the number and 2. Find the number.

18. Find the slope and the y-intercept of the line whose equation is $3x - 4y = 4$.

19. Find an equation of the line passing through $(-1, 5)$ with slope -2. Write the equation in standard form: $Ax + By = C$.

20. Graph: $y = x^2$

21. Find the domain and the range of the relation $\{(0, 2), (3, 3), (-1, 0), (3, -2)\}$.

22. Determine the number of solutions of the system.
$$\begin{cases} 3x - y = 4 \\ x + 2y = 8 \end{cases}$$

Solve each system.

23. $\begin{cases} x + 2y = 7 \\ 2x + 2y = 13 \end{cases}$

24. $\begin{cases} x + y = 7 \\ x - y = 5 \end{cases}$

25. Find two numbers whose sum is 37 and whose difference is 21.

Using a Variety of Mathematical Concepts

Carpenters make up the largest group of construction trade workers in the United States. There are nearly one million carpenters at work in this country. About one-third are self-employed.

Generally, carpenters learn their trade through on-the-job training, apprenticeships, or formal training programs. Because carpenters must be able to measure accurately, interpret blueprints, lay out building plans, and check that their work is level, straight, and square, they need to be able to work comfortably with many mathematical concepts. Many formal carpentry-training programs include courses in algebra, geometry, and trigonometry, or suggest these as prerequisites.

 For more information about a career as a carpenter, visit the Kentucky State District Council of Carpenters Website by first going to www.prenhall.com/martin-gay.

In the Spotlight on Decision Making feature on page 513, you will have the opportunity, as a carpenter, to decide if a deck is square.

ROOTS AND RADICALS

9.1 INTRODUCTION TO RADICALS

9.2 SIMPLIFYING RADICALS

9.3 ADDING AND SUBTRACTING RADICALS

9.4 MULTIPLYING AND DIVIDING RADICALS

9.5 SOLVING EQUATIONS CONTAINING RADICALS

9.6 RADICAL EQUATIONS AND PROBLEM SOLVING

9.7 RATIONAL EXPONENTS

Having spent the last chapter studying equations, we return now to algebraic expressions. We expand on your skills of operating on expressions—adding, subtracting, multiplying, dividing, and raising to powers—to include finding roots. Just as subtraction is defined by addition and division by multiplication, finding roots is defined by raising to powers. This chapter also includes working with equations that contain roots and solving problems that can be modeled by such equations.

9.1 INTRODUCTION TO RADICALS

▶ **OBJECTIVES**

1. Find square roots of perfect squares.
2. Approximate irrational square roots.
3. Simplify square roots containing variables.
4. Find higher roots.

1

In this section, we define finding the *root* of a number by its reverse operation, raising a number to a power. We begin with squares and square roots.

$$\text{The square of } 5 \text{ is } 5^2 = 25.$$
$$\text{The square of } -5 \text{ is } (-5)^2 = 25.$$
$$\text{The square of } \frac{1}{2} \text{ is } \left(\frac{1}{2}\right)^2 = \frac{1}{4}.$$

The reverse operation of squaring a number is finding the *square root* of a number. For example,

$$\text{A square root of } 25 \text{ is } 5, \text{ because } 5^2 = 25.$$
$$\text{A square root of } 25 \text{ is also } -5, \text{ because } (-5)^2 = 25.$$
$$\text{A square root of } \frac{1}{4} \text{ is } \frac{1}{2}, \text{ because } \left(\frac{1}{2}\right)^2 = \frac{1}{4}.$$

In general, a number b is a square root of a number a if $b^2 = a$.

Notice that both 5 and -5 are square roots of 25. The symbol $\sqrt{}$ is used to denote the **positive** or **principal square root** of a number. For example,

$$\sqrt{25} = 5 \text{ since } 5^2 = 25 \text{ and } 5 \text{ is positive.}$$

The symbol $-\sqrt{}$ is used to denote the **negative square root**. For example,

$$-\sqrt{25} = -5$$

SQUARE ROOT

The positive or principal square root of a positive number a is written as $\sqrt{a}$. The negative square root of a is written as $-\sqrt{a}$.

$$\sqrt{a} = b \qquad \text{only if} \qquad b^2 = a \text{ and } b > 0$$

Also, the square root of 0, written as $\sqrt{0}$, is 0.

The symbol $\sqrt{}$ is called a **radical** or **radical sign**. The expression within or under a radical sign is called the **radicand**. An expression containing a radical is called a **radical expression**.

◆ **Example 1**   Find each square root.

   a. $\sqrt{36}$ **b.** $\sqrt{64}$ **c.** $-\sqrt{25}$ **d.** $\sqrt{\dfrac{9}{100}}$ **e.** $\sqrt{0}$

Solution   **a.** $\sqrt{36} = 6$, because $6^2 = 36$ and 6 is positive.

   b. $\sqrt{64} = 8$, because $8^2 = 64$ and 8 is positive.

   c. $-\sqrt{25} = -5$. The negative sign in front of the radical indicates the negative square root of 25.

   d. $\sqrt{\dfrac{9}{100}} = \dfrac{3}{10}$ because $\left(\dfrac{3}{10}\right)^2 = \dfrac{9}{100}$ and $\dfrac{3}{10}$ is positive.

   e. $\sqrt{0} = 0$ because $0^2 = 0$.   ▬

Is the square root of a negative number a real number? For example, is $\sqrt{-4}$ a real number? To answer this question, we ask ourselves, is there a real number whose square is −4? Since there is no real number whose square is −4, we say that $\sqrt{-4}$ is not a real number. In general,

---

A square root of a negative number is not a real number.

We will discuss numbers such as $\sqrt{-4}$ in Chapter 10.

2   Recall that numbers such as 1, 4, 9, 25, and $\frac{4}{25}$ are called **perfect squares**, since $1^2 = 1$, $2^2 = 4$, $3^2 = 9$, $5^2 = 25$, and $\left(\frac{2}{5}\right)^2 = \frac{4}{25}$. Square roots of perfect square radicands simplify to rational numbers. What happens when we try to simplify a root such as $\sqrt{3}$? Since 3 is not a perfect square, $\sqrt{3}$ is not a rational number. It cannot be written as a quotient of integers. It is called an **irrational number** and we can find a decimal **approximation** of it. To find decimal approximations, use a calculator or Appendix E. (For calculator help, see the box at the end of this section.)

Example 2   Use a calculator or Appendix E to approximate $\sqrt{3}$ to three decimal places.

Solution   We may use Appendix E or a calculator to approximate $\sqrt{3}$. To use a calculator, find the square root key $\boxed{\sqrt{}}$.

$\sqrt{3} \approx 1.732050808$

To three decimal places, $\sqrt{3} \approx 1.732$.   ▬

3   Radicals can also contain variables. Since the square root of a negative number is not a real number, we want to make sure variables in the radicand do not have replacement values that would make the radicand negative. To avoid negative radicands, we assume for the rest of this chapter that **if a variable appears in the radicand of a radical expression, it represents positive numbers only**. Then

$$\sqrt{y^2} = y \qquad \text{Because } (y)^2 = y^2$$

$$\sqrt{x^8} = x^4 \qquad \text{Because } (x^4)^2 = x^8$$

$$\sqrt{9x^2} = 3x \qquad \text{Because } (3x)^2 = 9x^2$$

Example 3 Simplify the following expressions. Assume that each variable represents a positive number.

    a. $\sqrt{x^2}$ **b.** $\sqrt{x^6}$ **c.** $\sqrt{16x^{16}}$ **d.** $\sqrt{\dfrac{x^4}{25}}$

Solution **a.** $\sqrt{x^2} = x$ because x times itself equals x^2.
      b. $\sqrt{x^6} = x^3$ because $\left(x^3\right)^2 = x^6$.
      c. $\sqrt{16x^{16}} = 4x^8$ because $\left(4x^8\right)^2 = 16x^{16}$.

      d. $\sqrt{\dfrac{x^4}{25}} = \dfrac{x^2}{5}$ because $\left(\dfrac{x^2}{5}\right)^2 = \dfrac{x^4}{25}$.

---

4   We can find roots other than square roots. For example, since $2^3 = 8$, we call 2 the **cube root** of 8. In symbols, we write

$$\sqrt[3]{8} = 2 \qquad \text{The number 3 is called the \textbf{index}.}$$

Also,

$$\sqrt[3]{27} = 3 \qquad \text{\small Since } 3^3 = 27$$
$$\sqrt[3]{-64} = -4 \qquad \text{\small Since } (-4)^3 = -64$$

Notice that unlike the square root of a negative number, the cube root of a negative number is a real number. This is so because while we cannot find a real number whose **square** is negative, we **can** find a real number whose **cube** is negative. In fact, the cube of a negative number is a negative number. Therefore, the cube root of a negative number is a negative number.

Example 4 Find the cube roots.

    a. $\sqrt[3]{1}$ **b.** $\sqrt[3]{-27}$ **c.** $\sqrt[3]{\dfrac{1}{125}}$

Solution **a.** $\sqrt[3]{1} = 1$ because $1^3 = 1$.
      b. $\sqrt[3]{-27} = -3$ because $(-3)^3 = -27$.

      c. $\sqrt[3]{\dfrac{1}{125}} = \dfrac{1}{5}$ because $\left(\dfrac{1}{5}\right)^3 = \dfrac{1}{125}$.

Just as we can raise a real number to powers other than 2 or 3, we can find roots other than square roots and cube roots. In fact, we can take the *n*th root of a number where *n* is any natural number. An **_n_th root** of a number a is a number whose *n*th power is a. The natural number n is called the **index**.

    In symbols, the *n*th root of a is written as $\sqrt[n]{a}$. The index 2 is usually omitted for square roots.

> **HELPFUL HINT**
> If the index is even, such as $\sqrt{}, \sqrt[4]{}, \sqrt[6]{}$, and so on, the radicand must be nonnegative for the root to be a real number. For example,
>
> $$\sqrt[4]{16} = 2 \text{ but } \sqrt[4]{-16} \text{ is not a real number}$$
> $$\sqrt[6]{64} = 2 \text{ but } \sqrt[6]{-64} \text{ is not a real number}$$

Example 5 Simplify the following expressions.

    a. $\sqrt[4]{16}$ **b.** $\sqrt[5]{-32}$ **c.** $-\sqrt[3]{8}$ **d.** $\sqrt[4]{-81}$

Solution   **a.** $\sqrt[4]{16} = 2$ because $2^4 = 16$ and 2 is positive.

       b. $\sqrt[5]{-32} = -2$ because $(-2)^5 = -32$.

       c. $-\sqrt[3]{8} = -2$ since $\sqrt[3]{8} = 2$.

       d. $\sqrt[4]{-81}$ is not a real number since the index 4 is even and the radicand -81 is negative.

CALCULATOR EXPLORATIONS

To simplify or approximate square roots using a calculator, locate the key marked $\boxed{\sqrt{}}$.

To simplify $\sqrt{25}$ using a scientific calculator, press $\boxed{25}$ $\boxed{\sqrt{}}$. The display should read

$\boxed{5}$. To simplify $\sqrt{25}$ using a graphing calculator, press $\boxed{\sqrt{}}$ $\boxed{25}$ $\boxed{\text{ENTER}}$.

To approximate $\sqrt{30}$, press $\boxed{30}$ $\boxed{\sqrt{}}$ (or $\boxed{\sqrt{}}$ $\boxed{30}$ $\boxed{\text{ENTER}}$). The display should read

$\boxed{5.4772256}$. This is an approximation for $\sqrt{30}$. A three-decimal-place approximation is

$$\sqrt{30} \approx 5.477$$

Is this answer reasonable? Since 30 is between perfect squares 25 and 36, $\sqrt{30}$ is between $\sqrt{25} = 5$ and $\sqrt{36} = 6$. The calculator result is then reasonable since 5.4772256 is between 5 and 6.

Use a calculator to approximate each expression to three decimal places. Decide whether each result is reasonable.

  1. $\sqrt{7}$ **2.** $\sqrt{14}$ **3.** $\sqrt{11}$

  4. $\sqrt{200}$ **5.** $\sqrt{82}$ **6.** $\sqrt{46}$

Many scientific calculators have a key, such as $\boxed{\sqrt[x]{y}}$, that can be used to approximate roots other than square roots. To approximate these roots using a graphing calcuator, look under the $\boxed{\text{MATH}}$ menuor consult your manual.

Use a calculator to approximate each expression to three decimal places. Decide whether each result is reasonable.

  7. $\sqrt[3]{40}$ **8.** $\sqrt[3]{71}$ **9.** $\sqrt[4]{20}$

 10. $\sqrt[4]{15}$ **11.** $\sqrt[5]{18}$ **12.** $\sqrt[6]{2}$

SPOTLIGHT ON DECISION MAKING

Suppose you are a highway maintenance supervisor. You and your crew are heading out to post signs at a new cloverleaf exit ramp on the highway. One of the signs needed is a suggested ramp speed limit. Once you are at the ramp, you realize that you forgot to check with the highway engineers about which sign to post. You know that the formula $S = \sqrt{2.5r}$ can be used to estimate the maximum safe speed S, in miles per hour, at which a car can travel on a curved road with *radius of curvature*, r, in feet. Using a tape measure, your crew measures the radius of curvature as 400 feet. Which sign should you post? Explain your reasoning.

RAMP	RAMP	RAMP	RAMP
25	**30**	**35**	**40**
M. P. H.	M. P. H.	M. P. H.	M. P. H.

MENTAL MATH

Answer each exercise true or false.

1. $\sqrt{-16}$ simplifies to a real number.

2. $\sqrt{64} = 8$ while $\sqrt[3]{64} = 4$.

3. The number 9 has two square roots.

4. $\sqrt{0} = 0$ and $\sqrt{1} = 1$.

5. If x is a positive number, $\sqrt{x^{10}} = x^5$.

6. If x is a positive number, $\sqrt{x^{16}} = x^4$.

Exercise Set 9.1

Find each square root if it is a real number. See Example 1.

1. $\sqrt{16}$

2. $\sqrt{9}$

3. $\sqrt{81}$

4. $\sqrt{49}$

5. $\sqrt{\dfrac{1}{25}}$

6. $\sqrt{\dfrac{1}{64}}$

7. $-\sqrt{100}$

8. $-\sqrt{36}$

9. $\sqrt{-4}$

10. $\sqrt{-25}$

11. $-\sqrt{121}$

12. $-\sqrt{49}$

13. $\sqrt{\dfrac{9}{25}}$

14. $\sqrt{\dfrac{4}{81}}$

15. $\sqrt{144}$

16. $\sqrt{169}$

17. $\sqrt{\dfrac{49}{36}}$

18. $\sqrt{\dfrac{100}{121}}$

19. $-\sqrt{1}$

20. $-\sqrt{225}$

Approximate each square root to three decimal places. See Example 2.

21. $\sqrt{37}$

22. $\sqrt{27}$

23. $\sqrt{136}$

24. $\sqrt{8}$

△ **25.** A standard baseball diamond is a square with 90-foot sides connecting the bases. The distance from home plate to second base is $90 \cdot \sqrt{2}$ feet. Approximate $\sqrt{2}$ to two decimal places and use your result to approximate the distance $90 \cdot \sqrt{2}$ feet.

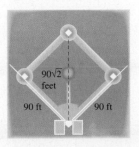

$90\sqrt{2}$ feet

90 ft   90 ft

△ **26.** The roof of the warehouse shown needs to be shingled. The total area of the roof is exactly $240 \cdot \sqrt{41}$ square feet. Approximate this area to the nearest whole number.

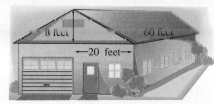

Find each root. Assume that all variables represent positive numbers. See Example 3.

27. $\sqrt{z^2}$ **28.** $\sqrt{y^{10}}$

29. $\sqrt{x^4}$ **30.** $\sqrt{x^6}$

31. $\sqrt{9x^8}$ **32.** $\sqrt{36x^{12}}$

33. $\sqrt{81x^2}$ **34.** $\sqrt{100z^4}$

35. $\sqrt{\dfrac{x^6}{36}}$ **36.** $\sqrt{\dfrac{y^8}{49}}$

37. $\sqrt{\dfrac{25y^2}{9}}$ **38.** $\sqrt{\dfrac{4x^2}{81}}$

Find each cube root. See Example 4.

39. $\sqrt[3]{125}$ **40.** $\sqrt[3]{64}$

41. $\sqrt[3]{-64}$ **42.** $\sqrt[3]{-27}$

43. $-\sqrt[3]{8}$ **44.** $-\sqrt[3]{27}$

45. $\sqrt[3]{\dfrac{1}{8}}$ **46.** $\sqrt[3]{\dfrac{1}{64}}$

47. $\sqrt[3]{-125}$ **48.** $-\sqrt[3]{64}$

49. $\sqrt[3]{-1000}$ **50.** $\sqrt[3]{-8}$

51. Explain why the square root of a negative number is not a real number.

52. Explain why the cube root of a negative number is a real number.

Find each root. See Example 5.

53. $\sqrt[5]{32}$ **54.** $\sqrt[4]{81}$

55. $\sqrt[4]{-16}$ **56.** $\sqrt{-9}$

57. $-\sqrt[4]{625}$ **58.** $-\sqrt[5]{32}$

59. $\sqrt[6]{1}$ **60.** $\sqrt[5]{1}$

61. $\sqrt[5]{-32}$ **62.** $-\sqrt[4]{256}$

63. $\sqrt[4]{256}$ **64.** $\sqrt[6]{64}$

△ **65.** If the amount of gold discovered by humankind could be assembled in one place, it would make a cube with a volume of 195,112 cubic feet. Each side of the cube would be $\sqrt[3]{195{,}112}$ feet long. How long would one side of the cube be? (*Source: Reader's Digest,* June 1996)

66. Graph $y = \sqrt{x}$. (*Hint:* Complete the table below, plot the ordered pair solutions, and draw a smooth curve through the points. Remember that since the radicand cannot be negative, this particular graph begins at the point with coordinates $(0, 0)$.)

x	y
0	0
1	
3	(approximate)
4	
9	

67. Graph $y = \sqrt[3]{x}$ (Complete the table below, plot the ordered pair solutions, and draw a smooth curve through the points.)

x	y
-8	
-2	(approximate)
-1	
0	
1	
2	(approximate)
8	

Simplify.

68. $\sqrt[3]{x^{12}}$ **69.** $\sqrt[3]{x^{21}}$

70. $\sqrt[4]{x^{12}}$ **71.** $\sqrt[4]{x^{20}}$

▦ *Use a grapher and graph each function. Observe the graph from left to right and give the ordered pair that corresponds to the "beginning" of the graph. Then tell why the graph starts at that point.*

72. $y = \sqrt{x - 2}$ **73.** $y = \sqrt{x + 3}$

74. $y = \sqrt{x + 4}$ **75.** $y = \sqrt{x - 5}$

REVIEW EXERCISES

Write each integer as a product of two integers such that one of the factors is a perfect square. For example, in $18 = 9 \cdot 2$, 9 is a perfect square.

76. 50   **77.** 8

78. 32   **79.** 75

80. 28   **81.** 44

82. 27   **83.** 90

9.2 SIMPLIFYING RADICALS

CD-ROM SSM

SSG Video

▶ **OBJECTIVES**

1. Use the product rule to simplify square roots.
2. Use the quotient rule to simplify square roots.
3. Simplify radicals containing variables.
4. Simplify higher roots.

1

A square root is simplified when the radicand contains no perfect square factors (other than 1). For example, $\sqrt{20}$ is not simplified because $\sqrt{20} = \sqrt{4 \cdot 5}$ and 4 is a perfect square.

To begin simplifying square roots, we notice the following pattern.

$$\sqrt{9 \cdot 16} = \sqrt{144} = 12$$
$$\sqrt{9} \cdot \sqrt{16} = 3 \cdot 4 = 12$$

Since both expressions simplify to 12, we can write

$$\sqrt{9 \cdot 16} = \sqrt{9} \cdot \sqrt{16}$$

This suggests the following product rule for square roots.

PRODUCT RULE FOR SQUARE ROOTS

If $\sqrt{a}$ and $\sqrt{b}$ are real numbers, then

$$\sqrt{a \cdot b} = \sqrt{a} \cdot \sqrt{b}$$

In other words, the square root of a product is equal to the product of the square roots.

To simplify $\sqrt{20}$, for example, we factor 20 so that one of its factors is a perfect square factor.

$$\sqrt{20} = \sqrt{4 \cdot 5} \qquad \text{Factor 20.}$$
$$= \sqrt{4} \cdot \sqrt{5} \qquad \text{Use the product rule.}$$
$$= 2\sqrt{5} \qquad \text{Write } \sqrt{4} \text{ as 2.}$$

The notation $2\sqrt{5}$ means $2 \cdot \sqrt{5}$. Since the radicand 5 has no perfect square factor other than 1, $2\sqrt{5}$ is in simplest form.

HELPFUL HINT

A radical expression in simplest form does *not mean* a decimal approximation. The simplest form of a radical expression is an exact form and may still contain a radical.

$$\underbrace{\sqrt{20} = 2\sqrt{5}}_{exact} \qquad\qquad \underbrace{\sqrt{20} \approx 4.47}_{decimal\ approximation}$$

Example 1 Simplify.

  a. $\sqrt{54}$ **b.** $\sqrt{12}$ **c.** $\sqrt{200}$ **d.** $\sqrt{35}$

Solution **a.** Try to factor 54 so that at least one of the factors is a perfect square. Since 9 is a perfect square and $54 = 9 \cdot 6$,

$$\begin{aligned} \sqrt{54} &= \sqrt{9 \cdot 6} && \text{Factor 54.} \\ &= \sqrt{9} \cdot \sqrt{6} && \text{Apply the product rule.} \\ &= 3\sqrt{6} && \text{Write } \sqrt{9} \text{ as 3.} \end{aligned}$$

b.
$$\begin{aligned} \sqrt{12} &= \sqrt{4 \cdot 3} && \text{Factor 12.} \\ &= \sqrt{4} \cdot \sqrt{3} && \text{Apply the product rule.} \\ &= 2\sqrt{3} && \text{Write } \sqrt{4} \text{ as 2.} \end{aligned}$$

c. The largest perfect square factor of 200 is 100.

$$\begin{aligned} \sqrt{200} &= \sqrt{100 \cdot 2} && \text{Factor 200.} \\ &= \sqrt{100} \cdot \sqrt{2} && \text{Apply the product rule.} \\ &= 10\sqrt{2} && \text{Write } \sqrt{100} \text{ as 10.} \end{aligned}$$

d. The radicand 35 contains no perfect square factors other than 1. Thus $\sqrt{35}$ is in simplest form.

In Example 1, part **(c)**, what happens if we don't use the largest perfect square factor of 200? Although using the largest perfect square factor saves time, the result is the same no matter what perfect square factor is used. For example, it is also true that $200 = 4 \cdot 50$. Then

$$\begin{aligned} \sqrt{200} &= \sqrt{4} \cdot \sqrt{50} \\ &= 2 \cdot \sqrt{50} \end{aligned}$$

Since $\sqrt{50}$ is not in simplest form, we continue.

$$\begin{aligned} \sqrt{200} &= 2 \cdot \sqrt{50} \\ &= 2 \cdot \sqrt{25} \cdot \sqrt{2} \\ &= 2 \cdot 5 \cdot \sqrt{2} \\ &= 10\sqrt{2} \end{aligned}$$

2 Next, let's examine the square root of a quotient.

$$\sqrt{\frac{16}{4}} = \sqrt{4} = 2$$

Also,

$$\frac{\sqrt{16}}{\sqrt{4}} = \frac{4}{2} = 2$$

Since both expressions equal 2, we can write

$$\sqrt{\frac{16}{4}} = \frac{\sqrt{16}}{\sqrt{4}}$$

This suggests the following quotient rule.

QUOTIENT RULE FOR SQUARE ROOTS

If $\sqrt{a}$ and $\sqrt{b}$ are real numbers and $b \neq 0$, then

$$\sqrt{\frac{a}{b}} = \frac{\sqrt{a}}{\sqrt{b}}$$

In other words, the square root of a quotient is equal to the quotient of the square roots.

Example 2   Simplify.

   a. $\sqrt{\dfrac{25}{36}}$ **b.** $\sqrt{\dfrac{3}{64}}$ **c.** $\sqrt{\dfrac{40}{81}}$

Solution   Use the quotient rule.

   a. $\sqrt{\dfrac{25}{36}} = \dfrac{\sqrt{25}}{\sqrt{36}} = \dfrac{5}{6}$ **b.** $\sqrt{\dfrac{3}{64}} = \dfrac{\sqrt{3}}{\sqrt{64}} = \dfrac{\sqrt{3}}{8}$

   c. $\sqrt{\dfrac{40}{81}} = \dfrac{\sqrt{40}}{\sqrt{81}}$ Use the quotient rule.

      $= \dfrac{\sqrt{4} \cdot \sqrt{10}}{9}$ Apply the product rule and write $\sqrt{81}$ as 9.

      $= \dfrac{2\sqrt{10}}{9}$ Write $\sqrt{4}$ as 2.

3   Recall that $\sqrt{x^6} = x^3$ because $\left(x^3\right)^2 = x^6$. If an odd exponent occurs, we write the exponential expression so that one factor is the greatest even power contained in the expression. Then we use the product rule to simplify.

Example 3   Simplify. Assume that variables represent positive numbers only.

   a. $\sqrt{x^5}$ **b.** $\sqrt{8y^2}$ **c.** $\sqrt{\dfrac{45}{x^6}}$

Solution   **a.** $\sqrt{x^5} = \sqrt{x^4 \cdot x} = \sqrt{x^4} \cdot \sqrt{x} = x^2\sqrt{x}$

      b. $\sqrt{8y^2} = \sqrt{4 \cdot 2 \cdot y^2} = \sqrt{4y^2 \cdot 2} = \sqrt{4y^2} \cdot \sqrt{2} = 2y\sqrt{2}$

      c. $\sqrt{\dfrac{45}{x^6}} = \dfrac{\sqrt{45}}{\sqrt{x^6}} = \dfrac{\sqrt{9 \cdot 5}}{x^3} = \dfrac{\sqrt{9} \cdot \sqrt{5}}{x^3} = \dfrac{3\sqrt{5}}{x^3}$

4 The product and quotient rules also apply to roots other than square roots. In general, we have the following product and quotient rules for radicals.

PRODUCT RULE FOR RADICALS

If $\sqrt[n]{a}$ and $\sqrt[n]{b}$ are real numbers, then

$$\sqrt[n]{a \cdot b} = \sqrt[n]{a} \cdot \sqrt[n]{b}$$

QUOTIENT RULE FOR RADICALS

If $\sqrt[n]{a}$ and $\sqrt[n]{b}$ are real numbers and $b \neq 0$, then

$$\sqrt[n]{\frac{a}{b}} = \frac{\sqrt[n]{a}}{\sqrt[n]{b}}$$

To simplify cube roots, look for perfect cube factors of the radicand. For example, 8 is a perfect cube, since $2^3 = 8$.

To simplify $\sqrt[3]{48}$, factor 48 as $8 \cdot 6$.

$$\begin{aligned}\sqrt[3]{48} &= \sqrt[3]{8 \cdot 6} & \text{Factor 48.}\\ &= \sqrt[3]{8} \cdot \sqrt[3]{6} & \text{Apply the product rule.}\\ &= 2\sqrt[3]{6} & \text{Write } \sqrt[3]{8} \text{ as 2.}\end{aligned}$$

$2\sqrt[3]{6}$ is in simplest form since the radicand 6 contains no perfect cube factors other than 1.

Example 4 Simplify.

  a. $\sqrt[3]{54}$ **b.** $\sqrt[3]{18}$ **c.** $\sqrt[3]{\frac{7}{8}}$ **d.** $\sqrt[3]{\frac{40}{27}}$

Solution   **a.** $\sqrt[3]{54} = \sqrt[3]{27 \cdot 2} = \sqrt[3]{27} \cdot \sqrt[3]{2} = 3\sqrt[3]{2}$

  b. The number 18 contains no perfect cube factors, so $\sqrt[3]{18}$ cannot be simplified further.

  c. $\sqrt[3]{\frac{7}{8}} = \frac{\sqrt[3]{7}}{\sqrt[3]{8}} = \frac{\sqrt[3]{7}}{2}$

  d. $\sqrt[3]{\frac{40}{27}} = \frac{\sqrt[3]{40}}{\sqrt[3]{27}} = \frac{\sqrt[3]{8 \cdot 5}}{3} = \frac{\sqrt[3]{8} \cdot \sqrt[3]{5}}{3} = \frac{2\sqrt[3]{5}}{3}$

To simplify fourth roots, look for perfect fourth powers of the radicand. For example, 16 is a perfect fourth power since $2^4 = 16$.

To simplify $\sqrt[4]{32}$, factor 32 as $16 \cdot 2$.

$$\begin{aligned}\sqrt[4]{32} &= \sqrt[4]{16 \cdot 2} & \text{Factor 32.}\\ &= \sqrt[4]{16} \cdot \sqrt[4]{2} & \text{Apply the product rule.}\\ &= 2\sqrt[4]{2} & \text{Write } \sqrt[4]{16} \text{ as 2.}\end{aligned}$$

Example 5   Simplify.

a. $\sqrt[4]{243}$ b. $\sqrt[4]{\dfrac{3}{16}}$ c. $\sqrt[5]{64}$

Solution   a. $\sqrt[4]{243} = \sqrt[4]{81 \cdot 3} = \sqrt[4]{81} \cdot \sqrt[4]{3} = 3\sqrt[4]{3}$

b. $\sqrt[4]{\dfrac{3}{16}} = \dfrac{\sqrt[4]{3}}{\sqrt[4]{16}} = \dfrac{\sqrt[4]{3}}{2}$

c. $\sqrt[5]{64} = \sqrt[5]{32 \cdot 2} = \sqrt[5]{32} \cdot \sqrt[5]{2} = 2\sqrt[5]{2}$

MENTAL MATH

Simplify each expression. Assume that all variables represent nonnegative real numbers.

1. $\sqrt{4 \cdot 9}$ 2. $\sqrt{9 \cdot 36}$ 3. $\sqrt{x^2}$ 4. $\sqrt{y^4}$
5. $\sqrt{0}$ 6. $\sqrt{1}$ 7. $\sqrt{25x^4}$ 8. $\sqrt{49x^2}$

Exercise Set 9.2

Use the product rule to simplify each radical. See Example 1.

1. $\sqrt{20}$ 2. $\sqrt{44}$

3. $\sqrt{18}$ 4. $\sqrt{45}$

5. $\sqrt{50}$ 6. $\sqrt{28}$

7. $\sqrt{33}$ 8. $\sqrt{98}$

9. $\sqrt{60}$ 10. $\sqrt{90}$

11. $\sqrt{180}$ 12. $\sqrt{150}$

13. $\sqrt{52}$ 14. $\sqrt{75}$

Use the quotient rule and the product rule to simplify each radical. See Example 2.

15. $\sqrt{\dfrac{8}{25}}$ 16. $\sqrt{\dfrac{63}{16}}$

17. $\sqrt{\dfrac{27}{121}}$ 18. $\sqrt{\dfrac{24}{169}}$

19. $\sqrt{\dfrac{9}{4}}$ 20. $\sqrt{\dfrac{100}{49}}$

21. $\sqrt{\dfrac{125}{9}}$ 22. $\sqrt{\dfrac{27}{100}}$

23. $\sqrt{\dfrac{11}{36}}$ 24. $\sqrt{\dfrac{30}{49}}$

25. $-\sqrt{\dfrac{27}{144}}$ 26. $-\sqrt{\dfrac{84}{121}}$

Simplify each radical. Assume that all variables represent positive numbers. See Example 3.

27. $\sqrt{x^7}$ 28. $\sqrt{y^3}$

29. $\sqrt{x^{13}}$ 30. $\sqrt{y^{17}}$

31. $\sqrt{75x^2}$ 32. $\sqrt{72y^2}$

33. $\sqrt{96x^4}$ 34. $\sqrt{40y^{10}}$

35. $\sqrt{\dfrac{12}{y^2}}$ 36. $\sqrt{\dfrac{63}{x^4}}$

37. $\sqrt{\dfrac{9x}{y^2}}$ 38. $\sqrt{\dfrac{6y^2}{x^4}}$

39. $\sqrt{\dfrac{88}{x^4}}$ 40. $\sqrt{\dfrac{x^{11}}{81}}$

Simplify each radical. See Example 4.

41. $\sqrt[3]{24}$ 42. $\sqrt[3]{81}$

43. $\sqrt[3]{250}$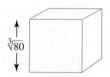

44. $\sqrt[3]{40}$

45. $\sqrt[3]{\dfrac{5}{64}}$

46. $\sqrt[3]{\dfrac{32}{125}}$

47. $\sqrt[3]{\dfrac{7}{8}}$

48. $\sqrt[3]{\dfrac{10}{27}}$

49. $\sqrt[3]{\dfrac{15}{64}}$

50. $\sqrt[3]{\dfrac{4}{27}}$

51. $\sqrt[3]{80}$

52. $\sqrt[3]{108}$

Simplify. See Example 5.

53. $\sqrt[4]{48}$

54. $\sqrt[4]{405}$

55. $\sqrt[4]{\dfrac{8}{81}}$

56. $\sqrt[4]{\dfrac{25}{256}}$

57. $\sqrt[5]{96}$

58. $\sqrt[5]{128}$

59. $\sqrt[5]{\dfrac{5}{32}}$

60. $\sqrt[5]{\dfrac{16}{243}}$

Simplify.

△ **61.** If a cube is to have a volume of 80 cubic inches, then each side must be $\sqrt[3]{80}$ inches long. Simplify the radical representing the side length.

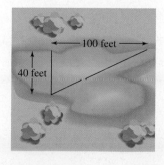

△ **62.** Jeannie Boswell is swimming across a 40-foot-wide river, trying to head straight across to the opposite shore. However, the current is strong enough to move her downstream 100 feet by the time she reaches land. (See the figure.) Because of the current, the actual distance she swam is $\sqrt{11,600}$ feet. Simplify this radical.

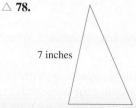

63. By using replacement values for a and b, show that $\sqrt{a^2 + b^2}$ does not equal $a + b$.

64. By using replacement values for a and b, show that $\sqrt{a + b}$ does not equal $\sqrt{a} + \sqrt{b}$.

The cost C in dollars per day to operate a small delivery service is given by $C = 100\sqrt[3]{n} + 700$ where n is the number of deliveries per day.

65. Find the cost if the number of deliveries is 1000.

66. Approximate the cost if the number of deliveries is 500.

Simplify.

67. $\sqrt[3]{-8x^6}$

68. $\sqrt[3]{y^{20}}$

69. $\sqrt[3]{\dfrac{2}{x^9}}$

70. $\sqrt[3]{\dfrac{48}{x^{12}}}$

REVIEW EXERCISES

Perform the following operations. See Sections 4.2 and 4.3.

71. $6x + 8x$

72. $(6x)(8x)$

73. $(2x + 3)(x - 5)$

74. $(2x + 3) + (x - 5)$

75. $9y^2 - 9y^2$

76. $(9y^2)(-8y^2)$

The following pairs of triangles are similar. Find the unknown lengths. See Section 6.7.

△ **77.**

6 cm

x

9 cm

12 cm

△ **78.**

7 inches

5 inches

x

3 inches

9.3 ADDING AND SUBTRACTING RADICALS

CD-ROM   SSM

SSG   Video

▶ **O B J E C T I V E S**

1. Add or subtract like radicals.
2. Simplify radical expressions, and then add or subtract any like radicals.

1   To combine like terms, we use the distributive property.

$$5x + 3x = (5 + 3)x = 8x$$

The distributive property can also be applied to expressions containing radicals. For example,

$$5\sqrt{2} + 3\sqrt{2} = (5 + 3)\sqrt{2} = 8\sqrt{2}$$

Also,

$$9\sqrt{5} - 6\sqrt{5} = (9 - 6)\sqrt{5} = 3\sqrt{5}$$

Radical terms $5\sqrt{2}$ and $3\sqrt{2}$ are **like radicals**, as are $9\sqrt{5}$ and $6\sqrt{5}$.

> **LIKE RADICALS**
>
> Like radicals are radical expressions that have the same index and the same radicand.

From the examples above, we can see that **only like radicals can be combined** in this way. For example, the expression $2\sqrt{3} + 3\sqrt{2}$ cannot be further simplified since the radicals are not like radicals. Also, the expression $4\sqrt{7} + 4\sqrt[3]{7}$ cannot be further simplified because the radicals are not like radicals since the indices are different.

Example 1   Simplify by combining like radical terms.

a. $4\sqrt{5} + 3\sqrt{5}$ **b.** $\sqrt{10} - 6\sqrt{10}$ **c.** $2\sqrt[3]{7} - 5\sqrt[3]{7} - 3\sqrt[3]{7}$ **d.** $2\sqrt{6} + 2\sqrt[3]{6}$

Solution   **a.** $4\sqrt{5} + 3\sqrt{5} = (4 + 3)\sqrt{5} = 7\sqrt{5}$

b. $\sqrt{10} - 6\sqrt{10} = 1\sqrt{10} - 6\sqrt{10} = (1 - 6)\sqrt{10} = -5\sqrt{10}$

c. $2\sqrt[3]{7} - 5\sqrt[3]{7} - 3\sqrt[3]{7} = (2 - 5 - 3)\sqrt[3]{7} = -6\sqrt[3]{7}$

d. $2\sqrt{6} + 2\sqrt[3]{6}$ cannot be simplified further since the indices are not the same. ■

2   At first glance, it appears that the expression $\sqrt{50} + \sqrt{8}$ cannot be simplified further because the radicands are different. However, the product rule can be used to simplify each radical, and then further simplification might be possible.

Example 2 Add or subtract by first simplifying each radical.

    a. $\sqrt{50} + \sqrt{8}$

    b. $7\sqrt{12} - \sqrt{75}$

    c. $\sqrt{25} - \sqrt{27} - 2\sqrt{18} - \sqrt{16}$

Solution   **a.** First simplify each radical.

$$\sqrt{50} + \sqrt{8} = \sqrt{25 \cdot 2} + \sqrt{4 \cdot 2}$$ Factor radicands.

$$= \sqrt{25} \cdot \sqrt{2} + \sqrt{4} \cdot \sqrt{2}$$ Apply the product rule.

$$= 5\sqrt{2} + 2\sqrt{2}$$ Simplify $\sqrt{25}$ and $\sqrt{4}$.

$$= 7\sqrt{2}$$ Add like radicals.

  b. $7\sqrt{12} - \sqrt{75} = 7\sqrt{4 \cdot 3} - \sqrt{25 \cdot 3}$ Factor radicands.

$$= 7\sqrt{4} \cdot \sqrt{3} - \sqrt{25} \cdot \sqrt{3}$$ Apply the product rule.

$$= 7 \cdot 2\sqrt{3} - 5\sqrt{3}$$ Simplify $\sqrt{4}$ and $\sqrt{25}$.

$$= 14\sqrt{3} - 5\sqrt{3}$$ Multiply.

$$= 9\sqrt{3}$$ Subtract like radicals.

  c. $\sqrt{25} - \sqrt{27} - 2\sqrt{18} - \sqrt{16}$

$$= 5 - \sqrt{9 \cdot 3} - 2\sqrt{9 \cdot 2} - 4$$ Factor radicands.

$$= 5 - \sqrt{9} \cdot \sqrt{3} - 2\sqrt{9} \cdot \sqrt{2} - 4$$ Apply the product rule.

$$= 5 - 3\sqrt{3} - 2 \cdot 3\sqrt{2} - 4$$ Simplify.

$$= 1 - 3\sqrt{3} - 6\sqrt{2}$$ Write $5 - 4$ as 1 and $2 \cdot 3$ as 6.

If radical expressions contain variables, we proceed in a similar way. Simplify radicals using the product and quotient rules. Then add or subtract any like radicals.

Example 3 Simplify: $2\sqrt{x^2} - \sqrt{25x} + \sqrt{x}$. Assume variables represent positive numbers.

Solution $2\sqrt{x^2} - \sqrt{25x} + \sqrt{x}$

$$= 2x - \sqrt{25} \cdot \sqrt{x} + \sqrt{x}$$ Write $\sqrt{x^2}$ as x and apply the product rule.

$$= 2x - 5\sqrt{x} + 1\sqrt{x}$$ Simplify.

$$= 2x - 4\sqrt{x}$$ Add like radicals.

Example 4 Add or subtract by first simplifying each radical.

$$2\sqrt[3]{27} - \sqrt[3]{54}$$

Solution $2\sqrt[3]{27} - \sqrt[3]{54} = 2 \cdot 3 - \sqrt[3]{27 \cdot 2}$ Simplify $\sqrt[3]{27}$ and factor 54.

$$= 6 - \sqrt[3]{27} \cdot \sqrt[3]{2}$$ Apply the product rule.

$$= 6 - 3\sqrt[3]{2}$$ Simplify $\sqrt[3]{27}$.

HELPFUL HINT
These two terms may not be combined. They are unlike terms.

MENTAL MATH

Simplify each expression by combining like radicals.

1. $3\sqrt{2} + 5\sqrt{2}$ **2.** $2\sqrt{3} + 7\sqrt{3}$ **3.** $5\sqrt{x} + 2\sqrt{x}$
4. $8\sqrt{x} + 3\sqrt{x}$ **5.** $5\sqrt{7} - 2\sqrt{7}$ **6.** $8\sqrt{6} - 5\sqrt{6}$

Exercise Set 9.3

Simplify each expression by combining like radicals where possible. See Example 1.

1. $4\sqrt{3} - 8\sqrt{3}$ **2.** $\sqrt{5} - 9\sqrt{5}$

3. $3\sqrt{6} + 8\sqrt{6} - 2\sqrt{6} - 5$

4. $12\sqrt{2} - 3\sqrt{2} + 8\sqrt{2} + 10$

5. $6\sqrt{5} - 5\sqrt{5} + \sqrt{2}$ **6.** $4\sqrt{3} + \sqrt{5} - 3\sqrt{3}$

7. $2\sqrt[3]{3} + 5\sqrt[3]{3} - \sqrt{3}$ **8.** $8\sqrt[3]{4} + 2\sqrt[3]{4} + 4$

9. $2\sqrt[3]{2} - 7\sqrt[3]{2} - 6$ **10.** $5\sqrt[3]{9} + 2 - 11\sqrt[3]{9}$

11. Find the perimeter of the rectangular picture frame.

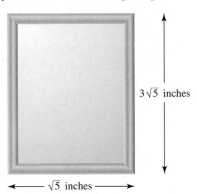

$3\sqrt{5}$ inches

$\sqrt{5}$ inches

12. Find the perimeter of the plot of land.

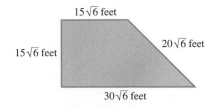

$15\sqrt{6}$ feet

$15\sqrt{6}$ feet

$20\sqrt{6}$ feet

$30\sqrt{6}$ feet

13. In your own words, describe like radicals.

14. In the expression $\sqrt{5} + 2 - 3\sqrt{5}$, explain why 2 and -3 cannot be combined.

Add or subtract by first simplifying each radical and then combining any like radical terms. Assume that all variables represent positive real numbers. See Examples 2 and 3.

15. $\sqrt{12} + \sqrt{27}$ **16.** $\sqrt{50} + \sqrt{18}$
17. $\sqrt{45} + 3\sqrt{20}$ **18.** $\sqrt{28} + \sqrt{63}$

19. $2\sqrt{54} - \sqrt{20} + \sqrt{45} - \sqrt{24}$
20. $2\sqrt{8} - \sqrt{128} + \sqrt{48} + \sqrt{18}$
21. $4x - 3\sqrt{x^2} + \sqrt{x}$ **22.** $x - 6\sqrt{x^2} + 2\sqrt{x}$
23. $\sqrt{25x} + \sqrt{36x} - 11\sqrt{x}$ **24.** $3\sqrt{x^3} - x\sqrt{4x}$
25. $\sqrt{16x} - \sqrt{x^3}$ **26.** $\sqrt{8x^3} - \sqrt{x^2}$
27. $12\sqrt{5} - \sqrt{5} - 4\sqrt{5}$ **28.** $7\sqrt{3} + 2\sqrt{3} - 13\sqrt{3}$
29. $\sqrt{5} + \sqrt[3]{5}$ **30.** $\sqrt{5} + \sqrt{5}$
31. $4 + 8\sqrt{2} - 9$ **32.** $6 - 2\sqrt{3} - \sqrt{3}$
33. $8 - \sqrt{2} - 5\sqrt{2}$ **34.** $\sqrt{75} + \sqrt{48}$
35. $5\sqrt{32} - \sqrt{72}$ **36.** $2\sqrt{80} - \sqrt{45}$
37. $\sqrt{8} + \sqrt{9} + \sqrt{18} + \sqrt{81}$
38. $\sqrt{6} + \sqrt{16} + \sqrt{24} + \sqrt{25}$

39. $\sqrt{\dfrac{5}{9}} + \sqrt{\dfrac{5}{81}}$ **40.** $\sqrt{\dfrac{3}{64}} + \sqrt{\dfrac{3}{16}}$

41. $\sqrt{\dfrac{3}{4}} - \sqrt{\dfrac{3}{64}}$ **42.** $\sqrt{\dfrac{7}{25}} - \sqrt{\dfrac{7}{100}}$

43. $2\sqrt{45} - 2\sqrt{20}$ **44.** $5\sqrt{18} + 2\sqrt{32}$
45. $\sqrt{35} - \sqrt{140}$ **46.** $\sqrt{6} - \sqrt{600}$
47. $5\sqrt{2x} + \sqrt{98x}$ **48.** $3\sqrt{9x} + 2\sqrt{x}$
49. $5\sqrt{x} + 4\sqrt{4x} - 13\sqrt{x}$ **50.** $\sqrt{9x} + \sqrt{81x} - 11\sqrt{x}$
51. $\sqrt{3x^3} + 3x\sqrt{x}$ **52.** $x\sqrt{4x} + \sqrt{9x^3}$

Add or subtract by first simplifying each radical and then combining any like radical terms. Assume that all variables represent positive real numbers. See Example 4.

53. $\sqrt[3]{81} + \sqrt[3]{24}$ **54.** $\sqrt[3]{32} - \sqrt[3]{4}$
55. $4\sqrt[3]{9} - \sqrt[3]{243}$ **56.** $7\sqrt[3]{6} - \sqrt[3]{48}$
57. $2\sqrt[3]{8} + 2\sqrt[3]{16}$ **58.** $3\sqrt[3]{27} + 3\sqrt[3]{81}$
59. $\sqrt[3]{8} + \sqrt[3]{54} - 5$ **60.** $\sqrt[3]{64} + \sqrt[3]{14} - 9$
61. $\sqrt{32x^2} + \sqrt[3]{32} + \sqrt{4x^2}$
62. $\sqrt{18x^2} + \sqrt[3]{24} + \sqrt{2x^2}$

63. $\sqrt{40x} + \sqrt[3]{40} - 2\sqrt{10x} - \sqrt[3]{5}$

64. $\sqrt{72x^2} + \sqrt[3]{54} - x\sqrt{50} - 3\sqrt[3]{2}$

65. A water trough is to be made of wood. Each of the two triangular end pieces has an area of $\dfrac{3\sqrt{27}}{4}$ square feet. The two side panels are both rectangular. In simplest radical form, find the total area of the wood needed.

66. Eight wooden braces are to be attached along the diagonals of the vertical sides of a storage bin. Each of four of these diagonals has a length of $\sqrt{52}$ feet, while each of the other four has a length of $\sqrt{80}$ feet. In simplest radical form, find the total length of the wood needed for these braces.

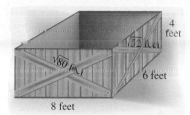

REVIEW EXERCISES

Square each binomial. See Section 4.4.

67. $(x + 6)^2$ **68.** $(3x + 2)^2$

69. $(2x - 1)^2$ **70.** $(x - 5)^2$

Solve each system of linear equations. See Section 8.2.

71. $\begin{cases} x = 2y \\ x + 5y = 14 \end{cases}$ **72.** $\begin{cases} y = -5x \\ x + y = 16 \end{cases}$

9.4 MULTIPLYING AND DIVIDING RADICALS

CD-ROM   SSM

SSG   Video

▶ **OBJECTIVES**

1. Multiply radicals.
2. Divide radicals.
3. Rationalize denominators.
4. Rationalize using conjugates.

1   In Section 9.2 we used the product and quotient rules for radicals to help us simplify radicals. In this section, we use these rules to simplify products and quotients of radicals.

PRODUCT RULE FOR RADICALS

If $\sqrt[n]{a}$ and $\sqrt[n]{b}$ are real numbers, then

$$\sqrt[n]{a} \cdot \sqrt[n]{b} = \sqrt[n]{a \cdot b}$$

This property says that the product of the *n*th roots of two numbers is the *n*th root of the product of the two numbers. For example,

$$\sqrt{3} \cdot \sqrt{2} = \sqrt{3 \cdot 2} = \sqrt{6}$$

Also,

$$\sqrt[3]{5} \cdot \sqrt[3]{7} = \sqrt[3]{5 \cdot 7} = \sqrt[3]{35}$$

Example 1   Multiply. Then simplify if possible.

a. $\sqrt{7} \cdot \sqrt{3}$ **b.** $\sqrt{3} \cdot \sqrt{15}$
c. $2\sqrt{6} \cdot 5\sqrt{2}$ **d.** $\left(3\sqrt{2}\right)^2$

Solution   **a.** $\sqrt{7} \cdot \sqrt{3} = \sqrt{7 \cdot 3} = \sqrt{21}$
b. $\sqrt{3} \cdot \sqrt{15} = \sqrt{45}$. Next, simplify $\sqrt{45}$.
$\qquad \sqrt{45} = \sqrt{9 \cdot 5} = \sqrt{9} \cdot \sqrt{5} = 3\sqrt{5}$
c. $2\sqrt{6} \cdot 5\sqrt{2} = 2 \cdot 5\sqrt{6 \cdot 2} = 10\sqrt{12}$. Next, simplify $\sqrt{12}$.
$\qquad 10\sqrt{12} = 10\sqrt{4 \cdot 3} = 10\sqrt{4} \cdot \sqrt{3} = 10 \cdot 2 \cdot \sqrt{3} = 20\sqrt{3}$
d. $\left(3\sqrt{2}\right)^2 = 3^2 \cdot \left(\sqrt{2}\right)^2 = 9 \cdot 2 = 18$

Example 2   Multiply: $\sqrt[3]{4} \cdot \sqrt[3]{18}$. Then simplify if possible.

Solution $\sqrt[3]{4} \cdot \sqrt[3]{18} = \sqrt[3]{4 \cdot 18} = \sqrt[3]{4 \cdot 2 \cdot 9} = \sqrt[3]{8 \cdot 9} = \sqrt[3]{8} \cdot \sqrt[3]{9} = 2\sqrt[3]{9}$

When multiplying radical expressions containing more than one term, use the same techniques we use to multiply other algebraic expressions with more than one term.

Example 3   Find the product and simplify.

a. $\sqrt{5}\left(\sqrt{5} - \sqrt{2}\right)$ **b.** $\left(\sqrt{x} + \sqrt{2}\right)\left(\sqrt{3} - \sqrt{2}\right)$

Solution   **a.** Using the distributive property, we have

$$\sqrt{5}\left(\sqrt{5} - \sqrt{2}\right) = \sqrt{5} \cdot \sqrt{5} - \sqrt{5} \cdot \sqrt{2}$$
$$= 5 - \sqrt{10}$$

b. Use the FOIL method of multiplication.

$$\left(\sqrt{x} + \sqrt{2}\right)\left(\sqrt{3} - \sqrt{2}\right) = \overset{F}{\sqrt{x} \cdot \sqrt{3}} - \overset{O}{\sqrt{x} \cdot \sqrt{2}} + \overset{I}{\sqrt{2} \cdot \sqrt{3}} - \overset{L}{\sqrt{2} \cdot \sqrt{2}}$$
$$= \sqrt{3x} - \sqrt{2x} + \sqrt{6} - \sqrt{4} \qquad \text{Apply the product rule.}$$
$$= \sqrt{3x} - \sqrt{2x} + \sqrt{6} - 2 \qquad \text{Simplify.}$$

Special products can be used to multiply expressions containing radicals.

Example 4   Find the product and simplify.

a. $\left(\sqrt{5} - 7\right)\left(\sqrt{5} + 7\right)$ **b.** $\left(\sqrt{7x} + 2\right)^2$

Solution   **a.** Recall from Chapter 4 that $(a - b)(a + b) = a^2 - b^2$. Then
$$\left(\sqrt{5} - 7\right)\left(\sqrt{5} + 7\right) = \left(\sqrt{5}\right)^2 - 7^2$$
$$= 5 - 49$$
$$= -44$$

b. Recall that $(a + b)^2 = a^2 + 2ab + b^2$. Then

$$(\sqrt{7x} + 2)^2 = (\sqrt{7x})^2 + 2(\sqrt{7x})(2) + (2)^2$$
$$= 7x + 4\sqrt{7x} + 4$$

2   To simplify quotients of radical expressions, we use the quotient rule.

QUOTIENT RULE FOR RADICALS

If $\sqrt[n]{a}$ and $\sqrt[n]{b}$ are real numbers and $b \neq 0$, then

$$\frac{\sqrt[n]{a}}{\sqrt[n]{b}} = \sqrt[n]{\frac{a}{b}}, \text{ providing } b \neq 0$$

Example 5   Divide. Then simplify if possible.

     a. $\dfrac{\sqrt{14}}{\sqrt{2}}$ **b.** $\dfrac{\sqrt{100}}{\sqrt{5}}$ **c.** $\dfrac{\sqrt{12x^3}}{\sqrt{3x}}$

Solution   Use the quotient rule and then simplify the resulting radicand.

     a. $\dfrac{\sqrt{14}}{\sqrt{2}} = \sqrt{\dfrac{14}{2}} = \sqrt{7}$

     b. $\dfrac{\sqrt{100}}{\sqrt{5}} = \sqrt{\dfrac{100}{5}} = \sqrt{20} = \sqrt{4 \cdot 5} = \sqrt{4} \cdot \sqrt{5} = 2\sqrt{5}$

     c. $\dfrac{\sqrt{12x^3}}{\sqrt{3x}} = \sqrt{\dfrac{12x^3}{3x}} = \sqrt{4x^2} = 2x$

Example 6   Divide: $\dfrac{\sqrt[3]{32}}{\sqrt[3]{4}}$. Then simplify if possible.

Solution $\dfrac{\sqrt[3]{32}}{\sqrt[3]{4}} = \sqrt[3]{\dfrac{32}{4}} = \sqrt[3]{8} = 2$

3   It is sometimes easier to work with radical expressions if the denominator does not contain a radical. To eliminate the radical in the denominator of a radical expression, we use the fact that we can multiply the numerator and the denominator of a fraction by the same nonzero number. This is equivalent to multiplying the fraction by 1. To eliminate the radical in the denominator of $\dfrac{\sqrt{5}}{\sqrt{2}}$, multiply the numerator and the denominator by $\sqrt{2}$. Then

$$\frac{\sqrt{5}}{\sqrt{2}} = \frac{\sqrt{5} \cdot \sqrt{2}}{\sqrt{2} \cdot \sqrt{2}} = \frac{\sqrt{10}}{2}$$

This process is called **rationalizing** the denominator.

Example 7 Rationalize each denominator.

a. $\dfrac{2}{\sqrt{7}}$ b. $\dfrac{\sqrt{5}}{\sqrt{12}}$ c. $\sqrt{\dfrac{1}{18x}}$

Solution a. To eliminate the radical in the denominator of $\dfrac{2}{\sqrt{7}}$, multiply the numerator and the denominator by $\sqrt{7}$.

$$\frac{2}{\sqrt{7}} = \frac{2 \cdot \sqrt{7}}{\sqrt{7} \cdot \sqrt{7}} = \frac{2\sqrt{7}}{7}$$

b. We can multiply the numerator and denominator by $\sqrt{12}$, but see what happens if we simplify first.

$$\frac{\sqrt{5}}{\sqrt{12}} = \frac{\sqrt{5}}{\sqrt{4 \cdot 3}} = \frac{\sqrt{5}}{2\sqrt{3}}$$

To rationalize the denominator now, multiply the numerator and the denominator by $\sqrt{3}$.

$$\frac{\sqrt{5}}{2\sqrt{3}} = \frac{\sqrt{5} \cdot \sqrt{3}}{2\sqrt{3} \cdot \sqrt{3}} = \frac{\sqrt{15}}{2 \cdot 3} = \frac{\sqrt{15}}{6}$$

c. $\sqrt{\dfrac{1}{18x}} = \dfrac{\sqrt{1}}{\sqrt{18x}} = \dfrac{1}{\sqrt{9} \cdot \sqrt{2x}} = \dfrac{1}{3\sqrt{2x}}$

To rationalize the denominator, multiply the numerator and denominator by $\sqrt{2x}$.

$$\frac{1}{3\sqrt{2x}} = \frac{1 \cdot \sqrt{2x}}{3\sqrt{2x} \cdot \sqrt{2x}} = \frac{\sqrt{2x}}{3 \cdot 2x} = \frac{\sqrt{2x}}{6x}$$ ▬

As a general rule, simplify a radical expression first and then rationalize the denominator.

Example 8 Rationalize each denominator.

a. $\dfrac{5}{\sqrt[3]{4}}$ b. $\dfrac{\sqrt[3]{7}}{\sqrt[3]{3}}$

Solution a. Since the denominator contains a cube root, we multiply the numerator and the denominator by a factor that gives the **cube root of a perfect cube** in the denominator. Recall that $\sqrt[3]{8} = 2$ and that the denominator $\sqrt[3]{4}$ multiplied by $\sqrt[3]{2}$ is $\sqrt[3]{4 \cdot 2}$ or $\sqrt[3]{8}$.

$$\frac{5}{\sqrt[3]{4}} = \frac{5 \cdot \sqrt[3]{2}}{\sqrt[3]{4} \cdot \sqrt[3]{2}} = \frac{5\sqrt[3]{2}}{\sqrt[3]{8}} = \frac{5\sqrt[3]{2}}{2}$$

b. Recall that $\sqrt[3]{27} = 3$. Multiply the denominator $\sqrt[3]{3}$ by $\sqrt[3]{9}$ and the result is $\sqrt[3]{3 \cdot 9}$ or $\sqrt[3]{27}$.

$$\frac{\sqrt[3]{7}}{\sqrt[3]{3}} = \frac{\sqrt[3]{7} \cdot \sqrt[3]{9}}{\sqrt[3]{3} \cdot \sqrt[3]{9}} = \frac{\sqrt[3]{63}}{\sqrt[3]{27}} = \frac{\sqrt[3]{63}}{3}$$ ▬

4 To rationalize a denominator that is a sum, such as the denominator in

$$\frac{2}{4 + \sqrt{3}}$$

we multiply the numerator and the denominator by $4 - \sqrt{3}$. The expressions $4 + \sqrt{3}$ and $4 - \sqrt{3}$ are called **conjugates** of each other. When a radical expression such as $4 + \sqrt{3}$ is multiplied by its conjugate $4 - \sqrt{3}$, the product simplifies to an expression that contains no radicals.

$$(a + b)(a - b) = a^2 - b^2$$
$$(4 + \sqrt{3})(4 - \sqrt{3}) = 4^2 - (\sqrt{3})^2 = 16 - 3 = 13$$

Then

$$\frac{2}{4 + \sqrt{3}} = \frac{2(4 - \sqrt{3})}{(4 + \sqrt{3})(4 - \sqrt{3})} = \frac{2(4 - \sqrt{3})}{13}$$

Example 9 Rationalize each denominator and simplify.

a. $\dfrac{2}{1 + \sqrt{3}}$

b. $\dfrac{\sqrt{5} + 4}{\sqrt{5} - 1}$

Solution **a.** Multiply the numerator and the denominator of this fraction by the conjugate of $1 + \sqrt{3}$, that is, by $1 - \sqrt{3}$.

$$\frac{2}{1 + \sqrt{3}} = \frac{2(1 - \sqrt{3})}{(1 + \sqrt{3})(1 - \sqrt{3})}$$

$$= \frac{2(1 - \sqrt{3})}{1^2 - (\sqrt{3})^2}$$

$$= \frac{2(1 - \sqrt{3})}{1 - 3}$$

$$= \frac{2(1 - \sqrt{3})}{-2}$$

$$= -\frac{2(1 - \sqrt{3})}{2} \qquad \frac{a}{-b} = -\frac{a}{b}$$

$$= -1(1 - \sqrt{3}) \qquad \text{Simplify.}$$

$$= -1 + \sqrt{3}$$

b. $\dfrac{\sqrt{5} + 4}{\sqrt{5} - 1} = \dfrac{(\sqrt{5} + 4)(\sqrt{5} + 1)}{(\sqrt{5} - 1)(\sqrt{5} + 1)}$ Multiply the numerator and denominator by $\sqrt{5} + 1$, the conjugate of $\sqrt{5} - 1$.

$$= \frac{5 + \sqrt{5} + 4\sqrt{5} + 4}{5 - 1} \qquad \text{Multiply.}$$

$$= \frac{9 + 5\sqrt{5}}{4} \qquad \text{Simplify.}$$

Example 10   Simplify $\dfrac{12 - \sqrt{18}}{9}$.

Solution   First simplify $\sqrt{18}$.

$$\frac{12 - \sqrt{18}}{9} = \frac{12 - \sqrt{9 \cdot 2}}{9} = \frac{12 - 3\sqrt{2}}{9}$$

Next, factor out a common factor of 3 from the terms in the numerator and the denominator and simplify.

$$\frac{12 - 3\sqrt{2}}{9} = \frac{3\left(4 - \sqrt{2}\right)}{3 \cdot 3} = \frac{4 - \sqrt{2}}{3}$$

MENTAL MATH

Find each product. Assume that variables represent nonnegative real numbers.

1. $\sqrt{2} \cdot \sqrt{3}$ **2.** $\sqrt{5} \cdot \sqrt{7}$ **3.** $\sqrt{1} \cdot \sqrt{6}$

4. $\sqrt{7} \cdot \sqrt{x}$ **5.** $\sqrt{10} \cdot \sqrt{y}$ **6.** $\sqrt{x} \cdot \sqrt{y}$

Exercise Set 9.4

Multiply and simplify. See Examples 1, 3, and 4.

1. $\sqrt{8} \cdot \sqrt{2}$ **2.** $\sqrt{3} \cdot \sqrt{12}$

3. $\sqrt{10} \cdot \sqrt{5}$ **4.** $3\sqrt{2} \cdot 5\sqrt{14}$

5. $\sqrt{10}\left(\sqrt{2} + \sqrt{5}\right)$ **6.** $\sqrt{6}\left(\sqrt{3} + \sqrt{2}\right)$

7. $\left(3\sqrt{5} - \sqrt{10}\right)\left(\sqrt{5} - 4\sqrt{3}\right)$

8. $\left(2\sqrt{3} - 6\right)\left(\sqrt{3} - 4\sqrt{2}\right)$

9. $\left(\sqrt{x} + 6\right)\left(\sqrt{x} - 6\right)$ **10.** $\left(2\sqrt{5} + 1\right)\left(2\sqrt{5} - 1\right)$

11. $\left(\sqrt{3} + 8\right)^2$ **12.** $\left(\sqrt{x} - 7\right)^2$

△ **13.** Find the area of a rectangular room whose length is $13\sqrt{2}$ meters and width is $5\sqrt{6}$ meters.

△ **14.** Find the volume of a microwave oven whose length is $\sqrt{3}$ feet, width is $\sqrt{2}$ feet, and height is $\sqrt{2}$ feet.

$\sqrt{2}$ feet

$\sqrt{3}$ feet

$\sqrt{2}$ feet

Divide and simplify. See Example 5.

15. $\dfrac{\sqrt{32}}{\sqrt{2}}$ **16.** $\dfrac{\sqrt{40}}{\sqrt{10}}$

17. $\dfrac{\sqrt{90}}{\sqrt{5}}$ **18.** $\dfrac{\sqrt{96}}{\sqrt{8}}$

19. $\dfrac{\sqrt{75y^5}}{\sqrt{3y}}$ **20.** $\dfrac{\sqrt{24x^7}}{\sqrt{6x}}$

Rationalize each denominator and simplify. See Example 7.

21. $\sqrt{\dfrac{3}{5}}$ **22.** $\sqrt{\dfrac{2}{3}}$

23. $\dfrac{1}{\sqrt{6y}}$ **24.** $\dfrac{1}{\sqrt{10z}}$

25. $\sqrt{\dfrac{5}{18}}$ **26.** $\sqrt{\dfrac{7}{12}}$

△ **27.** If a circle has area A, then the formula for the radius r of the circle is

$$r = \sqrt{\frac{A}{\pi}}$$

Simplify this expression by rationalizing the denominator.

△ **28.** If a round ball has volume V, then the formula for the radius r of the ball is

$$r = \sqrt[3]{\frac{3V}{4\pi}}$$

Simplify this expression by rationalizing the denominator.

✎ **29.** When rationalizing the denominator of $\dfrac{\sqrt{2}}{\sqrt{3}}$, explain why both the numerator and the denominator must be multiplied by $\sqrt{3}$.

✎ **30.** In your own words, explain why $\sqrt{6} + \sqrt{2}$ cannot be simplified further, but $\sqrt{6} \cdot \sqrt{2}$ can be.

Rationalize each denominator and simplify. See Example 9.

31. $\dfrac{3}{\sqrt{2} + 1}$

32. $\dfrac{6}{\sqrt{5} + 2}$

33. $\dfrac{2}{\sqrt{10} - 3}$

34. $\dfrac{4}{2 - \sqrt{3}}$

35. $\dfrac{\sqrt{5} + 1}{\sqrt{6} - \sqrt{5}}$

36. $\dfrac{\sqrt{3} + 1}{\sqrt{3} - \sqrt{2}}$

Simplify the following. See Example 10.

37. $\dfrac{6 + 2\sqrt{3}}{2}$

38. $\dfrac{9 + 6\sqrt{2}}{3}$

39. $\dfrac{18 - 12\sqrt{5}}{6}$

40. $\dfrac{8 - 20\sqrt{3}}{4}$

41. $\dfrac{15\sqrt{3} + 5}{5}$

42. $\dfrac{8 + 16\sqrt{2}}{8}$

Multiply or divide as indicated and simplify.

43. $2\sqrt{3} \cdot 4\sqrt{15}$

44. $3\sqrt{14} \cdot 4\sqrt{2}$

45. $\left(2\sqrt{5}\right)^2$

46. $\left(3\sqrt{10}\right)^2$

💿 **47.** $\left(6\sqrt{x}\right)^2$

48. $\left(8\sqrt{y}\right)^2$

💿 **49.** $\sqrt{6}\left(\sqrt{5} + \sqrt{7}\right)$

50. $\sqrt{10}\left(\sqrt{3} - \sqrt{7}\right)$

51. $4\sqrt{5x}\left(\sqrt{x} - 3\sqrt{5}\right)$

52. $3\sqrt{7y}\left(\sqrt{y} - 2\sqrt{7}\right)$

53. $\left(\sqrt{3} + \sqrt{5}\right)\left(\sqrt{2} - \sqrt{5}\right)$

54. $\left(\sqrt{6} + \sqrt{3}\right)\left(\sqrt{6} - \sqrt{3}\right)$

55. $\left(\sqrt{7} - 2\sqrt{3}\right)\left(\sqrt{7} + 2\sqrt{3}\right)$

56. $\left(\sqrt{2} - 4\sqrt{5}\right)\left(\sqrt{2} + 4\sqrt{5}\right)$

57. $\left(\sqrt{x} - 3\right)\left(\sqrt{x} + 3\right)$

58. $\left(2\sqrt{y} + 5\right)\left(2\sqrt{y} - 5\right)$

59. $\left(\sqrt{6} + 3\right)^2$

60. $\left(2 + \sqrt{7}\right)^2$

61. $\left(3\sqrt{x} - 5\right)^2$

62. $\left(2\sqrt{x} - 7\right)^2$

63. $\dfrac{\sqrt{150}}{\sqrt{2}}$

64. $\dfrac{\sqrt{120}}{\sqrt{3}}$

65. $\dfrac{\sqrt{72y^5}}{\sqrt{3y^3}}$

66. $\dfrac{\sqrt{54x^3}}{\sqrt{2x}}$

67. $\dfrac{\sqrt{24x^7y^4}}{\sqrt{2xy}}$

68. $\dfrac{\sqrt{96x^8y^2}}{\sqrt{3x^2y}}$

Rationalize each denominator and simplify.

69. $\sqrt{\dfrac{2}{15}}$ ___

70. $\sqrt{\dfrac{11}{14}}$ ___

71. $\sqrt{\dfrac{3}{20}}$ ___

72. $\sqrt{\dfrac{3}{50}}$ ___

💿 **73.** $\dfrac{3x}{\sqrt{2x}}$ ___

74. $\dfrac{5y}{\sqrt{3y}}$ ___

💿 **75.** $\dfrac{4}{2 - \sqrt{5}}$

76. $\dfrac{2}{1 - \sqrt{2}}$

77. $\dfrac{5}{3 + \sqrt{10}}$

78. $\dfrac{5}{\sqrt{6} + 2}$ ___

79. $\dfrac{2\sqrt{3}}{\sqrt{15} + 2}$ ___

80. $\dfrac{3\sqrt{2}}{\sqrt{10} + 2}$

81. $\dfrac{\sqrt{3} + 1}{\sqrt{2} - 1}$

82. $\dfrac{\sqrt{2} - 2}{2 - \sqrt{3}}$

Multiply or divide as indicated. See Examples 2 and 6.

83. $\sqrt[3]{12} \cdot \sqrt[3]{4}$

84. $\sqrt[3]{9} \cdot \sqrt[3]{6}$

85. $2\sqrt[3]{5} \cdot 6\sqrt[3]{2}$

86. $8\sqrt[3]{4} \cdot 7\sqrt[3]{7}$

87. $\sqrt[3]{15} \cdot \sqrt[3]{25}$

88. $\sqrt[3]{4} \cdot \sqrt[3]{4}$

89. $\dfrac{\sqrt[3]{54}}{\sqrt[3]{2}}$

90. $\dfrac{\sqrt[3]{80}}{\sqrt[3]{10}}$

91. $\dfrac{\sqrt[3]{120}}{\sqrt[3]{5}}$

92. $\dfrac{\sqrt[3]{270}}{\sqrt[3]{5}}$

Rationalize each denominator. See Example 8.

93. $\sqrt[3]{\dfrac{5}{4}}$

94. $\sqrt[3]{\dfrac{7}{9}}$

95. $\dfrac{6}{\sqrt[3]{2}}$

96. $\dfrac{3}{\sqrt[3]{5}}$

💿 **97.** $\sqrt[3]{\dfrac{1}{9}}$

98. $\sqrt[3]{\dfrac{8}{11}}$

99. $\sqrt[3]{\dfrac{2}{9}}$

100. $\sqrt[3]{\dfrac{3}{4}}$

101. When rationalizing the denominator of $\dfrac{\sqrt[3]{2}}{\sqrt[3]{3}}$, explain why both the numerator and the denominator must be multiplied by $\sqrt[3]{9}$.

It is often more convenient to work with a radical expression whose numerator is rationalized. Rationalize the numerator of each expression by multiplying numerator and denominator by the conjugate of the numerator.

102. $\dfrac{\sqrt{3} + 1}{\sqrt{2} - 1}$

103. $\dfrac{\sqrt{2} - 2}{2 - \sqrt{3}}$

REVIEW EXERCISES

Simplify the following expressions. See Section 6.1.

104. $\dfrac{3x + 12}{3}$

105. $\dfrac{12x + 8}{4}$

106. $\dfrac{6x^2 - 3x}{3x}$

107. $\dfrac{8y^2 - 2y}{2y}$

Solve each equation. See Sections 2.4 and 4.3.

108. $x + 5 = 7^2$

109. $2y - 1 = 3^2$

110. $4z^2 + 6z - 12 = (2z)^2$

111. $9x^2 + 5x + 4 = (3x + 1)^2$

9.5 SOLVING EQUATIONS CONTAINING RADICALS

CD-ROM   SSM

SSG   Video

▶ **OBJECTIVES**

1. Solve radical equations by using the squaring property of equality once.
2. Solve radical equations by using the squaring property of equality twice.

1   In this section, we solve **radical equations** such as

$$\sqrt{x + 3} = 5 \quad \text{and} \quad \sqrt{2x + 1} = \sqrt{3x}$$

Radical equations contain variables in the radicand. To solve these equations, we rely on the following squaring property.

THE SQUARING PROPERTY OF EQUALITY

If $a = b$, then $a^2 = b^2$

Unfortunately, this squaring property does not guarantee that all solutions of the new equation are solutions of the original equation. For example, if we square both sides of the equation

$$x = 2$$

we have

$$x^2 = 4$$

This new equation has two solutions, 2 and −2, while the original equation $x = 2$ has only one solution. For this reason, we must **always check proposed solutions of radical equations in the original equation**.

Example 1   Solve: $\sqrt{x + 3} = 5$

Solution   To solve this radical equation, we use the squaring property of equality and square both sides of the equation.

$$\sqrt{x + 3} = 5$$
$$(\sqrt{x + 3})^2 = 5^2 \qquad \text{Square both sides.}$$
$$x + 3 = 25 \qquad \text{Simplify.}$$
$$x = 22 \qquad \text{Subtract 3 from both sides.}$$

Check   We replace x with 22 in the original equation.

> **HELPFUL HINT**
> Don't forget to check the proposed solutions of radical equations in the original equation.

$$\sqrt{x + 3} = 5 \qquad \text{Original equation}$$
$$\sqrt{22 + 3} \stackrel{?}{=} 5 \qquad \text{Let } x = 22.$$
$$\sqrt{25} \stackrel{?}{=} 5$$
$$5 = 5 \qquad \text{True}$$

Since a true statement results, 22 is the solution.   ▬

Example 2   Solve: $\sqrt{x} + 6 = 4$

Solution   First we set the radical by itself on one side of the equation. Then we square both sides.

$$\sqrt{x} + 6 = 4$$
$$\sqrt{x} = -2 \qquad \text{Subtract 6 from both sides to get the radical by itself.}$$

Recall that $\sqrt{x}$ is the principal or nonnegative square root of x so that $\sqrt{x}$ cannot equal -2 and thus this equation has no solution. We arrive at the same conclusion if we continue by applying the squaring property.

$$\sqrt{x} = -2$$
$$(\sqrt{x})^2 = (-2)^2 \qquad \text{Square both sides.}$$
$$x = 4 \qquad \text{Simplify.}$$

Check   We replace x with 4 in the original equation.

$$\sqrt{x} + 6 = 4 \qquad \text{Original equation}$$
$$\sqrt{4} + 6 \stackrel{?}{=} 4 \qquad \text{Let } x = 4.$$
$$2 + 6 = 4 \qquad \text{False}$$

Since 4 *does not* satisfy the original equation, this equation has no solution.   ▬

Example 2 makes it very clear that we *must* check proposed solutions in the original equation to determine if they are truly solutions. If a proposed solution does not work, we say that the value is an **extraneous solution**.

The following steps can be used to solve radical equations containing square roots.

> **SOLVING A RADICAL EQUATION CONTAINING SQUARE ROOTS**
>
> **Step 1.** Arrange terms so that one radical is by itself on one side of the equation. That is, isolate a radical.
> **Step 2.** Square both sides of the equation.
> **Step 3.** Simplify both sides of the equation.
> **Step 4.** If the equation still contains a radical term, repeat steps 1 through 3.
> **Step 5.** Solve the equation.
> **Step 6.** Check all solutions in the original equation for extraneous solutions.

Example 3   Solve: $\sqrt{x} = \sqrt{5x - 2}$

Solution   Each of the radicals is already isolated, since each is by itself on one side of the equation. So we begin solving by squaring both sides.

$$\sqrt{x} = \sqrt{5x - 2} \qquad \text{Original equation}$$
$$(\sqrt{x})^2 = (\sqrt{5x - 2})^2 \qquad \text{Square both sides.}$$
$$x = 5x - 2 \qquad \text{Simplify.}$$
$$-4x = -2 \qquad \text{Subtract } 5x \text{ from both sides.}$$
$$x = \frac{-2}{-4} = \frac{1}{2} \qquad \text{Divide both sides by } -4 \text{ and simplify.}$$

Check   We replace x with $\frac{1}{2}$ in the original equation.

$$\sqrt{x} = \sqrt{5x - 2} \qquad \text{Original equation}$$
$$\sqrt{\frac{1}{2}} \stackrel{?}{=} \sqrt{5 \cdot \frac{1}{2} - 2} \qquad \text{Let } x = \frac{1}{2}.$$
$$\sqrt{\frac{1}{2}} \stackrel{?}{=} \sqrt{\frac{5}{2} - 2} \qquad \text{Multiply.}$$
$$\sqrt{\frac{1}{2}} \stackrel{?}{=} \sqrt{\frac{5}{2} - \frac{4}{2}} \qquad \text{Write 2 as } \frac{4}{2}.$$
$$\sqrt{\frac{1}{2}} = \sqrt{\frac{1}{2}} \qquad \text{True}$$

This statement is true, so the solution is $\frac{1}{2}$.

Example 4   Solve: $\sqrt{4y^2 + 5y - 15} = 2y$

Solution   The radical is already isolated, so we start by squaring both sides.

$$\sqrt{4y^2 + 5y - 15} = 2y$$
$$(\sqrt{4y^2 + 5y - 15})^2 = (2y)^2 \qquad \text{Square both sides.}$$
$$4y^2 + 5y - 15 = 4y^2 \qquad \text{Simplify.}$$
$$5y - 15 = 0 \qquad \text{Subtract } 4y^2 \text{ from both sides.}$$
$$5y = 15 \qquad \text{Add 15 to both sides.}$$
$$y = 3 \qquad \text{Divide both sides by 5.}$$

Check   We replace y with 3 in the original equation.

$$\sqrt{4y^2 + 5y - 15} = 2y \qquad \text{Original equation}$$
$$\sqrt{4 \cdot 3^2 + 5 \cdot 3 - 15} \stackrel{?}{=} 2 \cdot 3 \qquad \text{Let } y = 3.$$
$$\sqrt{4 \cdot 9 + 15 - 15} \stackrel{?}{=} 6 \qquad \text{Simplify.}$$
$$\sqrt{36} \stackrel{?}{=} 6$$
$$6 = 6 \qquad \text{True}$$

This statement is true, so the solution is 3.   ▬

Example 5   Solve: $\sqrt{x + 3} - x = -3$

Solution   First we isolate the radical by adding x to both sides. Then we square both sides.

$$\sqrt{x + 3} - x = -3$$
$$\sqrt{x + 3} = x - 3 \qquad \text{Add } x \text{ to both sides.}$$
$$(\sqrt{x + 3})^2 = (x - 3)^2 \qquad \text{Square both sides.}$$
$$x + 3 = \underbrace{x^2 - 6x + 9}$$

> **HELPFUL HINT**
> Don't forget that $(x - 3)^2 = (x - 3)(x - 3) = x^2 - 6x + 9$

To solve the resulting quadratic equation, we write the equation in standard form by subtracting x and 3 from both sides.

$$3 = x^2 - 7x + 9 \qquad \text{Subtract } x \text{ from both sides.}$$
$$0 = x^2 - 7x + 6 \qquad \text{Subtract 3 from both sides.}$$
$$0 = (x - 6)(x - 1) \qquad \text{Factor.}$$
$$0 = x - 6 \quad \text{or} \quad 0 = x - 1 \qquad \text{Set each factor equal to zero.}$$
$$6 = x \qquad\qquad 1 = x \qquad \text{Solve for } x.$$

Check   We replace x with 6 and then x with 1 in the original equation.

Let $x = 6$.   Let $x = 1$.

$$\sqrt{x + 3} - x = -3 \qquad\qquad \sqrt{x + 3} - x = -3$$
$$\sqrt{6 + 3} - 6 \stackrel{?}{=} -3 \qquad\qquad \sqrt{1 + 3} - 1 \stackrel{?}{=} -3$$
$$\sqrt{9} - 6 \stackrel{?}{=} -3 \qquad\qquad \sqrt{4} - 1 \stackrel{?}{=} -3$$
$$3 - 6 \stackrel{?}{=} -3 \qquad\qquad 2 - 1 \stackrel{?}{=} -3$$
$$-3 = -3 \quad \text{True} \qquad\qquad 1 = -3 \quad \text{False}$$

Since replacing x with 1 resulted in a false statement, 1 is an extraneous solution. The only solution is 6.   ▬

2   If a radical equation contains two radicals, we may need to use the squaring property twice.

Example 6   Solve: $\sqrt{x-4} = \sqrt{x} - 2$

Solution

$$\sqrt{x-4} = \sqrt{x} - 2$$

$$(\sqrt{x-4})^2 = (\sqrt{x}-2)^2 \qquad \text{Square both sides.}$$

$$x - 4 = \underbrace{x - 4\sqrt{x} + 4}$$

$$-8 = -4\sqrt{x}$$

$$2 = \sqrt{x} \qquad \text{Divide both sides by } -4.$$

$$4 = x \qquad \text{Square both sides again.}$$

> **HELPFUL HINT**
>
> $$(\sqrt{x}-2)^2 = (\sqrt{x}-2)(\sqrt{x}-2)$$
> $$= \sqrt{x} \cdot \sqrt{x} - 2\sqrt{x} - 2\sqrt{x} + 4$$
> $$= x - 4\sqrt{x} + 4$$

Check the proposed solution in the original equation. The solution is 4.

Exercise Set 9.5

Solve each equation. See Examples 1 through 5.

1. $\sqrt{x} = 9$

2. $\sqrt{x} = 4$

3. $\sqrt{x+5} = 2$

4. $\sqrt{x+12} = 3$

5. $\sqrt{2x+6} = 4$

6. $\sqrt{3x+7} = 5$

7. $\sqrt{x} - 2 = 5$

8. $4\sqrt{x} - 7 = 5$

9. $3\sqrt{x} + 5 = 2$

10. $3\sqrt{x} + 5 = 8$

11. $\sqrt{x+6} + 1 = 3$

12. $\sqrt{x+5} + 2 = 5$

13. $\sqrt{2x+1} + 3 = 5$

14. $\sqrt{3x-1} + 4 = 1$

15. $\sqrt{x} + 3 = 7$

16. $\sqrt{x} + 5 = 10$

17. $\sqrt{x+6} + 5 = 3$

18. $\sqrt{2x-1} + 7 = 1$

19. $\sqrt{4x-3} = \sqrt{x+3}$

20. $\sqrt{5x-4} = \sqrt{x+8}$

21. $\sqrt{x} = \sqrt{3x-8}$

22. $\sqrt{x} = \sqrt{4x-3}$

23. $\sqrt{4x} = \sqrt{2x+6}$

24. $\sqrt{5x+6} = \sqrt{8x}$

25. $\sqrt{9x^2 + 2x - 4} = 3x$

26. $\sqrt{4x^2 + 3x - 9} = 2x$

27. $\sqrt{16x^2 - 3x + 6} = 4x$

28. $\sqrt{9x^2 - 2x + 8} = 3x$

29. $\sqrt{16x^2 + 2x + 2} = 4x$

30. $\sqrt{4x^2 + 3x - 2} = 2x$

31. $\sqrt{2x^2 + 6x + 9} = 3$

32. $\sqrt{3x^2 + 6x + 4} = 2$

33. $\sqrt{x+7} = x + 5$

34. $\sqrt{x+5} = x - 1$

35. $\sqrt{x} = x - 6$

36. $\sqrt{x} = x + 6$

37. $\sqrt{2x+1} = x - 7$

38. $\sqrt{2x+5} = x - 5$

39. $x = \sqrt{2x-2} + 1$

40. $\sqrt{1-8x} + 2 = x$

41. $\sqrt{1-8x} - x = 4$

42. $\sqrt{3x+7} - x = 3$

43. $\sqrt{2x+5} - 1 = x$

44. $x = \sqrt{4x-7} + 1$

Solve each equation. See Example 6.

45. $\sqrt{x-7} = \sqrt{x} - 1$

46. $\sqrt{x} + 2 = \sqrt{x+24}$

47. $\sqrt{x} + 3 = \sqrt{x+15}$

48. $\sqrt{x-8} = \sqrt{x} - 2$

49. $\sqrt{x+8} = \sqrt{x} + 2$

50. $\sqrt{x} + 1 = \sqrt{x+15}$

Solve.

51. A number is 6 more than its principal square root. Find the number.

52. A number is 4 more than the principal square root of twice the number. Find the number.

53. The formula $b = \sqrt{\dfrac{V}{2}}$ can be used to determine the length b of a side of the base of a square-based pyramid with height 6 units and volume V cubic units.

a. Find the length of the side of the base that produces a pyramid with each volume. (Round to the nearest tenth of a unit.)

V	20	200	2000
b			

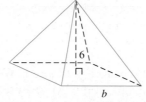

b. Notice in the table that volume V has been increased by a factor of 10 each time. Does the corresponding length b of a side increase by a factor of 10 each time also?

54. The formula $r = \sqrt{\dfrac{V}{2\pi}}$ can be used to determine the radius r of a cylinder with height 2 units and volume V cubic units.

a. Find the radius needed to manufacture a cylinder with each volume. (Round to the nearest tenth of a unit.)

V	10	100	1000
r			

b. Notice in the table that volume V has been increased by a factor of 10 each time. Does the corresponding radius increase by a factor of 10 each time also?

55. Explain why proposed solutions of radical equations must be checked in the original equation.

Graphing calculators can be used to solve equations. To solve $\sqrt{x-2} = x - 5$, for example, graph $y_1 = \sqrt{x-2}$ and $y_2 = x - 5$ on the same set of axes. Use the TRACE and ZOOM features or an INTERSECT feature to find the point of intersection of the graphs. The x-value of the point is the solution of the equation. Use a graphing calculator to solve the equations below. Approximate solutions to the nearest hundredth.

56. $\sqrt{x-2} = x - 5$

57. $\sqrt{x+1} = 2x - 3$

58. $-\sqrt{x+4} = 5x - 6$

59. $-\sqrt{x+5} = -7x + 1$

REVIEW EXERCISES

Translate each sentence into an equation and then solve. See Section 2.5.

60. If 8 is subtracted from the product of 3 and x, the result is 19. Find x.

61. If 3 more than x is subtracted from twice x, the result is 11. Find x.

62. The length of a rectangle is twice the width. The perimeter is 24 inches. Find the length.

63. The length of a rectangle is 2 inches longer than the width. The perimeter is 24 inches. Find the length.

9.6 RADICAL EQUATIONS AND PROBLEM SOLVING

CD-ROM   SSM

SSG   Video

▶ **OBJECTIVES**

1. Use the Pythagorean formula to solve problems.

2. Use the distance formula.

3. Solve problems using formulas containing radicals.

1 Applications of radicals can be found in geometry, finance, science, and other areas of technology. Our first application involves the Pythagorean theorem, giving a formula that relates the lengths of the three sides of a right triangle. We first studied the Pythagorean theorem in Chapter 4 and we review it here.

THE PYTHAGOREAN THEOREM

If a and b are lengths of the legs of a right triangle and c is the length of the hypotenuse, then $a^2 + b^2 = c^2$.

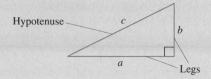

△ **Example 1**   Find the length of the hypotenuse of a right triangle whose legs are 6 inches and 8 inches long.

Solution   Because this is a right triangle, we use the Pythagorean theorem. We let $a = 6$ inches and $b = 8$ inches. Length c must be the length of the hypotenuse.

$$a^2 + b^2 = c^2 \qquad \text{Use the Pythagorean theorem.}$$
$$6^2 + 8^2 = c^2 \qquad \text{Substitute the lengths of the legs.}$$
$$36 + 64 = c^2 \qquad \text{Simplify.}$$
$$100 = c^2$$

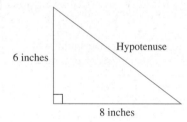

Since c represents a length, we know that c is positive and is the principal square root of 100.

$$100 = c^2$$
$$\sqrt{100} = c \qquad \text{Use the definition of principal square root.}$$
$$10 = c \qquad \text{Simplify.}$$

The hypotenuse has a length of 10 inches.

△ **Example 2**   Find the length of the leg of the right triangle shown. Give the exact length and a two-decimal-place approximation.

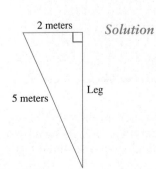

Solution   We let $a = 2$ meters and b be the unknown length of the other leg. The hypotenuse is $c = 5$ meters.

$$a^2 + b^2 = c^2 \qquad \text{Use the Pythagorean theorem.}$$
$$2^2 + b^2 = 5^2 \qquad \text{Let } a = 2 \text{ and } c = 5.$$
$$4 + b^2 = 25$$
$$b^2 = 21$$
$$b = \sqrt{21} \approx 4.58 \text{ meters}$$

The length of the leg is exactly $\sqrt{21}$ meters or approximately 4.58 meters.

△ **Example 3**   **FINDING A DISTANCE**

A surveyor must determine the distance across a lake at points P and Q as shown in the figure. To do this, she finds a third point R perpendicular to line PQ. If the length of $\overline{PR}$ is 320 feet and the length of $\overline{QR}$ is 240 feet, what is the distance across the lake? Approximate this distance to the nearest whole foot.

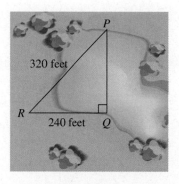

Solution

1. **UNDERSTAND.**   Read and reread the problem. We will set up the problem using the Pythagorean theorem. By creating a line perpendicular to line PQ, the surveyor deliberately constructed a right triangle. The hypotenuse, $\overline{PR}$, has a length of 320 feet, so we let $c = 320$ in the Pythagorean theorem. The side $\overline{QR}$ is one of the legs, so we let $a = 240$ and $b = $ the unknown length.

2. **TRANSLATE.**

$$a^2 + b^2 = c^2 \qquad \text{Use the Pythagorean theorem.}$$
$$240^2 + b^2 = 320^2 \qquad \text{Let } a = 240 \text{ and } c = 320.$$

3. **SOLVE.**

$$57{,}600 + b^2 = 102{,}400$$
$$b^2 = 44{,}800 \qquad \text{Subtract 57,600 from both sides.}$$
$$b = \sqrt{44{,}800} \qquad \text{Use the definition of principal square root.}$$

4. **INTERPRET.**

   Check: See that $240^2 + \left(\sqrt{44{,}800}\right)^2 = 320^2$.

   State: The distance across the lake is **exactly** $\sqrt{44{,}800}$ feet. The surveyor can now use a calculator to find that $\sqrt{44{,}800}$ feet is **approximately** 211.6601 feet, so the distance across the lake is roughly 212 feet.   ▬

2

A second important application of radicals is in finding the distance between two points in the plane. By using the Pythagorean theorem, the following formula can be derived.

DISTANCE FORMULA

The distance d between two points with coordinates (x_1, y_1) and (x_2, y_2) is given by

$$d = \sqrt{(x_2 - x_1)^2 + (y_2 - y_1)^2}$$

Example 4   Find the distance between $(-1, 9)$ and $(-3, -5)$.

Solution   Use the distance formula with $(x_1, y_1) = (-1, 9)$ and $(x_2, y_2) = (-3, -5)$.

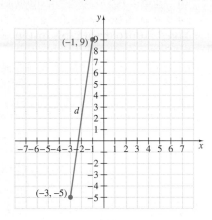

$$d = \sqrt{(x_2 - x_1)^2 + (y_2 - y_1)^2}$$ The distance formula.

$$= \sqrt{[-3 - (-1)]^2 + (-5 - 9)^2}$$ Substitute known values.

$$= \sqrt{(-2)^2 + (-14)^2}$$ Simplify.

$$= \sqrt{4 + 196}$$

$$= \sqrt{200} = 10\sqrt{2}$$ Simplify the radical.

The distance is **exactly** $10\sqrt{2}$ units or **approximately** 14.1 units.   ▬

3   The Pythagorean theorem is an extremely important result in mathematics and should be memorized. But there are other applications involving formulas containing radicals that are not quite as well known, such as the velocity formula used in the next example.

Example 5   **DETERMINING VELOCITY**

A formula used to determine the velocity v, in feet per second, of an object (neglecting air resistance) after it has fallen a certain height is $v = \sqrt{2gh}$, where g is the acceleration due to gravity, and h is the height the object has fallen. On Earth, the acceleration g due to gravity is approximately 32 feet per second per second. Find the velocity of a watermelon after it has fallen 5 feet.

Solution   We are told that $g = 32$ feet per second per second. To find the velocity v when $h = 5$ feet, we use the velocity formula.

$$v = \sqrt{2gh}$$ Use the velocity formula.

$$= \sqrt{2 \cdot 32 \cdot 5}$$ Substitute known values.

$$= \sqrt{320}$$

$$= 8\sqrt{5}$$ Simplify the radicand.

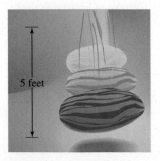

5 feet

The velocity of the watermelon after it falls 5 feet is **exactly** $8\sqrt{5}$ feet per second, or **approximately** 17.9 feet per second.   ▬

SPOTLIGHT ON DECISION MAKING

Suppose you are a carpenter. You are installing a 10-foot by 14-foot wooden deck attached to a client's house. Before sinking the deck posts into the ground, you lay out the dimensions and deck placement with stakes and string. It is very important that the string layout is "square"—that is, that the edges of the deck layout meet at right angles. If not, then the deck posts could be sunk in the wrong positions and portions of the deck will not line up properly.

From the Pythagorean theorem, you know that in a right triangle, $a^2 + b^2 = c^2$. It's also true that in a triangle, if $a^2 + b^2 = c^2$, you know that the triangle is a right triangle. This can be used to check that the two edges meet at right angles.

What should the diagonal of the string layout measure if everything is square? You measure one diagonal of the string layout as 17 feet, $2\frac{15}{32}$ inches. Should any adjustments be made to the string layout? Explain.

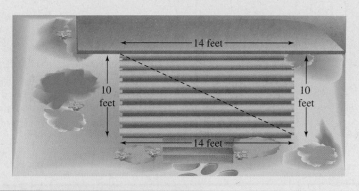

Exercise Set 9.6

Use the Pythagorean theorem to find the unknown side of each right triangle. See Examples 1 and 2.

1.
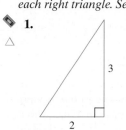

△ **2.**
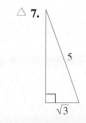

△ **3.**
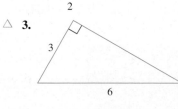

△ **4.**
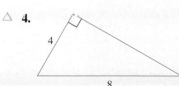

△ **5.**
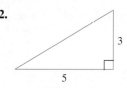

△ **6.**

△ **7.** △ **8.**
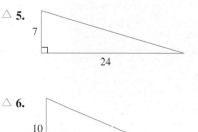
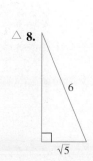

9.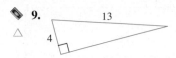

10.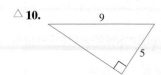

Find the length of the unknown side of each right triangle with sides a, b, and c, where c is the hypotenuse. See Examples 1 and 2.

11. $a = 4, b = 5$ **12.** $a = 2, b = 7$

13. $b = 2, c = 6$ **14.** $b = 1, c = 5$

15. $a = \sqrt{10}, c = 10$

16. $a = \sqrt{7}, c = \sqrt{35}$

Solve. See Examples 1 through 3.

17. Evan Saacks wants to determine the distance at certain points across a pond on his property. He is able to measure the distances shown on the following diagram. Find how wide the pond is to the nearest tenth of a foot.

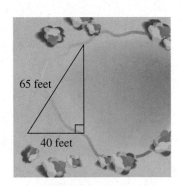

18. Use the formula from Example 5 and find the velocity of an object after it has fallen 20 feet.

19. A wire is used to anchor a 20-foot-high pole. One end of the wire is attached to the top of the pole. The other end is fastened to a stake five feet away from the bottom of the pole. Find the length of the wire, to the nearest tenth of a foot.

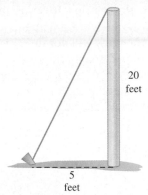

20. Jim Spivey needs to connect two underground pipelines, which are offset by 3 feet, as pictured in the diagram. Neglecting the joints needed to join the pipes, find the length of the shortest possible connecting pipe rounded to the nearest hundredth of a foot.

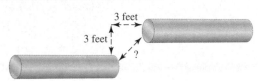

21. Robert Weisman needs to attach a diagonal brace to a rectangular frame in order to make it structurally sound. If the framework is 6 feet by 10 feet, find how long the brace needs to be to the nearest tenth of a foot.

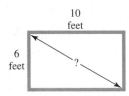

22. Elizabeth Kaster is flying a kite. She let out 80 feet of string and attached the string to a stake in the ground. The kite is now directly above her brother Mike, who is 32 feet away from Elizabeth. Find the height of the kite to the nearest foot.

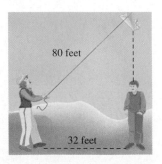

Use the distance formula to find the distance between the points given. See Example 4.

23. $(3, 6), (5, 11)$ **24.** $(2, 3), (9, 7)$

25. $(-3, 1), (5, -2)$ **26.** $(-2, 6), (3, -2)$

27. $(3, -2), (1, -8)$ **28.** $(-5, 8), (-2, 2)$

29. $\left(\frac{1}{2}, 2\right), (2, -1)$ **30.** $\left(\frac{1}{3}, 1\right), (1, -1)$

31. $(3, -2), (5, 7)$ **32.** $(-2, -3), (-1, 4)$

Solve each problem. See Example 5.

△ **33.** For a square-based pyramid, the formula $b = \sqrt{\dfrac{3V}{h}}$ describes the relationship between the length b of one side of the base, the volume V, and the height h. Find the volume if each side of the base is 6 feet long, and the pyramid is 2 feet high.

34. The formula $t = \dfrac{\sqrt{d}}{4}$ relates the distance d, in feet, that an object falls in t seconds, assuming that air resistance does not slow down the object. Find how long, to the nearest hundredth of a second, it takes an object to reach the ground from the top of the Sears Tower in Chicago, a distance of 1450 feet. (*Source: World Almanac and Book of Facts, 2000*)

35. Police use the formula $s = \sqrt{30fd}$ to estimate the speed s of a car in miles per hour. In this formula, d represents the distance the car skidded in feet and f represents the coefficient of friction. The value of f depends on the type of road surface, and for wet concrete f is 0.35. Find how fast a car was moving if it skidded 280 feet on wet concrete, to the nearest mile per hour.

36. The coefficient of friction of a certain dry road is 0.95. Use the formula in Exercise 35 to find how far a car will skid on this dry road if it is traveling at a rate of 60 mph. Round the length to the nearest foot.

37. The formula $v = \sqrt{2.5r}$ can be used to estimate the maximum safe velocity, v, in miles per hour, at which a car can travel if it is driven along a curved road with a **radius of curvature**, r, in feet. To the nearest whole number, find the maximum safe speed if a cloverleaf exit on an expressway has a radius of curvature of 300 feet.

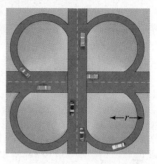

38. Use the formula from Exercise 37 to find the radius of curvature if the safe velocity is 30 mph.

△ **39.** The maximum distance d in kilometers that you can see from a height of h meters is given by $d = 3.5\sqrt{h}$. Find how far you can see from the top of the Texas Commerce Tower in Houston, a height of 305.4 meters. Round to the nearest tenth of a kilometer. (*Source: World Almanac and Book of Facts, 2000*)

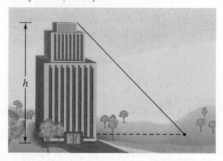

△ **40.** Use the formula from Exercise 39 to determine how high above the ground you need to be to see 40 kilometers. Round to the nearest tenth of a meter.

For each triangle, find the length of x.

△ **41.**

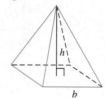

△ **42.**

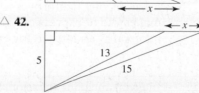

Solve.

△ **43.** Mike and Sandra Hallahan leave the seashore at the same time. Mike drives northward at a rate of 30 miles per hour, while Sandra drives west at 60 mph. Find how far apart they are after 3 hours to the nearest mile.

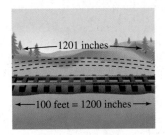

distance apart

30 mph
for 3 hours

60 mph
for 3 hours

△ **44.** Railroad tracks are invariably made up of relatively short sections of rail connected by expansion joints. To see why this construction is necessary, consider a single rail 100 feet long (or 1200 inches). On an extremely hot day, suppose it expands 1 inch in the hot sun to a new length of 1201 inches. Theoretically, the track would bow upward as pictured.

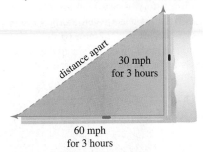

1201 inches

←—100 feet = 1200 inches —→

Let us approximate the bulge in the railroad this way.

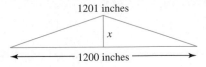

1201 inches

x

←———— 1200 inches ————→

Calculate the height of x of the bulge to the nearest tenth of an inch.

✎ **45.** Based on the results of Exercise 44, explain why rail-
△   roads use short sections of rail connected by expansion joints.

REVIEW EXERCISES

Simplify using rules for exponents. See Sections 4.1 and 4.2.

46. 2^5

47. $(-3)^3$

48. $\left(-\dfrac{1}{5}\right)^2$

49. $\left(\dfrac{2}{7}\right)^3$

50. $x^2 \cdot x^3$

51. $x^4 \cdot x^2$

52. $y^3 \cdot y$

53. $x \cdot x^7$

9.7   RATIONAL EXPONENTS

CD-ROM   SSM

SSG   Video

▶ **OBJECTIVES**

1. Evaluate exponential expressions of the form $a^{1/n}$.
2. Evaluate exponential expressions of the form $a^{m/n}$.
3. Evaluate exponential expressions of the form $a^{-m/n}$.
4. Use rules for exponents to simplify expressions containing fractional exponents.

1 Radical notation is widely used, as we've seen. In this section, we study an alternate notation, one that proves to be more efficient and compact. This alternate notation makes use of expressions containing an exponent that is a rational number but not necessarily an integer, for example,

$$3^{1/2}, 2^{-3/4}, \quad \text{and} \quad y^{5/6}$$

In giving meaning to rational exponents, keep in mind that we want the rules for operating with them to be the same as the rules for operating with integer exponents. For this to be true,

$$(3^{1/2})^2 = 3^{1/2 \cdot 2} = 3^1 = 3$$

Also, we know that

$$(\sqrt{3})^2 = 3$$

Since the square of both $3^{1/2}$ and $\sqrt{3}$ is 3, it would be reasonable to say that

$$3^{1/2} \text{ means } \sqrt{3}$$

In general, we have the following.

DEFINITION OF $a^{1/n}$

If n is a positive integer and $\sqrt[n]{a}$ is a real number, then

$$a^{1/n} = \sqrt[n]{a}$$

Notice that the denominator of the rational exponent is the same as the index of the corresponding radical.

◆ **Example 1**   Write in radical notation. Then simplify.

a. $25^{1/2}$ b. $8^{1/3}$ c. $-16^{1/4}$ d. $(-27)^{1/3}$ e. $\left(\dfrac{1}{9}\right)^{1/2}$

Solution   a. $25^{1/2} = \sqrt{25} = 5$
b. $8^{1/3} = \sqrt[3]{8} = 2$
c. In $-16^{1/4}$, the base of the exponent is 16. Thus **the negative sign is not affected by the exponent**; so $-16^{1/4} = -\sqrt[4]{16} = -2$.
d. The parentheses show that -27 is the base. $(-27)^{1/3} = \sqrt[3]{-27} = -3$.
e. $\left(\dfrac{1}{9}\right)^{1/2} = \sqrt{\dfrac{1}{9}} = \dfrac{1}{3}$

2   In Example 1, each rational exponent has a numerator of 1. What happens if the numerator is some other positive integer? Consider $8^{2/3}$. Since $\frac{2}{3}$ is the same $\frac{1}{3} \cdot 2$, we reason that

$$8^{2/3} = 8^{(1/3)2} = \left(8^{1/3}\right)^2 = \left(\sqrt[3]{8}\right)^2 = 2^2 = 4$$

The denominator 3 of the rational exponent is the same as the index of the radical. The numerator 2 of the fractional exponent indicates that the radical base is to be squared.

DEFINITION OF $a^{m/n}$

If m and n are integers with $n > 0$ and if a is a positive number, then

$$a^{m/n} = \left(a^{1/n}\right)^m = \left(\sqrt[n]{a}\right)^m$$

Also,

$$a^{m/n} = \left(a^m\right)^{1/n} = \sqrt[n]{a^m}$$

Example 2 Simplify each expression.

     a. $4^{3/2}$ **b.** $27^{2/3}$ **c.** $-16^{3/4}$

Solution   **a.** $4^{3/2} = \left(4^{1/2}\right)^3 = \left(\sqrt{4}\right)^3 = 2^3 = 8$

          b. $27^{2/3} = \left(27^{1/3}\right)^2 = \left(\sqrt[3]{27}\right)^2 = 3^2 = 9$

          c. The negative sign is **not** affected by the exponent since the base of the exponent is 16. $-16^{3/4} = -\left(16^{1/4}\right)^3 = -\left(\sqrt[4]{16}\right)^3 = -2^3 = -8$.

> **HELPFUL HINT**
> Recall that
>
> $$-3^2 = -(3 \cdot 3) = -9$$
>
> and
>
> $$(-3)^2 = (-3)(-3) = 9$$
>
> In other words, without parentheses the exponent 2 applies to the base 3, **not** -3. The same is true of rational exponents. For example,
>
> $$-16^{1/2} = -\sqrt{16} = -4$$
>
> and
>
> $$(-27)^{1/3} = \sqrt[3]{-27} = -3$$

---

3   If the exponent is a negative rational number, use the following definition.

> **DEFINITION OF $a^{-m/n}$**
>
> If $a^{-m/n}$ is a nonzero real number, then
>
> $$a^{-m/n} = \frac{1}{a^{m/n}}$$

Example 3 Write each expression with a positive exponent and then simplify.

     a. $36^{-1/2}$ **b.** $16^{-3/4}$ **c.** $-9^{1/2}$ **d.** $32^{-4/5}$

Solution   **a.** $36^{-1/2} = \dfrac{1}{36^{1/2}} = \dfrac{1}{\sqrt{36}} = \dfrac{1}{6}$

          b. $16^{-3/4} = \dfrac{1}{16^{3/4}} = \dfrac{1}{\left(\sqrt[4]{16}\right)^3} = \dfrac{1}{2^3} = \dfrac{1}{8}$

          c. $-9^{1/2} = -\sqrt{9} = -3$

          d. $32^{-4/5} = \dfrac{1}{32^{4/5}} = \dfrac{1}{\left(\sqrt[5]{32}\right)^4} = \dfrac{1}{2^4} = \dfrac{1}{16}$

4   It can be shown that the properties of integer exponents hold for rational exponents. By using these properties and definitions, we can now simplify products and quotients of expressions containing rational exponents.

Example 4   Simplify each expression. Write results with positive exponents only. Assume that all variables represent positive numbers.

a. $3^{1/2} \cdot 3^{3/2}$ **b.** $\dfrac{5^{1/3}}{5^{2/3}}$ **c.** $\left(x^{1/4}\right)^{12}$ **d.** $\dfrac{x^{1/5}}{x^{-4/5}}$ **e.** $\left(\dfrac{y^{3/5}}{z^{1/4}}\right)^2$

Solution   **a.** $3^{1/2} \cdot 3^{3/2} = 3^{(1/2)+(3/2)} = 3^{4/2} = 3^2 = 9$

b. $\dfrac{5^{1/3}}{5^{2/3}} = 5^{(1/3)-(2/3)} = 5^{-1/3} = \dfrac{1}{5^{1/3}}$

c. $\left(x^{1/4}\right)^{12} = x^{(1/4)12} = x^3$

d. $\dfrac{x^{1/5}}{x^{-4/5}} = x^{(1/5)-(-4/5)} = x^{5/5} = x^1$ or x

e. $\left(\dfrac{y^{3/5}}{z^{1/4}}\right)^2 = \dfrac{y^{(3/5)2}}{z^{(1/4)2}} = \dfrac{y^{6/5}}{z^{1/2}}$

Exercise Set 9.7

Simplify each expression. See Examples 1 and 2.

1. $8^{1/3}$ **2.** $16^{1/4}$

3. $9^{1/2}$ **4.** $16^{1/2}$

5. $16^{3/4}$ **6.** $27^{2/3}$

7. $32^{2/5}$ **8.** $64^{5/6}$

Simplify each expression. See Example 3.

9. $-16^{-1/4}$ **10.** $-8^{-1/3}$

11. $16^{-3/2}$ **12.** $27^{-4/3}$

13. $81^{-3/2}$ **14.** $32^{-2/5}$

15. $\left(\dfrac{4}{25}\right)^{-1/2}$ **16.** $\left(\dfrac{8}{27}\right)^{-1/3}$

17. Explain the meaning of the numbers 2, 3, and 4 in the exponential expression $4^{3/2}$.

18. Explain why $-4^{1/2}$ is a real number but $(-4)^{1/2}$ is not.

Simplify each expression. Write each answer with positive exponents. Assume that all variables represent positive numbers. See Example 4.

19. $2^{1/3} \cdot 2^{2/3}$ **20.** $4^{2/5} \cdot 4^{3/5}$

21. $\dfrac{4^{3/4}}{4^{1/4}}$ **22.** $\dfrac{9^{7/2}}{9^{3/2}}$

23. $\dfrac{x^{1/6}}{x^{5/6}}$ **24.** $\dfrac{x^{1/4}}{x^{3/4}}$

25. $\left(x^{1/2}\right)^6$ **26.** $\left(x^{1/3}\right)^6$

27. Explain how simplifying $x^{1/2} \cdot x^{1/3}$ is similar to simplifying $x^2 \cdot x^3$.

28. Explain how simplifying $\left(x^{1/2}\right)^{1/3}$ is similar to simplifying $\left(x^2\right)^3$.

Simplify each expression.

29. $81^{1/2}$ **30.** $(-27)^{1/3}$

31. $(-8)^{1/3}$ **32.** $36^{1/2}$

33. $-81^{1/4}$ **34.** $-64^{1/3}$

35. $\left(\dfrac{1}{81}\right)^{1/2}$ **36.** $\left(\dfrac{9}{16}\right)^{1/2}$

37. $\left(\dfrac{27}{64}\right)^{1/3}$

38. $\left(\dfrac{16}{81}\right)^{1/4}$

39. $9^{3/2}$

40. $16^{3/2}$

41. $64^{3/2}$

42. $64^{2/3}$

43. $-8^{2/3}$

44. $8^{2/3}$

45. $4^{5/2}$

46. $9^{4/2}$

47. $\left(\dfrac{4}{9}\right)^{3/2}$

48. $\left(\dfrac{8}{27}\right)^{2/3}$

49. $\left(\dfrac{1}{81}\right)^{3/4}$

50. $\left(\dfrac{1}{32}\right)^{3/5}$

51. $4^{-1/2}$

52. $9^{-1/2}$

53. $125^{-1/3}$

54. $216^{-1/3}$

55. $625^{-3/4}$

56. $256^{-5/8}$

Simplify each expression. Write each answer with positive exponents. Assume that all variables represent positive numbers.

57. $3^{4/3} \cdot 3^{2/3}$

58. $2^{5/4} \cdot 2^{3/4}$

59. $\dfrac{6^{2/3}}{6^{1/3}}$

60. $\dfrac{3^{3/5}}{3^{1/5}}$

61. $\left(x^{2/3}\right)^9$

62. $\left(x^6\right)^{3/4}$

63. $\dfrac{6^{1/3}}{6^{-5/3}}$

64. $\dfrac{2^{-3/4}}{2^{5/4}}$

65. $\dfrac{3^{-3/5}}{3^{2/5}}$

66. $\dfrac{5^{1/4}}{5^{-3/4}}$

67. $\left(\dfrac{x^{1/3}}{y^{3/4}}\right)^2$

68. $\left(\dfrac{x^{1/2}}{y^{2/3}}\right)^6$

69. $\left(\dfrac{x^{2/5}}{y^{3/4}}\right)^8$

70. $\left(\dfrac{x^{3/4}}{y^{1/6}}\right)^3$

71. If a population grows at a rate of 8% annually, the formula $P = P_O(1.08)^N$ can be used to estimate the total population P after N years have passed, assuming the original population is P_O. Find the population after $1\frac{1}{2}$ years if the original population of 10,000 people is growing at a rate of 8% annually.

72. Money grows in a certain savings account at a rate of 4% compounded annually. The amount of money A in the account at time t is given by the formula

$$A = P(1.04)^t$$

where P is the original amount deposited in the account. Find the amount of money in the account after $3\frac{3}{4}$ years if \$200 was initially deposited.

Use a calculator and approximate each to three decimal places.

73. $5^{3/4}$

74. $20^{1/8}$

75. $18^{3/5}$

76. $42^{3/10}$

REVIEW EXERCISES

Solve each system of linear inequalities by graphing on a single coordinate system. See Section 8.5.

77. $\begin{cases} x + y < 6 \\ \quad y \geq 2x \end{cases}$

78. $\begin{cases} 2x - y \geq 3 \\ \quad x < 5 \end{cases}$

Solve each quadratic equation. See Section 5.6.

79. $x^2 - 4 = 3x$

80. $x^2 + 2x = 8$

81. $2x^2 - 5x - 3 = 0$

82. $3x^2 + x - 2 = 0$

For additional Chapter Projects, visit the Real World Activities Website by going to http://www.prenhall.com/martin-gay.

CHAPTER PROJECT

Investigating the Dimensions of Cylinders

The volume V (in cubic units) of a cylinder is given by the formula $V = \pi r^2 h$, where r is the radius of the cylinder and h is its height. In this project, you will investigate the radii of several cylinders by completing the table on the next page.

For this project, you will need several empty cans of different sizes, a 2-cup (16-fluid-ounce) transparent measuring cup with metric markings (in milliliters), a metric ruler, and water. This project may be completed by working in groups or individually.

Can	Volume (ml)	Height (cm)	Calculated Radius (cm)	Measured Radius (cm)
A				
B				
C				
D				

△ 1. For each can, measure its volume by filling it with water and pouring the water into the measuring cup. Find the volume of the water in milliliters (ml). Record the volumes of the cans in the table. (Remember that $1 \text{ ml} = 1 \text{ cm}^3$.)

△ 2. Use a ruler to measure the height of each can in centimeters (cm). Record the heights in the table.

△ 3. Solve the formula $V = \pi r^2 h$ for the radius r.

△ 4. Use your formula from Question 3 to calculate an estimate of each can's radius based on the volume and height measurements recorded in the table. Record these calculated radii in the table.

△ 5. Try to measure the radius of each can and record these measured radii in the table. (Remember that radius $= \frac{1}{2}$ diameter.)

△ 6. How close are the values of the calculated radius and the measured radius of each can? What factors could account for the differences?

CHAPTER 9 VOCABULARY CHECK

Fill in each blank with one of the words or phrases listed below.

index   radicand   like radicals

rationalizing the denominator   conjugate

principal square root   radical

1. The expressions $5\sqrt{x}$ and $7\sqrt{x}$ are examples of _____.

2. In the expression $\sqrt[3]{45}$ the number 3 is the _____, the number 45 is the _____, and $\sqrt{}$ is called the _____ sign.

3. The _____ of $(a + b)$ is $(a - b)$.

4. The _____ of 25 is 5.

5. The process of eliminating the radical in the denominator of a radical expression is called _____.

CHAPTER 9 HIGHLIGHTS

DEFINITIONS AND CONCEPTS	EXAMPLES

Section 9.1 Introduction to Radicals

The **positive** or **principal square root** of a positive number a is written as $\sqrt{a}$. The **negative square root** of a is written as $-\sqrt{a}$.
$\sqrt{a} = b$ only if $b^2 = a$ and $b > 0$.

$$\sqrt{25} = 5 \qquad \sqrt{100} = 10$$
$$-\sqrt{9} = -3 \qquad \sqrt{\frac{4}{49}} = \frac{2}{7}$$
$$\sqrt{-4} \text{ is not a real number.}$$

A square root of a negative number is not a real number.
The **nth root** of a number a is written as $\sqrt[n]{a}$ and $\sqrt[n]{a} = b$ only if $b^n = a$.
The natural number n is called the **index**, the symbol $\sqrt{}$ is called a **radical**, and the expression within the radical is called the **radicand**.
(Note: If the index is even, the radicand must be nonnegative for the root to be a real number.)

$$\sqrt[3]{64} = 4 \qquad \sqrt[3]{-8} = -2$$
$$\sqrt[4]{81} = 3$$
$$\sqrt[5]{-32} = -2$$

Section 9.2 Simplifying Radicals

Product rule for radicals
If $\sqrt[n]{a}$ and $\sqrt[n]{b}$ are real numbers, then $\sqrt[n]{a} \cdot \sqrt[n]{b} = \sqrt[n]{a \cdot b}$.

$$\sqrt{2} \cdot \sqrt{3} = \sqrt{6}$$
$$\sqrt[3]{7} \cdot \sqrt[3]{2} = \sqrt[3]{14}$$

A square root is in **simplified form** if the radicand contains no perfect square factors other than 1. To simplify a square root, factor the radicand so that one of its factors is a perfect square factor.

$$\begin{aligned}\sqrt{45} &= \sqrt{9 \cdot 5}\\ &= \sqrt{9} \cdot \sqrt{5}\\ &= 3\sqrt{5}\end{aligned}$$
in simplest form.

To simplify cube roots, factor the radicand so that one of its factors is a perfect cube.

$$\begin{aligned}\sqrt[3]{48} &= \sqrt[3]{8 \cdot 6}\\ &= \sqrt[3]{8} \cdot \sqrt[3]{6}\\ &= 2\sqrt[3]{6}\end{aligned}$$

Quotient rule for radicals
If $\sqrt[n]{a}$ and $\sqrt[n]{b}$ are real numbers and $b \neq 0$, then

$$\sqrt[n]{\frac{a}{b}} = \frac{\sqrt[n]{a}}{\sqrt[n]{b}}$$

$$\sqrt{\frac{18}{x^6}} = \frac{\sqrt{9 \cdot 2}}{\sqrt{x^6}} = \frac{\sqrt{9} \cdot \sqrt{2}}{x^3} = \frac{3\sqrt{2}}{x^3}$$
$$\sqrt[3]{\frac{18}{x^6}} = \frac{\sqrt[3]{18}}{\sqrt[3]{x^6}} = \frac{\sqrt[3]{18}}{x^2}$$

Section 9.3 Adding and Subtracting Radicals

Like radicals are radical expressions that have the same index and the same radicand.

Like Radicals
$$5\sqrt{2}, -7\sqrt{2}, \sqrt{2}$$
$$-\sqrt[3]{11}, 3\sqrt[3]{11}$$

To combine like radicals, use the distributive property.

$$2\sqrt{7} - 13\sqrt{7} = (2 - 13)\sqrt{7} = -11\sqrt{7}$$
$$\begin{aligned}\sqrt[3]{24} &+ \sqrt[3]{8} + \sqrt[3]{81}\\ &= \sqrt[3]{8 \cdot 3} + 2 + \sqrt[3]{27 \cdot 3}\\ &= \sqrt[3]{8} \cdot \sqrt[3]{3} + 2 + \sqrt[3]{27} \cdot \sqrt[3]{3}\\ &= 2\sqrt[3]{3} + 2 + 3\sqrt[3]{3}\\ &= (2 + 3)\sqrt[3]{3} + 2\\ &= 5\sqrt[3]{3} + 2\end{aligned}$$

(continued)

DEFINITIONS AND CONCEPTS	EXAMPLES

Section 9.4 Multiplying and Dividing Radicals

The product and quotient rules for radicals may be used to simplify products and quotients of radicals.

Perform indicated operations and simplify.

$$(2\sqrt{5})^2 = 2^2 \cdot (\sqrt{5})^2 = 4 \cdot 5 = 20$$

Multiply.

$$(\sqrt{3x} + 1)(\sqrt{5} - \sqrt{3})$$
$$= \sqrt{15x} - \sqrt{9x} + \sqrt{5} - \sqrt{3}$$
$$= \sqrt{15x} - 3\sqrt{x} + \sqrt{5} - \sqrt{3}$$

$$\frac{\sqrt[3]{56x^4}}{\sqrt[3]{7x}} = \sqrt[3]{\frac{56x^4}{7x}} = \sqrt[3]{8x^3} = 2x$$

The process of eliminating the radical in the denominator of a radical expression is called **rationalizing the denominator**.

Rationalize the denominator.

$$\frac{5}{\sqrt{11}} = \frac{5 \cdot \sqrt{11}}{\sqrt{11} \cdot \sqrt{11}} = \frac{5\sqrt{11}}{11}$$

The **conjugate** of $a + b$ is $a - b$.

The conjugate of $2 + \sqrt{3}$ is $2 - \sqrt{3}$.

To rationalize a denominator that is a sum or difference of radicals, multiply the numerator and the denominator by the conjugate of the denominator.

Rationalize the denominator.

$$\frac{5}{6 - \sqrt{5}} = \frac{5(6 + \sqrt{5})}{(6 - \sqrt{5})(6 + \sqrt{5})}$$

$$= \frac{5(6 + \sqrt{5})}{36 + 6\sqrt{5} - 6\sqrt{5} - 5}$$

$$= \frac{5(6 + \sqrt{5})}{31}$$

Section 9.5 Solving Equations Containing Radicals

To solve a radical equation containing square roots

Step 1. Get one radical by itself on one side of the equation.

Step 2. Square both sides of the equation.

Step 3. Simplify both sides of the equation.

Step 4. If the equation still contains a radical term, repeat steps 1 through 3.

Step 5. Solve the equation.

Step 6. Check solutions in the original equation.

Solve: $\sqrt{2x - 1} - x = -2$.

$$\sqrt{2x - 1} = x - 2$$
$$(\sqrt{2x - 1})^2 = (x - 2)^2 \quad \text{Square both sides.}$$
$$2x - 1 = x^2 - 4x + 4$$
$$0 = x^2 - 6x + 5$$
$$0 = (x - 1)(x - 5) \qquad \text{Factor.}$$
$$x - 1 = 0 \quad \text{or} \quad x - 5 = 0$$
$$x = 1 \qquad \text{or} \quad x = 5 \quad \text{Solve.}$$

Check both proposed solutions in the original equation. 5 checks but 1 does not. The only solution is 5.

DEFINITIONS AND CONCEPTS	EXAMPLES

Section 9.6 Radical Equations and Problem Solving

Problem-solving steps

1. UNDERSTAND. Read and reread the problem.

A gutter is mounted on the eaves of a house 15 feet above the ground. A garden is adjacent to the house so that the closest a ladder can be placed to the house is 6 feet. How long a ladder is needed for installing the gutter? Let x = the length of the ladder.

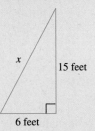

2. TRANSLATE.

Here, we use the Pythagorean theorem. The unknown length x is the hypotenuse.
In words:

$$\boxed{(\text{leg})^2} + \boxed{(\text{leg})^2} = \boxed{(\text{hypotenuse})^2}$$

3. SOLVE.

Translate:
$$6^2 + 15^2 = x^2$$
$$36 + 225 = x^2$$
$$261 = x^2$$
$$\sqrt{261} = x \quad \text{or} \quad x = 3\sqrt{29}$$

4. INTERPRET.

Check and state. The ladder needs to be $3\sqrt{29}$ feet or approximately 16.2 feet long.

Section 9.7 Rational Exponents

If n is a positive integer and $\sqrt[n]{a}$ is a real number, then $a^{1/n} = \sqrt[n]{a}$.

$$9^{1/2} = \sqrt{9} = 3$$
$$(-8)^{1/3} = \sqrt[3]{-8} = -2$$

If m and n are integers with $n > 0$ and a is positive, then
$$a^{m/n} = \left(a^{1/n}\right)^m = \left(\sqrt[n]{a}\right)^m$$
Also
$$a^{m/n} = \left(a^m\right)^{1/n} = \sqrt[n]{a^m}$$

$$25^{3/2} = \left(25^{1/2}\right)^3 = \left(\sqrt{25}\right)^3 = 5^3 \text{ or } 125$$

If $a^{m/n}$ is a nonzero real number, then
$$a^{-m/n} = \frac{1}{a^{m/n}}$$

Properties for integer exponents hold for rational exponents also.
$$a^m \cdot a^n = a^{m+n}$$
$$\left(a^m\right)^n = a^{mn}$$
$$\frac{a^m}{a^n} = a^{m-n}$$

$$81^{-3/4} = \frac{1}{81^{3/4}} = \frac{1}{\left(\sqrt[4]{81}\right)^3}$$
$$= \frac{1}{3^3} = \frac{1}{27}$$

$$x^{1/2} \cdot x^{1/4} = x^{(1/2)+(1/4)} = x^{(2/4)+(1/4)} = x^{3/4}$$
$$\left(x^{2/3}\right)^{1/5} = x^{(2/3)\cdot(1/5)} = x^{2/15}$$
$$\frac{x^{5/6}}{x^{1/6}} = x^{(5/6)-(1/6)} = x^{4/6} = x^{2/3}$$

CHAPTER 9 REVIEW

(9.1) *Find the root. Indicate if the expression is not a real number.*

1. $\sqrt{81}$

2. $-\sqrt{49}$

3. $\sqrt[3]{27}$

4. $\sqrt[4]{16}$

5. $-\sqrt{\dfrac{9}{64}}$

6. $\sqrt{\dfrac{36}{81}}$

7. $\sqrt[4]{-\dfrac{16}{81}}$

8. $\sqrt[3]{-\dfrac{27}{64}}$

Determine whether each of the following is rational or irrational. If rational, find the exact value. If irrational, use a calculator to find an approximation accurate to three decimal places.

9. $\sqrt{76}$

10. $\sqrt{576}$

Find the following roots. Assume that variables represent positive numbers only.

11. $\sqrt{x^{12}}$

12. $\sqrt{x^8}$

13. $\sqrt{9x^6}$

14. $\sqrt{25x^4}$

15. $\sqrt{\dfrac{16}{y^{10}}}$

16. $\sqrt{\dfrac{y^{12}}{49}}$

(9.2) *Simplify each expression using the product rule. Assume that variables represent nonnegative real numbers.*

17. $\sqrt{54}$

18. $\sqrt{88}$

19. $\sqrt{150x^3}$

20. $\sqrt{92y^5}$

21. $\sqrt[3]{54}$

22. $\sqrt[3]{88}$

23. $\sqrt[4]{48}$

24. $\sqrt[4]{162}$

Simplify each expression using the quotient rule. Assume that variables represent positive real numbers.

25. $\sqrt{\dfrac{18}{25}}$

26. $\sqrt{\dfrac{75}{64}}$

27. $\sqrt{\dfrac{45y^2}{4x^4}}$

28. $\sqrt{\dfrac{20x^5}{9x^2}}$

29. $\sqrt[4]{\dfrac{9}{16}}$

30. $\sqrt[3]{\dfrac{40}{27}}$

31. $\sqrt[3]{\dfrac{3}{8}}$

32. $\sqrt[4]{\dfrac{5}{81}}$

(9.3) *Add or subtract by combining like radicals.*

33. $3\sqrt[3]{2} + 2\sqrt[3]{3} - 4\sqrt[3]{2}$

34. $5\sqrt{2} + 2\sqrt[3]{2} - 8\sqrt{2}$

35. $\sqrt{6} + 2\sqrt[3]{6} - 4\sqrt[3]{6} + 5\sqrt{6}$

36. $3\sqrt{5} - \sqrt[3]{5} - 2\sqrt{5} + 3\sqrt[3]{5}$

Add or subtract by simplifying each radical and then combining like terms. Assume that variables represent nonnegative real numbers.

37. $\sqrt{28x} + \sqrt{63x} + \sqrt[3]{56}$

38. $\sqrt{75y} + \sqrt{48y} - \sqrt[4]{16}$

39. $\sqrt{\dfrac{5}{9}} - \sqrt{\dfrac{5}{36}}$

40. $\sqrt{\dfrac{11}{25}} + \sqrt{\dfrac{11}{16}}$

41. $2\sqrt[3]{125} - 5\sqrt[3]{8}$

42. $3\sqrt[3]{16} - 2\sqrt[3]{2}$

(9.4) *Find the product and simplify if possible.*

43. $3\sqrt{10} \cdot 2\sqrt{5}$

44. $2\sqrt[3]{4} \cdot 5\sqrt[3]{6}$

45. $\sqrt{3}(2\sqrt{6} - 3\sqrt{12})$

46. $4\sqrt{5}(2\sqrt{10} - 5\sqrt{5})$

47. $(\sqrt{3} + 2)(\sqrt{6} - 5)$

48. $(2\sqrt{5} + 1)(4\sqrt{5} - 3)$

Find the quotient and simplify if possible. Assume that variables represent positive real numbers.

49. $\dfrac{\sqrt{96}}{\sqrt{3}}$

50. $\dfrac{\sqrt{160}}{\sqrt{8}}$

51. $\dfrac{\sqrt{15x^6}}{\sqrt{12x^3}}$

52. $\dfrac{\sqrt{50y^8}}{\sqrt{72y^3}}$

Rationalize each denominator and simplify.

53. $\sqrt{\dfrac{5}{6}}$

54. $\sqrt{\dfrac{7}{10}}$

55. $\sqrt{\dfrac{3}{2x}}$

56. $\sqrt{\dfrac{6}{5y}}$

57. $\sqrt{\dfrac{7}{20y^2}}$

58. $\sqrt{\dfrac{5z}{12x^2}}$

59. $\sqrt[3]{\dfrac{7}{9}}$

60. $\sqrt[3]{\dfrac{3}{4}}$

61. $\sqrt[3]{\dfrac{3}{2}}$

62. $\sqrt[3]{\dfrac{5}{4}}$

63. $\dfrac{3}{\sqrt{5} - 2}$

64. $\dfrac{8}{\sqrt{10} - 3}$

65. $\dfrac{8}{\sqrt{6} + 2}$

66. $\dfrac{12}{\sqrt{15} - 3}$

67. $\dfrac{\sqrt{2}}{4 + \sqrt{2}}$

68. $\dfrac{\sqrt{3}}{5 + \sqrt{3}}$

69. $\dfrac{2\sqrt{3}}{\sqrt{3} - 5}$

70. $\dfrac{7\sqrt{2}}{\sqrt{2} - 4}$

(9.5) *Solve the following radical equations.*

71. $\sqrt{2x} = 6$

72. $\sqrt{x + 3} = 4$

73. $\sqrt{x} + 3 = 8$

74. $\sqrt{x} + 8 = 3$

75. $\sqrt{2x + 1} = x - 7$

76. $\sqrt{3x + 1} = x - 1$

77. $\sqrt{x+3} + x = 9$ **78.** $\sqrt{2x} + x = 4$

(9.6) *Use the Pythagorean theorem to find the length of the unknown side.*

△ **79.**

△ **80.**

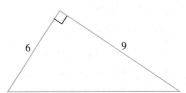

Solve.

△ **81.** Romeo is standing 20 feet away from the wall below Juliet's balcony during a school play. Juliet is on the balcony, 12 feet above the ground. Find how far apart Romeo and Juliet are.

△ **82.** The diagonal of a rectangle is 10 inches long. If the width of the rectangle is 5 inches, find the length of the rectangle.

Use the distance formula to find the distance between the points.

83. $(6, -2)$ and $(-3, 5)$ **84.** $(2, 8)$ and $(-6, 10)$

Use the formula $r = \sqrt{\dfrac{S}{4\pi}}$, *where r = the radius of a sphere and S = the surface area of the sphere, for Exercises 85 and 86.*

△ **85.** Find the radius of a sphere to the nearest tenth of an inch if the area is 72 square inches.

△ **86.** Find the exact surface area of a sphere if its radius is 6 inches. (Do not approximate π.)

(9.7) *Write each of the following with fractional exponents and simplify if possible. Assume that variables represent non-negative real numbers.*

87. $\sqrt{a^5}$ **88.** $\sqrt[5]{a^3}$
89. $\sqrt[6]{x^{15}}$ **90.** $\sqrt[4]{x^{12}}$

Simplify each of the following expressions.

91. $16^{1/2}$ **92.** $36^{1/2}$
93. $(-8)^{1/3}$ **94.** $(-32)^{1/5}$
95. $-64^{3/2}$ **96.** $-8^{2/3}$
97. $\left(\dfrac{16}{81}\right)^{3/4}$ **98.** $\left(\dfrac{9}{25}\right)^{3/2}$

99. $25^{-1/2}$ **100.** $64^{-2/3}$

Simplify each expression using positive exponents only. Assume that variables represent positive real numbers.

101. $8^{1/3} \cdot 8^{4/3}$ **102.** $4^{3/2} \cdot 4^{1/2}$
103. $\dfrac{3^{1/6}}{3^{5/6}}$ **104.** $\dfrac{2^{1/4}}{2^{-3/5}}$
105. $\left(x^{-1/3}\right)^6$ **106.** $\left(\dfrac{x^{1/2}}{y^{1/3}}\right)$

CHAPTER 9 TEST

Simplify the following. Indicate if the expression is not a real number.

1. $\sqrt{16}$ **2.** $\sqrt[3]{125}$

3. $16^{3/4}$ **4.** $\left(\dfrac{9}{16}\right)^{1/2}$

5. $\sqrt[4]{-81}$ **6.** $27^{-2/3}$

Simplify each radical expression. Assume that variables represent positive numbers only.

7. $\sqrt{54}$ **8.** $\sqrt{92}$
9. $\sqrt{3x^6}$ **10.** $\sqrt{8x^4y^7}$
11. $\sqrt{9x^9}$ **12.** $\sqrt[3]{40}$
13. $\sqrt[3]{8}$ **14.** $\sqrt{12} - 2\sqrt{75}$

15. $\sqrt{2x^2} + \sqrt[3]{54} - x\sqrt{18}$ **16.** $\sqrt{\dfrac{5}{16}}$

17. $\sqrt[3]{\dfrac{2}{27}}$ **18.** $3\sqrt{8x}$

Rationalize the denominator.

19. $\sqrt{\dfrac{2}{3}}$ **20.** $\sqrt[3]{\dfrac{5}{9}}$

21. $\sqrt{\dfrac{5}{12x^2}}$ **22.** $\dfrac{8}{\sqrt{6} + 2}$

23. $\dfrac{2\sqrt{3}}{\sqrt{3} - 3}$

Solve each of the following radical equations.

24. $\sqrt{x} + 8 = 11$ **25.** $\sqrt{3x - 6} = \sqrt{x + 4}$
26. $\sqrt{2x - 2} = x - 5$

△ **27.** Find the length of the unknown leg of a right triangle if the other leg is 8 inches long and the hypotenuse is 12 inches long.

28. Find the distance between $(-3, 6)$ and $(-2, 8)$.

Simplify each expression using positive exponents only.

29. $16^{-3/4} \cdot 16^{-1/4}$ **30.** $\left(x^{2/3}\right)^5$

CHAPTER 9 CUMULATIVE REVIEW

1. Simplify each expression.

a. $\dfrac{(-12)(-3) + 3}{-7 - (-2)}$

b. $\dfrac{2(-3)^2 - 20}{-5 + 4}$

2. Solve: $2x + 3x - 5 + 7 = 10x + 3 - 6x - 4$

3. Complete the table for the equation $y = 3x$.

x	y
-1	
	0
	-9

4. Combine like terms to simplify.

$3xy - 5y^2 + 7xy - 9x^2$

5. Factor: $x^2 + 5yx + 6y^2$

6. Simplify: $\dfrac{4 - x^2}{3x^2 - 5x - 2}$

7. Divide: $\dfrac{3x^3y^7}{40} \div \dfrac{4x^3}{y^2}$

8. Subtract: $\dfrac{2y}{2y - 7} - \dfrac{7}{2y - 7}$

9. Add: $\dfrac{2x}{x^2 + 2x + 1} + \dfrac{x}{x^2 - 1}$

10. Simplify: $\dfrac{\dfrac{1}{z} - \dfrac{1}{2}}{\dfrac{1}{3} - \dfrac{z}{6}}$

11. Solve: $\dfrac{x}{2} + \dfrac{8}{3} = \dfrac{1}{6}$

12. If the following two triangles are similar, find the missing length x.

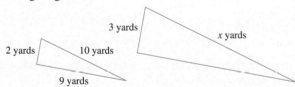

13. Find an equation of the line with y-intercept $(0, -3)$ and slope of $\frac{1}{4}$.

14. Find an equation of the line parallel to the line $y = 5$ and passing through $(-2, -3)$.

15. Which of the following linear equations are functions?

a. $y = x$

b. $y = 2x + 1$

c. $y = 5$

d. $x = -1$

16. Which of the following ordered pairs is a solution of the given system?

$$\begin{cases} 2x - 3y = 6 \\ x = 2y \end{cases}$$

a. $(12, 6)$

b. $(0, -2)$

17. Solve the system:

$$\begin{cases} 2x + y = 10 \\ x = y + 2 \end{cases}$$

18. Solve the system:

$$\begin{cases} -x - \dfrac{y}{2} = \dfrac{5}{2} \\ -\dfrac{x}{2} + \dfrac{y}{4} = 0 \end{cases}$$

19. Eric Daly, a chemistry teaching assistant, needs 10 liters of a 20% saline solution (salt water) for his 2 p.m. laboratory class. Unfortunately, the only mixtures on hand are a 5% saline solution and a 25% saline solution. How much of each solution should he mix to produce the 20% solution?

20. Graph the solution of the system

$$\begin{cases} 3x \ge y \\ x + 2y \le 8 \end{cases}$$

21. Simplify.

a. $\sqrt{54}$

b. $\sqrt{12}$

c. $\sqrt{200}$

d. $\sqrt{35}$

22. Find the product and simplify.

a. $(\sqrt{5} - 7)(\sqrt{5} + 7)$

b. $(\sqrt{7x} + 2)^2$

23. Solve: $\sqrt{x} + 6 = 4$

24. Find the length of the hypotenuse of a right triangle whose legs are 6 inches and 8 inches long.

25. Simplify each expression.

a. $4^{3/2}$

b. $27^{2/3}$

c. $-16^{3/4}$

A Team Player

Engineering technicians apply scientific and mathematical principles to solving practical problems. They frequently assist or work under the direction of engineers. Engineering technicians often specialize in civil, chemical, electrical, industrial, or mechanical branches of engineering; however, about one-third of all engineering technicians work in manufacturing settings.

Engineering technicians usually work as part of an engineering team, so they must work well with others. They should be good problem-solvers and communicators. Engineering technicians also need a solid background in both science and math to tackle such problems as planning workflow in a factory, checking electrical equipment, surveying land for construction of a new road, gauging product quality, or testing instrument accuracy.

 For more information about a career in engineering, visit the Junior Engineering Technical Society (JETS), Inc. Website by first going to www.prenhall.com/martin-gay.

In the Spotlight on Decision Making feature on page 544, you will have the opportunity, as an engineering technician, to decide whether a production line needs upgrading.

10

SOLVING QUADRATIC EQUATIONS

10.1 SOLVING QUADRATIC EQUATIONS BY THE SQUARE ROOT METHOD

10.2 SOLVING QUADRATIC EQUATIONS BY COMPLETING THE SQUARE

10.3 SOLVING QUADRATIC EQUATIONS BY THE QUADRATIC FORMULA

10.4 SUMMARY OF METHODS FOR SOLVING QUADRATIC EQUATIONS AND PROBLEM SOLVING

10.5 COMPLEX SOLUTIONS OF QUADRATIC EQUATIONS

10.6 GRAPHING QUADRATIC EQUATIONS

An important part of the study of algebra is learning to use methods for solving equations. In Chapter 2, we presented techniques for solving linear equations in one variable. In Chapter 5, we solved quadratic equations in one variable by factoring the quadratic expressions. We now present other methods for solving quadratic equations in one variable.

10.1 SOLVING QUADRATIC EQUATIONS BY THE SQUARE ROOT METHOD

CD-ROM   SSM

SSG   Video

▶ **O B J E C T I V E**

1. Use the square root property to solve quadratic equations.

1

Recall that a quadratic equation is an equation that can be written in the form

$$ax^2 + bx + c = 0$$

where a, b, and c are real numbers and $a \neq 0$.

To solve quadratic equations by factoring, use the **zero factor theorem**: If the product of two numbers is zero, then at least one of the two numbers is zero. For example, to solve $x^2 - 4 = 0$, we first factor the left side of the equation and then set each factor equal to 0.

$$x^2 - 4 = 0$$
$$(x + 2)(x - 2) = 0 \qquad \text{Factor.}$$
$$x + 2 = 0 \quad \text{or} \quad x - 2 = 0 \qquad \text{Apply the zero factor theorem.}$$
$$x = -2 \quad \text{or} \qquad x = 2 \qquad \text{Solve each equation.}$$

The solutions are -2 and 2.

Now let's solve $x^2 - 4 = 0$ another way. First, add 4 to both sides of the equation.

$$x^2 - 4 = 0$$
$$x^2 = 4 \qquad \text{Add 4 both sides.}$$

Now we see that the value for x must be a number whose square is 4. Therefore $x = \sqrt{4} = 2$ or $x = -\sqrt{4} = -2$. This reasoning is an example of the square root property.

SQUARE ROOT PROPERTY

If $x^2 = a$ for $a \geq 0$, then

$$x = \sqrt{a} \quad \text{or} \quad x = -\sqrt{a}$$

Example 1   Use the square root property to solve $x^2 - 9 = 0$.

Solution   First we solve for x^2 by adding 9 to both sides.

$$x^2 - 9 = 0$$
$$x^2 = 9 \qquad \text{Add 9 to both sides.}$$

Next we use the square root property.

$$x = \sqrt{9} \quad \text{or} \quad x = -\sqrt{9}$$
$$x = 3 \qquad\qquad x = -3$$

Check

$x^2 - 9 = 0$	*Original equation*	$x^2 - 9 = 0$	*Original equation*
$3^2 - 9 \stackrel{?}{=} 0$	*Let $x = 3$.*	$(-3)^2 - 9 \stackrel{?}{=} 0$	*Let $x = -3$.*
$0 = 0$	*True*	$0 = 0$	*True*

The solutions are 3 and −3.

Example 2   Use the square root property to solve $2x^2 = 7$.

Solution   First we solve for x^2 by dividing both sides by 2. Then we use the square root property.

$$2x^2 = 7$$

$$x^2 = \frac{7}{2} \qquad \text{Divide both sides by 2.}$$

$$x = \sqrt{\frac{7}{2}} \quad \text{or} \quad x = -\sqrt{\frac{7}{2}} \qquad \text{Use the square root property.}$$

If the denominators are rationalized, we have

$$x = \frac{\sqrt{7} \cdot \sqrt{2}}{\sqrt{2} \cdot \sqrt{2}} \qquad x = -\frac{\sqrt{7} \cdot \sqrt{2}}{\sqrt{2} \cdot \sqrt{2}} \qquad \text{Rationalize the denominator.}$$

$$x = \frac{\sqrt{14}}{2} \qquad x = -\frac{\sqrt{14}}{2} \qquad \text{Simplify.}$$

Remember to check both solutions in the original equation. The solutions are $\dfrac{\sqrt{14}}{2}$ and $-\dfrac{\sqrt{14}}{2}$.

Example 3   Use the square root property to solve $(x - 3)^2 = 16$.

Solution   Instead of x^2, here we have $(x - 3)^2$. But the square root property can still be used.

$$(x - 3)^2 = 16$$

$x - 3 = \sqrt{16}$	or	$x - 3 = -\sqrt{16}$	*Use the square root property.*
$x - 3 = 4$		$x - 3 = -4$	*Write $\sqrt{16}$ as 4 and $-\sqrt{16}$ as −4.*
$x = 7$		$x = -1$	*Solve.*

Check

$(x - 3)^2 = 16$	*Original equation*	$(x - 3)^2 = 16$	*Original equation*
$(7 - 3)^2 \stackrel{?}{=} 16$	*Let $x = 7$.*	$(-1 - 3)^2 \stackrel{?}{=} 16$	*Let $x = -1$.*
$4^2 \stackrel{?}{=} 16$	*Simplify.*	$(-4)^2 \stackrel{?}{=} 16$	*Simplify.*
$16 = 16$	*True*	$16 = 16$	*True*

Both 7 and −1 are solutions.

Example 4 Use the square root property to solve $(x + 1)^2 = 8$.

Solution $(x + 1)^2 = 8$

$$x + 1 = \sqrt{8} \quad \text{or} \quad x + 1 = -\sqrt{8} \qquad \text{Use the square root property.}$$
$$x + 1 = 2\sqrt{2} \qquad\qquad x + 1 = -2\sqrt{2} \qquad \text{Simplify the radical.}$$
$$x = -1 + 2\sqrt{2} \qquad\quad x = -1 - 2\sqrt{2} \qquad \text{Solve for } x.$$

Check both solutions in the original equation. The solutions are $-1 + 2\sqrt{2}$ and $-1 - 2\sqrt{2}$. This can be written compactly as $-1 \pm 2\sqrt{2}$. The notation $\pm$ is read as "plus or minus."

> **HELPFUL HINT**
>
> read "plus or minus"
>
> The notation $-1 \pm \sqrt{5}$, for example, is just a shorthand notation for both $-1 + \sqrt{5}$ and $-1 - \sqrt{5}$.

Example 5 Use the square root property to solve $(x - 1)^2 = -2$.

Solution This equation has no real solution because the square root of -2 is not a real number.

Example 6 Use the square root property to solve $(5x - 2)^2 = 10$.

Solution $(5x - 2)^2 = 10$

$$5x - 2 = \sqrt{10} \quad \text{or} \quad 5x - 2 = -\sqrt{10} \qquad \text{Use the square root property.}$$
$$5x = 2 + \sqrt{10} \qquad\qquad 5x = 2 - \sqrt{10} \qquad \text{Add 2 to both sides.}$$
$$x = \frac{2 + \sqrt{10}}{5} \qquad\qquad x = \frac{2 - \sqrt{10}}{5} \qquad \text{Divide both sides by 5.}$$

Check both solutions in the original equation. The solutions are $\dfrac{2 + \sqrt{10}}{5}$ and $\dfrac{2 - \sqrt{10}}{5}$, which can be written as $\dfrac{2 \pm \sqrt{10}}{5}$.

Exercise Set 10.1

Use the square root property to solve each quadratic equation. See Examples 1 and 2.

1. $x^2 = 64$

2. $x^2 = 121$

3. $x^2 = 21$

4. $x^2 = 22$

5. $x^2 = \dfrac{1}{25}$

6. $x^2 = \dfrac{1}{16}$

7. $x^2 = -4$

8. $x^2 = -25$

9. $3x^2 = 13$

10. $5x^2 = 2$

11. $7x^2 = 4$

12. $2x^2 = 9$

13. $x^2 - 2 = 0$

14. $x^2 - 15 = 0$

15. $2x^2 - 10 = 0$

16. $7x^2 - 21 = 0$

17. Explain why the equation $x^2 = -9$ has no real solution.

18. Explain why the equation $x^2 = 9$ has two solutions.

Use the square root property to solve each quadratic equation. See Examples 3 through 6.

19. $(x - 5)^2 = 49$

20. $(x + 2)^2 = 25$

21. $(x + 2)^2 = 7$

22. $(x - 7)^2 = 2$

23. $\left(m - \dfrac{1}{2}\right)^2 = \dfrac{1}{4}$

24. $\left(m + \dfrac{1}{3}\right)^2 = \dfrac{1}{9}$

25. $(p + 2)^2 = 10$

26. $(p - 7)^2 - 13$

27. $(3y + 2)^2 = 100$

28. $(4y - 3)^2 = 81$

29. $(z - 4)^2 = -9$

30. $(z + 7)^2 = -20$

31. $(2x - 11)^2 = 50$

32. $(3x - 17)^2 = 28$

33. $(3x - 7)^2 = 32$

34. $(5x - 11)^2 = 54$

35. $(2p - 5)^2 = 121$

36. $(3p - 1)^2 = 4$

Solve each quadratic equation by first factoring the perfect square trinomial on the left side. Then apply the square root property.

37. $x^2 + 4x + 4 = 16$

38. $z^2 - 6z + 9 = 25$

39. $y^2 - 10y + 25 = 11$

40. $x^2 + 14x + 49 = 31$

△ **41.** The area of a circle is found by the equation $A = \pi r^2$ If the area A of a certain circle is 36π square inches, find its radius r.

36π square inches

△ **42.** A 27-inch-square TV is advertised in the local paper. If 27 inches is the measure of the diagonal of the picture tube, use the Pythagorean theorem to find the measure of the side of the picture tube.

43. Neglecting air resistance, the distance d in feet that an object falls in t seconds is given by the equation $d = 16t^2$. If a sandblaster drops his goggles from a bridge 400 feet from the water below, find how long it takes for the goggles to hit the water.

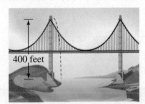

400 feet

44. The number of cattle y (in thousands) on farms in South Dakota from 1997 through 1999 is given by the equation $y = 150x^2 + 3700$. In this equation, $x = 0$ represents the year 1998. Assume that this trend continues and find the year in which there are 5050 thousand cattle on South Dakota farms. (*Hint:* Replace y with 5050 in the equation and solve for x.) (*Source:* Based on data from the U.S. Department of Agriculture)

45. The soybean yield y (in bushels per acre) in Nebraska from 1996 through 1998 is given by the equation $y = 3x^2 + 41$. In this equation, $x = 0$ represents the year 1997. Assume that this trend continues and predict the year in which the Nebraska soybean yield will be 116 bushels per acre. (*Hint:* Replace y with 116 in the equation and solve for x.) (*Source:* Based on data from the U.S. Department of Agriculture)

Solve each quadratic equation by using the square root property. Use a calculator and round each solution to the nearest hundredth.

46. $x^2 = 1.78$

47. $y^2 = 9.86$

48. $(x - 1.37)^2 = 5.71$

49. $(z + 10.68)^2 = 16.61$

50. $(2y + 1.58)^2 = 21.11$

51. $(5z - 5.95)^2 = 14.19$

REVIEW EXERCISES

Factor each perfect square trinomial. See Section 5.2.

52. $x^2 + 6x + 9$

53. $y^2 + 10y + 25$

54. $x^2 - 4x + 4$

55. $x^2 - 20x + 100$

The following graph shows the number of U.S. households with computers. Use this graph for Exercises 56 through 58. See Section 7.2.

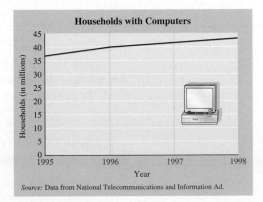

Households with Computers

Households (in millions)

Year

Source: Data from National Telecommunications and Information Ad.

56. Estimate the number of households with computers in 1996.

57. The growth of home offices has been almost linear since 1995. Approximate this growth with a linear equation. To do so, find an equation of the line through the ordered pairs (0, 36.6) and (3, 43) where x is the number of years since 1995 and y is the number of households (in millions) that have computers. Write the equation in slope-intercept form. (Approximate the slope to one decimal place.)

58. Use the equation found in Exercise 57 to predict the number of households with computers in 2005.

10.2 SOLVING QUADRATIC EQUATIONS BY COMPLETING THE SQUARE

CD-ROM   SSM

SSG   Video

▶ **OBJECTIVES**

1. Find perfect square trinomials.
2. Solve quadratic equations by completing the square.

---

1 In the last section, we used the square root property to solve equations such as

$$(x + 1)^2 = 8 \quad \text{and} \quad (5x - 2)^2 = 3$$

Notice that one side of each equation is a quantity squared and that the other side is a constant. To solve

$$x^2 + 2x = 4$$

notice that if we add 1 to both sides of the equation, the left side is a perfect square trinomial that can be factored.

$$x^2 + 2x + 1 = 4 + 1 \qquad \text{Add 1 to both sides.}$$
$$(x + 1)^2 = 5 \qquad \text{Factor.}$$

Now we can solve this equation as we did in the previous section by using the square root property.

$$x + 1 = \sqrt{5} \quad \text{or} \quad x + 1 = -\sqrt{5} \qquad \text{Use the square root property.}$$
$$x = -1 + \sqrt{5} \qquad x = -1 - \sqrt{5} \qquad \text{Solve.}$$

The solutions are $-1 \pm \sqrt{5}$.

Adding a number to $x^2 + 2x$ to form a perfect square trinomial is called **completing the square** on $x^2 + 2x$.

In general, we have the following.

COMPLETING THE SQUARE

To complete the square on $x^2 + bx$, add $\left(\dfrac{b}{2}\right)^2$. To find $\left(\dfrac{b}{2}\right)^2$, **find half the coefficient of x, then square the result.**

Example 1 Complete the square for each expression and then factor the resulting perfect square trinomial.

a. $x^2 + 10x$ **b.** $m^2 - 6m$ **c.** $x^2 + x$

Solution **a.** The coefficient of the x-term is 10. Half of 10 is 5, and $5^2 = 25$. Add 25.

$$x^2 + 10x + 25 = (x + 5)^2$$

b. Half the coefficient of m is -3, and $(-3)^2$ is 9. Add 9.

$$m^2 - 6m + 9 = (m - 3)^2$$

c. Half the coefficient of x is $\frac{1}{2}$ and $\left(\frac{1}{2}\right)^2 = \frac{1}{4}$. Add $\frac{1}{4}$.

$$x^2 + x + \frac{1}{4} = \left(x + \frac{1}{2}\right)^2$$

---

2 By completing the square, a quadratic equation can be solved using the square root property.

Example 2 Solve $x^2 + 6x + 3 = 0$ by completing the square.

Solution First we get the variable terms alone by subtracting 3 from both sides of the equation.

$$x^2 + 6x + 3 = 0$$
$$x^2 + 6x = -3 \qquad \text{Subtract 3 from both sides.}$$

Next we find half the coefficient of the x-term, then square it. Add this result to **both sides** of the equation. This will make the left side a perfect square trinomial. The coefficient of x is 6, and half of 6 is 3. So we add 3^2 or 9 to both sides.

$$x^2 + 6x + 9 = -3 + 9 \qquad \text{Complete the square.}$$
$$(x + 3)^2 = 6 \qquad \text{Factor the trinomial } x^2 + 6x + 9.$$
$$x + 3 = \sqrt{6} \quad \text{or} \quad x + 3 = -\sqrt{6} \qquad \text{Use the square root property.}$$
$$x = -3 + \sqrt{6} \qquad x = -3 - \sqrt{6} \qquad \text{Subtract 3 from both sides.}$$

Check by substituting $-3 + \sqrt{6}$ and $-3 - \sqrt{6}$ in the original equation. The solutions are $-3 \pm \sqrt{6}$.

HELPFUL HINT
Remember, when completing the square, add the number that completes the square to **both sides of the equation**. In Example 2, we added 9 to both sides to complete the square.

Example 3 Solve $x^2 - 10x = -14$ by completing the square.

Solution The variable terms are already alone on one side of the equation. The coefficient of x is -10. Half of -10 is -5, and $(-5)^2 = 25$. So we add 25 to both sides.

$$x^2 - 10x = -14$$

> **HELPFUL HINT**
> Add 25 to *both* sides of the equation.

$$x^2 - 10x + 25 = -14 + 25$$

$(x - 5)^2 = 11$ Factor the trinomial and simplify $-14 + 25$.

$x - 5 = \sqrt{11}$ or $x - 5 = -\sqrt{11}$ Use the square root property.

$x = 5 + \sqrt{11}$ $x = 5 - \sqrt{11}$ Add 5 to both sides.

The solutions are $5 \pm \sqrt{11}$.

The method of completing the square can be used to solve *any* quadratic equation whether the coefficient of the squared variable is 1 or not. When the coefficient of the squared variable is not 1, we first divide both sides of the equation by the coefficient of the squared variable so that the coefficient is 1. Then we complete the square.

Example 4 Solve $4x^2 - 8x - 5 = 0$ by completing the square.

Solution $4x^2 - 8x - 5 = 0$

$x^2 - 2x - \dfrac{5}{4} = 0$ Divide both sides by 4.

$x^2 - 2x = \dfrac{5}{4}$ Get the variable terms alone on one side of the equation.

The coefficient of x is -2. Half of -2 is -1, and $(-1)^2 = 1$. So we add 1 to both sides.

$x^2 - 2x + 1 = \dfrac{5}{4} + 1$

$(x - 1)^2 = \dfrac{9}{4}$ Factor $x^2 - 2x + 1$ and simplify $\dfrac{5}{4} + 1$.

$x - 1 = \sqrt{\dfrac{9}{4}}$ or $x - 1 = -\sqrt{\dfrac{9}{4}}$ Use the square root property.

$x = 1 + \dfrac{3}{2}$ $x = 1 - \dfrac{3}{2}$ Add 1 to both sides and simplify the radical.

$x = \dfrac{5}{2}$ $x = -\dfrac{1}{2}$ Simplify.

Both $\dfrac{5}{2}$ and $-\dfrac{1}{2}$ are solutions.

The following steps may be used to solve a quadratic equation in x by completing the square.

SOLVING A QUADRATIC EQUATION IN x BY COMPLETING THE SQUARE

Step 1. If the coefficient of x^2 is 1, go to Step 2. If not, divide both sides of the equation by the coefficient of x^2.

Step 2. Get all terms with variables on one side of the equation and constants on the other side.

Step 3. Find half the coefficient of x and then square the result. Add this number to both sides of the equation.

Step 4. Factor the resulting perfect square trinomial.

Step 5. Use the square root property to solve the equation.

Example 5   Solve $2x^2 + 6x = -7$ by completing the square.

Solution   The coefficient of x^2 is not 1. We divide both sides by 2, the coefficient of x^2.

$$2x^2 + 6x = -7$$

$$x^2 + 3x = -\frac{7}{2} \qquad \text{Divide both sides by 2.}$$

$$x^2 + 3x + \frac{9}{4} = -\frac{7}{2} + \frac{9}{4} \qquad \text{Add } \left(\frac{3}{2}\right)^2 \text{ or } \frac{9}{4} \text{ to both sides.}$$

$$\left(x + \frac{3}{2}\right)^2 = -\frac{5}{4} \qquad \text{Factor the left side and simplify the right.}$$

There is no real solution to this equation since the square root of a negative number is not a real number.

Example 6   Solve $2x^2 = 10x + 1$ by completing the square.

Solution   First we divide both sides of the equation by 2, the coefficient of x^2.

$$2x^2 = 10x + 1$$

$$x^2 = 5x + \frac{1}{2} \qquad \text{Divide both sides by 2.}$$

Next we get the variable terms alone by subtracting $5x$ from both sides.

$$x^2 - 5x = \frac{1}{2}$$

$$x^2 - 5x + \frac{25}{4} = \frac{1}{2} + \frac{25}{4} \qquad \text{Add } \left(-\frac{5}{2}\right)^2 \text{ or } \frac{25}{4} \text{ to both sides.}$$

$$\left(x - \frac{5}{2}\right)^2 = \frac{27}{4} \qquad \text{Factor the left side and simplify the right side.}$$

$$x - \frac{5}{2} = \sqrt{\frac{27}{4}} \quad \text{or} \quad x - \frac{5}{2} = -\sqrt{\frac{27}{4}} \qquad \text{Use the square root property.}$$

$$x - \frac{5}{2} = \frac{3\sqrt{3}}{2} \qquad\qquad x - \frac{5}{2} = -\frac{3\sqrt{3}}{2} \qquad \text{Simplify.}$$

$$x = \frac{5}{2} + \frac{3\sqrt{3}}{2} \qquad\qquad x = \frac{5}{2} - \frac{3\sqrt{3}}{2}$$

The solutions are $\dfrac{5 \pm 3\sqrt{3}}{2}$.

MENTAL MATH

Determine the number to add to make each expression a perfect square trinomial. See Example 1.

1. $p^2 + 8p$ **2.** $p^2 + 6p$ **3.** $x^2 + 20x$

4. $x^2 + 18x$ **5.** $y^2 + 14y$ **6.** $y^2 + 2y$

Exercise Set 10.2

Complete the square for each expression and then factor the resulting perfect square trinomial. See Example 1.

1. $x^2 + 4x$ **2.** $x^2 + 6x$

3. $k^2 - 12k$ **4.** $k^2 - 16k$

5. $x^2 - 3x$ **6.** $x^2 - 5x$

7. $m^2 - m$ **8.** $y^2 + y$

Solve each quadratic equation by completing the square. See Examples 2 and 3.

9. $x^2 - 6x = 0$ **10.** $y^2 + 4y = 0$

11. $x^2 + 8x = -12$ **12.** $x^2 - 10x = -24$

13. $x^2 + 2x - 5 = 0$ **14.** $z^2 + 6z - 9 = 0$

15. $x^2 + 6x - 25 = 0$ **16.** $x^2 - 6x + 7 = 0$

17. $z^2 + 5z = 7$ **18.** $x^2 - 7x = 5$

19. $x^2 - 2x - 1 = 0$ **20.** $x^2 - 4x + 2 = 0$

21. $y^2 + 5y + 4 = 0$ **22.** $y^2 - 5y + 6 = 0$

23. $x(x + 3) = 18$ **24.** $x(x - 3) = 18$

Solve each quadratic equation by completing the square. See Examples 4 through 6.

25. $4x^2 - 24x = 13$ **26.** $2x^2 + 8x = 10$

27. $5x^2 + 10x + 6 = 0$ **28.** $3x^2 - 12x + 14 = 0$

29. $2x^2 = 6x + 5$ **30.** $4x^2 = -20x + 3$

31. $3x^2 - 6x = 24$ **32.** $2x^2 + 18x = -40$

33. $2y^2 + 8y + 5 = 0$ **34.** $3z^2 + 6z + 4 = 0$

35. $2y^2 - 3y + 1 = 0$ **36.** $2y^2 - y - 1 = 0$

37. $3y^2 - 2y - 4 = 0$ **38.** $4y^2 - 2y - 3 = 0$

39. In your own words, describe a perfect square trinomial.

40. Describe how to find the number to add to $x^2 - 7x$ to make a perfect square trinomial.

41. Find a value of k that will make $x^2 + kx + 16$ a perfect square trinomial.

42. Find a value of k that will make $x^2 + kx + 25$ a perfect square trinomial.

Recall that a graphing calculator may be used to solve an equation. For example, to solve $x^2 + 8x = -12$ (Exercise 11), graph

$$\begin{array}{ll} y_1 = x^2 + 8x & \text{(left side of equation) and} \\ y_2 = -12 & \text{(right side of equation)} \end{array}$$

The x-coordinate of the point of intersection of the graphs is the solution. Use a graphing calculator and solve each equation. Round solutions to the nearest hundredth.

43. Exercise 11   **44.** Exercise 12

45. Exercise 29   **46.** Exercise 18

10.3 SOLVING QUADRATIC EQUATIONS BY THE QUADRATIC FORMULA

CD-ROM   SSM   SSG   Video

▶ **OBJECTIVES**

1. Use the quadratic formula to solve quadratic equations.
2. Determine the number of solutions of a quadratic equation by using the discriminant.

1   We can use the technique of completing the square to develop a formula to find solutions of any quadratic equation. We develop and use the **quadratic formula** in this section.

Recall that a quadratic equation in **standard form** is

$$ax^2 + bx + c = 0, \quad a \neq 0$$

To develop the quadratic formula, let's complete the square for this quadratic equation in standard form.

First we divide both sides of the equation by the coefficient of x^2 and then get the variable terms alone on one side of the equation.

$$x^2 + \frac{b}{a}x + \frac{c}{a} = 0 \qquad \text{Divide by } a; \text{ recall that } a \text{ cannot be 0.}$$

$$x^2 + \frac{b}{a}x = -\frac{c}{a} \qquad \text{Get the variable terms alone on one side of the equation.}$$

The coefficient of x is $\dfrac{b}{a}$. Half of $\dfrac{b}{a}$ is $\dfrac{b}{2a}$ and $\left(\dfrac{b}{2a}\right)^2 = \dfrac{b^2}{4a^2}$. So we add $\dfrac{b^2}{4a^2}$ to both sides of the equation.

$$x^2 + \frac{b}{a}x + \frac{b^2}{4a^2} = -\frac{c}{a} + \frac{b^2}{4a^2} \qquad \text{Add } \frac{b^2}{4a^2} \text{ to both sides.}$$

$$\left(x + \frac{b}{2a}\right)^2 = -\frac{c}{a} + \frac{b^2}{4a^2} \qquad \text{Factor the left side.}$$

$$\left(x + \frac{b}{2a}\right)^2 = -\frac{4ac}{4a^2} + \frac{b^2}{4a^2} \qquad \text{Multiply } -\frac{c}{a} \text{ by } \frac{4a}{4a} \text{ so that both terms on the right side have a common denominator.}$$

$$\left(x + \frac{b}{2a}\right)^2 = \frac{b^2 - 4ac}{4a^2} \qquad \text{Simplify the right side.}$$

Now we use the square root property.

$$x + \frac{b}{2a} = \sqrt{\frac{b^2 - 4ac}{4a^2}} \quad \text{or} \quad x + \frac{b}{2a} = -\sqrt{\frac{b^2 - 4ac}{4a^2}}$$ *Use the square root property.*

$$x + \frac{b}{2a} = \frac{\sqrt{b^2 - 4ac}}{2a} \qquad x + \frac{b}{2a} = -\frac{\sqrt{b^2 - 4ac}}{2a}$$ *Simplify the radical.*

$$x = -\frac{b}{2a} + \frac{\sqrt{b^2 - 4ac}}{2a} \qquad x = -\frac{b}{2a} - \frac{\sqrt{b^2 - 4ac}}{2a}$$ *Subtract $\frac{b}{2a}$ from both sides.*

$$x = \frac{-b + \sqrt{b^2 - 4ac}}{2a} \qquad x = \frac{-b - \sqrt{b^2 - 4ac}}{2a}$$ *Simplify.*

The solutions are $\dfrac{-b \pm \sqrt{b^2 - 4ac}}{2a}$. This final equation is called the **quadratic formula** and gives the solutions of any quadratic equation.

QUADRATIC FORMULA

If a, b, and c are real numbers and $a \neq 0$, a quadratic equation written in the form $ax^2 + bx + c = 0$ has solutions

$$x = \frac{-b \pm \sqrt{b^2 - 4ac}}{2a}$$

▼ HELPFUL HINT

Don't forget that to correctly identify a, b, and c in the quadratic formula, you should write the equation in standard form.

Quadratic Equations in Standard Form

$5x^2 - 6x + 2 = 0$	$a = 5, b = -6, c = 2$
$4y^2 - 9 = 0$	$a = 4, b = 0, c = -9$
$x^2 + x = 0$	$a = 1, b = 1, c = 0$
$\sqrt{2}x^2 + \sqrt{5}x + \sqrt{3} = 0$	$a = \sqrt{2}, b = \sqrt{5}, c = \sqrt{3}$

Example 1   Solve $3x^2 + x - 3 = 0$ using the quadratic formula.

Solution   This equation is in standard form with $a = 3$, $b = 1$, and $c = -3$. By the quadratic formula, we have

$$x = \frac{-b \pm \sqrt{b^2 - 4ac}}{2a}$$

$$x = \frac{-1 \pm \sqrt{1^2 - 4 \cdot 3 \cdot (-3)}}{2 \cdot 3}$$ *Let $a = 3, b = 1,$ and $c = -3$.*

$$= \frac{-1 \pm \sqrt{1 + 36}}{6}$$ *Simplify.*

$$= \frac{-1 \pm \sqrt{37}}{6}$$

Check both solutions in the original equation. The solutions are $\dfrac{-1 + \sqrt{37}}{6}$ and $\dfrac{-1 - \sqrt{37}}{6}$.

Example 2 Solve $2x^2 - 9x = 5$ using the quadratic formula.

Solution First we write the equation in standard form by subtracting 5 from both sides.

$$2x^2 - 9x = 5$$
$$2x^2 - 9x - 5 = 0$$

Next we note that $a = 2$, $b = -9$, and $c = -5$. We substitute these values into the quadratic formula.

$$x = \frac{-b \pm \sqrt{b^2 - 4ac}}{2a}$$

HELPFUL HINT
Notice that the fraction bar is under the entire numerator of $-b \pm \sqrt{b^2 - 4ac}$.

$$x = \frac{-(-9) \pm \sqrt{(-9)^2 - 4 \cdot 2 \cdot (-5)}}{2 \cdot 2}$$ Substitute in the formula.

$$= \frac{9 \pm \sqrt{81 + 40}}{4}$$ Simplify.

$$= \frac{9 \pm \sqrt{121}}{4} = \frac{9 \pm 11}{4}$$

Then,

$$x = \frac{9 - 11}{4} = -\frac{1}{2} \quad \text{or} \quad x = \frac{9 + 11}{4} = 5$$

Check $-\dfrac{1}{2}$ and 5 in the original equation. Both $-\dfrac{1}{2}$ and 5 are solutions.

The following steps may be useful when solving a quadratic equation by the quadratic formula.

SOLVING A QUADRATIC EQUATION BY THE QUADRATIC FORMULA

Step 1. Write the quadratic equation in standard form: $ax^2 + bx + c = 0$.
Step 2. If necessary, clear the equation of fractions to simplify calculations.
Step 3. Identify a, b, and c.
Step 4. Replace a, b, and c in the quadratic formula with the identified values, and simplify.

Example 3 Solve $7x^2 = 1$ using the quadratic formula.

Solution First we write the equation in standard form by subtracting 1 from both sides.

$$7x^2 = 1$$
$$7x^2 - 1 = 0$$

Next we replace a, b, and c with the identified values: $a = 7, b = 0, c = -1$.

$$x = \frac{0 \pm \sqrt{0^2 - 4 \cdot 7 \cdot (-1)}}{2 \cdot 7} \qquad \text{Substitute in the formula.}$$

$$= \frac{\pm\sqrt{28}}{14} \qquad \text{Simplify.}$$

$$= \frac{\pm 2\sqrt{7}}{14}$$

$$= \pm \frac{\sqrt{7}}{7}$$

The solutions are $\dfrac{\sqrt{7}}{7}$ and $-\dfrac{\sqrt{7}}{7}$.

Notice that the equation in Example 3, $7x^2 = 1$, could have been easily solved by dividing both sides by 7 and then using the square root property. We solved the equation by the quadratic formula to show that this formula can be used to solve any quadratic equation.

Example 4 Solve $x^2 = -x - 1$ using the quadratic formula.

Solution First we write the equation in standard form.

$$x^2 + x + 1 = 0$$

Next we replace a, b, and c in the quadratic formula with $a = 1, b = 1$, and $c = 1$.

$$x = \frac{-1 \pm \sqrt{1^2 - 4 \cdot 1 \cdot 1}}{2 \cdot 1} \qquad \text{Substitute in the formula.}$$

$$= \frac{-1 \pm \sqrt{-3}}{2} \qquad \text{Simplify.}$$

There is no real number solution because $\sqrt{-3}$ is not a real number.

Example 5 Solve $\dfrac{1}{2}x^2 - x = 2$ by using the quadratic formula.

Solution We write the equation in standard form and then clear the equation of fractions by multiplying both sides by the LCD, 2.

$$\frac{1}{2}x^2 - x = 2$$

$$\frac{1}{2}x^2 - x - 2 = 0 \qquad \text{Write in standard form.}$$

$$x^2 - 2x - 4 = 0 \qquad \text{Multiply both sides by 2.}$$

Here, $a = 1$, $b = -2$, and $c = -4$, so we substitute these values into the quadratic formula.

$$x = \frac{-(-2) \pm \sqrt{(-2)^2 - 4 \cdot 1 \cdot (-4)}}{2 \cdot 1}$$

$$= \frac{2 \pm \sqrt{20}}{2} = \frac{2 \pm 2\sqrt{5}}{2} \qquad \text{Simplify.}$$

$$= \frac{2\left(1 \pm \sqrt{5}\right)}{2} = 1 \pm \sqrt{5} \qquad \text{Factor and simplify.}$$

The solutions are $1 - \sqrt{5}$ and $1 + \sqrt{5}$. ▀

HELPFUL HINT

When simplifying expressions such as

$$\frac{3 \pm 6\sqrt{2}}{6}$$

first factor out a common factor from the terms of the numerator and then simplify.

$$\frac{3 \pm 6\sqrt{2}}{6} = \frac{3\left(1 \pm 2\sqrt{2}\right)}{2 \cdot 3} = \frac{1 \pm 2\sqrt{2}}{2}$$

---

2 In the quadratic formula, $x = \dfrac{-b \pm \sqrt{b^2 - 4ac}}{2a}$, the radicand $b^2 - 4ac$ is called the

discriminant because, by knowing its value, we can **discriminate** among the possible number and type of solutions of a quadratic equation. Possible values of the discriminant and their meanings are summarized next.

DISCRIMINANT

The following table corresponds the discriminant $b^2 - 4ac$ of a quadratic equation of the form $ax^2 + bx + c = 0$ with the number of solutions of the equation.

$b^2 - 4ac$	*Number of Solutions*
Positive	Two distinct real solutions
Zero	One real solution
Negative	No real solution*

---

* In this case, the quadratic equation will have two complex (but not real) solutions. See Section 10.5 for a discussion of complex numbers.

Example 6   Use the discriminant to determine the number of solutions of $3x^2 + x - 3 = 0$.

Solution   In $3x^2 + x - 3 = 0$, $a = 3$, $b = 1$, and $c = -3$. Then

$$b^2 - 4ac = (1)^2 - 4(3)(-3) = 1 + 36 = 37$$

Since the discriminant is 37, a positive number, this equation has two distinct real solutions.

We solved this equation in Example 1 of this section, and the solutions are $\dfrac{-1 + \sqrt{37}}{6}$ and $\dfrac{-1 - \sqrt{37}}{6}$, two distinct real solutions.

Example 7   Use the discriminant to determine the number of solutions of each quadratic equation.

a. $x^2 - 6x + 9 = 0$ **b.** $5x^2 + 4 = 0$

Solution   **a.** In $x^2 - 6x + 9 = 0$, $a = 1$, $b = -6$, and $c = 9$.

$$b^2 - 4ac = (-6)^2 - 4(1)(9) = 36 - 36 = 0$$

Since the discriminant is 0, this equation has one real solution.

b. In $5x^2 + 4 = 0$, $a = 5$, $b = 0$, and $c = 4$.

$$b^2 - 4ac = 0^2 - 4(5)(4) = 0 - 80 = -80$$

Since the discriminant is -80, a negative number, this equation has no real solution.

SPOTLIGHT ON DECISION MAKING

Suppose you are an engineering technician in a manufacturing plant. The engineering department has been asked to upgrade any production lines that produce less than 3000 items per hour. Production records show the following:

Production Lines A and B	
Lines A and B together:	produce 3000 items in 32 minutes
Line A alone:	takes 11.2 minutes longer than Line B to produce 3000 items

Decide whether either Line A or Line B should be upgraded. Explain your reasoning.

MENTAL MATH

Identify the value of a, b, and c in each quadratic equation.

1. $2x^2 + 5x + 3 = 0$

2. $5x^2 - 7x + 1 = 0$

3. $10x^2 - 13x - 2 = 0$

4. $x^2 + 3x - 7 = 0$

5. $x^2 - 6 = 0$

6. $9x^2 - 4 = 0$

Exercise Set 10.3

Simplify the following.

1. $\dfrac{-1 \pm \sqrt{1^2 - 4(1)(-2)}}{2(1)}$

2. $\dfrac{-(-5) \pm \sqrt{(-5)^2 - 4(2)(3)}}{2(2)}$

3. $\dfrac{-5 \pm \sqrt{5^2 - 4(1)(2)}}{2(1)}$

4. $\dfrac{-7 \pm \sqrt{7^2 - 4(2)(1)}}{2(2)}$

5. $\dfrac{-(-4) \pm \sqrt{(-4)^2 - 4(2)(1)}}{2(2)}$

6. $\dfrac{-6 \pm \sqrt{6^2 - 4(3)(1)}}{2(3)}$

23. $2a^2 - 7a + 3 = 0$

24. $3a^2 - 7a + 2 = 0$

25. $x^2 - 5x - 2 = 0$

26. $x^2 - 2x - 5 = 0$

27. $3x^2 - x - 14 = 0$

28. $5x^2 - 13x - 6 = 0$

29. $6x^2 + 9x = 2$

30. $3x^2 - 9x = 8$

31. $7p^2 + 2 = 8p$

32. $11p^2 + 2 = 10p$

33. $a^2 - 6a + 2 = 0$

34. $a^2 - 10a + 19 = 0$

35. $2x^2 - 6x + 3 = 0$

36. $5x^2 - 8x + 2 = 0$

37. $3x^2 = 1 - 2x$

38. $5y^2 = 4 - y$

39. $20y^2 = 3 - 11y$

40. $2z^2 = z + 3$

41. $x^2 + x + 1 = 0$

42. $k^2 + 2k + 5 = 0$

43. $4y^2 = 6y + 1$

44. $6z^2 + 3z + 2 = 0$

Use the quadratic formula to solve each quadratic equation. See Examples 1 through 4.

7. $x^2 - 3x + 2 = 0$

8. $x^2 - 5x - 6 = 0$

9. $3k^2 + 7k + 1 = 0$

10. $7k^2 + 3k - 1 = 0$

11. $49x^2 - 4 = 0$

12. $25x^2 - 15 = 0$

13. $5z^2 - 4z + 3 = 0$

14. $3z^2 + 2z + 1 = 0$

15. $y^2 = 7y + 30$

16. $y^2 = 5y + 36$

17. $2x^2 = 10$

18. $5x^2 = 15$

19. $m^2 - 12 = m$

20. $m^2 - 14 = 5m$

21. $3 - x^2 = 4x$

22. $10 - x^2 = 2x$

Use the quadratic formula to solve each quadratic equation. See Example 5.

45. $3p^2 - \dfrac{2}{3}p + 1 = 0$

46. $\dfrac{5}{2}p^2 - p + \dfrac{1}{2} = 0$

47. $\dfrac{m^2}{2} = m + \dfrac{1}{2}$

48. $\dfrac{m^2}{2} = 3m - 1$

49. $4p^2 + \dfrac{3}{2} = -5p$

50. $4p^2 + \dfrac{3}{2} = 5p$

51. $5x^2 = \dfrac{7}{2}x + 1$

52. $2x^2 = \dfrac{5}{2}x + \dfrac{7}{2}$

53. $28x^2 + 5x + \dfrac{11}{4} = 0$

54. $\dfrac{2}{3}x^2 - 2x - \dfrac{2}{3} = 0$

55. $5z^2 - 2z = \dfrac{1}{5}$

56. $9z^2 + 12z = -1$

57. $x^2 + 3\sqrt{2}x - 5 = 0$ **58.** $y^2 - 2\sqrt{5}y - 1 = 0$

Use the discriminant to determine the number of solutions of each quadratic equation.

59. $x^2 + 3x - 1 = 0$

60. $x^2 - 5x - 3 = 0$

61. $3x^2 + x + 5 = 0$

62. $2x^2 + x + 4 = 0$

63. $4x^2 + 4x = -1$

64. $7x^2 - x = 0$

65. $9x^2 + 2x = 0$

66. $x^2 + 10x = -25$

67. $5x^2 + 1 = 0$

68. $4x^2 + 9 = 12x$

69. $x^2 + 36 = -12x$

70. $10x^2 + 2 = 0$

71. For the quadratic equation $2x^2 - 5 = 7x$, if $a = 2$ and $c = -5$ in the quadratic formula, the value of b is

   a. $\dfrac{7}{2}$ **b.** 7   **c.** -5 **d.** -7

72. Explain how the quadratic formula is derived and why it is useful.

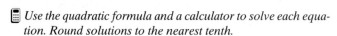 *Use the quadratic formula and a calculator to solve each equation. Round solutions to the nearest tenth.*

73. $x^2 + x = 15$ **74.** $y^2 - y = 11$

75. $1.2x^2 - 5.2x - 3.9 = 0$ **76.** $7.3z^2 + 5.4z - 1.1 = 0$

A rocket is launched from the top of an 80-foot cliff with an initial velocity of 120 feet per second. The height of the rocket h after t seconds is given by the equation

$$h = -16t^2 + 120t + 80$$

77. How long after the rocket is launched will it be 30 feet from the ground? Round to the nearest tenth of a second.

78. How long after the rocket is launched will it strike the ground? Round to the nearest tenth of a second. (*Hint: The rocket will strike the ground when its height $h = 0$.*)

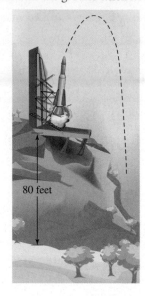

80 feet

REVIEW EXERCISES

Solve the following linear equations. See Section 2.4.

79. $\dfrac{7x}{2} = 3$

80. $\dfrac{5x}{3} = 1$

81. $\dfrac{5}{7}x - \dfrac{2}{3} = 0$

82. $\dfrac{6}{11}x + \dfrac{1}{5} = 0$

83. $\dfrac{3}{4}z + 3 = 0$

84. $\dfrac{5}{2}z + 10 = 0$

10.4 SUMMARY OF METHODS FOR SOLVING QUADRATIC EQUATIONS AND PROBLEM SOLVING

CD-ROM   SSM

SSG   Video

▶ **OBJECTIVES**

   1. Review methods for solving quadratic equations.

   2. Solve problems modeled by quadratic equations.

1

An important skill in mathematics is learning when to use one technique in favor of another. We now practice this by deciding which method to use when solving quadratic equations. Although both the quadratic formula and completing the square can be used to solve any quadratic equation, the quadratic formula is usually less tedious and thus preferred. The following steps may be used to solve a quadratic equation.

SOLVING A QUADRATIC EQUATION

Step 1. If the equation is in the form $(ax + b)^2 = c$, use the square root property and solve. If not, go to Step 2.

Step 2. Write the equation in standard form: $ax^2 + bx + c = 0$.

Step 3. Try to solve the equation by the factoring method. If not possible, go to Step 4.

Step 4. Solve the equation by the quadratic formula.

Example 1   Solve $m^2 - 2m - 7 = 0$.

Solution   The equation is in standard form, but the quadratic expression $m^2 - 2m - 7$ is not factorable, so use the quadratic formula with $a = 1$, $b = -2$, and $c = -7$.

$$m^2 - 2m - 7 = 0$$

$$m = \frac{-(-2) \pm \sqrt{(-2)^2 - 4 \cdot 1 \cdot (-7)}}{2 \cdot 1} = \frac{2 \pm \sqrt{32}}{2}$$

$$m = \frac{2 \pm 4\sqrt{2}}{2} = \frac{2(1 \pm 2\sqrt{2})}{2} = 1 \pm 2\sqrt{2}$$

The solutions are $1 - 2\sqrt{2}$ and $1 + 2\sqrt{2}$.

Example 2   Solve $(3x + 1)^2 = 20$.

Solution   This equation is in a form that makes the square root property easy to apply.

$$(3x + 1)^2 = 20$$

$$3x + 1 = \pm\sqrt{20} \qquad \text{Apply the square root property.}$$

$$3x + 1 = \pm 2\sqrt{5} \qquad \text{Simplify } \sqrt{20}.$$

$$3x = -1 \pm 2\sqrt{5}$$

$$x = \frac{-1 \pm 2\sqrt{5}}{3}$$

The solutions are $\dfrac{-1 - 2\sqrt{5}}{3}$ and $\dfrac{-1 + 2\sqrt{5}}{3}$.

Example 3   Solve $x^2 - \dfrac{11}{2}x = -\dfrac{5}{2}$.

Solution   The fractions make factoring more difficult and also complicate the calculations for using the quadratic formula. Clear the equation of fractions by multiplying both sides of the equation by the LCD 2.

$$x^2 - \frac{11}{2}x = -\frac{5}{2}$$

$$x^2 - \frac{11}{2}x + \frac{5}{2} = 0 \qquad \text{Write in standard form.}$$

$$2x^2 - 11x + 5 = 0 \qquad \text{Multiply both sides by 2.}$$

$$(2x - 1)(x - 5) = 0 \qquad \text{Factor.}$$

$$2x - 1 = 0 \quad \text{or} \quad x - 5 = 0 \qquad \text{Apply the zero factor theorem.}$$

$$2x = 1 \quad \text{or} \quad x = 5$$

$$x = \frac{1}{2} \quad \text{or} \quad x = 5$$

The solutions are $\frac{1}{2}$ and 5.

2   Many real-world applications are modeled by quadratic equations.

Example 4   **FINDING THE LENGTH OF TIME OF A DIVE**

The record for the highest dive into a lake was made by Harry Froboess of Switzerland. In 1936 he dove 394 feet from the airship Hindenburg into Lake Constance. To the nearest tenth of a second, how long did his dive take? (*Source: The Guiness Book of Records,* 1999)

Solution
1. UNDERSTAND. To approximate the time of the dive, we use the formula $h = 16t^2$* where t is time in seconds and s is the distance in feet, traveled by a free-falling body or object. For example, to find the distance traveled in 1 second, or 3 seconds, we let $t = 1$ and then $t = 3$.

$$\text{If } t = 1, h = 16(1)^2 = 16 \cdot 1 = 16 \text{ feet}$$
$$\text{If } t = 3, h = 16(3)^2 = 16 \cdot 9 = 144 \text{ feet}$$

Since a body travels 144 feet in 3 seconds, we now know the dive of 394 feet lasted longer than 3 seconds.

2. TRANSLATE. Use the formula $h = 16t^2$, let the distance $h = 394$, and we have the equation $394 = 16t^2$.

3. SOLVE. To solve $394 = 16t^2$ for t, we will use the square root property.

$$394 = 16t^2$$

$$\frac{394}{16} = t^2 \qquad \text{Divide both sides by 16.}$$

$$24.625 = t^2 \qquad \text{Simplify.}$$

$$\sqrt{24.625} = t \quad \text{or} \quad -\sqrt{24.625} = t \qquad \text{Use the square root property.}$$

$$5.0 \approx t \quad \text{or} \quad -5.0 \approx t \qquad \text{Approximate.}$$

4. INTERPRET.

Check: We reject the solution -5.0 since the length of the dive is not a negative number.

State: The dive lasted approximately 5 seconds.

*The formula $h = 16t^2$ does not take into account air resistance.

Exercise Set 10.4

Choose and use a method to solve each equation.

1. $5w^2 - 11w + 2 = 0$
2. $5w^2 + 13w - 6 = 0$
3. $x^2 - 1 = 2x$
4. $x^2 + 7 = 6x$
5. $a^2 = 20$
6. $a^2 = 72$
7. $x^2 - x + 4 = 0$
8. $x^2 - 2x + 7 = 0$
9. $3x^2 - 12x + 12 = 0$
10. $5x^2 - 30x + 45 = 0$
11. $9 - 6p + p^2 = 0$
12. $49 - 28p + 4p^2 = 0$
13. $4y^2 - 16 = 0$
14. $3y^2 - 27 = 0$
15. $x^4 - 3x^3 + 2x^2 = 0$
16. $x^3 + 7x^2 + 12x = 0$
17. $(2z + 5)^2 = 25$
18. $(3z - 4)^2 = 16$
19. $30x = 25x^2 + 2$
20. $12x = 4x^2 + 4$
21. $\dfrac{2}{3} m^2 - \dfrac{1}{3} m - 1 = 0$
22. $\dfrac{5}{8} m^2 + m - \dfrac{1}{2} = 0$
23. $x^2 - \dfrac{1}{2} x - \dfrac{1}{5} = 0$
24. $x^2 + \dfrac{1}{2} x - \dfrac{1}{8} = 0$
25. $4x^2 - 27x + 35 = 0$
26. $9x^2 - 16x + 7 = 0$
27. $(7 - 5x)^2 = 18$
28. $(5 - 4x)^2 = 75$
29. $3z^2 - 7z = 12$
30. $6z^2 + 7z = 6$
31. $x = x^2 - 110$
32. $x = 56 - x^2$
33. $\dfrac{3}{4} x^2 - \dfrac{5}{2} x - 2 = 0$
34. $x^2 - \dfrac{6}{5} x - \dfrac{8}{5} = 0$
35. $x^2 - 0.6x + 0.05 = 0$
36. $x^2 - 0.1x - 0.06 = 0$
37. $10x^2 - 11x + 2 = 0$
38. $20x^2 - 11x + 1 = 0$
39. $\dfrac{1}{2} z^2 - 2z + \dfrac{3}{4} = 0$
40. $\dfrac{1}{5} z^2 - \dfrac{1}{2} z - 2 = 0$

Solve. See Example 4. For exercises 41 through 44, use the formula from Example 4.

41. The highest regularly performed dives are made by professional divers from La Quebrada. This cliff in Acapulco has a height of 87.6 feet. To the nearest tenth of a second, determine the time of a dive. *(Source: The Guinness Book of Records, 1999).*

42. In 1988, Eddie Turner saved Frank Fanan, who became unconscious after an injury while jumping out of an airplane. Fanan fell 11,136 feet before Turner pulled his ripcord. To the nearest tenth of a second, determine the time of Fanan's unconscious free-fall.

43. In New Mexico, Joseph Kittinger fell 16 miles before opening his parachute on August 16, 1960. To the nearest tenth of a second, how long did Kittinger free-fall before opening his parachute? (*Hint:* First convert 16 miles to feet. Use 1 mile = 5280 feet.) *(Source: Guinness World Record, 2000).*

44. In the Ukraine, Elvira Fomitcheva fell 9 miles 1056 feet before opening her parachute on October 26, 1977. To the nearest tenth of a second, how long did she free-fall before opening her parachute. (Use the hint from Exercise 43.) *(Source: Guinness World Record, 2000).*

45. The number of Home Depot stores y operating in North America from 1994 through 1996 is given by the equation $y = 3x^2 + 80x + 340$. In this equation, x is the number of years after 1994. Assume that this trend continues and predict the year after 1994 in which the number of Home Depot stores will be 1172. *(Source: Based on data from The Home Depot, Inc.)*

46. The average price of tin y (in cents per pound) from 1993 through 1995 is given by the equation $y = 16x^2 + 3x + 350$. In this equation, x is the number of years after 1993. Assume that this trend continues and find the year after 1993 in which the price of tin will be 1155 cents per pound. *(Source: Based on data from the U.S. Bureau of Mines)*

47. The net income y (in millions of dollars) of Goodyear Tire and Rubber Company from 1994 through 1996 is given by the equation $y = 10x^2 + 34x + 567$, where $x = 0$ represents 1994. Assume that this trend continues and predict the year in which Goodyear's net income will be $1295 million. *(Source: Based on data from the Goodyear Tire and Rubber Company)*

48. The sales y (in billions of dollars) of Wal-Mart Stores from 1995 through 1997 is given by the equation $y = -0.5x^2 + 12.5x + 82$, where $x = 0$ represents 1995. Assume that this trend continues and predict the year in which Wal-Mart sales will first be $150 billion. (*Source:* Based on data from Wal-Mart Stores, Inc.)

△ **49.** If a line segment AB is divided by a point C into two segments, AC and CB, such that the proportion $\dfrac{AB}{AC} = \dfrac{AC}{CB}$ is true, this ratio $\dfrac{AB}{AC}$ (or $\dfrac{AC}{CB}$) is called the golden ratio.

If AC is 1 unit, find the length of AB. (*Hint:* Let x be the unknown length as shown and substitute into the given proportion.)

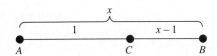

50. The formula $A = P(1 + r)^2$ is used to find the amount of money A in an account after P dollars have been invested in the account paying r annual interest rate for 2 years. Find the interest rate r if $1000 grows to $1690 in 2 years.

51. Explain how you will decide what method to use when solving quadratic equations.

REVIEW EXERCISES

Simplify each expression. See Section 9.2.

52. $\sqrt{48}$

53. $\sqrt{104}$

54. $\sqrt{50}$

55. $\sqrt{80}$

Solve the following. See Section 2.6.

△ **56.** The height of a triangle is 4 times the length of the base. The area of the triangle is 18 square feet. Find the height and base of the triangle.

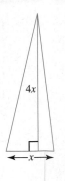

△ **57.** The length of a rectangle is 6 inches more than its width. The area of the rectangle is 391 square inches. Find the dimensions of the rectangle.

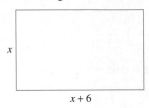

10.5 COMPLEX SOLUTIONS OF QUADRATIC EQUATIONS

CD-ROM   SSM

SSG   Video

▶ **OBJECTIVES**

1. Write complex numbers using i notation.
2. Add and subtract complex numbers.
3. Multiply complex numbers.
4. Divide complex numbers.
5. Solve quadratic equations that have complex solutions.

In Chapter 9, we learned that $\sqrt{-4}$, for example, is not a real number because there is no real number whose square is -4. However, our real number system can be extended to include numbers like $\sqrt{-4}$. This extended number system is called the **complex number** system. The complex number system includes the **imaginary unit i**, which is defined next.

IMAGINARY UNIT i

The imaginary unit, written i, is the number whose square is -1. That is,

$$i^2 = -1 \quad \text{and} \quad i = \sqrt{-1}$$

1 We use i to write numbers like $\sqrt{-6}$ as the product of a real number and i. Since $i = \sqrt{-1}$, we have

$$\sqrt{-6} = \sqrt{-1 \cdot 6} = \sqrt{-1} \cdot \sqrt{6} = i\sqrt{6}$$

Example 1 Write each radical as the product of a real number and i.

a. $\sqrt{-4}$ **b.** $\sqrt{-11}$ **c.** $\sqrt{-20}$

Solution Write each negative radicand as a product of a positive number and -1. Then write $\sqrt{-1}$ as i.

a. $\sqrt{-4} = \sqrt{-1 \cdot 4} = \sqrt{-1} \cdot \sqrt{4} = i \cdot 2 = 2i$
b. $\sqrt{-11} = \sqrt{-1 \cdot 11} = \sqrt{-1} \cdot \sqrt{11} = i\sqrt{11}$
c. $\sqrt{-20} = \sqrt{-1 \cdot 20} = \sqrt{-1} \cdot \sqrt{20} = i \cdot 2\sqrt{5} = 2i\sqrt{5}$

The numbers $2i$, $i\sqrt{11}$, and $2i\sqrt{5}$ are called **imaginary numbers**. Both real numbers and imaginary numbers are complex numbers.

COMPLEX NUMBERS AND IMAGINARY NUMBERS

A complex number is a number that can be written in the form

$$a + bi$$

where a and b are real numbers. A complex number that can be written in the form

$$0 + bi$$

$b \neq 0$, is also called an imaginary number.

A complex number written in the form $a + bi$ is in **standard form**. We call a the real part and bi the imaginary part of the complex number $a + bi$.

Example 2 Identify each number as a complex number by writing it in standard form $a + bi$.

a. 7 **b.** 0 **c.** $\sqrt{20}$ **d.** $\sqrt{-27}$ **e.** $2 + \sqrt{-4}$

Solution **a.** 7 is a complex number since $7 = 7 + 0i$.
b. 0 is a complex number since $0 = 0 + 0i$.
c. $\sqrt{20}$ is a complex number since $\sqrt{20} = 2\sqrt{5} = 2\sqrt{5} + 0i$.
d. $\sqrt{-27}$ is a complex number since $\sqrt{-27} = i \cdot 3\sqrt{3} = 0 + 3i\sqrt{3}$.
e. $2 + \sqrt{-4}$ is a complex number since $2 + \sqrt{-4} = 2 + 2i$.

2 We now present arithmetic operations—addition, subtraction, multiplication, and division—for the complex number system. Complex numbers are added and subtracted in the same way as we add and subtract polynomials.

Example 3   Simplify the sum or difference. Write the result in standard form.

   a. $(2 + 3i) + (-6 - i)$ **b.** $-i + (3 + 7i)$ **c.** $(5 - i) - 4$

Solution   Add the real parts and then add the imaginary parts.

   a. $(2 + 3i) + (-6 - i) = [2 + (-6)] + (3i - i) = -4 + 2i$
   b. $-i + (3 + 7i) = 3 + (-i + 7i) = 3 + 6i$
   c. $(5 - i) - 4 = (5 - 4) - i = 1 - i$

Example 4   Subtract $(11 - i)$ from $(1 + i)$.

Solution $(1 + i) - (11 - i) = 1 + i - 11 + i = (1 - 11) + (i + i) = -10 + 2i$

   3   Use the distributive property and the FOIL method to multiply complex numbers.

Example 5   Find the following products and write in standard form.

   a. $5i(2 - 1i)$ **b.** $(7 - 3i)(4 + 2i)$ **c.** $(2 + 3i)(2 - 3i)$

Solution   **a.** By the distributive property, we have

$5i(2 - 1i) = 5i \cdot 2 - 5i \cdot i$ Apply the distributive property.

$\qquad\qquad = 10i - 5i^2$

$\qquad\qquad = 10i - 5(-1)$ Write i^2 as -1.

$\qquad\qquad = 10i + 5$

$\qquad\qquad = 5 + 10i$ Write in standard form.

   b. $\qquad\qquad\qquad\quad$ F $\quad$ O $\quad$ I $\quad$ L

$(7 - 3i)(4 + 2i) = 28 + 14i - 12i - 6i^2$

$\qquad\qquad\qquad = 28 + 2i - 6(-1)$ Write i^2 as -1.

$\qquad\qquad\qquad = 28 + 2i + 6$

$\qquad\qquad\qquad = 34 + 2i$

   c. $(2 + 3i)(2 - 3i) = 4 - 6i + 6i - 9i^2$

$\qquad\qquad\qquad = 4 - 9(-1)$ Write i^2 as -1.

$\qquad\qquad\qquad = 13$

   The product in part (c) is the real number 13. Notice that one factor is the sum of 2 and $3i$, and the other factor is the difference of 2 and $3i$. When complex number factors are related as these two are, their product is a real number. In general,

$$(a + bi)(a - bi) = a^2 + b^2$$

$\qquad\qquad$ sum $\qquad$ difference   real number

The complex numbers $a + bi$ and $a - bi$ are called **complex conjugates** of each other. For example, $2 - 3i$ is the conjugate of $2 + 3i$, and $2 + 3i$ is the conjugate of $2 - 3i$. Also,

The conjugate of $3 - 10i$ is $3 + 10i$.
The conjugate of 5 is 5. (Note that $5 = 5 + 0i$ and its conjugate is $5 - 0i = 5$.)
The conjugate of $4i$ is $-4i$. ($0 - 4i$ is the conjugate of $0 + 4i$.)

4   The fact that the product of a complex number and its conjugate is a real number provides a method for dividing by a complex number and for simplifying fractions whose denominators are complex numbers.

Example 6   Write $\dfrac{4 + i}{3 - 4i}$ in standard form.

Solution   To write this quotient as a complex number in the standard form $a + bi$, we need to find an equivalent fraction whose denominator is a real number. By multiplying both numerator and denominator by the denominator's conjugate, we obtain a new fraction that is an equivalent fraction with a real number denominator.

$$\frac{4 + i}{3 - 4i} = \frac{(4 + i)}{(3 - 4i)} \cdot \frac{(3 + 4i)}{(3 + 4i)} \qquad \text{Multiply numerator and denominator by } 3 + 4i.$$

$$= \frac{12 + 16i + 3i + 4i^2}{9 - 16i^2}$$

$$= \frac{12 + 19i + 4(-1)}{9 - 16(-1)}$$

$$= \frac{12 + 19i - 4}{9 + 16} = \frac{8 + 19i}{25}$$

$$= \frac{8}{25} + \frac{19}{25}i \qquad \text{Write in standard form.}$$

Note that our last step was to write $\dfrac{4 + i}{3 - 4i}$ in standard form $a + bi$, where a and b are real numbers. ∎

5   Some quadratic equations have complex solutions.

Example 7   Solve $(x + 2)^2 = -25$ for x.

Solution   Begin by applying the square root property.

$$(x + 2)^2 = -25$$
$$x + 2 = \pm\sqrt{-25} \qquad \text{Apply the square root property.}$$
$$x + 2 = \pm 5i \qquad \text{Write } \sqrt{-25} \text{ as } 5i.$$
$$x = -2 \pm 5i$$

The solutions are $-2 + 5i$ and $-2 - 5i$. ∎

Example 8   Solve $m^2 = 4m - 5$.

Solution   Write the equation in standard form and use the quadratic formula to solve.

$$m^2 = 4m - 5$$

$$m^2 - 4m + 5 = 0 \qquad \text{Write the equation in standard form.}$$

Apply the quadratic formula with $a = 1$, $b = -4$, and $c = 5$.

$$m = \frac{4 \pm \sqrt{16 - 4 \cdot 1 \cdot 5}}{2 \cdot 1}$$

$$= \frac{4 \pm \sqrt{-4}}{2}$$

$$= \frac{4 \pm 2i}{2} \qquad \text{Write } \sqrt{-4} \text{ as } 2i.$$

$$= \frac{2\,(2 \pm i)}{2} = 2 \pm i$$

The solutions are $2 - i$ and $2 + i$.   ▪

Example 9   Solve $x^2 + x = -1$.

Solution

$$x^2 + x = -1$$

$$x^2 + x + 1 = 0 \qquad \text{Write in standard form.}$$

$$x = \frac{-1 \pm \sqrt{1 - 4 \cdot 1 \cdot 1}}{2 \cdot 1} \qquad \begin{array}{l}\text{Apply the quadratic formula} \\ \text{with } a = 1, b = 1, \text{ and } c = 1.\end{array}$$

$$= \frac{-1 \pm \sqrt{-3}}{2}$$

$$= \frac{-1 \pm i\sqrt{3}}{2}$$

The solutions are $\dfrac{-1 - i\sqrt{3}}{2}$ and $\dfrac{-1 + i\sqrt{3}}{2}$.   ▪

Exercise Set 10.5

Write each expression in i notation. See Example 1.

1. $\sqrt{-9}$ 2. $\sqrt{-64}$

3. $\sqrt{-100}$ 4. $\sqrt{-16}$

5. $\sqrt{-50}$ 6. $\sqrt{-98}$

7. $\sqrt{-63}$ 8. $\sqrt{-44}$

Add or subtract as indicated. See Examples 2 through 4.

9. $(2 - i) + (-5 + 10i)$ 10. $(-7 + 2i) + (5 - 3i)$

11. $(3 - 4i) - (2 - i)$ 12. $(-6 + i) - (3 + i)$

Multiply. See Example 5.

13. $4i(3 - 2i)$ 14. $-2i(5 + 4i)$

15. $(6 - 2i)(4 + i)$ 16. $(6 + 2i)(4 - i)$

17. Earlier in this text, we learned that $\sqrt{-4}$ is not a real number. Explain what that means and explain what type of number $\sqrt{-4}$ is.

18. Describe how to find the conjugate of a complex number.

Divide. Write each answer in standard form. See Example 6.

19. $\dfrac{8 - 12i}{4}$

20. $\dfrac{14 + 28i}{-7}$

21. $\dfrac{7 - i}{4 - 3i}$

22. $\dfrac{4 - 3i}{7 - i}$

Solve the following quadratic equations for complex solutions. See Example 7.

23. $(x + 1)^2 = -9$

24. $(y - 2)^2 = -25$

25. $(2z - 3)^2 = -12$

26. $(3p + 5)^2 = -18$

Solve the following quadratic equations for complex solutions. See Examples 8 and 9.

27. $y^2 + 6y + 13 = 0$

28. $y^2 - 2y + 5 = 0$

29. $4x^2 + 7x + 4 = 0$

30. $8x^2 - 7x + 2 = 0$

31. $2m^2 - 4m + 5 = 0$

32. $5m^2 - 6m + 7 = 0$

Perform the indicated operations. Write results in standard form.

33. $3 + (12 - 7i)$

34. $(-14 + 5i) + 3i$

35. $-9i(5i - 7)$

36. $10i(4i - 1)$

37. $(2 - i) - (3 - 4i)$

38. $(3 + i) - (-6 + i)$

39. $\dfrac{15 + 10i}{5i}$

40. $\dfrac{-18 + 12i}{-6i}$

41. Subtract $2 + 3i$ from $-5 + i$.

42. Subtract $-8 - i$ from $7 - 4i$.

43. $(4 - 3i)(4 + 3i)$

44. $(12 - 5i)(12 + 5i)$

45. $\dfrac{4 - i}{1 + 2i}$

46. $\dfrac{9 - 2i}{-3 + i}$

47. $(5 + 2i)^2$

48. $(9 - 7i)^2$

Solve the following quadratic equations for complex solutions.

49. $(y - 4)^2 = -64$

50. $(x + 7)^2 = -1$

51. $4x^2 = -100$

52. $7x^2 = -28$

53. $z^2 + 6z + 10 = 0$

54. $z^2 + 4z + 13 = 0$

55. $2a^2 - 5a + 9 = 0$

56. $4a^2 + 3a + 2 = 0$

57. $(2x + 8)^2 = -20$

58. $(6z - 4)^2 = -24$

59. $3m^2 + 108 = 0$

60. $5m^2 + 80 = 0$

61. $x^2 + 14x + 50 = 0$

62. $x^2 + 8x + 25 = 0$

Answer the following true or false.

63. Every real number is a complex number.

64. Every complex number is a real number.

65. If a complex number such as $2 + 3i$ is a solution of a quadratic equation, then its conjugate $2 - 3i$ is also a solution.

66. Some imaginary numbers are real numbers.

REVIEW EXERCISES

Graph the following linear equations in two variables. See Section 3.2.

67. $y = -3$

68. $x = 4$

69. $y = 3x - 2$

70. $y = 2x + 3$

Find the length of the unknown side of each triangle.

△ **71.**

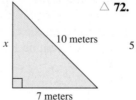

△ **72.**

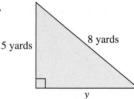

10.6 GRAPHING QUADRATIC EQUATIONS

CD-ROM SSM

SSG Video

▶ **OBJECTIVES**

1. Graph quadratic equations of the form $y = ax^2 + bx + c$.
2. Find the intercepts of a parabola.
3. Determine the vertex of a parabola.

---

1 Recall from Section 3.2 that the graph of a linear equation in two variables $Ax + By = C$ is a straight line. Also recall from Section 7.3 that the graph of a quadratic equation in two variables $y = ax^2 + bx + c$ is a parabola. In this section, we further investigate the graph of a quadratic equation.

To graph the quadratic equation $y = x^2$, select a few values for x and find the corresponding y-values. Make a table of values to keep track. Then plot the points corresponding to these solutions.

If $x = 0$, then $y = 0^2 = 0$.
If $x = -2$, then $y = (-2)^2 = 4$. And so on.

$y = x^2$

x	y
0	0
1	1
2	4
3	9
−1	1
−2	4
−3	9

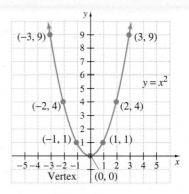

Clearly, these points are not on one straight line. As we saw in Chapter 7, the graph of $y = x^2$ is a smooth curve through the plotted points. This curve is called a **parabola**. The lowest point on a parabola opening upward is called the **vertex**. The vertex is $(0, 0)$ for the parabola $y = x^2$. If we fold the graph paper along the y-axis, the two pieces of the parabola match perfectly. For this reason, we say the graph is **symmetric about the y-axis**, and we call the y-axis the **axis of symmetry**.

Notice that the parabola that corresponds to the equation $y = x^2$ opens upward. This happens when the coefficient of x^2 is positive. In the equation $y = x^2$, the coefficient of x^2 is 1. Example 1 shows the graph of a quadratic equation whose coefficient of x^2 is negative.

Example 1   Graph $y = -2x^2$.

Solution   Select x-values and calculate the corresponding y-values. Plot the ordered pairs found. Then draw a smooth curve through those points. When the coefficient of x^2 is negative, the corresponding parabola opens downward. When a parabola opens downward, the vertex is the highest point of the parabola. The vertex of this parabola is $(0, 0)$ and the axis of symmetry is again the y-axis.

$y = -2x^2$

x	y
0	0
1	−2
2	−8
3	−18
−1	−2
−2	−8
−3	−18

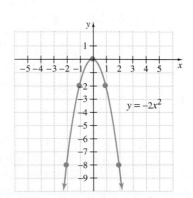

2 Just as for linear equations, we can use x- and y-intercepts to help graph quadratic equations. Recall from Chapter 3 that an x-intercept is the point where the graph intersects the x-axis. A y-intercept is the point where the graph intersects the y-axis.

> **HELPFUL HINT**
> Recall that:
> To find x-intercepts, let $y = 0$ and solve for x.
> To find y-intercepts, let $x = 0$ and solve for y.

Example 2 Graph $y = x^2 - 4$.

Solution First, find intercepts. To find the y-intercept, let $x = 0$. Then

$$y = 0^2 - 4 = -4$$

To find x-intercepts, we let $y = 0$.

$$0 = x^2 - 4$$
$$0 = (x - 2)(x + 2)$$
$$x - 2 = 0 \quad \text{or} \quad x + 2 = 0$$
$$x = 2 \qquad\qquad x = -2$$

Thus far, we have the y-intercept $(0, -4)$ and the x-intercepts $(2, 0)$ and $(-2, 0)$. Now we can select additional x-values, find the corresponding y-values, plot the points, and draw a smooth curve through the points.

$y = x^2 - 4$

x	y
0	−4
1	−3
2	0
3	5
−1	−3
−2	0
−3	5

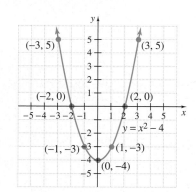

Notice that the vertex of this parabola is $(0, -4)$.

> **HELPFUL HINT**
> For the graph of $y = ax^2 + bx + c$,
> If a is positive, the parabola opens upward.
> If a is negative, the parabola opens downward.

3 Thus far, we have accidentally stumbled upon the vertex of each parabola that we have graphed. It would be helpful if we could first find the vertex of a parabola, next determine whether the parabola opens upward or downward, and finally calculate additional points such as x- and y-intercepts as needed. In fact, there is a formula that may be used to find the vertex of a parabola.

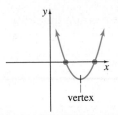

One way to develop this formula is to notice that the x-value of the vertex of the parabolas that we are considering lies halfway between its x-intercepts. We can use this fact to find a formula for the vertex.

Recall that the x-intercepts of a parabola may be found by solving $0 = ax^2 + bx + c$. These solutions, by the quadratic formula, are

$$x = \frac{-b - \sqrt{b^2 - 4ac}}{2a}, \, x = \frac{-b + \sqrt{b^2 - 4ac}}{2a}$$

The x-coordinate of the vertex of a parabola is halfway between its x-intercepts, so the x-value of the vertex may be found by computing the average, or $\frac{1}{2}$ of the sum of the intercepts.

$$x = \frac{1}{2}\left(\frac{-b - \sqrt{b^2 - 4ac}}{2a} + \frac{-b + \sqrt{b^2 - 4ac}}{2a} \right)$$

$$= \frac{1}{2}\left(\frac{-b - \sqrt{b^2 - 4ac} - b + \sqrt{b^2 - 4ac}}{2a} \right)$$

$$= \frac{1}{2}\left(\frac{-2b}{2a} \right)$$

$$= \frac{-b}{2a}$$

VERTEX FORMULA

The vertex of the parabola $y = ax^2 + bx + c$ has x-coordinate

$$\frac{-b}{2a}$$

The corresponding y-coordinate of the vertex is found by substituting the x-coordinate into the equation and evaluating y.

Example 3   Graph: $y = x^2 - 6x + 8$

Solution   In the equation $y = x^2 - 6x + 8$, $a = 1$ and $b = -6$. The x-coordinate of the vertex is

$$\frac{-b}{2a} = \frac{-(-6)}{2 \cdot 1} = 3 \qquad \text{Use the vertex formula, } \frac{-b}{2a}.$$

To find the corresponding y-coordinate, we let $x = 3$ in the original equation.

$$y = x^2 - 6x + 8 = 3^2 - 6 \cdot 3 + 8 = -1$$

The vertex is $(3, -1)$ and the parabola opens upward since a is positive. We now find and plot the intercepts.

To find the x-intercepts, we let $y = 0$.

$$0 = x^2 - 6x + 8$$

We factor the expression $x^2 - 6x + 8$ to find $(x - 4)(x - 2) = 0$. The x-intercepts are $(4, 0)$ and $(2, 0)$.

If we let $x = 0$ in the original equation, then $y = 8$ and the y-intercept is $(0, 8)$. Now we plot the vertex $(3, -1)$ and the intercepts $(4, 0)$, $(2, 0)$, and $(0, 8)$. Then we can sketch the parabola. These and two additional points are shown in the table.

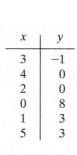

x	y
3	-1
4	0
2	0
0	8
1	3
5	3

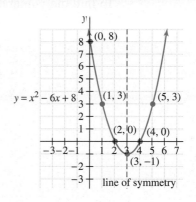

Example 4   Graph: $y = x^2 + 2x - 5$

Solution   In the equation $y = x^2 + 2x - 5$, $a = 1$ and $b = 2$. Using the vertex formula, we find that the x-coordinate of the vertex is

$$x = \frac{-b}{2a} = \frac{-2}{2 \cdot 1} = -1$$

The y-coordinate is

$$y = (-1)^2 + 2(-1) - 5 = -6$$

Thus the vertex is $(-1, -6)$.

To find the x-intercepts, we let $y = 0$.

$$0 = x^2 + 2x - 5$$

This cannot be solved by factoring, so we use the quadratic formula.

$$x = \frac{-2 \pm \sqrt{2^2 - 4(1)(-5)}}{2 \cdot 1} \qquad \text{Let } a = 1, b = 2, \text{ and } c = -5.$$

$$x = \frac{-2 \pm \sqrt{24}}{2}$$

$$x = \frac{-2 \pm 2\sqrt{6}}{2} \qquad \text{Simplify the radical.}$$

$$x = \frac{2(-1 \pm \sqrt{6})}{2} = -1 \pm \sqrt{6}$$

The x-intercepts are $(-1 + \sqrt{6}, 0)$ and $(-1 - \sqrt{6}, 0)$ We use a calculator to approximate these so that we can easily graph these intercepts.

$$-1 + \sqrt{6} \approx 1.4 \quad \text{and} \quad -1 - \sqrt{6} \approx -3.4$$

To find the y-intercept, we let $x = 0$ in the original equation and find that $y = -5$. Thus the y-intercept is $(0, -5)$.

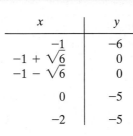

x	y
-1	-6
$-1 + \sqrt{6}$	0
$-1 - \sqrt{6}$	0
0	-5
-2	-5

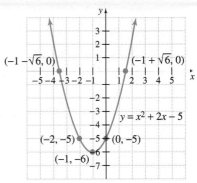

HELPFUL HINT

Notice that the number of x-intercepts of the graph of the parabola $y = ax^2 + bx + c$ is the same as the number of real solutions of $0 = ax^2 + bx + c$.

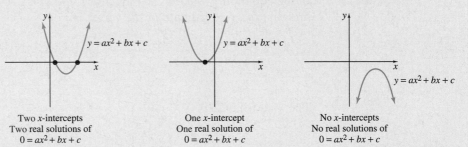

Two x-intercepts
Two real solutions of
$0 = ax^2 + bx + c$

One x-intercept
One real solution of
$0 = ax^2 + bx + c$

No x-intercepts
No real solutions of
$0 = ax^2 + bx + c$

GRAPHING CALCULATOR EXPLORATIONS

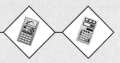

Recall that a graphing calculator may be used to solve quadratic equations. The x-intercepts of the graph of $y = ax^2 + bx + c$ are solutions of $0 = ax^2 + bx + c$. To solve $x^2 - 7x - 3 = 0$, for example, graph $y_1 = x^2 - 7x - 3$. The x-intercepts of the graph are the solutions of the equation.

Use a graphing calculator to solve each quadratic equation. Round solutions to two decimal places.

1. $x^2 - 7x - 3 = 0$
2. $2x^2 - 11x - 1 = 0$
3. $-1.7x^2 + 5.6x - 3.7 = 0$
4. $-5.8x^2 + 2.3x - 3.9 = 0$
5. $5.8x^2 - 2.6x - 1.9 = 0$
6. $7.5x^2 - 3.7x - 1.1 = 0$

SPOTLIGHT ON DECISION MAKING

Suppose you have some daffodil bulbs that you would like to plant in flower beds in your yard. It is recommended that daffodil bulbs be planted after the ground temperature has fallen below 50°F. You obtain the following graph of normal ground temperatures for your area from your county extension agent. When would you plant your daffodil bulbs? Explain.

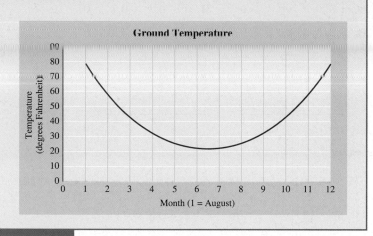

Exercise Set 10.6

Graph each quadratic equation by finding and plotting ordered pair solutions. See Example 1.

1. $y = 2x^2$ **2.** $y = 3x^2$

3. $y = -x^2$ **4.** $y = -4x^2$

5. $y = \dfrac{1}{3}x^2$ **6.** $y = -\dfrac{1}{2}x^2$

Sketch the graph of each equation. Identify the vertex and the intercepts. See Examples 2 through 4.

7. $y = x^2 - 1$ **8.** $y = x^2 - 16$

9. $y = x^2 + 4$ **10.** $y = x^2 + 9$

11. $y = x^2 + 6x$ **12.** $y = x^2 - 4x$

13. $y = x^2 + 2x - 8$ **14.** $y = x^2 - 2x - 3$

15. $y = -x^2 + x + 2$ **16.** $y = -x^2 - 2x - 1$

17. $y = x^2 + 5x + 4$ **18.** $y = x^2 + 7x + 10$

19. $y = -x^2 + 4x - 3$ **20.** $y = -x^2 + 6x - 8$

21. $y = x^2 + 2x - 2$ **22.** $y = x^2 - 4x - 3$

23. $y = x^2 - 3x + 1$ **24.** $y = x^2 - 2x - 5$

The graph of a quadratic equation that takes the form $y = ax^2 + bx + c$ is the graph of a function. Write the domain and the range of each of the functions graphed.

25.

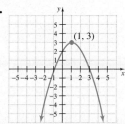

26.

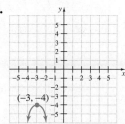

27.

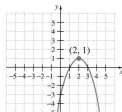

28.

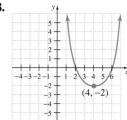

29. The height h of a fireball launched from a Roman candle with an initial velocity of 128 feet per second is given by the equation

$$h = -16t^2 + 128t$$

where t is time in seconds after launch.

Use the graph of this function to answer the questions.

a. Estimate the maximum height of the fireball.

b. Estimate the time when the fireball is at its maximum height.

c. Estimate the time when the fireball returns to the ground.

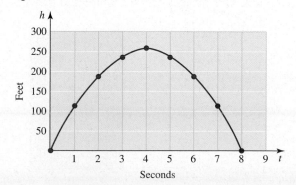

Match the values given with the correct graph of each quadratic equation of the form $y = a(x - h)^2 + k$.

30. $a > 0, h > 0, k > 0$ **31.** $a < 0, h > 0, k > 0$

32. $a > 0, h > 0, k < 0$ **33.** $a < 0, h > 0, k < 0$

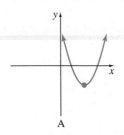

A

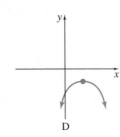

B

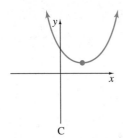

C

D

REVIEW EXERCISES

Simplify the following complex fractions. See Section 6.5.

34. $\dfrac{\dfrac{1}{7}}{\dfrac{2}{5}}$ **35.** $\dfrac{\dfrac{3}{8}}{\dfrac{1}{7}}$

36. $\dfrac{\dfrac{1}{x}}{\dfrac{2}{x^2}}$ **37.** $\dfrac{\dfrac{x}{5}}{\dfrac{2}{x}}$

38. $\dfrac{2x}{1 - \dfrac{1}{x}}$ **39.** $\dfrac{x}{x - \dfrac{1}{x}}$

40. $\dfrac{\dfrac{a - b}{2b}}{\dfrac{b - a}{8b^2}}$ **41.** $\dfrac{\dfrac{2a^2}{a - 3}}{\dfrac{a}{3 - a}}$

10 For additional Chapter Projects, visit the Real World Activities Website by going to http://www.prenhall.com/martin-gay.

CHAPTER PROJECT

Modeling a Physical Situation

When water comes out of a water fountain, it initially heads upward, but then gravity causes the water to fall. The curve formed by the stream of water can be modeled using a quadratic equation (parabola).

In this project, you will have the opportunity to model the parabolic path of water as it leaves a drinking fountain. This project may be completed by working in groups or individually.

1. Using Figure 2, collect data for the x-intercepts of the parabolic path. Let points A and B in Figure 2 be on the x-axis and let the coordinates of point A be $(0, 0)$. Use a ruler to measure the distance between points A and B **on Figure 2** to the nearest even one-tenth centimeter, and use this information to determine the coordinates of point B. Record this data in the data table. (*Hint:* If the distance from A to B measures 8 one-tenth centimeters, then the coordinates of point B are $(8, 0)$.)

DATA TABLE

	x	y
POINT A		
POINT B		
POINT V		

2. Next, collect data for the vertex V of the parabolic path. What is the relationship between the x-coordinate of the vertex and the x-intercepts found in Question 1? What is the line of symmetry? To locate point V in Figure 2, find the midpoint of the line segment joining points A and B and mark point V on the path of water directly above the midpoint. To approximate the y-coordinate of the vertex, use a ruler to measure its distance from the x-axis to the nearest one-tenth centimeter. Record this data in the data table.

3. Plot the points from the data table on a rectangular coordinate system. Sketch the parabola through your points A, B, and V.

4. Which of the following models best fits the data you collected? Explain your reasoning.

   A. $y = 16x + 18$
   B. $y = -13x^2 + 20x$
   C. $y = 0.13x^2 - 2.6x$
   D. $y = -0.13x^2 + 2.6x$

5. (Optional) Enter your data into a graphing calculator and use the quadratic curve-fitting feature to find a model for your data. How does the model compare with your selection from Question 4?

CHAPTER 10 VOCABULARY CHECK

Fill in each blank with one of the words listed below.

square root   complex   imaginary i
completing the square   quadratic   conjugate   vertex

1. If $x^2 = a$, then $x = \sqrt{a}$ or $x = -\sqrt{a}$. This property is called the _____ property.

2. A number that can be written in the form $a + bi$ is called a(n) _____ number.

3. The formula $\dfrac{-b}{2a}$ where $y = ax^2 + bx + c$ is called the _____ formula.

4. A complex number that can be written in the form $0 + bi$ is also called a(n) _____ number.

5. The _____ of $2 + 3i$ is $2 - 3i$.

6. $\sqrt{-1} =$

7. The process of solving a quadratic equation by writing it in the form $(x + a)^2 = c$ is called _____.

8. The formula $x = \dfrac{-b \pm \sqrt{b^2 - 4ac}}{2a}$ is called the _____ formula.

CHAPTER 10 HIGHLIGHTS

DEFINITIONS AND CONCEPTS	EXAMPLES

Section 10.1 Solving Quadratic Equations by the Square Root Method

Square root property
If $x^2 = a$ for $a \geq 0$, then $x = \pm\sqrt{a}$

Solve the equation.

$$(x - 1)^2 = 15$$

$$x - 1 = \pm\sqrt{15}$$

$$x = 1 \pm \sqrt{15}$$

Section 10.2 Solving Quadratic Equations by Completing the Square

To solve a quadratic equation by completing the square

Step 1. If the coefficient of x^2 is not 1, divide both sides of the equation by the coefficient.

Step 2. Isolate all terms with variables on one side.

Step 3. Complete the square by adding the square of half of the coefficient of x to both sides.

Step 4. Factor the perfect square trinomial.

Step 5. Apply the square root property to solve.

Solve $2x^2 + 12x - 10 = 0$ by completing the square.

$$\frac{2x^2}{2} + \frac{12x}{2} - \frac{10}{2} = \frac{0}{2} \quad \text{Divide by 2.}$$

$$x^2 + 6x - 5 = 0 \quad \text{Simplify.}$$

$$x^2 + 6x = 5 \quad \text{Add 5.}$$

The coefficient of x is 6. Half of 6 is 3 and $3^2 = 9$. Add 9 to both sides.

$$x^2 + 6x + 9 = 5 + 9$$

$$(x + 3)^2 = 14 \quad \text{Factor.}$$

$$x + 3 = \pm\sqrt{14}$$

$$x = -3 \pm \sqrt{14}$$

Section 10.3 Solving Quadratic Equations by the Quadratic Formula

Quadratic formula
If a, b, and c are real numbers and $a \neq 0$, the quadratic equation $ax^2 + bx + c = 0$ has solutions

$$x = \frac{-b \pm \sqrt{b^2 - 4ac}}{2a}$$

To solve a quadratic equation by the quadratic formula

Step 1. Write the equation in standard form:
$ax^2 + bx + c = 0$.

Step 2. If necessary, clear the equation of fractions.

Step 3. Identify a, b, and c.

Step 4. Replace a, b, and c in the quadratic formula by known values, and simplify.

Identify a, b, and c in the quadratic equation

$$4x^2 - 6x = 5$$

First, subtract 5 from both sides.

$$4x^2 - 6x - 5 = 0$$

$$a = 4, \quad b = -6, \quad \text{and} \quad c = -5$$

Solve: $3x^2 - 2x - 2 = 0$
In this equation, $a = 3$, $b = -2$, and $c = -2$.

$$x = \frac{-(-2) \pm \sqrt{(-2)^2 - 4(3)(-2)}}{2 \cdot 3}$$

$$= \frac{2 \pm \sqrt{4 - (-24)}}{6}$$

$$= \frac{2 \pm \sqrt{28}}{6} = \frac{2 \pm \sqrt{4 \cdot 7}}{6} = \frac{2 \pm 2\sqrt{7}}{6}$$

$$= \frac{2(1 \pm \sqrt{7})}{2 \cdot 3} = \frac{1 \pm \sqrt{7}}{3}$$

(continued)

DEFINITIONS AND CONCEPTS	EXAMPLES

Section 10.4 Summary of Methods for Solving Quadratic Equations and Problem Solving

To solve a quadratic equation

Step 1. If the equation is in the form $(ax + b)^2 = c$, use the square root property and solve. If not, go to step 2.

Step 2. Write the equation in standard form:
$ax^2 + bx + c = 0$.

Step 3. Try to solve by factoring. If not, go to step 4.

Step 4. Solve by the quadratic formula.

Solve $(3x - 1)^2 = 10$

$3x - 1 = \pm\sqrt{10}$ Square root property.

$3x = 1 \pm \sqrt{10}$ Add 1.

$x = \dfrac{1 \pm \sqrt{10}}{3}$ Divide by 3.

Solve $x(2x + 9) = 5$.
$$2x^2 + 9x - 5 = 0$$
$$(2x - 1)(x + 5) = 0$$
$$2x - 1 = 0 \quad \text{or} \quad x + 5 = 0$$
$$2x = 1 \qquad\qquad x = -5$$
$$x = \frac{1}{2}$$

Section 10.5 Complex Solutions of Quadratic Equations

The **imaginary unit**, written i, is the number whose square is -1. That is,

$$i^2 = -1 \quad \text{and} \quad i = \sqrt{-1}$$

A **complex number** is a number that can be written in the form
$$a + bi$$
where a and b are real numbers. A complex number that can be written in the form $0 + bi$, $b \neq 0$, is also called an **imaginary number**.

Complex numbers are added and subtracted in the same way as polynomials are added and subtracted.

Use the distributive property to multiply complex numbers.

The complex numbers $a + bi$ and $a - bi$ are called **complex conjugates**.

To write a quotient of complex numbers in standard form $a + bi$, multiply numerator and denominator by the denominator's conjugate.

Write $\sqrt{-10}$ as the product of a real number and i.
$$\sqrt{-10} = \sqrt{-1 \cdot 10} = \sqrt{-1} \cdot \sqrt{10} = i\sqrt{10}$$

Identify each number as a complex number by writing it in **standard form** $a + bi$.
$$7 = 7 + 0i$$
$$\sqrt{-5} = 0 + i\sqrt{5} \quad \text{Also an imaginary number}$$
$$1 - \sqrt{-9} = 1 - 3i$$

Simplify the sum or difference.
$$(2 + 3i) - (1 - 6i) = 2 + 3i - 1 + 6i$$
$$= 1 + 9i$$
$$2i + (5 - 3i) = 2i + 5 - 3i$$
$$= 5 - i$$

Multiply: $(4 - i)(-2 + 3i)$
$$= -8 + 12i + 2i - 3i^2$$
$$= -8 + 14i + 3$$
$$= -5 + 14i$$

The conjugate of $(5 - 6i)$ is $(5 + 6i)$.

Write $\dfrac{3 - 2i}{2 + 1}$ in standard form.

$$\frac{(3 - 2i)}{(2 + i)} \cdot \frac{(2 - i)}{(2 - i)} = \frac{6 - 7i + 2i^2}{4 - i^2}$$

$$= \frac{6 - 7i + 2(-1)}{4 - (-1)}$$

$$= \frac{4 - 7i}{5} \quad \text{or} \quad \frac{4}{5} - \frac{7}{5}i$$

(continued)

DEFINITIONS AND CONCEPTS	EXAMPLES

Section 10.5 Complex Solutions of Quadratic Equations

Some quadratic equations have complex solutions.

Solve $x^2 - 3x = -3$

$$x^2 - 3x + 3 = 0$$

$$a = 1, b = -3, c = 3$$

$$x = \frac{-(-3) \pm \sqrt{(-3)^2 - 4(1)(3)}}{2(1)}$$

$$= \frac{3 \pm \sqrt{-3}}{2} = \frac{3 \pm i\sqrt{3}}{2}$$

Section 10.6 Graphing Quadratic Equations

The graph of a quadratic equation $y = ax^2 + bx + c, a \neq 0$, is called a **parabola**. The lowest point on a parabola opening upward or the highest point on a parabola opening downward is called the **vertex**. The vertical line through the vertex is the **axis of symmetry**.

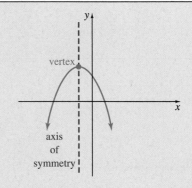

The vertex of the parabola $y = ax^2 + bx + c$ has x-value $\frac{-b}{2a}$.

Graph $y = 2x^2 - 6x + 4$.

The x-value of the vertex is

$$x = \frac{-b}{2a} = \frac{-(-6)}{2(2)} = \frac{6}{4} = \frac{3}{2}$$

The y-value is

$$y = 2\left(\frac{3}{2}\right)^2 - 6\left(\frac{3}{2}\right) + 4 = -\frac{1}{2}$$

The vertex is $\left(\frac{3}{2}, -\frac{1}{2}\right)$.

The y-intercept is

$$y = 2 \cdot 0^2 - 6 \cdot 0 + 4 = 4$$

The x-intercepts are found by

$$0 = 2x^2 - 6x + 4$$

$$0 = 2(x^2 - 3x + 2)$$

$$0 = 2(x - 2)(x - 1)$$

$$x - 2 = 0 \quad \text{or} \quad x - 1 = 0$$

$$x = 2 \quad \text{or} \quad x = 1$$

DEFINITIONS AND CONCEPTS	EXAMPLES
Section 10.6 Graphing Quadratic Equations	

Find more ordered pair solutions as needed.

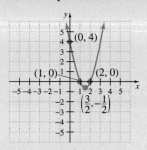

CHAPTER 10 REVIEW

(10.1) *Solve each quadratic equation by factoring or using the square root property.*

1. $(x - 4)(5x + 3) = 0$

2. $(x + 7)(3x + 4) = 0$

3. $3m^2 - 5m = 2$

4. $7m^2 + 2m = 5$

5. $k^2 = 50$

6. $k^2 = 45$

7. $(x - 5)(x - 1) = 12$

8. $(x - 3)(x + 2) = 6$

9. $(x - 11)^2 = 49$

10. $(x + 3)^2 = 100$

11. $6x^3 - 54x = 0$

12. $2x^2 - 8 = 0$

13. $(4p + 2)^2 = 100$

14. $(3p + 6)^2 = 81$

(10.2) *Complete the square for the following expressions and then factor the resulting perfect square trinomial.*

15. $x^2 - 10x$

16. $x^2 + 16x$

17. $a^2 + 4a$

18. $a^2 - 12a$

19. $m^2 - 3m$

20. $m^2 + 5m$

Solve each quadratic equation by completing the square.

21. $x^2 - 6x + 7 = 0$

22. $x^2 + 6x + 7 = 0$

23. $2y^2 + y - 1 = 0$

24. $y^2 + 3y - 1 = 0$

(10.3) *Solve each quadratic equation by using the quadratic formula.*

25. $x^2 - 10x + 7 = 0$

26. $x^2 + 4x - 7 = 0$

27. $2x^2 + x - 1 = 0$

28. $x^2 + 3x - 1 = 0$

29. $9x^2 + 30x + 25 = 0$

30. $16x^2 - 72x + 81 = 0$

31. $15x^2 + 2 = 11x$

32. $15x^2 + 2 = 13x$

33. $2x^2 + x + 5 = 0$

34. $7x^2 - 3x + 1 = 0$

Use the discriminant to determine the number of solutions of each quadratic equation.

35. $x^2 - 7x - 1 = 0$

36. $x^2 + x + 5 = 0$

37. $9x^2 + 1 = 6x$

38. $x^2 + 6x = 5$

39. $5x^2 + 4 = 0$

40. $x^2 + 25 = 10x$

(10.4) *Solve the following equations by using the most appropriate method.*

41. $5z^2 + z - 1 = 0$

42. $4z^2 + 7z - 1 = 0$

43. $4x^4 = x^2$

44. $9x^3 = x$

45. $2x^2 - 15x + 7 = 0$

46. $x^2 - 6x - 7 = 0$

47. $(3x - 1)^2 = 0$

48. $(2x - 3)^2 = 0$

49. $x^2 = 6x - 9$

50. $x^2 = 10x - 25$

51. $\left(\dfrac{1}{2}x - 3\right)^2 = 64$

52. $\left(\dfrac{1}{3}x + 1\right)^2 = 49$

53. $x^2 - 0.3x + 0.01 = 0$

54. $x^2 + 0.6x - 0.16 = 0$

55. $\frac{1}{10}x^2 + x - \frac{1}{2} = 0$ **56.** $\frac{1}{12}x^2 - \frac{1}{2}x + \frac{1}{3} = 0$

Solve. For Exercises 57–58, use the formula $h = 16t^2$.

57. If Kara Washington dives from a height of 100 feet, how long before she hits the water?

58. How long does a 5-mile free-fall take? Round your result to the nearest tenth of a second.

59. The average price of silver (in cents per ounce) from 1996 to 1998 is given by the equation $y = 25x^2 - 54x + 519$. In this equation, x is the number of years since 1996. Assume that this trend continues and find the year after 1996 in which the price of silver will be 1687 cents per ounce. (*Source:* U.S. Bureau of Mines)

60. The average price of platinum (in dollars per ounce) from 1996 to 1998 is given by the equation $y = 5x^2 - 6x + 398$. In this equation, x is the number of years since 1996. Assume that this trend continues and find the year after 1996 in which the price of platinum will be 670 dollars per ounce. (*Source:* U.S. Bureau of Mines)

(10.5) Perform the indicated operations. Write the resulting complex number in standard form.

61. $\sqrt{-144}$ **62.** $\sqrt{-36}$

63. $\sqrt{-108}$ **64.** $\sqrt{-500}$

65. $(7 - i) + (14 - 9i)$ **66.** $(10 - 4i) + (9 - 21i)$

67. $3 - (11 + 2i)$ **68.** $(-4 - 3i) + 5i$

69. $(2 - 3i)(3 - 2i)$ **70.** $(2 + 5i)(5 - i)$

71. $(3 - 4i)(3 + 4i)$ **72.** $(7 - 2i)(7 - 2i)$

73. $\frac{2 - 6i}{4i}$ **74.** $\frac{5 - i}{2i}$

75. $\frac{4 - i}{1 + 2i}$ **76.** $\frac{1 + 3i}{2 - 7i}$

Solve each quadratic equation.

77. $3x^2 = -48$ **78.** $5x^2 = -125$

79. $x^2 - 4x + 13 = 0$

80. $x^2 + 4x + 11 = 0$

(10.6) Identify the vertex, axis of symmetry, and whether the parabola opens upward or downward for the given quadratic equation.

81. $y = -3x^2$ **82.** $y = -\frac{1}{2}x^2$

83. $y = (x - 3)^2$ **84.** $y = (x - 5)^2$

85. $y = 3x^2 - 7$ **86.** $y = -2x^2 + 25$

87. $y = -5(x - 72)^2 + 14$ **88.** $y = 2(x - 35)^2 - 21$

Graph the following quadratic equations. Label the vertex and the intercepts with their coordinates.

89. $y = -x^2$ **90.** $y = 4x^2$

91. $y = \frac{1}{2}x^2$ **92.** $y = \frac{1}{4}x^2$

93. $y = x^2 + 5x + 6$ **94.** $y = x^2 - 4x - 8$

95. $y = 2x^2 - 11x - 6$ **96.** $y = 3x^2 - x - 2$

Quadratic equations $y = ax^2 + bx + c$ are graphed below. Determine the number of real solutions for the related equation $0 = ax^2 + bx + c$ from each graph. List the solutions.

97.

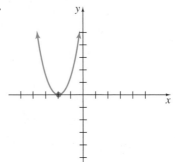

98.

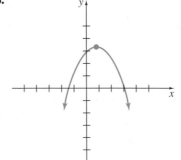

99.

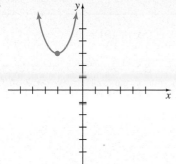

100.

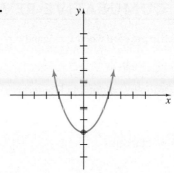

CHAPTER 10 TEST

Solve by factoring.

1. $2x^2 - 11x = 21$ **2.** $x^4 + x^3 - 2x^2 = 0$

Solve using the square root property.

3. $5k^2 = 80$ **4.** $(3m - 5)^2 = 8$

Solve by completing the square.

5. $x^2 - 26x + 160 = 0$ **6.** $5x^2 + 9x = 2$

Solve using the quadratic formula.

7. $x^2 - 3x - 10 = 0$ **8.** $p^2 - \dfrac{5}{3}p - \dfrac{1}{3} = 0$

Solve by the most appropriate method.

9. $(3x - 5)(x + 2) = -6$ **10.** $(3x - 1)^2 = 16$

11. $3x^2 - 7x - 2 = 0$ **12.** $x^2 - 4x + 5 = 0$

13. $3x^2 - 7x + 2 = 0$ **14.** $2x^2 - 6x + 1 = 0$

15. $2x^5 + 5x^4 - 3x^3 = 0$ **16.** $9x^3 = x$

Perform the indicated operations. Write the resulting complex number in standard form.

17. $\sqrt{-25}$ **18.** $\sqrt{-200}$

19. $(3 + 2i) + (5 - i)$ **20.** $(4 - i) - (-3 + 5i)$

21. $(3 + 2i) - (3 - 2i)$ **22.** $(3 + 2i) + (3 - 2i)$

23. $(3 + 2i)(3 - 2i)$ **24.** $\dfrac{3 - i}{1 + 2i}$

Graph the quadratic equations. Label the vertex and the intercept points with their coordinates.

25. $y = -3x^2$ **26.** $y = \dfrac{1}{4}x^2$

27. $y = x^2 - 7x + 10$

Solve.

28. The highest dive from a diving board by a woman was made by Lucy Wardle of the United States. She dove from a height of 120.75 feet at Ocean Park, Hong Kong, in 1985. To the nearest tenth of a second, how long did the dive take? Use the formula $h = 16t^2$.

29. The value of Washington State's mineral production y (in millions of dollars) from 1995 through 1997 is given by the equation $y = 28x^2 + 555$. In this equation, $x = 0$ represents the year 1996. Assume that this trend continues and find the year in which the value of mineral production is $1003 million. (*Source:* U.S. Bureau of Mines)

CHAPTER 10 CUMULATIVE REVIEW

1. Find the value of each expression when $x = 2$ and $y = -5$.

   a. $\dfrac{x - y}{12 + x}$ **b.** $x^2 - y$

2. Simplify each expression by combining like terms.

   a. $2x + 3x + 5 + 2$ **b.** $-5a - 3 + a + 2$

   c. $4y - 3y^2$ **d.** $2.3x + 5x - 6$

3. Identify the x- and y- intercepts.

a.

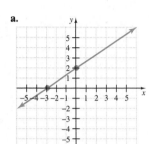

b.

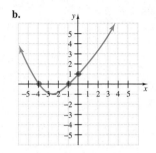

c.

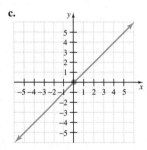

d.

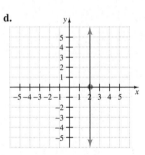

e.

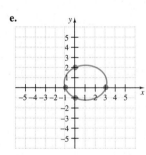

4. Simplify the following.

   a. $\left(\dfrac{-5x^2}{y^3}\right)^2$ **b.** $\dfrac{(x^3)^4 x}{x^7}$

   c. $\dfrac{(2x)^5}{x^3}$ **d.** $\dfrac{(a^2 b)^3}{a^3 b^2}$

5. Solve: $(5x - 1)(2x^2 + 15x + 18) = 0$

6. Solve for x: $\dfrac{x - 5}{3} = \dfrac{x + 2}{5}$

7. Determine whether the graphs of $y = -\dfrac{1}{5}x + 1$ and $2x + 10y = 30$ are parallel lines, perpendicular lines, or neither.

8. Graph the equation $y = |x| - 3$.

9. Solve the following system of equations by graphing.

$$\begin{cases} 2x + y = 7 \\ 2y = -4x \end{cases}$$

10. Solve the system

$$\begin{cases} 7x - 3y = -14 \\ -3x + y = 6 \end{cases}$$

11. Solve the system:

$$\begin{cases} 3x - 2y = 2 \\ -9x + 6y = -6 \end{cases}$$

12. Albert and Louis live 15 miles away from each other. They decide to meet one day by walking toward one another. After 2 hours they meet. If Louis walks one mile per hour faster than Albert, find both walking speeds.

13. Graph the solution of the system

$$\begin{cases} -3x + 4y < 12 \\ x \geq 2 \end{cases}$$

14. Simplify the following expressions.

   a. $\sqrt[4]{16}$ **b.** $\sqrt[5]{-32}$

   c. $-\sqrt[3]{8}$ **d.** $\sqrt[4]{-81}$

15. Simplify the following expressions.

   a. $\sqrt{\dfrac{25}{36}}$ **b.** $\sqrt{\dfrac{3}{64}}$

   c. $\sqrt{\dfrac{4}{81}}$

16. Add or subtract by first simplifying each radical.

   a. $\sqrt{50} + \sqrt{8}$ **b.** $7\sqrt{12} - \sqrt{75}$

   c. $\sqrt{25} - \sqrt{27} - 2\sqrt{18} - \sqrt{16}$

17. Multiply. Then simplify if possible.

   a. $\sqrt{7} \cdot \sqrt{3}$ **b.** $\sqrt{3} \cdot \sqrt{15}$

   c. $2\sqrt{6} \cdot 5\sqrt{2}$ **d.** $(3\sqrt{2})^2$

18. Solve: $\sqrt{x} = \sqrt{5x - 2}$

19. A surveyor must determine the distance across a lake at points P and Q as shown in the figure. To do this, she finds a third point R perpendicular to line PQ. If the length of $\overline{PR}$ is 320 feet and the length of $\overline{QR}$ is 240 feet, what is the distance across the lake? Approximate this distance to the nearest whole foot.

20. Write in radical notation. Then simplify.

    a. $25^{1/2}$ **b.** $8^{1/3}$

    c. $16^{1/4}$ **d.** $(-27)^{1/3}$

    e. $\left(\dfrac{1}{9}\right)^{1/2}$

21. Use the square root property to solve $2x^2 = 7$.

22. Solve $x^2 - 10x = -14$ by completing the square.

23. Solve $2x^2 - 9x = 5$ using the quadratic formula.

24. Write each radical as the product of a real number and i.

    a. $\sqrt{-4}$ **b.** $\sqrt{-11}$

    c. $\sqrt{-20}$

25. Graph $y = x^2 - 4$.

OPERATIONS ON DECIMALS

To **add** or **subtract** decimals, write the numbers vertically with decimal points lined up. Add or subtract as with whole numbers and place the decimal point in the answer directly below the decimal points in the problem.

Example 1   Add $5.87 + 23.279 + 0.003$.

Solution
$$
\begin{array}{r}
5.87 \\
23.279 \\
+\,0.003 \\
\hline
29.152
\end{array}
$$

Example 2   Subtract $32.15 - 11.237$.

Solution
$$
\begin{array}{r}
3 \quad \overset{1}{\cancel{2}} \, . \, \overset{11}{\cancel{1}} \quad \overset{4}{\cancel{5}} \quad \overset{10}{\cancel{0}} \\
-\; 1 \quad 1 \, . \, 2 \quad 3 \quad 7 \\
\hline
2 \quad 0 \, . \, 9 \quad 1 \quad 3
\end{array}
$$

To **multiply** decimals, multiply the numbers as if they were whole numbers. The decimal point in the product is placed so that the number of decimal places in the product is the same as the sum of the number of decimal places in the factors.

Example 3   Multiply 0.072×3.5.

Solution
$$
\begin{array}{r}
0.072 \\
\times \quad 3.5 \\
\hline
360 \\
216 \quad\;\; \\
\hline
0.2520
\end{array}
$$

       0.072   *3 decimal places*
       ×   3.5   *1 decimal place*
       0.2520   *4 decimal places*

To **divide** decimals, move the decimal point in the divisor to the right of the last digit. Move the decimal point in the dividend the same number of places that the decimal point in the divisor was moved. The decimal point in the quotient lies directly above the decimal point in the dividend.

Example 4 Divide $9.46 \div 0.04$.

Solution

$$
\begin{array}{r}
236.5 \\
04.\overline{)946.0} \\
-8 \\
\hline
14 \\
-12 \\
\hline
26 \\
-24 \\
\hline
20 \\
-20 \\
\hline
\end{array}
$$

Appendix A Exercise Set

Perform the indicated operations.

1. $9.076 + 8.004$

2. $\begin{array}{r} 6.3 \\ \times\, 0.05 \\ \hline \end{array}$

3. $\begin{array}{r} 27.004 \\ -14.2 \\ \hline \end{array}$

4. $\begin{array}{r} 0.0036 \\ 7.12 \\ 32.502 \\ +0.05 \\ \hline \end{array}$

5. $\begin{array}{r} 107.92 \\ +3.04 \\ \hline \end{array}$

6. $7.2 \div 4$

7. $10 - 7.6$

8. $40 \div 0.25$

9. $126.32 - 97.89$

10. $\begin{array}{r} 3.62 \\ 7.11 \\ 12.36 \\ 4.15 \\ +2.29 \\ \hline \end{array}$

11. $\begin{array}{r} 3.25 \\ \times\, 70 \\ \hline \end{array}$

12. $\begin{array}{r} 26.014 \\ -\, 7.8 \\ \hline \end{array}$

13. $8.1 \div 3$

14. $\begin{array}{r} 1.2366 \\ 0.005 \\ 15.17 \\ +\, 0.97 \\ \hline \end{array}$

15. $55.405 - 6.1711$

16. $8.09 + 0.22$

17. $60 \div 0.75$

18. $20 - 12.29$

19. $7.612 \div 100$

20. $\begin{array}{r} 8.72 \\ 1.12 \\ 14.86 \\ 3.98 \\ +\, 1.99 \\ \hline \end{array}$

21. $12.312 \div 2.7$

22. $0.443 \div 100$

23. $\begin{array}{r} 569.2 \\ 71.25 \\ +\, 8.01 \\ \hline \end{array}$

24. $3.706 - 2.91$

25. $768 - 0.17$

26. $63 \div 0.28$

27. $12 + 0.062$

28. $0.42 + 18$

29. $76 - 14.52$

30. $1.1092 \div 0.47$

31. $3.311 \div 0.43$

32. $7.61 + 0.0004$

33. $\begin{array}{r} 762.12 \\ 89.7 \\ +\, 11.55 \\ \hline \end{array}$

34. $444 \div 0.6$

35. $23.4 - 0.821$

36. $3.7 + 5.6$

37. $476.12 - 112.97$

38. $19.872 \div 0.54$

39. $0.007 + 7$

40. $\begin{array}{r} 51.77 \\ +\, 3.6 \\ \hline \end{array}$

REVIEW OF ANGLES, LINES, AND SPECIAL TRIANGLES

The word **geometry** is formed from the Greek words, **geo**, meaning earth, and **metron**, meaning measure. Geometry literally means to measure the earth.

This section contains a review of some basic geometric ideas. It will be assumed that fundamental ideas of geometry such as point, line, ray, and angle are known. In this appendix, the notation ∠1 is read "angle 1" and the notation $m\angle 1$ is read "the measure of angle 1."

We first review types of angles.

ANGLES

A **right angle** is an angle whose measure is 90°. A right angle can be indicated by a square drawn at the vertex of the angle, as shown below.
An angle whose measure is more than 0° but less than 90° is called an **acute angle**.
An angle whose measure is greater than 90° but less than 180° is called an **obtuse angle**.
An angle whose measure is 180° is called a **straight angle**.
Two angles are said to be **complementary** if the sum of their measures is 90°. Each angle is called the **complement** of the other.
Two angles are said to be **supplementary** if the sum of their measures is 180°. Each angle is called the **supplement** of the other.

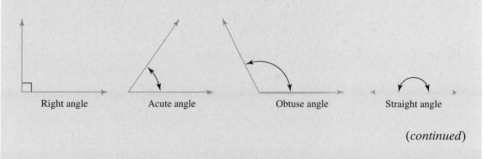

| Right angle | Acute angle | Obtuse angle | Straight angle |

(continued)

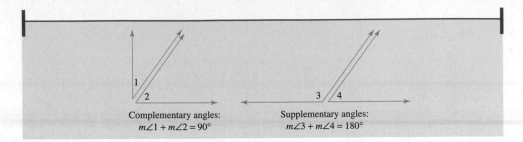

Complementary angles:
$m\angle 1 + m\angle 2 = 90°$

Supplementary angles:
$m\angle 3 + m\angle 4 = 180°$

Example 1   If an angle measures 28°, find its complement.

Solution   Two angles are complementary if the sum of their measures is 90°. The complement of a 28° angle is an angle whose measure is $90° - 28° = 62°$. To check, notice that $28° + 62° = 90°$.

Plane is an undefined term that we will describe. A plane can be thought of as a flat surface with infinite length and width, but no thickness. A plane is two dimensional. The arrows in the following diagram indicate that a plane extends indefinitely and has no boundaries.

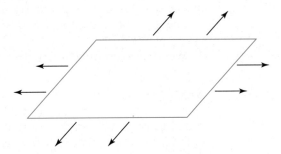

Figures that lie on a plane are called **plane figures**. (See the description of common plane figures in Appendix C.) Lines that lie in the same plane are called **coplanar**.

LINES

Two lines are **parallel** if they lie in the same plane but never meet.
Intersecting lines meet or cross in one point.
Two lines that form right angles when they intersect are said to be **perpendicular**.

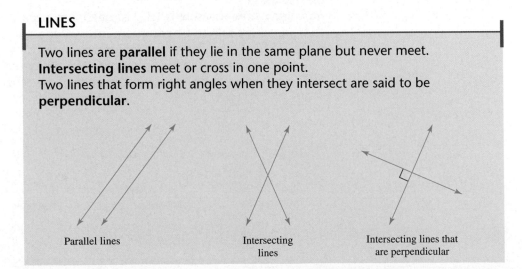

Parallel lines

Intersecting lines

Intersecting lines that are perpendicular

Two intersecting lines form **vertical angles**. Angles 1 and 3 are vertical angles. Also angles 2 and 4 are vertical angles. It can be shown that **vertical angles have equal measures**.

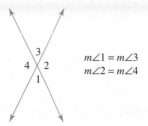

$$m\angle 1 = m\angle 3$$
$$m\angle 2 = m\angle 4$$

Adjacent angles have the same vertex and share a side. Angles 1 and 2 are adjacent angles. Other pairs of adjacent angles are angles 2 and 3, angles 3 and 4, and angles 4 and 1.

A **transversal** is a line that intersects two or more lines in the same plane. Line l is a transversal that intersects lines m and n. The eight angles formed are numbered and certain pairs of these angles are given special names.

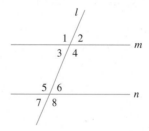

Corresponding angles: $\angle 1$ and $\angle 5$, $\angle 3$ and $\angle 7$, $\angle 2$ and $\angle 6$, and $\angle 4$ and $\angle 8$.

Exterior angles: $\angle 1, \angle 2, \angle 7$, and $\angle 8$.

Interior angles: $\angle 3, \angle 4, \angle 5$, and $\angle 6$.

Alternate interior angles: $\angle 3$ and $\angle 6$, $\angle 4$ and $\angle 5$.

These angles and parallel lines are related in the following manner.

PARALLEL LINES CUT BY A TRANSVERSAL

1. If two parallel lines are cut by a transversal, then
   a. **corresponding angles are equal** and
   b. **alternate interior angles are equal.**

2. If corresponding angles formed by two lines and a transversal are equal, then the lines are parallel.

3. If alternate interior angles formed by two lines and a transversal are equal, then the lines are parallel.

Example 2 Given that lines *m* and *n* are parallel and that the measure of angle 1 is 100°, find the measures of angles 2, 3, and 4.

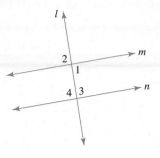

Solution $m\angle 2 = 100°$, since angles 1 and 2 are vertical angles.
$m\angle 4 = 100°$, since angles 1 and 4 are alternate interior angles.
$m\angle 3 = 180° - 100° = 80°$, since angles 4 and 3 are supplementary angles.

A **polygon** is the union of three or more coplanar line segments that intersect each other only at each end point, with each end point shared by exactly two segments.
A **triangle** is a polygon with three sides. The sum of the measures of the three angles of a triangle is 180°. In the following figure, $m\angle 1 + m\angle 2 + m\angle 3 = 180°$.

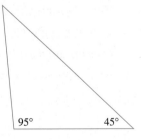

Example 3 Find the measure of the third angle of the triangle shown.

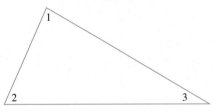

Solution The sum of the measures of the angles of a triangle is 180°. Since one angle measures 45° and the other angle measures 95°, the third angle measures $180° - 45° - 95° = 40°$.

Two triangles are **congruent** if they have the same size and the same shape. In congruent triangles, the measures of corresponding angles are equal and the lengths of corresponding sides are equal. The following triangles are congruent.

Corresponding angles are equal: $m\angle 1 = m\angle 4$, $m\angle 2 = m\angle 5$, and $m\angle 3 = m\angle 6$. Also, lengths of corresponding sides are equal: $a = x$, $b = y$, and $c = z$.

Any one of the following may be used to determine whether two triangles are congruent.

CONGRUENT TRIANGLES

1. If the measures of two angles of a triangle equal the measures of two angles of another triangle and the lengths of the sides between each pair of angles are equal, the triangles are congruent.

$m\angle 1 = m\angle 3$
$m\angle 2 = m\angle 4$
and
$a = x$

2. If the lengths of the three sides of a triangle equal the lengths of corresponding sides of another triangle, the triangles are congruent.

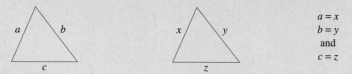

$a = x$
$b = y$
and
$c = z$

3. If the lengths of two sides of a triangle equal the lengths of corresponding sides of another triangle, and the measures of the angles between each pair of sides are equal, the triangles are congruent.

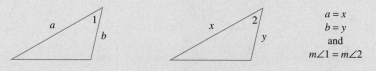

$a = x$
$b = y$
and
$m\angle 1 = m\angle 2$

Two triangles are **similar** if they have the same shape. In similar triangles, the measures of corresponding angles are equal and corresponding sides are in proportion. The following triangles are similar. (All similar triangles drawn in this appendix will be oriented the same.)

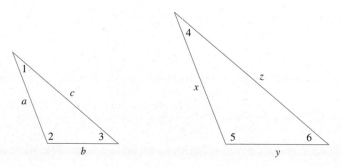

Corresponding angles are equal: $m\angle 1 = m\angle 4$, $m\angle 2 = m\angle 5$, and $m\angle 3 = m\angle 6$. Also, corresponding sides are proportional: $\dfrac{a}{x} = \dfrac{b}{y} = \dfrac{c}{z}$.

Any one of the following may be used to determine whether two triangles are similar.

SIMILAR TRIANGLES

1. If the measures of two angles of a triangle equal the measures of two angles of another triangle, the triangles are similar.

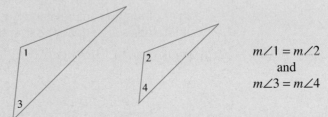

$$m\angle 1 = m\angle 2$$
and
$$m\angle 3 = m\angle 4$$

2. If three sides of one triangle are proportional to three sides of another triangle, the triangles are similar.

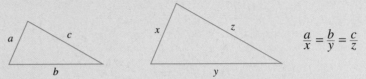

$$\frac{a}{x} = \frac{b}{y} = \frac{c}{z}$$

3. If two sides of a triangle are proportional to two sides of another triangle and the measures of the included angles are equal, the triangles are similar.

$$m\angle 1 = m\angle 2$$
and
$$\frac{a}{x} = \frac{b}{y}$$

Example 4   Given that the following triangles are similar, find the missing length x.

Solution   Since the triangles are similar, corresponding sides are in proportion. Thus, $\frac{2}{3} = \frac{10}{x}$. To solve this equation for x, we multiply both sides by the LCD, $3x$.

$$3x\left(\frac{2}{3}\right) = 3x\left(\frac{10}{x}\right)$$
$$2x = 30$$
$$x = 15$$

The missing length is 15 units.

A **right triangle** contains a right angle. The side opposite the right angle is called the **hypotenuse**, and the other two sides are called the **legs**. The **Pythagorean theorem** gives a formula that relates the lengths of the three sides of a right triangle.

THE PYTHAGOREAN THEOREM

If a and b are the lengths of the legs of a right triangle, and c is the length of the hypotenuse, then $a^2 + b^2 = c^2$.

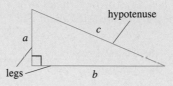

Example 5   Find the length of the hypotenuse of a right triangle whose legs have lengths of 3 centimeters and 4 centimeters.

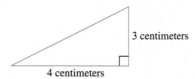

Solution   Because we have a right triangle, we use the Pythagorean theorem. The legs are 3 centimeters and 4 centimeters, so let $a = 3$ and $b = 4$ in the formula.

$$a^2 + b^2 = c^2$$
$$3^2 + 4^2 = c^2$$
$$9 + 16 = c^2$$
$$25 = c^2$$

Since c represents a length, we assume that c is positive. Thus, if c^2 is 25, c must be 5. The hypotenuse has a length of 5 centimeters.

Appendix B Exercise Set

Find the complement of each angle. See Example 1.

1. $19°$

2. $65°$

3. $70.8°$

4. $45\frac{2}{3}°$

5. $11\frac{1}{4}°$

6. $19.6°$

Find the supplement of each angle.

7. $150°$

8. $90°$

9. $30.2°$

10. $81.9°$

11. $79\frac{1}{2}°$

12. $165\frac{8}{9}°$

13. If lines m and n are parallel, find the measures of angles 1 through 7. See Example 2.

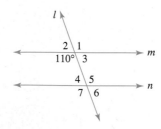

14. If lines *m* and *n* are parallel, find the measures of angles 1 through 5. See Example 2.

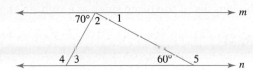

30.

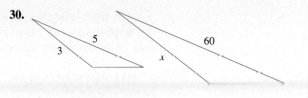

In each of the following, the measures of two angles of a triangle are given. Find the measure of the third angle. See Example 3.

15. 11°, 79°   **16.** 8°, 102°

17. 25°, 65°   **18.** 44°, 19°

19. 30°, 60°   **20.** 67°, 23°

In each of the following, the measure of one angle of a right triangle is given. Find the measures of the other two angles.

21. 45°   **22.** 60°

23. 17°   **24.** 30°

25. $39\frac{3}{4}°$ **26.** 72.6°

Given that each of the following pairs of triangles is similar, find the missing lengths. See Example 4.

27.

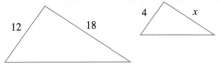

28.

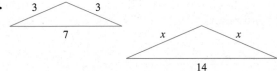

29.

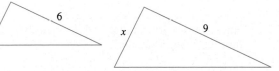

Use the Pythagorean theorem to find the missing lengths in the right triangles. See Example 5.

31.

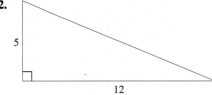

32.

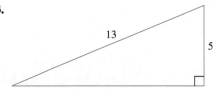

33.

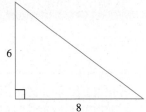

34.

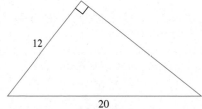

REVIEW OF GEOMETRIC FIGURES

Plane figures have length and width but no thickness or depth.

Name	Description	Figure
POLYGON	Union of three or more coplanar line segments that intersect with each other only at each end point, with each end point shared by two segments.	
TRIANGLE	Polygon with three sides (sum of measures of three angles is 180°).	
SCALENE TRIANGLE	Triangle with no sides of equal length.	
ISOSCELES TRIANGLE	Triangle with two sides of equal length.	
EQUILATERAL TRIANGLE	Triangle with all sides of equal length.	
RIGHT TRIANGLE	Triangle that contains a right angle.	leg, hypotenuse, leg

Plane figures have length and width but no thickness or depth.

Name	Description	Figure
QUADRILATERAL	Polygon with four sides (sum of measures of four angles is 360°).	
TRAPEZOID	Quadrilateral with exactly one pair of opposite sides parallel.	
ISOSCELES TRAPEZOID	Trapezoid with legs of equal length.	
PARALLELOGRAM	Quadrilateral with both pairs of opposite sides parallel and equal in length.	
RHOMBUS	Parallelogram with all sides of equal length.	
RECTANGLE	Parallelogram with four right angles.	
SQUARE	Rectangle with all sides of equal length.	
CIRCLE	All points in a plane the same distance from a fixed point called the **center**.	

Solids have length, width, and depth.

Name	Description	Figure
RECTANGULAR SOLID	A solid with six sides, all of which are rectangles.	
CUBE	A rectangular solid whose six sides are squares.	
SPHERE	All points the same distance from a fixed point, called the center.	
RIGHT CIRCULAR CYLINDER	A cylinder consisting of two circular bases that are perpendicular to its altitude.	
RIGHT CIRCULAR CONE	A cone with a circular base that is perpendicular to its altitude.	

MEAN, MEDIAN, AND MODE

Appendix **D**

It is sometimes desirable to be able to describe a set of data, or a set of numbers, by a single "middle" number. Three such **measures of central tendency** are the mean, the median, and the mode.

The most common measure of central tendency is the mean (sometimes called the arithmetic mean or the average). The **mean** of a set of data items, denoted by $\bar{x}$, is the sum of the items divided by the number of items.

Example 1 Seven students in a psychology class conducted an experiment on mazes. Each student was given a pencil and asked to successfully complete the same maze. The timed results are below.

STUDENT	Ann	Thanh	Carlos	Jesse	Melinda	Ramzi	Dayni
TIME (SECONDS)	13.2	11.8	10.7	16.2	15.9	13.8	18.5

a. Who completed the maze in the shortest time? Who completed the maze in the longest time?
b. Find the mean.
c. How many students took longer than the mean time? How many students took shorter than the mean time?

Solution   **a.** Carlos completed the maze in 10.7 seconds, the shortest time. Dayni completed the maze in 18.5 seconds, the longest time.
b. To find the mean, $\bar{x}$, find the sum of the data items and divide by 7, the number of items.

$$\bar{x} = \frac{13.2 + 11.8 + 10.7 + 16.2 + 15.9 + 13.8 + 18.5}{7} = \frac{100.1}{7} = 14.3$$

c. Three students, Jesse, Melinda, and Dayni, had times longer than the mean time. Four students, Ann, Thanh, Carlos, and Ramzi, had times shorter than the mean time.

Two other measures of central tendency are the median and the mode.

The **median** of an ordered set of numbers is the middle number. If the number of items is even, the median is the mean of the two middle numbers. The **mode** of a set of numbers is the number that occurs most often. It is possible for a data set to have no mode or more than one mode.

Example 2   Find the median and the mode of the following list of numbers. These numbers were high temperatures for fourteen consecutive days in a city in Montana.

$$76, 80, 85, 86, 89, 87, 82, 77, 76, 79, 82, 89, 89, 92$$

Solution   First, write the numbers in order.

$$76, 76, 77, 79, 80, 82, 82, 85, 86, 87, 89, 89, 89, 92$$

two middle numbers   mode

Since there are an even number of items, the median is the mean of the two middle numbers.

$$\text{median} = \frac{82 + 85}{2} = 83.5$$

The mode is 89, since 89 occurs most often.

Appendix D Exercise Set

For each of the following data sets, find the mean, the median, and the mode. If necessary, round the mean to one decimal place.

1. 21, 28, 16, 42, 38

2. 42, 35, 36, 40, 50

3. 7.6, 8.2, 8.2, 9.6, 5.7, 9.1

4. 4.9, 7.1, 6.8, 6.8, 5.3, 4.9

5. 0.2, 0.3, 0.5, 0.6, 0.6, 0.9, 0.2, 0.7, 1.1

6. 0.6, 0.6, 0.8, 0.4, 0.5, 0.3, 0.7, 0.8, 0.1

7. 231, 543, 601, 293, 588, 109, 334, 268

8. 451, 356, 478, 776, 892, 500, 467, 780

Ten tall buildings in the United States are listed below. Use this table for Exercises 9–12.

Building	Height (feet)
Sears Tower, Chicago, IL	1454
One World Trade Center (1972), New York, NY	1368
One World Trade Center (1973), New York, NY	1362
Empire State, New York, NY	1250
Amoco, Chicago, IL	1136
John Hancock Center, Chicago, IL	1127
First Interstate World Center, Los Angeles, CA	1107
Chrysler, New York, NY	1046
NationsBank Tower, Atlanta, GA	1023
Texas Commerce Tower, Houston, TX	1002

9. Find the mean height for the five tallest buildings.

10. Find the median height for the five tallest buildings.

11. Find the median height for the ten tallest buildings.

12. Find the mean height for the ten tallest buildings.

During an experiment, the following times (in seconds) were recorded: 7.8, 6.9, 7.5, 4.7, 6.9, 7.0.

13. Find the mean. Round to the nearest tenth.

14. Find the median.   **15.** Find the mode.

In a mathematics class, the following test scores were recorded for a student: 86, 95, 91, 74, 77, 85.

16. Find the mean. Round to the nearest hundredth.

17. Find the median.   **18.** Find the mode.

The following pulse rates were recorded for a group of fifteen students: 78, 80, 66, 68, 71, 64, 82, 71, 70, 65, 70, 75, 77, 86, 72.

19. Find the mean.   **20.** Find the median.

21. Find the mode.

22. How many rates were higher than the mean?

23. How many rates were lower than the mean?

24. Have each student in your algebra class take his/her pulse rate. Record the data and find the mean, the median, and the mode.

Find the missing numbers in each list of numbers. (These numbers are not necessarily in numerical order)

25. __, __, 16, 18, __

The mode is 21.

The mean is 20.

26. __, __, __, __, 40

The mode is 35.

The median is 37.

The mean is 38.

$\Lambda ppendix$ E

TABLE OF SQUARES AND SQUARE ROOTS

n	n^2	$\sqrt{n}$	n	n^2	$\sqrt{n}$
1	1	1.000	51	2,601	7.141
2	4	1.414	52	2,704	7.211
3	9	1.732	53	2,809	7.280
4	16	2.000	54	2,916	7.348
5	25	2.236	55	3,025	7.416
6	36	2.449	56	3,136	7.483
7	49	2.646	57	3,249	7.550
8	64	2.828	58	3,364	7.616
9	81	3.000	59	3,481	7.681
10	100	3.162	60	3,600	7.746
11	121	3.317	61	3,721	7.810
12	144	3.464	62	3,844	7.874
13	169	3.606	63	3,969	7.937
14	196	3.742	64	4,096	8.000
15	225	3.873	65	4,225	8.062
16	256	4.000	66	4,356	8.124
17	289	4.123	67	4,489	8.185
18	324	4.243	68	4,624	8.246
19	361	4.359	69	4,761	8.307
20	400	4.472	70	4,900	8.367
21	441	4.583	71	5,041	8.426
22	484	4.690	72	5,184	8.485
23	529	4.796	73	5,329	8.544
24	576	4.899	74	5,476	8.602
25	625	5.000	75	5,625	8.660
26	676	5.099	76	5,776	8.718
27	729	5.196	77	5,929	8.775
28	784	5.292	78	6,084	8.832
29	841	5.385	79	6,241	8.888
30	900	5.477	80	6,400	8.944

n	n^2	$\sqrt{n}$	n	n^2	$\sqrt{n}$
31	961	5.568	81	6,561	9.000
32	1,024	5.657	82	6,724	9.055
33	1,089	5.745	83	6,889	9.110
34	1,156	5.831	84	7,056	9.165
35	1,225	5.916	85	7,225	9.220
36	1,296	6.000	86	7,396	9.274
37	1,369	6.083	87	7,569	9.327
38	1,444	6.164	88	7,744	9.381
39	1,521	6.245	89	7,921	9.434
40	1,600	6.325	90	8,100	9.487
41	1,681	6.403	91	8,281	9.539
42	1,764	6.481	92	8,464	9.592
43	1,849	6.557	93	8,649	9.644
44	1,936	6.633	94	8,836	9.695
45	2,025	6.708	95	9,025	9.747
46	2,116	6.782	96	9,216	9.798
47	2,209	6.856	97	9,409	9.849
48	2,304	6.928	98	9,604	9.899
49	2,401	7.000	99	9,801	9.950
50	2,500	7.071	100	10,000	10.000

Appendix F

REVIEW OF VOLUME AND SURFACE AREA

A **convex solid** is a set of points, *S*, not all in one plane, such that for any two points *A* and *B* in *S*, all points between *A* and *B* are also in *S*. In this appendix, we will find the volume and surface area of special types of solids called polyhedrons. A solid formed by the intersection of a finite number of planes is called a **polyhedron**. The box below is an example of a polyhedron.

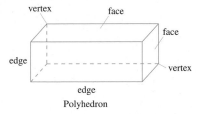

Polyhedron

Each of the plane regions of the polyhedron is called a **face** of the polyhedron. If the intersection of two faces is a line segment, this line segment is an **edge** of the polyhedron. The intersections of the edges are the **vertices** of the polyhedron.

   Volume is a measure of the space of a solid. The volume of a box or can, for example, is the amount of space inside. Volume can be used to describe the amount of juice in a pitcher or the amount of concrete needed to pour a foundation for a house.

   The volume of a solid is the number of **cubic units** in the solid. A cubic centimeter and a cubic inch are illustrated.

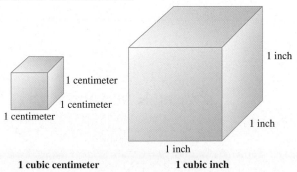

1 cubic centimeter   **1 cubic inch**

   The **surface area** of a polyhedron is the sum of the areas of the faces of the polyhedron. For example, each face of the cube to the left above has an area of 1 square centimeter. Since there are 6 faces of the cube, the sum of the areas of the faces is 6 square centimeters. Surface area can be used to describe the amount of material needed to cover or form a solid. Surface area is measured in square units.

Formulas for finding the volumes, V, and surface areas, SA, of some common solids are given next.

VOLUME AND SURFACE AREA FORMULAS OF COMMON SOLIDS

Solid	*Formulas*

RECTANGULAR SOLID

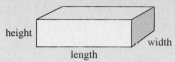

$V = lwh$
$SA = 2lh + 2wh + 2lw$
where h = height, w = width, l = length

CUBE

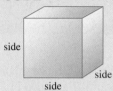

$V = s^3$
$SA = 6s^2$
where s = side

SPHERE

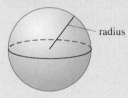

$V = \dfrac{4}{3}\pi r^3$
$SA = 4\pi r^2$
where r = radius

CIRCULAR CYLINDER

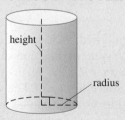

$V = \pi r^2 h$
$SA = 2\pi rh + 2\pi r^2$
where h = height, r = radius

CONE

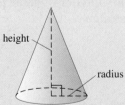

$V = \dfrac{1}{3}\pi r^2 h$

$SA = \pi r\sqrt{r^2 + h^2} + \pi r^2$
where h = height, r = radius

SQUARE-BASED PYRAMID

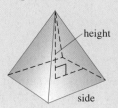

$V = \dfrac{1}{3}s^2 h$

$SA = B + \dfrac{1}{2}pl$

where B = area of base; p = perimeter of base, h = height, s = side, l = slant height

> **HELPFUL HINT**
> Volume is measured in cubic units. Surface area is measured in square units.

Example 1   Find the volume and surface area of a rectangular box that is 12 inches long, 6 inches wide, and 3 inches high.

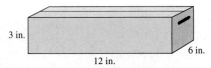

3 in.

12 in.

6 in.

Solution   Let $h = 3$ in., $l = 12$ in., and $w = 6$ in.

$V = lwh$

$V = 12$ inches $\cdot$ 6 inches $\cdot$ 3 inches $= 216$ cubic inches

The volume of the rectangular box is 216 cubic inches.

$SA = 2lh + 2wh + 2lw$

$\quad = 2(12 \text{ in.})(3 \text{ in.}) + 2(6 \text{ in.})(3 \text{ in.}) + 2(12 \text{ in.})(6 \text{ in.})$

$\quad = 72 \text{ sq. in.} + 36 \text{ sq. in.} + 144 \text{ sq. in.}$

$\quad = 252 \text{ sq. in.}$

The surface area of rectangular box is 252 square inches.

Example 2   Find the volume and surface area of a ball of radius 2 inches. Give the exact volume and surface area and then use the approximation $\dfrac{22}{7}$ for π.

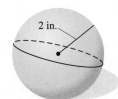

2 in.

Solution

$$V = \frac{4}{3}\pi r^3 \qquad \text{Formula for volume of a sphere.}$$

$$V = \frac{4}{3} \cdot \pi (2 \text{ in.})^3 \qquad \text{Let } r = 2 \text{ inches.}$$

$$= \frac{32}{3}\pi \text{ cu. in.} \qquad \text{Simplify.}$$

$$\approx \frac{32}{3} \cdot \frac{22}{7} \text{ cu. in.} \qquad \text{Approximate } \pi \text{ with } \frac{22}{7}.$$

$$= \frac{704}{21} \text{ or } 33\frac{11}{21} \text{ cu. in.}$$

The volume of the sphere is exactly $\dfrac{32}{3}\pi$ cubic inches or approximately $33\dfrac{11}{21}$ cubic inches.

$$SA = 4\pi r^2 \qquad \text{Formula for surface area.}$$

$$SA = 4 \cdot \pi (2 \text{ in.})^2 \qquad \text{Let } r = 2 \text{ inches.}$$

$$= 16\pi \text{ sq. in.} \qquad \text{Simplify.}$$

$$\approx 16 \cdot \frac{22}{7} \text{ sq. in.} \qquad \text{Approximate } \pi \text{ with } \frac{22}{7}.$$

$$= \frac{352}{7} \text{ or } 50\frac{2}{7} \text{ sq. in.}$$

The surface area of the sphere is exactly 16π square inches or approximately $50\dfrac{2}{7}$ square inches.

Appendix F Exercise Set

Find the volume and surface area of each solid. See Examples 1 through 3. For formulas that contain π, give an exact answer and then approximate using $\frac{22}{7}$ for π.

1.

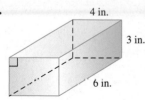

4 in.
3 in.
6 in.

2.

3 mi

3.

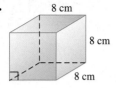

8 cm
8 cm
8 cm

4.

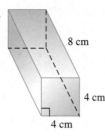

8 cm
4 cm
4 cm

5. (For surface area, use 3.14 for π.)

3 yd
2 yd

6.

10 ft
6 ft

7.

10 in.

8. Find the volume only.

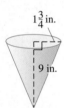

$1\frac{3}{4}$ in.
9 in.

9.

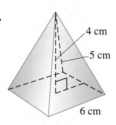

4 cm
5 cm
6 cm

10.

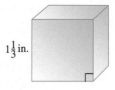

1 ft

Solve.

11. Find the volume of a cube with edges of $1\frac{1}{3}$ inches.

$1\frac{1}{3}$ in.

12. A water storage tank is in the shape of a cone with the pointed end down. If the radius is 14 ft and the depth of the tank is 15 ft, approximate the volume of the tank in cubic feet. Use $\frac{22}{7}$ for π.

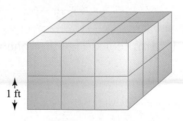

14 ft
15 ft

13. Find the surface area of a rectangular box 2 ft by 1.4 ft by 3 ft.

14. Find the surface area of a box in the shape of a cube that is 5 ft on each side.

15. Find the volume of a pyramid with a square base 5 in. on a side and a height of 1.3 in.

16. Approximate to the nearest hundredth the volume of a sphere with a radius of 2 cm. Use 3.14 for π.

17. A paperweight is in the shape of a square-based pyramid 20 cm tall. If an edge of the base is 12 cm, find the volume of the paperweight.

18. A bird bath is made in the shape of a hemisphere (half-sphere). If its radius is 10 in., approximate the volume. Use $\dfrac{22}{7}$ for π.

19. Find the exact surface area of a sphere with a radius of 7 in.

20. A tank is in the shape of a cylinder 8 ft tall and 3 ft in radius. Find the exact surface area of the tank.

21. Find the volume of a rectangular block of ice 2 ft by $2\dfrac{1}{2}$ ft by $1\dfrac{1}{2}$ ft.

22. Find the capacity (volume in cubic feet) of a rectangular ice chest with inside measurements of 3 ft by $1\dfrac{1}{2}$ ft by $1\dfrac{3}{4}$ ft.

23. An ice cream cone with a 4-cm diameter and 3-cm depth is filled exactly level with the top of the cone. Approximate how much ice cream (in cubic centimeters) is in the cone. Use $\dfrac{22}{7}$ for π.

24. A child's toy is in the shape of a square-based pyramid 10 in. tall. If an edge of the base is 7 in., find the volume of the toy.

ANSWERS TO SELECTED EXERCISES

■ CHAPTER 1   REVIEW OF REAL NUMBERS

Exercise Set 1.2 **1.** < **3.** > **5.** = **7.** < **9.** 32 < 212 **11.** 44,300 > 34,611 **13.** true **15.** false **17.** false
19. true **21.** 30 ≤ 45 **23.** 8 < 12 **25.** 5 ≥ 4 **27.** 15 ≠ −2 **29.** 535; −8 **31.** −433,853 **33.** 350; −126 **35.** 1988
37. 1988, 1989, 1990, 1991 **39.** 6068 ≥ 2649 **41.** whole, integers, rational, real **43.** integers, rational, real
45. natural, whole, integers, rational, real **47.** rational, real **49.** irrational, real **51.** false **53.** true **55.** true
57. true **59.** false **61.** > **63.** > **65.** < **67.** < **69.** > **71.** = **73.** < **75.** < **77.** −0.04 > −26.7 **79.** Sun
81. Sun **83.** 20 ≤ 25 **85.** 6 > 0 **87.** −12 < −10 **89.** Answers may vary.

Mental Math **1.** $\dfrac{3}{8}$ **3.** $\dfrac{5}{7}$ **5.** numerator; denominator

Exercise Set 1.3 **1.** 3·11 **3.** 2·7·7 **5.** 2·2·5 **7.** 3·5·5 **9.** 3·3·5 **11.** $\dfrac{1}{2}$ **13.** $\dfrac{2}{3}$ **15.** $\dfrac{3}{7}$ **17.** $\dfrac{3}{5}$ **19.** $\dfrac{3}{8}$ **21.** $\dfrac{1}{2}$

23. $\dfrac{6}{7}$ **25.** 15 **27.** $\dfrac{1}{6}$ **29.** $\dfrac{25}{27}$ **31.** $\dfrac{11}{20}$ sq. mi **33.** $\dfrac{3}{5}$ **35.** 1 **37.** $\dfrac{1}{3}$ **39.** $\dfrac{9}{35}$ **41.** $\dfrac{21}{30}$ **43.** $\dfrac{4}{18}$ **45.** $\dfrac{16}{20}$ **47.** $\dfrac{23}{21}$ **49.** $1\dfrac{2}{3}$

51. $\dfrac{5}{66}$ **53.** $\dfrac{7}{5}$ **55.** $\dfrac{1}{5}$ **57.** $\dfrac{3}{8}$ **59.** $\dfrac{1}{9}$ **61.** $\dfrac{5}{7}$ **63.** $\dfrac{65}{21}$ **65.** $\dfrac{2}{5}$ **67.** $\dfrac{9}{7}$ **69.** $\dfrac{3}{4}$ **71.** $\dfrac{17}{3}$ **73.** $\dfrac{7}{26}$ **75.** 1 **77.** $\dfrac{1}{5}$ **79.** $5\dfrac{1}{6}$

81. $\dfrac{17}{18}$ **83.** $55\dfrac{1}{4}$ ft **85.** $5\dfrac{8}{25}$ m **87.** Answers may vary. **89.** $3\dfrac{3}{8}$ mi **91.** $\dfrac{3}{4}$ **93.** $\dfrac{49}{200}$

Calculator Explorations **1.** 125 **3.** 59,049 **5.** 30 **7.** 9857 **9.** 2376

Mental Math **1.** multiply **3.** subtract

Exercise Set 1.4 **1.** 243 **3.** 27 **5.** 1 **7.** 5 **9.** $\dfrac{1}{125}$ **11.** $\dfrac{16}{81}$ **13.** 49 **15.** 16 **17.** 1.44 **19.** 17 **21.** 20 **23.** 10

25. 21 **27.** 45 **29.** 0 **31.** $\dfrac{2}{7}$ **33.** 30 **35.** 2 **37.** $\dfrac{7}{18}$ **39.** $\dfrac{27}{10}$ **41.** $\dfrac{7}{5}$ **43.** no **45. a.** 64 **b.** 43 **c.** 19 **d.** 22 **47.** 9

49. 1 **51.** 1 **53.** 11 **55.** 45 **57.** 27 **59.** 132 **61.** $\dfrac{37}{18}$ **63.** 16, 64, 144, 256 **65.** solution **67.** not a solution
69. not a solution **71.** solution **73.** not a solution **75.** $x + 15$ **77.** $x − 5$ **79.** $3x + 22$ **81.** $1 + 2 = 9 ÷ 3$
83. $3 ≠ 4 ÷ 2$ **85.** $5 + x = 20$ **87.** $13 − 3x = 13$ **89.** $\dfrac{12}{x} = \dfrac{1}{2}$ **91.** Answers may vary. **93.** 28 m **95.** 12,000 sq. ft
97. 6.5% **99.** $27.75

Mental Math **1.** negative **3.** 0 **5.** negative

Exercise Set 1.5 **1.** 9 **3.** −14 **5.** 1 **7.** −12 **9.** −5 **11.** −12 **13.** −4 **15.** 7 **17.** −2 **19.** 0 **21.** −19 **23.** 31
25. −47 **27.** −2.1 **29.** −8 **31.** 38 **33.** −13.1 **35.** $\dfrac{2}{8} = \dfrac{1}{4}$ **37.** $-\dfrac{3}{16}$ **39.** $-\dfrac{13}{10}$ **41.** −8 **43.** −59 **45.** −9 **47.** 5

49. 11 **51.** −18 **53.** 19 **55.** −0.7 **57.** Tues **59.** 7° **61.** 1° **63.** −6° **65.** −654 ft **67.** −$218.8 million **69.** −17

71. −6 **73.** 2 **75.** 0 **77.** −6 **79.** Answers may vary. **81.** −2 **83.** 0 **85.** $-\dfrac{2}{3}$ **87.** Answers may vary.

89. negative **91.** positive **93.** yes **95.** no

Exercise Set 1.6 **1.** −10 **3.** −5 **5.** 19 **7.** $\dfrac{1}{6}$ **9.** 2 **11.** −11 **13.** 11 **15.** 5 **17.** 37 **19.** −6.4 **21.** −71 **23.** 0

25. 4.1 **27.** $\dfrac{2}{11}$ **29.** $-\dfrac{11}{12}$ **31.** 8.92 **33.** 13 **35.** −5 **37.** −1 **39.** −23 **41.** Answers may vary. **43.** −26 **45.** −24

47. 3 **49.** −45 **51.** −4 **53.** 13 **55.** 6 **57.** 9 **59.** −9 **61.** −7 **63.** $\dfrac{7}{5}$ **65.** 21 **67.** $\dfrac{1}{4}$ **69.** January, −22° **71.** −12°

73. 49° **75.** 100° **77.** 23 yd loss **79.** 384 B.C. **81.** $-2\dfrac{3}{8}$ points **83.** 22,965 ft **85.** 130° **87.** 30° **89.** not a solution

91. not a solution **93.** solution **95.** true **97.** false **99.** negative, −30,387

Calculator Explorations **1.** 38 **3.** −441 **5.** 163.$\overline{3}$ **7.** 54,499 **9.** 15,625

Mental Math **1.** positive **3.** negative **5.** positive

Exercise Set 1.7 **1.** −24 **3.** −2 **5.** 50 **7.** −12 **9.** 42 **11.** −18 **13.** $\dfrac{3}{10}$ **15.** $\dfrac{2}{3}$ **17.** −7 **19.** 0.14 **21.** −800

23. −28 **25.** 25 **27.** $-\dfrac{8}{27}$ **29.** −121 **31.** $-\dfrac{1}{4}$ **33.** 0.84 **35.** −30 **37.** 90 **39.** true **41.** false **43.** 16 **45.** −1

47. 25 **49.** −49 **51.** $\dfrac{1}{9}$ **53.** $\dfrac{3}{2}$ **55.** $-\dfrac{1}{14}$ **57.** $-\dfrac{11}{3}$ **59.** $\dfrac{1}{0.2}$ **61.** −6.3 **63.** −9 **65.** 4 **67.** −4 **69.** 0 **71.** −5

73. undefined **75.** 3 **77.** −15 **79.** $-\dfrac{18}{7}$ **81.** $\dfrac{20}{27}$ **83.** −1 **85.** $-\dfrac{9}{2}$ **87.** −4 **89.** 16 **91.** −3 **93.** $-\dfrac{16}{7}$ **95.** 2

97. $\dfrac{6}{5}$ **99.** −5 **101.** $\dfrac{3}{2}$ **103.** −21 **105.** 41 **107.** −134 **109.** 3 **111.** −1 **113.** −$498 million **115.** Answers may vary.

117. 1, −1 **119.** positive **121.** not possible **123.** negative **125.** $-2 + -\dfrac{15}{3}; -7$ **127.** $2(-5 + (-3)); -16$ **129.** yes

131. yes **133.** yes

Exercise Set 1.8 **1.** $16 + x$ **3.** $y \cdot (-4)$ **5.** yx **7.** $13 + 2x$ **9.** $x \cdot (yz)$ **11.** $(2 + a) + b$ **13.** $(4a) \cdot b$
15. $a + (b + c)$ **17.** $17 + b$ **19.** $24y$ **21.** y **23.** $26 + a$ **25.** $-72x$ **27.** s **29.** Answers may vary. **31.** $4x + 4y$
33. $9x - 54$ **35.** $6x + 10$ **37.** $28x - 21$ **39.** $18 + 3x$ **41.** $-2y + 2z$ **43.** $-21y - 35$ **45.** $5x + 20m + 10$
47. $-4 + 8m - 4n$ **49.** $-5x - 2$ **51.** $-r + 3 + 7p$ **53.** $3x + 4$ **55.** $-x + 3y$ **57.** $6r + 8$ **59.** $-36x - 70$
61. $-16x - 25$ **63.** $4(1 + y)$ **65.** $11(x + y)$ **67.** $-1(5 + x)$ **69.** $30(a + b)$ **71.** commutative property of
multiplication **73.** associative property of addition **75.** distributive property **77.** associative property of multiplication
79. identity property of addition **81.** distributive property **83.** commutative and associative properties of multiplication
85. $-8; \dfrac{1}{8}$ **87.** $-x; \dfrac{1}{x}$ **89.** $-2x; \dfrac{1}{2x}$ **91.** no **93.** yes **95.** Answers may vary.

Exercise Set 1.9 **1.** approx. 7.8 million **3.** 2002 **5.** PGA/LPGA tours **7.** Major League Baseball, NBA
9. approx. 15 million **11.** France **13.** France, United States, Spain, Italy **15.** 34 million **17.** approx. 142 million
19. *Snow White and the Seven Dwarfs* **21.** Answers may vary. **23.** 1994 **25.** 1989, 1990 **27.** approx. 54%
29. 1994 **31.** approx. 59 beats per minute **33.** approx. 26 beats per minute **35.** 20 students **37.** 1985
39. Answers may vary. **41.** 4 million **43.** 69 million **45.** 1960 **47.** 12 million **49.** 30° north, 90° west
51. Answers may vary.

Chapter 1 Review **1.** < **3.** > **5.** < **7.** = **9.** > **11.** $4 \geq -3$ **13.** $0.03 < 0.3$ **15. a.** $\{1, 3\}$ **b.** $\{0, 1, 3\}$

c. $\{-6, 0, 1, 3\}$ **d.** $\left\{-6, 0, 1, 1\dfrac{1}{2}, 3, 9.62\right\}$ **e.** $\{\pi\}$ **f.** $\left\{-6, 0, 1, 1\dfrac{1}{2}, 3, \pi, 9.62\right\}$ **17.** Friday **19.** $2 \cdot 2 \cdot 3 \cdot 3$ **21.** $\dfrac{12}{25}$

23. $\dfrac{13}{10}$ **25.** $9\dfrac{3}{8}$ **27.** 15 **29.** $\dfrac{7}{12}$ **31.** $A = \dfrac{34}{121}$ sq. in.; $P = 2\dfrac{4}{11}$ in. **33.** $2\dfrac{15}{16}$ lb **35.** $11\dfrac{5}{16}$ lb **37.** Odera **39.** $3\dfrac{7}{8}$ lb

41. 16 **43.** $\dfrac{4}{49}$ **45.** 70 **47.** 37 **49.** $\dfrac{18}{7}$ **51.** $20 - 12 = 2 \cdot 4$ **53.** 18 **55.** 5 **57.** 63° **59.** no **61.** $\dfrac{-2}{3}$ **63.** 7

65. -17 **67.** -5 **69.** 3.9 **71.** -14 **73.** 5 **75.** -19 **77.** 15 **79.** $\$51$ **81.** $-\dfrac{1}{6}$ **83.** -48 **85.** 3 **87.** -36

89. undefined   **91.** undefined   **93.** -5 **95.** commutative property of addition   **97.** distributive property
99. associative property of addition   **101.** distributive property   **103.** multiplicative inverse
105. commutative property of addition   **107.** 7 million   **109.** number of subscribers is increasing   **111.** Miami, 1.8%
113. New York, Cleveland, Houston, Chicago, Boston, Dallas/Fort Worth, San Francisco

Chapter 1 Test   **1.** $|-7| > 5$ **2.** $(9 + 5) \geq 4$ **3.** -5 **4.** -11 **5.** -14 **6.** -39 **7.** 12 **8.** -2 **9.** undefined   **10.** -8
11. $-\dfrac{1}{3}$ **12.** $4\dfrac{5}{8}$ **13.** $\dfrac{51}{40}$ **14.** -32 **15.** -48 **16.** 3 **17.** 0 **18.** $>$ **19.** $>$ **20.** $>$ **21.** $=$ **22.** $2221 < 10{,}993$

23. a. $\{1, 7\}$ **b.** $\{0, 1, 7\}$ **c.** $\{-5, -1, 0, 1, 7\}$ **d.** $\left\{-5, -1, 0, \dfrac{1}{4}, 1, 7, 11.6\right\}$ **e.** $\{\sqrt{7}, 3\pi\}$

f. $\left\{-5, -1, 0, \dfrac{1}{4}, 1, 7, 11.6, \sqrt{7}, 3\pi\right\}$ **24.** 40 **25.** 12 **26.** 22 **27.** -1 **28.** associative property

29. commutative property   **30.** distributive property   **31.** multiplicative inverse   **32.** 9 **33.** -3 **34.** second down
35. yes   **36.** $17°$ **37.** 650 million   **38.** loss of $\$420$ **39.** $\$8$ billion   **40.** $\$25$ billion   **41.** $\$5.5$ billion   **42.** 1996
43. Indiana, 25.2 million tons   **44.** Texas, 5 million tons   **45.** 16 million tons   **46.** 3 million tons

■ CHAPTER 2   EQUATIONS, INEQUALITIES, AND PROBLEM SOLVING

Exercise Set 2.1   **1.** $15y$ **3.** $13w$ **5.** $-7b - 9$ **7.** $-m - 6$ **9.** $5y - 20$ **11.** $7d - 11$ **13.** $-3x + 2y - 1$
15. $2x + 14$ **17.** Answers may vary.   **19.** $10x - 3$ **21.** $-4x - 9$ **23.** $5x^2$ **25.** $4x - 3$ **27.** $8x - 53$ **29.** -8
31. $7.2x - 5.2$ **33.** $k - 6$ **35.** $0.9m + 1$ **37.** $-12y + 16$ **39.** $x + 5$ **41.** -11 **43.** $1.3x + 3.5$ **45.** $x + 2$

47. $-15x + 18$ **49.** $2k + 10$ **51.** $-3x + 5$ **53.** $2x - 4$ **55.** $\dfrac{3}{4}x + 12$ **57.** $-2 + 12x$ **59.** $-4m - 3$ **61.** $8(x + 6)$

63. $x - 10$ **65.** $\dfrac{7x}{6}$ **67.** $7x - 7$ **69.** $(18x - 2)$ ft   **71.** balanced   **73.** balanced   **75.** $(15x + 23)$ in.   **77.** 2
79. -23 **81.** -25 **83.** $5b^2c^3 + b^3c^2$ **85.** $5x^2 + 9x$ **87.** $-7x^2y$

Mental Math   **1.** 2 **3.** 12 **5.** 17

Exercise Set 2.2   **1.** 3 **3.** -2 **5.** -14 **7.** 0.5 **9.** $\dfrac{5}{12}$ **11.** -0.7 **13.** 3 **15.** Answers may vary.   **17.** -3 **19.** -10

21. 11 **23.** $\dfrac{2}{3}$ **25.** -9 **27.** 13 **29.** -17.9 **31.** $-\dfrac{1}{2}$ **33.** 11 **35.** -30 **37.** -7 **39.** 2 **41.** $-\dfrac{3}{4}$ **43.** 21 **45.** 25
47. 0 **49.** 1.83 **51.** $20 - p$ **53.** $(10 - x)$ ft   **55.** $(180 - x)°$ **57.** $n + 284$ **59.** $(m - 60)$ ft   **61.** $(173 - 3x)°$

63. 250 ml   **65.** Answers may vary.   **67.** solution   **69.** not a solution   **71.** $\dfrac{6}{7}$ **73.** $\dfrac{1}{5}$ **75.** $-\dfrac{5}{3}$ **77.** y **79.** r **81.** x

Mental Math   **1.** 9 **3.** 2 **5.** -5

Exercise Set 2.3   **1.** -4 **3.** 0 **5.** 12 **7.** -12 **9.** 3 **11.** 2 **13.** 0 **15.** 6.3 **17.** 6 **19.** -5.5 **21.** $\dfrac{14}{3}$ **23.** -9

25. 10 **27.** -20 **29.** 0 **31.** -5 **33.** 0 **35.** $-\dfrac{3}{2}$ **37.** -21 **39.** $\dfrac{11}{2}$ **41.** 1 **43.** $-\dfrac{1}{4}$ **45.** -30 **47.** $\dfrac{9}{10}$ **49.** -30

51. 2 **53.** -2 **55.** Answers may vary.   **57.** Answers may vary.   **59.** $2x + 2$ **61.** $2x + 2$ **63.** $\dfrac{700}{3}$ mg   **65.** -6.3
67. -1.23 **69.** $-8y - 3$ **71.** $-a - 3$ **73.** $15z - 49$ **75.** $>$ **77.** $=$ **79.** $=$

Calculator Explorations   **1.** solution   **3.** not a solution   **5.** solution

Exercise Set 2.4   **1.** 1 **3.** $\dfrac{9}{2}$ **5.** $\dfrac{3}{2}$ **7.** 0 **9.** 2 **11.** -5 **13.** 10 **15.** 18 **17.** 1 **19.** 50 **21.** 0.2 **23.** all real numbers
25. no solution   **27.** no solution   **29.** Answers may vary.   **31.** Answers may vary.   **33.** 4 **35.** -4 **37.** 3 **39.** -2
41. 4 **43.** $\dfrac{7}{3}$ **45.** no solution   **47.** $\dfrac{9}{5}$ **49.** $\dfrac{4}{19}$ **51.** 1 **53.** no solution   **55.** $\dfrac{7}{2}$ **57.** -17 **59.** $\dfrac{19}{6}$ **61.** all real numbers

63. 3 **65.** 13 **67.** 15.3 **69.** -0.2 **71.** $2x + \dfrac{1}{5} = 3x - \dfrac{4}{5}; 1$ **73.** $2x + 7 = x + 6; -1$ **75.** $3x - 6 = 2x + 8; 14$

77. $\frac{1}{3}x = \frac{5}{6}; \frac{5}{2}$ **79.** $x - 4 = 2x; -4$ **81.** $\frac{x}{4} + \frac{1}{2} = \frac{3}{4}; 1$ **83.** $x = 4$ cm, $2x = 8$ cm   **85.** $\frac{5}{4}$ **87.** Midway   **89.** 145

91. -1 **93.** $\frac{1}{5}$ **95.** $(6x - 8)$ m   **97.** $-\frac{7}{8}$ **99.** no solution   **101.** 0

Exercise Set 2.5   **1.** 1   **3.** -25 **5.** $-\frac{3}{4}$ **7.** -16 **9.** governor of Nebraska: \$65,000; governor of Washington: \$130,000

11. 1st piece: 5 in.; 2nd piece: 10 in.; 3rd piece: 25 in.   **13.** 172 mi   **15.** 1st angle: 37.5°; 2nd angle: 37.5°; 3rd angle: 105°
17. Brown: 66,362; Randall: 53,074   **19.** 45°, 135°   **21.** Belgium: 32; France: 33; Spain: 34   **23.** 0.2 in./year

25. height: 34 in.; diameter: 49 in.   **27.** Spurs: 78; Knicks: 77   **29.** $2\frac{1}{2}$ **31.** 17 Democratic governors; 31 Republican governors

33. 5 ft, 12 ft   **35.** 58°, 60°, 62°   **37.** 350 mi   **39.** Answers may vary.   **41.** Texas and Florida
43. Hawaii: \$37.9 million; Pennsylvania: \$23 million   **45.** 15 ft by 24 ft   **47.** Answers may vary.   **49.** Answers may vary.
51. c   **53.** -10 **55.** -9 **57.** -15 **59.** $5(-x) = x + 60$ **61.** $50 - (x + 9) = 0$

Exercise Set 2.6   **1.** $h = 3$ **3.** $h = 3$ **5.** $h = 20$ **7.** $c = 12$ **9.** $r = 2.5$ **11.** $T = 3$ **13.** $h = 15$ **15.** $h = \frac{f}{5g}$

17. $W = \frac{V}{LH}$ **19.** $y = 7 - 3x$ **21.** $R = \frac{A - P}{PT}$ **23.** $A = \frac{3V}{h}$ **25.** $a = P - b - c$ **27.** $\frac{S - 2\pi r^2}{2\pi r}$ **29.** 131 ft

31. 7.5 hrs   **33.** $-10°$C   **35.** 96 piranhas   **37.** 137.5 mi   **39.** 2 bags   **41.** 6.25 hrs   **43.** 800 cu. ft   **45.** one 16-in. pizza

47. 4.65 min   **49.** 2.25 hrs   **51.** $-109.3°$F   **53.** 500 sec or $8\frac{1}{3}$ min   **55.** 33,493,333,333 cu. mi   **57.** 10.7   **59.** 44.3 sec

61. $-40°$ **63.** It multiplies the volume by 8.   **65.** $\frac{9}{x + 5}$ **67.** $3(x + 4)$ **69.** $2(10 + 4x)$ **71.** $3(x - 12)$

Mental Math   **1.** no   **3.** yes

Exercise Set 2.7   **1.** 1.2   **3.** 0.225   **5.** 0.0012   **7.** 75%   **9.** 200%   **11.** 12.5%   **13.** 38%   **15.** 54%   **17.** 136.8°
19. Answers may vary.   **21.** 4%   **23.** 14.4%   **25.** 49,950   **27.** 11.2   **29.** 55%   **31.** 180   **33.** 4.6   **35.** 50   **37.** 30%
39. \$39 decrease; \$117 sale price   **41.** 243   **43.** 55.40%   **45.** 54 people   **47.** No, many people use several medications.
49. 31 men   **51.** 27%, 5%, 35%, 2%   **53.** 0.15%   **55.** 23.6 million   **57.** 75% increase   **59.** 24%; no   **61.** 166.567%
63. 16%   **65.** 12   **67.** Answers may vary.   **69.** 11%   **71.** 7.7%   **73.** 19.3%   **75.** Answers may vary.   **77.** 6   **79.** 208
81. -55

Exercise Set 2.8   **1.** length: 78 ft; width: 52 ft   **3.** 18 ft, 36 ft, 48 ft   **5.** $666\frac{2}{3}$ mi   **7.** 160 mi   **9.** 2 gal   **11.** $6\frac{2}{3}$ lbs   **13.** no

15. \$11,500 @ 8%; \$13,500 @ 9%   **17.** \$7000 @ 11% profit; \$3000 @ 4% loss   **19.** 13 in.   **21.** \$30,000 @ 8%; \$24,000 @ 10%
23. \$5000 @ 12%; \$15,000 @ 4%   **25.** \$4500   **27.** \$4500 @ 9%; \$9000 @ 10%; \$13,500 @ 11%   **29.** 3 adult tickets

31. 2.2 mph; 3.3 mph   **33.** 27.5 mi   **35.** 500; \$30,000   **37.** 25 monitors   **39.** -4 **41.** $\frac{9}{16}$ **43.** -4 **45.** $>$ **47.** $=$

Mental Math   **1.** $x > 2$ **3.** $x \geq 8$

Exercise Set 2.9   **1.** [number line: closed circle at -1]   **3.** [number line: open circle at $\frac{1}{2}$]   **5.** [number line: open circles at -1, 3]   **7.** [number line: closed circles at 0, 2]

9. $x < -3$, [number line: open circle at -3]   **11.** $x \geq -5$, [number line: closed circle at -5]   **13.** $x \geq -2$, [number line: closed circle at -2]   **15.** $x > -3$, [number line: open circle at -3]

17. $x \leq 1$, [number line: closed circle at 1]   **19.** $x > -5$, [number line: open circle at -5]   **21.** $x \leq -2$, [number line: closed circle at -2]   **23.** $x \leq -8$, [number line: closed circle at -8]

25. $x > 4$, [number line: open circle at 4]   **27.** $-1 < x < 2$, [number line: open circles at -1, 2]   **29.** $4 \leq x \leq 5$, [number line: closed circles at 4, 5]

31. $1 < x < 5$, [number line: open circles at 1, 5]   **33.** Answers may vary.   **35.** $x \geq 20$, [number line: closed circle at 20]   **37.** $x > 16$, [number line: open circle at 16]

39. $x > -3$, [number line: open circle at -3]   **41.** $x \geq -\frac{2}{3}$, [number line: closed circle at $-\frac{2}{3}$]   **43.** $x > \frac{8}{3}$, [number line: open circle at $\frac{8}{3}$]   **45.** $x > -13$, [number line: open circle at -13]

47. $x > 0$, [number line: open circle at 0]   **49.** $x \geq 0$, [number line: closed circle at 0]   **51.** $1 < x < 4$, [number line: open circles at 1, 4]   **53.** $x > 3$, [number line: open circle at 3]

55. $x \leq 0$,   **57.** $0 < x < \dfrac{14}{3}$,   **59.** $x < 200$ recommended; $200 \leq x \leq 240$ borderline;

$x > 240$ high   **61.** $x > -10$ **63.** 86 people   **65.** 35 cm   **67.** $-38\dfrac{1}{5}°$F to $113°$F   **69.** $0.924 \leq d \leq 0.987$

71. $-3 < x < 3$ **73.** 10%   **75.** 193   **77.** 8   **79.** 1   **81.** $\dfrac{16}{49}$ **83.** 51 million   **85.** 1996   **87.** $x > 1$,

89. $x < \dfrac{5}{8}$,   **91.** $x \leq 0$,

Chapter 2 Review   **1.** $6x$ **3.** $4x - 2$ **5.** $3n - 18$ **7.** $-6x + 7$ **9.** $3x - 7$ **11.** 4   **13.** -6 **15.** -9 **17.** 5; 5

19. $10 - x$ **21.** $(175 - x)°$ **23.** 4   **25.** -1 **27.** -1 **29.** 6   **31.** 2   **33.** no solution   **35.** $\dfrac{3}{4}$ **37.** 20   **39.** 0   **41.** $\dfrac{20}{7}$

43. $\dfrac{23}{7}$ **45.** 102   **47.** 3   **49.** 1052 ft   **51.** 307; 955   **53.** $w = 9$ **55.** $m = \dfrac{y - b}{x}$ **57.** $x = \dfrac{2y - 7}{5}$ **59.** $\pi = \dfrac{C}{D}$

61. 15 m   **63.** 1 hr and 20 min   **65.** 93.5   **67.** 70%   **69.** 1280   **71.** 6%   **73.** 120 travelers   **75.** 14.3%
77. 80 nickels   **79.** 48 mi   **81.** $x > 0$,   **83.** $0.5 \leq y < 1.5$,   **85.** $x < -4$,

87. $x \leq 4$,   **89.** $-\dfrac{1}{2} < x < \dfrac{3}{4}$,   **91.** $x \leq \dfrac{19}{3}$,   **93.** score must be less than 83

Chapter 2 Test   **1.** $y - 10$ **2.** $5.9x + 1.2$ **3.** $-2x + 10$ **4.** $-15y + 1$ **5.** -5 **6.** 8   **7.** $\dfrac{7}{10}$ **8.** 0   **9.** 27   **10.** $-\dfrac{19}{6}$

11. 3   **12.** $\dfrac{3}{11}$ **13.** 0.25   **14.** $\dfrac{25}{7}$ **15.** 21   **16.** 7 gal   **17.** \$8500 @ 10%; \$17,000 @ 12%   **18.** $2\dfrac{1}{2}$ hrs   **19.** $x = 6$

20. $h = \dfrac{V}{\pi r^2}$ **21.** $y = \dfrac{3x - 10}{4}$ **22.** $x < -2$,   **23.** $x < 4$,

24. $-1 < x < \dfrac{7}{3}$,   **25.** $\dfrac{7}{4} < x < 4$,   **26.** $x > \dfrac{2}{5}$,   **27.** 81.3%

28. \$5.9314 billion   **29.** $24.12°$ **30.** 17%   **31.** 13%   **32.** about 5980

Chapter 2 Cumulative Review   **1. a.** $11, 112$ **b.** $0, 11, 112$ **c.** $-3, -2, 0, 11, 112$ **d.** $-3, -2, 0, \dfrac{1}{4}, 11, 112$ **e.** $\sqrt{2}$
f. all numbers in the given set; Sec. 1.2, Ex. 5   **2. a.** 4   **b.** 5   **c.** 0; Sec. 1.2, Ex. 7

3. a. $2 \cdot 2 \cdot 2 \cdot 5$ **b.** $3 \cdot 3 \cdot 7$; Sec. 1.3, Ex. 1   **4.** $\dfrac{8}{20}$; Sec. 1.3, Ex. 6   **5.** 54; Sec. 1.4, Ex. 4   **6.** 2 is a solution; Sec. 1.4, Ex. 7

7. -3; Sec. 1.5, Ex. 2   **8.** 2; Sec. 1.5, Ex. 4   **9. a.** 10   **b.** $\dfrac{1}{2}$ **c.** $2x$ **d.** -6; Sec. 1.5, Ex. 10   **10. a.** 9.9   **b.** $-\dfrac{4}{5}$

c. $\dfrac{2}{15}$; Sec. 1.6, Ex. 2   **11. a.** $52°$ **b.** $118°$; Sec. 1.6, Ex. 8   **12. a.** -0.06 **b.** $-\dfrac{7}{15}$; Sec. 1.7, Ex. 3   **13. a.** 6   **b.** -12

c. $-\dfrac{8}{15}$; Sec. 1.7, Ex. 7   **14. a.** $5 + x$ **b.** $x \cdot 3$; Sec. 1.8, Ex. 1   **15. a.** $8(2 + x)$ **b.** $7(s + t)$; Sec. 1.8, Ex. 5

16. $-2x - 1$; Sec. 2.1, Ex. 7   **17.** $\dfrac{5}{4}$; Sec. 2.2, Ex. 3   **18.** 19; Sec. 2.2, Ex. 6   **19.** 140; Sec. 2.3, Ex. 3   **20.** 2; Sec. 2.4, Ex. 1

21. 10; Sec. 2.5, Ex. 1   **22.** $\dfrac{V}{wh} = l$; Sec. 2.6, Ex. 4   **23. a.** 0.35   **b.** 0.895   **c.** 1.5; Sec. 2.7, Ex. 1
24. 87.5%; Sec. 2.7, Ex. 5   **25.** $\{x | x \leq -10\}$; Sec. 2.9, Ex. 3

■ CHAPTER 3   GRAPHING

Mental Math   **1.** Answers may vary.; Ex. $(5, 5), (7, 3)$ **3.** Answers may vary.; Ex. $(3, 5), (3, 0)$

Exercise Set 3.1   **1.** quadrant I   **3.** no quadrant, x-axis   **5.** quadrant IV   **7.** no quadrant, x-axis

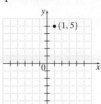
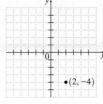
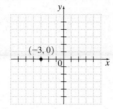
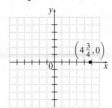

9. no quadrant, origin   **11.** no quadrant, y-axis   **13.** $A(0, 0); B\left(3\dfrac{1}{2}, 0\right); C(3, 2); D(-1, 3); E(-2, -2);$
$F(0, -1); G(2, -1)$

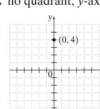

15. 26 units

17. a. $(1991, 1.14), (1992, 1.13), (1993, 1.11),$
$(1994, 1.11), (1995, 1.15), (1996, 1.23),$
$(1997, 1.23), (1998, 1.06), (1999, 1.17)$

b.
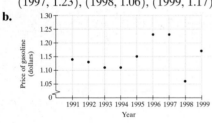

19. a. $(1994, 578), (1995, 613), (1996, 654), (1997, 675),$
$(1998, 717)$

b.
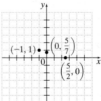

c. Average monthly mortgage payment increases
each year.

21. yes; no; yes   **23.** no; yes; yes   **25.** no; yes; yes   **27.** yes; no   **29.** no; no   **31.** yes; yes   **33.** yes; yes
35. $(-4, -2), (4, 0)$ **37.** $(0, 9), (3, 0)$ **39.** $(11, -7)$; Answers may vary. Ex. $(2, -7)$

41. $(0, 2), (6, 0), (3, 1)$ **43.** $(0, -12), (5, -2), (-3, -18)$ **45.** $\left(0, \dfrac{5}{7}\right), \left(\dfrac{5}{2}, 0\right), (-1, 1)$

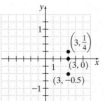
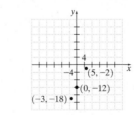
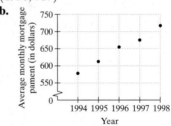

47. $(3, 0), (3, -0.5), \left(3, \dfrac{1}{4}\right)$ **49.** $(0, 0), (-5, 1), (10, -2)$ **51.** Answers may vary.

53. a. 13,000; 21,000; 29,000

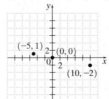

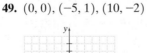

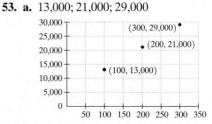

b. 45 desks

55. a. 29.219; 45.599; 54.699 **b.** 1977 **57.** In 1995, there were 670 Target stores. **59.** year 6: 66 stores; year 7: 60 stores; year 8: 55 stores **61.** $a = b$ **63.** quadrant IV **65.** quadrants II or III **67.** $y = 5 - x$

69. $y - -\dfrac{1}{2}x + \dfrac{5}{4}$ **71.** $y = 2x$ **73.** $y = \dfrac{1}{3}x - 2$

Calculator Explorations 1.

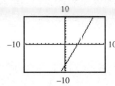

3.

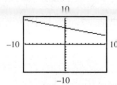

5

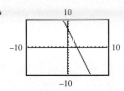

Exercise Set 3.2 1. yes **3.** yes **5.** no **7.** yes

9.

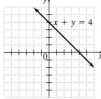

11.

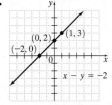

13.

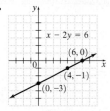

15.

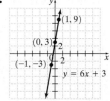

17.

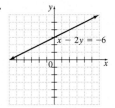

19.

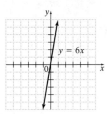

21.

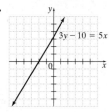

23.

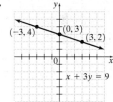

25.

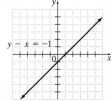

27.

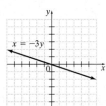

29.

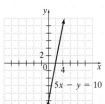

31.

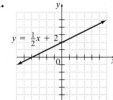

33.

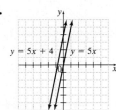

35.

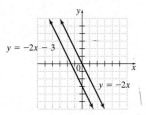

37.

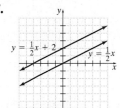

39. c **41.** d

43.

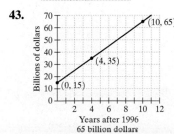

65 billion dollars

45.

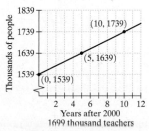

1699 thousand teachers

47. $y = x + 5$

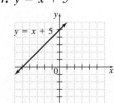

49. $2x + 3y = 6$

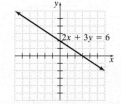

51. Answers may vary. **53.** Answers may vary.

55. $0, 1, 1, 4, 4$ **57.** $(4, -1)$ **59.** -5 **61.** $-\dfrac{1}{10}$ **63.** $(0, 3), (-3, 0)$ **65.** $(0, 0), (0, 0)$

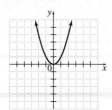

Calculator Explorations **1.**

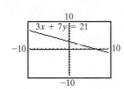

3.

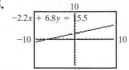

5.

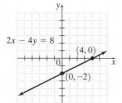

Mental Math **1.** false **3.** true

Exercise Set 3.3 **1.** $(-1, 0); (0, 1)$ **3.** $(-2, 0)$ **5.** $(-1, 0); (1, 0); (0, 1); (0, -2)$ **7.** infinite **9.** 0

11.

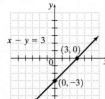

13.

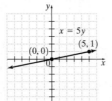

15.

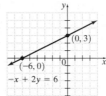

17.

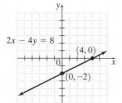

19.

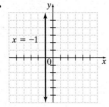

21.

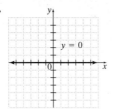

23.

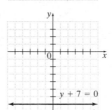

25.

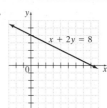

27.

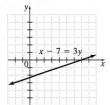

29.

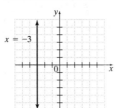

31.

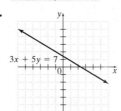

33.

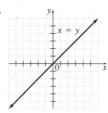

35.

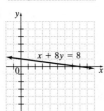

37.

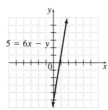

39.

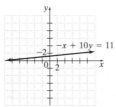

41.

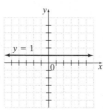

43.

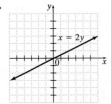

45.

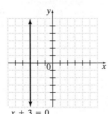

47.

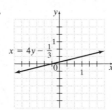

49.

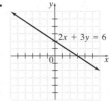

51. C **53.** E **55.** B

57. a. (0, 200); Answers may vary.   **59. a.** (20.8, 0) **b.** About 20.8 years after 1987, 0 music cassettes will be shipped.
  b. (400, 0); Answers may vary.   **61.** Answers may vary. **63.** Answers may vary.

c.

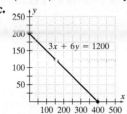

65.

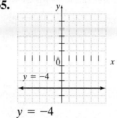

67. 1 **69.** 9 **71.** 0

d. 300 chairs

Calculator Explorations **1.**

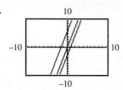

3.

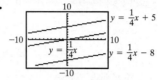

Mental Math **1.** upward **3.** horizontal

Exercise Set 3.4 **1.** $m = -\dfrac{4}{3}$ **3.** undefined slope **5.** $m = \dfrac{5}{2}$ **7.** $\dfrac{8}{7}$ **9.** -1 **11.** $-\dfrac{1}{4}$ **13.** undefined **15.** $-\dfrac{2}{3}$
17. undefined **19.** 0 **21.** line 1 **23.** line 2 **25.** D **27.** B **29.** E **31.** no slope **33.** $m = 0$ **35.** no slope
37. $m = 0$ **39. a.** 1 **b.** -1 **41. a.** $\dfrac{9}{11}$ **b.** $-\dfrac{11}{9}$ **43.** parallel **45.** perpendicular **47.** parallel **49.** neither
51. perpendicular **53.** $-\dfrac{1}{5}$ **55.** $\dfrac{1}{3}$ **57.** 1 **59.** $\dfrac{3}{5}$ **61.** 12.5% **63.** 40% **65.** 0.02 **67.** Every 1 year, there are/should
be 15 million more internet users. **69.** It costs $0.36 per 1 mile to own and operate a compact car. **71.** 18.7 mpg
73. 3.8 mpg **75.** from 1989 to 1990 **77.** 1990 **79.** $x = 6$ **81.** Answers may vary. **83.** -0.25 **85.** 0.875
87.

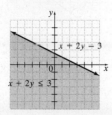

The line becomes steeper.

89. $x < 16$ **91.** $x \geq -2$

93.

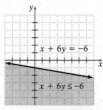

95.

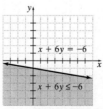

97.

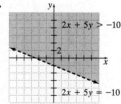

Mental Math **1.** yes **3.** yes **5.** yes **7.** no

Exercise Set 3.5 **1.** no; yes **3.** no; no **5.** no; yes

7.

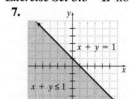

9.

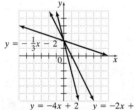

11.

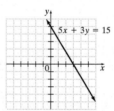

13.

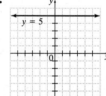

15.

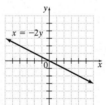

17.

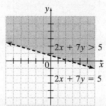

19.

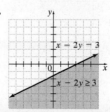

21.

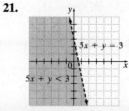

23.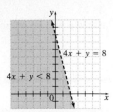

25.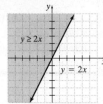

27.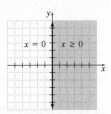

29.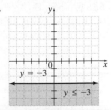

31.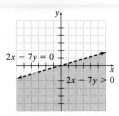

33.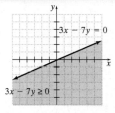

35.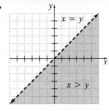

37.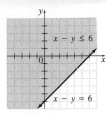

39.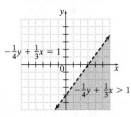

41.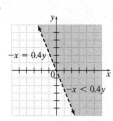

43. e   **45.** c   **47.** f   **49.** $x + y \geq 13$

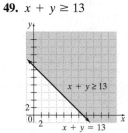

51. Answers may vary.   **53.** 8   **55.** -32 **57.** 48   **59.** 25   **61.** -2

Chapter 3 Review   **1.**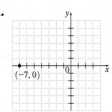

3.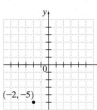

5.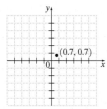

7. a. $(8.00, 1)$; $(7.50, 10)$; $(6.50, 25)$; $(5.00, 50)$; $(2.00, 100)$

9. no; yes   **11.** yes; yes   **13.** $(7, 44)$

15. $(-3, 0)$; $(1, 3)$; $(9, 9)$ **17.** $(0, 0)$; $(10, 5)$; $(-10, -5)$

b.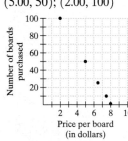

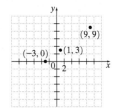

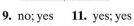

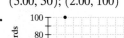

19.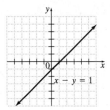

21.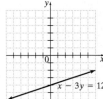

23.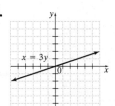

25.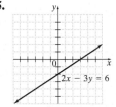

27. $506 billion

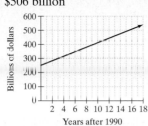

29. $y = -3; (0, -3)$ **31.** $x = -1; x = 2; x = 3; y = -2; (-1, 0); (2, 0); (3, 0); (0, -2)$

33.

35.

37.

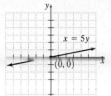

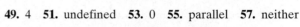

39.

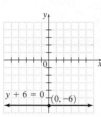

41. $m = \dfrac{1}{5}$ **43.** b   **45.** a   **47.** $\dfrac{3}{4}$ **49.** 4   **51.** undefined   **53.** 0   **55.** parallel   **57.** neither

59. Every 1 year, 1.24 million more persons have a bachelors degree or higher.

61.

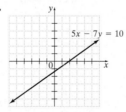

63.

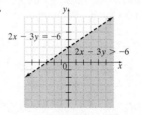

65.

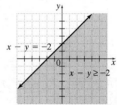

67.

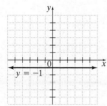

Chapter 3 Test   **1.** no   **2.** no   **3.** $(1, 1)$ **4.** $(-4, 17)$ **5.** $\dfrac{2}{5}$ **6.** 0   **7.** -1 **8.** -7 **9.** 3   **10.** undefined

11. parallel   **12.** neither   **13.**

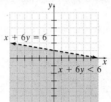

14.

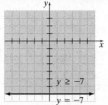

15.

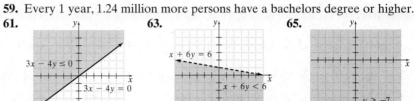

16.

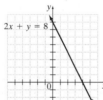

17.

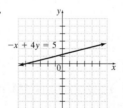

18.

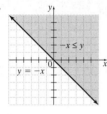

19.

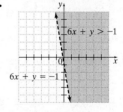

20.

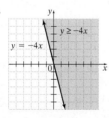

21.

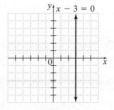

22.

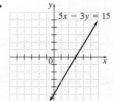

23. $y = 2x + 1$

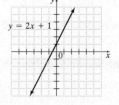

24. $x + 4y < -4$ **25.** $x + 2y = 21; x = 5$ m   **26. a.** (1986, 38); (1988, 44); (1990, 50); (1992, 53); (1994, 57); (1996, 62); (1997, 64)

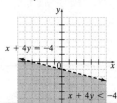

b.

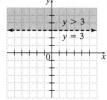

27. Every 1 year, 71 million more movie tickets are sold.

Chapter 3 Cumulative Review   **1. a.** $<$ **b.** $>$ **c.** $>$; Sec. 1.2, Ex. 1   **2.** $\frac{2}{39}$; Sec. 1.3, Ex. 3   **3.** $\frac{8}{3}$; Sec. 1.4, Ex. 3

4. a. -19 **b.** 30   **c.** -0.5 **d.** $-\frac{4}{5}$ **e.** 6.7   **f.** $\frac{1}{40}$; Sec. 1.5, Ex. 6   **5.** -6; Sec. 1.6, Ex. 4   **6. a.** -6 **b.** -24

c. $\frac{3}{4}$; Sec. 1.7, Ex. 10   **7. a.** $22 + x$ **b.** $-21x$; Sec. 1.8, Ex. 3   **8. a.** -3 **b.** 22   **c.** 1   **d.** -1 **e.** $\frac{1}{7}$; Sec. 2.1, Ex. 1

9. -1.6; Sec. 2.2, Ex. 2   **10.** $\frac{15}{4}$; Sec. 2.3, Ex. 5   **11.** $3x + 3$; Sec. 2.3, Ex. 8   **12.** 0; Sec. 2.4, Ex. 4

13. 46 Democratic senators; 54 Republican senators; Sec. 2.5, Ex. 3   **14.** 40 ft; Sec. 2.6, Ex. 2   **15.** $\frac{y - b}{m} = x$; Sec. 2.6, Ex. 5

16. a. 73%   **b.** 139%   **c.** 25%; Sec. 2.7, Ex. 2   **17.** 800; Sec. 2.7, Ex. 6   **18.** 40% solution: 81; 70% solution: 41; Sec. 2.8, Ex. 3

19. ⟵———•———⟶; Sec. 2.9, Ex. 1   **20.** $\{x | 1 \le x < 4\}$, ⟵———•———◦———⟶; Sec. 2.9, Ex. 10

21. a. solution   **b.** not a solution   **c.** solution; Sec. 3.1, Ex. 3
22. a. linear   **b.** linear   **c.** not linear   **d.** linear; Sec. 3.2, Ex. 1   **23.** 0; Sec. 3.4, Ex. 3

24. $\frac{3}{5}$; Sec. 3.4, Ex. 6   **25.** ; Sec. 3.5, Ex. 5

■ CHAPTER 4   EXPONENTS AND POLYNOMIALS

Mental Math   **1.** base: 3; exponent: 2   **3.** base: -3; exponent: 6   **5.** base: 4; exponent: 2
7. base: 5; exponent: 1; base: 3; exponent: 4   **9.** base: 5; exponent: 1; base: x; exponent: 2

Exercise Set 4.1   **1.** 49   **3.** -5 **5.** -16 **7.** 16   **9.** $\frac{1}{27}$ **11.** 112   **13.** Answers may vary.   **15.** 4   **17.** 135   **19.** 150

21. $\frac{32}{5}$ **23.** 343 cu. m   **25.** volume   **27.** x^7 **29.** $(-3)^{12}$ **31.** $15y^5$ **33.** $-24z^{20}$ **35.** p^7q^7 **37.** $\frac{m^9}{n^9}$ **39.** $x^{10}y^{15}$

41. $\frac{4x^2z^2}{y^{10}}$ **43.** x^2 **45.** 4   **47.** p^6q^5 **49.** $\frac{y^3}{2}$ **51.** 1   **53.** -2 **55.** 2   **57.** $\frac{-27a^6}{b^9}$ **59.** x^{39} **61.** $\frac{z^{14}}{625}$ **63.** $7776m^4n^3$

65. -25 **67.** $\frac{1}{64}$ **69.** $81x^2y^2$ **71.** 1   **73.** 40   **75.** b^6 **77.** a^9 **79.** $-16x^7$ **81.** $64a^3$ **83.** $36x^2y^2z^6$ **85.** $\frac{y^{15}}{8x^{12}}$ **87.** x

89. $2x^2y$ **91.** $243x^7y^{24}$ **93.** Answers may vary.   **95.** $20x^5$ sq. ft   **97.** $25\pi y^2$ sq. cm   **99.** $27y^{12}$ cu. ft   **101.** $-2x + 7$

103. $2y - 10$ **105.** $-x - 4$ **107.** $-x + 5$ **109.** x^{9a} **111.** a^{5b} **113.** x^{5a} **115.** $x^{5a^2}y^{5ab}z^{5ac}$

Mental Math   **1.** $-14y$ **3.** $7y^3$ **5.** $7x$

Exercise Set 4.2   **1.** 1; binomial   **3.** 3; none of these   **5.** 6; trinomial   **7.** 2; binomial   **9.** 3   **11.** 2
13. Answers may vary.   **15.** Answers may vary.   **17. (a)** 6   **(b)** 5   **19. (a)** -2 **(b)** 4   **21. (a)** -15 **(b)** -16
23. -45.24 ft; the object has reached the ground.   **25.** $23x^2$ **27.** $12x^2 - y$ **29.** $7s$ **31.** $-1.1y^2 + 4.8$

33. $-\dfrac{7}{12}x^3 + \dfrac{7}{5}x^2 + 6$ **35.** $5a^2 - 9ab + 16b^2$ **37.** $12x + 12$ **39.** $-3x^2 + 10$ **41.** $-x^2 + 14$ **43.** $-2x + 9$
45. $2x^2 + 7x - 16$ **47.** $(x^2 + 7x + 4)$ ft   **49.** $(3y^2 + 4y + 11)$ m   **51.** $8t^2 - 4$ **53.** $-2z^2 - 16z + 6$
55. $2x^3 - 2x^2 + 7x + 2$ **57.** $62x^2 + 5$ **59.** $12x + 2$ **61.** $-11x$ **63.** $7x - 13$ **65.** $-y^2 - 3y - 1$ **67.** $2x^2 + 11x$
69. $-16x^4 + 8x + 9$ **71.** $7x^2 + 14x + 18$ **73.** $3x - 3$ **75.** $7x^2 - 2x + 2$ **77.** $4y^2 + 12y + 19$
79. $6x^2 - 5x + 21$ **81.** $-6.6x^2 - 1.8x - 1.8$ **83.** $4x^2 + 7x + x^2 + 5x; 5x^2 + 12x$
85. a. 184 ft   **b.** 600 ft   **c.** 595.84 ft   **d.** 362.56 ft   **87.** 12.5 sec   **89.** $-2a - b + 1$ **91.** $3x^2 + 5$ **93.** $6x^2 - 2xy + 19y^2$
95. $8r^2s + 16rs - 8 + 7r^2s^2$ **97.** 11.7 million   **99.** $928x^2 - 858x + 43{,}288$ **101.** $6x^2$ **103.** $-12x^8$ **105.** $200x^3y^2$

Mental Math   **1.** $10xy$ **3.** x^7 **5.** $18x^3$

Exercise Set 4.3   **1.** $4a^2 - 8a$ **3.** $7x^3 + 14x^2 - 7x$ **5.** $6x^4 - 3x^3$ **7.** $x^2 + 3x$ **9.** $a^2 + 5a - 14$
11. $4y^2 - 16y + 16$ **13.** $30x^2 - 79xy + 45y^2$ **15.** $4x^4 - 20x^2 + 25$ **17.** $x^2 + 5x + 6$ **19.** $x^3 - 5x^2 + 13x - 14$
21. $x^4 + 5x^3 - 3x^2 - 11x + 20$ **23.** $10a^3 - 27a^2 + 26a - 12$ **25.** $x^3 + 6x^2 + 12x + 8$ **27.** $8y^3 - 36y^2 + 54y - 27$
29. $2x^3 + 10x^2 + 11x - 3$ **31.** $x^5 - 4x^3 - 7x^2 - 45x + 63$ **33. a.** 25; 13   **b.** 324; 164   **c.** no; Answers may vary.
35. $2a^2 + 8a$ **37.** $6x^3 - 9x^2 + 12x$ **39.** $15x^2 + 37xy + 18y^2$ **41.** $x^3 + 7x^2 + 16x + 12$ **43.** $49x^2 + 56x + 16$
45. $-6a^4 + 4a^3 - 6a^2$ **47.** $x^3 + 10x^2 + 33x + 36$ **49.** $a^3 + 3a^2 + 3a + 1$ **51.** $x^2 + 2xy + y^2$ **53.** $x^2 - 13x + 42$
55. $3a^3 + 6a$ **57.** $-4y^3 - 12y^2 + 44y$ **59.** $25x^2 - 1$ **61.** $5x^3 - x^2 + 16x + 16$ **63.** $8x^3 - 60x^2 + 150x - 125$
65. $32x^3 + 48x^2 - 6x - 20$ **67.** $49x^2y^2 - 14xy^2 + y^2$ **69.** $5y^4 - 16y^3 - 4y^2 - 7y - 6$
71. $6x^4 - 8x^3 - 7x^2 + 22x - 12$ **73.** $(4x^2 - 25)$ sq. yds   **75.** $(6x^2 - 4x)$ sq. in.   **77.** $(x^2 + 6x + 5)$ sq. units
79. a. $a^2 - b^2$ **b.** $4x^2 - 9y^2$ **c.** $16x^2 - 49$ **d.** Answers may vary.   **81.** $16p^2$ **83.** $49m^4$ **85.** \$3500   **87.** \$500
89. There is a loss in value each year.

Mental Math   **1.** false   **3.** false

Exercise Set 4.4   **1.** $x^2 + 7x + 12$ **3.** $x^2 + 5x - 50$ **5.** $5x^2 + 4x - 12$ **7.** $4y^2 - 25y + 6$ **9.** $6x^2 + 13x - 5$
11. $x^2 - 4x + 4$ **13.** $4x^2 - 4x + 1$ **15.** $9a^2 - 30a + 25$ **17.** $25x^2 + 90x + 81$ **19.** Answers may vary.

21. $a^2 - 49$ **23.** $9x^2 - 1$ **25.** $9x^2 - \dfrac{1}{4}$ **27.** $81x^2 - y^2$ **29.** $(4x^2 + 4x + 1)$ sq. ft   **31.** $a^2 + 9a + 20$

33. $a^2 + 14a + 49$ **35.** $12a^2 - a - 1$ **37.** $x^2 - 4$ **39.** $9a^2 + 6a + 1$ **41.** $4x^2 + 3xy - y^2$ **43.** $x^3 - 3x^2 - 17x + 3$

45. $4a^2 - 12a + 9$ **47.** $25x^2 - 36z^2$ **49.** $x^2 - 8x + 15$ **51.** $x^2 - \dfrac{1}{9}$ **53.** $a^2 + 8a - 33$ **55.** $x^2 - 4x + 4$

57. $6b^2 - b - 35$ **59.** $49p^2 - 64$ **61.** $\dfrac{1}{9}a^4 - 49$ **63.** $15x^4 - 5x^3 + 10x^2$ **65.** $4r^2 - 9s^2$ **67.** $9x^2 - 42xy + 49y^2$

69. $16x^2 - 25$ **71.** $x^2 + 8x + 16$ **73.** $a^2 - \dfrac{1}{4}y^2$ **75.** $\dfrac{1}{25}x^2 - y^2$ **77.** $3a^3 + 2a^2 + 1$ **79.** $(3x^2 - 9)$ sq. units

81. $(24x^2 - 32x + 8)$ sq. m   **83.** Answers may vary.   **85.** x^2y^4 **87.** $-2a^4$ **89.** $-\dfrac{3b^3}{2}$ **91.** $-\dfrac{1}{2}$ **93.** $-\dfrac{3}{2}$

95. $a^2 + 2ac + c^2 - 25$ **97.** $x^2 - 4x + 4 - y^2$

Calculator Explorations   **1.** 5.31 EE 3   **3.** 6.6 EE −9   **5.** 1.5×10^{13} **7.** 8.15×10^{19}

Mental Math   **1.** $\dfrac{5}{x^2}$ **3.** y^6 **5.** $4y^3$

Exercise Set 4.5   **1.** $\dfrac{1}{64}$ **3.** $\dfrac{7}{x^3}$ **5.** -64 **7.** $\dfrac{5}{6}$ **9.** p^3 **11.** $\dfrac{q^4}{p^5}$ **13.** $\dfrac{1}{x^3}$ **15.** z^3 **17.** $\dfrac{4}{3}$ **19.** $\dfrac{1}{9}$ **21.** $-p^4$ **23.** -2

25. x^4 **27.** p^4 **29.** m^{11} **31.** r^6 **33.** $\dfrac{1}{x^{15}y^9}$ **35.** $\dfrac{1}{x^4}$ **37.** $\dfrac{1}{a^2}$ **39.** $4k^3$ **41.** $3m$ **43.** $-\dfrac{4a^5}{b}$ **45.** $-\dfrac{6x}{7y^2}$ **47.** $\dfrac{a^{30}}{b^{12}}$

49. $\dfrac{1}{x^{10}y^6}$ **51.** $\dfrac{z^2}{4}$ **53.** $\dfrac{1}{32x^5}$ **55.** $\dfrac{49a^4}{b^6}$ **57.** $a^{24}b^8$ **59.** y^9y^{19} **61.** $-\dfrac{y^8}{8x^2}$ **63.** $\dfrac{27}{x^6z^3}$ cu. in.   **65.** 7.8×10^4
67. 1.67×10^{-6} **69.** 6.35×10^{-3} **71.** 1.16×10^6 **73.** 2.0×10^7 **75.** 9.3×10^7 **77.** 1.2×10^8
79. 0.0000000008673   **81.** 0.033   **83.** 20,320   **85.** 6,250,000,000,000,000,000   **87.** 9,460,000,000,000   **89.** 0.000036
91. 0.0000000000000000028   **93.** 0.0000005   **95.** 200,000   **97.** 1.512×10^{10} cu. ft   **99.** Answers may vary.   **101.** 100
103. -394.5 **105.** 1.3 sec   **107.** $\dfrac{5x^3}{3}$ **109.** $\dfrac{5z^3y^2}{7}$ **111.** $5y - 6 + \dfrac{5}{y}$ **113.** $20x^3 - 12x^2 + 2x$ **115.** a^m **117.** $27y^{6z}$

119. y^{5a} **121.** $\dfrac{1}{z^{6a+4}}$

Mental Math   **1.** a^2 **3.** a^2 **5.** k^3 **7.** p^5 **9.** k^2

Exercise Set 4.6   **1.** $5p^2 + 6p$ **3.** $-\dfrac{3}{2x} + 3$ **5.** $-3x^2 + x - \dfrac{4}{x^3}$ **7.** $-1 + \dfrac{3}{2x} - \dfrac{7}{4x^4}$ **9.** $5x^3 - 3x + \dfrac{1}{x^2}$

11. $(3x^3 + x - 4)$ ft   **13.** $x + 1$ **15.** $2x + 3$ **17.** $2x + 1 + \dfrac{7}{x-4}$ **19.** $4x + 9$ **21.** $3a^2 - 3a + 1 + \dfrac{2}{3a+2}$

23. $2b^2 + b + 2 - \dfrac{12}{b+4}$ **25.** Answers may vary.   **27.** $(2x + 5)$ m   **29.** $\dfrac{4}{x} + \dfrac{1}{x^2} + \dfrac{9}{5x^3}$ **31.** $5x - 2 + \dfrac{2}{x+6}$

33. $x^2 - \dfrac{12x}{5} - 1$ **35.** $6x - 1 - \dfrac{1}{x+3}$ **37.** $4x^2 + 1$ **39.** $2x^2 + 6x - 5 - \dfrac{2}{x-2}$ **41.** $6x - 1$ **43.** $-x^3 + 3x^2 - \dfrac{4}{x}$

45. $4x + 3 - \dfrac{2}{2x+1}$ **47.** $2x + 9$ **49.** $2x^2 + 3x - 4$ **51.** $x^2 + 3x + 9$ **53.** $x^2 - x + 1$ **55.** $-3x + 6 - \dfrac{11}{x+2}$

57. $2b - 1 - \dfrac{6}{2b-1}$ **59.** $2a^3 + 2a$ **61.** $2x^3 + 14x^2 - 10x$ **63.** $-3x^2y^3 - 21x^3y^2 - 24xy$

65. $9a^2b^3c + 36ab^2c - 72ab$ **67.** The Rolling Stones (1994)   **69.** \$80 million

Chapter 4 Review   **1.** base: 3; exponent: 2   **3.** base: 5; exponent: 4   **5.** 36   **7.** -65 **9.** 1   **11.** 8   **13.** $-10x^5$ **15.** $\dfrac{b^4}{16}$

17. $\dfrac{x^6y^6}{4}$ **19.** $40a^{19}$ **21.** $\dfrac{3}{64}$ **23.** $\dfrac{1}{x}$ **25.** 5   **27.** 1   **29.** $6a^6b^9$ **31.** 7   **33.** 8   **35.** 5   **37.** 5

39. a. $3, -2, 1, -1, -6; 3, 2, 2, 2, 0$ **b.** 3   **41.** $22; 78; 154.02; 400$ **43.** $22x^2y^3 + 3xy + 6$ **45.** cannot be combined
47. $2s^5 + 3s^4 + 4s^3 + s^2 - 7s - 6$ **49.** $8r^2 + 11rs + 7s^2$ **51.** $4x - 13y$ **53.** \$99.44   **55.** $-56xz^2$ **57.** $-12x^2a^2y^4$
59. $9x - 63$ **61.** $54a - 27$ **63.** $-32y^3 + 48y$ **65.** $-3a^3b - 3a^2b - 3ab^2$ **67.** $42b^4 - 28b^2 + 14b$
69. $6x^2 - 11x - 10$ **71.** $42a^2 + 11a - 3$ **73.** $x^6 + 2x^5 + x^2 + 3x + 2$ **75.** $x^6 + 8x^4 + 16x^2 - 16$
77. $8x^3 - 60x^2 + 150x - 125$ **79.** $15y^3 - 3y^2 + 6y$ **81.** $12a^2 + 25a - 7$ **83.** $x^2 - 10x + 25$ **85.** $16x^2 + 16x + 4$

87. $x^3 - 3x^2 + 4$ **89.** $25x^2 - 1$ **91.** $a^2 - 4b^2$ **93.** $16a^4 - 4b^2$ **95.** $-\dfrac{1}{49}$ **97.** $\dfrac{1}{16x^4}$ **99.** $\dfrac{9}{4}$ **101.** $\dfrac{1}{42}$ **103.** $-q^3r^3$

105. $\dfrac{r^5}{s^3}$ **107.** $-\dfrac{s^4}{4}$ **109.** $\dfrac{y^6}{3}$ **111.** $\dfrac{x^6y^9}{64z^3}$ **113.** $\dfrac{y^8}{49}$ **115.** $\dfrac{x^3}{y^3}$ **117.** $\dfrac{a^{10}}{b^{10}}$ **119.** $-\dfrac{4}{27}$ **121.** x^{10+3h} **123.** a^{2m+5}

125. 8.868×10^{-1} **127.** -8.68×10^5 **129.** 4.0×10^3 **131.** 0.00386 **133.** $893{,}600$

135. $0.00000000000000000000000003$ **137.** $400{,}000{,}000{,}000$ **139.** $-a^2 + 3b - 4$ **141.** $4x + \dfrac{7}{x+5}$

143. $3b^2 - 4b - \dfrac{1}{3b-2}$ **145.** $-x^2 - 16x - 117 - \dfrac{684}{x-6}$

Chapter 4 Test   **1.** 32   **2.** 81   **3.** -81 **4.** $\dfrac{1}{64}$ **5.** $-15x^{11}$ **6.** y^5 **7.** $\dfrac{1}{r^5}$ **8.** $\dfrac{y^{14}}{x^2}$ **9.** $\dfrac{1}{6xy^8}$ **10.** 5.63×10^5

11. 8.63×10^{-5} **12.** 0.0015 **13.** $62{,}300$ **14.** 0.036 **15. a.** $4, 7, 1, -2; 3, 3, 4, 0$ **b.** 4   **16.** $-2x^2 + 12xy + 11$
17. $16x^3 + 7x^2 - 3x - 13$ **18.** $-3x^3 + 5x^2 + 4x + 5$ **19.** $x^3 + 8x^2 + 3x - 5$ **20.** $3x^3 + 22x^2 + 41x + 14$
21. $6x^4 - 9x^3 + 21x^2$ **22.** $3x^2 + 16x - 35$ **23.** $9x^2 - 49$ **24.** $16x^2 - 16x + 4$ **25.** $64x^2 + 48x + 9$

26. $x^4 - 81b^2$ **27.** 1001 ft; 985 ft; 857 ft; 601 ft   **28.** $\dfrac{2}{x^2yz}$ **29.** $\dfrac{x}{2y} + \dfrac{1}{4} - \dfrac{7}{8y}$ **30.** $x + 2$ **31.** $9x^2 - 6x + 4 - \dfrac{16}{3x+2}$

32. 2917 thousand cases

Chapter 4 Cumulative Review   **1. a.** true   **b.** true   **c.** false   **d.** true; Sec. 1.2, Ex. 2   **2. a.** $\dfrac{64}{25}$ **b.** $\dfrac{1}{20}$

c. $\dfrac{5}{4}$; Sec. 1.3, Ex. 4   **3. a.** 9   **b.** 125   **c.** 16   **d.** 7   **e.** $\dfrac{9}{49}$; Sec. 1.4, Ex. 1   **4. a.** -10 **b.** -21 **c.** -12; Sec. 1.5, Ex. 3

5. -12; Sec. 1.6, Ex. 3   **6. a.** $\dfrac{1}{22}$ **b.** $\dfrac{16}{3}$ **c.** $-\dfrac{1}{10}$ **d.** $-\dfrac{13}{9}$; Sec. 1.7, Ex. 5   **7. a.** $(5 + 4) + 6$ **b.** $-1 \cdot (2 \cdot 5)$; Sec. 1.8, Ex. 2

8. a. Alaska Village Electric   **b.** American Electric Power (Kentucky)   **c.** American Electric Power: 5¢ per kilowatt-hour;
Green Mountain Power: 11¢ per kilowatt-hour; Montana Power Co.: 6¢ per kilowatt-hour;
Alaska Village Electric: 42¢ per kilowatt-hour   **d.** 37¢; Sec. 1.9, Ex. 1
9. a. $5x + 10$ **b.** $-2y - 0.6z + 2$ **c** $-x - y + 2z - 6$; Sec. 2.1, Ex. 5   **10.** 17; Sec. 2.2, Ex. 1   **11.** 6; Sec. 2.3, Ex. 1

12. -10; Sec. 2.4, Ex. 8   **13.** $\dfrac{5F - 160}{9} = C$; Sec. 2.6, Ex. 7   **15.** width: 4 ft; length: 10 ft; Sec. 2.8, Ex. 1

16. 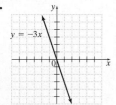; Sec. 2.9, Ex. 2   **17. a.** $(0, 12)$ **18.** 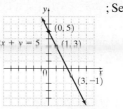 ; Sec. 3.2, Ex. 2
b. $(2, 6)$
c. $(-1, 15)$; Sec. 3.1, Ex. 4

19. 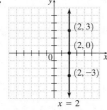 ; Sec. 3.3, Ex. 5   **20.** undefined slope; Sec. 3.4, Ex. 4   **21.** 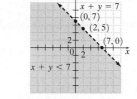 ; Sec. 3.5, Ex. 1

22. $-6x^7$; Sec. 4.1, Ex. 4   **23.** $-4x^2 + 6x + 2$; Sec. 4.2, Ex. 8   **24.** $4x^2 - 4xy + y^2$; Sec. 4.3, Ex. 3
25. $3m + 1$; Sec. 4.6, Ex. 1

■ CHAPTER 5   FACTORING POLYNOMIALS

Mental Math   **1.** $2 \cdot 7$ **3.** $2 \cdot 5$ **5.** 3   **7.** 3

Exercise Set 5.1   **1.** 4   **3.** 6   **5.** y^2 **7.** xy^2 **9.** 4   **11.** $4y^3$ **13.** $3x^3$ **15.** $9x^2y$ **17.** $15(2x - 1)$ **19.** $6cd(4d^2 - 3c)$
21. $-6a^3x(4a - 3)$ **23.** $4x(3x^2 + 4x - 2)$ **25.** $5xy(x^2 - 3x + 2)$ **27.** Answers may vary.   **29.** $(x + 2)(y + 3)$
31. $(y - 3)(x - 4)$ **33.** $(x + y)(2x - 1)$ **35.** $(x + 3)(5 + y)$ **37.** $(y - 4)(2 + x)$ **39.** $(y - 2)(3x + 8)$
41. $(y + 3)(y^2 + 1)$ **43.** $12x^3 - 2x; 2x(6x^2 - 1)$ **45.** $200x + 25\pi; 25(8x + \pi)$ **47.** $3(x - 2)$ **49.** $2x(16y - 9x)$
51. $4(x - 2y + 1)$ **53.** $(x + 2)(8 - y)$ **55.** $-8x^8y^5(5y + 2x)$ **57.** $-3(x - 4)$ **59.** $6x^3y^2(3y - 2 + x^2)$
61. $(x - 2)(y^2 + 1)$ **63.** $(y + 3)(5x + 6)$ **65.** $(x - 2y)(4x - 3)$ **67.** $42yz(3x^3 + 5y^3z^2)$ **69.** $(3 - x)(5 + y)$
71. $2(3x^2 - 1)(2y - 7)$ **73.** Answers may vary.   **75.** factored   **77.** not factored   **79.** $(n^3 - 6)$ units
81. a. 850 million   **b.** 1855 million   **c.** $5(12x^2 - 17x + 156)$ **83** $x^2 + 7x + 10$ **85.** $a^2 - 15a + 56$ **87.** 2, 6
89. $-1, -8$ **91.** $-2, 5$ **93.** $-8, 3$

Mental Math   **1.** $+5$ **3.** -3 **5.** $+2$

Exercise Set 5.2   **1.** $(x + 6)(x + 1)$ **3.** $(x + 5)(x + 4)$ **5.** $(x - 5)(x - 3)$ **7.** $(x - 9)(x - 1)$ **9.** prime
11. $(x - 6)(x + 3)$ **13.** prime   **15.** $(x + 5y)(x + 3y)$ **17.** $(x - y)(x - y)$ **19.** $(x - 4y)(x + y)$
21. $2(z + 8)(z + 2)$ **23.** $2x(x - 5)(x - 4)$ **25.** $7(x + 3y)(x - y)$ **27.** product; sum   **29.** $(x + 12)(x + 3)$
31. $(x - 2)(x + 1)$ **33.** $(r - 12)(r - 4)$ **35.** $(x - 7)(x + 3)$ **37.** $(x + 5y)(x + 2y)$ **39.** prime
41. $2(t + 8)(t + 4)$ **43.** $x(x - 6)(x + 4)$ **45.** $(x - 9)(x - 7)$ **47.** $(x + 2y)(x - y)$ **49.** $3(x - 18)(x - 2)$
51. $(x - 24)(x + 6)$ **53.** $6x(x + 4)(x + 5)$ **55.** $2t^3(t - 4)(t - 3)$ **57.** $5xy(x - 8y)(x + 3y)$ **59.** $4y(x^2 + x - 3)$
61. $2b(a - 7b)(a - 3b)$ **63.** 8; 16   **65.** 6; 26   **67.** 5; 8; 9   **69.** 3; 4   **71.** Answers may vary.   **73.** $2x^2 + 11x + 5$
75. $15y^2 - 17y + 4$ **77.** $9a^2 + 23a - 12$
79. 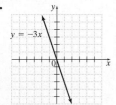   **81.** 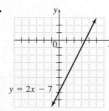   **83.** $2y(x + 5)(x + 10)$
85. $-12y^3(x^2 + 2x + 3)$
87. $(x + 1)(y - 5)(y + 3)$

Mental Math   **1.** yes   **3.** no   **5.** yes

Exercise Set 5.3   **1.** $(2x + 3)(x + 5)$ **3.** $(2x + 1)(x - 5)$ **5.** $(2y + 3)(y - 2)$ **7.** $(4a - 3)^2$ **9.** $(9r - 8)(4r + 3)$
11. $(5x + 1)(2x + 3)$ **13.** $3(7x + 5)(x - 3)$ **15.** $2(2x - 3)(3x + 1)$ **17.** $x(4x + 3)(x - 3)$ **19.** $(x + 11)^2$
21. $(x - 8)^2$ **23.** $(4y - 5)^2$ **25.** $(xy - 5)^2$ **27.** Answers may vary.   **29.** $(2x + 11)(x - 9)$ **31.** $(2x - 7)(2x + 3)$

33. $(6x - 7)(5x - 3)$ **35.** $(4x - 9)(6x - 1)$ **37.** $(3x - 4y)^2$ **39.** $(x - 7y)^2$ **41.** $(2x + 5)(x + 1)$ **43.** prime
45. $(5 - 2y)(2 + y)$ **47.** $(4x + 3y)^2$ **49.** $2y(4x - 7)(x + 6)$ **51.** $(3x - 2)(x + 1)$ **53.** $(xy + 2)^2$ **55.** $(7y + 3x)^2$
57. $3(x^2 - 14x + 21)$ **59.** $(7a - 6)(6a - 1)$ **61.** $(6x - 7)(3x + 2)$ **63.** $(5p - 7q)^2$ **65.** $(5x + 3)(3x - 5)$
67. $(7t + 1)(t - 4)$ **69.** $a^2 + 2ab + b^2$ **71.** $2; 14$ **73.** $5; 13$ **75.** 2 **77.** $4; 5$ **79.** $x^2 - 4$ **81.** $a^3 + 27$
83. $y^3 - 125$ **85.** \$75,000 and above   **87.** Answers may vary.   **89.** $-3xy^2(4x - 5)(x + 1)$
91. $-2pq(p - 3q)(15p + q)$ **93.** $(y - 1)^2(4x^2 + 10x + 25)$

Calculator Explorations

16	14	16
16	14	16
2.89	0.89	2.89
171.61	169.61	171.61
1	−1	1

Mental Math   1. 1^2 **3.** 9^2 **5.** 3^2 **7.** 1^3 **9.** 2^3

Exercise Set 5.4   1. $(x + 2)(x - 2)$ **3.** $(y + 7)(y - 7)$ **5.** $(5y - 3)(5y + 3)$ **7.** $(11 - 10x)(11 + 10x)$
9. $3(2x - 3)(2x + 3)$ **11.** $(13a - 7b)(13a + 7b)$ **13.** $(xy - 1)(xy + 1)$ **15.** $(x^2 + 3)(x^2 - 3)$
17. $(7a^2 + 4)(7a^2 - 4)$ **19.** $(x^2 + y^5)(x^2 - y^5)$ **21.** $(x + 6)$ **23.** $(a + 3)(a^2 - 3a + 9)$ **25.** $(2a + 1)(4a^2 - 2a + 1)$
27. $5(k + 2)(k^2 - 2k + 4)$ **29.** $(xy - 4)(x^2y^2 + 4xy + 16)$ **31.** $(x + 5)(x^2 - 5x + 25)$
33. $3x(2x - 3y)(4x^2 + 6xy + 9y^2)$ **35.** $(2x + y)$ **37.** $(x - 2)(x + 2)$ **39.** $(9 - p)(9 + p)$ **41.** $(2r - 1)(2r + 1)$
43. $(3x - 4)(3x + 4)$ **45.** prime   **47.** $(3 - t)(9 + 3t + t^2)$ **49.** $8(r - 2)(r^2 + 2r + 4)$ **51.** $(t - 7)(t^2 + 7t + 49)$
53. $(x - 13y)(x + 13y)$ **55.** $(xy - z)(xy + z)$ **57.** $(xy + 1)(x^2y^2 - xy + 1)$ **59.** $(s - 4t)(s^2 + 4st + 16t^2)$
61. $2(3r - 2)(3r + 2)$ **63.** $x(3y - 2)(3y + 2)$ **65.** $25y^2(y - 2)(y + 2)$ **67.** $xy(x - 2y)(x + 2y)$
69. $4s^3t^3(2s^3 + 25t^3)$ **71.** $xy^2(27xy - 1)$ **73. a.** 777 ft   **b.** 441 ft   **c.** 7 sec   **d.** $(29 + 4t)(29 - 4t)$

75. Answers may vary.   **77.** $4x^3 + 2x^2 - 1 + \dfrac{3}{x}$ **79.** $2x + 1$ **81.** $3x + 4 - \dfrac{2}{x + 3}$ **83.** $(a - 2 - b)(a + 2 + b)$
85. $(x - 2)^2(x + 1)(x + 3)$

Exercise Set 5.5   1. $(a + b)^2$ **3.** $(a - 3)(a + 4)$ **5.** $(a + 2)(a - 3)$ **7.** $(x + 1)^2$ **9.** $(x + 1)(x + 3)$
11. $(x + 3)(x + 4)$ **13.** $(x + 4)(x - 1)$ **15.** $(x + 5)(x - 3)$ **17.** $(x - 6)(x + 5)$ **19.** $2(x - 7)(x + 7)$
21. $(x + 3)(x + y)$ **23.** $(x + 8)(x - 2)$ **25.** $4x(x + 7)(x - 2)$ **27.** $2(3x + 4)(2x + 3)$ **29.** $(2a - b)(2a + b)$
31. $(5 - 2x)(4 + x)$ **33.** prime   **35.** $(4x - 5)(x + 1)$ **37.** $4(t^2 + 9)$ **39.** $(x + 1)(a + 2)$ **41.** $4a(3a^2 - 6a + 1)$
43. prime   **45.** $(5p - 7q)^2$ **47.** $(5 - 2y)(25 + 10y + 4y^2)$ **49.** $(5 - x)(6 + x)$ **51.** $(7 - x)(2 + x)$
53. $3x^2y(x + 6)(x - 4)$ **55.** $5xy^2(x - 7y)(x - y)$ **57.** $3xy(4x^2 + 81)$ **59.** $(x - y - z)(x - y + z)$
61. $(s + 4)(3r - 1)$ **63.** $(4x - 3)(x - 2y)$ **65.** $6(x + 2y)(x + y)$ **67.** $(x + 3)(y - 2)(y + 2)$
69. $(5 + x)(x + y)$ **71.** $(7t - 1)(2t - 1)$ **73.** $(3x + 5)(x - 1)$ **75.** $(x + 12y)(x - 3y)$ **77.** $(1 - 10ab)(1 + 2ab)$
79. $(x - 3)(x + 3)(x - 1)(x + 1)$ **81.** $(x - 4)(x + 4)(x^2 + 2)$ **83.** $(x - 15)(x - 8)$ **85.** $2x(3x - 2)(x - 4)$
87. $(3x - 5y)(9x^2 + 15xy + 25y^2)$ **89.** $(xy + 2z)(x^2y^2 - 2xyz + 4z^2)$ **91.** $2xy(1 - 6x)(1 + 6x)$
93. $(x - 2)(x + 2)(x + 6)$ **95.** $2a^2(3a + 5)$ **97.** $(a^2 + 2)(a + 2)$ **99.** $(x - 2)(x + 2)(x + 7)$ **101.** Answers may vary.
103. 6   **105.** −2   **107.** $\dfrac{1}{5}$ **109.** 8 in.   **111.** $(-2, 0); (4, 0); (0, 2); (0, -2)$ **113.** $(2, 0); (4, 0); (0, 4)$

Calculator Explorations   1. $-0.9, 2.2$ **3.** no real solution   **5.** $-1.8, 2.8$

Mental Math   1. $3, 7$ **3.** $-8, -6$ **5.** $-1, 3$

Exercise Set 5.6   1. $2, -1$ **3.** $0, -6$ **5.** $-\dfrac{3}{2}, \dfrac{5}{4}$ **7.** $\dfrac{7}{2}, -\dfrac{2}{7}$ **9.** $(x - 6)(x + 1) = 0$ **11.** $9, 4$ **13.** $-4, 2$ **15.** $8, -4$

17. $\dfrac{7}{3}, -2$ **19.** $\dfrac{8}{3}, -9$ **21.** $x^2 - 12x + 35 = 0$ **23.** $0, 8, 4$ **25.** $\dfrac{3}{4}$ **27.** $0, \dfrac{1}{2}, -\dfrac{1}{2}$ **29.** $0, \dfrac{1}{2}, -\dfrac{3}{8}$ **31.** $0, -7$ **33.** $-5, 4$

35. $-5, 6$ **37.** $-\dfrac{4}{3}, 5$ **39.** $-\dfrac{3}{2}, -\dfrac{1}{2}, 3$ **41.** $-5, 3$ **43.** $0, 16$ **45.** $-\dfrac{9}{2}, \dfrac{8}{3}$ **47.** $\dfrac{5}{4}, \dfrac{11}{3}$ **49.** $-\dfrac{2}{3}, \dfrac{1}{6}$ **51.** $-\dfrac{5}{3}, \dfrac{1}{2}$ **53.** $\dfrac{17}{2}$

55. $2, -\dfrac{4}{5}$ **57.** $y-4, 3$ **59.** $\dfrac{1}{2}, -\dfrac{1}{2}$ **61.** $-2, -11$ **63.** 3 **65.** -3 **67.** $\dfrac{3}{4}, -\dfrac{4}{3}$ **69.** $0, \dfrac{5}{2}, -1$ **71.** $-\dfrac{4}{3}, 1$ **73.** $-2, 5$

75. $-6, \dfrac{1}{2}$ **77.** E   **79.** B   **81.** C

83. a. 300; 304; 276; 216; 124; 0; −156
b. 5 sec   **c.** 304 ft

85. $\dfrac{47}{45}$ **87.** $\dfrac{17}{60}$ **89.** $\dfrac{15}{8}$ **91.** $\dfrac{7}{10}$ **93.** $0, \dfrac{1}{2}$

95. 0, −15   **97.** 0, 13

d. 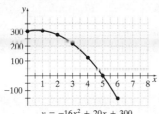 ; Answers may vary.

$$y = -16x^2 + 20x + 300$$

Exercise Set 5.7   1. width $= x$; length $= x + 4$ **3.** x and $x + 2$ if x is an odd integer   **5.** base $= x$; height $= 4x + 1$
7. 11 units   **9.** 15 cm, 13 cm, 70 cm, 22 cm   **11.** base $= 16$ mi; height $= 6$ mi   **13.** 5 sec   **15.** length $= 5$ cm; width $= 6$ cm
17. 54 diagonals   **19.** 10 sides   **21.** −12 or 11   **23.** slow boat: 8 mph; fast boat: 15 mph   **25.**1 13 and 7   **27.** 5 in.
29. 12 mm, 16 mm, 20 mm   **31.** 10 km   **33.** 36 ft   **35.** 6.25 sec   **37.** 20%   **39.** length: 15 mi; width: 8 mi   **41.** 105 units
43. boom length: 36 ft; height of mainsail: 100 ft   **45.** 435 acres   **47.** 2 million   **49.** Answers may vary.   **51.** $\dfrac{4}{7}$ **53.** $\dfrac{3}{2}$ **55.** $\dfrac{1}{3}$

Chapter 5 Review   1. $2x - 5$ **3.** $4x(5x + 3)$ **5.** $-2x^2y(4x - 3y)$ **7.** $(x + 1)(5x - 1)$ **9.** $(2x - 1)(3x + 5)$
11. $(x + 4)(x + 2)$ **13.** prime   **15.** $(x + 4)(x - 2)$ **17.** $(x + 5y)(x + 3y)$ **19.** $2(3 - x)(12 + x)$
21. $(2x - 1)(x + 6)$ **23.** $(2x + 3)(2x - 1)$ **25.** $(6x - y)(x - 4y)$ **27.** $(2x + 3y)(x - 13y)$
29. $(6x + 5y)(3x - 4y)$ **31.** $(2x - 3)(2x + 3)$ **33.** prime   **35.** $(2x + 3)(4x^2 - 6x + 9)$
37. $2(3 - xy)(9 + 3xy + x^2y^2)$ **39.** $(2x - 1)(2x + 1)(4x^2 + 1)$ **41.** $(2x - 3)(x + 4)$ **43.** $(x - 1)(x + 3)$
45. $2xy(2x - 3y)$ **47.** $(5x + 3)(25x^2 - 15x + 9)$ **49.** $(x + 7 - y)(x + 7 + y)$ **51.** $-6, 2$
53. $-\dfrac{1}{5}, -3$ **55.** $-4, 6$ **57.** 2, 8   **59.** $-\dfrac{2}{7}, \dfrac{3}{8}$ **61.** $-\dfrac{2}{5}$ **63.** 3   **65.** $0, -\dfrac{7}{4}, 3$ **67.** 36 yd

69. a. 17.5 sec and 10 sec; The rocket reaches a height of 2800 ft on its way up and on its way back down.   **b.** 27.5 sec

Chapter 5 Test   1. $3x(3x + 1)(x + 4)$ **2.** prime   **3.** prime   **4.** $(y - 12)(y + 4)$ **5.** $(3a - 7)(a + b)$
6. $(3x - 2)(x - 1)$ **7.** prime   **8.** $(x + 12y)(x + 2y)$ **9.** $x^4(26x^2 - 1)$ **10.** $5x(10x^2 + 2x - 7)$
11. $5(6 - x)(6 + x)$ **12.** $(4x - 1)(16x^2 + 4x + 1)$ **13.** $(6t + 5)(t - 1)$ **14.** $(y - 2)(y + 2)(x - 7)$
15. $x(1 - x)(1 + x)(1 + x^2)$ **16.** $-xy(y^2 + x^2)$ **17.** $-7, 2$ **18.** $-7, 1$ **19.** $0, \dfrac{3}{2}, -\dfrac{4}{3}$ **20.** $0, 3, -3$ **21.** $0, -4$
22. $-3, 5$ **23.** $-3, 8$ **24.** $0, \dfrac{5}{2}$ **25.** width: 6 ft; length: 11 ft   **26.** 17 ft   **27.** 8 and 9   **28.** 7 sec

Chapter 5 Cumulative Review   1. a. $9 \le 11$ **b.** $8 > 1$ **c.** $3 \ne 4$; Sec. 1.2, Ex. 3   **2. a.** $\dfrac{6}{7}$ **b.** $\dfrac{11}{27}$ **c.** $\dfrac{22}{5}$; Sec. 1.3, Ex. 2
3. $\dfrac{14}{3}$; Sec. 1.4, Ex. 5   **4. a.** −12   **b.** −9; Sec. 1.5, Ex. 7   **5. a.** −24   **b.** −2   **c.** 50; Sec. 1.7, Ex. 1
6. a. $4x$ **b.** $11y^2$ **c.** $8x^2 - x$; Sec. 2.1, Ex. 3   **7.** −4; Sec. 2.2, Ex. 7   **8.** −11; Sec. 2.3, Ex. 2   **9.** $\dfrac{16}{3}$; Sec. 2.4, Ex. 2
10. shorter: 2 ft; longer: 8 ft; Sec. 2.5, Ex. 2
11. 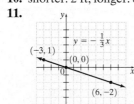 ; Sec. 3.2, Ex. 5   **12.** yes; Sec. 3.4, Ex. 5   **13. a.** 250   **b.** 1; Sec. 4.1, Ex. 2
14. a. 2   **b.** 5   **c.** 0; Sec. 4.2, Ex. 1   **15.** $9x^2 - 6x - 1$; Sec. 4.2, Ex. 12
16. $6x^2 - 11x - 10$; Sec. 4.3, Ex. 2   **17.** $9y^2 + 6y + 1$; Sec. 4.4, Ex. 4
18. a. $\dfrac{1}{9}$ **b.** $\dfrac{2}{x^3}$ **c.** $\dfrac{3}{4}$ **d.** $\dfrac{1}{16}$; Sec. 4.5, Ex. 1
19. a. 3.67×10^6 **b.** 3.0×10^{-6} **c.** 2.052×10^{10} **d.** 8.5×10^{-4}; Sec. 4.5, Ex. 5
20. $x + 4$; Sec. 4.6, Ex. 4   **21. a.** x^3 **b.** y; Sec. 5.1, Ex. 2
22. $(x + 3)(x + 4)$; Sec. 5.2, Ex. 1   **23.** $(4x - 1)(2x - 5)$; Sec. 5.3, Ex. 2   **24.** $(5a + 3b)(5a - 3b)$; Sec. 5.4, Ex. 2b
25. 3, −1; Sec. 5.6, Ex. 1

■ CHAPTER 6   RATIONAL EXPRESSIONS

Exercise Set 6.1   **1.** $\dfrac{7}{4}$ **3.** $\dfrac{13}{3}$ **5.** $-\dfrac{11}{2}$ **7.** $\dfrac{7}{4}$ **9.** $-\dfrac{8}{3}$ **11.** $x = -2$ **13.** $x = 4$ **15.** $x = -2$ **17.** none

19. Answers may vary.   **21.** $\dfrac{2}{x^4}$ **23.** $\dfrac{5}{x + 1}$ **25.** -5 **27.** $\dfrac{1}{x - 9}$ **29.** $5x + 1$ **31.** $\dfrac{x + 4}{2x + 1}$ **33.** Answers may vary.

35. -1 **37.** $-\dfrac{y}{2}$ **39.** $\dfrac{2 - x}{x + 2}$ **41.** $-\dfrac{3y^5}{x^4}$ **43.** $\dfrac{x - 2}{5}$ **45.** -6 **47.** $\dfrac{x + 2}{2}$ **49.** $\dfrac{11x}{6}$ **51.** $\dfrac{1}{x - 2}$ **53.** $x + 2$

55. $\dfrac{x + 1}{x - 1}$ **57.** $\dfrac{m - 3}{m + 3}$ **59.** $\dfrac{2(a - 3)}{a + 3}$ **61.** -1 **63.** $-x - 1$ **65.** $\dfrac{x + 5}{x - 5}$ **67.** $\dfrac{x + 2}{x + 4}$ **69.** 400 mg   **71.** no; $B \approx 24$

73. a. \$37.5 million   **b.** \$85.7 million   **c.** \$48.2 million   **75.** a, c; Answers may vary.   **77.** $a + b$ **79.** $\dfrac{x + 2}{y}$

81. $x^2 - 4x + 16$ **83.** $-\dfrac{1}{x^2 + 3x + 9}$ **85.**

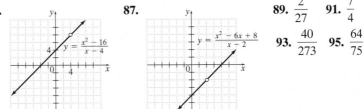

87.   **89.** $\dfrac{2}{27}$ **91.** $\dfrac{7}{4}$

93. $\dfrac{40}{273}$ **95.** $\dfrac{64}{75}$

Mental Math   **1.** $\dfrac{2x}{3y}$ **3.** $\dfrac{5y^2}{7x^2}$ **5.** $\dfrac{9}{5}$

Exercise Set 6.2   **1.** $\dfrac{21}{4y}$ **3.** x^4 **5.** $-\dfrac{b^2}{6}$ **7.** $\dfrac{x^2}{10}$ **9.** $\dfrac{1}{3}$ **11.** 1 **13.** $\dfrac{x + 5}{x}$ **15.** $\dfrac{2}{9x^2(x - 5)}$ sq. ft   **17.** x^4 **19.** $\dfrac{12}{y^6}$

21. $x(x + 4)$ **23.** $\dfrac{3(x + 1)}{x^3(x - 1)}$ **25.** $m^2 - n^2$ **27.** $-\dfrac{x + 2}{x - 3}$ **29.** $-\dfrac{x + 2}{x - 3}$ **31.** Answers may vary.   **33.** $\dfrac{1}{6b^4}$ **35.** $\dfrac{9}{7x^2y^7}$

37. $\dfrac{5}{6}$ **39.** $\dfrac{3x}{8}$ **41.** $\dfrac{3}{2}$ **43.** $\dfrac{3x + 4y}{2(x + 2y)}$ **45.** $-2(x + 3)$ **47.** $\dfrac{2(x + 2)}{x - 2}$ **49.** $\dfrac{(a + 5)(a + 3)}{(a + 2)(a + 1)}$ **51.** $-\dfrac{1}{x}$ **53.** $\dfrac{2(x + 3)}{x - 4}$

55. $-(x^2 + 2)$ **57.** -1 **59.** $4x^3(x - 3)$ **61.** 1440 **63.** $411{,}972$ sq. yd   **65.** 73 **67.** 3424.8 mph   **69.** $\dfrac{(a + b)^2}{a - b}$

71. $\dfrac{3x + 5}{x^2 + 4}$ **73.** $\dfrac{4}{x - 2}$ **75.** $\dfrac{a - b}{6(a^2 + ab + b^2)}$ **77.** 1 **79.** $-\dfrac{10}{9}$ **81.** $-\dfrac{1}{5}$

83.   **85.** $\dfrac{x}{2}$ **87.** $\dfrac{5a(2a + b)(3a - 2b)}{b^2(a - b)(a + 2b)}$

Mental Math   **1.** 1 **3.** $\dfrac{7x}{9}$ **5.** $\dfrac{1}{9}$ **7.** $\dfrac{7 - 10y}{5}$

Exercise Set 6.3   **1.** $\dfrac{a + 9}{13}$ **3.** $\dfrac{y + 10}{3 + y}$ **5.** $\dfrac{3m}{n}$ **7.** $\dfrac{5x + 7}{x - 3}$ **9.** $\dfrac{1}{2}$ **11.** 4 **13.** $x + 5$ **15.** $x + 4$ **17.** 1 **19.** $\dfrac{12}{x^3}$

21. $-\dfrac{5}{x + 4}$ **23.** $\dfrac{x - 2}{x + y}$ **25.** 4 **27.** 3 **29.** $\dfrac{1}{a + 5}$ **31.** $\dfrac{1}{x - 6}$ **33.** $\dfrac{20}{x - 2}$ m   **35.** Answers may vary.   **37.** 33

39. $4x^3$ **41.** $8x(x + 2)$ **43.** $6(x + 1)^2$ **45.** $8 - x$ or $x - 8$ **47.** $40x^3(x - 1)^2$ **49.** $(2x + 1)(2x - 1)$

51. $(2x - 1)(x + 4)(x + 3)$ **53.** Answers may vary.   **55.** $\dfrac{6x}{4x^2}$ **57.** $\dfrac{24b^2}{12ab}$ **59.** $\dfrac{18}{2(x + 3)}$ **61.** $\dfrac{9ab + 2b}{5b(a + 2)}$

63. $\dfrac{x^2 + x}{(x + 4)(x + 2)(x + 1)}$ **65.** $\dfrac{18y - 2}{30x^2 - 60}$ **67.** $\dfrac{15x(x - 7)}{3x(2x + 1)(x - 7)(x - 5)}$ **69.** $-\dfrac{5}{x - 2}$ **71.** $\dfrac{7 + x}{x - 2}$

73. $95{,}304$ Earth days   **75.** Answers may vary.   **77.** $0, -5$ **79.** $1, 5$ **81.** $\dfrac{3}{10}$ **83.** $\dfrac{58}{45}$

Mental Math   **1.** D   **3.** A

Exercise Set 6.4   **1.** $\dfrac{5}{x}$ **3.** $\dfrac{75a + 6b^2}{5b}$ **5.** $\dfrac{6x + 5}{2x^2}$ **7.** $\dfrac{21}{2(x + 1)}$ **9.** $\dfrac{17x + 30}{2(x - 2)(x + 2)}$ **11.** $\dfrac{35x - 6}{4x(x - 2)}$

13. $\dfrac{5 + 10y - y^3}{y^2(2y + 1)}$ **15.** Answers may vary.   **17.** $-\dfrac{2}{x - 3}$ **19.** $-\dfrac{1}{x^2 + 1}$ **21.** $\dfrac{1}{x - 2}$ **23.** $\dfrac{5 + 2x}{r}$ **25.** $\dfrac{6x - 7}{x - 2}$

27. $-\dfrac{y + 4}{y + 3}$ **29.** $\left(\dfrac{90x - 40}{x}\right)^\circ$ **31.** 2   **33.** $3x^3 - 4$ **35.** $\dfrac{x + 2}{(x + 3)^2}$ **37.** $\dfrac{9b - 4}{5b(b - 1)}$ **39.** $\dfrac{2 + m}{m}$ **41.** $\dfrac{10}{1 - 2x}$

43. $\dfrac{15x - 1}{(x + 1)^2(x - 1)}$ **45.** $\dfrac{x^2 - 3x - 2}{(x - 1)^2(x + 1)}$ **47.** $\dfrac{a + 2}{2(a + 3)}$ **49.** $\dfrac{x - 10}{2(x - 2)}$ **51.** $\dfrac{-3 - 2y}{(y - 1)(y - 2)}$ **53.** $\dfrac{-5x + 23}{(x - 3)(x - 2)}$

55. $\dfrac{12x - 32}{(x + 2)(x - 2)(x - 3)}$ **57.** $\dfrac{6x + 9}{(3x - 2)(3x + 2)}$ **59.** $\dfrac{2x^2 - 2x - 46}{(x + 1)(x - 6)(x - 5)}$ **61.** $\dfrac{2x - 16}{(x - 4)(x + 4)}$ in.   **63.** C

65. B   **67.** 10   **69.** 2   **71.** $\dfrac{25a}{9(a - 2)}$ **73.** $\dfrac{x + 4}{(x - 2)(x - 1)}$ **75.** Answers may vary.   **77.** $\dfrac{1}{2}$ **79.** $\dfrac{3}{7}$ **81.** $m = 2$

83. $m = -\dfrac{1}{3}$ **85.** $\dfrac{4x^2 - 15x + 6}{(x - 2)^2(x + 2)(x - 3)}$ **87.** $\dfrac{2(x + 1)}{(x - 3)(x^2 + 3x + 9)}$

Mental Math   **1.** $\dfrac{y}{5x}$ **3.** $\dfrac{3x}{5}$

Exercise Set 6.5   **1.** $\dfrac{2}{3}$ **3.** $\dfrac{2}{3}$ **5.** $\dfrac{1}{2}$ **7.** $-\dfrac{21}{5}$ **9.** $\dfrac{27}{16}$ **11.** $\dfrac{4}{3}$ **13.** $\dfrac{1}{21}$ **15.** $-\dfrac{4x}{15}$ **17.** $\dfrac{m - n}{m + n}$ **19.** $\dfrac{2x(x - 5)}{7x^2 + 10}$

21. $\dfrac{1}{y - 1}$ **23.** $\dfrac{1}{6}$ **25.** $\dfrac{x + y}{x - y}$ **27.** $\dfrac{3}{7}$ **29.** $\dfrac{a}{x + b}$ **31.** $\dfrac{7(y - 3)}{8 + y}$ **33.** $\dfrac{3x}{x - 4}$ **35.** $-\dfrac{x + 8}{x - 2}$ **37.** $\dfrac{s^2 + r^2}{s^2 - r^2}$

39. Answers may vary.   **41.** $\dfrac{13}{24}$ **43.** $\dfrac{R_1 R_2}{R_2 + R_1}$ **45.** 12 hr   **47.** $\dfrac{2}{3}$ **49.** $-\dfrac{1}{2}, 1$ **51.** 1   **53.** -1 **55.** $\dfrac{2x}{2 - x}$

57. $\dfrac{xy + 1}{x}$ **59.** $\dfrac{1}{y^2 - 1}$

Calculator Explorations   **1.**   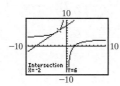   **2.**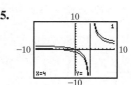

3.   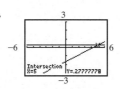   **4.**   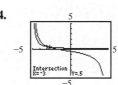   **5.**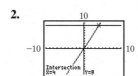

6.

Mental Math   **1.** $x = 10$ **3.** $z = 36$

Exercise Set 6.6   **1.** 30   **3.** 0   **5.** $-5, 2$ **7.** 5   **9.** 3   **11.** $100°, 80°$ **13.** $22.5°, 67.5°$ **15.** $\dfrac{1}{4}$ **17.** no solution   **19.** $6, -4$

21. 5   **23.** $-\dfrac{10}{9}$ **25.** $\dfrac{11}{14}$ **27.** no solution   **29.** expression; $\dfrac{3 + 2x}{3x}$ **31.** equation; $x = 3$ **33.** expression; $\dfrac{x - 1}{x(x + 1)}$

35. equation; no solution   **37.** Answers may vary.   **39.** $-\dfrac{21}{11}$ **41.** 1   **43.** $\dfrac{9}{2}$ **45.** no solution   **47.** $-\dfrac{9}{4}$ **49.** $-\dfrac{3}{2}, 4$

51. -2 **53.** $\dfrac{12}{5}$ **55.** $-2, 8$ **57.** $\dfrac{1}{5}$ **59.** $R = \dfrac{D}{T}$ **61.** $y = \dfrac{3x + 6}{5x}$ **63.** $b = -\dfrac{3a^2 + 2a - 4}{6}$ **65.** $B = \dfrac{2A}{H}$

67. $r = \dfrac{C}{2\pi}$ **69.** $a = \dfrac{bc}{c + b}$ **71.** $n = \dfrac{m^2 - 3p}{2}$ **73.** $\dfrac{17}{4}$ **75.** $(2, 0), (0, -2)$ **77.** $(-4, 0), (-2, 0), (3, 0), (0, 4)$

Exercise Set 6.7   **1.** $\dfrac{2}{15}$ **3.** $\dfrac{5}{6}$ **5.** $\dfrac{5}{12}$ **7.** $\dfrac{1}{10}$ **9.** $\dfrac{7}{20}$ **11.** $\dfrac{19}{18}$ **13.** Answers may vary.   **15.** $x = 4$ **17.** $x = \dfrac{50}{9}$

19. $x = \dfrac{21}{4}$ **21.** $a = 30$ **23.** $x = 7$ **25.** $x = -\dfrac{1}{3}$ **27.** $x = -3$ **29.** $x = \dfrac{14}{9}$ **31.** $x = 5$ **33.** 123 lb

35. 165 calories   **37.** 3833 women   **39.** 9 gal   **41.** 7800 people   **43.** $182\dfrac{6}{7}$ cal   **45.** 110 oz for \$5.79   **47.** 8 oz for \$0.90

49. 4-pack   **51.** gallon   **53.** 530 megawatts   **55.** 1,484,000 people   **57.** yes; Answers may vary.   **59.** a

61. $m = 3$; upward   **63.** $m = -\dfrac{9}{5}$; downward   **65.** $m = 0$; horizontal

Exercise Set 6.8   **1.** 2   **3.** -3 **5.** $2\dfrac{2}{9}$ hr   **7.** $1\dfrac{1}{2}$ min   **9.** trip to park rate: r; to park time: $\dfrac{12}{r}$; return trip rate: r;

return time: $\dfrac{18}{r}$; $r = 6$ mph   **11.** 1st portion: 10 mph; cooldown: 8 mph   **13.** $x = 6$ **15.** $x = 5$ **17.** 2   **19.** \$108.00

21. 63 mph   **23.** $y = 21.25$ **25.** $y = 5\dfrac{5}{7}$ ft   **27.** $37\dfrac{1}{2}$ ft   **29.** 5   **31.** 217 mph   **33.** 8 mph   **35.** 3 hr   **37.** 20 hr

39. $26\dfrac{2}{3}$ ft   **41.** $1\dfrac{1}{5}$ hr   **43.** $5\dfrac{1}{4}$ hr   **45.** first pump: 28 min; second pump: 84 min   **47.** 35 yr; 42 yr

49. Answers may vary.   **51.** 3.75 min   **53.**   **55.**   **57.**

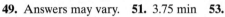

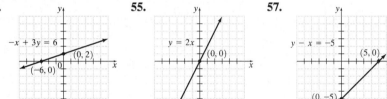

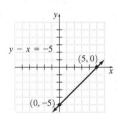

Chapter 6 Review   **1.** $x = 2, x = -2$ **3.** $\dfrac{2}{11}$ **5.** $\dfrac{1}{x - 5}$ **7.** $\dfrac{x(x - 2)}{x + 1}$ **9.** $\dfrac{x - 3}{x - 5}$ **11.** $\dfrac{x + 1}{2x + 1}$ **13.** $\dfrac{x + 5}{x - 3}$

15. $-\dfrac{1}{x^2 + 4x + 16}$ **17.** $-\dfrac{9x^2}{8}$ **19.** $-\dfrac{2x(2x + 5)}{(x - 6)^2}$ **21.** $\dfrac{4x}{3y}$ **23.** $\dfrac{2}{3}$ **25.** $\dfrac{x}{x + 6}$ **27.** $\dfrac{3(x + 2)}{3x + y}$ **29.** $-\dfrac{2(2x + 3)}{y - 2}$

31. $\dfrac{5x + 2}{3x - 1}$ **33.** $\dfrac{2x + 1}{2x^2}$ **35.** $(x - 8)(x + 8)(x + 3)$ **37.** $\dfrac{3x^2 + 4x - 15}{(x + 2)^2(x + 3)}$ **39.** $\dfrac{-2x + 10}{(x - 3)(x - 1)}$ **41.** $\dfrac{-2x - 2}{x + 3}$

43. $\dfrac{x - 4}{3x}$ **45.** $\dfrac{x^2 + 2x - 3}{(x + 2)^2}$ **47.** $\dfrac{29x}{12(x - 1)}; \dfrac{3xy}{5(x - 1)}$ **49.** $\dfrac{2x}{x - 3}$ **51.** $\dfrac{5}{3a^2}$ **53.** $\dfrac{2x^2 + 1}{x + 2}$ **55.** $-\dfrac{7 + 2x}{2x}$ **57.** 30

59. $3, -4$ **61.** no solution   **63.** 5   **65.** $-6, 1$ **67.** $y = \dfrac{560 - 8x}{7}$ **69.** $\dfrac{2}{3}$ **71.** $x = 500$ **73.** $c = 50$ **75.** no solution

77. no solution   **79.** 15 oz for \$1.63   **81.** \$33.75   **83.** 3   **85.** 30 mph; 20 mph   **87.** $17\dfrac{1}{2}$ hr   **89.** $x = 15$ **91.** $x = 15$

Chapter 6 Test   **1.** $x = -1, x = -3$ **2. a.** \$115   **b.** \$103   **3.** $\dfrac{3}{5}$ **4.** $\dfrac{1}{x - 10}$ **5.** $\dfrac{1}{x + 6}$ **6.** $\dfrac{1}{x^2 - 3x + 9}$

7. $\dfrac{2m(m + 2)}{m - 2}$ **8.** $\dfrac{a + 2}{a + 5}$ **9.** $-\dfrac{1}{x + y}$ **10.** $\dfrac{(x - 6)(x - 7)}{(x + 7)(x + 2)}$ **11.** 15   **12.** $\dfrac{y - 2}{4}$ **13.** $-\dfrac{1}{2x + 5}$

14. $\dfrac{3a - 4}{(a - 3)(a + 2)}$ **15.** $\dfrac{3}{x - 1}$ **16.** $\dfrac{2(x + 5)}{x(y + 5)}$ **17.** $\dfrac{x^2 + 2x + 35}{(x + 9)(x + 2)(x - 5)}$ **18.** $\dfrac{4y^2 + 13y - 15}{(y + 4)(y + 5)(y + 1)}$

19. $\dfrac{30}{11}$ **20.** -6 **21.** no solution   **22.** no solution   **23.** $\dfrac{xz}{2y}$ **24.** $\dfrac{b^2 - a^2}{2b^2}$ **25.** $\dfrac{5y^2 - 1}{y + 2}$

26. 18 bulbs   **27.** 5 or 1   **28.** 30 mph   **29.** $6\dfrac{2}{3}$ hr   **30.** 6 oz for \$1.19   **31.** $x = 12$

Chapter 6 Cumulative Review   **1. a.** $\dfrac{15}{x} = 4$ **b.** $12 - 3 = x$ **c.** $4x + 17 = 21$; Sec. 1.4, Ex. 9

2. amount at 7%: \$12,500; amount at 9%: \$7500; Sec. 2.8, Ex. 4

3. 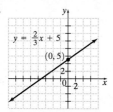 ; Sec. 3.3, Ex. 2   **4. a.** 4^7 **b.** x^7 **c.** y^4 **d.** y^{12} **e.** $(-5)^{15}$; Sec. 4.1, Ex. 3

5. $12z + 16$; Sec. 4.2, Ex. 14   **6.** $27a^3 + 27a^2b + 9ab^2 + b^3$; Sec. 4.3, Ex. 5

7. a. $t^2 + 4t + 4$ **b.** $p^2 - 2qp + q^2$ **c.** $4x^2 + 20x + 25$

d. $x^4 - 14x^2y + 49y^2$; Sec. 4.4, Ex. 5   **8. a.** $\dfrac{27}{8}$ **b.** r^3 **c.** $\dfrac{q^9}{p^4}$; Sec. 4.5, Ex. 2

9. $4x^2 - 4x + 6 + \dfrac{-11}{2x + 3}$; Sec. 4.6, Ex. 6   **10. a.** 4   **b.** 1   **c.** 3; Sec. 5.1, Ex. 1

11. $-3a(3a^4 - 6a + 1)$; Sec. 5.1, Ex. 5   **12.** $3(m + 2)(m - 10)$; Sec. 5.2, Ex. 7   **13.** $(3x + 2)(x + 3)$; Sec. 5.3, Ex. 1

14. $(x + 6)^2$; Sec. 5.3, Ex. 7   **15.** prime polynomial; Sec. 5.4, Ex. 5   **16.** $(x + 2)(x^2 - 2x + 4)$; Sec. 5.4, Ex. 7

17. $(2x + 3)(x + 1)(x - 1)$; Sec. 5.5, Ex. 2   **18.** $3(2m + n)(2m - n)$; Sec. 5.5, Ex. 3   **19.** $-\dfrac{1}{2}, 4$; Sec. 5.6, Ex. 3

20. $(1, 0), (4, 0)$; Sec. 5.6, Ex. 8   **21.** base: 6 m; height: 10 m; Sec. 5.7, Ex. 3   **22.** $\dfrac{5}{x^2}$; Sec. 6.1, Ex. 5

23. $\dfrac{2}{x(x + 1)}$; Sec. 6.2, Ex. 6   **24.** $\dfrac{x + 1}{x + 2y}$; Sec. 6.5, Ex. 5   **25.** 14-ounce box; Sec. 6.7, Ex. 5

■ CHAPTER 7   FURTHER GRAPHING

Calculator Explorations   **1.** 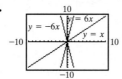   **3.** 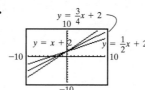   **5.**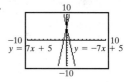

Mental Math   **1.** $m = 2; (0, -1)$ **3.** $m = 1; \left(0, \dfrac{1}{3}\right)$ **5.** $m = \dfrac{5}{7}; (0, -4)$

Exercise Set 7.1   **1.** $m = -2; (0, 4)$ **3.** $m = -\dfrac{1}{9}; \left(0, \dfrac{1}{9}\right)$ **5.** $m = \dfrac{4}{3}; (0, -4)$ **7.** $m = -1; (0, 0)$ **9.** $m = 0; (0, -3)$

11. $m = \dfrac{1}{5}; (0, 4)$ **13.** B   **15.** D   **17.** neither   **19.** neither   **21.** perpendicular   **23.** parallel   **25.** Answers may vary.

27. $y = -x + 1$ **29.** $y = 2x + \dfrac{3}{4}$ **31.** $y = \dfrac{2}{7}x$

33. 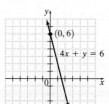   **35.** 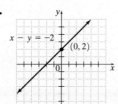   **37.** 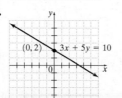   **39.**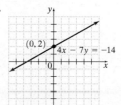

41. 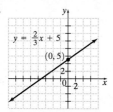   **43.** 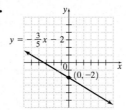   **45.** 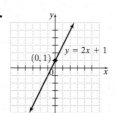   **47.**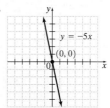

49. a. $(0, 28); (5, 22)$ **b.** $y = -1.2x + 28$ **c.** 17.2 heads/sq. ft   **51. a.** The temperature 100° Celsius is equivalent to 212° Fahrenheit.   **b.** 68°F   **c.** 27°C   **d.** $F = \dfrac{9}{5}C + 32$ **53.** Answers may vary.

55. 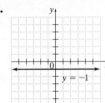   **57.** 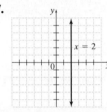   **59.** $3x - y = 16$ **61.** $6x + y = -10$

Mental Math **1.** $m = 3$; Answers may vary. Ex. $(4, 8)$ **3.** $m = -2$; Answers may vary. Ex. $(10, -3)$
5. $m = \dfrac{2}{5}$; Answers may vary. Ex. $(-1, 0)$

Exercise Set 7.2 **1.** $6x - y = 10$ **3.** $8x + y = -13$ **5.** $x - 2y = 17$ **7.** $2x - y = 4$ **9.** $8x - y = -11$
11. $4x - 3y = -1$ **13.** $x = 0$ **15.** $y = 3$ **17.** $x = -\dfrac{7}{3}$ **19.** $y = 2$ **21.** $y = 5$ **23.** $x = 6$ **25.** $3x + 6y = 10$
27. $x - y = -16$ **29.** $x + y = 17$ **31.** $y = 7$ **33.** $4x + 7y = -18$ **35.** $x + 8y = 0$ **37.** $3x - y = 0$
39. $x - y = 0$ **41.** $5x + y = 7$ **43.** $11x + y = -6$ **45.** $x = -\dfrac{3}{4}$ **47.** $y = -3$ **49.** $7x - y = 4$
51. a. $y = 740x + 3280$ **b.** 9940 vehicles   **53. a.** $s = 32t$ or $y = 32x$ **b.** 128 ft/sec   **55. a.** $y = -2.5x + 150$
b. 105 thousand apparel and accessory stores   **57.** $31x - 5y = -5$ **59. a.** $V = -200t + 3000$ **b.** $1000
61. Answers may vary.   **63. a.** $3x - y = -5$ **b.** $x + 3y = 5$ **65. a.** $3x + 2y = -1$ **b.** $2x - 3y = 21$
67. 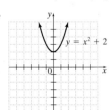   **69.** 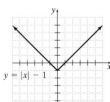 $x + 3y = 5$ **71.** 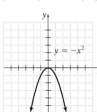 $y = -2$

Calculator Explorations **1.** $(0.56, 0), (-3.56, 0)$ **3.** $(-0.87, 0), (2.79, 0)$ **5.** $(-0.65, 0), (0.65, 0)$ **7.** $(-1.51, 0)$

Exercise Set 7.3 **1.** 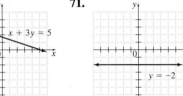   **3.**   **5.**   **7.**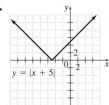

9. C   **11.** D   **13.** A   **15.** $(0, 1), (0, -1), (-2, 0), (2, 0)$ **17.** $(2, 0), \left(\dfrac{2}{3}, -2\right)$ **19.** $(2, \text{any real number})$
21. There is no such point.   **23.** $(2, -1)$
25. linear   **27.** linear   **29.** linear   **31.** not linear

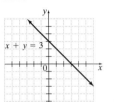 $x + y = 3$ 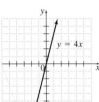 $y = 4x$ 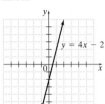 $y = 4x - 2$ 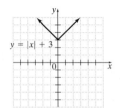 $y = |x| + 3$

33. linear 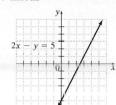 **35.** not linear 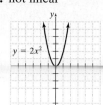 **37.** not linear 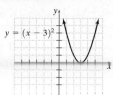 **39.** linear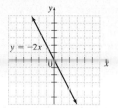

41. linear 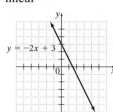 **43.** not linear 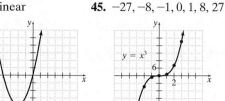 **45.** $-27, -8, -1, 0, 1, 8, 27$

47. Answers may vary.
49. Answers may vary.
51. 11
53. 19
55. no
57. yes

Exercise Set 7.4 **1.** domain: $\{-7, 0, 2, 10\}$; range: $\{-7, 0, 4, 10\}$ **3.** domain: $\{0, 1, 5\}$; range: $\{-2\}$ **5.** yes **7.** no
9. no **11.** yes **13.** yes **15.** no **17.** no **19.** yes **21.** yes **23.** yes **25.** yes **27.** no **29.** no **31.** 5:20 A.M.
33. Answers may vary. **35.** $4.75 per hour **37.** 2002 **39.** Answers may vary. **41.** $-9, -5, 1$ **43.** $6, 2, 11$
45. $-8, 0, 27$ **47.** $2, 0, 3$ **49.** $-5, 0, 20$ **51.** $5, 3, 35$ **53.** $4, 3, -21$ **55.** $6, 6, 6$ **57.** all real numbers
59. all real numbers except -5 **61.** all real numbers **63.** domain: all real numbers; range: $y \geq -4$
65. domain: all real numbers; range: all real numbers **67.** domain: all real numbers; range: $y = 2$
69. a. 166.38 cm **b.** 148.25 cm **71.** Answers may vary. **73.** $f(x) = x + 7$ **75.** $(3, 0)$ **77.** $(-3, -3)$
79. a. $-3s + 12$ **b.** $-3r + 12$ **81. a.** 132 **b.** $a^2 - 12$

Chapter 7 Review **1.** $m = -3; (0, 7)$ **3.** $m = 0; (0, 2)$ **5.** perpendicular **7.** neither **9.** $y = \dfrac{2}{3}x + 6$

11. 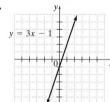 **13.** 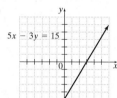 **15.** C **17.** B **19.** $3x + y = -5$ **21.** $y = -3$
23. $6x + y = 11$ **25.** $x + y = 6$ **27.** $x = 5$ **29.** $x = 6$
31. a. $3x + y = 15$ **b.** $x - 3y = 5$

33. 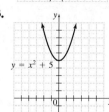 **35.** 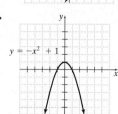 **37.** 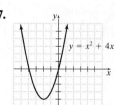 **39.** $(-1, -3)$ **41.** $(6, -1)$
43. yes **45.** yes **47.** yes **49.** no
51. a. 6 **b.** 10 **c.** 5
53. a. 45 **b.** -35 **c.** 0
55. all real numbers

57. domain: $-3 \leq x \leq 5$; range: $-4 \leq y \leq 2$ **59.** domain: $x = 3$; range: all real numbers

Chapter 7 Test **1.** $m = \dfrac{7}{3}; \left(0, -\dfrac{2}{3}\right)$ **2.** neither **3.** $x + 4y = 10$ **4.** $7x + 6y = 0$ **5.** $8x + y = 11$ **6.** $x = -5$
7. $x - 8y = -96$ **8.** 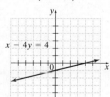 **9.** 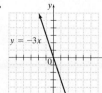 **10.** 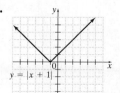 **11.**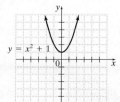

12. yes **13.** yes **14.** yes **15.** no **16.** no **17. a.** -8 **b.** -3.6 **c.** -4 **18. a.** 0 **b.** 0 **c.** 60 **19. a.** 6 **b.** 6 **c.** 6

20. all real numbers except -1 **21.** domain: all real numbers; range: $y \leq 4$ **22.** domain: all real numbers; range: all real numbers **23.** 9 P.M. **24.** 4 P.M. **25.** January 1st and December 1st **26.** June 1st and end of July **27.** yes; it passes the vertical line test **28.** yes; every location has exactly 1 sunset time per day.

Chapter 7 Cumulative Review **1. a.** commutative property of multiplication **b.** associative property of addition **c.** identity element for addition **d.** commutative property of multiplication **e.** multiplication inverse property **f.** additive inverse property **g.** commutative and associative properties of multiplication; Sec. 1.8, Ex. 6 **2.** 12; Sec. 2.4, Ex. 3

3. 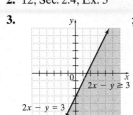 ; Sec. 3.5, Ex. 2 **4. a.** $\dfrac{m^7}{n^7}$ **b.** $\dfrac{x^{12}}{81y^{20}}$; Sec. 4.1, Ex. 7 **5. a.** $4x$ **b.** $13x^2 - 2$

c. $\dfrac{1}{2}x^4 + \dfrac{1}{2}x^3 - x^2$; Sec. 4.2, Ex. 6 **6.** $3t^3 + 2t^2 - 6t + 4$; Sec. 4.3, Ex. 4

7. $x^2 + x - 12$; Sec. 4.4, Ex. 1 **8. a.** 0.056 **b.** 200,000; Sec. 4.5, Ex. 7

9. $3x^3 - 4 + \dfrac{1}{x}$; Sec. 4.6, Ex. 2 **10.** $(x + 3)(5 + y)$; Sec. 5.1, Ex. 7

11. $(x - 2)(x + 6)$; Sec. 5.2, Ex. 3 **12.** $(2x - 3y)(5x + y)$; Sec. 5.3, Ex. 4 **13.** $(x + 5)(x - 5)$; Sec. 5.4, Ex. 1 **14.** $5a^2b(2b - 1)(3b + 7)$; Sec. 5.5, Ex. 5 **15.** $0, -2, 2$; Sec. 5.6, Ex. 5 **16.** 4 sec; Sec. 5.7, Ex. 1

17. a. 1 **b.** -1; Sec. 6.1, Ex. 7 **18.** $\dfrac{-3(x + 1)}{5x(2x - 3)}$; Sec. 6.2, Ex. 3 **19.** $3x - 5$; Sec. 6.3, Ex. 3

20. $\dfrac{17t + 2}{3t(t + 1)}$; Sec. 6.4, Ex. 3 **21.** $\dfrac{2x^2 + 3y}{x^2y + 2xy^2}$; Sec. 6.5, Ex. 6 **22.** 5; Sec. 6.6, Ex. 2 **23.** $\dfrac{3}{4}$; Sec. 7.1, Ex. 1 **24.** $x = -1$; Sec. 7.2, Ex. 3 **25. a.** function **b.** not a function; Sec. 7.4, Ex. 2

■ CHAPTER 8 SOLVING SYSTEMS OF LINEAR EQUATIONS

Calculator Explorations **1.** $(0.37, 0.23)$ **3.** $(0.03, -1.89)$

Mental Math **1.** 1 solution, $(-1, 3)$ **3.** infinite number of solutions **5.** no solution **7.** 1 solution, $(3, 2)$

Exercise Set 8.1 **1. a.** no **b.** yes **c.** no **3. a.** no **b.** yes **c.** no **5. a.** yes **b.** yes **c.** yes **7.** Answers may vary.

9. $(2, 3)$; consistent; independent

11. $(1, -2)$; consistent; independent

13. $(-2, 1)$; consistent; independent

15. $(4, 2)$; consistent; independent

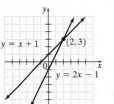
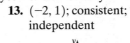
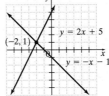
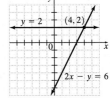

17. no solution; inconsistent; independent

19. infinite number of solutions; consistent; dependent

21. $(0, -1)$; consistent; independent

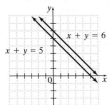
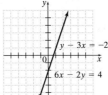
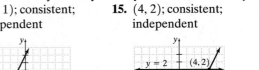

23. $(4, -3)$; consistent; independent

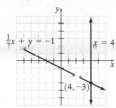

25. $(-5, -7)$; consistent; independent

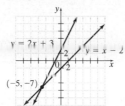

27. $(5, 2)$; consistent; independent

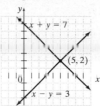

29. Answers may vary.
31. intersecting, one solution
33. parallel, no solution
35. identical lines, infinite number of solutions
37. intersecting, one solution
39. intersecting, one solution
41. identical lines, infinite number of solutions

43. parallel, no solution
45. Answers may vary.
47. 1984, 1988

49. a. $(4, 9)$ **b.**

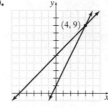

; yes   **51.** -1 **53.** 3   **55.** -7

Exercise Set 8.2   **1.** $(2, 1)$ **3.** $(-3, 9)$ **5.** $(4, 2)$ **7.** $(10, 5)$ **9.** $(2, 7)$ **11.** $(-2, 4)$ **13.** $(-2, -1)$ **15.** no solution

17. infinite number of solutions   **19.** $(3, -1)$ **21.** $(3, 5)$ **23.** $\left(\dfrac{2}{3}, -\dfrac{1}{3}\right)$ **25.** $(-1, -4)$ **27.** $(-6, 2)$ **29.** $(2, 1)$

31. no solution   **33.** $\left(-\dfrac{1}{5}, \dfrac{43}{5}\right)$ **35.** Answers may vary.   **37.** $(1, -3)$

39. a. $(21, 18)$ **b.** Answers may vary.   **c.**

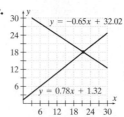

41. $(-2.6, 1.3)$
43. $(3.28, 2.11)$
45. $-6x - 4y = -12$
47. $-12x + 3y = 9$
49. $5n$
51. $-15b$

Exercise Set 8.3   **1.** $(1, 2)$ **3.** $(2, -3)$ **5.** $(6, 0)$ **7.** no solution   **9.** $\left(2, -\dfrac{1}{2}\right)$ **11.** $(6, -2)$

13. infinite number of solutions   **15.** $(-2, -5)$ **17.** $(5, -2)$ **19.** $(-7, 5)$ **21.** $\left(\dfrac{12}{11}, -\dfrac{4}{11}\right)$ **23.** no solution   **25.** $\left(\dfrac{3}{2}, 3\right)$

27. $(1, 6)$ **29.** infinite number of solutions   **31.** $(-2, 0)$ **33.** Answers may vary.   **35.** $(2, 5)$ **37.** $(-3, 2)$ **39.** $(0, 3)$

41. $(5, 7)$ **43.** $\left(\dfrac{1}{3}, 1\right)$ **45.** infinite number of solutions   **47.** $(-8.9, 10.6)$ **49. a.** $(12, 555)$ **b.** Answers may vary.

c. 1993 to 1997   **51. a.** $b = 15$ **b.** any real number except 15   **53. a.** Answers may vary.   **b.** Answers may vary.
55. $2x + 6 = x - 3$ **57.** $20 - 3x = 2$ **59.** $4(n + 6) = 2n$

Exercise Set 8.4   **1.** c   **3.** b   **5.** a   **7.** $\begin{cases} x + y = 15 \\ x - y = 7 \end{cases}$ **9.** $\begin{cases} x + y = 6500 \\ x = y + 800 \end{cases}$ **11.** 33 and 50   **13.** 14 and -3

15. Cooper: 686 points; Swoopes: 585 points   **17.** child's ticket: $18; adult's ticket: $29   **19.** quarters: 53; nickels: 27
21. IBM: $43.51; GA Financial: $107.76   **23.** still water: 6.5 mph; current: 2.5 mph   **25.** still air: 455 mph; wind: 65 mph
27. 12% solution: $7\dfrac{1}{2}$ oz; 4% solution: $4\dfrac{1}{2}$ oz   **29.** $4.95 beans: 113 lbs; $2.65 beans: 87 lbs   **31.** $60°, 30°$ **33.** $20°, 70°$
35. 20% solution: 10 l; 70% solution: 40 l   **37.** number sold at $9.50: 23; number sold at $7.50: 67

39. width: 9 ft; length: 15 ft   **41.** $4\dfrac{1}{2}$ hr   **43.** westbound: 80 mph; eastbound: 40 mph   **45.** B

47.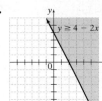

$y \geq 4 - 2x$

49.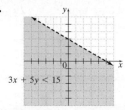

$3x + 5y < 15$

Exercise Set 8.5 **1.**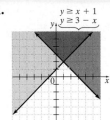

$y \geq x + 1$
$y \geq 3 - x$

3.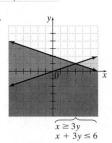

$\begin{cases} y < 3x - 4 \\ y \leq x + 2 \end{cases}$

5.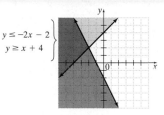

$\left. \begin{array}{l} y \leq -2x - 2 \\ y \geq x + 4 \end{array} \right\}$

7.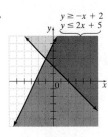

$y \geq -x + 2$
$y \leq 2x + 5$

9.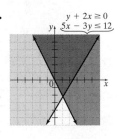

$x \geq 3y$
$x + 3y \leq 6$

11.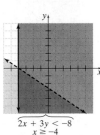

$y + 2x \geq 0$
$5x - 3y \leq 12$

13.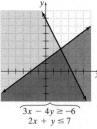

$3x - 4y \geq -6$
$2x + y \leq 7$

15. $x \leq 2$
$y \geq -3$
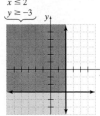

17. $y \geq 1$
$x < -3$
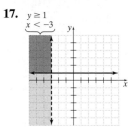

19.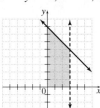

$2x + 3y < -8$
$x \geq -4$

21.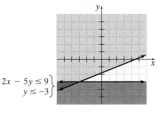

$\left. \begin{array}{l} 2x - 5y \leq 9 \\ y \leq -3 \end{array} \right.$

23.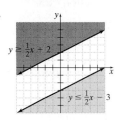

$y \geq \frac{1}{2}x + 2$

$y \leq \frac{1}{2}x - 3$

25. C   **27.** D   **29. a.** $x + y \leq 8$; $x < 3$; $x \geq 0$; $y \geq 0$

b.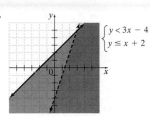

31. Answers may vary.
33. 16
35. $36x^2$
37. $100y^6$

Chapter 8 Review   **1. a.** no **b.** yes **c.** no   **3. a.** no **b.** no **c.** yes

5. $(3, -1)$ **7.** $(-3, -2)$ **9.** $\left(\dfrac{1}{2}, \dfrac{1}{2} \right)$

11. intersecting, one solution

13. identical, infinite number of solutions

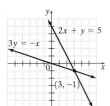

$2x + y = 5$
$3y = -x$
$(3, -1)$

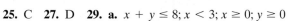

$y - 2x = 4$
$(-3, -2)$
$x + y = -5$

$3x + y = 2$
$3x - 6 = -9y$
$\left(\frac{1}{2}, \frac{1}{2} \right)$

15. $(-1, 4)$ **17.** $(3, -2)$
19. no solution; inconsistent
21. $(3, 1)$ **23.** $(8, -6)$
25. $(-6, 2)$ **27.** $(3, 7)$

29. infinite number of solutions; dependent   **31.** $\left(2, -2\frac{1}{2}\right)$ **33.** $(-6, 15)$ **35.** $(-3, 1)$ **37.** -6 and 22

39. ship: 21.1 mph; current: 3.2 mph   **41.** width: 1.15 ft; length: 1.85 ft   **43.** one egg: \$0.40; one strip of bacon: \$0.65

45. 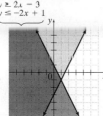   **47.** 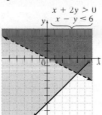   **49.** 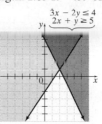   **51.**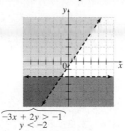

Chapter 8 Test   **1.** false   **2.** false   **3.** true   **4.** false   **5.** no   **6.** yes

7. $(-4, 2)$ 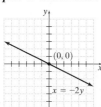   **8.** $(-4, 1)$ **9.** $\left(\frac{1}{2}, -2\right)$ **10.** $(4, -2)$ **11.** $\left(2, \frac{1}{2}\right)$ **12.** $(4, -5)$ **13.** $(7, 2)$ **14.** $(5, -2)$

15. 20 \$1.00 bills; 42 \$5.00 bills   **16.** \$1225 at 5%; \$2775 at 9%

17. Texas: 226 thousand; Missouri: 110 thousand

18. 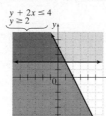   **19.**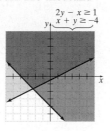

Chapter 8 Cumulative Review   **1. a.** $<$ **b.** $=$ **c.** $>$; Sec. 1.2, Ex. 6   **2.** -7; Sec. 2.2, Ex. 4

3. 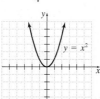 ; Sec. 3.3, Ex. 3   **4.** $5x^2 - 17x + 14$; Sec. 4.4, Ex. 2   **5.** $(x - 3)(x - 5)$; Sec. 5.2, Ex. 2

6. $(2m - 1)^2$; Sec. 5.3, Ex. 9   **7.** $2(3a - 2b)(9a^2 + 6ab + 4b^2)$; Sec. 5.4, Ex. 9

8. $(2t - 3)(5t - 1)$; Sec. 5.5, Ex. 1   **9.** 4, 5; Sec. 5.6, Ex. 2

10. a. $\frac{9}{7}$ **b.** $-\frac{2}{7}$; Sec. 6.1, Ex. 1   **11.** $\frac{2}{5}$; Sec. 6.2, Ex. 2

12. LCD $= x - 2$ or LCD $= 2 - x$; Sec. 6.3, Ex. 8   **13.** $\frac{3}{x - 2}$; Sec. 6.4, Ex. 2

14. $\frac{15}{16}$; Sec. 6.5, Ex. 1   **15.** no solution; Sec. 6.6, Ex. 5   **16.** 63; Sec. 6.7, Ex. 2   **17.** -5; Sec. 6.8, Ex. 1

18. slope: $\frac{3}{4}$; y-intercept: -1; Sec. 7.1, Ex. 3   **19.** $2x + y = 3$; Sec. 7.2, Ex. 1

20. ; Sec. 7.3, Ex. 1   **21.** domain: $\{-1, 0, 3\}$; range: $\{-2, 0, 2, 3\}$; Sec. 7.4, Ex. 1

22. one solution; Sec. 8.1, Ex. 6

23. $\left(6, \frac{1}{2}\right)$; Sec. 8.2, Ex. 2

24. $(6, 1)$; Sec. 8.3, Ex. 1

25. 29 and 8; Sec. 8.4, Ex. 1

■ CHAPTER 9   ROOTS AND RADICALS

Calculator Explorations   **1.** 2.646; yes   **3.** 3.317; yes   **5.** 9.055; yes   **7.** 3.420; yes   **9.** 2.115; yes   **11.** 1.783; yes

Mental Math   **1.** false   **3.** true   **5.** true

Exercise Set 9.1   **1.** 4   **3.** 9   **5.** $\frac{1}{5}$ **7.** -10 **9.** not a real number   **11.** -11 **13.** $\frac{3}{5}$ **15.** 12   **17.** $\frac{7}{6}$ **19.** -1

21. 6.083   **23.** 11.662   **25.** $\sqrt{2} \approx 1.41$; 126.90 ft   **27.** z **29.** x^2 **31.** $3x^4$ **33.** $9x$ **35.** $\frac{x^3}{6}$ **37.** $\frac{5y}{3}$ **39.** 5   **41.** -4

43. -2 **45.** $\dfrac{1}{2}$ **47.** -5 **49.** -10 **51.** Answers may vary. **53.** 2 **55.** not a real number **57.** -5 **59.** 1 **61.** -2

63. 4 **65.** 58 ft **67.** $-2, -1.3, -1, 0, 1, 1.3, 2$ **69.** x^7 **71.** x^5 **73.** $(-3, 0)$ **75.** $(5, 0)$ **77.** $4 \cdot 2$ **79.** $25 \cdot 3$ **81.** $4 \cdot 11$

83. $9 \cdot 10$

Mental Math **1.** 6 **3.** x **5.** 0 **7.** $5x^2$

Exercise Set 9.2 **1.** $2\sqrt{5}$ **3.** $3\sqrt{2}$ **5.** $5\sqrt{2}$ **7.** $\sqrt{33}$ **9.** $2\sqrt{15}$ **11.** $6\sqrt{5}$ **13.** $2\sqrt{13}$ **15.** $\dfrac{2\sqrt{2}}{5}$ **17.** $\dfrac{3\sqrt{3}}{11}$ **19.** $\dfrac{3}{2}$

21. $\dfrac{5\sqrt{5}}{3}$ **23.** $\dfrac{\sqrt{11}}{6}$ **25.** $-\dfrac{\sqrt{3}}{4}$ **27.** $x^3\sqrt{x}$ **29.** $x^6\sqrt{x}$ **31.** $5x\sqrt{3}$ **33.** $4x^2\sqrt{6}$ **35.** $\dfrac{2\sqrt{3}}{y}$ **37.** $\dfrac{3\sqrt{x}}{y}$ **39.** $\dfrac{2\sqrt{22}}{x^2}$

41. $2\sqrt[3]{3}$ **43.** $5\sqrt[3]{2}$ **45.** $\dfrac{\sqrt[3]{5}}{4}$ **47.** $\dfrac{\sqrt[3]{7}}{2}$ **49.** $\dfrac{\sqrt[3]{15}}{4}$ **51.** $2\sqrt[3]{10}$ **53.** $2\sqrt[4]{10}$ **55.** $\dfrac{\sqrt[4]{8}}{3}$ **57.** $2\sqrt[5]{3}$ **59.** $\dfrac{\sqrt[5]{5}}{2}$ **61.** $2\sqrt[3]{10}$

63. Answers may vary. **65.** $1700 **67.** $-2x^2$ **69.** $\dfrac{\sqrt[3]{2}}{x^3}$ **71.** $14x$ **73.** $2x^2 - 7x - 15$ **75.** 0 **77.** 8 cm

Mental Math **1.** $8\sqrt{2}$ **3.** $7\sqrt{x}$ **5.** $3\sqrt{7}$

Exercise Set 9.3 **1.** $-4\sqrt{3}$ **3.** $9\sqrt{6} - 5$ **5.** $\sqrt{5} + \sqrt{2}$ **7.** $7\sqrt[3]{3} - \sqrt{3}$ **9.** $-5\sqrt[3]{2} - 6$ **11.** $8\sqrt{5}$ in.

13. Answers may vary. **15.** $5\sqrt{3}$ **17.** $9\sqrt{5}$ **19.** $4\sqrt{6} + \sqrt{5}$ **21.** $x + \sqrt{x}$ **23.** 0 **25.** $4\sqrt{x} - x\sqrt{x}$ **27.** $7\sqrt{5}$

29. $\sqrt{5} + \sqrt[3]{5}$ **31.** $-5 + 8\sqrt{2}$ **33.** $8 - 6\sqrt{2}$ **35.** $14\sqrt{2}$ **37.** $5\sqrt{2} + 12$ **39.** $\dfrac{4\sqrt{5}}{9}$ **41.** $\dfrac{3\sqrt{3}}{8}$ **43.** $2\sqrt{5}$

45. $-\sqrt{35}$ **47.** $12\sqrt{2x}$ **49.** 0 **51.** $x\sqrt{3x} + 3x\sqrt{x}$ **53.** $5\sqrt[3]{3}$ **55.** $\sqrt[3]{9}$ **57.** $4 + 4\sqrt[3]{2}$ **59.** $-3 + 3\sqrt[3]{2}$

61. $4x\sqrt{2} + 2\sqrt[3]{4} + 2x$ **63.** $\sqrt[3]{5}$ **65.** $\left(48 + \dfrac{9\sqrt{3}}{2}\right)$ sq. ft **67.** $x^2 + 12x + 36$ **69.** $4x^2 - 4x + 1$ **71.** $(4, 2)$

Mental Math **1.** $\sqrt{6}$ **3.** $\sqrt{6}$ **5.** $\sqrt{10y}$

Exercise Set 9.4 **1.** 4 **3.** $5\sqrt{2}$ **5.** $2\sqrt{5} + 5\sqrt{2}$ **7.** $15 - 12\sqrt{15} - 5\sqrt{2} + 4\sqrt{30}$ **9.** $x - 36$ **11.** $67 + 16\sqrt{3}$

13. $130\sqrt{3}$ sq. m **15.** 4 **17.** $3\sqrt{2}$ **19.** $5y^2$ **21.** $\dfrac{\sqrt{15}}{5}$ **23.** $\dfrac{\sqrt{6y}}{6y}$ **25.** $\dfrac{\sqrt{10}}{6}$ **27.** $\dfrac{\sqrt{A\pi}}{\pi}$ **29.** Answers may vary.

31. $3\sqrt{2} - 3$ **33.** $2\sqrt{10} + 6$ **35.** $\sqrt{30} + 5 + \sqrt{6} + \sqrt{5}$ **37.** $3 + \sqrt{3}$ **39.** $3 - 2\sqrt{5}$ **41.** $3\sqrt{3} + 1$ **43.** $24\sqrt{5}$

45. 20 **47.** $36x$ **49.** $\sqrt{30} + \sqrt{42}$ **51.** $4x\sqrt{5} - 60\sqrt{x}$ **53.** $\sqrt{6} - \sqrt{15} + \sqrt{10} - 5$ **55.** -5 **57.** $x - 9$

59. $15 + 6\sqrt{6}$ **61.** $9x - 30\sqrt{x} + 25$ **63.** $5\sqrt{3}$ **65.** $2y\sqrt{6}$ **67.** $2xy\sqrt{3y}$ **69.** $\dfrac{\sqrt{30}}{15}$ **71.** $\dfrac{\sqrt{15}}{10}$ **73.** $\dfrac{3\sqrt{2x}}{2}$

75. $-8 - 4\sqrt{5}$ **77.** $5\sqrt{10} - 15$ **79.** $\dfrac{6\sqrt{5} - 4\sqrt{3}}{11}$ **81.** $\sqrt{6} + \sqrt{3} + \sqrt{2} + 1$ **83.** $2\sqrt[3]{6}$ **85.** $12\sqrt[3]{10}$ **87.** $5\sqrt[3]{3}$

89. 3 **91.** $2\sqrt[3]{3}$ **93.** $\dfrac{\sqrt[3]{10}}{2}$ **95.** $3\sqrt[3]{4}$ **97.** $\dfrac{\sqrt[3]{3}}{3}$ **99.** $\dfrac{\sqrt[3]{6}}{3}$ **101.** Answers may vary. **103.** $\dfrac{2}{-2\sqrt{2} - 4 + \sqrt{6} + 2\sqrt{3}}$

105. $3x + 2$ **107.** $4y - 1$ **109.** 5 **111.** 3

Exercise Set 9.5 **1.** 81 **3.** -1 **5.** 5 **7.** 49 **9.** no solution **11.** -2 **13.** $\dfrac{3}{2}$ **15.** 16 **17.** no solution **19.** 2 **21.** 4

23. 3 **25.** 2 **27.** 2 **29.** no solution **31.** $0, -3$ **33.** -3 **35.** 9 **37.** 12 **39.** 3, 1 **41.** -1 **43.** 2 **45.** 16 **47.** 1

49. 1 **51.** 9 **53. a.** $3.2, 10, 31.6$ **b.** no **55.** Answers may vary. **57.** 2.43 **59.** 0.48 **61.** $2x - (x + 3) = 11; x = 14$

63. $2x + 2(x + 2) = 24$; length: 7 in.

Exercise Set 9.6 **1.** $\sqrt{13}$ **3.** $3\sqrt{3}$ **5.** 25 **7.** $\sqrt{22}$ **9.** $3\sqrt{17}$ **11.** $\sqrt{41}$ **13.** $4\sqrt{2}$ **15.** $3\sqrt{10}$ **17.** 51.2 ft

19. 20.6 ft **21.** 11.7 ft **23.** $\sqrt{29}$ **25.** $\sqrt{73}$ **27.** $2\sqrt{10}$ **29.** $\dfrac{3\sqrt{5}}{2}$ **31.** $\sqrt{85}$ **33.** 24 cu. ft **35.** 54 mph **37.** 27 mph

39. 61.2 km **41.** $2\sqrt{10}$ **43.** 201 mi **45.** Answers may vary. **47.** -27 **49.** $\dfrac{8}{343}$ **51.** x^6 **53.** x^8

Exercise Set 9.7   **1.** 2   **3.** 3   **5.** 8   **7.** 4   **9.** $-\dfrac{1}{2}$ **11.** $\dfrac{1}{64}$ **13.** $\dfrac{1}{729}$ **15.** $\dfrac{5}{2}$ **17.** Answers may vary.   **19.** 2

21. 2   **23.** $\dfrac{1}{x^{2/3}}$ **25.** x^3 **27.** Answers may vary.   **29.** 9   **31.** -3 **33.** -3 **35.** $\dfrac{1}{9}$ **37.** $\dfrac{3}{4}$ **39.** 27   **41.** 512   **43.** -4

45. 32   **47.** $\dfrac{8}{27}$ **49.** $\dfrac{1}{27}$ **51.** $\dfrac{1}{2}$ **53.** $\dfrac{1}{5}$ **55.** $\dfrac{1}{125}$ **57.** 9   **59.** $6^{1/3}$ **61.** x^6 **63.** 36   **65.** $\dfrac{1}{3}$ **67.** $\dfrac{x^{2/3}}{y^{3/2}}$ **69.** $\dfrac{x^{16/5}}{y^6}$

71. 11,224 people   **73.** 3.344   **75.** 3.665   **77.**   **79.** $-1, 4$ **81.** $-\dfrac{1}{2}, 3$

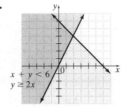

$x + y < 6$
$y \geq 2x$

Chapter 9 Review   **1.** 9   **3.** 3   **5.** $-\dfrac{3}{8}$ **7.** not a real number   **9.** irrational, 8.718   **11.** x^6 **13.** $3x^3$ **15.** $\dfrac{4}{y^5}$ **17.** $3\sqrt{6}$

19. $5x\sqrt{6x}$ **21.** $3\sqrt[3]{2}$ **23.** $2\sqrt[4]{3}$ **25.** $\dfrac{3\sqrt{2}}{5}$ **27.** $\dfrac{3y\sqrt{5}}{2x^2}$ **29.** $\dfrac{\sqrt[4]{9}}{2}$ **31.** $\dfrac{\sqrt[3]{3}}{2}$ **33.** $2\sqrt[3]{3} - \sqrt[3]{2}$ **35.** $6\sqrt{6} - 2\sqrt[3]{6}$

37. $5\sqrt{7x} + 2\sqrt[3]{7}$ **39.** $\dfrac{\sqrt{5}}{6}$ **41.** 0   **43.** $30\sqrt{2}$ **45.** $6\sqrt{2} - 18$ **47.** $3\sqrt{2} - 5\sqrt{3} + 2\sqrt{6} - 10$ **49.** $4\sqrt{2}$ **51.** $\dfrac{x\sqrt{5x}}{2}$

53. $\dfrac{\sqrt{30}}{6}$ **55.** $\dfrac{\sqrt{6x}}{2x}$ **57.** $\dfrac{\sqrt{35}}{10y}$ **59.** $\dfrac{\sqrt[3]{21}}{3}$ **61.** $\dfrac{\sqrt[3]{12}}{2}$ **63.** $3\sqrt{5} + 6$ **65.** $4\sqrt{6} - 8$ **67.** $\dfrac{2\sqrt{2} - 1}{6}$ **69.** $-\dfrac{3 + 5\sqrt{3}}{11}$
71. 18   **73.** 25   **75.** 12   **77.** 6   **79.** $2\sqrt{14}$ **81.** $4\sqrt{34}$ ft   **83.** $\sqrt{130}$ **85.** 2.4 in.   **87.** $a^{5/2}$ **89.** $x^{5/2}$ **91.** 4   **93.** -2

95. -512 **97.** $\dfrac{8}{27}$ **99.** $\dfrac{1}{5}$ **101.** 32   **103.** $\dfrac{1}{3^{2/3}}$ **105.** $\dfrac{1}{x^2}$

Chapter 9 Test   **1.** 4   **2.** 5   **3.** 8   **4.** $\dfrac{3}{4}$ **5.** not a real number   **6.** $\dfrac{1}{9}$ **7.** $3\sqrt{6}$ **8.** $2\sqrt{23}$ **9.** $x^3\sqrt{3}$ **10.** $2x^2y^3\sqrt{2y}$

11. $3x^4\sqrt{x}$ **12.** $2\sqrt[3]{5}$ **13.** 2   **14.** $-8\sqrt{3}$ **15.** $3\sqrt[3]{2} - 2x\sqrt{2}$ **16.** $\dfrac{\sqrt{5}}{4}$ **17.** $\dfrac{\sqrt[3]{2}}{3}$ **18.** $6\sqrt{2x}$ **19.** $\dfrac{\sqrt{6}}{3}$ **20.** $\dfrac{\sqrt[3]{15}}{3}$

21. $\dfrac{\sqrt{15}}{6x}$ **22.** $4\sqrt{6} - 8$ **23.** $-1 - \sqrt{3}$ **24.** 9   **25.** 5   **26.** 9   **27.** $4\sqrt{5}$ in.   **28.** $\sqrt{5}$ **29.** $\dfrac{1}{16}$ **30.** $x^{10/3}$

Chapter 9 Cumulative Review   **1. a.** $-\dfrac{39}{5}$ **b.** 2; Sec. 1.7, Ex. 9   **2.** -3; Sec. 2.2, Ex. 5   **3.**

x	y
-1	-3
0	0
-3	-9

; Sec 3.1, Ex. 5

4. $10xy - 5y^2 - 9x^2$; Sec. 4.2, Ex. 7   **5.** $(x + 2y)(x + 3y)$; Sec. 5.2, Ex. 6   **6.** $\dfrac{-2 - x}{3x + 1}$; Sec. 6.1, Ex. 8

7. $\dfrac{3y^9}{160}$; Sec. 6.2, Ex. 4   **8.** 1; Sec. 6.3, Ex. 2   **9.** $\dfrac{x(3x - 1)}{(x + 1)^2(x - 1)}$; Sec. 6.4, Ex. 7   **10.** $\dfrac{3}{z}$; Sec. 6.5, Ex. 3

11. -5; Sec. 6.6, Ex. 1   **12.** 15 yd; Sec. 6.8, Ex. 4   **13.** $y = \dfrac{1}{4}x - 3$; (Sec. 7.1, Ex. 5)   **14.** $y = -3$; Sec. 7.2, Ex. 4

15. a, b, c; Sec. 7.4, Ex. 5   **16. a.** solution   **b.** not a solution; Sec. 8.1, Ex. 1   **17.** $(4, 2)$; Sec. 8.2, Ex. 1

18. $\left(-\dfrac{5}{4}, -\dfrac{5}{2}\right)$; Sec. 8.3, Ex. 6   **19.** 5% saline solution: 2.5 L; 25% saline solution: 7.5 L; Sec. 8.4, Ex. 4

20. 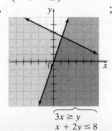   ; Sec. 8.5, Ex. 1   **21. a.** $3\sqrt{6}$ **b.** $2\sqrt{3}$ **c.** $10\sqrt{2}$ **d.** $\sqrt{35}$; Sec. 9.2, Ex. 1

$3x \geq y$
$x + 2y \leq 8$

22. a. -44 **b.** $7x + 4\sqrt{7x} + 4$; Sec. 9.4, Ex. 4
23. no solution; Sec. 9.5, Ex. 2
24. 10 in.; Sec. 9.6, Ex. 1
25. a. 8   **b.** 9   **c.** -8; Sec. 9.7, Ex. 2

■ CHAPTER 10   SOLVING QUADRATIC EQUATIONS

Exercise Set 10.1   **1.** ± 8 **3.** $\pm\sqrt{21}$ **5.** $\pm\dfrac{1}{5}$ **7.** no real solution   **9.** $\pm\dfrac{\sqrt{39}}{3}$ **11.** $\pm\dfrac{2\sqrt{7}}{7}$ **13.** $\pm\sqrt{2}$ **15.** $\pm\sqrt{5}$

17. Answers may vary.   **19.** $12, -2$ **21.** $-2 \pm \sqrt{7}$ **23.** $1, 0$ **25.** $-2 \pm \sqrt{10}$ **27.** $\dfrac{8}{3}, -4$ **29.** no real solution

31. $\dfrac{11 \pm 5\sqrt{2}}{2}$ **33.** $\dfrac{7 \pm 4\sqrt{2}}{3}$ **35.** $8, -3$ **37.** $2, -6$ **39.** $5 \pm \sqrt{11}$ **41.** $r = 6$ in.   **43.** 5 sec   **45.** 2002   **47.** ± 3.14

49. $-14.76, -6.60$ **51.** $0.44, 1.94$ **53.** $(y + 5)^2$ **55.** $(x - 10)^2$ **57.** $y = 2.1x + 36.6$

Mental Math   **1.** 16   **3.** 100   **5.** 49

Exercise Set 10.2   **1.** $(x + 2)^2$ **3.** $(k - 6)^2$ **5.** $\left(x - \dfrac{3}{2}\right)^2$ **7.** $\left(m - \dfrac{1}{2}\right)^2$ **9.** $0, 6$ **11.** $-6, -2$ **13.** $-1 \pm \sqrt{6}$

15. $-3 \pm \sqrt{34}$ **17.** $\dfrac{-5 \pm \sqrt{53}}{2}$ **19.** $1 \pm \sqrt{2}$ **21.** $-4, -1$ **23.** $-6, 3$ **25.** $-\dfrac{1}{2}, \dfrac{13}{2}$ **27.** no real solution

29. $\dfrac{3 \pm \sqrt{19}}{2}$ **31.** $-2, 4$ **33.** $\dfrac{-4 \pm \sqrt{6}}{2}$ **35.** $\dfrac{1}{2}, 1$ **37.** $\dfrac{1 \pm \sqrt{13}}{3}$ **39.** Answers may vary.   **41.** $k = 8$ or $k = -8$

43. $x = -6, -2$ **45.** $x \approx -0.68, 3.68$ **47.** $-\dfrac{1}{2}$ **49.** -1 **51.** $3 + 2\sqrt{5}$ **53.** $\dfrac{1 - 3\sqrt{2}}{2}$

Mental Math   **1.** $a = 2, b = 5, c = 3$ **3.** $a = 10, b = -13, c = -2$ **5.** $a = 1, b = 0, c = -6$

Exercise Set 10.3   **1.** $-2, 1$ **3.** $\dfrac{-5 \pm \sqrt{17}}{2}$ **5.** $\dfrac{2 \pm \sqrt{2}}{2}$ **7.** $1, 2$ **9.** $\dfrac{-7 \pm \sqrt{37}}{6}$ **11.** $\pm\dfrac{2}{7}$ **13.** no real solution

15. $-3, 10$ **17.** $\pm\sqrt{5}$ **19.** $-3, 4$ **21.** $-2 \pm \sqrt{7}$ **23.** $\dfrac{1}{2}, 3$ **25.** $\dfrac{5 \pm \sqrt{33}}{2}$ **27.** $-2, \dfrac{7}{3}$ **29.** $\dfrac{-9 \pm \sqrt{129}}{12}$ **31.** $\dfrac{4 \pm \sqrt{2}}{7}$

33. $3 \pm \sqrt{7}$ **35.** $\dfrac{3 \pm \sqrt{3}}{2}$ **37.** $-1, \dfrac{1}{3}$ **39.** $-\dfrac{3}{4}, \dfrac{1}{5}$ **41.** no real solution   **43.** $\dfrac{3 \pm \sqrt{13}}{4}$ **45.** no real solution

47. $1 \pm \sqrt{2}$ **49.** $-\dfrac{3}{4}, -\dfrac{1}{2}$ **51.** $\dfrac{7 \pm \sqrt{129}}{20}$ **53.** no real solution   **55.** $\dfrac{1 \pm \sqrt{2}}{5}$ **57.** $\dfrac{-3\sqrt{2} \pm \sqrt{38}}{2}$

59. 2 real solutions   **61.** no real solutions   **63.** 1 real solution   **65.** 2 real solutions   **67.** no real solutions

69. 1 real solution   **71.** d   **73.** $-4.4, 3.4$ **75.** $-0.7, 5.0$ **77.** 7.9 sec   **79.** $\dfrac{6}{7}$ **81.** $\dfrac{14}{15}$ **83.** -4

Exercise Set 10.4   **1.** $\dfrac{1}{5}, 2$ **3.** $1 \pm \sqrt{2}$ **5.** $\pm 2\sqrt{5}$ **7.** no real solution   **9.** 2   **11.** 3   **13.** ± 2 **15.** $0, 1, 2$ **17.** $-5, 0$

19. $\dfrac{3 \pm \sqrt{7}}{5}$ **21.** $-1, \dfrac{3}{2}$ **23.** $\dfrac{5 \pm \sqrt{105}}{20}$ **25.** $\dfrac{7}{4}, 5$ **27.** $\dfrac{7 \pm 3\sqrt{2}}{5}$ **29.** $\dfrac{7 \pm \sqrt{193}}{6}$ **31.** $-10, 11$ **33.** $-\dfrac{2}{3}, 4$

35. $0.1, 0.5$ **37.** $\dfrac{11 \pm \sqrt{41}}{20}$ **39.** $\dfrac{4 \pm \sqrt{10}}{2}$ **41.** 2.3 sec   **43.** 72.7 sec   **45.** 2002   **47.** 2001   **49.** $\dfrac{1 + \sqrt{5}}{2}$

51. Answers may vary.   **53.** $2\sqrt{26}$ **55.** $4\sqrt{5}$ **57.** width: 17 in.; length: 23 in.

Exercise Set 10.5   **1.** $3i$ **3.** $10i$ **5.** $5i\sqrt{2}$ **7.** $3i\sqrt{7}$ **9.** $-3 + 9i$ **11.** $1 - 3i$ **13.** $8 + 12i$ **15.** $26 - 2i$

17. Answers may vary.   **19.** $2 - 3i$ **21.** $\dfrac{31}{25} + \dfrac{17}{25}i$ **23.** $-1 \pm 3i$ **25.** $\dfrac{3 \pm 2i\sqrt{3}}{2}$ **27.** $-3 \pm 2i$ **29.** $\dfrac{-7 \pm i\sqrt{15}}{8}$

31. $\dfrac{2 \pm i\sqrt{6}}{2}$ **33.** $15 - 7i$ **35.** $45 + 63i$ **37.** $-1 + 3i$ **39.** $2 - 3i$ **41.** $-7 - 2i$ **43.** 25   **45.** $\dfrac{2}{5} - \dfrac{9}{5}i$

47. $21 + 20i$ **49.** $4 \pm 8i$ **51.** $\pm 5i$ **53.** $-3 \pm i$ **55.** $\dfrac{5 \pm i\sqrt{47}}{4}$ **57.** $-4 \pm i\sqrt{5}$ **59.** $\pm 6i$ **61.** $-7 \pm i$ **63.** true

65. true   **67.**   **69.**   **71.** $x = \sqrt{51}$ m

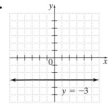

Calculator Explorations   **1.** $x = -0.41, 7.41$ **3.** $x = 0.91, 2.38$ **5.** $x = -0.39, 0.84$

Exercise Set 10.6   **1.** 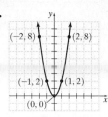   **3.** 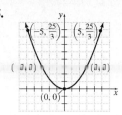   **5.** 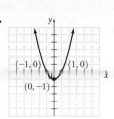   **7.**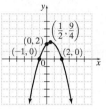

9. 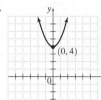   **11.** 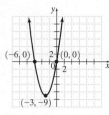   **13.** 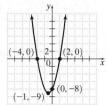   **15.**

17. 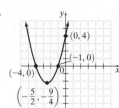   **19.** 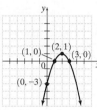   **21.** 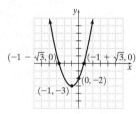   **23.**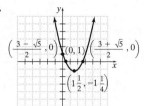

25. domain: all real numbers; range: $y \le 3$ **27.** domain: all real numbers; range: $y \le 1$ **29. a.** 256 ft   **b.** $t = 4$ sec
c. $t = 8$ sec   **31.** B   **33.** D   **35.** $\dfrac{21}{8}$ **37.** $\dfrac{x^2}{10}$ **39.** $\dfrac{x^2}{x^2 - 1}$ **41.** $-2a$

Chapter 10 Review   **1.** $-\dfrac{3}{5}, 4$ **3.** $-\dfrac{1}{3}, 2$ **5.** $\pm 5\sqrt{2}$ **7.** $-1, 7$ **9.** $4, 18$ **11.** $0, \pm 3$ **13.** $-3, 2$

15. $x^2 - 10x + 25 = (x - 5)^2$ **17.** $a^2 + 4a + 4 = (a + 2)^2$ **19.** $m^2 - 3m + \dfrac{9}{4} = \left(m - \dfrac{3}{2}\right)^2$ **21.** $3 \pm \sqrt{2}$

23. $-1, \dfrac{1}{2}$ **25.** $5 \pm 3\sqrt{2}$ **27.** $-1, \dfrac{1}{2}$ **29.** $-\dfrac{5}{3}$ **31.** $\dfrac{2}{5}, \dfrac{1}{3}$ **33.** no real solution   **35.** 2 real solutions   **37.** 1 real solution

39. no real solutions   **41.** $\dfrac{-1 \pm \sqrt{21}}{10}$ **43.** $0, \pm\dfrac{1}{2}$ **45.** $\dfrac{1}{2}, 7$ **47.** $\dfrac{1}{3}$ **49.** 3   **51.** $-10, 22$ **53.** $\dfrac{3 \pm \sqrt{5}}{20}$ **55.** $-5 \pm \sqrt{30}$

57. 2.5 sec   **59.** 2004   **61.** $12i$ **63.** $6i\sqrt{3}$ **65.** $21 - 10i$ **67.** $-8 - 2i$ **69.** $-13i$ **71.** 25   **73.** $-\dfrac{3}{2} - \dfrac{1}{2}i$ **75.** $\dfrac{2}{5} - \dfrac{9}{5}i$

77. $\pm 4i$ **79.** $2 \pm 3i$ **81.** vertex: $(0, 0)$; axis of symmetry: $x = 0$; opens downward
83. vertex: $(3, 0)$; axis of symmetry: $x = 3$; opens upward   **85.** vertex: $(0, -7)$; axis of symmetry: $x = 0$; opens upward
87. vertex: $(72, 14)$; axis of symmetry: $x = 72$; opens downward

89. 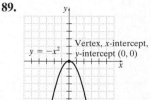   **91.** 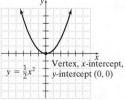   **93.** 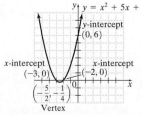   **95.**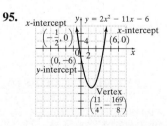

97. 1 solution; $x = -2$ **99.** no real solutions

Chapter 10 Test   **1.** $-\dfrac{3}{2}, 7$ **2.** $-2, 0, 1$ **3.** ± 4 **4.** $\dfrac{5 \pm 2\sqrt{2}}{3}$ **5.** $10, 16$ **6.** $-2, \dfrac{1}{5}$ **7.** $-2, 5$ **8.** $\dfrac{5 \pm \sqrt{37}}{6}$ **9.** $-\dfrac{4}{3}, 1$

10. $-1, \dfrac{5}{3}$ **11.** $\dfrac{7 \pm \sqrt{73}}{6}$ **12.** $2 \pm i$ **13.** $\dfrac{1}{3}, 2$ **14.** $\dfrac{3 \pm \sqrt{7}}{2}$ **15.** $-3, 0, \dfrac{1}{2}$ **16.** $0, \pm\dfrac{1}{3}$ **17.** $5i$ **18.** $10i\sqrt{2}$ **19.** $8 + i$

20. $7 - 6i$ **21.** $4i$ **22.** 6   **23.** 13   **24.** $\dfrac{1}{5} - \dfrac{7}{5}i$

25. 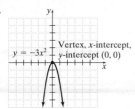   **26.** 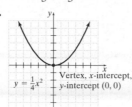   **27.** 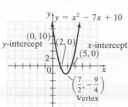   **28.** 27 sec   **29.** 2002

Chapter 10 Cumulative Review   **1. a.** $\dfrac{1}{2}$ **b.** 9; Sec. 1.6, Ex. 6   **2. a.** $5x + 7$ **b.** $-4a - 1$ **c.** $4y - 3y^2$

d. $7.3x - 6$; Sec. 2.1, Ex. 4   **3. a.** x-int: $(-3, 0)$; y-int: $(0, 2)$ **b.** x-int: $(-4, 0)$, $(-1, 0)$; y-int: $(0, 1)$

c. x-int and y-int: $(0, 0)$ **d.** x-int: $(2, 0)$; y-int: none   **e.** x-int: $(-1, 0)$, $(3, 0)$; y-int: $(0, -1)$, $(0, 2)$; Sec. 3.3, Ex. 1

4. a. $\dfrac{25x^4}{y^6}$ **b.** x^6 **c.** $32x^2$ **d.** a^3b; Sec. 4.1, Ex. 10   **5.** $-6, -\dfrac{3}{2}, \dfrac{1}{5}$; Sec. 5.6, Ex. 6   **6.** $\dfrac{31}{2}$; Sec. 6.7, Ex. 3

7. parallel; Sec. 7.1, Ex. 4

8. 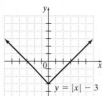 ; Sec. 7.3, Ex. 6   **9.** no solution; Sec. 8.1, Ex. 3   **10.** $(-2, 0)$; Sec. 8.2, Ex. 3

11. infinite number of solutions; Sec. 8.3, Ex. 4

12. Albert: 3.25 mph; Louis: 4.25 mph; Sec. 8.4, Ex. 3

13. 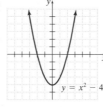 ; Sec. 8.5, Ex. 3   **14. a.** 2   **b.** -2 **c.** -2 **d.** not a real number; Sec. 9.1, Ex. 5

15. a. $\dfrac{5}{6}$ **b.** $\dfrac{\sqrt{3}}{8}$ **c.** $\dfrac{2}{9}$; Sec. 9.2, Ex. 2

16. a. $7\sqrt{2}$ **b.** $9\sqrt{3}$ **c.** $1 - 3\sqrt{3} - 6\sqrt{2}$; Sec. 9.3, Ex. 2

17. a. $\sqrt{21}$ **b.** $3\sqrt{5}$ **c.** $20\sqrt{3}$ **d.** 18; Sec. 9.4, Ex. 1

18. $\dfrac{1}{2}$; Sec. 9.5, Ex. 3   **19.** $\sqrt{44,800} \approx 212$ ft; Sec. 9.6, Ex. 3

20. a. $\sqrt{25} = 5$ **b.** $\sqrt[3]{8} = 2$ **c.** $-\sqrt[4]{16} = -2$ **d.** $\sqrt[3]{-27} = -3$ **e.** $\sqrt{\dfrac{1}{9}} = \dfrac{1}{3}$; Sec. 9.7, Ex. 1

21. $\dfrac{\sqrt{14}}{2}, -\dfrac{\sqrt{14}}{2}$; Sec. 10.1, Ex. 2   **22.** $5 \pm \sqrt{11}$; Sec. 10.2, Ex. 3   **23.** $-\dfrac{1}{2}, 5$; Sec. 10.3, Ex. 2

24. a. $2i$ **b.** $i\sqrt{11}$ **c.** $2i\sqrt{5}$; Sec. 10.5, Ex. 1   **25.** ; Sec. 10.6, Ex. 2

■ APPENDIX A   OPERATIONS ON DECIMALS

Appendix A Exercise Set   **1.** 17.08   **3.** 12.804   **5.** 110.96   **7.** 2.4   **9.** 28.43   **11.** 227.5   **13.** 2.7   **15.** 49.2339   **17.** 80
19. 0.07612   **21.** 4.56   **23.** 648.46   **25.** 767.83   **27.** 12.062   **29.** 61.48   **31.** 7.7   **33.** 863.37   **35.** 22.579   **37.** 363.15
39. 7.007

■ APPENDIX B   REVIEW OF ANGLES, LINES, AND SPECIAL TRIANGLES

Appendix B Exercise Set   **1.** 71°   **3.** 19.2°   **5.** $78\frac{3}{4}°$ **7.** 30°   **9.** 149.8°   **11.** $100\frac{1}{2}°$

13. $m\angle 1 = m\angle 5 = m\angle 7 = 110°$; $m\angle 2 = m\angle 3 = m\angle 4 = m\angle 6 = 70°$ **15.** 90°   **17.** 90°   **19.** 90°   **21.** 45°, 90°

23. 73°, 90°   **25.** $50\frac{1}{4}°, 90°$ **27.** $x = 6$ **29.** $x = 4.5$ **31.** 10   **33.** 12

■ APPENDIX D   MEAN, MEDIAN, AND MODE

Appendix D Exercise Set   **1.** mean: 29, median: 28, no mode   **3.** mean: 8.1, median: 8.2, mode: 8.2
5. mean: 0.6, median: 0.6, mode: 0.2 and 0.6   **7.** mean: 370.9, median: 313.5, no mode   **9.** 1314 ft   **11.** 1131.5 ft   **13.** 6.8
15. 6.9   **17.** 85.5   **19.** 73   **21.** 70 and 71   **23.** 9   **25.** 21, 21, 24

■ APPENDIX F   REVIEW OF VOLUME AND SURFACE AREA

Appendix F Exercise Set   **1.** $V = 72$ cu. in.; $SA = 108$ sq. in.   **3.** $V = 512$ cu. cm; $SA = 384$ sq. cm

5. $V = 4\pi$ cu. yd $\approx 12\frac{4}{7}$ cu. yd; $SA = (2\sqrt{13}\pi + 4\pi)$ sq. yd ≈ 35.20 sq. yd

7. $V = \frac{500}{3}\pi$ cu. in. $\approx 523\frac{17}{21}$ cu. in.; $SA = 100\pi$ sq. in. $\approx 314\frac{2}{7}$ sq. in.   **9.** $V = 48$ cu. cm; $SA = 96$ sq. cm

11. $2\frac{10}{27}$ cu. in.   **13.** 26 sq. ft   **15.** $10\frac{5}{6}$ cu. in.   **17.** 960 cu. cm   **19.** 196π sq. in.   **21.** $7\frac{1}{2}$ cu. ft   **23.** $12\frac{4}{7}$ cu. cm

SUBJECT INDEX

A

Absolute value, 14, 35, 36, 69
Addition
  associative property of, 56, 72
  commutative property of, 56, 72
  of complex numbers, 551–52, 565
  of exponential expressions, 219
  of fractions with same denominator, 20–22, 70
  of fractions with unlike denominators, 348–49
  identity for, 59
  of like radicals, 522
  of polynomials, 232, 264
  of radicals, 494–95
  of rational expressions with common denominators,
    347, 390
  of rational expressions with different denominators,
    353–56, 390
  of real numbers, 34–39, 71
  of two numbers with different signs, 36–37, 71
  of two numbers with same sign, 35, 71
Addition (or elimination) method, 469
  solving systems of linear equations by, 453–57, 475
Addition property of equality, 87–90, 151, 453
Addition property of inequality, 142, 155
Additive inverses (or opposites), 38–39, 59, 71, 72
Algebra, generality with, 3
Algebraic expressions, 28, 70, 97
  difference between equations and, 31
  evaluating, 52–53
  phrases translated into, 30
  simplifying, 80–84
  word phrases written as, 83
American Physical Therapy Association website, 398
Angles
  complementary, 45
  supplementary, 45, 476
Approximations, or irrational numbers, 483
Associative property of addition, 56, 72
Associative property of multiplication, 57, 72
Axis of symmetry, 556, 566

B

Bar graphs, reading, 61–63
Base, 25, 70, 218
Binomials, 228, 229, 264
  factoring, 294–97, 324
  special products for multiplying, 241–42
  squaring, 243–44, 265
Boundary, 203
Boundary lines, 469, 470
Braces, 26
Brackets, 26, 43
Broken line graph, 64
Budget development, 150
Building options choices, 322

C

Calculator explorations
  checking equations, 104
  exponents and order of operations, 31
  operations with real numbers, 53
  quadratic equations, 560
  rational expressions, 370
  scientific notation, 253
  square roots, 485
Carpenters, 480
Celsius
  Fahrenheit converted to, 118–19
Chapter projects
  budget development, 150
  building options, 322
  creating and interpreting graphs, 68
  cylinder dimensions, 520-21
  dosage formulas, 387–88
  financial analysis, 208
  matching descriptions of linear data to
    equations/graphs, 430–31
  modeling physical situation, 562–62
Chapter projects
  modeling with polynomials, 262–63
  ship courses, 472-73
Circle graph, 126

Coefficient, 228, 264, 455
Colon notation, 392
Combining like terms, 81, 82, 90, 151
    for multiplying polynomials, 237
    simplifying polynomials with like terms by, 231
Common base, 219
Common factors, factoring out, 289
Commutative property of addition, 56, 72
Commutative property of multiplication, 56, 72
Complementary angles, 45
Completing the square, 546
    quadratic equations solved by, 534–38, 564
Complex conjugates, 553, 565
Complex fractions, 359
    simplifying, 359–62, 391
Complex numbers
    adding and subtracting, 551–53, 565
    division of, 553
    and imaginary numbers, 551
    multiplying, 552–53, 565
    system of, 550, 551
Complex rational expressions, 359
Compound inequalities, 142, 155
    solving, 146, 156
Conjugates, 523
    rationalizing using, 501
Consistent system, 441, 443, 444, 474
Constant, 228, 229
Corner point, 469
Cowling's Rule for dosage formulas, 387
Credit card offers, 303
Criminalistics, 330
Cross products, 374–75, 392
Cube roots, 484
    of perfect cube in denominator, 500
    simplifying, 491, 522
Cubes, sum or difference of two, 296–97, 324
Cylinder dimensions, 520–21

D

Decimals
    clearing equation of, 102
    percents written as, 125, 153
    rational/irrational numbers written as, 12
Decision making. *See* Spotlight on decision making
Degree of polynomial or term, 229, 264
Denominator, 17, 69
    cube root of perfect cube in, 500
Dependent equations, 442, 443, 474
Descending powers of x, polynomial written in, 228
Difference
    of squares, 244, 292, 295, 324
    of two cubes, 296–97, 324
Discriminant, 543
Distance
    finding, 511

formula for, 116, 117, 511
Distributive property, 72
    for combining like radicals, 522
    combining like terms by applying, 231
    for combining numerical coefficients of like terms, 82
    and expressions containing radicals, 494
    of multiplication over addition, 57–58
    and multiplying complex numbers, 552, 565
    and multiplying polynomials, 237
    for removing parentheses, 82, 83, 90, 100, 102, 151
Dividend, zero as, 51
Diving length of time, finding, 548
Division
    of complex numbers, 553
    of exponential expressions, 219
    of fractions, 20, 70
    of polynomials, 256–60, 266
    of radicals, 523
    of rational expressions, 339–41, 389
    of real numbers, 51–53, 71–72
Divisor, zero as, 51
Domain
    of function, 426
    of relation, 421, 433
Dropped objects, height of, 230

E

Element, of set, 8, 69
Elimination method, 453
Ellipsis, 9
Engine displacement, 225
Engineering technicians, 528
Equal symbol, 9, 29
Equations, 29, 71, 79, 86 (*See also* Linear equations; Quadratic equations; Systems of linear equations)
    checking on calculator, 104
    difference between expressions and, 31
    with infinitely many solutions, 103
    with no solution, 102, 103, 368
    sentences translated into, 30–31
    solving for specified variable, 119, 370
    solving those containing radicals, 504–8, 523
    solving those containing rational expressions, 365–70, 391
    of vertical and horizontal lines, 409
Equations in two variables
    ordered pairs as solution of, 165, 210
    and problem solving, 459–60
    table of values and solutions of, 167
Equivalent equations, 86, 87, 88, 151
Equivalent fractions, 20–21, 70
Equivalent inequalities, 143
Evaluating algebraic expressions, 28, 70
Evaluating expressions, for given replacement values, 43
Exponential expressions, 25, 70, 217, 218, 219
    base of, 50

evaluating, 27
form *a-m/n* and evaluation of, 518, 524
form *am/n* and evaluation of, 517–18, 524
form *a1/n* and evaluation of, 516–17, 524
multiplying, 237
raising to powers, 220
rules for exponents for simplifying those containing fractional exponents, 519
Exponential notation, 25
Exponents, 25, 70, 217, 218–25, 264
on calculator, 31
power of a product rule for, 221, 250, 264
power of a quotient rule for, 222
power rule for, 221, 250, 264
product rule for, 219–20, 250, 264
quotient rule for, 223, 250, 264
summary of rules for, 250
Expressions, simplifying, 70 (*See also* Algebraic expressions; Rational expressions)
Extraneous solution, 505

F

Factoring, 273, 323
binomials, 294–97, 324
completely, 284
by grouping, 323
polynomials, 274
quadratic equations solved by, 305-10, 325
strategy for, 300-302, 323-25
trinomials of form $ax^2 + bx + c$, 286–92, 323–24
trinomials of form $x^2 + bx + c$, 280–84
Factoring out the GCF, 276–77
Factors, 18, 49, 69, 274
Fahrenheit, converted to Celsius, 118–19
Financial analysis, 208
FOIL method, 265, 281
of multiplication, 498
to multiply complex numbers, 552
to multiply two binomials, 241-42
squaring binomial with, 243–44
Forensic science, 330
Formulas
and problem solving, 116–21, 153
solving for specified variable, 119
Four-term polynomials, 278, 291
Fractional notation, 392
Fraction bar, 26, 27
Fractions, 17–22, 69
adding and subtracting those with same denominator, 20–22, 70
adding and subtracting those with unlike denominators, 348–49
clearing equation of, 101
complex, 359
dividing, 20, 70
fundamental principle of, 18

in lowest terms, 70, 334
multiplying, 19, 70
simplifying, 18
subtracting those with same denominator, 70
unit, 343
Fujita, T. Theodore, 148
Fujita Scale (F-Scale), 148
Function notation, 425, 433
Functions, 397, 421–27, 433
defined, 421
and function notation, 425–26
graphs of, 433
relations, domains, and ranges, 421
and vertical line test, 422–25
Fundamental principle of fractions, 18

G

Graphing
of inequalities, 141, 142
linear equations, 173–79, 210
linear inequalities, 202–6, 212
linear inequalities in two variables, 202–5
nonlinear equations, 414–18, 432
ordered pair, 209
quadratic equations, 555–60, 566
solutions of system of linear inequalities, 470
solving systems of linear equations by, 440–44, 474
system of inequalities, 468–69
system of linear inequalities, 476
Graphing calculator explorations
approximating solutions of systems of equations, 444
evaluating expressions at given replacement values, 298
graphing equations, 404
Graphing calculator explorations
graphing equations on same set of axes, 198
negative numbers entered on, 53
solutions of quadratic equation, 310
solving equation for *y*, 187
windows, 179
x-intercepts of graph of equation, 418
Graphs
creating and interpreting, 68
of functions, 433
of horizontal line, 186
of inequality, 141
intercepts of, 182–86
reading, 61–65, 72
of vertical line, 186
x-intercept of, for quadratic equations in two variables, 309
Greater than or equal to symbol, 10
Greatest common factor (GCF)
factoring out, 274-75, 278, 284
of list of integers, 323
of list of terms, 275
Grouping, 277–78

Grouping (*cont.*)
    factoring by, 323
    symbols, 25–26

H

Half-planes, 203
Horizontal axis, 162
Horizontal lines, 186, 211, 410
    equations of, 409
    slope of, 193
Hypotenuse, 316, 510

I

Identities, 72, 103
Identity properties, 59–60
Imaginary numbers, 551, 565
Imaginary unit (*i*), 550, 565
Improper fraction, mixed number written as, 22
Inconsistent system, 441, 443, 444, 474
Independent equations, 442, 443, 474
Index, 484, 522
Inequalities, graphing, 141
Inequality symbols, 9, 10, 140
Information technology (IT) professions, 2
Integer exponents, 519, 523
Integers, 11, 69
    GCF of list of, 274
Intercepts, 182–86, 210–11
Inverses, 72
Irrational numbers, 12, 69, 483

J

Junior Engineering Technical Society (JETS) website, 528

K

Kentucky State District Council of Carpenters website, 480

L

LAN. *See* Local area network
Landscaping industry, 272
Least common denominator, 21
    finding, 348, 390
Legs, of right triangle, 316
Less than or equal to symbol, 10
Like radicals, 494, 495, 522
Like terms, 81, 151
Linear equations (*See also* Point-slope form; Systems of
    linear equations)
    forms of, 410
    graphing, 173–79
    slope-intercept from for graphing, 402-3
    solving, 152
    solving systems of, 439–76
    solving systems of, by graphing, 440–44
Linear equations in one variable, 86, 99, 151
Linear equations in two variables, 210, 414

Linear inequalities
    graphing, 202–6
    solving, 140–47, 155, 156
Linear inequalities in one variable, 141, 144, 155
Linear inequalities in two variables, graphing, 202–5, 212
Linear programming, 468, 469
Line graphs, reading, 63–65
Lines
    and graph of linear equation in two variables, 173
    horizontal, 410
    parallel, 194, 195, 410, 442, 444
    perpendicular, 410
    point-slope form of equation of, 407
    slope of, 191
    vertical, 410
List of variables raised to powers, greatest common factor
    raised to, 275
Local area network (LAN), 2
Local area network (LAN) administrators, 2, 32
Long division, and polynomials, 258-60, 266
Lowest common denominator, and solving linear inequal-
    ities in one variable, 144
Lowest terms, fractions in, 18, 70

M

Mathematical statements, sentences translated into, 10
Mathematics
    flexibility with, 3
    tips for success in, 4–7
Measurements, formulas for, 116
Member, of set, 8
Mixed numbers, 22
Mixture problems, 135
Monomials, 228, 229, 264
    dividing polynomial by, 256
    dividing polynomial by polynomial other than, 266
    multiplying, 237
Multiplication
    associative property of, 57, 72
    commutative property of, 56, 72
    of complex numbers, 552–53
    of exponential expressions, 219
    of fractions, 19, 70
    identity for, 59
    of polynomials, 237–39, 265
    of radicals, 523
    of rational expressions, 339–41, 389
    of real numbers, 48–51, 71–72
    of same factor, 25
    of sum and difference of two terms, 244, 265
Multiplication property of equality, 94–96, 152, 365
Multiplication property of inequality, 143, 155
Multiplicative inverses (or reciprocals), 50, 59, 72

N

National Education Association, 78

National Federation of Paralegal Associations website, 216
National League for Nursing website, 438
Natural numbers, 8, 69
Negative constant, in trinomials, 284
Negative exponents, 248, 250, 265
Negative integers, 11
Negative numbers, 12, 37, 48
  cube roots of, 484
  entering on calculators, 53
  multiplying, 49
  and variation in elevation, 43, 44
Negative reciprocals, 195, 211
Negative slope, 192, 194
Negative square root, 482, 522
Nonlinear equations
  graphing, 414–18, 432
  identifying, 414
  writing ordered pairs from graphs of, 416–18
Nonvertical parallel lines, slopes of, 401–2
Numbers (*See also* Irrational numbers; Natural numbers;
   Rational numbers; Real numbers; Whole numbers)
  common sets of, 13
  nth root of, 484, 522
  on number line, 9, 69
  in scientific notation, 251, 265
  in standard form, 252
Numerator, 17, 69
Numerical coefficients, 80, 151, 264
  of term, 228
  of trinomials, 286–87

O

On-line shopping, 169
Opposites (or additive inverses), 38–39, 59, 71, 72
Ordered pairs, 162, 163
  graphing, 209
  and quadratic equations, 566
  as solution of equation in two variables, 165, 210
  as solution of inequality, 202
  as solutions of system of two equations in two variables, 440
  writing from graphs of nonlinear equations, 416–18
Order of operations, 26
  on calculator, 31
Order property for real numbers, 13, 69
Origin, 162, 177, 209
Original inequality, 205

P

Paired data, 164
Parabolas, 309, 415, 556, 566
  finding vertex of, 557–58
  locating intercept points of, 557-60, 566
  vertex of, 556, 566
Paralegal profession, 216

Parallel lines, 194, 410, 442, 444
  and slope-intercept form, 401–2
  slope of, 211
Parentheses, 26, 43
  distributive property and removal of, 82, 83, 90, 100, 102, 151
  in exponential expressions, 218
Partial product, 238, 239
Percent, 124, 153
  decimal written as, 153
  and problem solving, 154
  written as decimal, 125
Percent decrease, 127
Percent increase, 127
Perfect fourth powers, 491
Perfect square factors, 488, 489
Perfect squares, 483
Perfect square trinomials, 290, 324
  finding, 534–35
Perpendicular lines, 410
  and slope-intercept form, 401–2
  slope of, 194–95, 196, 211
Physical therapists, 398
Pie chart, 126
Plotting, ordered pair, 209
Point-slope form, 406–410, 432
  and equations of vertical and horizontal lines, 408–9
  and equation of line given slope and point of line, 406–7
  and equation of line given two points of line, 407–8
  of linear equation, 410
  for solving problems, 409–410
Polynomials, 217, 264
  addition of, 228-33, 264
  degree of, 229
  dividing by monomial, 256–57, 266
Polynomials
  dividing by polynomial other than monomial, 258, 266
  division of, 256–60, 266
  factoring, 274, 300-302
  modeling with, 262–63
  multiplication of, 237–39, 265
  subtraction of, 264
  in x, 228
Positive constant, in trinomials, 284
Positive integers, 11
Positive numbers, 12, 37, 48
  multiplying, 49
  and variation in elevation, 43, 44
Positive (or principal) square root, 482, 522
Positive slope, 192, 194, 397
Power of a product rule, 221, 250, 264
Power of a quotient rule, 222, 250, 264
Power rule for exponents, 221, 250, 264
Powers, raising to, 218, 220
Prices, solving problem about, 461–62
Prime numbers, 18

Prime polynomial, 283
Problem solving (*See also* Spotlight on decision making)
    and formulas, 116–21, 153
    and percent, 154
    point-slope form for, 409–410
    for problems written in words, 103–4
    with proportions, 375–77
    and quadratic equations, 313-17, 325–26, 548
    and radical equations, 509–512, 523
    and rational equations, 379–83, 392
    steps for, 152–53
    strategy for, 107, 133
    and systems of linear equations, 459–65, 476
Product, 18, 48, 69, 72, 278
Product rule for exponents, 219–20, 250, 264
Product rule for radicals, 491, 497, 522
Product rule for square roots, 488–89
Proportion, 373–74, 392
    problem solving with, 375–21
Pythagorean theorem, 316
    and problem solving, 509-513, 523

Q

Quadrants, 162
Quadratic equations, 415, 507, 565
    complex solutions of, 550–54, 565–66
    defined, 305, 325
    discriminants and number of solutions of, 543–44
    graphing, 555–60, 566
    graphing form $y = ax^2 + bx + c$, 555–56, 560
    methods for solving, 546–48
    and problem solving, 313–17, 325–26
    solving by completing the square, 534–38, 564
    solving by factoring, 305–310
    solving by quadratic formula, 539–44, 564
    solving by square root method, 530–32, 564
    solving problems modeled by, 548
    in standard form, 539, 540
Quadratic equations in two variables, 415, 432
Quadratic formula, 540, 546, 559
    quadratic equations solved by, 539–44, 564
Quotient rule for exponents, 223, 247, 250, 264
Quotient rule for radicals, 491, 499, 522
Quotient rule for square roots, 489–90
Quotients
    raising to a power, 222
    of two real numbers, 50-51
    and zero, 72

R

Radical equations
    and problem solving, 509–512, 523
    solving, 504, 505-6
Radical expressions, 482
    simplifying, 494–95, 499
Radical notation, 516

Radicals, 482, 522
    adding and subtracting, 494–95
    dividing, 499
    multiplying, 497–98, 523
    product rule for, 491, 497, 522
    quotient rule for, 491, 522
    simplifying, 488–92, 522
    solving equations containing, 504–8, 523
    variables in, 483
Radical sign, 482
Radicand, 482, 522
    perfect cube factors of, 491
Radius, of cylinder, 520–21
Radius of curvature, 486, 515
Range
    of function, 426
    of relation, 421, 433
Rates, finding, 462–64
Ratio, 373, 392
Rational equations, and problem solving, 379–83, 392
Rational exponents, 516–19, 523
Rational expressions, 257, 331–92
    adding and subtracting those with different denominators, 353–56
    adding and subtracting those with same denominator, 346–48, 390
    adding and subtracting those with unlike denominators, 390
    complex, 359
    defined, 332, 389
    dividing, 339–41, 389
    as equivalent expression with given denominator, 350
    fundamental principle of, 334
    least common denominator of, 348
    multiplying, 339–41, 389
    simplifying, 332, 334, 389
    solving equations containing, 365–70, 391
    undefined, 366, 367
    value of, given replacement number, 332–33
    values for which expression is undefined, 333
Rationalizing the denominator, 499–500, 523
Rational numbers, 11, 12, 69
Real numbers, 69, 551
    absolute value of, 14
    adding, 34–39, 71
    dividing, 51–53, 71-72
    multiplying, 48–51, 71-72
    operations with, on calculator, 53
    order property for, 13
    properties of, 56–60, 72
    relationships among sets of, 13
    set of, 12
    subtracting, 41–44, 71
Reciprocals (or multiplicative inverses), 19, 50, 59, 70, 72
Rectangle
    area formula for, 116

perimeter formula for, 118
Rectangular coordinate system, 162–69, 209
Rectangular solid, volume formula for, 116, 120
Registered nurses, 438
Relations, 421, 433
Right triangle, 316
    hypotenuse of, 510
Rise, between points, 190
ROOT feature, on graphers, 310, 418
Root (or solution) of equation, 71, 86
Roots, 481
Run, between points, 190

S

Scatter diagrams, 164–65
Scientific calculator, negative numbers entered on, 53
Scientific notation, 250–53, 265
Sentences, writing as percent equations, 126–27
Sets, 8, 69
Signed decimals/fractions, multiplying, 49
Signed numbers, 12, 35
Sign patterns
    with sums or differences of cubes, 297
    with trinomials, 284
Similar triangles, 382–83
Simple inequalities, 142, 146
Simple interest formula, 116
Simplifying complex fractions
    method 1, 359–61
    method 2, 359–61
Slope, 189–97, 211, 397
    of horizontal line, 193
    of line, 191–92
    negative, 192, 194
    of parallel lines, 194, 195, 211
    of perpendicular lines, 194–95, 196, 211
    positive, 192, 194
    summary of, 194
    of vertical lines, 193–94
    of zero, 194
Slope-intercept form, 400–403, 432, 442, 443
    for determining parallel or perpendicular lines, 401–2
    to find slope/y-intercept of line, 400–401
    to graph linear equation, 402–3
    of linear equation, 410
    to write equation of line, 402
Small businesses, 160
Solution amounts, finding, 464–65
Solution (or root) of equation, 71, 86
Solutions
    of compound inequalities, 142
    of equations, 29
    of inequalities, 141
    of system of linear inequalities, 468, 470
    of system of two equations in two variables, 474
Solving equation for variable, 29

Solving formulas for specified variable, 153
Solving the equation for the variable, 86
Special products, 241–46, 265
    and expressions containing radicals, 498
Spotlight on decision making (*See also* Problem solving)
    air conditioning installation, 427
    blood pressure, 471
    child care centers, 377
    credit card offers, 303
    dimensions for deck installation, 513
    engine displacement, 225
    extra credit projects, 91
    fishing record, 22
    forensics, 336
    ground temperature, 561
    gum tissue pocket depth, 44
    homeowner's insurance, 138
    income tax preparation, 128
    landscaping public park, 318
    local area network administrator, 32
    mail-order business, 169
    paralegal investigation, 253
    patient's progress on treadmill, 411
    production line analysis, 544
    quality control engineering, 15
Spotlight on decision making
    ramp speed limit, 486
    real estate agency choices, 197
    sales associate positions, 457
    softball team coaching, 383
    stock trade, 39
Square root property, 547
    quadratic equations solved by, 530–32, 564
Square roots, 482–83
    product rule for, 488–89
    quotient rule for, 489–90
    simplifying, 522
    simplifying those containing variables, 483–84
    solving radical equations containing, 505–6
Squares, 482
Squaring a binomial, 243–44, 265
Squaring property of equality
    using once, 504–5
    using twice, 507–8
Standard form, 174, 210
    complex number in, 551, 565
    of linear equation, 407, 410
    quadratic equation in, 306, 325, 539, 540
    scientific notation number written in, 252
Standard window, on graphing calculator, 179
Substitution method, 469
    solving systems of linear equations by, 447–51, 475
Subtraction
    of complex numbers, 551–52, 565
    of exponential expressions, 219
    of fractions with same denominator, 20–22, 70

Subtraction (*cont.*)
  of fractions with unlike denominators, 348–49
  of like radicals, 522
  of polynomials, 232–33, 264
  of radicals, 494–95
  of rational expressions with common denominators, 347, 390
  of rational expressions with different denominators, 353–56, 390
  of real numbers, 41–44, 71
Sum, 278
  of cubes, 296, 324
Supplementary angles, 45, 476
Surveys, 68
Symbols
  brace, 8
  cube roots, 484
  equal, 9, 29
  function notation, 425
  greater than or equal to, 10
  grouping, 25–26
  inequality, 9, 10, 140
  less than or equal to, 10
  nth root of number, 484
  for percent, 125, 153
  radicals, 482, 522
  and sets of numbers, 69
  square roots, 482
Symmetric about the y-axis, 556
System of equations, 439
  with no solution, 474
  with one solution, 474
System of two linear equations, solving by addition method, 456
Systems of linear equations
  with infinite number of solutions, 450, 455, 474
  with no solution, 451, 455
  and problem solving, 459–65, 476
  solving by addition method, 453–57
  solving by graphing, 440-44, 474
  solving by substitution, 447–51, 475
Systems of linear inequalities, 468-70, 476

T

Table of values, 167
Teachers and teacher aides, 78
Temperature formula, 116, 118–19
Terms, 80, 151, 264
  degree of, 229
  GCF of list of, 275, 323
Tests
  preparing for, 6
  taking, 7
Time management, 7

TRACE feature, on graphers, 310, 419
Triangles
  lengths of, 383
  perimeter formula for, 116
Trinomials, 228, 229, 264
  factoring form $ax^2 + bx + c$, 286–92, 323–324
  factoring form $x^2 + bx + c$, 280–84
  perfect square, 290-91, 324
  sign patterns with, 284

U

Undefined quotient, 51
Undefined rational expressions, 366, 367
Undefined slope, 194, 211
Unicom Institute of Technology, 2
Unit fractions, 343
Units of measurements, converting, 343
Unknown numbers, finding, 107–8, 379–80, 460-61
Unknown values, finding, 166
Unlike denominators, adding and subtracting rational expressions with, 390
Unlike terms, 81
U.S. Bureau of Labor Statistics, 216, 438

V

Variables, 28, 70, 79
  eliminating, when adding two equations, 455
  in radicals, 483
  replacement values for, 230
Vehicle speeds, finding, 381–82
Velocity formula, 512
Vertex, 556
  of parabola, 566
  formula, 558
Vertical axis, 162
Vertical lines, 186, 211, 410
  equation of, 409
  slope of, 193–94
Vertical line test, 422–25, 433
Volume, of cylinder, 520

W

Whole numbers, 9, 69
Windows, on graphing calculator, 179
Work rates, finding, 380–81

X

x-axis, 163, 209
x-coordinate, 163
  of vertex, 558, 559
x-intercepts, 210
  finding/identifying, 182–85, 211, 558–59, 560, 566
  of graph of quadratic equation in two variables, 309–10
  to graph quadratic equations, 557
  of parabola, 558

63
rtex, 558, 566

axis, 163, 209
    and axis of symmetry, 556
y-coordinate, 163
    of vertex, 558, 559
y-intercept, 210
    equation of line given slope and, 402
    finding, 183-85, 211, 566
    for graphing quadratic equations, 557
    identifying, 182
    slope-intercept form for finding, 400–401
Young's Rule for dosage formulas, 387

y-value, 163
    on graphs of nonlinear equations, 414, 415, 416, 417
    of vertex, 566

Z

Zero
    as divisor or dividend, 51
    as factor, 49
    products and quotients involving, 72
    slope of, 194
Zero exponent, 224, 250, 264
Zero factor theorem, 306, 325, 530
Zero slope, 211
ZOOM feature, on grapher, 310, 418, 419

Photo Credits

Chapter 1 CO Bob Daemmrich/The Image Works, (p.5) Frank Johnston/PhotoDisc, Inc., (p.11) Tim Davis/Photo Researchers, Inc., (p.17) John Chumack/Photo Researchers, Inc., (p.28) Didier Klein/Vandystadt/Allsport Photography (USA), Inc., (p.40) © Richard T. Nowitz/Corbis, (p.40) Emory Kristof/NGS Image Collection, (p.47) N. Gellette/Liaison Agency, Inc., (p.55) David Young-Wolfe/PhotoEdit

Chapter 2 CO PhotoEdit, (p.92) AP/Wide World Photos, (p.109) Bruce Hoertel/Liaison Agency, Inc., (p.111) Brian K. Diggs/ AP/Wide World Photos, (p.113) Nasa/ Science Source/Photo Researchers, Inc., (p.113) Mark Lennihan/ AP/Wide World Photos, (p.115) Bachmann/Stock Boston, (p.117) Sean Reid/Alaska Stock, (p.123) Mt. Stromlo and Siding Spring Observations, Australian National University/Science Photo Library/Photo Researchers, Inc., (p.123) Norbert Wu/Stone, (p.124) John Elk III/Stock Boston, (p.128) Jonathan Nourok/PhotoEdit, (p.157) Jean-Claude LeJeune/Stock Boston

Chapter 3 CO Photo Researchers, Inc. (p.178) Michael Newman/PhotoEdit, (p.181) Amy C. Etra/PhotoEdit, (p.196) John Neubauer/PhotoEdit

Chapter 4 CO Michael Newman/PhotoEdit, (p.255) Reuters/Jack Newton/Archive Photos

Chapter 5 CO Jonathan Nourok/PhotoEdit

Chapter 6 CO Carl J. Single/The Image Works, (p.338) Chuck Keeler/The Image Works, (p.343) Gary J. Thibeault, (p.345) Rich Pedroncelli/AP/Wide World Photos, (p.345) UPI/Corbis, (p.380) © Charles O'Rear/CORBIS, (p.395) Keith Brofsky/PhotoDisc, Inc.

Chapter 7 CO The Image Works, (p.405) Orlin Wagner/AP/Wide World Photos, (p.410) Doug Menuez/PhotoDisc, Inc., (p.412) M. K. Denny/PhotoEdit, (p.412) Spencer Grant/PhotoEdit, (p.413) Dwight Cendrowski/Focusing Group Photography, (p.413) Lawrence Migdale/Stock Boston

Chapter 8 CO Nicola Sutton/PhotoDisc, Inc., (p.458) David Young-Wolfe/PhotoEdit, (p.466) Elsa Hasch/Allsport Photography (USA), Inc., (p.466) Doug Densinger/Allsport Photography (USA), Inc., (p.467) J. Scott Applewhite/AP/Wide World Photos, (p.479) Donna McWilliam/AP/Wide World Photos,

Chapter 9 CO Alexandria Town Talk/Katharine Ganter/AP/Wide World Photos, (p.493) Peter Poulides/Stone

Chapter 10 CO N. R. Rodman/The Image Works, (p.533) Deborah Davis/PhotoEdit, (p.548) Hulton Getty/Liaison Agency, Inc., (p.549) Amy C. Etra/PhotoEdit, (p.549) Joe Cavaretta/AP/Wide World Photos, (p.568) © Jeremy Horner/CORBIS, (p.562) Lawrence Migdale/Photo Researchers, Inc.

COMMON GRAPHS

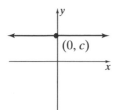

Horizontal Line;
Zero Slope
$y = c$

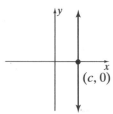

Vertical Line;
Undefined Slope
$x = c$

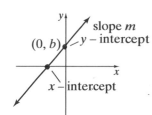

Linear Equation;
Positive Slope
$y = mx + b; m > 0$

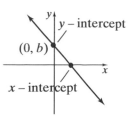

Linear Equation;
Negative Slope
$y = mx + b; m < 0$

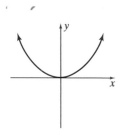

$y = x$

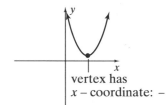

$y = x^2$

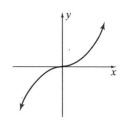

$y = x^3$

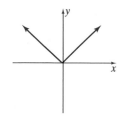

$y = |x|$

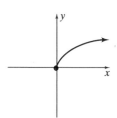

$y = \sqrt{x}; x \geq 0$

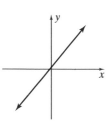

Quadratic Equation
$y = ax^2 + bx + c; a \neq 0$
Parabola opens upward if $a > 0$
Parabola opens downward if $a < 0$

vertex has
x – coordinate: $-\dfrac{b}{2a}$.

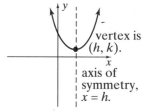

vertex is
(h, k).

axis of
symmetry,
$x = h$.

Quadratic Equation
$y = a(x - h)^2 + k; a \neq 0$
Parabola opens upward if $a > 0$
Parabola opens downward if $a < 0$

SYSTEMS OF LINEAR EQUATIONS

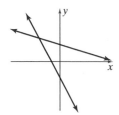

Independent and
consistent; one solution

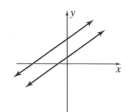

Independent and
inconsistent, no solution

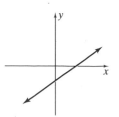

Dependent and
consistent; infinitely many solutions